Fundamentals of
Environmental
Chemistry

Stanley E. Manahan

LEWIS PUBLISHERS
Boca Raton   Ann Arbor   London   Tokyo

Library of Congress Cataloging-in-Publication Data

Manahan, Stanley E.
  Fundamentals of environmental chemistry / Stanley E. Manahan.
    p. cm.
  Includes bibliographical references and index.
  1. Environmental chemistry. I. Title.
TD193.M38   1993
540—dc20   93-10356
ISBN 0-87371-587-X (acid-free)

LEWIS PUBLISHERS
121 South Main Street, Chelsea, Michigan 48118

PRINTED IN THE UNITED STATES OF AMERICA   3 4 5 6 7 8 9 0

Printed on acid-free paper

Stanley E. Manahan is Professor of Chemistry at the University of Missouri–Columbia, where he has been on the faculty since 1965. He received his A.B. in chemistry from Emporia State University in 1960 and his Ph.D. in analytical chemistry from the University of Kansas in 1965. Since 1968 his primary research and professional activities have been in environmental chemistry and have included development of methods for the chemical analysis of pollutant species, environmental aspects of coal conversion processes, development of coal products useful for pollutant control, hazardous waste treatment, and toxicological chemistry. He teaches courses on environmental chemistry, hazardous wastes, toxicological chemistry, and analytical chemistry and has lectured on these topics throughout the U.S. as an American Chemical Society Local Section tour speaker. He is also President of ChemChar Research, Inc., a firm working with the development of nonincinerative thermochemical and electrothermochemical treatment of mixed hazardous substances containing refractory organic compounds and heavy metals.

Professor Manahan has written books on hazardous wastes (*Hazardous Waste Chemistry, Toxicology and Treatment*, 1990, Lewis Publishers, Inc.), toxicological chemistry (*Toxicological Chemistry*, 2nd ed., 1992, Lewis Publishers, Inc.), and environmental chemistry (*Environmental Chemistry*, 5th ed., 1991, Lewis Publishers, Inc.), applied chemistry, and quantitative chemical analysis. He has been the author or co-author of approximately 75 research articles.

Preface

Fundamentals of Environmental Chemistry is written to provide a reader having little or no background in chemistry with the fundamentals of chemistry and environmental chemistry needed for a trade, profession, or curriculum of study that requires a basic knowledge of these topics. It is designed to serve as a general reference and textbook for a broad spectrum of applications.

Virtually everyone needs some knowledge of chemistry. Unfortunately, this vital, interesting discipline "turns off" many of the very people who need a rudimentary knowledge of it. There are many reasons why this is so. "Chemophobia," a fear of insidious contamination of food, water, and air with chemicals at undetectable levels that may cause cancer and other maladies is widespread among the general population. Chemistry courses have been misused, especially at the beginning college level, to "weed out" the unworthy (in the view of some who teach the courses); those who fail such courses or who decide that further study of chemistry would not be in their best interests often go into nonscientific professions, such as law, journalism, and politics, where they make decisions that come back to haunt the chemical profession, which they have come to dislike. The language of chemistry is often made too complex so that those who try to learn it retreat from concepts such as moles, orbitals, electronic configurations, chemical bonds, and molecular structure before coming to realize that these ideas are comprehensible and even interesting and useful.

Just as everyone needs to know some chemistry, any citizen of Planet Earth must have some knowledge of environmental sciences. Here chemistry is often regarded as the villain, producing substances and using processes that cause environmental damage. However, it is through understanding the chemical processes that cause environmental harm and by being aware of chemical processes that are essential for preventing environmental pollution that most environmental problems can be solved.

With the above points in mind, *Fundamentals of Environmental Chemistry* has been written to meet two objectives: (1) to provide those who need it with a rudimentary knowledge of chemistry and (2) to present this knowledge within a framework of environmental science; specifically, environmental chemistry. This book draws on the author's extensive experience in writing books on related areas of environmental chemistry, general chemistry, toxicological chemistry, hazardous wastes chemistry, and chemical analysis. It is designed to be simple, understandable, and interesting. Without being overly simplistic or misleading, it seeks to present chemical principles in ways so that even a reader with a minimal background in, or no particular aptitude for science and mathematics can master the material in it and apply it to a trade, profession, or course of study.

One of the ways in which *Fundamentals of Environmental Chemistry* presents chemistry in a "reader-friendly" manner is through an organizational structure that the author has found to be useful in a previous book, *General Applied Chemistry* and in teaching courses in preparation for general chemistry. In the first few pages of Chapter 1, the reader is presented with a "mini-course" in chemistry that consists of the most basic concepts and terms needed to really begin to understand chemistry. In order to study chemistry it is necessary to know a few essential things—what an atom is, what is meant by elements, chemical formulas, chemical bonds, molecular mass. With these terms defined in very basic ways it is possible to go into more detail on chemical concepts without having to assume—as many introductory chemistry books do very awkwardly—that the reader knows nothing of the meaning of these terms.

Chapter 2 discusses matter largely on the basis of its physical nature and behavior, introducing physical and chemical properties, states of matter, the mole as a quantity of matter, and other ideas required to visualize chemical substances as physical entities. Chapters 3–5 cover the core of chemical knowledge constructed as a language in which elements and the atoms of which they are composed (Chapter 3) are presented as letters of an alphabet, the compounds made up of elements (Chapter 4) are analogous to words, the reactions by which compounds are synthesized and changed (Chapter 5) are like sentences in the chemical language, and the mathematical aspects hold it all together quantitatively. Chapters 6–8 constitute the remainder of material that is usually regarded as essential material in general chemistry. Chapter 9 presents a basic coverage of organic chemistry. Although this topic is often ignored at the beginning chemistry level, those who deal with the real world of environmental pollution, hazardous wastes, agricultural science, and other applied areas quickly realize that a rudimentary understanding of organic chemistry is essential. Chapter 10 covers biological chemistry, an area essential to understanding later material dealing with environmental and toxicological chemistry.

Chapters 11–18 cover the core of environmental chemistry—water, soil, and air. These chapters are based upon material and concepts, condensed, simplified, and updated, from the author's book, *Environmental Chemistry*, 5th edition. Chapters 19 and 20 deal with hazardous substances and hazardous wastes, and address their sources, environmental chemistry, reduction, treatment, and disposal. Chapter 21 uses toxicological chemistry as the basis for discussing occupational health and human exposure to pollutants and toxicants. Chapter 22 discusses resources and energy with emphasis upon the environmental considerations involved with their extraction and utilization.

Acknowledgments

The author would like to acknowledge the assistance of Vivian Collier, Kathy Walters, and the rest of the staff of Lewis Publishers who have been outstanding to work with in preparing this book. Their unfailing helpfulness, courtesy, and professionalism have contributed markedly to making the production of this book a rewarding experience. The outstanding efforts of Brian Lewis, Publisher, in recognizing the need and opportunity for works in the environmental sciences are very much appreciated. The author would also like to acknowledge the careful and constructive reviews provided by Professor Edward J. Baum of Grand Valley State University (Michigan) and by Frankie L. Elqué.

The author appreciates the efforts of those reviewers who have carefully and constructively critiqued the book and would appreciate receiving copies of their reviews. Those most qualified to review a book are, of course, the individuals who have used it. Feedback in the form of comments and suggestions from users are welcomed and may be directed to the author at 123 Chemistry Building, University of Missouri, Columbia, Missouri 65211.

CONTENTS

Fundamentals of

Environmental Chemistry

1 INTRODUCTION TO CHEMISTRY

1.1 CHEMISTRY AND ENVIRONMENTAL CHEMISTRY

Chemistry is defined as the science of matter. Therefore, it deals with the air we breathe, the water we drink, the soil that grows our food, and vital life substances and processes. Our own bodies contain a vast variety of chemical substances and are tremendously sophisticated chemical factories that carry out an incredible number of complex chemical processes.

There is a deep concern today about the uses — and particularly the misuses — of chemistry as it relates to the environment. Ongoing events serve as constant reminders of threats to the environment ranging from individual exposures to toxicants to phenomena on a global scale that may cause massive, perhaps catastrophic, alterations in climate. These include, for example, a five-year drought in California during the late 1980s and early 1990s that may strongly affect the economy and social structure of the state, massive petroleum fires in Kuwait in 1991, and air quality in Mexico City so bad that it threatens human health. Furthermore, increasing numbers of employees are involved in dealing with hazardous substances and wastes in laboratories and the workplace. All such matters involve environmental chemistry for understanding of the problems and for arriving at solutions to them.

What is environmental chemistry? **Environmental chemistry** is that branch of chemistry that deals with the origins, transport, reactions, effects, and fates of chemical species in the water, air, terrestrial, and living environments.[1] A related discipline, **toxicological chemistry**, is the chemistry of toxic substances with emphasis on their interaction with biologic tissue and living systems.[2] In addition to its being an essential, vital discipline in its own right, environmental chemistry provides an excellent framework for the study of chemistry. This is because it cuts across all subdivisions of chemistry. In addition to what is commonly regarded as "general chemistry," environmental chemistry must deal with organic chemistry, chemical analysis, physical chemistry, photochemistry, geochemistry, and biological chemistry. By necessity it breaks down the barriers that tend to compartmentalize chemistry as it is conventionally addressed. Therefore, this book is written with two major goals — to provide an overview of chemical science within an environmental chemistry framework and to provide the basics of environmental chemistry for those who need to know about this essential topic for their profession or for their overall education.

1.2. A MINI-COURSE IN CHEMISTRY

It is much easier to learn chemistry if one already knows some chemistry! That is, in order to go into any detail on any chemical topic, it is extremely helpful to have some rudimentary knowledge of chemistry as a whole. For example, a crucial part of chemistry is an understanding of the nature of chemical compounds, the chemical formulas used to describe them, and the chemical bonds that hold them together; these are topics addressed in Chapter 3 of this book. However, to understand these concepts, it is very helpful to know some things about the chemical reactions by which chemical compounds are formed, as addressed in Chapter 4. To work around this problem, Chapter 1 provides a highly condensed, simplified, but meaningful overview of chemistry to give the reader the essential concepts and terminology required to understand more advanced chemical material.

1.3.   THE BUILDING BLOCKS OF MATTER

All matter is composed of only about a hundred fundamental kinds of matter called **elements**. Each element is made up of very small entities called **atoms**; all atoms of the same element behave identically chemically. The study of chemistry, therefore, can logically begin with elements and the atoms of which they are composed.

Subatomic Particles and Atoms

Figure 1.1 represents an atom of deuterium, a form of the element hydrogen. It is seen that such an atom is made up of even smaller **subatomic particles**—

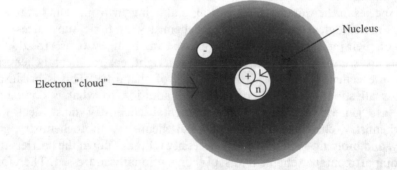

Figure 1.1.   Representation of a deuterium atom. The nucleus contains one proton (+) and one neutron (n). The electron (-) is in constant, rapid motion around the nucleus forming a cloud of negative electrical charge, the density of which drops off with increasing distance from the nucleus.

positively charged **protons**, negatively charged **electrons**, and uncharged (neutral) **neutrons**.

Protons and neutrons have relatively high masses compared to electrons and are contained in the positively charged **nucleus** of the atom. The nucleus has essentially all of the mass, but occupies virtually none of the volume, of the atom. An uncharged atom has the same number of electrons as protons. The electrons in an atom are contained in a cloud of negative charge around the nucleus that occupies most of the volume of the atom.

Atoms and Elements

All of the literally millions of different substances are composed of only around 100 elements. Each atom of a particular element is chemically identical to every other atom and contains the same number of protons in its nucleus. This number of protons in the nucleus of each atom of an element is the **atomic number** of the element. Atomic numbers are integers ranging from 1 to more than 100, each of which denotes a particular element. In addition to atomic numbers, each element has a name and a **chemical symbol**, such as carbon, C; potassium, K (for its Latin name kalium); or cadmium, Cd. In addition to atomic number, name, and chemical symbol, each element has an **atomic mass** (atomic weight). The atomic mass of each element is the average mass of all atoms of the element, including the various isotopes of which it consists. The **atomic mass unit, u** (also called the **dalton**), is used to express masses of individual atoms and molecules (aggregates of atoms). These terms are summarized in Figure 1.2.

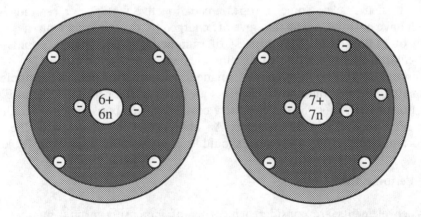

An atom of carbon, symbol C. Each C atom has 6 protons (+) in its nucleus, so the atomic number of C is 6. The atomic mass of C is 12.

An atom of nitrogen, symbol N. Each N atom has 7 protons (+) in its nucleus, so the atomic number of N is 7. The atomic mass of N is 14.

Figure 1.2. Atoms of carbon and nitrogen

Table 1.1.   List of Some of the More Important Common Elements

Element	Symbol	Atomic Number	Atomic Mass (relative to carbon-12)
Aluminum	Al	13	26.9815
Argon	Ar	18	39.948
Bromine	Br	35	79.904
Calcium	Ca	20	40.078
Carbon	C	6	12.01115
Chlorine	Cl	17	35.453
Copper	Cu	29	63.546
Fluorine	F	9	18.998403
Helium	He	2	4.00260
Hydrogen	H	1	1.0080
Iron	Fe	26	55.847
Magnesium	Mg	12	24.305
Mercury	Hg	80	200.59
Neon	Ne	10	20.180
Nitrogen	N	7	14.0067
Oxygen	O	8	15.9994
Potassium	K	19	39.0983
Silicon	Si	14	28.0855
Sodium	Na	11	22.9898
Sulfur	S	16	32.06

Although atoms of the same element are *chemically* identical, atoms of most elements consist of two or more **isotopes** that have different numbers of neutrons in their nuclei. Some isotopes are **radioactive isotopes** or **radionuclides**, which have unstable nuclei that give off charged particles and gamma rays in the form of **radioactivity**. This process of **radioactive decay** changes atoms of a particular element to atoms of another element.

Throughout this book reference is made to various elements. The symbols and atomic numbers of the known elements are given in the periodic table in Figure 1.3. Fortunately, most of the chemistry covered in this book requires familiarity with only about 25 or 30 elements. An abbreviated list of a few of the most important elements that the reader should learn at this point is given in Table 1.1.

The Periodic Table

When elements are considered in order of increasing atomic number, it is observed that their properties are repeated in a periodic manner. For example, elements with atomic numbers 2, 10, and 18 are gases that do not undergo chemical reactions and consist of individual molecules, whereas those with atomic numbers larger by one—3, 11, and 19—are unstable, highly reactive

Active metals

Nonmetals

Transition metals

1A 1	2A 2	3B 3	4B 4	5B 5	6B 6	7B 7	8B 8	8B 9	8B 10	1B 11	2B 12	3A 13	4A 14	5A 15	6A 16	7A 17	8A 18
1 H 1.0079																	2 He 4.00260
3 Li 6.941	4 Be 9.01218											5 B 10.81	6 C 12.011	7 N 14.0067	8 O 15.9994	9 F 18.998403	10 Ne 20.179
11 Na 22.98977	12 Mg 24.305											13 Al 26.98154	14 Si 28.0855	15 P 30.97376	16 S 32.06	17 Cl 35.453	18 Ar 39.948
19 K 39.0983	20 Ca 40.078	21 Sc 44.9559	22 Ti 47.88	23 V 50.9415	24 Cr 51.996	25 Mn 54.9380	26 Fe 55.847	27 Co 58.9332	28 Ni 58.69	29 Cu 63.546	30 Zn 65.38	31 Ga 69.72	32 Ge 72.61	33 As 74.9216	34 Se 78.96	35 Br 79.904	36 Kr 83.80
37 Rb 85.4678	38 Sr 87.62	39 Y 88.9059	40 Zr 91.22	41 Nb 92.9064	42 Mo 95.94	43 Tc (98)	44 Ru 101.07	45 Rh 102.9055	46 Pd 106.42	47 Ag 107.8682	48 Cd 112.41	49 In 114.82	50 Sn 118.69	51 Sb 121.75	52 Te 127.60	53 I 126.9045	54 Xe 131.29
55 Cs 132.9054	56 Ba 137.33	57 *La 138.9055	72 Hf 178.49	73 Ta 180.9479	74 W 183.85	75 Re 186.207	76 Os 190.2	77 Ir 192.22	78 Pt 195.08	79 Au 196.9665	80 Hg 200.59	81 Tl 204.383	82 Pb 207.2	83 Bi 208.9804	84 Po (209)	85 At (210)	86 Rn (222)
87 Fr (223)	88 Ra 226.0254	89 +Ac 227.0278	104 Rf (261)	105 Ha (262)	106 Unh (263)	107 Uns (262)		109 Une (266)									

*Lanthanide series

58 Ce 140.12	59 Pr 140.9077	60 Nd 144.24	61 Pm (145)	62 Sm 150.36	63 Eu 151.96	64 Gd 157.25	65 Tb 158.9254	66 Dy 162.50	67 Ho 164.9304	68 Er 167.26	69 Tm 168.9342	70 Yb 173.04	71 Lu 174.967

+Actinide series

90 Th 232.0381	91 Pa 231.0359	92 U 238.0289	93 Np (237)	94 Pu (244)	95 Am (243)	96 Cm (247)	97 Bk (247)	98 Cf (251)	99 Es (252)	100 Fm (257)	101 Md (258)	102 No (259)	103 Lr (260)

The larger and smaller labels reflect two different numbering schemes in common usage.

Figure 1.3.   The periodic table of the elements.

metals. An arrangement of the elements in a manner that reflects this recurring behavior is known as the **periodic table** (Figure 1.3). The periodic table is extremely useful in understanding chemistry and predicting chemical behavior. As shown in Figure 1.3, the entry for each element in the periodic table gives the element's atomic number, symbol, and atomic mass. More detailed versions of the table include other information as well.

Features of the Periodic Table

The periodic table gets its name from the fact that the properties of elements are repeated periodically in going from left to right across a horizontal row of elements. The table is arranged so that an element has properties similar to those of other elements above or below it in the table. Elements with similar chemical properties are called **groups** of elements and are contained in vertical columns in the periodic table.

1.4.   CHEMICAL BONDS AND COMPOUNDS

Only a few elements, particularly the noble gases, exist as individual atoms; most atoms are joined by chemical bonds to other atoms. This can be illustrated very simply by elemental hydrogen, which exists as **molecules**, each consisting of 2 H atoms linked by a **chemical bond** as shown in Figure 1.4. Because hydrogen molecules contain 2 H atoms, they are said to be diatomic and are denoted by the **chemical formula**, H_2. The H atoms in the H_2 molecule are held together by a **covalent bond** made up of 2 electrons, each contributed by one of the H atoms, and shared between the atoms.

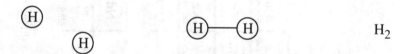

The H atoms in elemental hydrogen are held together by chemical bonds in molecules that have the chemical formula H_2.

Figure 1.4.   Molecule of H_2.

Chemical Compounds

Most substances consist of two or more elements joined by chemical bonds. As an example, consider the chemical combination of the elements hydrogen and oxygen shown in Figure 1.5. Oxygen, chemical symbol O, has an atomic number of 8 and an atomic mass of 16.00 and exists in the elemental form as diatomic molecules of O_2. Hydrogen atoms combine with oxygen atoms to form molecules in which 2 H atoms are bonded to 1 O atom in a substance with a

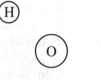

H_2O

Hydrogen atoms and oxygen atoms bond together to form molecules in
which 2 H atoms are attached to 1 O atom. The chemical formula of the
resulting compound, water, is H_2O.

**Figure 1.5.   A molecule of water, H_2O, formed from 2 H atoms and 1 O atom held
together by chemical bonds.**

chemical formula of H_2O (water). A substance such as H_2O that consists of a
chemically bonded combination of two or more elements is called a **chemical
compound**. (A chemical compound is a substance that consists of atoms of two
or more different elements bonded together.) In the chemical formula for water
the letters H and O are the chemical symbols of the two elements in the compound, and the subscript 2 indicates that there are 2 H atoms per O atom. (The
absence of a subscript after the O denotes the presence of just 1 O atom in the
molecule.) Each of the chemical bonds holding a hydrogen atom to the oxygen
atom in the water molecule is composed of two electrons shared between the
hydrogen and oxygen atoms.

Ionic Bonds

As shown in Figure 1.6, the transfer of electrons from one atom to another
produces charged species called **ions**. Positively charged ions are called **cations**
and negatively charged ions are called **anions**. Ions that make up a solid compound are held together by **ionic bonds** in a **crystalline lattice** consisting of an

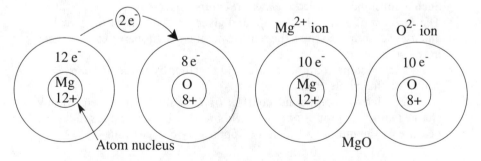

The transfer of two electrons from an atom of Mg to an O atom yields an ion
of Mg^{2+} and one of O^{2-} in the compound MgO.

**Figure 1.6. Ionic bonds are formed by the transfer of electrons and the mutual
attraction of oppositely charged ions in a crystalline lattice.**

ordered arrangement of the ions in which each cation is largely surrounded by anions and each anion by cations. The attracting forces of the oppositely charged ions in the crystalline lattice constitute the ionic bonds in the compound.

The formation of magnesium oxide is shown in Figure 1.6. In naming this compound, the cation is simply given the name of the element from which it was formed, magnesium. However, the ending of the name of the anion, o*xide*, is different from that of the element from which it was formed, o*xygen*.

Rather than individual atoms that have lost or gained electrons, many ions are groups of atoms bonded together covalently and having a net charge. A common example of such an ion is the ammonium ion, NH_4^+,

$$
\begin{array}{c}
H \\
| \\
H - N - H^+ \\
| \\
H \qquad \text{Ammonium ion, } NH_4^+
\end{array}
$$

which consists of 4 hydrogen atoms covalently bonded to a single nitrogen (N) atom and having a net electrical charge of $+1$ for the whole cation, as shown by the formula above.

Summary of Chemical Compounds and the Ionic Bond

The preceding several pages have just covered some material on chemical compounds and bonds that is essential to understanding chemistry. To summarize, these essentials are:

- Atoms of two or more different elements can form *chemical bonds* with each other to yield a product that is entirely different from the elements.
- Such a substance is called a *chemical compound*.
- The *formula* of a chemical compound gives the symbols of the elements and uses subscripts to show the relative numbers of atoms of each element in the compound.
- *Molecules* of some compounds are held together by *covalent bonds* consisting of shared electrons.
- Another kind of compound consists of *ions* made up of electrically charged atoms or groups of atoms held together by *ionic bonds* that exist because of the mutual attraction of oppositely charged ions.

Molecular Mass

The average mass of all molecules of a compound is its **molecular mass** (formerly called molecular weight). The molecular mass of a compound is calculated by multiplying the atomic mass of each element by the relative number of atoms of the element, then adding all the values obtained for each element in the compound. For example, the molecular mass of NH_3 is $14.0 + 3 \times 1.0 = 17.0$. As another example consider the following calculation of the molecular mass of ethylene, C_2H_4.

1. The chemical formula of the compound is C_2H_4.
2. Each molecule of C_2H_4 consists of 2 C atoms and 4 H atoms.
3. From the periodic table or Table 1.1, the atomic mass of C is 12.0 and that of H is 1.0.
4. Therefore, the molecular mass of C_2H_4 is
$$\underbrace{12.0 + 12.0}_{\text{From 2 C atoms}} + \underbrace{1.0 + 1.0 + 1.0 + 1.0}_{\text{From 4 H atoms}} = 28.0.$$

1.5.   CHEMICAL REACTIONS AND EQUATIONS

Chemical reactions occur when substances are changed to other substances through the breaking and formation of chemical bonds. For example, water is produced by the chemical reaction of hydrogen and oxygen:

Hydrogen plus oxygen yields water

Chemical reactions are written as **chemical equations**. The chemical reaction between hydrogen and water is written as the **balanced chemical equation**

$$2H_2 + O_2 \rightarrow 2H_2O \qquad (1.5.1)$$

in which the arrow is read as "yields" and separates the hydrogen and oxygen **reactants** from the water **product**. Note that because elemental hydrogen and elemental oxygen occur as *diatomic molecules* of H_2 and O_2, respectively, it is necessary to write the equation in a way that reflects these correct chemical formulas of the elemental form. All correctly written chemical equations are **balanced**, in that *they must show the same number of each kind of atom on both sides of the equation*. The equation above is balanced because of the following:

On the left

- There are 2 H_2 *molecules* each containing 2 H *atoms* for a total of 4 H atoms on the left.
- There is 1 O_2 *molecule* containing 2 O *atoms* for a total of 2 O atoms on the left.

On the right

• There are 2 H_2O *molecules* each containing 2 H *atoms* and 1 O atom for a total of 4 H atoms and 2 O atoms on the right.

The process of balancing chemical equations is relatively straightforward for simple equations. It is discussed in Chapter 5.

1.6.   NUMBERS IN CHEMISTRY: EXPONENTIAL NOTATION

An essential skill in chemistry is the ability to handle numbers, including very large and very small numbers. An example of the former is Avogadro's number, which is discussed further in Chapters 2 and 5. Avogadro's number is a way of expressing quantities of entities such as atoms or molecules and is equal to 602,000,000,000,000,000,000,000. A number written in this decimal form is very cumbersome to express and very difficult to handle in calculations. It can be expressed much more conveniently in exponential notation. Avogadro's number in exponential notation is 6.02×10^{23}. It is put into decimal form by moving the decimal in 6.02 to the right by 23 places. Exponential notation works equally well to express very small numbers, such as 0.000,000,000,000,000,087. In exponential notation this number is 8.7×10^{-17}. To convert this number back to decimal form, the decimal point in 8.7 is simply moved 17 places to the left.

A number in exponential notation consists of a *digital number* equal to or greater than exactly 1 and less than exactly 10 (examples are 1.00000, 4.3, 6.913, 8.005, 9.99999) multiplied by a *power of 10* (10^{-17}, 10^{13}, 10^{-5}, 10^{3}, 10^{23}). Some examples of numbers expressed in exponential notation are given in Table 1.2. As seen in the second column of the table, a positive power of 10 shows the number of times that the digital number is multiplied by ten and a negative power of 10 shows the number of times that the digital number is divided by 10.

Table 1.2.   Numbers in Exponential and Decimal Form

Exponential Form of Number	Places Decimal Moved for Decimal Form	Decimal Form
$1.37 \times 10^5 = 1.37 \times 10 \times 10 \times 10 \times 10 \times 10$	→ 5 places	137,000
$7.19 \times 10^7 = 7.19 \times 10 \times 10 \times 10 \times 10 \times 10$ $\times 10 \times 10$	→ 7 places	71,900,000
$3.25 \times 10^{-2} = 3.25/(10 \times 10)$	← 2 places	0.0325
$2.6 \times 10^{-6} = 3.25/(10 \times 10 \times 10 \times 10 \times 10 \times 10)$	← 6 places	0.000 0026
$5.39 \times 10^{-5} = 3.25/(10 \times 10 \times 10 \times 10 \times 10)$	← 5 places	0.000 0539

Addition and Subtraction of Exponential Numbers

Everyone now uses a calculator or computer to do mathematical operations. Properly used, a calculator keeps track of exponents automatically and with total accuracy. For example, getting the sum $7.13 \times 10^3 + 3.26 \times 10^4$ on a calculator simply involves the following sequence:

$$\boxed{7.13 \text{ EE3}} \quad \boxed{+} \quad \boxed{3.26 \text{ EE4}} \quad \boxed{=} \quad \boxed{3.97 \text{ EE4}}$$

where 3.97 EE4 stands for 3.97×10^4. To do such a sum manually, the largest number in the sum should be set up in the standard exponential notation form and each of the other numbers should be taken to the same power of 10 as that of the largest number as shown below for the calculation of $3.07 \times 10^{-2} - 6.22 \times 10^{-3} + 4.14 \times 10^{-4}$.

$$
\begin{array}{ll}
3.07 \times 10^{-2} & \text{(largest number, digital portion between 1 and 10)} \\
-0.622 \times 10^{-2} & \text{(same as } 6.22 \times 10^{-3}\text{)} \\
+0.041 \times 10^{-2} & \text{(same as } 4.1 \times 10^{-4}\text{)} \\
\hline
\end{array}
$$

Answer: 2.49×10^{-2}

Multiplication and Division of Exponential Numbers

As with addition and subtraction, multiplication and division of exponential numbers on a calculator or computer is simply a matter of (correctly) pushing buttons. For example, to solve

$$\frac{1.39 \times 10^{-2} \times 9.05 \times 10^8}{3.11 \times 10^4}$$

on a calculator, the sequence below is followed:

$$\boxed{1.39 \text{ EE-2}} \quad \boxed{\times} \quad \boxed{9.05 \text{ EE8}} \quad \boxed{\div} \quad \boxed{3.11 \text{ EE4}} \quad = 4.04 \text{ EE2 (same as } 4.04 \times 10^2\text{)}$$

In multiplication and division of exponential numbers, the digital portions of the numbers are handled conventionally. For the powers of 10, in multiplication, exponents are added algebraically, whereas in division the exponents are subtracted algebraically. Therefore, in the preceding example,

$$\frac{1.39 \times 10^{-2} \times 9.05 \times 10^{8}}{3.11 \times 10^{4}}$$

the digital portion is

$$\frac{1.39 \times 9.05}{3.11} = 4.04$$

and the exponential portion is,

$$\frac{10^{-2} \times 10^{8}}{10^{4}} = 10^{2} \text{ (The exponent is } -2 + 8 - 4)$$

So the answer is 4.04×10^{2}.

Example: Solve

$$\frac{7.39 \times 10^{-2} \times 4.09 \times 10^{5}}{2.22 \times 10^{4} \times 1.03 \times 10^{-3}}$$

without using exponential notation on the calculator.

Answer: Exponent of answer = $\underbrace{-2 + 5}$ - $\underbrace{(4 - 3)}$ = 2

Algebraic addition	Algebraic subtraction
of exponents	of exponents
in the numerator	in the denominator

$$\frac{7.39 \times 4.09}{2.22 \times 1.03} = 13.2 \quad \text{The answer is } 13.2 \times 10^{2} = 1.32 \times 10^{3}$$

Example: Solve

$$\frac{3.49 \times 10^{3}}{3.26 \times 10^{18} \times 7.47 \times 10^{-5} \times 6.18 \times 10^{-8}}$$

Answer: 2.32×10^{-4}

1.7. SIGNIFICANT FIGURES AND UNCERTAINTIES IN NUMBERS

The preceding section illustrated how to handle very large and very small numbers with *exponential notation*. This section considers **uncertainties** in numbers, taking into account the fact that numbers are known only to a certain degree of **accuracy**. The accuracy of a number is shown by how many **significant figures** or **significant digits** it contains. This may be illustrated by considering the atomic masses of elemental boron and sodium. The atomic mass of boron is given as 10.81. Written in this way, the number expressing the atomic mass of boron contains four significant digits — the 1, the 0, the 8, and the 1. It is understood to have an uncertainty of + or – 1 in the last digit, meaning that it is really 10.81 ± 0.01. The atomic mass of sodium is given as 22.98977, a number with seven significant digits understood to mean 22.98977 ± 0.00001. Therefore, the atomic mass of sodium is known with more *certainty* than that of boron.

It is important to express numbers to the correct number of significant digits in chemical calculations and in the laboratory. Too many digits imply an accuracy in the number that does not exist, and too few do not express the number to the degree of accuracy to which it is known. The rules for expressing significant digits are summarized in Table 1.3.

Exercise: Referring to Table 1.3, give the number of significant digits and the rule(s) upon which they are based for each of the following numbers:

(a) 17.000	(b) 9.5378	(c) 7.001
(d) $50	(e) 0.00300	(f) 7400
(g) 6.207×10^{-7}	(h) 13.5269184	(i) 0.05029

Answers: (a) 5, Rule 4; (b) 5, Rule 1; (c) 4, Rule 2; (d) exact number; (e) 3, Rules 3 and 4; (f) uncertain, Rule 5; (g) 4, Rule 6; (h) 9, Rule 1; (i) 4 Rules 2 and 3

Significant Figures in Calculations

After numbers are obtained by a laboratory measurement, they are normally subjected to mathematical operations to get the desired final result. It is important that the answer have the correct number of significant figures. It should not have so few that accuracy is sacrificed or so many that an unjustified degree of accuracy is implied. The two major rules that apply, one for addition/subtraction, the other for multiplication/division, are the following:

1. In addition and subtraction, the number of digits retained to the right of the decimal point should be the same as that in the number in the calculation with the fewest such digits.

   Example: 273.591 + 1.00327 + 229.13 = 503.7247 is rounded to 503.72 because 229.13 has only two significant digits beyond the decimal.

   Example: 313.4 + 11.0785 + 229.13 = 553.6085 is rounded to 553.6 because 313.4 has only one significant digit beyond the decimal.

Table 1.3.   Rules for Use of Significant Digits

Example Number	Number of Significant Digits	Rule
11.397	5	1. Nonzero digits in a number are always significant. The 1,1,3,9, and 7 in this number are each significant.
140.039	6	2. Zeros between nonzero digits are significant. The 1, 4, 0, 0, 3, and 9 in this number are each significant.
0.00329	3	3. Zeros on the left of the first nonzero digit are not significant because they are used only to locate the decimal point. Only 3, 2, and 9 in this number are significant.
70.00	4	4. Zeros to the right of a decimal point that are preceded by a significant figure are significant. All three 0s, as well as the 7, are significant.
32 000	Uncertain	5. The number of significant digits in a number with zeros to the left, but not to the right of a decimal point (1700, 110 000) may be uncertain. Such numbers should be written in exponential notation.
3.20×10^3	3	6. The number of significant digits in a number written in exponential notation is equal to the number of significant digits in the decimal portion.
Exactly 50 dollars	Unlimited	7. Some numbers, such as the amount of money that one expects to receive when cashing a check or the number of children claimed for income tax exemptions, are defined as **exact numbers** without any uncertainty.

2. The number of significant figures in the result of multiplication/ division should be the same as that in the number in the calculation having the fewest significant figures.

Example: $\dfrac{3.7218 \times 4.019 \times 10^{-3}}{1.48} = 1.0106699 \times 10^{-2}$ is rounded to

1.01×10^{-2} (3 significant figures because 1.48 has only 3 significant figures)

Example: $\dfrac{5.27821 \times 10^7 \times 7.245 \times 10^{-5}}{1.00732} = 3.7962744 \times 10^3$ is rounded to 3.796×10^3 (4 significant figures because 7.245 has only 4 significant figures)

It should be noted that an exact number is treated in calculations as though it has an unlimited number of significant figures.

Exercise: Express each of the following to the correct number of significant figures:

(a) $13.1 + 394.0000 + 8.1937$ (b) $1.57 \times 10^{-4} \times 7.198 \times 10^{-2}$
(c) $189.2003 - 13.47 - 2.563$ (d) $221.9 \times 54.2 \times 123.008$
(e) $\dfrac{603.9 \times 21.7 \times 0.039217}{87}$ (f) $\dfrac{3.1789 \times 10^{-3} \times 7.000032 \times 10^4}{27.130921}$

(g) $100 \times 7.428 \times 7.000032 \times 10^4$ (where 100 is an exact number)

Answers: (a) 415.3, (b) 1.13×10^{-5}, (c) 173.17, (d) 1.48×10^6, (e) 5.9, (f) 8.2019, (g) 5.074×10^7

Rounding Numbers

With an electronic calculator it is easy to obtain a long string of digits that must be rounded to the correct number of significant figures. The rules for doing this are the following:

1. If the digit to be dropped is 0, 1, 2, 3, or 4, leave the last digit unchanged

   Example: Round 4.17821 to 4 significant figures
   Answer: 4.178
          Last retained digit   Digit to be dropped

2. If the digit to be dropped is 5, 6, 7, 8 or 9, increase the last retained digit by 1

   Example: Round 4.17821 to 3 significant figures
   Answer: 4.18
          Last retained digit   Digit to be dropped

Use of Three Significant Digits

It is possible to become thoroughly confused about how many significant figures to retain in an answer. In such a case it is often permissible to use 3 significant figures. Generally, this gives sufficient accuracy without doing grievous harm to the concept of significant figures.

1.8.   MEASUREMENTS AND SYSTEMS OF MEASUREMENT

The development of chemistry has depended strongly upon careful measurements. Historically, measurements of the quantities of substances reacting and produced in chemical reactions have allowed the explanation of the fundamental nature of chemistry. Exact measurements continue to be of the utmost importance in chemistry and are facilitated by increasingly sophisticated instrumentation. For example, atmospheric chemists can determine a small degree of stratospheric ozone depletion by measuring minute amounts of ultraviolet radiation absorbed by ozone with satellite-mounted instruments. Determinations of a part per trillion or less of a toxic substance in water may serve to trace the source of a hazardous pollutant. This section discusses the basic measurements commonly made in chemistry and environmental chemistry.

SI Units of Measurement

It is necessary to deal with several systems of measurement in chemistry and environmental chemistry. The most systematic of these is the **International System of Units**, abbreviated **SI**, a self-consistent set of units based upon the metric system. SI was recommended in 1960 by the General Conference of Weights and Measures to simplify and make more logical the proliferation of units in the scientific and engineering community. Table 1.4 gives the seven base SI units from which all others are derived.

Multiples of Units

Quantities expressed in science often range over many orders of magnitude (many factors of 10). For example, a mole of molecular nitrogen contains 6.02×10^{23} molecules of N_2 and very small particles in the atmosphere may be only about 1×10^{-6} meters in diameter. A convenient means of expressing very large or very small multiples is by means of **prefixes** that express the number of times that the basic unit is multiplied. Each prefix has a name and an abbreviation. The ones that are used in this book, or that are most commonly encountered, are given in Table 1.5.

Metric and English Systems of Measurement

The **metric** system has long been the standard system for scientific measurement and is the one most commonly used in this book. It was the first to use multiples of 10 to designate units that differ by orders of magnitude from a basic unit. The **English** system is still employed for many measurements encountered in normal everyday activities in the United States, including some environmental engineering measurements. Bathroom scales are still calibrated in pounds, well

Table 1.4.   Units of the International System of Units (SI)

Physical Quantity Measured	Unit Name	Unit Symbol	Definition
Base units			
Length	metre	m	Distance travelled by light in a vacuum in $\dfrac{1}{299\ 792\ 458}$ second
Mass	kilogram	kg	Mass of a platinum-iridium block located at the International Bureau of Weights and Measures at Sevres, France
Time	second	s	9 192 631 770 periods of a specific line in the microwave spectrum of the cesium-133 isotope
Temperature	kelvin	K	1/273.16 the temperature interval between absolute zero and the triple point of water at 273.16 K (0.01°C)
Amount of substance	mole	mol	Amount of substance containing as many entities (atoms, molecules) as there are atoms in exactly 0.012 kilograms of the carbon-12 isotope
Electric current	ampere	A	—
Luminous intensity	candela	cd	—
Examples of derived units			
Force	newton	N	Force required to impart an acceleration of 1 m/s^2 to a mass of 1 kg
Energy (heat)	joule	J	Work performed by 1 newton acting over a distance of 1 meter
Pressure	pascal	Pa	Force of 1 newton acting on an area of 1 square meter

Table 1.5.   Prefixes Commonly Used to Designate Multiples of Units

Prefix	Basic Unit is Multiplied By	Abbreviation
Mega	1 000 000(10^6)	M
Kilo	1 000(10^3)	k
Hecto	100(10^2)	h
Deka	10(10)	da
Deci	0.1 (10^{-1})	d
Centi	0.01 (10^{-2})	c
Milli	0.001 (10^{-3})	m
Micro	0.000 001 (10^{-6})	μ
Nano	0.000 000 001 (10^{-9})	n
Pico	0.000 000 000 001 (10^{-12})	p

depths may be given in feet, and quantities of liquid wastes are frequently expressed as gallons or barrels. Furthermore, English units of pounds, tons, and gallons are still commonly used in commerce, even in the chemical industry. Therefore, it is still necessary to have some familiarity with this system; conversion factors between it and metric units are given in this book.

1.9.   UNITS OF MASS

Mass expresses the degree to which an object resists a change in its state of rest or motion and is proportional to the amount of matter in the object. **Weight** is the gravitational force acting upon an object and is proportional to mass. An object weighs much less in the gravitational force on the Moon's surface than on Earth, but the object's mass is the same in both places. Figure 1.7 illustrates this effect. Although mass and weight are not usually distinguished from each other

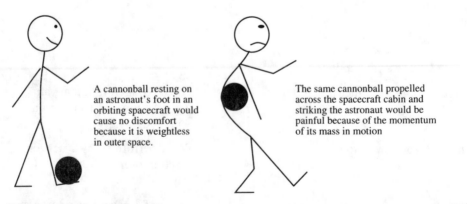

A cannonball resting on an astronaut's foot in an orbiting spacecraft would cause no discomfort because it is weightless in outer space.

The same cannonball propelled across the spacecraft cabin and striking the astronaut would be painful because of the momentum of its mass in motion

Figure 1.7.   An object maintains its mass even in the weightless surroundings of outer space.

Table 1.6. Metric Units of Mass

Unit of Mass	Abbreviation	Number of Grams	Example of Use for Measurement
Megagram or metric ton	Mg	10^6	Quantities of industrial chemicals (1 Mg = 1.102 short tons)
Kilogram	kg	10^3	Body weight and other quantities for which the pound has been commonly used (1 kg = 2.2046 lb)
Gram	g	1	Mass of laboratory chemicals (1 ounce = 28.35 g and 1 lb = 453.6 g)
Milligram	mg	10^{-3}	Small quantities of chemicals
Microgram	µg	10^{-6}	Quantities of toxic pollutants

in everyday activities, it is important for the science student to be aware of the differences between them.

The **gram** (g) with a mass equal to 1/1000 that of the SI kilogram (see Table 1.4) is the fundamental unit of mass in the metric system. Although the gram is a convenient unit for many laboratory-scale operations, other units that are multiples of the gram are often more useful for expressing mass. The names of these are obtained by affixing the appropriate prefixes from Table 1.5 to "gram." Global burdens of atmospheric pollutants may be given in units of teragrams, each equal to 1×10^{12} grams. Significant quantities of toxic water pollutants may be measured in micrograms (1×10^{-6} grams). Large-scale industrial chemicals are marketed in units of megagrams (Mg). This quantity is also known as a metric ton, or tonne, and is somewhat larger (2205 lb) than the 2000-lb short ton still used in commerce in the United States. Table 1.6 summarizes some of the more commonly used metric units of mass and their relationship to some English units.

1.10. UNITS OF LENGTH

Length in the metric system is expressed in units based upon the **meter**, m (SI spelling *metre*, Table 1.4). A meter is 39.37 inches long, slightly longer than a yard.

A kilometer (km) is equal to 1000 m and, like the mile, is used to measure relatively great distances. A centimeter (cm) equal to 0.01 m is often convenient to designate lengths such as the dimensions of laboratory instruments. There are 2.540 cm per inch, and the cm is employed to express lengths that would be given

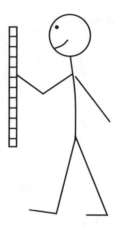

Figure 1.8. The meter stick is a common tool for measuring length.

in inches in the English system. The micrometer (μm) is about the same length as that of a typical bacterial cell. The μm is also used to express wavelengths of infrared radiation by which Earth reradiates solar energy back to outer space. The nanometer (nm), equal to 10^{-9} m, is a convenient unit for the wavelength of visible light, which ranges from 400 to 800 nm. Atoms are even smaller than 1 nm; their dimensions used to be expressed in angstrom units (1 Å = 0.1 nm), but are now commonly given in picometers (pm, 10^{-12} m). Table 1.7 lists the common metric units of length, along with some examples of their use and some related English units.

Table 1.7. Metric Units of Length

Unit of Length	Abbreviation	Number of Meters	Example of Use for Measurement
Kilometer	km	10^3	Geographic distance (1 mile = 1.609 km)
Meter	m	1	Standard metric unit of length (1 m = 1.094 yards)
Centimeter	cm	10^{-2}	Used in place of inches (1 inch = 2.54 cm)
Millimeter	mm	10^{-3}	Same order of magnitude as sizes of letters on this page
Micrometer	μm	10^{-6}	Size of typical bacteria
Nanometer	nm	10^{-9}	Measurement of light wavelength

1.11.   UNITS OF VOLUME

The basic metric unit of **volume** is the **liter**. This volume is defined in terms of metric units of length. A liter is the volume of a decimeter cubed, that is, 1 L = 1 dm³ (a decimeter is 0.1 meter, about 4 inches). A milliliter (mL) is the same volume as a centimeter cubed (cm³ or cc), and a liter is 1000 cm³. A kiloliter, usually designated as a cubic meter (m³), is a common unit of measurement for the volume of air. For example, standards for human exposure to toxic substances in the workplace are frequently given in units of $\mu g/m^3$. Table 1.8 gives some common metric units of volume. The measurement of volume is one of the more frequently performed routine laboratory measurements; Figure 1.10 shows some of the more common tools for laboratory volume measurement of liquids.

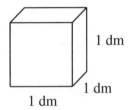

1 dm

1 dm

1 dm

Figure 1.9. A cube that is 1 decimeter to the side has a volume of 1 liter.

Table 1.8.   Metric Units of Volume

Unit of Volume	Abbreviation	Number of Liters	Example of Use for Measurement
Kiloliter or cubic meter	kL	10^3	Volumes of air in air pollution studies
Liter	L	1	Basic metric unit of volume (1 liter = 1 dm³ = 1.057 quarts; 1 cubic foot = 28.32 L)
Milliliter	mL	10^{-3}	Equal to 1 cm³. Convenient unit for laboratory volume measurements
Microliter	μL	10^{-6}	Used to measure very small volumes for chemical analysis

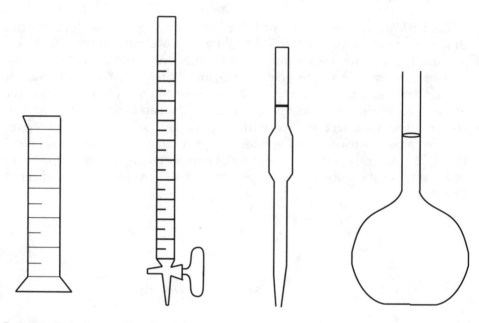

Graduated cylinder for
approximate measure-
ment of volume

Buret for accurate
measurement of
varying volumes

Pipet for quantitative
transfer of
solution volumes

Volumetric flask
containing a specific,
accurately known volume

Figure 1.10.   Glassware for volume measurement in the laboratory.

1.12.   TEMPERATURE, HEAT, AND ENERGY

Temperature Scales

In chemistry, temperatures are usually expressed in metric units of **Celsius degrees**, °C. On this scale, water freezes at 0°C and boils at 100°C. The **Fahrenheit** temperature scale still used for some nonscientific temperature measurements in the U.S. defines the freezing temperature of water at 32 degrees Fahrenheit (°F) and boiling at 212°F, a range of 180°F. Therefore, a span of 100 Celsius degrees is equivalent to one of 180 Fahrenheit degrees, so that each °C is equivalent to 1.8°F.

The most fundamental temperature scale is the **Kelvin** or **absolute** scale, for which zero is the lowest attainable temperature. A unit of temperature on this scale is equal to a Celsius degree, but it is called a **kelvin**, abbreviated K, not a degree. Kelvin temperatures are designated as K, not °K. The value of absolute zero on the Kelvin scale is –273.15°C, so that the kelvin temperature is always a number 273.15 (usually rounded to 273) higher than the Celsius temperature.

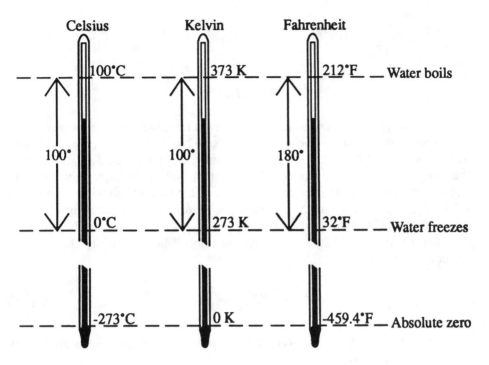

Figure 1.11. Comparison of temperature scales.

Thus water boils at 373 K and freezes at 273 K. The relationships among Kelvin, Celsius, and Fahrenheit temperatures are illustrated in Figure 1.11.

Converting from Fahrenheit to Celsius

With Figure 1.11 in mind, it is easy to convert from one temperature scale to another. Examples of how this is done are given below:

Example: What is the Celsius temperature equivalent to room temperature of 70°F?

Answer: Step 1. Subtract 32 Fahrenheit degrees from 70 Fahrenheit degrees to get the number of Fahrenheit degrees above zero. This is done because 0 on the Celsius scale is at the freezing point of water.

Step 2. Multiply the number of Fahrenheit degrees above the freezing point of water obtained above by the number of Celsius degrees per Fahrenheit degree.

$$°C = \frac{1.00°C}{1.80°F} \times (70°F - 32°F) = \frac{1.00°C}{1.80°F} \times 38°F = 21.1°C \qquad (1.12.1)$$

↑   ↖

Factor for conversion   Number of °F
from °F to °C   above freezing

In working the above example it is first noted (as is obvious from Figure 1.11) that zero on the Celsius scale corresponds to 32°F on the Fahrenheit scale. So 32°F is subtracted from 70°F to give the number of Fahrenheit degrees by which the temperature is above the freezing point of water. These are converted to Celsius degrees above the freezing point of water by multiplying by the factor 1.00°C/1.80°F. The origin of this factor is readily seen by referring to Figure 1.11 and observing that there are 100°C between the freezing and boiling temperatures of water and 180°F over the same range. Mathematically, the equation for converting from °F to °C is

$$°C = \frac{1.00°C}{1.80°F} \times (°F - 32) \qquad (1.12.2)$$

Example: What is the Celsius temperature corresponding to normal body temperature of 98.6°F?

Answer: From Equation 1.12.2

$$°C = \frac{1.00°C}{1.80°F} \times (98.6°F - 32°F) = 37.0°C \qquad (1.12.3)$$

Example: What is the Celsius temperature corresponding to –5°F?

Answer: From Equation 1.12.2

$$°C = \frac{1.00°C}{1.80°F} \times (-5°F - 32°F) = °C = -20.6°C \qquad (1.12.4)$$

Converting from Celsius to Fahrenheit

To convert from Celsius to Fahrenheit first requires multiplying the Celsius temperature by 1.80°F/1.00°C to get the number of Fahrenheit degrees above the freezing temperature of 32°F, then adding 32°F.

Example: What is the Fahrenheit temperature equivalent to 10°C?

Answer: Step 1. Multiply 10°C by 1.80°F/1.00°C to get the number of Fahrenheit degrees above the freezing point of water.

Step 2. Since the freezing point of water is 32°F, add 32°F to the result of Step 1.

$$°F = \frac{1.80°F}{1.00°C} \times °C + 32°F = \frac{1.80°F}{1.00°C} \times 10°C + 32°F = 50°F \quad (1.12.5)$$

The formula for converting °C to °F is simply

$$°F = \frac{1.80°F}{1.00°C} \times °C + 32°F \quad\quad (1.12.6)$$

To convert from °C to K, simply add 273 to the Celsius temperature. To convert from K to °C, subtract 273 from K. All of the conversions discussed here can be deduced without memorizing any equations by remembering that the freezing point of water is 0°C, 273 K, and 32°F, whereas the boiling point is 100°C, 373 K, and 212°F.

Melting Point and Boiling Point

In the preceding discussion, the melting and boiling points of water were both used in defining temperature scales. These are important thermal properties of any substance. For the present, **melting temperature** may be defined as the temperature at which a substance changes from a solid to a liquid. **Boiling temperature** is defined as the temperature at which a substance changes from a liquid to a gas. More exacting definitions of these terms, particularly boiling temperature, are given later in the book.

Heat and Energy

As illustrated in Figure 1.12, when two objects at different temperatures are placed in contact with each other, the warmer object becomes cooler and the

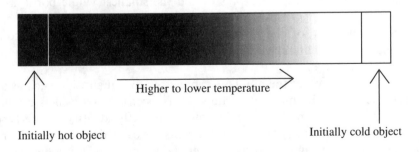

Higher to lower temperature

Initially hot object

Initially cold object

Figure 1.12. Heat energy flow from a hot to a colder object.

cooler one warmer until they reach the same temperature. This occurs because of a flow of energy between the objects. Such a flow is called **heat**.

The SI unit of heat is the **joule** (J, see Table 1.4). The kilojoule (1 kJ = 1000 J) is a convenient unit to use to express energy values in laboratory studies. The metric unit of energy is the **calorie** (cal), equal to 4.184 J. Throughout the liquid range of water, essentially 1 calorie of heat energy is required to raise the temperature of 1 g of water by 1°C. The **calories** most people hear about are those used to express energy values of foods and are actually kilocalories (1 kcal = 4.184 kJ).

1.13. PRESSURE

Pressure is force per unit area. The SI unit of pressure is the pascal (Pa), defined in Table 1.4. The kilopascal (1 kPa = 1000 Pa) is often a more convenient unit of pressure to use than is the pascal.

Like many other quantities, pressure has been plagued with a large number of different kinds of units. One of the more meaningful of these is the **atmosphere** (atm), which is the average pressure exerted by air at sea level. One atmosphere is equal to 101.3 kPa or 14.7 lb/in². The latter means that an evacuated cube, 1 inch to the side, has a force of 14.70 lb exerted on each side due to atmospheric pressure. It is also the pressure that will hold up a column of liquid mercury metal 760 mm long, as shown in Figure 1.13. Such a device used to measure atmospheric pressure is called a **barometer**, and the mercury barometer was the first instrument used to measure pressures with a high degree of accuracy. Consequently, the practice developed of expressing pressure in units of **millimeters of mercury** (mm Hg), where 1 mm of mercury is a unit called the **torr**.

Pressure is an especially important variable describing gas behavior. The temperature/pressure/volume relationships of gases are discussed in Chapters 2 and 18.

1.14. UNITS AND THEIR USE IN CALCULATIONS

Most numbers used in chemistry are accompanied by a **unit** that tells the type of quantity that the number expresses and the smallest whole portion of that quantity. For example, "36 liters" denotes that a volume is expressed and the smallest whole unit of the volume is 1 liter. The same quantity could be expressed as 360 deciliters, where the number is multiplied by 10 because the unit is only 1/10 as large.

Except in cases where the numbers express relative quantities, such as atomic masses relative to the mass of carbon-12 or specific gravity, it is essential to include units with numbers. In addition to correctly identifying the type and magnitude of the quantity expressed, the units are carried through mathematical operations. The wrong unit in the answer shows that something has been done wrong in the calculation and it must be checked.

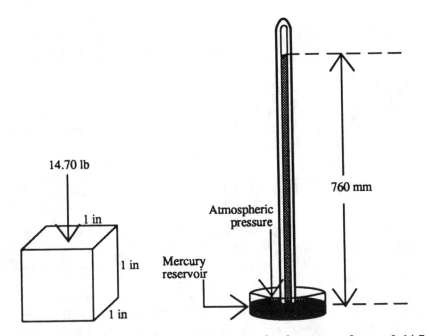

14.70 lb

1 in

1 in

1 in

760 mm

Atmospheric
pressure

Mercury
reservoir

Figure 1.13.   Average atmospheric pressure at sea level exerts a force of 14.7 pounds on an inch-square surface. This corresponds to a pressure sufficient to hold up a 760 mm column of mercury.

Unit Conversion Factors

Most chemical calculations involve calculating one type of quantity, given another, or converting from one unit of measurement to another. For example, in the chemical reaction

$$2H_2 + O_2 \rightarrow 2H_2O$$

someone might want to calculate the number of grams of H_2O produced when 3 g of H_2 react or they might want to convert the number of grams of H_2 to ounces. These kinds of calculations are carried out with **unit conversion factors**. Suppose for example that the mass of a 160 lb person is to be expressed in kilograms; the person doing the calculation does not know the factor to convert from lb to kg, but does know that a 551 lb motorcycle has a mass of 250 kg. From this information the needed unit conversion factor can be derived and the calculation completed as follows:

Mass of person in kg = 160 lb × unit conversion factor
(problem to be solved)   (1.14.1)

250 kg = 551 lb (known relationship between lb and kg)   (1.14.2)

$$\frac{250 \text{ kg}}{551 \text{ lb}} = \frac{551 \text{ lb}}{551 \text{ lb}} = 1 \quad \text{(The unit of kg is left on top because it} \quad (1.14.3)$$
$$\text{is the unit needed; division is by 551 lb.)}$$

$$\frac{0.454 \text{ kg}}{1.00 \text{ lb}} = 1 \text{ (The unit conversion factor expressed in the form} \quad (1.14.4)$$
$$250 \text{ kg/551 lb. could have been used, but dividing}$$
$$250 \text{ by } 551 \text{ gives the unit conversion factor}$$
$$\text{in a more concise form.)}$$

$$\text{Mass of person} = 160 \text{ lb} \times \frac{0.454 \text{ kg}}{1.00 \text{ lb}} = 72.6 \text{ kg} \quad (1.14.5)$$

It is permissible to multiply 160 lb by 0.454 kg/1.00 lb because, as shown by Equation 1.14.4, this unit conversion factor has a value of exactly 1. Any quantity can be multiplied by 1 without changing the quantity itself, only the units in which it is expressed.

As another example of the use of a unit conversion factor, calculate the number of liters of gasoline required to fill a 12-gallon fuel tank, given that there are 4 gallons in a quart and that a volume of 1 liter is equal to that of 1.057 quarts. This problem can be worked by first converting gallons to quarts, then quarts to liters. For the first step the unit conversion factor is

$$1 \text{ gal} = 4 \text{ qt} \quad (1.14.6)$$

$$\frac{1 \text{ gal}}{1 \text{ gal}} = \frac{4 \text{ qt}}{1 \text{ gal}} = 1 \quad \text{(Enables conversion from gallons to quarts)} \quad (1.14.7)$$

$$1.057 \text{ qt} = 1 \text{ L} \quad (1.14.8)$$

$$\frac{1.057 \text{ qt}}{1.057 \text{ qt}} = \frac{1 \text{ L}}{1.057 \text{ qt}} = 1 \text{ (Enables conversion from quarts to liters) } (1.14.9)$$

Both unit conversion factors are used to calculate the capacity of the tank in liters:

$$\text{Tank capacity} = 12 \text{ gal} \times \frac{4 \text{ qt}}{1 \text{ gal}} \times \frac{1 \text{ L}}{1.057 \text{ qt}} = 45.4 \text{ L} \quad (1.14.10)$$

Cancellation of Units

The preceding examples show that units are cancelled in mathematical operations, just as numbers may be. When the same unit appears both above and below the line in a mathematical operation such as lb in

$$160 \text{ lb} \times \frac{0.454 \text{ kg}}{1.00 \text{ lb}}$$

the unit simply cancels.

Calculation of Some Unit Conversion Factors

In Tables 1.6 to 1.8 several values of units are given that enable conversion between metric and English units. For example, Table 1.6 states that a mega-gram (Mg, metric ton) is equal to 1.102 short tons (T). By using this equality to give the correct unit conversion factors it is easy to calculate the number of metric tons in a given number of short tons of material or vice versa. To do this, first write the known equality:

$$1 \text{ Mg} = 1.102 \text{ T} \qquad (1.14.11)$$

If the number of Mg is to be calculated given a mass in T, the unit conversion factor needed is

$$\frac{1 \text{ Mg}}{1.102 \text{ T}} = \frac{1.102 \text{ T}}{1.102 \text{ T}} = 1 \qquad (1.14.12)$$

leaving Mg on top. Suppose, for example, that the problem is to calculate the mass in Mg of a 3521 T shipment of industrial soda ash. The calculation is simply

$$3521 \text{ T} \times \frac{1 \text{ Mg}}{1.102 \text{ T}} = 3195 \text{ Mg} \qquad (1.14.13)$$

If the problem had been to calculate the number of T in 789 Mg of copper ore, the following steps would be followed:

$$1.102 \text{ T} = 1 \text{ Mg}, \frac{1.102 \text{ T}}{1 \text{ Mg}} = \frac{1 \text{ Mg}}{1 \text{ Mg}} = 1, \qquad (1.14.14)$$

$$789 \text{ Mg} \times \frac{1.102 \text{ T}}{1 \text{ Mg}} = 869 \text{ T copper ore} \qquad (1.14.15)$$

Table 1.9 gives some unit conversion factors calculated from the information given in Tables 1.6 to 1.8 and in preceding parts of this chapter. Note that in each case, two unit conversion factors are calculated; the one that is used depends on the units that are required for the answer.

Table 1.9.   Examples of Some Unit Conversion Factors

Equality	Conversion Factors	
1 kg = 2.2046 lb	$\dfrac{1 \text{ kg}}{2.2046 \text{ lb}} = 1$	$\dfrac{2.2046 \text{ lb}}{1 \text{ kg}} = 1$
1 oz = 28.35 g	$\dfrac{1 \text{ oz}}{28.35 \text{ g}} = 1$	$\dfrac{28.35 \text{ g}}{1 \text{ oz}} = 1$
1 mi = 1.609 km	$\dfrac{1 \text{ mi}}{1.609 \text{ km}} = 1$	$\dfrac{1.609 \text{ km}}{1 \text{ mi}} = 1$
1 in = 2.54 cm	$\dfrac{1 \text{ in}}{2.54 \text{ cm}} = 1$	$\dfrac{2.54 \text{ cm}}{1 \text{ in}} = 1$
1 L = 1.057 qt	$\dfrac{1 \text{ L}}{1.057 \text{ qt}} = 1$	$\dfrac{1.057 \text{ qt}}{1 \text{ L}} = 1$
1 cal = 4.184 J	$\dfrac{1 \text{ cal}}{4.184 \text{ J}} = 1$	$\dfrac{4.184 \text{ J}}{1 \text{ cal}} = 1$
1 atm = 101.4 kPa	$\dfrac{1 \text{ atm}}{101.4 \text{ kPa}} = 1$	$\dfrac{101.4 \text{ kPa}}{1 \text{ atm}} = 1$

CHAPTER SUMMARY

The chapter summary below is presented in a programmed format to review the main points covered in this chapter. It is used most effectively by filling in the blanks, referring back to the chapter as necessary. The correct answers are given at the end of the summary.

Chemistry is defined as (1)_____. Environmental chemistry is (2)_____

_____

_____

Toxicological chemistry is defined as (3)_____

_____

All matter is composed of only about a hundred fundamental kinds of matter called (4)_____, each composed of very small entities called (5)_____. The three major subatomic particles and their charges are (6)_____

_____.

Of these, the two that have relatively high masses are contained in the

(7)_____ of the atom. The subatomic particles with a relatively low mass are contained in (8)_____ in the atom.

The number of protons in the nucleus of each atom of an element is the (9)_____ of the element. Each element is represented by an abbreviation called a (10)_____. In addition to atomic number, name, and chemical symbol, each element has a characteristic (11)_____ _____. Atoms of most elements consist of two or more isotopes that have different (12)_____. An arrangement of the elements in a manner that reflects their recurring behavior with increasing atomic number is the (13)_____ in which elements with similar chemical properties are called (14)_____ and are contained in (15)_____ in the periodic table.

Instead of existing as atoms, elemental hydrogen consists of (16)_____ _____, each consisting of (17)_____ linked by a (18)_____. Water is not an element, but is a (19)_____, for which the (20)_____ is H_2O. Species consisting of electrically charged atoms or groups of atoms are called (21)_____. Those with positive charges are called (22)_____ and those with negative charges are (23)_____. Compounds made of these kinds of entities are held together by (24)_____.

The average mass of all molecules of a compound is its (25)_____, which is calculated by (26) _____

_____

_____

(27)_____ occur when substances are changed to other substances through the breaking and formation of chemical bonds and are written as (28)_____. To be correct, these must be (29)_____. In them, the arrow is read as (30)_____ and separates the (31)_____ from the (32)_____.

Very large or small numbers are conveniently expressed in (33)_____ _____, which is the product of a (34)_____ _____ with a value equal to or greater than (35)_____ and less than (36)_____ multiplied times a (37)_____. In such a notation, 3,790,000 is expressed as (38)_____ and 0.000 000 057 is expressed as (39)_____.

The accuracy of a number is shown by how many (40)_____ _____ it contains. Nonzero digits in a number are always (41)_____. Zeros between nonzero digits are (42)_____. Zeros on the left of the first nonzero digit are (43)_____. Zeros to the right of a decimal point that are preceded by a significant figure are (44)_____. The number of significant digits in a number written in exponential notation is equal to (45)_____. Some numbers, such as the amount of money that one expects to receive when cashing a check are defined as (46)_____. In addition and subtraction, the number of digits retained to the right of the decimal point should be (47)_____.

_____ The number of significant figures in the result of multiplication/division should be (48)_____ _____.

In rounding numbers, if the digit to be dropped is 0, 1, 2, 3, or 4, (49)_____, whereas if the digit to be dropped is 5, 6, 7, 8, or 9, (50)_____.

A self-consistent set of units based upon the metric system is the (51)_____. (52)_____ is proportional to the amount of matter in the object, the metric unit for which is the (53)_____. Length in the metric system is expressed in units based upon the (54)_____. The basic metric unit of volume is the (55)_____. On the Celsius temperature scale water

freezes at (56)_____ and boils at (57)_____. On the Fahrenheit temperature scale water freezes at (58)_____ and boils at (59)_____. On the Kelvin temperature scale water freezes at (60)_____ and boils at (61)_____. Melting temperature may be defined as (62)_____. Boiling temperature is defined as (63)_____. Energy that flows from a warmer to a colder object is called (64)_____, commonly expressed in units of (65)_____ or (66)_____. (67)_____ is force per unit area, some of the common units for which are (68)_____
_____.

A unit after a number tells the (69)_____ that the number expresses and the smallest (70)_____. The quantity 0.454 kg/1 lb is an example of (71)_____.

Answers

1. the science of matter
2. that branch of chemistry which deals with the origins, transport, reactions, effects, and fates of chemical species in the water, air, terrestrial, and living environments
3. the chemistry of toxic substances with emphasis upon their interaction with biologic tissue and living systems
4. elements
5. atoms
6. positively charged protons, negatively charged electrons, and uncharged (neutral) neutrons
7. nucleus
8. a cloud of negative charge
9. atomic number
10. chemical symbol
11. atomic mass
12. numbers of neutrons in their nuclei
13. periodic table
14. groups of elements
15. vertical columns
16. molecules
17. 2 H atoms
18. chemical bond

19. chemical compound
20. chemical formula
21. ions
22. cations
23. anions
24. ionic bonds
25. molecular mass
26. multiplying the atomic mass of each element by the relative number of atoms of the element, then adding all the values obtained for each element in the compound
27. Chemical reactions
28. chemical equations
29. balanced
30. yields
31. reactants
32. products
33. exponential notation
34. digital number
35. exactly 1
36. exactly 10
37. power of 10
38. 3.79×10^6
39. 5.7×10^{-8}
40. significant figures or significant digits
41. significant
42. significant
43. not significant
44. significant
45. the number of significant digits in the decimal portion
46. exact numbers
47. the same as that in the number in the calculation with the fewest such digits
48. the same as that in the number in the calculation having the fewest significant figures
49. leave the last digit unchanged
50. increase the last retained digit by 1
51. International System of Units, SI
52. Mass
53. gram
54. meter
55. liter
56. 0°C
57. 100°C
58. 32°F
59. 212°F

60. 273 K
61. 373 K
62. the temperature at which a substance changes from a solid to a liquid
63. the temperature at which a substance changes from a liquid to a gas
64. heat
65. joules
66. calories
67. Pressure
68. atmosphere, torr, mm Hg, lb/in^2
69. type of quantity
70. whole portion of that quantity
71. a unit conversion factor

QUESTIONS AND PROBLEMS

1. Consider the following atom:

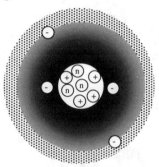

   How many electrons, protons, and neutrons does it have? What is its atomic number? Give the name and chemical symbol of the element of which it is composed.

2. What distinguishes a radioactive isotope from a "normal" stable isotope?

3. Why is the periodic table so named?

4. Match the following:
   1. O_2 (a) Element consisting of individual atoms
   2. NH_3 (b) Element consisting of chemically bonded atoms
   3. Ar   (c) Ionic compound
   4. NaCl   (d) Covalently bound compound

5. After examining Figure 1.6, consider what might happen when an atom of sodium (Na), atomic number 11, loses an electron to an atom of fluorine, (F), atomic number 9. What kinds of particles are formed by this transfer of

a negatively charged electron? Is a chemical compound formed? What is it called?

6. Give the chemical formula and molecular mass of the molecule represented below:

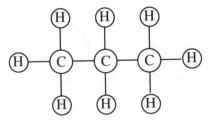

7. Calculate the molecular masses of (a) C_2H_2, (b) N_2H_2, (c) Na_2O, (d) O_3 (ozone), (e) PH_3, (f) CO_2.

8. Is the equation, $H_2 + O_2 \rightarrow H_2O$, a balanced chemical equation? Explain. Point out the reactants and products in the equation.

9. Write each of the following in exponential form to three significant figures: (a) 321,000, (b) 0.000 005 29, (c) 5170, (d) 000 000 000 000 784, (e) 86,300,000,000,000.

10. Write each of the following in decimal form: (a) 7.49×10^3, (b) 9.6×10^{-5}, (c) 1.16×10^{21}, (d) 4.47×10^{-17}, (e) 2.93×10^{13}.

11. Without using a calculator, calculate the sum $4.13 \times 10^3 + 8.76 \times 10^2 + 1.22 \times 10^4$ expressed to three significant figures in the correct exponential notation.

12. Without using a calculator, calculate the sum $4.13 \times 10^{-4} + 8.76 \times 10^{-3} + 1.22 \times 10^{-2}$ expressed to three significant figures in the correct exponential notation.

13. Without using a calculator for the exponential portions, calculate $1.39 \times 10^{-2} \times 9.05 \times 10^8 \times 3.11 \times 10^4$.

14. Without using a calculator for the exponential portions, calculate

$$\frac{9.05 \times 10^{-6} \times 3.19 \times 10^3}{4.02 \times 10^5 \times 1.93 \times 10^{-7}}$$

15. Match the following numbers, with the significant figures for each given in parentheses with the rule for assigning significant figures that applies to each:

    1. 0.00027(2)   (a) Nonzero digits in a number are always significant.
    2. 7.28139(6)   (b) Zeros between nonzero digits are significant.
    3. 7.4×10^3(2)   (c) Zeros on the left of the first nonzero digit are not significant.
    4. $50(infinite)   (d) Zeros to the right of a decimal point that are preceded
    5. 81.000   by a significant figure are significant.
    6. 40.007(5)   (e) The number of significant digits in a number written in exponential notation is equal to the number of significant digits in the decimal portion.
      (f) Some numbers are defined as exact numbers without any uncertainty.

16. Using the appropriate rules round each of the following to the correct number of significant digits: (a) 923.527 + 3.02891 + 729.29, (b) 273.591 + 12.72489 + 0.1324, (c) 473 + 9.3827 + 349.17, (d) 693.59102 + 9.00327 + 229.461853.

17. Using the appropriate rules, round each of the following to the correct number of significant digits:
    (a) $3.52 \times 8.02891 \times 729$, (b) $4.52 \times 10^3 \times 8.021 \times 0.5779$,

    (c) $\dfrac{7.7218 \times 10^7 \times 4.019 \times 10^{-3}}{1.48 \times 10^{-5}}$, (d) $\dfrac{7.8 \times 6.028 \times 10^{-3}}{4.183 \times 10^{-5} \times 2.19 \times 10^5}$

18. Round each of the following properly: (a) 7.32987 to 3 places, (b) 1.193528 to 4 places, (c) 7.1382×10^3 to 2 places (d) 9.04557×10^{-17} to 4 places, (e) 71235.801 to 3 places, (f) 5.8092355 to 3 places.

19. Match each of the following units on the left with its description from the column on the right, below:
    1. Mole   (a) 1/273.16 the temperature interval between absolute zero and the triple point of water at 273.16 K (0.01°C)
    2. Metre
    3. Gram   (b) Metric unit of volume
    4. Kelvin   (c) SI unit for amount of substance
    5. Liter   (d) Metric unit of mass
      (e) Distance travelled by light in a vacuum
    in $\dfrac{1}{299\ 792\ 458}$ second

20. Match the following:
    1. Centi   (a) $\times 10^{-3}$

2. Kilo   (b) × 10^{-6}
3. Deka   (c) × 1000
4. Micro   (d) × 10
5. Milli   (e) × 0.01

21. Denote each of the following as characteristic of mass (m) or characteristic of weight (w):
(   ) Varies with gravity
(   ) Degree to which an object resists a change in its state of rest or motion
(   ) Direct measure of the amount of matter in the object
(   ) Different on the Moon's surface than on Earth

22. Match the following units on the left, below, with the quantity they are most likely to be used to express:
1. Mg   (a) Quantities of toxic pollutants
2. μg   (b) Quantities of large-scale industrial chemicals
3. kg   (c) Quantities of laboratory chemicals
4. g   (d) Global burdens of atmospheric pollutants
5. Teragrams   (e) Body mass

23. Calculate (a) the number of grams in 1.56 pounds, (b) the number of kilograms in a 2000-pound ton, (c) the number of g in 2.14 kg, (d) the number of atmospheric dust particles, each weighing an average of 2.56 μg, to make up an ounce of dust particles

24. Distinguish between a meter and a metre.

25. How tall is a 6-foot person in cm?

26. Estimate approximately how many bacterial cells would have to be laid end-to-end to reach an inch.

27. Match the following units on the left, below, with the quantity they are most likely to be used to measure:
1. km   (a) Distance between this line and the line directly below
2. m   (b) Distance run by an athlete in 5 seconds
3. Nanometer   (c) Distance travelled by an automobile in 1 hour
4. Centimeter   (d) Dimensions of this book
5. mm   (e) Wavelength of visible light

28. Explain how metric units of volume may be defined in terms of length.

29. Recalling the appropriate formula from elementary geometry, what is the volume in liters of a round tank with a radius of 39.0 cm and a depth of 15.0 cm?

30. Consider gasoline at a price of $1.45 per gallon. What is its equivalent price in dollars per liter ($/L)?

31. Match the following units on the left, below, with the quantity they are most likely to be used to measure:
    1. Milliliter   (a) Volume of milk
    2. Kiloliter   (b) Volume of a laboratory chemical
    3. Microliter   (c) Volume of air in air pollution studies
    4. Liter   (d) Volume of chemical reagent in a syringe for chemical analysis

32. Convert each of the following Fahrenheit temperatures to Celsius and Kelvin: (a) 237°F, (b) 105°F, (c) 17°F, (d) 2°F, (e) –32°F, (f) –5°F, (g) 31.2°F.

33. Convert each of the following Celsius temperatures to Fahrenheit: (a) 237°C, (b) 75°C, (c) 17°C, (d) 100°C, (e) –32°C, (f) –40°C, (g) –11°C.

34. Convert each of the following Celsius temperatures to kelvin: (a) 237°C, (b) 75°C, (c) 48°C, (d) 100°C, (e) –0°C, (f) –40°C, (g) –200°C.

35. Calculate the value of temperature in °F that is numerically equal to the temperature in °C.

36. The number 1.8 can be used in making conversions between Fahrenheit and Celsius temperatures. Explain how it is used and where it comes from in this application.

37. Calculate how many calories there are in 1 joule.

38. Match each of the following pertaining to units of pressure:
    1. pascal   (a) Based on a column of liquid
    2. atm   (b) Takes 14.7 to equal 1 atm
    3. mm Hg   (c) SI unit
    4. lb/in^2 (d) Essentially 1 for air at sea level.

39. Try to explain why pressure is a more important variable for gases than for liquids.

40. The pressure in a typical automobile tire is supposed to be 35 lb/in^2 (above normal atmospheric pressure). Calculate the equivalent pressure in (a) pascal, (b) atm, and (c) torr.

41. Knowing that there are 12 inches per foot, calculate the normal pressure of the atmosphere in lb/ft^2.

42. Atmospheric pressure readings on weather reports in the U.S. are typically given as 29–30 inches. Speculate regarding what such a reading may mean.

43. An analyst reported some titration data as 34.52 mL. What two things are stated by this expression?

44. Why is it not wrong to give the atomic mass of aluminum as 26.98, even though a unit is not specified for the mass?

45. Explain what is meant by a unit conversion factor. How is such a factor used? Why may quantities be multiplied by a proper unit conversion factor without concern about changing the magnitude of the quantity?

46. Consider only the following information: *An object with a mass of 1 kg also has a mass of 2.2046 lb. A pound contains 16 ounces. A piece of chalk 2 inches long is also 5.08 cm long. There are 36 inches in 1 yard. A 1-liter volumetric flask contains 1.057 quarts. A cubic centimeter is the same vol-ume as a milliliter.* From this information show how to calculate unit conversion factors to convert the following: (a) From yards to meters, (b) from ounces to grams, (c) from quarts to deciliters, (d) from cubic inches to cubic centimeters.

47. Using unit conversion factors, calculate the following:
    (a) A pressure in inches of mercury equivalent to 1 atm pressure
    (b) The mass in metric tons of 760,000 tons of contaminated soil
    (c) The number of cubic meters in a cubic mile of atmospheric air
    (d) The cost of 100 liters of gasoline priced at $1.33/gallon.
    (e) The number of kilograms of cheese in 200 ounces of this food
    (f) The pressure in kPa equivalent to 5.00 atm pressure

LITERATURE CITED

1. Manahan, S. E., *Environmental Chemistry*, 5th ed. (Chelsea, MI: Lewis Publishers, Inc., 1991).
2. Manahan, S. E., *Toxicological Chemistry*, 2nd ed. (Chelsea, MI: Lewis Publishers, Inc., 1992).

2   MATTER AND PROPERTIES OF MATTER

2.1.   WHAT IS MATTER?

In Chapter 1, chemistry was defined as the science of **matter,** *anything that has mass and occupies space*. This chapter deals specifically with matter—what it is, how it acts, what its properties are. Most of this book is concerned with chemical processes—those in which chemical bonds are broken and formed to produce different substances—and the *chemical properties* that they define. However, before going into chemical phenomena in any detail, it is helpful to consider matter in its bulk form, aside from its chemical behavior. Nonchemical aspects include physical state—whether a substance is a solid, liquid, or gas—color, hardness, extent to which matter dissolves in water, melting temperature,

Warm, moist air tends to rise, causing cloud formation and influencing atmospheric chemistry

A small amount of matter consumed
in a nuclear reaction can release an
awesome amount of energy

A layer of ice — matter in the solid
state — can be hard on pedestrians

Figure 2.1. **Different kinds of matter have a vast variety of properties that determine what matter does and how it is used.**

and boiling temperature. These kinds of *physical properties* are essential in describing the nature of matter and the chemical changes that it undergoes. They are also essential in understanding and describing its environmental chemical and biological behavior. For example, as discussed in Chapter 12, the temperature-density relationships of water result in stratification of bodies of water such that much different chemical and biochemical processes occur at different depths in a pond, lake, or reservoir. An analogous stratification of the atmosphere profoundly influences atmospheric chemistry.

2.2. CLASSIFICATION OF MATTER

Matter exists in the form of either *elements* or *compounds* (see Sections 1.3 and 1.4, respectively). Recall that compounds consist of atoms of two or more elements bonded together. Chemical changes in which chemical bonds are broken or formed are involved when a compound is produced from two or more elements or from other compounds, or when one or more elements are isolated from a compound. These kinds of changes, which are illustrated by the examples in Table 2.1, profoundly affect the properties of matter. For example, the first reaction in the table shows that sulfur (S), a yellow, crumbly solid, reacts with oxygen, a colorless, odorless gas that is essential for life, to produce sulfur dioxide, a toxic gas with a choking odor. In the second reaction in Table 2.1, the hydrogen atoms in methane, CH_4, a highly flammable, light gas, are replaced by chlorine atoms to produce carbon tetrachloride, CCl_4, a dense liquid so nonflammable that it has been used as a fire extinguisher.

Table 2.1. Chemical Processes Involving Elements and Compounds

Type of Process	Chemical Reaction
Two elements combining to form a chemical compound, sulfur plus oxygen yield sulfur dioxide	$S + O_2 \rightarrow SO_2$
A compound reacting with an element to form two different compounds, methane plus chlorine yield carbon tetrachloride plus hydrogen chloride	$CH_4 + 4Cl_2 \rightarrow$ $CCl_4 + 4HCl$
Two compounds reacting to form a different compound, calcium oxide plus water yield calcium hydroxide	$CaO + H_2O \rightarrow$ $Ca(OH)_2$
A compound plus an element reacting to produce another compound and another pure element, iron oxide plus carbon yield carbon monoxide plus elemental iron (Fe)	$Fe_2O_3 + 3C \rightarrow$ $3CO + 2Fe$
A compound breaking down to its constituent elements, passing an electrical current through water yields hydrogen and oxygen	$2H_2O \rightarrow 2H_2 + O_2$

Some General Types of Matter

In discussing matter, some general terms are employed which are encountered frequently enough that they should be mentioned here. More exact definitions are given later in the text.

Elements are divided between metals and nonmetals. **Metals** are elements that are generally solid, shiny in appearance, electrically conducting, and malleable — that is, they can be pounded into flat sheets without disintegrating. Examples of metals are iron, copper, and silver. Most metallic objects that are commonly encountered are not composed of just one kind of elemental metal, but are alloys consisting of homogeneous mixtures of two or more metals. **Nonmetals** often have a dull appearance, are not at all malleable, and are frequently present as gases or liquids. Colorless oxygen gas, green chlorine gas (transported and stored as a liquid under pressure), and brown bromine liquid are common nonmetals. In a very general sense, we tend to regard as nonmetals substances that actually contain metal in a chemically combined form. One would classify table salt, sodium chloride, as a nonmetal even though it contains chemically-bound sodium metal.

Another general classification of matter is in the categories of inorganic and organic substances. **Organic substances** consist of virtually all compounds that contain carbon, including substances made by life processes (wood, flesh, cotton, wool), petroleum, natural gas (methane), solvents (dry cleaning fluids), synthetic fibers, and plastics. All of the rest of the chemical kingdom is composed of **inorganic substances** made up of virtually all substances that do not contain carbon. These include metals, rocks, table salt, water, sand, and concrete.

Mixtures and Pure Substances

Matter consisting of only one compound or of only one form of an element is a **pure substance**; all other matter is a **mixture** of two or more substances. The compound water with nothing dissolved in it is a pure substance. Highly purified helium gas is an elemental pure substance. The composition of a pure substance is defined and constant and its properties are always the same under specified conditions. Therefore, pure water at $-1°C$ is always a solid (ice), whereas a mixture of water and salt might be either a liquid or a solid at that temperature, depending upon the amount of salt dissolved in the water. Air is a mixture of elemental gases and compounds, predominantly nitrogen, oxygen, argon, carbon dioxide, and water vapor. Drinking water is a mixture containing calcium ion (Ca^{2+}, see ions in Section 1.4), hydrogen carbonate ion, (bicarbonate, HCO_3^-), nitrogen gas, carbon dioxide gas, and other substances dissolved in the water. Mixtures can be separated into their constituent pure substances by **physical processes**. Liquefied air is distilled to isolate pure oxygen (used in welding, industrial processes, for some specialized wastewater treatment processes, and for breathing by people with emphysema), liquid nitrogen (used for quick-freezing frozen foods), and argon (used in specialized types of welding because of its chemically nonreactive nature).

Homogeneous and Heterogeneous Mixtures

A **heterogeneous mixture** is one that is not uniform throughout and possesses readily distinguishable constituents (matter in different phases, see Section 2.5) that can be isolated by means as simple as mechanical separation. Concrete (see Figure 2.2) is such a mixture; individual grains of sand and pieces of gravel may be separated from concrete with a pick or other tool. **Homogeneous mixtures** are uniform throughout; to observe different constituents would require going down to molecular levels. Homogeneous mixtures may have varying compositions and consist of two or more chemically distinct constituents, so they are not pure substances. Air, well filtered to remove particles of dust, pollen, or smoke is a homogeneous mixture. Homogeneous mixtures are also called **solutions**, a term that is usually applied to mixtures composed of gases, solids, and other liquids dissolved in a liquid. A hazardous waste leachate is a solution containing — in addition to water — contaminants such as dissolved acids, iron, heavy metals, and toxic organic compounds. Whereas mechanical means including centrifugation and filtration can be used to separate the solids from the liquids in a heterogeneous mixture of solids and liquids, the isolation of components of homogeneous mixtures requires processes such as distillation and freezing.

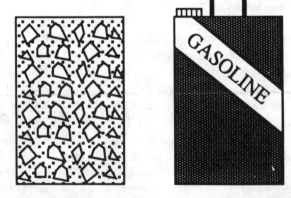

Figure 2.2. Concrete (left) is a heterogeneous mixture with readily visible grains of sand, pieces of gravel, and cement dust. Gasoline is a homogeneous mixture consisting of a solution of petroleum hydrocarbon liquids, dye, and additives such as ethyl alcohol to improve its fuel qualities.

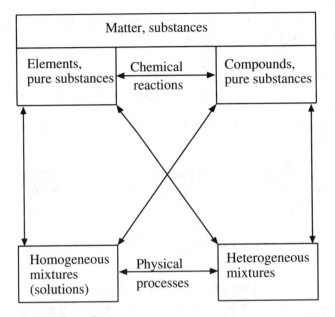

Figure 2.3. Classification of matter.

Summary of the Classification of Matter

As discussed above, all matter consists of compounds or elements. These may exist as pure substances or mixtures; the latter may be either homogeneous or heterogeneous. These relationships are summarized in Figure 2.3.

2.3.   QUANTITY OF MATTER: THE MOLE

One of the most fundamental characteristics of a specific mass of matter is the quantity of it. In discussing the quantitative chemical characteristics of matter it is essential to have a way of expressing quantity in a way that is proportional to the number of individual entities of the substance—that is, atoms, molecules, or ions—in numbers that are readily related to the properties of the atoms or molecules of the substance. The simplest way to do this would be as individual atoms, molecules, or ions, but for laboratory quantities these number in the order of 10^{23}, far too large to be used routinely in expressing quantities of matter. Instead, such quantities are readily expressed as moles of substance. A mole is defined in terms of specific entities, such as atoms of Ar, molecules of H_2O, or Na^+ and Cl^- ions, each pair of which composes a "molecule" of NaCl. A **mole** is defined as *the quantity of substance that contains the same number of specified entities as there are atoms of C in exactly 0.012 kg (12 g) of carbon-12.* It is easier to specify the quantity of a substance equivalent to its number of moles than it is to define the mole. To do so, simply state the atomic mass (of an element) or the molecular mass (of a compound) and affix "mole" to it as shown by the examples below:

- *A mole of argon, which always exists as individual Ar atoms:* The atomic mass of Ar is 40.0. Therefore exactly one mole of Ar is 40.0 grams of argon.
- *A mole of molecular elemental hydrogen, H_2:* The atomic mass of H is 1.0, the molecular mass of H_2 is, therefore, 2.0, and a mole of H_2 is 2.0 g of H_2.
- *A mole of methane, CH_4:* The atomic mass of H is 1.0 and that of C is 12.0, so the molecular mass of CH_4 is 16.0. Therefore a mole of methane has a mass of 16.0 g.

The Mole and Avogadro's Number

In Section 1.6, Avogadro's number was mentioned as an example of a huge number, 6.02×10^{23}. **Avogadro's number** *is the number of specified entities in a mole of substance.* The "specified entities" may consist of atoms or molecules or they may be groups of ions making up the smallest possible unit of an ionic compound, such as 2 Na^+ ions and one S^{2-} ion in Na_2S. (It is not really correct to refer to Na_2S as a molecule because the compound consists of ions arranged in a crystalline structure so that there are 2 Na^+ ions for each S^{2-} ion.) A general term that covers all these possibilities is the **formula unit**. The average mass of a formula unit is called the **formula mass**. Examples of the terms defined in this section are given in Table 2.2.

Table 2.2. Relationships Involving Moles of Substance

Substance	Formula Unit	Formula Mass	Mass of 1 mole	Numbers and Kinds of Individual Entities in 1 mole
Helium	He atom	4.003*	4.003 g	6.02×10^{23} (Avogadro's number) of He atoms
Fluorine gas	F_2 molecule	38.00**	38.00 g	6.02×10^{23} F_2 molecules $2 \times 6.02 \times 10^{23}$ F atoms
Methane	CH_4 molecules	16.04**	16.04 g	6.02×10^{23} CH_4 molecules 6.02×10^{23} C atoms $4 \times 6.02 \times 10^{23}$ H atoms
Sodium oxide	Na_2O	62.00†	62.00 g	6.02×10^{23} Na_2O formula units $2 \times 6.02 \times 10^{23}$ Na^+ ions 6.02×10^{23} O^{2-} atoms

*Specifically, atomic mass.
**Specifically, molecular mass.
†Reference should be made to formula mass for this compound because it consists of Na^+ and O^{2-} ions in a ratio of 2/1.

2.4. PHYSICAL PROPERTIES OF MATTER

Physical properties of matter are those that can be measured without altering the chemical composition of the matter. A typical physical property is color, which can be observed without changing matter at all. Malleability of metals, the degree to which they can be pounded into thin sheets, certainly alters the shape of an object but does not change it chemically. On the other hand, observation of a *chemical property*, such as whether or not sugar burns when ignited in air, potentially involves a complete change in the chemical composition of the substance tested.

Physical properties are important in describing and identifying particular kinds of matter. For example, if a substance is liquid at room temperature, has a lustrous metallic color, conducts electricity well, and has a very high density (mass per unit volume) of 13.6 g/cm^3, it is doubtless elemental mercury metal. Physical properties are very useful in assessing the hazards and predicting the fates of environmental pollutants. An organic substance that readily forms a vapor (volatile organic compound) will tend to enter the atmosphere or to pose an inhalation hazard. A brightly-colored water soluble pollutant may cause deterioration of water quality by adding water "color." Much of the health hazard of asbestos is its tendency to form extremely small diameter fibers that are readily carried far into the lungs and puncture individual cells.

Several important physical properties of matter are discussed in this section. The most commonly considered of these are density, color, and solubility. Thermal properties are addressed separately in Section 2.9.

Density

Density (d) is defined as mass per unit volume and is expressed by the formula

$$d = \frac{mass}{volume} \tag{2.4.1}$$

Density is useful for identifying and characterizing pure substances and mixtures. For example, the density of a mixture of automobile system antifreeze and water, used to prevent the engine coolant from freezing in winter or boiling in summer, varies with the composition of the mixture. Its composition and, therefore, the degree of protection that it offers against freezing may be estimated by measuring its density and relating that through a table to the freezing temperature of the mixture.

Density may be expressed in any units of mass or volume. The densities of liquids and solids are normally given in units of grams per cubic centimeter (g/cm^3, the same as grams per milliliter, g/mL). These values are convenient because they are of the order of 1 g/cm^3 for common liquids and solids.

The volume of a given mass of substance varies with temperature, so the density is a function of temperature. This variation is relatively small for solids, greater for liquids, and very high for gases. Densities of gases vary a great deal with pressure, as well. The temperature-dependent variation of density is an important property of water and results in stratification of bodies of water, which greatly affects the environmental chemistry that occurs in lakes and reservoirs. The density of liquid water has a maximum value of 1.0000 g/mL at 4°C, is 0.9998 g/mL at 0°C, and is 0.9970 g/mL at 25°C. The combined temperature/pressure relationship for the density of air causes air to become stratified into layers, particularly the troposphere near the surface and the stratosphere from about 13 to 50 kilometers altitude.

Whereas liquids and solids have densities of the order of 1 to several g/cm^3, gases at atmospheric temperature and pressure have densities of only about one one-thousandth as much (see Table 2.3 and Figure 2.4).

Table 2.3.   Densities of Some Solids, Liquids, and Gases (at 1 atm Pressure and 0°C)

Solids	d, g/cm^3	Liquids	d, g/cm^3	Gases	d. g/L
Sugar	1.59	Benzene	0.879	Helium	0.178
Sodium chloride	2.16	Water (4°C)	1.0000	Methane	0.714
Iron	7.86	Carbon		Air	1.293
Lead	11.34	tetrachloride	1.595	Chlorine	3.17
		Sulfuric acid	1.84		

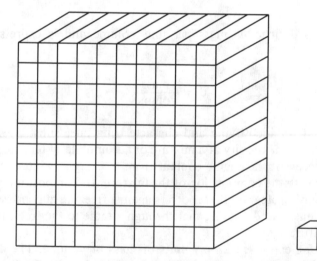

Figure 2.4.   A quantity of air that occupies a volume of 1000 cm³ of air at room temperature and atmospheric pressure (left) has about the same mass as 1 cm³ of liquid water (right).

Specific Gravity

Often densities are expressed by means of **specific gravity**, defined as the ratio of the density of a substance to that of a standard substance. For solids and liquids, the standard substance is usually water; for gases it is usually air. For example, the density of ethanol (ethyl alcohol) at 20°C is 0.7895 g/mL. The specific gravity of ethanol at 20°C referred to water at 4°C is given by

$$\text{Specific gravity} = \frac{\text{density of ethanol}}{\text{density of water}} = \frac{0.7895 \text{ g/mL}}{1.0000 \text{ g/mL}} = 0.7895 \qquad (2.4.2)$$

For an exact value of specific gravity, the temperatures of the substances should be specified. In this case the notation of "specific gravity of ethanol at 20°/4°C" shows that the specific gravity is the ratio of the density of ethanol at 20°C to that of water at 4°C.

Color

Color is one of the more useful properties for identifying substances without doing any chemical or physical tests. A violet vapor, for example, is characteristic of iodine. A red/brown gas is likely to be bromine or nitrogen dioxide (NO_2); a practiced eye can distinguish the two. Just a small amount of potassium permanganate, $KMnO_4$, in solution provides an intense purple color. A characteristic yellow/brown color in water may be indicative of organically-bound iron.

The human eye responds to colors of electromagnetic radiation ranging in wavelength from about 400 nanometers (nm) to somewhat over 700 nm. Within this wavelength range, humans see **light**; immediately below 400 nm is **ultraviolet radiation**, and somewhat above 700 nm is **infrared radiation** (see Figure 2.5). Light with a mixture of wavelengths throughout the visible region, such as sunlight, appears white to the eye. Light over narrower wavelength regions has the following colors: 400–450 nm, blue; 490–550 nm, green; 550–580 nm, yellow; 580–650 nm, orange; 650 nm-upper limit of visible region, red. Solutions are colored because of the light they absorb. Red, orange, and yellow solutions absorb violet and blue light; purple solutions absorb green and yellow light; and blue and green solutions absorb orange and red light. Solutions that do not absorb light are colorless (clear); solids that do not absorb light are white.

2.5.   STATES OF MATTER

Figure 2.6 illustrates the three **states of matter** in which matter may exist. **Solids** have a definite shape and volume. **Liquids** have an indefinite shape and take on the shape of the container in which they are contained. Solids and

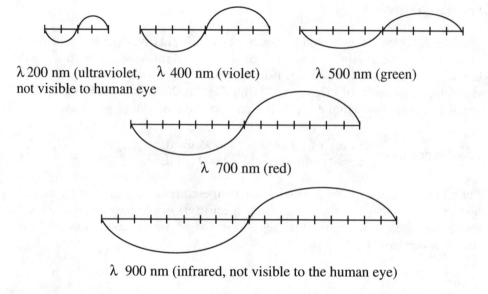

λ 200 nm (ultraviolet, λ 400 nm (violet) λ 500 nm (green)
not visible to human eye

λ 700 nm (red)

λ 900 nm (infrared, not visible to the human eye)

Figure 2.5. Electromagnetic radiation of different wavelengths. Each division represents 50 nm.

liquids are not significantly compressible, which means that a specific quantity of a substance has a definite volume and cannot be squeezed into a significantly smaller volume. **Gases** take on both the shape and volume of their containers. A quantity of gas may be compressed to a very small volume and will expand to occupy the volume of any container into which it is introduced.

Each of these separate phases is discussed in separate sections in this chapter. Everyone is familiar with the three states of matter for water. These are the following:

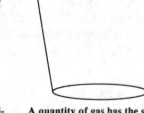

A solid has a definite shape and volume regardless of the container into which it is placed.

A quantity of liquid has a definite volume, but takes on the shape of its container.

A quantity of gas has the shape and volume of the container it occupies.

Figure 2.6.   Representations of the three states of matter.

- Gas: Water vapor in a humid atmosphere, steam
- Liquid: Water in a lake, groundwater
- Solid: Ice in polar ice caps, snow in snowpack

Changes in matter from one phase to another are very important in the environment. For example, water vapor changing from the gas phase to liquid results in cloud formation or precipitation. Water is desalinated by producing water vapor from sea water, leaving the solid salt behind, and recondensing the pure water vapor as a salt-free liquid. Some organic pollutants are extracted from water for chemical analysis by transferring them from the water to another organic phase that is immiscible with water. The condensation of organic pollutants from gaseous materials to solid products in the atmosphere is responsible for the visibility-reducing particulate matter in photochemical smog. Additional examples of phase changes are discussed in Section 2.10 and some of the energy relationships involved are covered in Section 2.9.

2.6.   GASES

We live at the bottom of a "sea" of gas—the earth's atmosphere. It is composed of a mixture of gases, the most abundant of which are nitrogen, oxygen, argon, carbon dioxide, and water vapor. Although these gases have neither color nor odor, so that we are not aware of their presence, they are crucial to the well-being of life on Earth. Deprived of oxygen, an animal loses consciousness and dies within a short time. Nitrogen extracted from the air is converted to chemically bound forms that are crucial to plant growth. Plants require carbon dioxide, and water vapor condenses to produce rain.

Gases and the atmosphere are addressed in more detail in Chapter 16. At this point, however, it is important to have a basic understanding of the nature and behavior of gases. Physically, gases are the "loosest" form of matter. A quantity of gas has neither a definite shape nor a definite volume so that it takes on the shape and volume of the container in which it is held. The reason for this behavior is that gas molecules move independently and at random, bouncing off each other as they do so. They move very rapidly; at 0°C the average molecule of hydrogen gas moves at 3600 miles per hour, 6 to 7 times as fast as a passenger jet airplane. Gas molecules colliding with container walls exert **pressure**. An under-inflated automobile tire has a low pressure because there are relatively few gas molecules in the tire to collide with the container walls. As more air is added the pressure is increased and may become so high that it causes the tire wall to burst.

A quantity of a gas is mostly empty space, which is reflected in the high volume of gas compared to the same amount of material in a liquid or solid. For example, at 100°C a mole (18 g) of liquid water occupies a little more than 18 mL of volume, equivalent to just a few teaspoonsful. When enough heat energy is added to the water to convert it all to gaseous steam at 100°C, the volume becomes 30,600 mL, which is 1,700 times the original volume. This huge increase in volume occurs because of the large distances separating gas molecules compared to their own diameters. It is this great distance that allows gas to be compressed (pushed together).

The rapid, constant motion of gas molecules explains the phenomenon of gas **diffusion** in which gases move large distances from their sources. Diffusion is responsible for much of the hazard of volatile, flammable liquids, such as gasoline. Molecules of gasoline evaporated from an open container of this liquid can spread from their source. If the gaseous gasoline reaches an ignition source, such as an open flame, it may ignite and cause a fire or explosion.

The Gas Laws

To describe the physical and chemical behavior of gases, it is essential to have a rudimentary understanding of the relationships among the following: Quantity of gas in numbers of moles (n, see Section 2.3), volume (V), temperature (T), and pressure (P). These relationships are expressed by the **gas laws.** The gas laws as discussed here apply to **ideal gases**, for which it is assumed that the molecules of gas have negligible volume, that there are no forces of attraction between them, and that they collide with each other in a perfectly elastic manner. Real gases such as H_2, O_2, He, and Ar behave in a nearly ideal fashion at moderate pressures and all but very low temperatures. The gas laws were derived from observations of the effects on gas volume of changing pressure, temperature, and quantity of gas, as stated by Boyle's law, Charles' law, and Avogadro's law, respectively. These three laws are defined below:

Boyle's Law

As illustrated in Figure 2.7, doubling the pressure on a quantity of a gas halves its volume. This illustrates **Boyle's law**, according to which *at constant temperature, the volume of a fixed quantity of gas is inversely proportional to the pressure of the gas.* Boyle's law may be stated mathematically as

$$V = \text{(a constant)} \times \frac{1}{P} \tag{2.6.1}$$

This inverse relationship between the volume of a gas and its pressure is a consequence of the compressibility of matter in the gas phase (recall that solids and liquids are not significantly compressible).

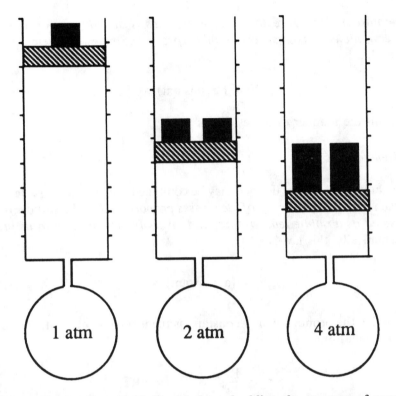

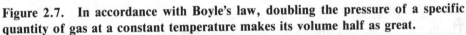

Figure 2.7.   In accordance with Boyle's law, doubling the pressure of a specific quantity of gas at a constant temperature makes its volume half as great.

Charles' Law

Charles' law gives the relationship between gas volume and temperature and states that *the volume of a fixed quantity of gas is directly proportional to the absolute temperature (°C + 273) at constant pressure*. This law may be stated mathematically as

$$V = (a \text{ constant}) \times T \tag{2.6.2}$$

where T is the temperature in K (see Section 1.12). According to this relationship, doubling the absolute temperature of a fixed quantity of gas at constant pressure doubles the volume.

Avogadro's Law

The third of three fundamentally important gas laws is **Avogadro's law** relating the volume of gas to its quantity in moles. This law states that *at constant* *temperature and pressure the volume of a gas is directly proportional to the number of molecules of gas, commonly expressed as moles.* Mathematically, this relationship is

$$V = (a\ constant) \times n \tag{2.6.3}$$

where n is the number of moles of gas.

The General Gas Law

The three gas laws just defined may be combined into a **general gas law** stating that *the volume of a quantity of ideal gas is proportional to the number of moles of gas and its absolute temperature, and inversely proportional to its pressure.* Mathematically, this law is

$$V = (a\ constant) \times \frac{nT}{P} \tag{2.6.4}$$

Designating the proportionality constant as the **ideal gas constant, R**, yields the **ideal gas equation**:

$$V = \frac{RnT}{P} \text{ or } PV = nRT \tag{2.6.5}$$

The units of R depend upon the way in which the ideal gas equation is used. For calculations involving volume in liters and pressures in atmospheres, the value of R is 0.0821 L-atm/deg-mol.

The ideal gas equation shows that at a chosen temperature and pressure a mole of any gas should occupy the same volume. A temperature of 0°C (273.15 K) and 1 atm pressure have been chosen as **standard temperature and pressure (STP)**. *At STP the volume of 1 mole of ideal gas is 22.4 L.* This volume is called the **molar volume of a gas**. It should be remembered that at a temperature of 273 K and a pressure of 1 atm the volume of 1 mole of ideal gas is 22.4 L. Knowing these values, it is always possible to calculate the value of R as follows:

$$R = \frac{PV}{nT} = \frac{1\ atm \times 22.4\ L}{1\ mole \times 273\ K} = 0.0821 \frac{L \times atm}{deg \times mol} \tag{2.6.6}$$

Gas Law Calculations

A common calculation is that of the volume of a gas, starting with a particular volume and changing temperature and/or pressure. This kind of calculation follows logically from the ideal gas equation, but can also be reasoned out knowing that an increase in temperature causes an increase in volume, whereas an increase in pressure causes a decrease in volume. Therefore, a second volume, V_2, is calculated from an initial volume, V_1, by the relationship

$$V_2 = V_1 \times \boxed{} \times \boxed{}$$

(2.6.7)

↑   ↑

Ratio of temperatures   Ratio of pressures

This relationship requires only that one remember the following:

- If the temperature increases (if T_2 is greater than T_1), the volume increases. Therefore, T_2 is always placed over T_1 in the ratio of temperatures.
- If the pressure increases (if P_2 is greater than P_1), the volume decreases. Therefore, P_1 is always placed over P_2 in the ratio of pressures.
- If there is no change in temperature or pressure, the corresponding ratio remains 1.

These relationships can be illustrated by several examples.

Charles' Law Calculation

When the temperature of a quantity of gas changes while the pressure stays the same, the resulting calculation is a *Charles' law calculation*. As an example of such a calculation, calculate the volume of a gas with an initial volume of 10.0 L when the temperature changes from –11.0°C to 95.0°C at constant pressure. The first step in solving any gas law problem involving a temperature change is to convert Celsius temperatures to kelvin:

$$T_1 = 273 + (-11) = 262 \text{ K} \qquad T_2 = 273 + 95 = 368 \text{ K}$$

Substitution into Equation 2.6.7 gives

$$V_2 = V_1 \times \frac{368K}{262K} \times 1 \text{ (factor for pressure} = 1 \text{ because } P_2 = P_1) \quad (2.6.8)$$

$$V_2 = 10.0 \text{ L} \times \frac{368 \text{ K}}{262 \text{ K}} = 14.1 \text{ L}$$

Note that in this calculation it is seen that the temperature increases; this increases the volume, so the higher temperature is placed over the lower. Furthermore, since $P_2 = P_1$, the factor for the pressure ratio simply drops out.

Calculate next the volume of a gas with an initial volume of 10.0 L after the temperature decreases from 111.0°C to 2.0°C at constant pressure:

$$T_1 = 273 + (111) = 384 \text{ K} \qquad T_2 = 273 + 2 = 275 \text{ K}$$

Substitution into Equation 2.6.7 gives

$$V_2 = V_1 \times \frac{275 \text{ K}}{384 \text{ K}} = 10.0 \text{ L} \times \frac{275 \text{ K}}{384 \text{ K}} \tag{2.6.9}$$

$$= 7.16 \text{ L}$$

In this calculation the temperature decreases; this decreases the volume, so the lower temperature is placed over the higher (mathematically it is still T_2 over T_1 regardless of which way the temperature changes).

Boyle's Law Calculation

When the pressure of a quantity of gas changes while the temperature stays the same, the resulting calculation is a *Boyle's law calculation*. As an example, calculate the new volume that results when the pressure of a quantity of gas initially occupying 12.0 L is changed from 0.856 atm to 1.27 atm at constant temperature. In this case, the pressure has increased; this decreases the volume of the gas so that V_2 is given by the following:

$$V_2 = V_1 \times \frac{0.856 \text{ atm}}{1.27 \text{ atm}} \times 1(\text{because } T_2 = T_1) = 12.0 \text{ L} \times \frac{0.856 \text{ atm}}{1.27 \text{ atm}} \tag{2.6.10}$$

$$= 8.09 \text{ L}$$

Calculate next the volume of a gas with an initial volume of 15.0 L when the pressure decreases from 1.71 atm to 1.07 atm at constant temperature. In this case the pressure decreases; this increases the volume, so the appropriate factor to multiply by is 1.71/1.07:

$$V_2 = V_1 \times \frac{1.71 \text{ atm}}{1.07 \text{ atm}} = 15.0 \text{ L} \times \frac{1.71 \text{ atm}}{1.07 \text{ atm}} = 24.0 \text{ L} \tag{2.6.11}$$

Calculations Using the Ideal Gas Law:

The ideal gas law equation

$$PV = nRT \qquad\qquad (2.6.5)$$

may be used to calculate any of the variables in it if the others are known. As an example, calculate the volume of 0.333 moles of gas at 300 K under a pressure of 0.950 atm:

$$V = \frac{nRT}{P} = \frac{0.333 \text{ mol} \times 0.0821 \text{ L atm/K mol} \times 300 \text{ K}}{0.950 \text{ atm}} = 8.63 \text{ L} \quad (2.6.12)$$

As another example calculate the temperature of 2.50 mol of gas that occupies a volume of 52.6 L under a pressure of 1.15 atm:

$$T = \frac{PV}{nR} = \frac{1.15 \text{ atm} \times 52.6 \text{ L}}{2.50 \text{ mol} \times 0.0821 \text{ L atm/K mol}} = 295 \text{ K} \qquad (2.6.13)$$

Other examples of gas law calculations are given in the problem section at the end of this chapter.

The changes in a gas resulting from variations in one or more of the parameters of P, V, n, or T can be calculated by the relationship,

$$\frac{P_2 V_2}{n_2 T_2} = \frac{P_1 V_1}{n_1 T_1} \qquad\qquad (2.6.14)$$

derived from the fact that R is a constant, where the subscripts 1 and 2 denote values of the parameters designated before and after these are changed. By holding the appropriate parameters constant, this equation can be used to derive the mathematical relationships used in calculations involving Boyle's law, Charles' law, or Avogadro's law, as well as any combination of these laws.

Exercise: Derive an equation that will give the volume of a gas after all three of the parameters P, T, and n have been changed.

Answer: Rearrangement of Equation 2.6.14 gives the relationship needed:

$$V_2 = V_1 \frac{P_1 n_2 T_2}{P_2 n_1 T_1} \qquad\qquad (2.6.15)$$

by holding the appropriate parameters constant, this equation can be used to give mathematical expressions for Boyle's law, Charles' law, or Avogadro's law.

2.7.   LIQUIDS AND SOLUTIONS

Whereas the molecules in gases are relatively far apart and have minimal attractive forces between them (none in the case of ideal gases, which are discussed in the preceding section), molecules of liquids are close enough together that they may be regarded as "touching" and are strongly attracted to each other. Like those of gases, however, the molecules of liquids move freely relative to each other. These characteristics give rise to several significant properties of liquids. Whereas gases are mostly empty space, most of the volume of a liquid is occupied by molecules; therefore, the densities of liquids are much higher than those of gases. Because there is little unoccupied space between molecules of liquid, liquids are only slightly *compressible*; doubling the pressure on a liquid causes a barely perceptible change in volume, whereas it halves the volume of an ideal gas. Liquids do not expand to fill their containers as do gases. Essentially, therefore, a given quantity of liquid occupies a fixed volume. Because of the free movement of molecules relative to each other in a liquid, it takes on the shape of that portion of the container that it occupies.

Evaporation and Condensation of Liquids

Consider the surface of a body of water as illustrated in Figure 2.8. The molecules of liquid water are in constant motion relative to each other. They have a distribution of energies such that there are more higher energy molecules at higher temperatures. Some molecules are energetic enough that they escape the attractive forces of the other molecules in the mass of liquid and enter the gas phase. This phenomenon is called **evaporation**.

Molecules of liquid in the gas phase, such as water vapor molecules above a body of water, may come together or strike the surface of the liquid and re-enter

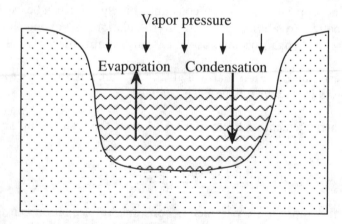

Figure 2.8.   Evaporation and condensation of liquid.

the liquid phase. This process is called **condensation**. It is what happens, for example, when water vapor molecules in the atmosphere strike the surface of very small water droplets in clouds and enter the liquid water phase, causing the droplets to grow to sufficient size to fall as precipitation.

Vapor Pressure

In a confined area above a liquid, equilibrium is established between the evaporation and condensation of molecules from the liquid. For a given temperature, this results in a steady-state level of the vapor, which can be described as a pressure. Such a pressure is called the **vapor pressure** of the liquid (see Figure 2.8).

Vapor pressure is very important in determining the fates and effects of substances, including hazardous substances, in the environment. Loss of a liquid by evaporation to the atmosphere increases with increasing vapor pressure and is an important mechanism by which volatile pollutants enter the atmosphere. High vapor pressure of a flammable liquid can result in the formation of explosive mixtures of vapor above a liquid, such as in a storage tank.

Solutions

Solutions were mentioned in Section 2.2 as homogeneous mixtures. Here solutions are regarded as liquids in which quantities of gases, solids, or other liquids are dispersed as individual ions or molecules. Rainwater, for example, is a solution containing molecules of N_2, O_2 and CO_2 from air dispersed among the water molecules. Rainwater from polluted air may contain harmful substances such as sulfuric acid (H_2SO_4) in small quantities among the water molecules. The predominant liquid constituent of a solution is called the **solvent**. A substance dispersed in it is said to be **dissolved** in the solvent and is called the **solute**.

Solubility

Mixing and stirring some solids, such as sugar, with water results in a rapid and obvious change in which all or part of the substance seems to disappear as it dissolves. When this happens, the substance is said to *dissolve*, and one that does so to a significant extent is said to be **soluble** in the solvent. Substances that do not dissolve to a perceptible degree are called **insoluble**. *Solubility* is a significant physical property. It is a relative term. For example, Teflon is insoluble in water, limestone is slightly soluble, and ethanol (ethyl alcohol) is infinitely so in that it may be mixed with water in any proportion ranging from pure water to pure ethanol with only a single liquid phase consisting of an ethanol/water solution being observed.

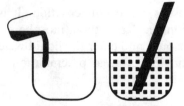

Whether or not a solid dissolves
significantly in water may be
determined in the laboratory by
stirring the solid with water
in a beaker.

An insoluble substance will   An appreciably soluble sub-   If the dissolved substance absorbs
show no evidence of dis-   stance will seem to disappear, light, the solution formed will be
solving and will remain as   leaving only liquid   colored; the color may be useful
a residue in the beaker   to identify the substance.

Figure 2.9.   Illustration of whether or not a substance dissolves in a liquid.

Solution Concentration

Solutions are addressed specifically and in more detail in Chapter 7. At this
point, however, it is useful to have some understanding of the quantitative
expression of the amount of solute dissolved in a specified amount of solution.
In a relative sense, a solution that has comparatively little solute dissolved per
unit volume of solution is called a **dilute solution**, whereas one with an amount
of solute of the same order of magnitude as that of the solvent is a **concentrated
solution**. Water saturated with air at 25°C contains only about 8 milligrams of
oxygen dissolved in 1 liter of water; this composes a *dilute solution* of oxygen. A
typical engine coolant solution contains about as much ethylene glycol (antifreeze) as it does water, so it is a *concentrated solution*. A solution that is at
equilibrium with excess solute so that it contains the maximum amount of solute
that it can dissolve is called a **saturated solution**. One that can still dissolve more
solute is called an **unsaturated solution**.

For the chemist, the most useful way to express the concentration of a solution is in terms of the number of moles of solute dissolved per liter of solution.
The **molar concentration, M,** of a solution is *the number of moles of solute
dissolved per liter of solution.* Mathematically,

$$M = \frac{\text{moles of solute}}{\text{number of liters of solution}} \qquad (2.7.1)$$

Example: Exactly 34.0 g of ammonia, NH_3, were dissolved in water and the solution was made up to a volume of exactly 0.500 L. What was the molar concentration, M, of ammonia in the resulting solution?

Answer: The molar mass of NH_3 is 17.0 g/mole. Therefore

$$\text{Number of moles of } NH_3 = \frac{34.0 \text{ g}}{17.0 \text{ g/mole}} = 2.00 \text{ mol} \qquad (2.7.2)$$

The molar concentration of the solution is calculated from Equation 2.7.1:

$$M = \frac{2.00 \text{ mol}}{0.500 \text{ L}} = 4.00 \text{ mol/L} \qquad (2.7.3)$$

Equations 2.7.1 and 2.7.2 are useful in doing calculations that relate mass of solute to solution volume and solution concentration in chemistry (see stoichiometry in Chapter 5).

2.8. SOLIDS

Matter in the form of *solids* is said to be in the **solid state**. The solid state is the most organized form of matter in that the atoms, molecules, and ions in it are in essentially fixed relative positions and are highly attracted to each other. Therefore, solids have a definite shape, maintain a constant volume, are virtually noncompressible under pressure, and expand and contract only slightly with changes in temperature. Like liquids, solids have generally very high densities relative to gases. (Some solids such as those made from styrofoam appear to have very low densities, but that is because such materials are composed mostly of bubbles of air in a solid matrix.) Because of the strong attraction of the atoms, molecules, and ions of solids for each other, solids do not enter the vapor phase very readily at all; the phenomenon by which this happens to a limited extent is called **sublimation**.

Some solids (quartz, sodium chloride) have very well defined geometric shapes and form characteristic crystals. These are called **crystalline solids** and occur because the molecules or ions of which they are composed assume well defined, specific positions relative to each other. Other solids have indefinite shapes because their constituents are arranged at random; glass is such a solid. These are **amorphous solids**.

2.9. THERMAL PROPERTIES

The behavior of a substance when heated or cooled defines several important physical properties of it. These include the temperature at which it melts or vaporizes. Also included are the amounts of heat required by a given mass of the substance to raise its temperature by a unit of temperature, to melt it, or to vaporize it.

Melting Point

The **melting point** of a pure substance is the temperature at which the substance changes from a solid to a liquid. At the melting temperature pure solid and pure liquid composed of the substance may be present together in a state of equilibrium. An impure substance does not have a single melting temperature. Instead, as the substance is melted by application of heat, it begins to melt at a temperature below the melting temperature of the pure substance; melting proceeds as the temperature increases until no solid remains. Therefore, melting temperature measurements serve two purposes in characterizing a substance. The melting point of a pure substance is indicative of the identity of the substance. The melting behavior — whether melting occurs at a single temperature or over a temperature range — is a measure of substance purity.

Boiling Point

Boiling occurs when a liquid is heated to a temperature so that bubbles of vapor of the substance are evolved. The boiling temperature depends upon pressure. When the surface of a pure liquid substance is in contact with the pure vapor of the substance at 1 atm pressure, boiling occurs at a temperature called the **normal boiling point** (see Figure 2.10). Whereas a pure liquid remains at a constant temperature during boiling until it has all turned to vapor, the temperature of an impure liquid increases during boiling. As with melting point, the boiling point is useful for identifying liquids. The extent to which the boiling temperature is constant as the liquid is converted to vapor is a measure of the purity of the liquid.

Specific Heat

As the temperature of a substance is raised, energy must be put into it to enable the molecules of the substance to move more rapidly relative to each other and to overcome the attractive forces between them. The amount of heat energy required to raise the temperature of a unit mass of a solid or liquid substance by a degree of temperature varies with the substance. Of the common liquids, the most heat is required to raise the temperature of water. This is

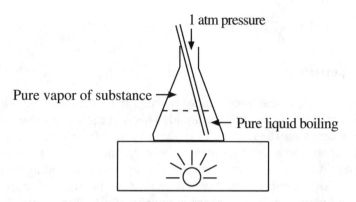

1 atm pressure

Pure vapor of substance

Pure liquid boiling

Figure 2.10.   Illustration of normal boiling liquid.

because of the molecular structure of water, the molecules of which are strongly attracted to each other by electrical charges between the molecules (ea·h of which has a relatively positive and a relatively more negative end). Furthermore, water molecules tend to be held in molecular networks by special kinds of chemical bonds called hydrogen bonds, in which hydrogen atoms form bridges between O atoms in different water molecules. The very high amount of energy required to increase the temperature of water has some important environmental implications because it stabilizes the temperatures of bodies of water and of geographic areas close to bodies of water. In contrast to water, a hydrocarbon liquid such as those in gasoline requires relatively little heat for warming because the molecules interact much less with each other than do those of water.

The heat required to raise the temperature of a substance is called the **specific heat** defined as the amount of heat energy required to raise the temperature of a gram of substance by 1 degree Celsius. It can be expressed by the equation

$$\text{Specific heat} = \frac{\text{heat energy absorbed, J}}{(\text{mass, g})(\text{increase in temperature, }°C)} \qquad (2.9.1)$$

where the heat energy is in units of joule, J. The most important value of specific heat is that of water, 4.18 J/g°C. From Equation 2.9.1, the amount of heat, q, required to raise the mass, m, of a particular substance over a temperature range of ΔT is

$$q = (\text{specific heat}) \times m \times \Delta T \qquad (2.9.2)$$

This equation can be used to calculate quantities of heat used in increasing water temperature or released when water temperature decreases. As an example, consider the amount of heat required to raise the temperature of 11.6 g of liquid water from 14.3°C to 21.8°C:

$$q = 4.18 \frac{J}{g°C} \times 11.6 \text{ g} \times (21.8°C - 14.3°C) = 364 \text{ J} \qquad (2.9.3)$$

Heat of Vaporization

Large amounts of heat energy may be required to convert liquid to vapor compared to the specific heat of the liquid. This is because changing a liquid to a gas requires breaking the molecules of liquid away from each other, not just increasing the rates at which they move relative to one another. The vaporization of a pure liquid occurs at a constant temperature (normal boiling temperature, see above). For example, the temperature of a quantity of pure water is increased by heating to 100°C where it remains until all the water has evaporated. The **heat of vaporization** is the quantity of heat taken up in converting a unit mass of liquid entirely to vapor at a constant temperature. The heat of vaporization of water is 2,260 J/g (2.26 kJ/g) for water boiling at 100°C at 1 atm pressure. This amount of heat energy is about 540 times that required to raise the temperature of 1 gram of liquid water by 1°C. The fact that water's heat of vaporization is the largest of any common liquid has significant environmental effects. It means that enormous amounts of heat are required to produce water vapor from liquid water in bodies of water. This means that the temperature of a body of water is relatively stable when it is exposed to high temperatures because so much heat is dissipated when a fraction of the water evaporates. When water vapor condenses, similar enormous amounts of heat energy called **heat of condensation** are released. This occurs when water vapor forms precipitation in storm clouds and is the driving force behind the tremendous releases that occur in thunderstorms and hurricanes.

The heat of vaporization of water can be used to calculate the heat required to evaporate a quantity of water. For example, the heat, q, required to evaporate 2.50 g of liquid water is

$$q = (\text{heat of vaporization})(\text{mass of water}) =$$
$$2.26 \text{ kJ/g} \times 2.50 \text{ g} = 5.65 \text{ kJ} \qquad (2.9.4)$$

and that released when 5.00 g of water vapor condenses is

$$q = -(\text{heat of vaporization})(\text{mass of water}) =$$
$$-2.26 \text{ kJ/g} \times 5.00 \text{ g} = -11.3 \text{ kJ} \qquad (2.9.5)$$

the latter value is negative to express the fact that heat is <u>released</u>.

Heat of Fusion

Heat of fusion is the quantity of heat taken up in converting a unit mass of solid entirely to liquid at a constant temperature. The heat of fusion of water is

330 J/g for ice melting at 0°C. This amount of heat energy is 80 times that required to raise the temperature of 1 gram of liquid water by 1°C. When liquid water freezes, an exactly equal amount of energy per unit mass is released, but is denoted as a negative value.

The heat of fusion of water can be used to calculate the heat required to melt a quantity of ice. For example, the heat, q, required to melt 2.50 g of ice is

$$q = \text{(heat of fusion)(mass of water)} =$$
$$330 \text{ J/g} \times 2.50 \text{ g} = 825 \text{ J} \qquad (2.9.6)$$

and that released when 5.00 g of liquid water freezes is

$$q = -\text{(heat of fusion)(mass of water)} =$$
$$-330 \text{J/g} \times 5.00 \text{ g} = -1,650 \text{ J} \qquad (2.9.7)$$

2.10. SEPARATION AND CHARACTERIZATION OF MATTER

A very important aspect of the understanding of matter is the separation of mixtures of matter into their constituent pure substances. Consider, for example, the treatment of an uncharacterized waste sludge found improperly disposed in barrels. Typically, the sludge will appear as a black goo with an obnoxious odor from several constituents. In order to do a proper analysis of this material, it often needs to be separated into its constituent components, such as by extracting water-soluble substances into water, extracting organic materials into an organic solvent, and distilling off volatile constituents. The materials thus isolated can be analyzed to determine what is present. With this knowledge and with information about the separation itself a strategy can be devised to treat the waste in an effective, economical manner.

Many kinds of processes are used for separations, and it is beyond the scope of this chapter to go into detail about them here. However, several of the more important separation operations are discussed briefly below.

Distillation

Distillation consists of evaporating a liquid by heating and cooling the vapor so that it condenses back to the liquid in a different container, as shown for water in Figure 2.11. Less volatile constituents such as solids are left behind in the distillation flask; more volatile impurities such as volatile organic compound pollutants in water will distill off first and can be placed in different containers before the major liquid constituent is collected.

The apparatus shown in Figure 2.11 could be used, for example, to prepare fresh water from seawater, which may be considered as a solution of sodium

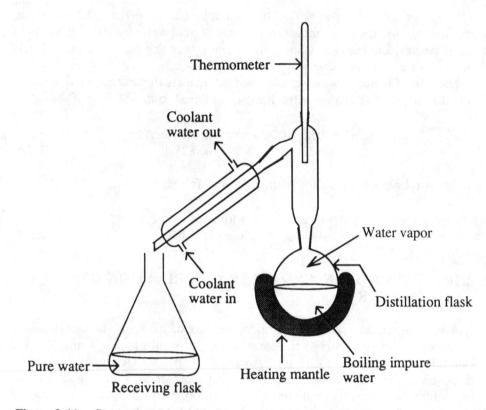

Figure 2.11.   Separation by distillation.

chloride in water. The seawater to be purified is placed in a round-bottom distillation flask heated with an electrically powered heating mantle. The seawater is boiled and pure water vapor in the gaseous state flows into the condenser where it is cooled and condensed back to a purified salt-free product in the receiving flask. When most of the seawater has been evaporated, the distillation flask and its contents are cooled and part of the sodium chloride separates out as crystals that can be removed. Sophisticated versions of this distillation process are used in some arid regions of the world to produce potable (drinking) water from seawater.

Distillation is widely used in the petroleum and chemical industries. A sophisticated distillation apparatus is used to separate the numerous organic components of crude oil, including petroleum ether, gasoline, kerosene, and jet fuel fractions. This requires **fractional distillation** in which part of the vapor recondenses in a **fractionating column** which would be mounted vertically on top of the distillation flask in the apparatus shown in Figure 2.11. The most volatile components enter into the condenser and are condensed back to liquid products;

less volatile liquid constituents return to the distillation flask. The net effect is that of numerous distillations, which gives a very efficient separation.

In some cases the residues left from distillation, **distillation bottoms** ("still bottoms") are waste materials. Several important categories of hazardous wastes consist of distillation bottoms, which pose disposal problems.

Separation in Waste Treatment

In addition to distillation, several other kinds of separation procedures are used for waste treatment. These are addressed in more detail in Chapter 20 and are mentioned here as examples of separation processes.

Phase Transitions

Distillation is an example of a separation in which one of the constituents being separated undergoes a **phase transition** from one phase to another, then often back again to the same phase. In distillation, for example, a liquid is converted to the vapor state, then recondensed as a liquid. Several other kinds of separations by phase transition as applied to waste treatment are the following:

Evaporation is usually employed to remove water from an aqueous waste to concentrate it. A special case of this technique is **thin-film evaporation** in which volatile constituents are removed by heating a thin layer of liquid or sludge waste spread on a heated surface.

Drying — removal of solvent or water from a solid or semisolid (sludge) or the removal of solvent from a liquid or suspension — is a very important operation because water is often the major constituent of waste products, such as sludges. In **freeze drying**, the solvent, usually water, is sublimed from a frozen material. Hazardous waste solids and sludges are dried to reduce the quantity of waste, to remove solvent or water that might interfere with subsequent treatment processes, and to remove hazardous volatile constituents.

Stripping is a means of separating volatile components from less volatile ones in a liquid mixture by the partitioning of the more volatile materials to a gas phase of air or steam (steam stripping). The gas phase is introduced into the aqueous solution or suspension containing the waste in a stripping tower that is equipped with trays or packed to provide maximum turbulence and contact between the liquid and gas phases. The two major products are condensed vapor and a stripped bottoms residue. Examples of two volatile components that can be removed from water by air stripping are the organic solvents benzene and dichloromethane.

Physical precipitation is used here as a term to describe processes in which a solid forms from a solute in solution as a result of a physical change in the solution. The major changes that can cause physical precipitation are cooling the solution, evaporation of solvent, or alteration of solvent composition. The most common type of physical precipitation by alteration of solvent composition occurs when a water-miscible organic solvent is added to an aqueous (water) solution of a salt so that the solubility of a salt is lowered below its concentration in the solution.

Phase Transfer

Phase transfer consists of the transfer of a solute in a mixture from one phase to another. An important type of phase transfer process is **solvent extraction** (Figure 2.12), a process in which a substance is transferred from solution in one solvent (usually water) to another (usually an organic solvent) without any chemical change taking place. When solvents are used to leach substances from solids or sludges, the process is called **leaching**. One of the more promising approaches to solvent extraction and leaching of hazardous wastes is the use of **supercritical fluids**, most commonly CO_2, as extraction solvents. A supercritical fluid is one that has characteristics of both liquid and gas and consists of a substance above its supercritical temperature and pressure ($31.1°C$ and 73.8 atm, respectively, for CO_2). After a substance has been extracted from a waste into a supercritical fluid at high pressure, the pressure can be released, resulting in separation of the substance extracted. The fluid can then be compressed again and recirculated through the extraction system. Some possibilities for treatment of hazardous wastes by extraction with supercritical CO_2 include removal of organic contaminants from wastewater, extraction of organohalide pesticides from soil, extraction of oil from emulsions used in aluminum and steel processing, and regeneration of spent activated carbon.

Transfer of a substance from a solution to a solid phase is called **sorption**. The most important sorbent used in waste treatment is **activated carbon**. It can also

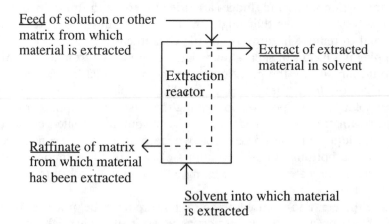

Figure 2.12.   Outline of solvent extraction/leaching process with important terms underlined.

be applied to pretreatment of waste streams going into other processes to improve treatment efficiency and reduce fouling. Activated carbon sorption is most effective for removing from water organic materials that are poorly water soluble and that have high molecular masses.

Molecular Separation

A third major class of physical separation is **molecular separation**, often based upon **membrane processes** in which dissolved contaminants or solvent pass through a size-selective membrane under pressure. **Reverse osmosis** is the most widely used of the membrane techniques. It operates by virtue of a membrane that is selectively permeable to water and excludes ionic solutes. Reverse osmosis uses high pressures to force permeate through the membrane, producing a concentrate containing high levels of dissolved salts.

CHAPTER SUMMARY

The chapter summary below is presented in a programmed format to review the main points covered in this chapter. It is used most effectively by filling in the blanks, referring back to the chapter as necessary. The correct answers are given at the end of the summary.

Matter is defined as (1)_____

_____. Chemical properties of matter are

those that are observed (2)_____

_____, whereas physical properties are those that are

observed (3)_____

_____. Matter consists of only a few (4)_____ and a large

number of substances formed from them called (5)_____. Some

general properties of metals are (6)_____

_____. Some common proper-

ties of nonmetals are (7)_____

_____

(8)_____ substances consist of virtually all compounds that

contain carbon, whereas (9)_____ make up virtually all

matter that does not contain carbon. Matter consisting of (10)_____

_____ is a pure substance; any other form of matter is a (11)_____ of two or more substances. A (12)_____ is not uniform throughout and possesses readily distinguishable constituents, whereas (13)_____ are uniform throughout.

A mole is defined as the quantity of substance that (14)_____ _____. A mole of H_2O contains (15)_____ grams of this compound. Avogadro's number is the number of specified entities in (16)_____. A general term that includes atoms and molecules as well as groups of ions making up the smallest possible quantity of an ionic compound is the (17)____ _____. The average mass of a formula unit is called the (18)_____.

Density is defined as (19)_____ and expressed by the formula (20)_____. Relative densities are expressed by means of a number called (21)_____ defined as (22)_____ _____. Some solutions are colored because they (23)_____ _____.

Solids, liquids, and gases are the three (24)_____ of matter. A solid has a definite (25)_____ and (26)_____ regardless of the container in which it is placed, a liquid has a definite (27)_____, but takes on the (28)_____ of its container, and a gas has the same (29)_____ as its container.

Gas molecules move (30)_____, colliding with container walls to exert (31)_____. A quantity of a gas has a (32)_____ compared to the same amount of material in a liquid or solid. (33)_____ enables gas molecules to move large distances from their sources. The gas laws relate the four parameters of (34)_____

_____ of gas. Boyle's law states that (35)_____

_____

_____. Charles' law states that (36)_____

_____

_____. Avogadro's law states that (37)_____

_____

_____.

The general gas law states that (38)_____

_____

_____. The ideal gas law is given mathematically as

(39)_____. At a temperature of 0°C and 1 atm pressure

known as (40)_____ the volume

of 1 mole of ideal gas is (41)_____.

   Molecules of liquids move (42)_____ relative to each other.
Because there is little unoccupied space between molecules of liquid, liquids are
only slightly (43)_____. The phenomenon by which molecules of
liquid enter the gas phase is called (44)_____, whereas the opposite phenomenon is termed (45)_____. A steady-state
level of the vapor of a liquid above the liquid surface is called the
(46)_____ of the liquid. Solutions are regarded as liquids in
which (47)_____

_____. The predominant
liquid constituent of a solution is called the (48)_____,
a substance dispersed in it is said to be (49)_____
in the solvent and is called the (50)_____. A substance that dissolves
to a significant extent is said to be (51)_____ in the
solvent. Substances that do not dissolve to a perceptible degree are classified as
(52)_____. A solution that has comparatively
little solute dissolved per unit volume of solution is called a (53)_____

solution, whereas one with an amount of solute of the same order of magnitude as that of the solvent is a (54)_____ solution. The molar concentration, M, of a solution is defined as (55)_____ and is given by formula (56)_____.

The solid state is the most (57)_____ form of matter because the atoms, molecules, and ions in it are (58)_____ _____.The phenomenon by which solids enter the gaseous state directly is called (59)_____. Solids that form well-defined crystals are called (60)_____. and those that have indefinite shapes because their constituents are arranged at random are called (61)_____.

The melting point of a pure substance is the temperature at which the substance changes (62)_____. At this temperature pure solid and pure liquid composed of the substance (63)_____ _____. An impure substance does not have (64)_____ _____. The two purposes served by melting temperature measurements in characterizing a substance are (65)_____ _____ (66) _____ _____ _____. When a pure liquid is heated so that the surface of the liquid is in contact with the pure vapor of the substance at 1 atm pressure, boiling occurs at a temperature called the (67)_____ _____. The extent to which the boiling temperature is constant as the liquid is converted to vapor is a measure of (68)_____.

The amount of heat energy required to raise the temperature of a unit mass of a solid or liquid substance by a degree of temperature is called the (69)_____ _____ of the substance and can be expressed by the equation

(70)_____. The heat of vaporiza-

tion is defined as (71)_____

_____.

Mathematically, the heat, q, required to evaporate a specified mass of liquid

water is (72)_____. Heat of

fusion is defined as (73)_____

_____.

and is given by the formula (74)_____.

   Distillation consists of (75)_____

_____. Fractional

distillation is a separation process in which part of the vapor (76)_____.

Residues left from distillation are called (77)_____.

   A process in which a solvent is sublimed from a frozen material is called

(78)_____. Stripping is a means of separating volatile components

from less volatile ones in a liquid mixture by (79)_____

_____.

Solvent extraction is a process in which (80)_____

_____

_____. The most important sorbent used in waste treatment is (81)_____.

A water purification process called (82)_____ uses high pres-

sures to force permeate through a membrane that is selectively permeable to

water and excludes ionic solutes.

Answers

1. anything that has mass and occupies space
2. by chemical change or the potential for chemical change
3. without any chemical change
4. elements
5. compounds
6. generally solid, shiny in appearance, electrically conducting, and malleable
7. a dull appearance, not at all malleable, and frequently present as gases or liquids
8. Organic

9. inorganic substances
10. only one compound or of only one form of an element
11. mixture
12. heterogeneous mixture
13. homogeneous mixtures
14. contains the same number of specified entities as there are atoms of C in exactly 0.012 kg (12 g) of carbon-12
15. 18
16. a mole of substance
17. formula unit
18. formula mass
19. mass per unit volume
20. density $= \dfrac{\text{mass}}{\text{volume}}$
21. specific gravity
22. the ratio of the density of a substance to that of a standard substance
23. absorb light
24. states
25. shape
26. volume
27. volume
28. shape
29. shape and volume
30. independently and at random
31. pressure
32. high volume
33. Diffusion
34. moles (n), volume (V), temperature (T), and pressure (P)
35. at constant temperature, the volume of a fixed quantity of gas is inversely proportional to the pressure of the gas
36. the volume of a fixed quantity of gas is directly proportional to the absolute temperature at constant pressure
37. at constant temperature and pressure the volume of a gas is directly proportional to the number of molecules of gas, commonly expressed as moles
38. the volume of a quantity of ideal gas is proportional to the number of moles of gas and its absolute temperature and inversely proportional to its pressure
39. $PV = nRT$
40. standard temperature and pressure (STP)
41. 22.4 L
42. freely
43. compressible
44. evaporation
45. condensation

46. vapor pressure
47. quantities of gases, solids, or other liquids are dispersed as individual ions or molecules
48. solvent
49. dissolved
50. solute
51. soluble
52. insoluble
53. dilute
54. concentrated
55. the number of moles of solute dissolved per liter of solution
56. $M = \dfrac{\text{moles of solute}}{\text{number of liters of solution}}$
57. organized
58. in essentially fixed relative positions and are highly attracted to each other
59. sublimation
60. crystalline solids
61. amorphous solids
62. from a solid to a liquid
63. may be present together in a state of equilibrium
64. a single melting temperature
65. the melting point of a pure substance is indicative of the identity of the substance
66. the melting behavior—whether melting occurs at a single temperature or over a temperature range—is a measure of substance purity
67. normal boiling point
68. the purity of the liquid
69. specific heat
70. Specific heat $= \dfrac{\text{heat energy absorbed, J}}{\text{(mass, g)(increase in temperature, }°\text{C)}}$
71. the quantity of heat taken up in converting a unit mass of liquid entirely to vapor at a constant temperature
72. $q =$ (heat of vaporization)(mass of water)
73. the quantity of heat taken up in converting a unit mass of solid entirely to liquid at a constant temperature
74. $q =$ (heat of fusion)(mass of substance)
75. evaporating a liquid by heating and cooling the vapor so that it condenses back to the liquid in a different container
76. recondenses in a fractionating column
77. distillation bottoms
78. freeze drying
79. the partitioning of the more volatile materials to a gas phase of air or steam
80. a substance is transferred from solution in one solvent to another without any chemical change taking place

81. activated carbon
82. reverse osmosis

QUESTIONS AND PROBLEMS

1. What is the definition of matter?

2. Classify each of the following as chemical processes (c) or physical processes (p):

   ( ) Ice melts
   ( ) $CaO + H_2O \rightarrow Ca(OH)_2$
   ( ) A small silver sphere pounded into a thin sheet
   ( ) A copper-covered roof turns green in the atmosphere
   ( ) Wood burns
   ( ) Oxygen isolated by the distillation of liquid air

3. List (a) three common characteristics of metals and (b) three common characteristics of nonmetals.

4. Classify each of the following as organic substances (o) or inorganic substances (i):

   ( ) Crude oil
   ( ) A dime coin
   ( ) A shirt
   ( ) Asphalt isolated as a residue of petroleum distillation
   ( ) A compound composed of silicon, oxygen, and aluminum
   ( ) Concrete

5. Sandy sediment from the bottom of a river was filtered to isolate a clear liquid, which was distilled to give a liquid compound that could not be broken down further by nonchemical means. Explain how these operations illustrate heterogeneous mixtures, homogeneous mixtures, and pure substances.

6. Explain why a solution is defined as a homogeneous mixture.

7. Justify the statement that "exactly 12 g of carbon-12 contain 6.02×10^{23} atoms of the isotope of carbon that contains 6 protons and 6 neutrons in its nucleus." What two terms relating to quantity of matter are illustrated by this statement?

8. Where X is a number equal to the formula mass of a substance, complete the following formula: Mass of a mole of a substance = _____.

9. Fill in the blanks in the table below:

Substance	Formula unit	Formula mass	Mass of 1 mole	Numbers and kinds of individual entities in 1 mole
Neon	Ne atom	_____	_____ g	_____Ne atoms
Nitrogen gas	N_2 molecule	_____	_____ g	_____N_2 molecules
				_____N atoms
Ethene	C_2H_4 molecules	_____	_____ g	_____C_2H_4 molecules
				_____C atoms
				_____H atoms

10. Calculate the density of each of the following substances, for which the mass of a specified volume is given:

(a) 93.6 g occupying 15 mL
(b) 0.992 g occupying 1,005 mL
(c) 13.7 g occupying 11.4 mL
(d) 16.8 g occupying 3.19 mL

11. Calculate the volume in mL occupied by 156.3 g of water at 4°C.

12. Given the densities in Table 2.3, calculate the following:

(a) Volume of 105 g of iron
(b) Mass of 157 cm³ of benzene
(c) Volume of 1,932 g of air
(d) Mass of 0.525 cm³ of carbon tetrachloride
(e) Volume of 10.0 g of helium
(f) Mass of 13.2 cm³ of sugar

13. From information given in Table 2.3, calculate the specific gravities of (a) benzene, (b) carbon tetrachloride, and (c) sulfuric acid relative to water at 4°C.

14. From information given in this chapter, calculate the density of pure nitrogen gas, N_2, at STP. To do so will require some knowledge of the mole and of the gas laws.

15. Different colors are due to different _____ of light.

16. Explain how the various forms of water observed in the environment illustrate the three states of matter.

17. How might an individual molecule of a gas describe its surroundings. What might it see, feel, and do?

18. Explain pressure and diffusion in terms of the behavior of gas molecules.

19. Use the gas law to calculate whether or not the volume of 1 mole of water vapor at 100°C and 1 atm pressure is 30,600 mL as stated in this chapter.

20. Match each gas law stated on the left, below, with the phenomenon that it explains on the right:

( ) Charles' law   1. The volume of a mole of ideal gas at STP is 22.4 L.

( ) General gas law   2. Tripling the number of moles of gas at constant temperature and pressure triples the volume.

( ) Boyle's law   3. Raising the absolute temperature of a quantity of gas at constant pressure from 300 K to 450 K increases the gas volume by 50%.

( ) Avogadro's law   4. Doubling the pressure on a quantity of gas at constant temperature halves the volume.

21. Using the appropriate gas laws, calculate the quantities denoted by the blanks in the table below:

Conditions before				Conditions after			
n_1, mol	V_1, L	P_1, atm	T_1, K	n_2, mol	V_2, L	P_2, atm	T_2, K
1.25	13.2	1.27	298	1.25	(a)___	0.987	298
1.25	13.2	1.27	298	1.25	(b)___	1.27	407
1.25	13.2	1.27	298	4.00	(c)___	1.27	298
1.25	13.2	1.27	298	1.36	(d)___	1.06	372
1.25	13.2	1.27	298	1.25	30.5	(e)___	298

22. In terms of distances between molecules and movement of molecules relative to each other, distinguish between liquids and gases and liquids and solids. Compare and explain shape and compressibility.

23. Define (a) evaporation, (b) condensation, (c) vapor pressure as related to liquids.

24. Define (a) solvent, (b) solute, (c) solubility, (d) saturated solution, (e) unsaturated solution, (f) insoluble substance.

25. From information given in Section 2.7, calculate the molar concentration of O_2 in water saturated with air at 25°C

26. Calculate the molar concentration of solute in each of the following solutions:

(a) Exactly 7.59 g of NH_3 dissolved in water and made up to a volume of 1.500 L.

(b) Exactly 0.291 g of H_2SO_4 dissolved in water and made up to a volume of 8.65 L.

(c) Exactly 85.2 g of NaOH dissolved in water and made up to a volume of 5.00 L.

(d) Exactly 31.25 g of NaCl dissolved in water and made up to a volume of 2.00 L.

(e) Exactly 6.00 mg of atmospheric N_2 dissolved in water in a total solution volume of 2.00 L.

(f) Exactly 1,000 g of ethylene glycol, $C_2H_6O_2$ dissolved in water and made up to a volume of 2.00 L.

27. Why is the solid state called the most organized form of matter?

28. As related to solids, define (a) sublimation, (b) crystalline solids, and (c) amorphous solids.

29. Experimentally, how could a chemist be sure that (a) a melting point temperature was being observed, (c) the melting point observed was that of a pure substance, and (c) the melting point observed was that of a mixture.

30. Experimentally, how could a chemist be sure that (a) a boiling point temperature was being observed, (c) the boiling point observed was that of a pure substance, and (c) the boiling point observed was that of a mixture.

31. Calculate the specific heat of substances for which

(a) 22.5 J of heat energy was required to raise the temperature of 2.60 g of the substance from 11.5 to 16.2°C

(b) 45.6 J of heat energy was required to raise the temperature of 1.90 g of the substance from 8.6 to 19.3°C

(c) 1.20 J of heat energy was required to raise the temperature of 3.00 g of the substance from 25.013 to 25.362°C

32. Calculate the amount of heat required to raise the temperature of 7.25 g of liquid water from 22.7°C to 29.2°C.

33. Distinguish between specific heat and (a) heat of vaporization and (b) heat of fusion.

34. Calculate the heat, q, (a) required to evaporate 5.20 g of liquid water and (b) released when 6.50 g of water vapor condenses.

35. Calculate the heat, q, (a) required to melt 5.20 g of ice and (b) released when 6.50 g of liquid water freezes.

36. Suppose that salt were removed from seawater by distillation. In reference to the apparatus shown in Figure 2.11, explain where the pure water product would be isolated and where the salt would be found.

37. How does fractional distillation differ from simple distillation?

38. What are distillation bottoms and what do they have to do in some cases with wastes?

39. Describe solvent extraction as an example of a phase transfer process.

40. For what kind of treatment process is activated carbon used?

41. Name and describe a purification process that operates by virtue of "a membrane that is selectively permeable to water and excludes ionic solutes."

3   ATOMS AND ELEMENTS

3.1. ATOMS AND ELEMENTS

At the very beginning of this book chemistry was defined as the science of matter. Chapter 2 examined the nature of matter, largely at a macroscopic level and primarily through its physical properties. Today the learning of chemistry is simplified by taking advantage of what is known about matter at its most microscopic and fundamental level—atoms and molecules. Although atoms have been viewed so far in this book as simple, indivisible particles, they are in fact complicated bodies. Subtle differences in the arrangements and energies of electrons that make up most of the volume of atoms determine the chemical characteristics of the atoms. In particular, the electronic structures of atoms cause the periodic behavior of the elements, as described briefly in Section 1.3 and summarized in the periodic table in Figure 1.3. This chapter addresses atomic theory and atomic structure in more detail to provide a basis for the understanding of chemistry.

3.2. THE ATOMIC THEORY

The nature of atoms in relation to chemical behavior is summarized in the **atomic theory**. This theory in essentially its modern form was advanced in 1808 by John Dalton, an English schoolteacher, taking advantage of a substantial body of chemical knowledge and the contributions of others. It has done more than any other concept to place chemistry on a sound, systematic theoretical foundation. Those parts of the atomic theory that are consistent with current understanding of atoms are summarized in Figure 3.1.

Laws That Are Explained by Dalton's Atomic Theory

The atomic theory outlined in Figure 3.1 explains the following three laws that are of fundamental importance in chemistry:

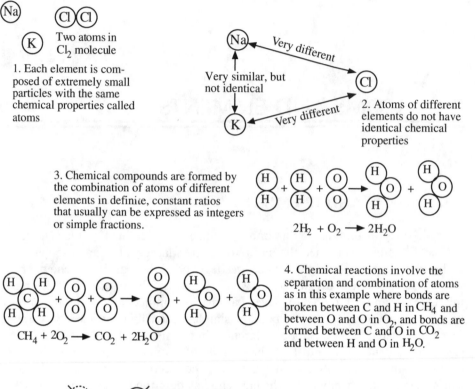

3. Chemical compounds are formed by the combination of atoms of different elements in definite, constant ratios that usually can be expressed as integers or simple fractions.

$$2H_2 + O_2 \longrightarrow 2H_2O$$

$$CH_4 + 2O_2 \longrightarrow CO_2 + 2H_2O$$

4. Chemical reactions involve the separation and combination of atoms as in this example where bonds are broken between C and H in CH_4 and between O and O in O_2, and bonds are formed between C and O in CO_2 and between H and O in H_2O.

(Atoms don't come from nothing.)

(Atoms aren't destroyed.)

(Atoms of one kind aren't changed to atoms of another kind.)

5. During the course of ordinary chemical reactions, the phenomena illustrated above do not occur: Atoms are not created, destroyed, or changed to atoms of other elements.

Figure 3.1.   Illustration of Dalton's atomic theory.

1. **Law of Conservation of Mass:** *There is no detectable change in mass in an ordinary chemical reaction.* (This law was first stated in 1798 by "the father of chemistry," the Frenchman Antoine Lavoisier. Since, as shown in Item 5 of Figure 3.1, no atoms are lost, gained, or changed in chemical reactions, mass is conserved.)

2. **Law of Constant Composition:** *A specific chemical compound always contains the same elements in the same proportions by mass.* (If atoms always combine in definite, constant ratios to form a particular chemical compound as implied in Item 3 of Figure 3.1, the elemental composition of the compound by mass always remains the same.)

3. **Law of Multiple Proportions:** *When two elements combine to form two or more compounds, the masses of one combining with a fixed mass of the other are in ratios of small whole numbers.* (This law is illustrated for two compounds composed only of carbon and hydrogen below.)

For CH_4: Relative mass of hydrogen $= 4 \times 1.0 = 4.0$ (because there are 4 atoms of H, atomic mass 1.0)

Relative mass of carbon $= 1 \times 12.0 = 12.0$ (because there is 1 atom of C, atomic mass 12.0)

$$\text{C/H ratio for } CH_4 = \frac{\text{Mass of C}}{\text{Mass of H}} = \frac{12.0}{4.0} = 3.0$$

For C_2H_6: Relative mass of hydrogen $= 6 \times 1.0 = 6.0$ (because there are 6 atoms of H, atomic mass 1.0)

Relative mass of carbon $= 2 \times 12.0 = 24.0$ (because there are 2 atoms of C, atomic mass 12.0)

$$\text{C/H ratio for } C_2H_6 = \frac{\text{Mass of C}}{\text{Mass of H}} = \frac{24.0}{6.0} = 4.0$$

$$\text{Comparing } C_2H_6 \text{ and } CH_4: \quad \frac{\text{C/H ratio for } C_2H_6}{\text{C/H ratio for } CH_4} = \frac{4.0}{3.0} = 4/3$$

Note that this is a ratio of small, whole numbers.

Small Size of Atoms

It is difficult to imagine just how small an individual atom is. An especially small unit of mass, the *atomic mass unit, u,* is used to express the masses of atoms. This unit was defined in Section 1.3 as a mass equal to exactly 1/12 that of the carbon-12 isotope. An atomic mass unit is only 1.66×10^{-24} g. An average atom of hydrogen, the lightest element, has a mass of only 1.0079 u. The average mass of an atom of uranium, the heaviest naturally-occurring element, is 238.03 u. To place these values in perspective, consider that a signature written by ballpoint pen on a piece of paper typically has a mass of 0.1 mg (1×10^{-4} g). This mass is equal to 6×10^{19} u. It would take almost 60 000 000 000 000 000 000 hydrogen atoms to equal the mass of ink in such a signature!

Atoms visualized as spheres have diameters of around $1\text{–}3 \times 10^{-10}$ meters (100–300 picometers, pm). By way of comparison, a small marble has a diameter of around 1 cm (1×10^{-2} m), which is about 100 000 000 times that of a typical atom.

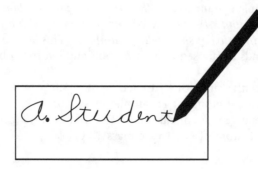

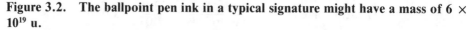

Figure 3.2. **The ballpoint pen ink in a typical signature might have a mass of 6 $\times$ 10^{19} u.**

Atomic Mass

As defined in Section 1.3, the *atomic mass* of an element is the average mass of all atoms of the element relative to carbon-12 taken as exactly 12. Since atomic masses are *relative* quantities, they can be expressed without units. Or atomic masses may be given in atomic mass units. For example, an atomic mass of 14.0067 for nitrogen means that the average mass of all nitrogen atoms is 14.0067/12 as great as the mass of the carbon-12 isotope and is also 14.0067 u.

3.3. SUBATOMIC PARTICLES

Small as atoms are, they in turn consist of even smaller entities called **subatomic particles**. Although physicists have found several dozen of these, chemists need consider only three—*protons, neutrons*, and *electrons* (these were introduced briefly in Section 1.3). These subatomic particles differ in mass and charge. Like the atom, their masses are expressed in atomic mass units.

The proton, *p*, has a mass of 1.007277 u and a unit charge of + 1. This charge is equal to 1.6022×10^{-19} coulombs, where a coulomb is the amount of electrical charge involved in a flow of electrical current of 1 ampere for 1 second.

The neutron, *n*, has no electrical charge and a mass of 1.009665 u. The proton and neutron each have a mass of essentially 1 u and are said to have a *mass number* of 1. (Mass number is a useful concept expressing the total number of protons and neutrons, as well as the approximate mass, of a nucleus or subatomic particle.)

The electron, *e*, has an electrical charge of –1. It is very light, however, with a mass of only 0.00054859 u, about 1/1840 that of the proton or neutron. Its mass number is 0. The properties of protons, neutrons, and electrons are summarized in Table 3.1.

Table 3.1.   Properties of Protons, Neutrons, and Electrons

Subatomic Particle	Symbol	Unit Charge	Mass Number[a]	Mass in u	Mass in grams
Proton	p	+1	1	1.007277	1.6726×10^{-24}
Neutron	n	0	1	1.008665	1.6749×10^{-24}
Electron	e	−1	0	0.000549	9.1096×10^{-28}

[a]The mass number and charge of each of these kinds of particles may be indicated by a superscript and subscript, respectively in the symbols $^1_1 p$, $^1_0 n$, and $^0_{-1} e$.

Although it is convenient to think of the proton and neutron as having the same mass, and each is assigned a mass number of 1, it is seen in Table 3.1 that their exact masses differ slightly from each other. Furthermore, the mass of an atom differs slightly from the sum of the masses of subatomic particles composing the atom. This is because of the energy relationships involved in holding the subatomic particles together in atoms so that the masses of the atom's constituent subatomic particles do not add up to exactly the mass of the atom.

3.4.   THE BASIC STRUCTURE OF THE ATOM

As mentioned in Section 1.3, protons and neutrons are located in the *nucleus* of an atom; the remainder of the atom consists of a cloud of rapidly moving electrons. Since protons and neutrons have much higher masses than electrons, essentially all the mass of an atom is in its nucleus. However, the electron cloud makes up virtually all of the volume of the atom, and the nucleus is very small.

Atomic Number, Isotopes, and Mass Number of Isotopes

All atoms of the same element have the same number of protons and electrons equal to the *atomic number* of the element. (When reference is made to atoms here it is understood that they are electrically neutral atoms and not ions consisting of atoms that have lost or gained 1 or more electrons.) Thus all helium atoms have 2 protons in their nuclei, and all nitrogen atoms have 7 protons.

In Section 1.3, *isotopes* were defined as atoms of the same element that differ in the number of neutrons in their nuclei. It was noted in Section 3.3 that both the proton and neutron have a *mass number* of exactly 1. Mass number is commonly used to denote isotopes. The **mass number of an isotope** *is the sum of the number of protons and neutrons in the nucleus of the isotope.* Since atoms with the same number of protons—that is, atoms of the same element—may have different numbers of neutrons, several isotopes may exist of a particular element. The naturally-occurring forms of some elements consist of only one isotope, but isotopes of these elements can be made artificially, usually by exposing the elements to neutrons produced by a nuclear reactor.

It is convenient to have a symbol that clearly designates an isotope of an element. Borrowing from nuclear science, a superscript in front of the symbol for the element is used to show the mass number and a subscript before the symbol designates the number of protons in the nucleus (atomic number). Using this notation denotes carbon-12 as $^{12}_{6}C$.

Exercise: Fill in the blanks designated with letters in the table below:

Element	Atomic Number	Mass Number of Isotope	Number of Neutrons in Nucleus	Isotope Symbol
Nitrogen	7	14	7	$^{14}_{7}N$
Chlorine	17	35	(a)_____	$^{35}_{17}Cl$
Chlorine	(b)_____	37	(c)_____	(d)_____
(e)_____	6	(f)_____	7	(g)_____
(h)_____	(i)_____	(j)_____	(k)_____	$^{11}_{5}B$

Answers: (a) 18, (b) 17, (c) 20, (d) $^{37}_{17}Cl$, (e) carbon, (f) 13, (g) $^{13}_{6}C$, (h) boron, (i) 5, (j) 11, (k) 6

Electrons in Atoms

In Section 1.3 electrons around the nucleus of an atom were depicted as forming a cloud of negative charge. The energy levels, orientations in space, and behavior of electrons varies with the number of them contained in an atom. In a general sense, the arrangements of electrons in atoms are described by *electron configuration*, a term discussed in some detail later in this chapter.

Attraction Between Electrons and the Nucleus

The electrons in an atom are held around the nucleus by the attraction between their negative charges and the positive charges of the protons in the nucleus. Opposite electrical charges attract, and like charges repel. The forces of attraction and repulsion are expressed quantitatively by **Coulomb's law**,

$$F = \frac{Q_1 Q_2}{d^2}$$ (3.4.1)

where F is the force of interaction, Q_1 and Q_2 are the electrical charges on the bodies involved, and d is the distance between the bodies.

Coulomb's law explains the attraction between negatively charged electrons and the positively charged nucleus in an atom. However, the law does not explain why electrons move around the nucleus, rather than coming to rest on it. The behavior of electrons in atoms — as well as the numbers, types, and strengths of chemical bonds formed between atoms — is explained by *quantum theory*, a concept that is discussed in some detail in Sections 3.11 to 3.17.

3.5.   DEVELOPMENT OF THE PERIODIC TABLE

The *periodic table*, which was mentioned briefly in connection with the elements in Section 1.3, was first described by the Russian chemist Dmitry Mendeleyev in 1867. It was based upon Mendeleyev's observations of periodicity in chemical behavior without any knowledge of atomic structure. The periodic table is discussed in more detail in this chapter along with the development of the concepts of atomic structure. Elements are listed in the periodic table in an ordered, systematic way that correlates with their electron structures. The elements are placed in rows or **periods** of the periodic table in order of increasing atomic number so that there is a periodic repetition of elemental properties across the periods. The rows are arranged so that elements in vertical columns called **groups** have similar chemical properties reflecting similar arrangements of the outermost electrons in their atoms.

With some knowledge of the ways that electrons behave in atoms it is much easier to develop the concept of the periodic table. This is done for the first 20 elements in the following sections. After these elements are discussed, they are placed in an abbreviated 20-element version of the periodic table as shown in Figure 3.9.

3.6.   HYDROGEN, THE SIMPLEST ATOM

The simplest atom is that of **hydrogen**, having only one positively charged proton in its nucleus, which is surrounded by a cloud of negative charge formed by only one electron. By far the most abundant kind of hydrogen atom has no neutrons in its nucleus. Having only one proton with a mass number (see Section 3.4) of 1, and 1 electron with a mass number of zero, the mass number of this form of hydrogen is $1 + 0 = 1$. There are, however, two other forms of hydrogen atoms having, in addition to the proton, 1 and 2 neutrons, respectively, in their nuclei. The three different forms of elemental hydrogen are *isotopes* of hydrogen that all have the same number of protons, but different numbers of neutrons in their nuclei. Only about 1 of 7000 hydrogen atoms is ^2_1H, deuterium, mass number 2 (from 1 proton + 1 neutron). The mass number of *tritium*, which has 2 neutrons, is 3 (from 1 proton + 2 neutrons) = 3. These three forms of hydrogen atoms may be designated as ^1_1H, ^2_1H, and ^3_1H. The meaning of this notation is reviewed below:

Mass number (number
of protons plus neutrons
in the nucleus)   —— Symbol of element

$$^{3}_{1}\text{H}$$

Atomic number (number
of protons in the nucleus)

Designation of Hydrogen in the Periodic Table

The atomic mass of hydrogen is 1.0079. This means that the average atom of hydrogen has a mass of 1.0079 u (atomic mass units); hydrogen's atomic mass is simply 1.0079 relative to the carbon-12 isotope taken as exactly 12. With this information it is possible to place hydrogen in the periodic table with the designation shown in Figure 3.3.

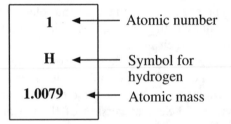

1	←—— Atomic number
H	←—— Symbol for hydrogen
1.0079	←—— Atomic mass

Figure 3.3.   Designation of hydrogen in the periodic table.

Showing Electrons in Hydrogen Atoms and Molecules

It is useful to have some simple way of showing the hydrogen atom's single electron in chemical symbols and formulas. This is accomplished with **electron-dot symbols** or **Lewis symbols** (after G. N. Lewis), which use dots around the symbol of an element to show outer electrons (those that may become involved in chemical bonds). The Lewis symbol for hydrogen is

H•

As mentioned in Section 1.4, elemental hydrogen consists of molecules made up of 2 H atoms held together by a chemical bond consisting of two shared electrons. Just as it is useful to show electrons in atoms with a Lewis symbol, it is helpful to visualize molecules and the electrons in them with **electron-dot formulas** or **Lewis formulas** as shown for the H_2 molecule in Figure 3.4.

$$H\cdot \longrightarrow \overleftarrow{} \cdot H$$

$$H_2$$

$$H:H$$

Elemental hydrogen does not exist as individual H atoms.

Instead, it exists as molecules made up to 2 H atoms with the chemical formula H_2.

The covalent bond holding the two H atoms together consists of 2 shared electrons shown in the Lewis formula for H_2 above.

Figure 3.4. Lewis formula for hydrogen molecules.

Properties of Elemental Hydrogen

Pure hydrogen is colorless and odorless. It is a gas at all but very low temperatures; liquid hydrogen boils at $-253\,^\circ$C and solidifies at $-259\,^\circ$C. Hydrogen gas has the lowest density of any pure substance. It reacts chemically with a large number of elements, and hydrogen-filled balloons that float readily in the atmosphere also explode convincingly when touched with a flame because of the following very rapid reaction with oxygen in the air:

$$2H_2 + O_2 \rightarrow 2H_2O \tag{3.6.1}$$

Production and Uses of Elemental Hydrogen

Elemental hydrogen is one of the more widely produced industrial chemicals. On an industrial scale, it is commonly made by **steam reforming** of methane (natural gas, CH_4) under high-temperature, high-pressure conditions:

$$CH_4 + H_2O \xrightarrow[\text{30 atm P}]{\text{800}^\circ\text{C T}} CO + 3H_2 \tag{3.6.2}$$

Hydrogen is used to manufacture a number of chemicals. One of the most important of these is ammonia, NH_3. Methanol (methyl alcohol, CH_3OH) is a widely used industrial chemical and solvent synthesized by the following reaction between carbon monoxide and hydrogen:

$$CO + 2H_2 \rightarrow CH_3OH \tag{3.6.3}$$

Methanol made by this process can be blended with gasoline to yield a fuel that produces relatively less pollutant carbon monoxide; such "oxygenated gasoline additives" are now required for use in some cities during winter months. Gasoline is upgraded by the chemical addition of hydrogen to some petroleum fractions. Synthetic petroleum can be made by the addition of hydrogen to coal at high temperatures and pressures. A widely used process in the food industry is the addition of hydrogen to unsaturated vegetable oils to give margarine and other hydrogenated fats and oils.

3.7. HELIUM, THE FIRST ATOM WITH A FILLED ELECTRON SHELL

The *electron configurations* of atoms determine their chemical behavior. Electrons in atoms occupy distinct *energy levels*. At this point it is useful to introduce the concept of the **electron shell** to help explain electron energy levels and their influence on chemical behavior. Each electron shell can hold a maximum number of electrons. An atom with a **filled electron shell** is especially content in a chemical sense, with little or no tendency to lose, gain, or share electrons. Elements with these characteristics exist as gas-phase atoms and are called *noble gases*. The high stability of the noble gas electron configuration is explained in more detail in Section 3.9.

Examined in order of increasing atomic number, the first element consisting of atoms with filled electron shells is helium, He, atomic number 2. All helium atoms contain 2 protons and 2 electrons. Virtually all helium atoms are $_2^4He$ that contain 2 neutrons in their nuclei; the $_2^3He$ isotope containing only 1 neutron in its nucleus occurs to a very small extent. The atomic mass of helium is 4.00260.

The two electrons in the helium atom are shown by the Lewis symbol illustrated in Figure 3.5. These electrons constitute a *filled electron shell*, so that helium is a *noble gas* composed of individual helium atoms that have no tendency to form chemical bonds with other atoms. Helium gas has a very low density of only 0.164 g/L at 25°C and 1 atm pressure.

Occurrence and Uses of Helium

Helium is extracted from some natural gas sources that contain up to 10% helium by volume. It has many uses that depend upon its unique properties. Because of its very low density compared to air, helium is used to fill weather balloons and airships. Helium is nontoxic, odorless, tasteless, and colorless. Because of these properties and its low solubility in blood, helium is mixed with

He:

A helium atom has a filled electron
shell containing 2 electrons.

It can be represented by the Lewis
symbol above.

Figure 3.5.   Two representations of the helium atom having a filled electron shell

oxygen for breathing by deep sea divers and persons with some respiratory ailments. Use of helium by divers avoids the very painful condition called "the bends" caused by bubbles of nitrogen forming from nitrogen gas dissolved in blood.

Liquid helium, which boils at a temperature of only 4.2 K above absolute zero is especially useful in the growing science of **cryogenics**, which deals with very low temperatures. Some metals are superconductors at such temperatures so that helium is used to cool electromagnets that develop very powerful magnetic fields for a relatively small magnet. Such magnets are components of the very useful chemical tool known as nuclear magnetic resonance (NMR). The same kind of instrument modified for clinical applications and called MRI is used as a medical diagnostic tool.

3.8.   LITHIUM, THE FIRST ATOM WITH BOTH INNER AND OUTER ELECTRONS

The third element in the periodic table is lithium (Li), atomic number 3, atomic mass 6.941. The most abundant lithium isotope has 4 neutrons in its nucleus, so it has a mass number of 7 and is designated ^7_3Li. A less common isotope, ^6_3Li, has only 3 neutrons.

Lithium is the first element in the periodic table that is a *metal*. Mentioned in Section 2.2, metals tend to have the following properties:

- Characteristic **luster** (like freshly-polished silverware or a new penny)
- **Malleable** (can be pounded or pressed into various shapes without breaking)
- **Conduct electricity**
- Chemically, tend to **lose electrons** and form cations (see Section 1.4) with charges of +1 to +3.

Lithium is the lightest metal, with a density of only 0.531 g/cm^3.

Uses of Lithium

Lithium compounds have a number of important uses in industry and medicine. Lithium carbonate, Li_2CO_3, is one of the most important lithium compounds and is used as the starting material for the manufacture of many other lithium compounds. It is an ingredient of specialty glasses, enamels, and specialty ceramic ware having low thermal expansion coefficients (minimum expansion when heated). Lithium carbonate is widely prescribed as a drug to treat acute mania in manic-depressive and schizo-affective mental disorders. Lithium hydroxide, LiOH, is an ingredient in the manufacture of some lubricant greases and in some long-life alkaline storage batteries.

The lithium atom has three electrons. As shown in Figure 3.6, lithium has both **inner electrons** — in this case 2 contained in an **inner shell** — as in the immediately preceding noble gas helium, and an **outer electron** that is farther from, and less strongly attracted to, the nucleus. The outer electron is said to be in the atom's **outer shell**. The inner electrons are, on the average, closer to the nucleus than is the outer electron, are very difficult to remove from the atom, and do not become involved in chemical bonds. Lithium's outer electron is relatively easy to remove from the atom, which is what happens when ionic bonds involving Li^+ ion are formed (see Section 1.4). The distinction between inner and outer electrons is developed to a greater extent later in this chapter.

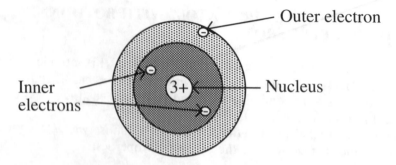

Figure 3.6.   An atom of lithium, Li, has 2 inner electrons and 1 outer electron. The latter can be lost to another atom to produce the Li^+ ion, which is present in ionic compounds (see Section 1.4).

In atoms such as lithium that have both outer and inner electrons, the Lewis symbol shows only the outer electrons. Therefore, the Lewis symbol of lithium is

$$Li\cdot$$

A lithium atom's loss of its single outer electron is shown by the **half-reaction** (one in which there is a net number of electrons on either the reactant or product side) in Figure 3.7. The Li^+ product of this reaction has the very stable helium core of 2 electrons. The Li^+ ion is a constituent of ionic lithium compounds in which it is held by attraction for negatively charged anions (such as Cl^-) in the crystalline lattice of the ionic compound. The tendency to lose its outer electron and to be stabilized in ionic compounds determines lithium's chemical behavior.

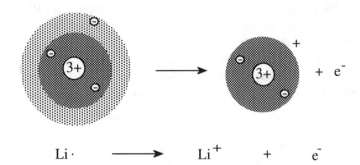

Li · ⟶ Li⁺ + e⁻

A neutral lithium atom loses an electron to form a positively charged lithium ion and an electron that is acquired by an atom of some other element.

Figure 3.7.   Half-reaction showing the formation of Li⁺ from an Li atom. The Li⁺ ion has the especially stable helium core of just 2 electrons. The atom to which the electron is lost is not shown, so this is a half-reaction.

3.9.   THE SECOND PERIOD, ELEMENTS 4–10

In this section elements 4–10 will be discussed and placed in the periodic table to complete a period in the table.

Beryllium, Atomic Number 4

Each atom of **beryllium** — atomic number 4, atomic mass 9.01218 — contains 4 protons and 5 neutrons in its nucleus. The beryllium atom has two inner electrons and two outer electrons, the latter designated by the two dots in the Lewis symbol below:

$$\text{Be:}$$

Beryllium can react chemically by losing 2 electrons from the beryllium atom. This occurs according to the half reaction

$$\text{Be:} \rightarrow \text{Be}^{2+} + 2\text{e}^- \text{ (lost to another atom)} \qquad (3.9.1)$$

in which the beryllium atom, Lewis symbol Be:, loses two e⁻ to form a beryllium ion with a charge of +2. The loss of these two outer electrons gives the beryllium atom the same stable helium core as that of the Li⁺ ion discussed in the preceding section.

Beryllium is melted together with certain other metals to give homogeneous mixtures of metals called **alloys**. The most important beryllium alloys are hard, corrosion-resistant, nonsparking, and good conductors of electricity. They are used to make such things as springs, switches, and small electrical contacts. A very high melting temperature of about 1290°C combined with good heat absorption and conduction properties has led to the use of beryllium metal in aircraft brake components.

Beryllium is an environmentally and toxicologically important element because it causes **berylliosis,** a disease marked by lung deterioration. Inhalation is particularly hazardous, and atmospheric standards have been set at very low levels.

Boron, Atomic Number 5

Boron, B, has an atomic number of 5 and an atomic mass of 10.81. Most boron atoms have 6 neutrons in addition to 5 protons in their nuclei; a less common isotope has 5 neutrons. Two of boron's 5 electrons are in a helium core and 3 are outer electrons, as shown by the Lewis symbol

$$\dot{B}:$$

Boron—along with silicon, germanium, arsenic, antimony, and tellurium—is one of a few elements called **metalloids** with properties intermediate between those of metals and nonmetals. Although they have a luster like metals, metalloids do not form positively-charged ions (cations). The melting temperature of boron is very high, 2190°C. Boron is added to copper, aluminum, and steel to improve their properties. It is used in control rods of nuclear reactors because of the good neutron-absorbing properties of the $^{10}_{5}B$ isotope. Some chemical compounds of boron, especially boron nitride, BN, are noted for their hardness. Boric acid, H_3BO_3, is used as a flame retardant in cellulose insulation in houses. The oxide of boron, B_2O_3, is an ingredient of fiberglass used in textiles and insulation.

Carbon, Atomic Number 6

Atoms of **carbon,** C, have 2 inner and 4 outer electrons, the latter shown by the Lewis symbol

The carbon-12 isotope with 6 protons and 6 neutrons in its nucleus, $^{12}_{6}C$, constitutes 98.9% of all naturally occurring carbon. The $^{13}_{6}C$ isotope makes up 1.1% of all carbon atoms. Radioactive carbon-14, $^{14}_{6}C$, is present in some carbon sources.

Carbon is an extremely important element with unique chemical properties, without which life could not exist. All of organic chemistry (Chapter 9) is based upon compounds of carbon, and it is an essential element in life molecules (studied as part of biochemistry, Chapter 10). Carbon atoms are able to bond to each other to form long straight chains, branched chains, rings, and three-dimensional structures. As a result of its self-bonding abilities, carbon exists in several elemental forms. These include powdery carbon black, very hard, clear diamonds, and graphite so soft that it is used as a lubricant. Activated carbon prepared by treating carbon with air, carbon dioxide, or steam at high temperatures is widely used to absorb undesirable pollutant substances from air and water. Carbon fiber has been developed as a structural material in the form of composites consisting of strong strands of carbon bonded together with special plastics and epoxy resins.

Nitrogen, Atomic Number 7

Nitrogen, N, composes 78% by volume of air in the form of N_2 molecules. The atomic mass of nitrogen is 14.0067, and the nuclei of nitrogen atoms contain 7 protons and 7 neutrons. Nitrogen has 5 outer electrons, so its Lewis symbol is

$$\cdot \ddot{\underset{\cdot}{N}} \colon$$

Like carbon, nitrogen is a nonmetal. Pure N_2 is prepared by distilling liquified air, and it has a number of uses. Since nitrogen gas is not very chemically reactive, it is used as an inert atmosphere in some industrial applications, particularly where fire or chemical reactivity may be a hazard. People have been killed by accidentally entering chambers filled with nitrogen gas, which acts as a simple asphyxiant with no warning odor of its presence. Liquid nitrogen boils at a very cold -190°C. It is widely used to maintain very low temperatures in the laboratory, for quick-freezing foods, and in freeze-drying processes. Freeze-drying is used to isolate fragile biochemical compounds from water solution, for the concentration of environmental samples to be analyzed for pollutants, and for the preparation of instant coffee and other dehydrated foods. It has potential applications in the concentration and isolation of hazardous waste substances.

Like carbon, nitrogen is an essential element for life processes. Nitrogen is an ingredient of all of the amino acids found in proteins. Nitrogen compounds are fertilizers essential for the growth of plants. The **nitrogen cycle**, which involves incorporation of N_2 from the atmosphere into living matter and chemically-bound nitrogen in soil and water, then back into the atmosphere again, is one of nature's fundamental cycles. Nitrogen compounds, particularly ammonia (NH_3), and nitric acid (HNO_3), are widely used industrial chemicals.

Oxygen, Atomic Number 8

Like carbon and nitrogen, oxygen, atomic number 8, is a major component of living organisms. Oxygen is a nonmetal existing as molecules of O_2 in the elemental gas state, and air is 21% oxygen by volume. Like all animals, humans require oxygen to breathe and to maintain their life processes. The nuclei of oxygen atoms contain 8 protons and 8 neutrons, and the atomic mass of oxygen is 15.9994. The oxygen atom has 6 outer electrons as shown by its Lewis symbol below:

$$\cdot \ddot{\underset{\cdot}{O}} \colon$$

In addition to O_2, there are two other important elemental oxygen species in the atmosphere. These are atomic oxygen, O, and ozone, O_3. These species are normal constituents of the stratosphere, a region of the atmosphere that extends from about 11 kilometers to about 50 km in altitude. Oxygen atoms are formed when high-energy ultraviolet radiation strikes oxygen molecules high in the stratosphere:

$$O_2 \quad \xrightarrow[\text{radiation}]{\text{Ultraviolet}} \quad O + O \qquad (3.9.2)$$

The oxygen atoms formed by the above reaction combine with O_2 molecules,

$$O + O_2 \rightarrow O_3 \qquad (3.9.3)$$

to form ozone molecules. These molecules make up the **ozone layer** in the stratosphere and effectively absorb additional high-energy ultraviolet radiation. If it weren't for this phenomenon, the ultraviolet radiation would reach the Earth's surface and cause painful sunburn and skin cancer in exposed people. However, ozone produced in photochemical smog at ground level is toxic to animals and plants.

The most notable chemical characteristic of elemental oxygen is its tendency to combine with other elements in energy-yielding reactions. Such reactions provide the energy that propels automobiles, heats buildings, and keeps body processes going. One of the most widely used chemical reactions of oxygen is that with hydrocarbons, particularly those from petroleum and natural gas. For example, butane (C_4H_{10}, a liquifiable gaseous hydrocarbon fuel) burns in oxygen from the atmosphere,

$$2C_4H_{10} + 13O_2 \rightarrow 8CO_2 + 10H_2O \qquad\qquad (3.9.4)$$

a reaction that provides heat in home furnaces, water heaters, and other applications.

Fluorine, Atomic Number 9

Fluorine, F, has 7 outer electrons, so its Lewis symbol is

Under ordinary conditions elemental fluorine is a greenish-yellow gas consisting of F_2 molecules.

Fluorine compounds have many uses. One of the most notable of these is the manufacture of chlorofluorocarbon compounds known by the trade name Freons. These are chemical combinations of chlorine, fluorine, and carbon, an example of which is dichlorodifluoromethane, Cl_2CF_2. These compounds until recently were widely used as refrigerant fluids and blowing agents to make foam plastics; they were once widely employed as propellants in aerosol spray cans. Uses of chlorofluorocarbons are now being phased out because of their role in destroying stratospheric ozone (discussed with oxygen, above).

Neon, Atomic Number 10

The last element in the period of the periodic table under discussion is **neon**. Air is about 2 parts per thousand neon by volume, and neon is obtained by the distillation of liquid air. Neon is especially noted for its use in illuminated signs that consist of glass tubes containing neon, through which an electrical current is passed causing the neon gas to emit a characteristic glow.

In addition to 10 protons, most neon atoms have 10 neutrons in their nuclei, although some have 12, and a very small percentage have 11. As shown by its Lewis symbol,

the neon atom has 8 outer electrons. These 8 electrons constitute a *filled electron shell*, just as the 2 electrons in helium give it a filled electron shell. Because of this "satisfied" outer shell, the neon atom has no tendency to acquire, give away, or share electrons. Therefore, neon is a *noble gas*, like helium, and consists of individual neon atoms.

Stability of the Neon Noble Gas Electron Octet

In going through the rest of the periodic table it can be seen that all other atoms with 8 outer electrons like neon are also noted for a high degree of chemical stability. In addition to neon, these noble gases are argon (atomic number 18), krypton (atomic number 36), xenon (atomic number 54), and radon (atomic number 86). Each of these may be represented by the Lewis symbol

where X is the chemical symbol of the noble gas. It is seen that these atoms each have 8 outer electrons, a group known as an **octet** of electrons. In many cases atoms that do not have an octet of outer electrons acquire one by losing, gaining, or sharing electrons in chemical combination with other atoms; that is, they acquire a **noble gas outer electron configuration**. For all noble gases except helium, which has only 2 electrons, the noble gas outer electron configuration consists of eight electrons. The tendency of elements to acquire an 8-electron outer electron configuration, which is very useful in predicting the nature of chemical bonding and the formulas of compounds that result, is called the **octet rule**. Although the use of the octet rule to explain and predict bonding is discussed in some detail in Chapter 4, at this point it is useful to show how it explains bonding between hydrogen and carbon in methane:

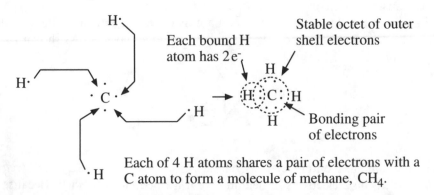

Figure 3.8. Illustration of the octet rule in methane.

3.10.   ELEMENTS 11–20, AND BEYOND

The abbreviated version of the periodic table will be finished with elements 11 through 20. The names, symbols, electron configurations, and other pertinent information about these elements are given in Table 3.2. An abbreviated periodic table with these elements in place is shown in Figure 3.9. This table shows the Lewis symbols of all of the elements to emphasize their orderly variation across periods and similarity in groups of the periodic table.

The first 20 elements in the periodic table are very important. They include the three most abundant elements on the earth's surface (oxygen, silicon, aluminum); all elements of any appreciable significance in the atmosphere (hydrogen in H_2O vapor, N_2, O_2, carbon in CO_2, argon, and neon); the elements making up most of living plant and animal matter (hydrogen, oxygen, carbon, nitrogen, phosphorus, and sulfur); and elements such as sodium, magnesium, potassium, calcium, and chlorine that are essential for life processes. The chemistry of these elements is relatively straightforward and reasonably easy to relate to their atomic structures. Therefore, emphasis is placed on them in the earlier chapters of this book. It is helpful to remember their names, symbols, atomic numbers, atomic masses, and Lewis symbols.

As mentioned in Section 1.3, the vertical columns of the table contain *groups* of elements that have similar chemical structures. One exception is hydrogen, H, which has unique chemical properties. All elements other than hydrogen in the first column of the abbreviated table are **alkali metals**—lithium, sodium, and potassium. These are generally soft silvery-white metals of low density that react violently with water to produce hydroxides (LiOH, NaOH, KOH) and with chlorine to produce chlorides (LiCl, NaCl, KCl). The **alkaline earth** metals—beryllium, magnesium, calcium—are in the second column of the table. When

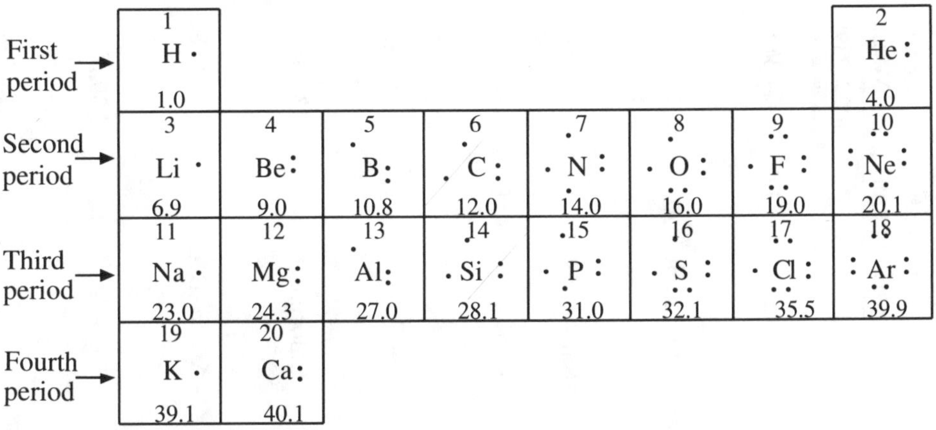

	1								2
First period →	H · 1.0								He : 4.0
Second period →	3 Li · 6.9	4 Be: 9.0	5 B: 10.8	6 · C: 12.0	7 · N: 14.0	8 · O: 16.0	9 · F: 19.0	10 : Ne: 20.1
Third period →	11 Na · 23.0	12 Mg: 24.3	13 Al: 27.0	14 · Si: 28.1	15 · P: 31.0	16 · S: 32.1	17 · Cl: 35.5	18 : Ar: 39.9
Fourth period →	19 K · 39.1	20 Ca: 40.1						

Figure 3.9.   Abbreviated 20-element version of the periodic table showing Lewis symbols of the elements.

Table 3.2.   Elements 11–20

Atomic Number	Name and Lewis Symbol	Atomic Mass	Number of Outer e^-	Major Properties and Uses
11	Sodium (Na·)	22.9898	1	Soft, chemically very reactive metal. Nuclei contain 11 p and 12 n.
12	Magnesium (Mg:)	24.312	2	Lightweight metal used in aircraft components, extension ladders, portable tools. Chemically very reactive. Three isotopes with 12, 13, 14 n.
13	Aluminum (A̤l:)	26.9815	3	Lightweight metal used in aircraft, automobiles, electrical transmission line. Chemically reactive, but forms self-protective coating.
14	Silicon (·Si:)	28.086	4	Nonmetal, 2nd most abundant metal in Earth's crust. Rock constituent. Used in semiconductors.
15	Phosphorus (·P̤:)	30.9738	5	Chemically very reactive nonmetal. Highly toxic as elemental white phosphorus. Component of bones and teeth, genetic material (DNA), fertilizers, insecticides.
16	Sulfur (·S̤:)	32.064	6	Brittle, generally yellow nonmetal. Essential nutrient for plants and animals, occurring in amino acids. Used to manufacture sulfuric acid. Present in pollutant sulfur dioxide, SO_2.
17	Chlorine (·C̤l:)	35.453	7	Greenish-yellow toxic gas composed of molecules of Cl_2. Manufactured in large quantities to disinfect water and to manufacture plastics and solvents.
18	Argon (:A̤r·)	39.948	8	Noble gas used to fill light bulbs and as a plasma medium in inductively coupled plasma atomic emission analysis of elemental pollutants.
19	Potassium (K·)	39.098	1	Chemically reactive alkali metal very similar to sodium in chemical and physical properties. Essential fertilizer for plant growth as K^+ ion.
20	Calcium (Ca:)	40.078	2	Chemically reactive alkaline earth metal with properties similar to those of magnesium.

freshly cut, these metals have a grayish-white luster. They are chemically reactive and have a strong tendency to form doubly-charged cations (Be^{2+}, Mg^{2+}, Ca^{2+}) by losing two electrons from each atom. Another group notable for the very close similarities of the elements in it consists of the **noble gases** in the far right column of the table. Each of these — helium, neon, argon — is a monatomic gas that does not react chemically.

The Elements Beyond Calcium

The electron structures of elements beyond atomic number 20 are more complicated than those of the first 20 elements just discussed. As shown in the complete periodic table in Figure 1.3, included among these heavier elements are transition metals, the lanthanides, and the actinides. The transition metals include many metals that are important in industry and in life processes. Among the transition metals are chromium, manganese, iron, cobalt, nickel, and copper. The actinides contain thorium, uranium, and plutonium — familiar names to those concerned with nuclear energy, nuclear warfare, and related issues.

3.11.   A MORE DETAILED LOOK AT ATOMIC STRUCTURE

So far, this chapter has covered some important aspects of atoms. These include the facts that an atom is made of three major subatomic particles, and consists of a very small, very dense, positively charged nucleus surrounded by a cloud of negatively charged electrons in constant, rapid motion. The first 20 elements have been discussed in some detail and placed in an abbreviated version of the periodic table. Important concepts introduced so far in this chapter include:

- Dalton's atomic theory
- Electron shells
- Inner shell electrons
- Octet rule

- Lewis symbols to represent outer e^-
- Significance of filled electron shells
- Outer shell electrons
- Abbreviated periodic table

The information presented about atoms so far in this chapter is adequate to meet the needs of many readers. These readers may choose to forego the details of atomic structure presented in the rest of this chapter without major harm to their understanding of chemistry. However, for those who wish to go into more detail, or who do not have a choice, the remainder of this chapter discusses in more detail the electronic structures of atoms as related to their chemical behavior and introduces the quantum theory of electrons in atoms.

Electromagnetic Radiation

The **quantum theory** explains the unique behavior of charged particles that are as small and move as rapidly as electrons. Because of its close relationship to electromagnetic radiation, an appreciation of quantum theory requires an understanding of the following important points related to electromagnetic radiation:

- Energy can be carried through space at the speed of light, 3.00×10^8 meters per second (m/s) in a vacuum, by **electromagnetic radiation**, which includes visible light, ultraviolet radiation, infrared radiation, microwaves, and radio waves.
- Electromagnetic radiation has a **wave character**. The waves move at the speed of light, c, and have characteristics of **wavelength** (λ), amplitude, and **frequency** (ν, Greek "nu") as illustrated below:

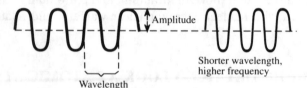

- The wavelength is the distance required for one complete cycle and the frequency is the number of cycles per unit time. They are related by the following equation:

$$\nu \lambda = c$$

where ν is in units of cycles per second (s^{-1}, a unit called the **hertz**, Hz) and λ is in meters (m).
- In addition to behaving as a wave, electromagnetic radiation also has characteristics of particles.
- The dual wave/particle nature of electromagnetic radiation is the basis of the **quantum theory** of electromagnetic radiation, which states that radiant energy may be absorbed or emitted only in discrete packets called **quanta** or **photons**. The energy, E, of each photon is given by

$$E = h\nu$$

where h is Planck's constant, 6.63×10^{-34} J-s (joule × second).
- From the preceding, it is seen that *the energy of a photon is higher when the frequency of the associated wave is higher* (and the wavelength shorter).

3.12. QUANTUM AND WAVE MECHANICAL MODELS OF ELECTRONS IN ATOMS

The *quantum theory* introduced in the preceding section provided the key concepts needed to explain the energies and behavior of electrons in atoms. One of the best clues to this behavior, and one that ties the nature of electrons in atoms to the properties of electromagnetic radiation, is the emission of light by energized atoms. This is easiest to explain for the simplest atom of all, that of hydrogen, which consists of only one electron moving around a nucleus with a single positive charge. Energy added to hydrogen atoms, such as by an electrical discharge through hydrogen gas, is re-emitted in the form of light at very specific wavelengths (656, 486, 434, 410 nm in the visible region). The highly energized atoms that can emit this light are said to be "excited" by the excess energy originally put into them and to be in an **excited state**. The reason for this is that the electrons in the excited atoms are forced farther from the nuclei of the atoms and, when they return to a lower energy state, energy is emitted in the form of light. The fact that very specific wavelengths of light are emitted in this process means that electrons can be present only in specified states at highly specific energy levels. Therefore, the transition from one energy state to a lower one involves the emission of a specific energy of electromagnetic radiation (light). Consider the equation

$$E = h\nu \tag{3.12.1}$$

that relates energy to frequency, ν, of electromagnetic radiation. If a transition of an electron from one excited state to a lower one involves a specific amount of energy, E, a corresponding value of ν is observed. This is reflected by a specific wavelength of light according to the following relationship:

$$\lambda = \frac{c}{\nu} \tag{3.12.2}$$

The first accepted explanation of the behavior outlined above was the **Bohr theory** advanced by the Danish physicist Niels Bohr in 1913. Although this theory has been shown to be too simplistic, it had some features that are still pertinent to atomic structure. The Bohr theory visualized an electron orbiting the nucleus (a proton) of the hydrogen atom in orbits. Only specific orbits called **quantum states** were allowed. When energy was added to a hydrogen atom, its electron could jump to a higher orbit. When the atom lost its energy as the electron returned to a lower orbit, the energy lost was emitted in the form of electromagnetic radiation as shown in Figure 3.10. Because the two energy levels are of a definite magnitude according to quantum theory, the energy lost by the electron must also be of a definite energy, $E = h\nu$. Therefore, the electromagnetic radiation (light) emitted is of a specific frequency and wavelength.

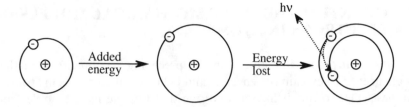

Figure 3.10. According to the Bohr model, adding energy to the hydrogen atom promotes an electron to a higher energy level. When the electron falls back to a lower energy level, excess energy is emitted in the form of electromagnetic radiation of a specific energy, $E = h\nu$.

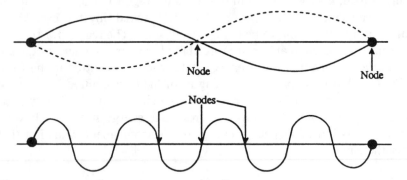

Figure 3.11. A string anchored at both ends (such as on a stringed musical instrument) can only vibrate at multiples of waves or half-waves. The illustration above shows one wave (top) and four waves (bottom). Each point at which a wave intersects the horizontal lines is called a node.

The Wave Mechanical Model of Atomic Structure

Though shown to have some serious flaws and long since abandoned, the Bohr model laid the groundwork for the more sophisticated theories of atomic structure that are accepted today and introduced the all-important concept that *only specific energy states are allowed for an electron in an atom*. Like electro-magetic radiation, electrons in atoms are now visualized as having a dual wave/particle nature. They are treated theoretically by the **wave mechanical model** as **standing waves** around the nucleus of an atom. The idea of a standing wave can be visualized for the string of a musical instrument as represented in Figure 3.11. Such a wave does not move along the length of a string because both ends are anchored, which is why it is called a *standing wave*. Each wave has **nodes**, which are points of zero displacement. Because there must be a node on each end where the string is anchored, the standing waves can only exist as multiples of *half-wavelengths*.

According to the *wave mechanical* or *quantum mechanical* model of electrons in atoms, the known *quantization* of electron energy in atoms occurs because only specific multiples of the standing wave associated with an electron's movement are allowed. Such a phenomenon is treated mathematically with the **Schrödinger equation** resulting from the work of Erwin Schrödinger, first published in 1926. Even for the one-electron hydrogen atom, the mathematics is quite complicated, and no attempt will be made to go into it here. The Schrödinger equation is represented as

$$H\Psi = E\Psi \tag{3.12.3}$$

where Ψ is the Greek psi. Specifically, Ψ is the **wave function**, a function of the electron's energy and the coordinates in space where it may be found. The term H is an **operator** consisting of a set of mathematical instructions, and E is the sum of the kinetic energy due to the motion of the electron and the potential energy of the mutual attraction between the electron and the nucleus.

Solutions to the Schrödinger Equation, Electrons in Orbitals

With its movement governed by the laws of quantum mechanics, it is not possible to know where an electron is relative to an atom's nucleus at any specific instant. However, the Schrödinger equation permits calculation of the probability of an electron being in a specified region. Solution of the equation gives numerous wave functions, each of which corresponds to a definite energy level and to the probability of finding an electron at various locations and distances relative to the nucleus (Figure 3.12). The wave function describes **orbitals**, *each of which has a characteristic energy and region around the nucleus where the electron has certain probabilities of being found.* The term orbital is used because quantum mechanical calculations do not give specific orbits consisting of defined paths for electrons, like planets going around the sun.

The square of the wave function, Ψ^2, calculated and integrated for a small segment of volume around a nucleus is proportional to the probability of finding an electron in that small volume, which may be regarded as part of an *electron*

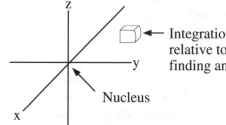 Integration of ψ^2 for any small volume segment relative to the nucleus gives the probability of finding an electron in that volume segment.

Figure 3.12. The xyz coordinate system used to describe the orientations in space of orbitals around the nucleus of an atom.

cloud. With this view, the electron is more likely to be found in a region where the cloud is relatively more dense. The cloud has no definite outer limits, but fades away with increasing distance from the nucleus to regions in which there is essentially no probability of finding the electron.

Multielectron Atoms and Quantum Numbers

Wave mechanical calculations describe various **energy levels** for electrons in an atom. Each energy level has at least one orbital. *A single orbital may contain a maximum of only two electrons*. Although the Schrödinger approach was originally applied to the simplest atom, hydrogen, it has been extended to atoms with many electrons (multielectron atoms). Electrons in atoms are described by four **quantum numbers**, which are defined and described briefly below. With these quantum numbers and a knowledge of the rules governing their use, it is possible to specify the orbitals allowed in a particular atom. *An electron in an atom has its own unique set of quantum numbers; no two electrons may have exactly identical quantum numbers.*

The Principal Quantum Number, n

Main energy levels corresponding to *electron shells* discussed earlier in this chapter are designated by a **principal quantum number, n**. Both the size of orbitals and the magnitude of the average energy of electrons contained therein increase with increasing n. Permitted values of n are 1, 2, 3, . . ., extending through 7 for the known elements.

The Azimuthal Quantum Number, l

Within each main energy level (shell) represented by a principal quantum number, there are **sublevels** (subshells). Each sublevel is denoted by an **azimuthal quantum number**, *l*. For any shell with a principal quantum number of n, the possible values of *l* are 0, 1, 2, 3, . . .,(n − 1). This gives the following:

- For n = 1, there is only 1 possible subshell, *l* = 0.
- For n = 2, there are 2 possible subshells, *l* = 0, 1.
- For n = 3, there are 3 possible subshells, *l* = 0, 1, 2.
- For n = 4, there are 4 possible subshells, *l* = 0, 1, 2, 3.

From the above it is seen that the maximum number of sublevels within a principal energy level designated by n is equal to n. Within a main energy level, sublevels denoted by different values of *l* have slightly different energies. Furthermore, *l* designates *different shapes of orbitals*.

The italicized letters *s, p, d,* and *f,* corresponding to *l* values of 0, 1, 2, and 3, respectively, are frequently used to designate sublevels. The number of electrons present in a given sublevel is limited to 2, 6, 10, and 14 for *s, p, d,* and *f* sublevels, respectively. It follows that, since each orbital may be occupied by a maximum of 2 electrons, there is only 1 orbital in the *s* sublevel, 3 orbitals in the *p* sublevel, 5 in the *d* and 7 in the *f*. The value of the principal quantum number and the letter designating the azimuthal quantum number are written in sequence to designate both the shell and subshell. For example, 4*d* represents the *d* subshell of the fourth shell.

The Magnetic Quantum Number, m_l

The **magnetic quantum number**, m_l, is also known as the **orientational quantum number**. It designates the orientation of orbitals in space relative to each other and distinguishes orbitals within a subshell from each other. It is called the magnetic quantum number because the presence of a magnetic field can result in the appearance of additional lines among those emitted by electronically excited atoms (that is, in atomic emission spectra). The possible values of m_l in a subshell with azimuthal quantum number *l* are given by $m_l = +l, +(l-1), \ldots ,0, \ldots, -(l-1), -l$. As examples, for $l = 0$, the only possible value of m_l is 0, and for $l = 3$, m_l, may have values of 3, 2, 1, 0, –1, –2, –3.

Spin Quantum Number, m_s

The fourth and final quantum number to be considered for electrons in atoms is the **spin quantum number** m_s, which may have values of only $+1/2$ or $-1/2$. It results from the fact that an electron spins in either of two directions and generates a tiny magnetic field with an associated magnetic moment. *Two electrons may occupy the same orbital only if they have opposite spins so that their magnetic moments cancel each other.*

Quantum Numbers Summarized

The information given by each quantum number is summarized below:

- The value of the principal quantum number, n, gives the shell and main energy level of the electron.
- The value of the azimuthal quantum number, *l*, specifies sublevels with somewhat different energies within the main energy levels and describes the shapes of orbitals.
- The magnetic quantum number, m_l, distinguishes the orientations in space of orbitals in a subshell and may provide additional distinctions of orbital shapes.
- The spin quantum number, m_s, with possible values of only $+1/2$ and $-1/2$, accounts for the fact that each orbital may be occupied by a maximum number of only 2 electrons with opposing spins.

Specification of the values of 4 quantum numbers describes each electron in an atom, as shown in Table 3.3.

For each electron in an atom, the orbital that it occupies and the direction of its spin are specified by values of n, l, m_l, and m_s, assigned to it. These values are unique for each electron; *no two electrons in the same atom may have identical values of all four quantum numbers.* This rule is known as the **Pauli exclusion principle**.

3.13.   ENERGY LEVELS OF ATOMIC ORBITALS

Electrons in atoms occupy the lowest energy levels available to them. Figure 3.13 is an energy level diagram that shows the relative energy levels of electrons in various atomic orbitals. Each dash, $-$, in the figure represents <u>one</u> orbital that is a potential slot for <u>two</u> electrons with opposing spins.

The electron energies represented in Figure 3.13 can be visualized as those required to remove an electron from a particular orbital to a location completely away from the atom. As indicated by its lowest position in the diagram, the most energy would be needed to remove an electron in the 1s orbital; comparatively little energy is required to remove electrons from orbitals having higher n values, such as 5, 6, or 7. Furthermore, the diagram shows decreasing separation of the energy levels (values of n) with increasing n. It also shows that some sublevels with a particular value of n have lower energies than sublevels for which the principal quantum number is n – 1. This is first seen from the placement of the 4s sublevel below that of the 3d sublevel. As a consequence, with increasing atomic number, the 4s orbital acquires 2 electrons before any electrons are placed in the 3d orbitals.

The value of the principal quantum number, n, of an orbital is a measure of the relative distance of the maximum electron density from the nucleus. An electron with a lower value of n, being on the average closer to the nucleus, is more strongly attracted to the nucleus than an electron with a higher n value that is at a relatively greater distance from the nucleus.

Hund's Rule of Maximum Multiplicity

Orbitals that are in the same sublevel, but that have different values of m_l, have the same energy. This is first seen in the 2p sublevel where the three orbitals with m_l of –1, 0, +1 (shown as three dashes on the same level in Figure 3.13) all have the same energies. The order in which electrons go into such a sublevel follows **Hund's rule of maximum multiplicity**, which states that *electrons in a sublevel are distributed to give the maximum number of unpaired electrons* (that is, those with parallel spins having the same sign of m_s). Therefore, the first three electrons to be placed in the 2p sublevel would occupy the three separate available orbitals and would have the same spins. Not until the fourth electron out of

Table 3.3. Quantum Numbers for Electrons in Atoms

n, Principal Quantum Number Denoting an Energy Level, or Shell	l, Azimuthal Quantum Number of a Sublevel or Subshell[a]		m_l, Magnetic Quantum Number Designating Orbital (m_s in parentheses)[b]	
1	0	(1s)	$0(m_s = +1/2$ or $-1/2)$ } 1 orbital	
2	0	(2s)	$0(m_s = +1/2$ or $-1/2)$ } 1 orbital	
	1	(2p)	$-1(m_s = +1/2$ or $-1/2)$ $0(m_s = +1/2$ or $-1/2)$ $+1(m_s = +1/2$ or $-1/2)$	3 orbitals
3	0	(3s)	$0(m_s = +1/2$ or $-1/2)$ } 1 orbital	
	1	(3p)	$-1(m_s = +1/2$ or $-1/2)$ $0(m_s = +1/2$ or $-1/2)$ $+1(m_s = +1/2$ or $-1/2)$	3 orbitals
	2	(3d)	$-2(m_s = +1/2$ or $-1/2)$ $-1(m_s = +1/2$ or $-1/2)$ $0(m_s = +1/2$ or $-1/2)$ $+1(m_s = +1/2$ or $-1/2)$ $+2(m_s = +1/2$ or $-1/2)$	5 orbitals
4	0	(4s)	$0(m_s = +1/2$ or $-1/2)$ } 1 orbital	
	1	(4p)	$-1(m_s = +1/2$ or $-1/2)$ $0(m_s = +1/2$ or $-1/2)$ $+1(m_s = +1/2$ or $-1/2)$	3 orbitals
	2	(4d)	$-2(m_s = +1/2$ or $-1/2)$ $-1(m_s = +1/2$ or $-1/2)$ $0(m_s = +1/2$ or $-1/2)$ $+1(m_s = +1/2$ or $-1/2)$ $+2(m_s = +1/2$ or $-1/2)$	5 orbitals
	3	(4f)	$-3(m_s = +1/2$ or $-1/2)$ $-2(m_s = +1/2$ or $-1/2)$ $-1(m_s = +1/2$ or $-1/2)$ $0(m_s = +1/2$ or $-1/2)$ $+1(m_s = +1/2$ or $-1/2)$ $+2(m_s = +1/2$ or $-1/2)$ $+3(m_s = +1/2$ or $-1/2)$	7 orbitals

[a]Subshell designation in parentheses.
[b]Each entry in this column designates an orbital capable of holding 2 electrons with opposing spins.

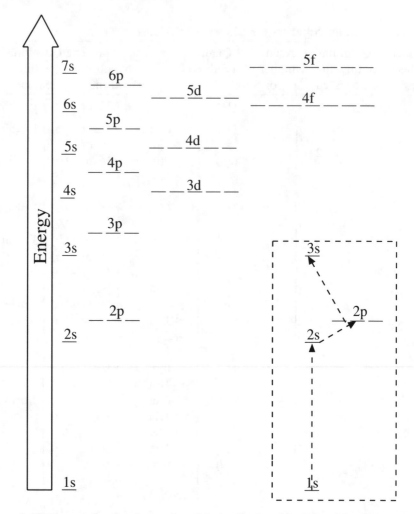

Figure 3.13. Energy levels of atomic orbitals. Each orbital capable of containing 2 electrons is shown with a dash, −. The order of placing electrons in orbitals is from the lowest-lying orbitals up, as shown in the inset.

a maximum number of 6 is added to this sublevel are two electrons placed in the same orbital.

3.14.   SHAPES OF ATOMIC ORBITALS

In trying to represent different shapes of orbitals, it is important to keep in mind that they do not contain electrons within finite volumes, because there is no outer boundary of an orbital at which the probability of finding an electron drops to exactly 0. However, an orbital can be drawn as a figure ("fuzzy cloud") around the atom nucleus within which there is a relatively high probability (typically 90%) of finding an electron within the orbital. These figures are called **contour representations of orbitals** and are as close as one can come to visualizing the shapes of orbitals. Each point on the surface of a contour representation of an orbital has the same value of Ψ^2 (square of the wave function of the Schrödinger equation, Section 3.12), and the entire surface encloses the volume within which an electron spends 90% of its time. The two most important aspects of an orbital are its size and shape, both of which are rather well illustrated by a contour representation.

Figure 3.14 shows the contour representations of the first three s orbitals. These are seen to be spherically shaped and to increase markedly in size with increasing principal quantum number, reflecting increased average distance of the electron from the nucleus.

Figure 3.15 shows contour representations of the three $2p$ orbitals. These are seen to have different orientations in space; they have directional properties. In fact, s orbitals are the only ones that are spherically symmetrical. As discussed further in Chapter 4, the shapes of orbitals are involved in molecular geometry and help to determine the shapes of molecules. The shapes of d and f orbitals are

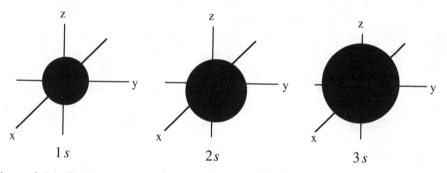

Figure 3.14. Contour representations of s orbitals for the first three main energy levels.

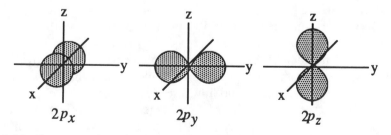

Figure 3.15. Contour representations of 2p orbitals.

more complex and variable than those of p orbitals and are not discussed in this book.

3.15.   ELECTRON CONFIGURATION

Electron configuration is a means of stating which kinds of orbitals contain electrons and the numbers of electrons in each kind of orbital of an atom. It is expressed by the number and letter representing each kind of orbital and superscript numbers telling how many electrons are in each sublevel. The hydrogen atom's one electron in the $1s$ orbital is shown by the following notation:

Sublevel notation
↓
$1s^1$
↑
Number of electrons in sublevel

Nitrogen, atomic number 7, has 7 electrons, of which 2 are paired in the $1s$ orbital, 2 are paired in the $2s$ orbital and 3 occupy singly each of the 3 available $2p$ orbitals. This electron configuration is designated as $1s^2 2s^2 2p^3$.

The **orbital diagram** is an alternative way of expressing electron configurations in which each separate orbital is represented by a box. Individual electrons in the orbitals are shown as arrows pointing up or down to represent opposing spins ($m_s = +1/2$ or $-1/2$). The orbital diagram for nitrogen is the following:

$$\boxed{\uparrow\downarrow}\quad \boxed{\uparrow\downarrow}\quad \boxed{\uparrow}\,\boxed{\uparrow}\,\boxed{\uparrow}$$

$1s$ $2s$ $2p$

This gives all the information contained in the notation $1s^2 2s^2 2p^3$, but emphasizes that the three electrons in the three available $2p$ orbitals each occupy separate orbitals, a condition of Hund's rule of maximum multiplicity (Section 3.13).

Most of the remainder of this chapter is devoted to a discussion of the placement of electrons in atoms with increasing atomic number, how this affects the chemical behavior of the atoms, and how it leads to a systematic organization of the elements in the periodic table. This placement of electrons is in the order shown in increasing energy levels from bottom to top in Figure 3.13 as illustrated for nitrogen in Figure 3.16.

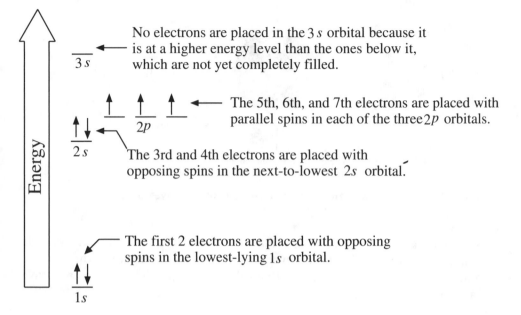

Figure 3.16. Placement of electrons in orbitals for the 7-electron nitrogen atom according to the energy level diagram.

3.16. ELECTRONS IN THE FIRST 20 ELEMENTS

The most meaningful way to place electrons in the orbitals of atoms is on the basis of the periodic table. This enables relating electron configurations to chemical properties and the properties of elements in groups and periods of the periodic table. In this section, electron configurations are deduced for the first 20 elements and given in an abbreviated version of the periodic table.

Electron Configuration of Hydrogen

The 1 electron in the hydrogen atom goes into its lowest-lying 1s orbital. Figure 3.17 summarizes all the information available about this electron and its configuration.

Electron Configuration of Helium

An atom of helium, atomic number 2, has 2 electrons, both contained in the 1s orbital and having quantum numbers $n = 1$, $l = 0$, $m_l = 0$, and $m_s = +1/2$ and $-1/2$. Both electrons have the same set of quantum numbers except for m_s. The electron configuration of helium can be represented as $1s^2$, showing that there are 2 electrons in the 1s orbital. Two is the maximum number of electrons that can be contained in the first principal energy level; additional electrons in

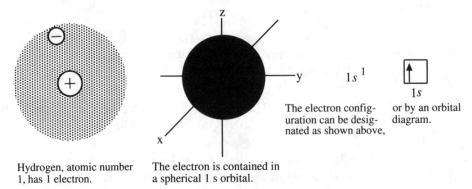

Hydrogen, atomic number 1, has 1 electron.

The electron is contained in a spherical 1 s orbital.

The electron configuration can be designated as shown above,

or by an orbital diagram.

Figure 3.17. Representation of the electron on the hydrogen atom.

atoms with atomic number greater than 2 must go into principal energy levels with n greater than 1. There are several other noble gas elements in the periodic table, but helium is the only one with a filled shell of 2 electrons—the rest have stable outer shells of 8 electrons.

Electron Configurations of Elements 2–20

The electron configurations of elements with atomic numbers through 20 are very straightforward. Electrons are placed in order of orbitals with increasing energy as shown in Figure 3.13. This order is

$$1s^2 2s^2 2p^6 3s^2 3p^6 4s^2$$

It should be noted that in this configuration the electrons go into the $4s$ orbital before the $3d$ orbital, which lies at a slightly higher energy level. In filling the p orbitals, it should also be kept in mind that 1 electron goes into each of three p orbitals before pairing occurs. In order to follow the discussion of electron configurations for elements through 20, it is useful to refer to the abbreviated periodic table in Figure 3.20.

Lithium

For lithium, atomic number 3, two electrons are placed in the $1s$ orbital, leaving the third electron for the $2s$ orbital. This gives an electron configuration of $1s^2 2s^1$. The two $1s$ electrons in lithium are in the stable noble gas electron configuration of helium and are very difficult to remove. These are lithium's *inner electrons* and, along with the nucleus, constitute the **core** of the lithium atom. The electron configuration *of the core* of the lithium atom is $1s^2$, the same as that of helium. Therefore, the lithium atom is said to have a *helium core. The core of any atom consists of its nucleus plus its inner electrons, those with the* *same electron configuration as the noble gas immediately preceding the element in the periodic table.*

Valence Electrons

Lithium's lone 2s electron is an outer electron contained in the *outer shell* of the atom. Outer shell electrons are also called **valence electrons,** and are the electrons that can be shared in covalent bonding or lost to form cations in ionic compounds. Lithium's valence electron is illustrated in Figure 3.18.

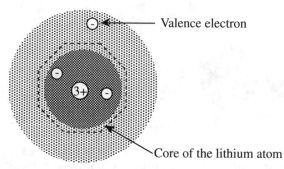

Figure 3.18. Core (kernel) and valence electrons of the lithium atom.

Beryllium

Beryllium, atomic number 4, has 2 inner electrons in the 1s orbital and 2 outer electrons in the 2s orbital. Therefore, beryllium has a *helium core*, plus 2 *valence electrons*. Its electron configuration is $1s^2 2s^2$.

Filling the 2p Orbitals

Boron, atomic number 5, is the first element containing an electron in a 2p orbital. Its electron cofiguration is $1s^2 2s^2 2p^1$. Two of the five electrons in the boron atom are contained within the spherical orbital closest to the nucleus, and 2 more are in the larger spherical 2s orbital. The lone 2p electron is in an approximately dumbbell-shaped orbital in which the average distance of the electron from the nucleus is about the same as that of the 2s electrons. This electron could have any one of the three orientations in space shown for *p* orbitals in Figure 3.15.

The electron configuration of carbon, atomic number 6, is $1s^2 2s^2 2p^2$. The four outer, valence electrons are shown by the Lewis symbol

Two of the four valence electrons are represented by a pair of dots, :, to indicate that these electrons are paired in the same orbital. These are the two $2s$ electrons. The other two are shown as individual dots to represent two unpaired $2p$ electrons in separate orbitals.

The 7 electrons in nitrogen, N, are in a $1s^2 2s^2 2p^3$ electron configuration. Nitrogen is the first element to have at least one electron in each of 3 possible p orbitals (see Figure 3.19).

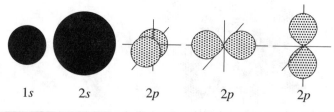

Figure 3.19. The N atom contains 2 electrons in the $1s$ orbital, 2 in the $2s$ orbital, and 1 in each of three separate $2p$ orbitals oriented in different directions in space.

The next element to be considered is oxygen, atomic number 8. Its electron configuration is $1s^2 2s^2 2p^4$. It is the first element in which it is necessary for 2 electrons to occupy the same p orbital as shown by the following orbital diagrams:

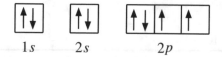

The electron configuration of fluorine, atomic number 9, is $1s^2 2s^2 2p^5$. Its Lewis symbol,

shows that it has only one unpaired electron in its valence shell. In its chemical reactions, fluorine seeks to obtain another electron to give a stable *octet*.

Neon, atomic number 10, is at the end of the second period of the periodic table and is a noble gas as shown by its Lewis symbol,

denoting a filled octet of electrons. Its electron configuration is $1s^22s^22p^6$. In this configuration the $2s^22p^6$ portion stands for the outer electrons. Like neon, all other elements with the outer electron configuration $\mathbf{n}s^2\mathbf{n}p^6$ are noble gases located in the far right of the periodic table (for example, argon, with the electron configuration $1s^22s^22p^63s^23p^6$).

Filling the 3s, 3p, and 4s Orbitals

The 3s and 3p orbitals are filled in going across the third period of the periodic table from sodium through argon. Atoms of these elements, like all atoms beyond neon, have their 10 innermost electrons in the neon electron configuration of $1s^22s^22p^6$. Therefore, these atoms have a *neon core*, which may be designated {Ne}.

$$\{Ne\} \text{ stands for } 1s^22s^22p^6$$

With this notation, the electron configuration of element number 11, sodium, may be shown as {Ne}$3s^1$, which is an abbreviation for $1s^22s^22p^63s^1$. The former notation has some advantage in simplicity, while showing the outer electrons specifically. In the example just cited it is easy to see that sodium has 1 outer shell 3s electron, which it can lose to form the Na$^+$ ion with its stable noble gas neon electron configuration.

At the end of the third period is located the noble gas argon, atomic number 18, with the electron configuration $1s^22s^22p^63s^23p^6$. For elements beyond argon, that portion of the electron configuration identical to argon's may be represented simply as {Ar}. Therefore, the electron configuration of potassium, atomic number 19 is $1s^22s^22p^63s^23p^64s^1$, abbreviated {Ar}$4s^1$, and that of calcium, atomic number 20 is $1s^22s^22p^63s^23p^64s^2$, abbreviated {Ar}$4s^2$.

3.17.   ELECTRON CONFIGURATIONS AND THE PERIODIC TABLE

There are several learning devices to assist expression of the order in which atomic orbitals are filled (electron configuration for each element). However, it is of little significance to express electron configurations without an understand-

1							2
H							He
$1s^1$							$1s^2$

3	4	5	6	7	8	9	10
Li	Be	B	C	N	O	F	Ne
{He}$2s^1$	{He}$2s^2$	{He}$2s^22p^1$	{He}$2s^22p^2$	{He}$2s^22p^3$	{He}$2s^22p^4$	{He}$2s^22p^5$	{He}$2s^22p^6$

11	12	13	14	15	16	17	18
Na	Mg	Al	Si	P	S	Cl	Ar
{Ne}$3s^1$	{Ne}$3s^2$	{Ne}$3s^23p^1$	{Ne}$3s^23p^2$	{Ne}$3s^23p^3$	{Ne}$3s^23p^4$	{Ne}$3s^23p^5$	{Ne}$3s^23p^6$

19	20
K	Ca
{Ar}$4s^1$	{Ar}$4s^2$

Figure 3.20. Abbreviated periodic table showing the electron configurations of the first 20 elements.

ing of the meaning of the configurations. By far the most meaningful way to understand this important aspect of chemistry is within the context of the periodic table as outlined in Figure 3.21. To avoid clutter, only atomic numbers of key elements are shown in this table. The double-pointed arrows drawn horizontally across the periods are labeled with the kind of orbital being filled in that period.

The first step in using the periodic table to figure out electron configurations is to note that the periods are numbered 1 through 7 from top to bottom along the left side of the table. These numbers correspond to the principal quantum numbers (n values) for the orbitals that become filled across the period for both s and p orbitals. Therefore, in the first group of elements — those with atomic numbers 1, 3, 11, 19, 37, 55, and 87 — the last electron added is in the ns orbital; for example, 5s for Rb, atomic number 37. The last electron added to each of the second group of elements — those with atomic numbers 4, 12, 20, 38, 56, and 88 — is the second electron going into the ns orbital. For example, in the third period, Mg, atomic number 12, has a filled 3s orbital containing 2 electrons. For the group of elements in which the p orbitals start to be filled — those in the column with atomic numbers 5, 13, 31, 49, 81 — the last electron added to each atom is the first one to enter an np orbital. For example for element 31, Ga, which is contained in the 4th period, the outermost electron is in the 4p orbital.

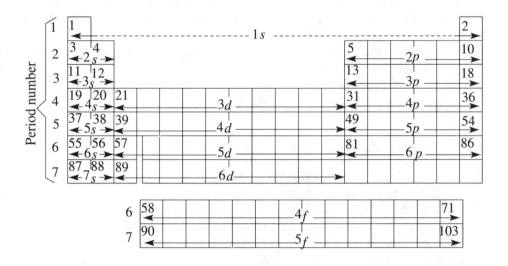

Figure 3.21. Outline of the periodic table labeled to show the filling of atomic orbitals. The type of element being filled with increasing atomic number is shown by the labeled horizontal lines across the periods. The left point of each arrow, ←, marks an element in which the filling of a new kind of orbital starts, and the right point of each arrow, →, marks an element for which the filling of a new kind of orbital is completed. The atomic numbers are given for each element at which the filling of a kind of orbital begins or is completed.

In going to the right across each period in this part of the periodic table, the outermost electrons are, successively, np^2, np^3, np^4, np^5, and np^6. Therefore, for elements with atomic numbers 6, 7, 8, 9, and 10 in the 2nd period, the outermost electrons are $2p^2$, $2p^3$, $2p^4$, $2p^5$, and $2p^6$. Each of the noble gases beyond helium has filled np orbitals with a total of 6 outermost p electrons in the three np orbitals.

Each set of five d orbitals becomes filled for the *transition metals* in the three horizontal periods beginning with atomic numbers 21, 39, and 57 and ending with, successively, atomic numbers 30, 48, and 80. *For each of these orbitals the value of n is 1 less than the period number in which the orbitals become filled.* The first d orbitals to become filled are the lowest-lying ones possible, the $3d$ orbitals, which become filled in the fourth period. Across the 5th period, where the $5s$ and $5p$ orbitals become filled for elements 37–38 and 49–54, respectively, the $4d$ orbitals (*n 1 less than the period number*) become filled for the transition metals, atomic numbers 39–48.

Shown below the main body of the periodic table, the **inner transition elements** consist of two rows of elements that are actually parts of the 6th and 7th periods, respectively. The principal quantum numbers of their f orbitals are 2 *less than* their period numbers. Figure 3.21 shows that the first f orbitals begin to fill with the first of the *lanthanides*, element number 58. These are $4f$ orbitals, of which there are 7 that are filled completely with element number 71. The $4f$ orbitals are filled in the *6th* period. The $5f$ orbitals are filled with the actinide elements, atomic numbers 90–103.

With Figure 3.21 in mind, it is possible to write the expected electron configurations of any of the elements. Consider the following examples:

- Atomic No. 16, S: 3rd period, 16 electrons, Ne core, outermost electrons $3p$, electron configuration $\{Ne\}3s^23p^4$
- Atomic No. 23, V: 4th period, 23 electrons, Ar core, outermost electrons $3d$, electron configuration $\{Ar\}4s^23d^3$
- Atomic No. 35, Br: 4th period, 35 electrons, Ar core, all $3d$ orbitals filled, outermost electrons $4p$, electron configuration $\{Ne\}4s^23d^{10}4p^5$
- Atomic No. 38, Sr: 5th period, 38 electrons, Kr core, outermost electrons $5s$, electron configuration $\{Kr\}5s^2$
- Atomic No. 46, Cd: 5th period, 48 electrons, Kr core, outermost electrons $4d$, electron configuration $\{Kr\}5s^24d^{10}$
- Atomic No. 77, Ir: 6th period, 77 electrons, Xe core, all $4f$ orbitals filled in the 6th period, outermost electrons $5d$, electron configuration $\{Xe\}6s^24f^{14}5d^7$

The actual electron configurations are given in Table 3.4. In some cases these vary slightly from those calculated according to the rules outlined above. These exceptions occur because of the relatively higher stabilities of two half-filled sets of outermost orbitals, or orbitals in which one is half-filled and one entirely filled. The examples below illustrate this point:

- Cr, atomic number 24. Rules predict $\{Ar\}4s^23d^4$. However the actual electron configuration is $\{Ar\}4s^13d^5$ because this gives the slightly more stable electron configuration with *half-filled* $4s$ and $3d$ orbitals.
- Cu, atomic number 29. Rules predict $\{Ar\}4s^23d^9$. However the actual electron configuration is $\{Ar\}4s^13d^{10}$ because this gives *half-filled* $4s$ and *filled* $3d$ orbitals.

CHAPTER SUMMARY

The chapter summary below is presented in a programmed format to review the main points covered in this chapter. It is used most effectively by filling in the blanks, referring back to the chapter as necessary. The correct answers are given at the end of the summary.

Table 3.4. Electron Configurations of the Elements in the Ground (Unexcited) State

Atomic Number	Symbol	Configuration of Ground State Atoms	Atomic Number	Symbol	Configuration of Ground State Atoms
1	H	$1s^1$	55	Cs	$\{Xe\}6s^1$
2	He	$1s^2$	56	Ba	$\{Xe\}6s^2$
3	Li	$\{He\}2s^1$	57	La	$\{Xe\}6s^25d^1$
4	Be	$\{He\}2s^2$	58	Ce	$\{Xe\}6s^24f^15d^1$
5	B	$\{He\}2s^22p^1$	59	Pr	$\{Xe\}6s^24f^3$
6	C	$\{He\}2s^22p^2$	60	Nd	$\{Xe\}6s^24s^4$
7	N	$\{He\}2s^22p^3$	61	Pm	$\{Xe\}6s^24f^5$
8	O	$\{He\}2s^22p^4$	62	Sm	$\{Xe\}6s^24f^6$
9	F	$\{He\}2s^22p^5$	63	Eu	$\{Xe\}6s^24f^7$
10	Ne	$\{He\}2s^22p^6$	64	Gd	$\{Xe\}6s^24f^75d^1$
11	Na	$\{Ne\}3s^1$	65	Tb	$\{Xe\}6s^24f^9$
12	Mg	$\{Ne\}3s^2$	66	Dy	$\{Xe\}6s^24f^{10}$
13	Al	$\{Ne\}3s^23p^1$	67	Ho	$\{Xe\}6s^24f^{11}$
14	Si	$\{Ne\}3s^23p^2$	68	Er	$\{Xe\}6s^24f^{12}$
15	P	$\{Ne\}3s^23p^3$	69	Tm	$\{Xe\}6s^24f^{13}$
16	S	$\{Ne\}3s^23p^4$	70	Yb	$\{Xe\}6s^24f^{14}$
17	Cl	$\{Ne\}3s^23p^5$	71	Lu	$\{Xe\}6s^24f^{14}5d^1$
18	Ar	$\{Ne\}3s^23p^6$	72	Hf	$\{Xe\}6s^24f^{14}5d^2$
19	K	$\{Ar\}4s^1$	73	Ta	$\{Xe\}6s^24f^{14}5d^3$
20	Ca	$\{Ar\}4s^2$	74	W	$\{Xe\}6s^24f^{14}5d^4$
21	Sc	$\{Ar\}4s^23d^1$	75	Re	$\{Xe\}6s^24f^{14}5d^5$
22	Ti	$\{Ar\}4s^23d^2$	76	Os	$\{Xe\}6s^24f^{14}5d^6$
23	V	$\{Ar\}4s^23d^3$	77	Ir	$\{Xe\}6s^24f^{14}5d^7$
24	Cr	$\{Ar\}4s^13d^5$	78	Pt	$\{Xe\}6s^14f^{14}5d^9$
25	Mn	$\{Ar\}4s^23d^5$	79	Au	$\{Xe\}6s^14f^{14}5d^{10}$
26	Fe	$\{Ar\}4s^23d^6$	80	Hg	$\{Xe\}6s^24f^{14}5d^{10}$
27	Co	$\{Ar\}4s^23d^7$	81	Tl	$\{Xe\}6s^24f^{14}5d^{10}6p^1$
28	Ni	$\{Ar\}4s^23d^8$	82	Pb	$\{Xe\}6s^24f^{14}5d^{10}6p^2$
29	Cu	$\{Ar\}4s^13d^{10}$	83	Bi	$\{Xe\}6s^24f^{14}5d^{10}6p^3$
30	Zn	$\{Ar\}4s^23d^{10}$	84	Po	$\{Xe\}6s^24f^{14}5d^{10}6p^4$
31	Ga	$\{Ar\}4s^23d^{10}4p^1$	85	At	$\{Xe\}6s^24f^{14}5d^{10}6p^5$
32	Ge	$\{Ar\}4s^23d^{10}4p^2$	86	Rn	$\{Xe\}6s^24f^{14}5d^{10}6p^6$
33	As	$\{Ar\}4s^23d^{10}4p^3$	87	Fr	$\{Rn\}7s^1$
34	Se	$\{Ar\}4s^23d^{10}4p^4$	88	Ra	$\{Rn\}7s^2$
35	Br	$\{Ar\}4s^23d^{10}4p^5$	89	Ac	$\{Rn\}7s^26d^1$
36	Kr	$\{Ar\}4s^23d^{10}4p^6$	90	Th	$\{Rn\}7s^26d^2$
37	Rb	$\{Kr\}5s^1$	91	Pa	$\{Rn\}7s^25f^26d^1$
38	Sr	$\{Kr\}5s^2$	92	U	$\{Rn\}7s^25f^36d^1$
39	Y	$\{Kr\}5s^24d^1$	93	Np	$\{Rn\}7s^25f^46d^1$
40	Zr	$\{Kr\}5s^24d^2$	94	Pu	$\{Rn\}7s^25f^6$
41	Nb	$\{Kr\}5s^14d^4$	95	Am	$\{Rn\}7s^25f^7$
42	Mo	$\{Kr\}5s^14d^5$	96	Cm	$\{Rn\}7s^25f^76d^1$
43	Tc	$\{Kr\}5s^24d^5$	97	Bk	$\{Rn\}7s^25f^9$
44	Ru	$\{Kr\}5s^14d^7$	98	Cf	$\{Rn\}7s^25f^{10}$
45	Rh	$\{Kr\}5s^14d^8$	99	Es	$\{Rn\}7s^25f^{11}$
46	Pd	$\{Kr\}4d^{10}$	100	Fm	$\{Rn\}7s^25f^{12}$
47	Ag	$\{Kr\}5s^14d^{10}$	101	Md	$\{Rn\}7s^25f^{13}$
48	Cd	$\{Kr\}5s^24d^{10}$	102	No	$\{Rn\}7s^25f^{14}$
49	In	$\{Kr\}5s^24d^{10}5p^1$	103	Lr	$\{Rn\}7s^25f^{14}6d^1$
50	Sn	$\{Kr\}5s^24d^{10}5p^2$	104	Rf	$\{Rn\}7s^25f^{14}6d^2$
51	Sb	$\{Kr\}5s^24d^{10}5p^3$	105	Ha	$\{Rn\}7s^25f^{14}6d^3$
52	Te	$\{Kr\}5s^24d^{10}5p^4$	106	Unh	$\{Rn\}7s^25f^{14}6d^4$
53	I	$\{Kr\}5s^24d^{10}5p^5$	107	Uns	$\{Rn\}7s^25f^{14}6d^5$
54	Xe	$\{Kr\}5s^24d^{10}5p^6$	109	Une	$\{Rn\}7s^25f^{14}6d^7$

Briefly, the basic parts of the atomic theory are (1)_____

_____,

(2)_____

_____,

(3)_____

_____,

(4)_____

_____,

(5)_____

_____.

Three fundamental laws that are explained by the atomic theory are (6)_____

_____, (7)_____

_____, and (8)_____.

The atomic mass unit is used to (9)_____

and is defined as (10)_____.

Most of the volume of an atom is composed of (11)_____

_____. The three subatomic particles of concern to

chemists, their charges, and mass numbers are (12)_____

_____.

The atomic number and mass number of $^{14}_{7}N$ are (13)_____.

    A systematic arrangement of elements that places those with similar chemical

properties and electron configurations in the same groups is the (14)_____

_____. Most hydrogen atoms have a nucleus consisting of (15)___

_____. The notation Ca: is an example of (16)_____

_____ and H:H is an example of (17)_____

_____. Helium is the first element with atoms that have a

(18)_____. Lithium has both (19)_____

and (20)_____ electrons. Four general charac-

teristics of metals are (21)_____

_____.

The Lewis symbols of elements 4–10 are (22)_____
_____.

The octet rule is (23)_____

_____.

The names of the elements in the third period of the periodic table are (24)___

_____

_____. The noble gases in the first 20 elements

are (25)_____.

   The unique behavior of charged particles that are as small and move as rapidly
as electrons is explained by (26)_____. Electromag-
netic radiation has a characteristic (27)_____ and (28)_____
related by the equation (29)_____. According to the
quantum theory, radiant energy may be absorbed or emitted only in discrete
packets called (30)_____, the energy of
which is given by the equation (31)_____. The Bohr model
introduced the all-important concept that (32)_____

_____.

The wave mechanical model of electrons treats them as (33)_____
_____ around the nucleus of an atom. The wave mechanical
model of electrons in atoms is treated by the (34)_____
equation expressed mathematically as (35)_____. In this
equation Ψ is the (36)_____ and is a function of the
electron's (37)_____

_____. According to the wave
mechanical model electrons occupy (38)_____ each of which has
(39)_____.

A single orbital may contain a maximum of (40)_____ electrons. An electron
in an atom is described by four (41)_____ which may not be
(42)_____ for any two electrons in an atom. The symbol n represents the
(43)_____ which may have values of (44)_____

_____. The symbol l represents the (45)_____ quantum number, which may have values of (46)_____.
The symbol m_l represents (47)_____ with possible values of (48)_____.
The symbol m_s is the (49)_____ which may have values of only (50)_____. Using standard notation for electron configuration (starting $1s^22s^22p^6$) the order of filling of orbitals and the maximum number of electrons in each is (51)_____.
The orbital diagram for the p electrons in nitrogen,

↑	↑	↑

illustrates the rule that (52)_____

_____.

The entire surface of a contour representation of an orbital encloses (53)_____

_____.

The contour representation of an s orbital is that of (54)_____ whereas that of a p orbital is shaped like a (55)_____.

(56)_____ is a means of stating which kinds of orbitals contain electrons and the numbers of electrons in each kind of orbital of an atom. It is expressed by the number and letter representing (57)_____ and superscript numbers telling (58)_____.
The electron configurations of phosphorus (P), potassium (K), and arsenic (As) are, respectively, (59)_____.
The part of an atom consisting of its nucleus and the electrons in it equivalent to those of the noble gas immediately preceding the element in the periodic table is the (60)_____ of the atom. Electrons that can be shared in covalent bonding or lost to form cations in ionic compounds are called (61)_____.
Other than helium, all elements with the outer electron configuration ns^2np^6 are (62)_____

_____.

In respect to period number in the periodic table, the principal quantum number of s electrons is (63)_____, and of p electrons is (64)_____, that of d electrons is (65)_____, and that of f electrons is (66)_____. The types of elements in which d orbitals become filled are (67)_____. The electron configuration of Cr, atomic number 24, is $\{Ar\}4s^13d^5$, which appears to deviate slightly from the rules because it gives (68)_____ _____.

Answers

1. Elements are composed of small objects called atoms.
2. Atoms of different elements do not have identical chemical properties.
3. Chemical compounds are formed by combination of atoms of different elements in definite ratios.
4. Chemical reactions involve the separation and combination of atoms.
5. During the course of ordinary chemical reactions, atoms are not created, destroyed, or changed to atoms of other elements.
6. law of conservation of mass
7. law of constant composition
8. law of multiple proportions
9. express masses of atoms
10. exactly 1/12 mass of carbon-12 isotope
11. electrons around the nucleus
12. proton $(+1,1)$ electron $(-1,0)$, neutron $(0,1)$
13. 7 and 14
14. periodic table
15. 1 proton
16. an electron-dot symbol or Lewis symbol
17. an electron-dot formula or Lewis formula
18. filled electron shell
19. inner
20. outer
21. luster, malleable, conduct electricity, tend to lose electrons to form cations
22.

   3 Li ·, 4 Be:, 5 ·B:, 6 · C:, 7 · N:, 8 · O:, 9 · F:, 10, :Ne:

23. The tendency of elements to acquire an 8-electron outer electron configuration in chemical compounds.

24. sodium, magnesium, aluminum, silicon, phosphorus, sulfur, chlorine, and argon
25. helium, neon, argon
26. the quantum theory
27. wavelength
28. frequency
29. $\nu\lambda = c$
30. quanta or photons
31. $E = h\nu$
32. only specific energy states are allowed for an electron in an atom
33. standing waves
34. Schrödinger
35. $H\Psi = E\Psi$
36. wave function
37. energy and the coordinates in space where it may be found
38. orbitals
39. characteristic energy and region around the nucleus where the electron has certain probabilities of being found
40. two
41. quantum numbers
42. identical
43. principal quantum number
44. 0, 1, 2, 3, 4, 5, 6, 7 . . .
45. azimuthal
46. 0, 1, 2, 3, . . . , (n − 1)
47. magnetic quantum number
48. $+l, +(l-1), \ldots, 0, \ldots, -(l-1), -l$
49. spin quantum number
50. $+1/2$ or $-1/2$
51. $1s^2 2s^2 2p^6 3s^2 3p^6 4s^2 3d^{10} 4p^6 5s^2 4d^{10} 5p^6 6s^2 4f^{14} 5d^{10} 6p^6 7s^2$
52. electrons in a sublevel are distributed to give the maximum number of unpaired electrons
53. the volume within which an electron spends 90% of its time
54. a sphere
55. dumbbell, or two spheres touching at the nucleus
56. Electron configuration
57. each kind of orbital
58. how many electrons are in each sublevel
59. $1s^2 2s^2 2p^6 3s^2 3p^3$, $1s^2 2s^2 2p^6 3s^2 3p^6 4s^1$,
    $1s^2 2s^2 2p^6 3s^2 3p^6 4s^2 3d^{10} 4p^3$
60. core
61. valence electrons
62. noble gases located in the far right of the periodic table
63. the same as the period number

64. the same as the period number
65. one less than the period number
66. two less than the period numberr
67. transition metals
68. slightly more stable electron configuration with half-filled $4s$ and $3d$ orbitals

QUESTIONS AND PROBLEMS

1. Match the law or observation on the left below with the portion of Dalton's atomic theory that explains it from the column on the right:

   1. Law of Conservation of Mass
   2. Law of Constant Composition
   3. Law of Multiple Proportions
   4. The reaction of C with O_2 does not produce SO_2.

   A. Illustrated by groups of compounds such as $CHCl_3$, CH_2Cl_2, or CH_3Cl.
   B. Chemical compounds are formed by the combination of atoms of different elements in definite constant ratios that usually can be expressed as integers or simple fractions.
   C. During the course of ordinary chemical reactions, atoms are not created or destroyed.
   D. During the course of ordinary chemical reactions, atoms are not changed to atoms of other elements.

2. Particles of pollutant fly ash may be very small. Estimate the number of atoms in such a small particle assumed to have the shape of a cube that is 1 micrometer (μm) to the side. Assume also that an atom is shaped like a cube 100 picometers (pm) on a side.

3. Explain why it is incorrect to say that atomic mass is the mass of any atom of an element. How is atomic mass defined?

4. The $^{12}_{6}C$ isotope has a mass of exactly 12 u. Compare this to the sum of the masses of the subatomic particles that compose this isotope. Is it correct to say that the mass of an isotope is exactly equal to the sum of the masses of its subatomic particles? Is it close to the sum?

5. What is the distinction between the mass of a subatomic particle and its mass number?

6. Add up the masses of all the subatomic particles in an isotope of carbon-12. Is the sum exactly 12? Should it be?

7. Fill in the blanks in the table below

Subatomic Particle	Symbol	Unit Charge	Mass Number	Mass in u	Mass in grams
Proton	(a)___	(b)_____	(c)_____	(d)____	(e)_____
(f)____	n	0	1	(g)____	(h)_____
Electron	e	(i)_____	(j)_____	(k)____	9.1096×10^{-28}

8. Define what is meant by x, y, and A in the notation ${}_x^y A$.

9. Describe what happens to the magnitude and direction of the forces between charged particles (electrons, protons, nuclei) of (a) like-charge and (b) unlike charge with distance and magnitude of charge.

10. What is the Lewis symbol of hydrogen and what does it show? What is the Lewis formula of H_2 and what does it show?

11. In many respects the properties of elemental hydrogen are unique. List some of these properties and some of the major uses of H_2.

12. Give the Lewis symbol of helium and explain what it has to do with (a) electron shell, (b) filled electron shell, and (c) noble gases.

13. Where is helium found, and for what purpose is it used?

14. Using dots to show all of its electrons, give the Lewis symbol of Li. Explain how this symbol shows (a) inner and outer electrons and electron shells, (b) valence electrons, (c) and how Li^+ ion is formed.

15. Discuss the chemical and physical properties of lithium that indicate that it is a metal.

16. What is a particular health concern with beryllium?

17. Based upon its electronic structure, suggest why boron behaves like a metalloid; that is, neither a metal nor a nonmetal.

18. Carbon has two isotopes that are of particular importance. What are they and why are they important?

19. Why might carbon be classified as a "life element"?

20. What two species other than O_2 are possible for elemental oxygen, particularly in the stratosphere?

21. What do particular kinds of fluorine compounds have to do with atmospheric ozone?

22. In reference to neon define and explain the significance of (a) noble gas, (b) octet of outer shell electrons, (c) noble gas outer electron configuration, (d) octet rule.

23. To which class of elements do lithium, sodium, and potassium belong? What are their elemental properties?

24. Define and explain (a) electromagnetic radiation, (b) wave character of electromagnetic radiation, (c) quanta (photons).

25. What is the significance of the fact that very specific wavelengths of light are emitted when atoms in an excited state revert back to a lower energy state (ground state)?

26. What are the major accomplishments and shortcomings of the Bohr theory?

27. How are electrons visualized in the wave mechanical model of the atom?

28. What is the fundamental equation for the wave mechanical or quantum mechanical model of electrons in atoms? What is the significance of Ψ in this equation?

29. Why are orbitals not simply called "orbits" in the quantum mechanical model of atoms?

30. Name and define the four quantum numbers used to describe electrons in atoms.

31. What does the Pauli exclusion principle say about electrons in atoms?

32. Use the notation employed for electron configurations to denote the order in which electrons are placed in orbitals through the $4f$ orbitals.

33. What is a statement and significance of Hund's Rule of Maximum Multiplicity?

34. Discuss shapes and sizes of s orbitals with increasing principal quantum number.

35. Complete the figure below for contour representations of the three $2p$ orbitals.

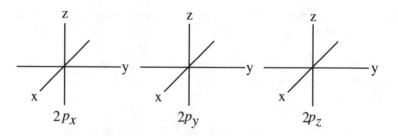

$2p_x$ $2p_y$ $2p_z$

36. An atom has an electron configuration of $1s^2 2s^2 2p^3$. Give the electron configuration of an atom with 5 more electrons and also give the orbital diagram of such an atom.

37. Give the electron configurations of carbon, neon, aluminum, phosphorus, and calcium.

38. Give the symbols of the elements with the following electron configurations: (a) $1s^2 2s^3$, (b) $1s^2 2s^2 2p^6 3s^1$, (c) $1s^2 2s^2 2p^6 3s^2 3p^4$, and (d) $1s^2 2s^2 2p^6 3s^2 3p^6 4s^1$.

39. What is the core of an atom? Specifically, what is the neon core?

40. What are valence electrons? What are the electron configurations of the valence electrons in aluminum?

41. What is wrong with the orbital diagram below?

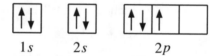

$1s$ $2s$ $2p$

42. State the rule that gives which orbitals are filled for transition elements in relation to the period number of the elements involved. Do the same for the inner transition elements.

43. Explain why the electron configuration of Cr is $\{Ar\}4s^1 3d^5$ and why that of Cu is $\{Ar\}4s^1 3d^{10}$.

4 CHEMICAL BONDS, MOLECULES, AND COMPOUNDS

4.1. CHEMICAL BONDS AND COMPOUND FORMATION

Chemical compounds and chemical bonds have already been mentioned several times in this book. It is not possible to even begin the study of chemistry without some knowledge of compounds and bonding because they are the essence of chemical science. The first three chapters of this book have provided an overview of chemistry, a discussion of the properties of matter, and an explanation of atoms, and elements. With this background it is possible to discuss chemical bonds, molecules, and compounds in more detail.

By now the reader should have gained some appreciation of the diversity of types of chemical compounds, all of which are formed from a limited number of elements. The variety and large numbers of chemical compounds are possible because atoms of the elements are joined together through chemical bonds. The chemical behavior of the elements can only be understood with a knowledge of their abilities to form chemical bonds. Shapes and structures of molecules are determined by the orientation of chemical bonds attached to the atoms composing the elements. The strengths of chemical bonds in a compound largely determine its stability and resistance to chemical change. For example, photons (see Section 3.11) of sunlight in the ultraviolet and shorter wavelength regions of the visible light spectrum can be absorbed by a molecule of nitrogen dioxide, NO_2, and break a chemical bond between an N and O atom,

$$NO_2 + NO + O \qquad (4.1.1)$$

yielding very reactive atoms of oxygen. These atoms attack hydrocarbons (compounds composed of C and H) introduced as pollutants into the atmosphere from incompletely burned automobile exhaust gases and other sources. This sets off a series of reactions—chain reactions—that lead to the production of atmospheric ozone, formaldehyde, additional NO_2, and other products known as photochemical smog (see Figure 4.1).

Figure 4.1. Breakage of a chemical bond between N and O in NO_2 starts the process involving a series of many reactions leading to the formation of unhealthy pollutant photochemical smog.

Chemical Bonds and Valence Electrons

Chemical bonds are normally formed by the transfer or sharing of *valence electrons*, those that are in the *outermost shell* of the atom (see Section 3.8). To visualize how this is done, it is helpful to consider the first 20 elements in the periodic table, their Lewis symbols, and their electron configurations given in Figure 4.2.

Recall from Chapter 3 that in forming chemical bonds atoms often attain the same valence electron configuration as the noble gas closest to the element in the periodic table. This provides the basis of the *octet rule* discussed in Section 3.9. According to this rule atoms attain a stable outer shell of 8 electrons in chemical compounds. This is the same as the outer shell electrons of the noble gases other than helium, such as neon and argon in the periodic table shown in Figure 1.3. Atoms can attain a stable octet of outer shell electrons by losing electrons to become *cations*, gaining electrons to become anions, or sharing electrons with other atoms in *covalent bonds*.

Figure 4.2. Abbreviated version of the periodic table showing outer shell (valence) electrons as dots. Electron configurations are also given.

4.2.   CHEMICAL BONDING AND THE OCTET RULE

The octet rule for chemical bonding was mentioned briefly in Chapter 3. At this point it will be useful to examine the octet rule in more detail. Although there are many exceptions to it, the octet rule remains a valuable concept for an introduction to chemical bonding.

The Octet Rule for Some Diatomic Gases

Recall that some elemental gases — hydrogen, nitrogen, oxygen, and fluorine — do not consist of individual atoms, but of *diatomic molecules* of H_2, N_2, O_2, and F_2, respectively. This can be explained by the tendencies of these elements to attain a noble gas outer electron configuration. The bonds in molecules of F_2 are shown by the following:

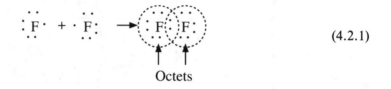

(4.2.1)

It is seen that the F_2 molecule is held together by a *single covalent bond* consisting of two shared electrons. As shown by the circles around them, each atom in the molecule has eight outer electrons, some of them shared. These 8 electrons constitute an *octet* of outer electrons. Such an octet is possessed by the neon atom, which is the noble gas nearest to F in the periodic table.

The Octet Rule for Chemical Compounds

The octet rule illustrated for molecules of diatomic elemental fluorine above applies to molecules of *chemical compounds*. Recall from Section 1.4 that a *chemical compound* is formed from atoms or ions of two or more elements joined by chemical bonds. In many cases, compound formation enables elements to attain a noble gas electron configuration, usually an octet of outer shell electrons.

The most straightforward way for atoms to gain a stable octet is through loss and gain of electrons to form ions. This may be illustrated by the ionic compound sodium chloride as shown in Figure 4.3. By considering the Lewis symbols of Na and Cl atoms in Figure 4.2, it is easy to see that the octet of the Na^+ ion is formed by the loss of an electron from an Na atom and that of the Cl^- ion is produced by a Cl atom gaining 1 electron.

It is difficult for atoms of an element in the middle of a period of the periodic table, such as those of carbon or nitrogen, to either gain or lose enough electrons

Figure 4.3.   In the ionic compound, NaCl, both the Na⁺ ion and the Cl⁻ ion have octets of outer shell electrons. The Na⁺ ion is formed by the loss of one electron from a sodium atom (see Figure 4.2) and the Cl⁻ ion is formed by a Cl atom gaining an electron.

to attain a noble gas outer electron configuration as ions. Normally, atoms of these elements share electrons to form *covalent bonds*. As an example of co-valent bonding, consider the chemical combination of C with H in which each of 4 H atoms shares an electron with 1 C atom to form a molecule of methane, CH_4. This sharing enables each C atom to have an 8-electron outer shell (octet) like neon. Each H atom has 2 electrons, both shared with C, which provides a shell of 2 electrons like that in the noble gas helium.

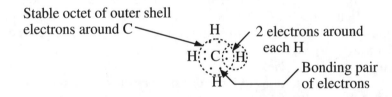

Figure 4.4.   Stable outer electron shells from covalent bonding in compounds.

4.3.   IONIC BONDING

Ionic compounds and ionic bonding were introduced in Chapter 1 and in the preceding section. In this section ionic bonding and the nature of ions are discussed in more detail.

An **ion** consists of an atom or group of atoms having an unequal number of electrons and protons and, therefore, a net electrical charge. A **cation** has a positive charge and an anion has a negative charge. The simplest kind of ion to visualize is one formed by the loss or gain by an atom of one or more electrons. As examples, loss of 2 electrons from a calcium atom gives the Ca^{2+} cation, and the gain of an electron by a Cl atom gives a Cl⁻ anion. An **ionic compound** is one that contains cations and anions. Such a compound is held together by **ionic bonds** that result from the mutual attraction of positively charged cations and negatively charged anions.

Table 4.1.   Ions with Neon and Argon Electron Configurations

Element	Atomic Number	Electron Configuration	Electrons Gained/Lost	Ion	Ion Electron Configuration	
N	7	$1s^22s^22p^3$	Gain 3 e^-	N^{3-}	$1s^22s^22p^6$	⎫
O	8	$1s^22s^22p^4$	Gain 2 e^-	O^{2-}	$1s^22s^22p^6$	⎬ neon
F	9	$1s^22s^22p^5$	Gain 1 e^-	F^-	$1s^22s^22p^6$	⎪
Na	11	$1s^22s^22p^63s^1$	Lose 1 e^-	Na^+	$1s^22s^22p^6$	⎪
Mg	12	$1s^22s^22p^63s^2$	Lose 2 e^-	Mg^{2+}	$1s^22s^22p^6$	⎭
S	16	$1s^22s^22p^63s^23p^4$	Gain 2 e^-	S^{2-}	$1s^22s^22p^63s^23p^6$	⎫
Cl	17	$1s^22s^22p^63s^23p^5$	Gain 1 e^-	Cl^-	$1s^22s^22p^63s^23p^6$	⎬ argon
K	19	$1s^22s^22p^63s^23p^64s^1$	Lose 1 e^-	K^+	$1s^22s^22p^63s^23p^6$	⎪
Ca	20	$1s^22s^22p^63s^23p^64s^2$	Lose 2 e^-	Ca^{2+}	$1s^22s^22p^63s^23p^6$	⎭

Electron Configurations of Ions from a Single Atom

Electron configurations of atoms were discussed in some detail in Chapter 3. Emphasis has been placed on the stability of a stable *octet* of outer shell electrons, which is characteristic of a *noble gas*. Such an octet is in the ns^2np^6 electron configuration. The first element to have this configuration is neon, electron configuration $1s^22s^22p^6$. (Recall that helium is also a noble gas, but with only two electrons, its electron configuration is simply $1s^2$). The next noble gas beyond neon is argon, which has the electron configuration $1s^22s^22p^63s^23p^6$, commonly abbreviated $\{Ne\}3s^23p^6$.

The electron configurations of ions consisting of only one atom can be shown in the same way as illustrated for atoms above. For the lighter elements the electron configurations are those of the nearest noble gas. Therefore, in forming ions an atom of an element that is just before a particular noble gas in the periodic table <u>gains</u> enough electrons to get the configuration of that noble gas, whereas an atom of an element just beyond a noble gas <u>loses</u> enough electrons to attain the electron configuration of the nearest noble gas. This is best seen in reference to the periodic table showing electron configurations as seen in Figure 4.2.

As an example of electron configurations of ions, consider an anion, F^-, and a cation, Mg^{2+}, both near neon in the periodic table. From Figure 4.2, it is seen that the electron configuration of the neutral fluoride atom is $1s^22s^22p^5$. To get the stable noble gas electron configuration of neon, the fluoride atom may gain 1 electron to become F^- ion, electron configuration $1s^22s^22p^6$. The electron configuration of magnesium, atomic number 12, is $1s^22s^22p^63s^2$. A magnesium atom can attain the stable electron configuration of neon by losing its 2 outer shell $3s$ electrons to become Mg^{2+} ion, electron configuration $1s^22s^22p^6$. The electron configurations of some other ions near neon and argon in the periodic table are given in Table 4.1.

Sodium Chloride as an Ionic Compound

The formation of sodium chloride, NaCl, is often cited as an example of the production of an ionic compound and its constituent ions, and the stable octets of Na^+ and Cl^- ions were shown in Figure 4.3. Here the nature of sodium chloride as an ionic compound is considered in more detail. The electron configurations of both Na^+ ion and Cl^- ion are given in Table 4.1. The reaction between a neutral elemental sodium atom and a neutral elemental chlorine atom to form NaCl consisting of Na^+ and Cl^- ions is as follows:

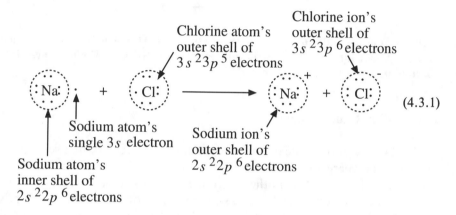

Chlorine atom's outer shell of $3s^2 3p^5$ electrons

Chlorine ion's outer shell of $3s^2 3p^6$ electrons

Sodium atom's single $3s$ electron

Sodium atom's inner shell of $2s^2 2p^6$ electrons

Sodium ion's outer shell of $2s^2 2p^6$ electrons

(4.3.1)

A compound such as NaCl that is composed of ions is an ionic compound. The ionic bonds in such a compound exist because of the electrostatic attraction between ions of opposite charge. An ionic compound possesses a crystal structure such that a particular ion is located as close as possible to attracting ions of opposite charge and as far as possible from repelling ions with the same charge. Figure 4.5 shows such a structure for NaCl. Visualized in three dimensions, this figure shows how the six nearest neighbors of each ion are ions of opposite charge. Consider the Cl^- ion with the arrow pointing to it in the center of the structure. Every line intersecting the sphere that represents the Cl^- ion leads to a nearest neighbor ion, each of which is an Na^+ ion. The closest Cl^- ions are actually farther away than any of the six nearest neighbor Na^+ ions. The crystal structure is such that each Na^+ ion is similarly surrounded by six nearest neighbor Cl^- ions. Therefore, every ion in the crystal is closest to ions of opposite charge resulting in forces of attraction that account for the stability of ionic bonds. Depending upon the relative numbers of cations and anions in a compound and upon their sizes, there are various crystalline structures that maximize the closeness of ions of opposite charge and, therefore, the strengths of ionic bonds. Although the ball and stick model in Figure 4.5 shows the relative positions of the ions, the ions are actually considered to be of a size such that they fill the spaces in the structure and touch each other (the Cl^- ion is larger than the Na^+ ion, as discussed later and shown in Figure 4.7).

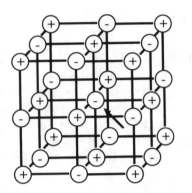

Figure 4.5. Representation of the structure of NaCl. This structure can be visualized in three dimensions as "balls" with interconnecting "sticks." The balls are Na⁺ and Cl⁻ ions with the charges designated.

Energetics in Ionic Bonding

The stability of the ionic bond is the result of the interaction of several factors involving the energetics of the attraction of atoms for electrons and of oppositely charged ions for each other. To understand some of the factors involved, consider the reaction of isolated gas-phase atoms of Na and Cl to form solid NaCl. This is a somewhat simplified view in that it does not take account of the fact that sodium is a solid under normal conditions and that elemental chlorine exists as diatomic molecules of Cl_2; when solid Na and gaseous Cl_2 react, energy is involved in removing Na atoms from the elemental solid, as well as in breaking the covalent bond between Cl atoms in Cl_2. The reaction to be considered is the following:

$$\text{Na·}(g) + \text{·}\overset{..}{\underset{..}{\text{Cl}}}\text{:}(g) \rightarrow \text{Na}^+ \quad + \quad \text{:}\overset{..}{\underset{..}{\text{Cl}}}\text{:}^-(g) \rightarrow \text{Na}^+\text{:}\overset{..}{\underset{..}{\text{Cl}}}\text{:}^- (s) \ (4.3.2)$$

When gas-phase sodium and chlorine atoms react, *ionization energy* is involved in the removal of an electron from a sodium atom to form Na⁺, *electron affinity* is involved in the addition of an electron to a chlorine atom to produce Cl⁻, and *lattice energy* is released by oppositely charged ions coming together in the sodium chloride crystal.

Three energy changes are listed for the process outlined in the above equation. These are the following:

- Even though the sodium ion has been represented as being "willing" to lose its valence electron to attain a noble gas electron configuration, some energy is required. This is called the **ionization energy**. In the case of sodium, the ionization energy for the removal of the single outer shell electron is + 490 kJ/mole. The positive value indicates that energy must be put into the system to remove the electron from Na. For each mole of Na atoms, the amount of energy is 490 kilojoules.

• When an electron is added to a Cl atom, the energy is expressed as **electron affinity**, with a value of −349 kJ/mole. The negative sign denotes that energy is <u>released</u>; for the addition of 1 electron to each atom in a mole of Cl atoms, 349 kJ of energy is evolved.

• A very large amount of energy is released when a mole of solid NaCl is formed by a mole of Na^+ and a mole of Cl^- ions coming together to form a **crystal lattice** of ions arranged in a crystalline structure. The energy change for this process is −785 kJ/mole.

The net energy change for Reaction 4.3.2 is given by the following:

Energy change	= Ionization energy for Na(g)	+ Electron affinity of Cl(g)	+ Energy released by Na^+ and Cl^- coming together to form NaCl(s)	(4.3.3)
	= 490 kJ	−349 kJ	−785kJ	
	= −644 kJ			

This calculation shows that when 1 mole of solid NaCl is formed from 1 mole each of gas-phase Na and Cl atoms, the energy change is −644 kilojoules; the negative sign shows that energy is released, so that the process is energetically favored. Under normal conditions of temperature and pressure, however, sodium exists as the solid metal, Na(s), and chlorine is the diatomic gas, Cl_2(s). The reaction of 1 mole of solid sodium with one mole of chlorine atoms contained in gaseous Cl_2 to yield 1 mole of NaCl is represented by

$$Na(s) + \tfrac{1}{2}\, Cl_2(g) \rightarrow NaCl(g) \qquad (4.3.4)$$

The energy change for this reaction is −411 kJ/mole, representing a significantly lower release of energy from that calculated in Equation 4.3.3. This difference is due to the energy used to vaporize solid Na and to break the Cl-Cl bonds in diatomic Cl_2. The total process for the formation of solid NaCl from solid Na and gaseous Cl_2 is outlined in Figure 4.6, a form of the Born-Haber cycle.

Energy of Ion Attraction

In the case of NaCl, it is seen that the largest component of the energy change that occurs in the formation of the ionic solid is that released when the ions come together to form the solid. This may be understood in light of the potential energy of interaction, E, between two charged bodies given by the equation

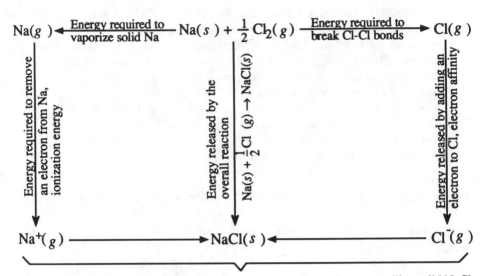

Lattice energy released when gas-phase ions come together to form crystalline solid NaCl

Figure 4.6. Energy changes involved in the formation of an ionic compound from its constituent elements. The overall process is shown by the vertical arrow in the middle and the constituent parts by the arrows around the periphery.

$$E = k \frac{Q_1 Q_2}{d} \tag{4.3.5}$$

where Q_1 and Q_2 are the charges on the two bodies in coulombs (C), d is the distance between their centers in meters, and k is a constant with a value of 8.99×10^9 J-m/C^2. If one of the bodies has a + and the other a – charge, as is the case with oppositely charged ions, the sign of E is negative, consistent with the fact that energy is released when oppositely charged bodies come together. The magnitude of E is greater for larger values of Q_1 and Q_2 (higher ionic charges, such as the 2+ for Ca^{2+} ion compared to the 1+ for Na^+ ion) and greater for smaller values of d (smaller radii of ions allowing them to approach more closely).

Lattice Energy

Having considered the energy of interaction between charged particles, next consider the normally large amount of energy involved with the packing of ions into a crystalline lattice. The lattice energy for a crystalline ionic compound is the energy required to separate all of the ions in the compound and remove them a sufficient distance from each other so that there is no interaction between them. Since energy input is required to do this for ionic compounds, lattice energy has a positive value; it is conventionally expressed in kilojoules per mole of compound. For the example being considered, the lattice energy of NaCl is $+785$ kJ/mole. However, in considering the production of an ionic compound, as shown in Reaction 4.3.3, the opposite process was considered, which is the assembling of a mole of Na^+ ions and a mole of Cl^- ions to produce a mole of solid NaCl. This releases the lattice energy, so that its contribution to the energy change involved was given a minus sign, that is, -785 kJ/mole.

Ion Size

Equation 4.3.5 shows that the distance between the centers of charged bodies, d, affects their energy of interaction. For ionic compounds this is a function of ionic size; smaller ions can get closer and therefore, their energies of interaction as measured by lattice energy are greater. Because of the removal of their outer shell electrons, cations are smaller than the atoms from which they are formed; because of the addition of outer shell electrons, anions are larger than their corresponding neutral atoms. Some ion sizes are shown in Figure 4.7.

Formation of Some Example Ionic Compounds

Consider next several other ionic compounds that contain some of the ions shown in Figure 4.7. To better illustrate the transfer of electrons, these are shown for the formation of ions from single atoms, even in cases where diatomic gases are involved. Calcium and chlorine react

$$\text{Ca:} + \begin{array}{c} \cdot \ddot{\text{Cl}} : \\ \cdot \ddot{\text{Cl}} : \end{array} \rightarrow : \ddot{\text{Cl}} :^- \ Ca^{2+} : \ddot{\text{Cl}} :^- \tag{4.3.6}$$

to form calcium chloride, $CaCl_2$. This compound is a by-product of some industrial processes and is used to salt "icy" streets, which causes the ice to melt and, unfortunately, automobiles to rust. The reaction of magnesium and oxygen

$$Mg: + \cdot \ddot{O} : \rightarrow Mg^{2+} \ \ddot{O} :^{2-} \tag{4.3.7}$$

yields magnesium oxide, MgO. This compound is used in heat-resistant refractory materials (such as those used to line industrial furnaces), insulation, cement, and paper manufacture. Sodium and sulfur react,

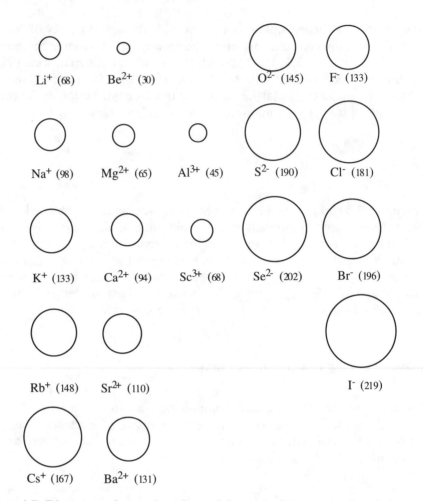

Figure 4.7. Diameters of some ions formed from single atoms, units of picometers (1 Ångstrom = 100 pm)

$$\begin{matrix} Na \cdot \longrightarrow \cdot \\ Na \cdot \longrightarrow \end{matrix} + \overset{..}{:} S \overset{..}{:} \longrightarrow Na^+ \overset{..}{:} \overset{..}{S} \overset{..}{:} {}^{2-} Na^+ \qquad (4.3.8)$$

to yield sodium sulfide, Na_2S. Among its other uses, this ionic compound is an intermediate in one of the processes used in wood pulping for paper manufacture. As a final example, consider the following reaction of aluminum and oxygen:

$$\text{(4.3.9)}$$

The product, aluminum oxide, Al_2O_3, is the most widely used aluminum compound. It is commonly called alumina. Among the products that it is used to make are abrasives, ceramics, antacids, and antiperspirants.

Exercise: Fill in the table below for the ionic products of the reaction of the metal and nonmetal indicated.

Metal	Nonmetal	Cation formed	Anion formed	Ionic compound
Li	Cl	(a)_____	(b)_____	(c)_____
K	S	(d)_____	(e)_____	(f)_____
Ca	F	(g)_____	(h)_____	(i)_____
Ba	O	(j)_____	(k)_____	(l)_____
Ca	N	(m)_____	(n)_____	(o)_____

Answers: (a) Li^+, (b) Cl^-, (c) LiCl, (d) K^+, (e) S^{2-}, (f) K_2S, (g) Ca^{2+}, (h) F^-, (i) CaF_2, (j) Ba^{2+}, (k) O^{2-}, (l) BaO, (m) Ca^{2+}, (n) N^{3-}, (o) Ca_3N_2

4.4. FUNDAMENTALS OF COVALENT BONDING

Chemical Bonds and Energy

The energy changes involved in the formation of ions and ionic compounds were mentioned prominently in the preceding discussion of ionic compounds. In fact, bonds form largely as the result of the tendency of atoms to attain minimum energy. Several kinds of bonds are possible between atoms, and energy considerations largely determine the type of bonds that join atoms together. It was shown in the preceding section that *ionization energy, electron affinity*, and *lattice energy* are all strongly involved in forming ionic bonds. Ionic compounds form when the sum of the energy released in forming anions (electron affinity) and in the cations and anions coming together (lattice energy) is sufficient to remove electrons from a neutral atom to produce cations (ionization energy). It was seen that for simple monatomic ions conditions are most likely to be favorable for the formation of ionic bonds when the cations are derived from elements on the left side of the periodic table and the anions from those near the right side. In addition, the transition elements can lose 1–3 electrons to form simple monatomic cations, such as Cu^+, Zn^{2+}, and Fe^{3+}.

In many cases, however, the criteria outlined above are not energetically favorable for the formation of ionic bonds. In such cases, bonding can be accomplished by the sharing of electrons in covalent bonds. Whereas oppositely charged ions attract each other, neutral atoms have a tendency to repel each other because of the forces of repulsion between their positively charged nuclei and the negatively charged clouds of electrons around the nuclei. Sharing electrons in covalent bonds between atoms enables these forces of repulsion to be overcome and results in energetically favored bonding between atoms. Covalent bonding is discussed in this and following sections.

Covalent Bonding

A **covalent bond** is one that joins two atoms through the sharing of 1 or more pairs of electrons between them. Covalent bonds were discussed briefly in Chapter 1. The simplest such bond to visualize is that which forms between two hydrogen atoms, to yield a diatomic molecule of hydrogen, H_2:

$$H\cdot \quad + \quad \cdot H \quad \longrightarrow \quad \text{(H}\colon\text{H)} \qquad\qquad (4.4.1)$$

Figures 4.8 and 4.9 illustrate some important characteristics of the H-H bond.

As shown in Figure 4.8, there is an electron cloud between the two positively charged H atom nuclei. This is the covalent bond consisting of a pair of shared electrons that hold the two atoms together. Figure 4.9 is an energy diagram for the H_2 molecule. The two hydrogen nuclei are shown at the distance that results in minimum energy, the bond length. Forcing the nuclei close together results in rapidly increasing energy because of the repulsive forces between the nuclei. The energy increases more gradually with distances farther apart than the bond length, approaching zero energy as complete separation is approached. The bond energy of 435 kJ/mol means that a total energy of 435 kilojoules is

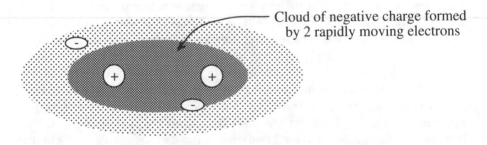

Cloud of negative charge formed by 2 rapidly moving electrons

Figure 4.8.   Two hydrogen nuclei, +, covalently bonded together by sharing two electrons, –. The negatively charged cloud of electrons is concentrated between the two nuclei, reducing their natural tendency to repel each other.

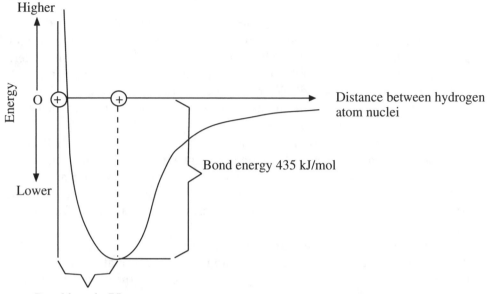

Figure 4.9.   Relationship between energy and distance between hydrogen nuclei
(energy diagram of H₂). The nuclei are shown at the most energetically favored
distance apart, i.e., the bond length of the H₂ atom.

required to break all the bonds in a mole of H_2 molecules and totally separate the
resulting H atoms.

4.5.   COVALENT BONDS IN COMPOUNDS

Consider next some example covalent bonds between atoms of some of the
lighter elements. These are best understood in reference to Figure 4.2, the abbreviated version of the periodic table showing the Lewis symbols (outer shell
valence electrons) of the first 20 elements. As is the case with ions, atoms that
are covalently bonded in molecules often have an arrangement of outer shell
electrons like that of the noble gas with an atomic number closest to the element
in question. It was just seen that covalently bonded H atoms in molecules of H_2
have 2 outer shell electrons like the nearby noble gas helium. For atoms of many
other elements, the tendency is to acquire 8 outer shell electrons — an octet — in
sharing electrons through covalent bonds. This tendency forms the basis of the
octet rule discussed in Section 4.2. In illustrating the application of the octet rule
to covalent bonding consider first the bonding of atoms of hydrogen to atoms of
elements with atomic numbers 6 through 9 in the second period of the periodic
table. These elements are close to the noble gas neon and tend to attain a "neon-
like" octet of outer shell electrons when they form covalently bonded molecules.

As shown in Figure 4.10 (p. 147), when each of the 4 H atoms shares an electron with 1 C atom to form a molecule of methane, CH_4, each C atom attains an 8-electron outer shell (octet) like neon. Every hydrogen atom attains an outer shell of 2 electrons through the arrangement of shared electrons. There is a total of 4 covalent bonds, 1 per H attached to the C atom, and each composed of a shared pair of electrons. Each of the hydrogen atoms has two electrons like the noble gas helium and carbon has an octet of outer shell electrons like the noble gas neon.

Also shown in Figure 4.10, nitrogen bonds with hydrogen such that 1 N atom shares an electron with each of 3 H atoms to form a molecule of ammonia, NH_3. This arrangement gives the N atom 8 outer shell electrons, of which 6 are shared with H atoms and 2 constitute an **unshared pair** of electrons. In both CH_4 and NH_3, each H atom has 2 shared electrons, which provides a shell of 2 electrons like that in the noble gas helium.

In discussing covalent bonds and molecules it is sometimes convenient to use the term **central atom** in reference to an atom to which several other atoms are bonded. In the case of CH_4 that was just discussed, C is the central atom. For NH_3 nitrogen, N, is the central atom.

The use of Lewis symbols and formulas in the preceding examples readily shows how many atoms are bound together in each of the compounds and the types of bonds in each. The products of each reaction are shown in two ways, one in which all the valence electrons are represented as dots, and the other in which each pair of valence electrons in a chemical bond is shown as a dash, and the unbonded valence electrons as dots. In cases where all that is needed is to show which atoms are bonded together and the types of bonds (one dash for a single covalent bond, two for a double bond of 2 shared electron pairs, and three for a triple bond of 3 shared electron pairs), the dots representing unshared electrons may be omitted.

Figure 4.11 shows covalent bonding in the formation of hydrogen compounds of O and F. In the case of oxygen, two H atoms are combined with one O atom having 6 valence electrons, sharing electrons such that each H atom has 2 electrons and the O atom has 8 outer shell electrons, 4 of which are in 2 shared pairs. To form HF, only 1 H atom is required to share its electron with an atom of F having 7 outer shell electrons, leading to a compound in which the F atom has 8 outer shell electrons, 2 of which are shared with H.

Of the compounds whose formation is shown above, CH_4 is methane, the simplest of the hydrocarbon compounds and the major component of natural gas. The compound formed with nitrogen and hydrogen is ammonia, NH_3. It is the second most widely produced synthetic chemical, a pungent-smelling gas with many uses in the manufacture of fertilizer, explosives, and synthetic chemicals. The product of the reaction between hydrogen and oxygen is, of course, water, H_2O. (Another compound composed only of hydrogen and oxygen is hydrogen peroxide, H_2O_2, a reactive compound widely used as a bleaching agent). Hydrogen fluoride, HF, is the product of the chemical combination of H

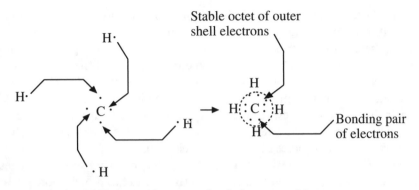

Each of 4 H atoms shares a pair of electrons with a
C atom to form a molecule of methane, CH_4.

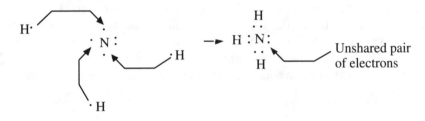

Each of 3 H atoms shares a pair of electrons with an atom
of N to form a molecule of ammonia, NH_3.

**Figure 4.10.   Formation of stable outer electron shells by covalent bonding in
compounds.**

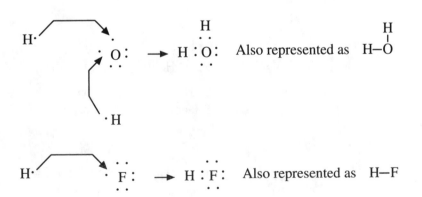

**Figure 4.11.   Covalent bonding of H to O and F to produce H_2O and HF,
respectively.**

and F. It is a toxic corrosive gas (boiling point 19.5°C) used as a raw material to manufacture Freon refrigerants and Teflon plastics.

Having seen how hydrogen combines with carbon, nitrogen, oxygen, and fluorine, it is easy to predict the hydrogen compounds with the elements directly below C, N, O, and F in the third period of the periodic table. These elements are silicon, phosphorus, sulfur, and chlorine, atomic numbers 14 through 17. As shown in Figure 4.12, these elements combined with hydrogen have covalent bonds such that every hydrogen atom has 2 outer shell electrons and each of the other elements has an octet of outer shell electrons.

Silane, SH_4, is a colorless gas. **Phosphine**, PH_3 is also a gas; it catches fire spontaneously when exposed to air, is very toxic, and has a bad odor. **Hydrogen sulfide**, H_2S, is a toxic gas with a foul, rotten, egg odor. Natural gas contaminated with hydrogen sulfide is said to be sour, and the hydrogen sulfide must be removed before the natural gas may be used as a fuel. Fortunately, H_2S is readily converted to elemental sulfur or to sulfuric acid, for which there are ready markets. **Hydrogen chloride** gas has a sharp odor and is very soluble in water. Solutions of hydrogen chloride in water are called **hydrochloric acid**. About 2.5 million tons of hydrochloric acid are produced for industrial applications in the U.S. each year.

4.6.   SOME OTHER ASPECTS OF COVALENT BONDING

Multiple Bonds and Bond Order

Several examples of single bonds consisting of a pair of shared electrons have just been seen. Two other types of covalent bonds are the *double bond* consisting of two shared pairs of electrons (4 electrons total) and the *triple bond* made up of three shared pairs of electrons (6 electrons total). These are both examples

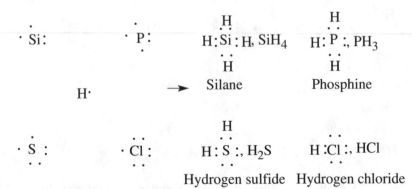

Figure 4.12.   Chemical compounds of Si, P, S, and Cl with H.

of **multiple bonds**. Examples of a double bond and of a triple bond are shown in Figure 4.13.

Atoms of carbon, oxygen, and nitrogen bonded to each other are the most likely to form multiple bonds. In some cases sulfur forms multiple bonds.

Lengths and Strengths of Multiple Bonds

In various compounds, bonds between the same two atoms (either of the same or different elements) may be single, double, or triple. This is seen below for the three types of carbon-carbon bonds illustrated in Figure 4.13:

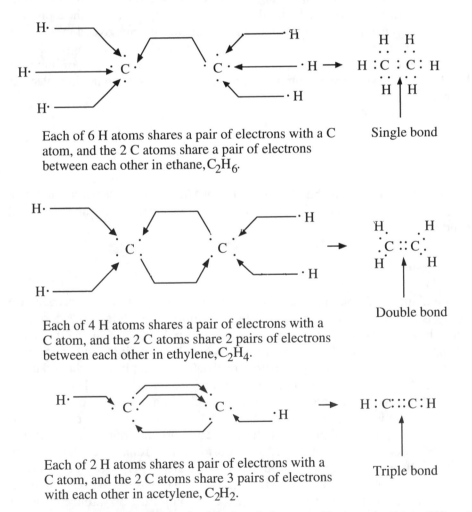

Each of 6 H atoms shares a pair of electrons with a C atom, and the 2 C atoms share a pair of electrons between each other in ethane, C_2H_6.

Single bond

Each of 4 H atoms shares a pair of electrons with a C atom, and the 2 C atoms share 2 pairs of electrons between each other in ethylene, C_2H_4.

Double bond

Each of 2 H atoms shares a pair of electrons with a C atom, and the 2 C atoms share 3 pairs of electrons with each other in acetylene, C_2H_2.

Triple bond

Figure 4.13. Single, double, and triple bonds between C atoms in three different hydrocarbon molecules, each of which contains 2 C atoms.

Bond type:	Single in C_2H_6	Double in C_2H_4	Triple in C_2H_2
Bond length:	154 pm	134 pm	120 pm

(pm stands for picometers)

Bond strength: 348 kJ/mol   614 kJ/mol   839 kJ/mol

The above shows that as the bond multiplicity increases, the bonds become shorter. Furthermore, the bond strength, expressed as the energy in kilojoules needed to break a mole of the bonds, increases. These trends are reasonable if a covalent bond is viewed as consisting of electrons located between the two atoms that are bonded together, such that the positively charged nuclei of the atoms are attracted to the electrons in the bond. Increasing multiplicity gives a more dense cloud of electrons enabling the atoms to approach closer resulting in shorter bond length. The same reasoning explains the greater strength of multiple bonds. One of the strongest common bonds is the triple $N \equiv N$ bond in molecular nitrogen, N_2:

$$(4.6.1)$$

The bond energy of this bond is 941 kJ/mol. As a result, elemental diatomic nitrogen is very stable, reacting with other elements only under special conditions.

Bond Order

The number of bonding electron pairs that make up the bond between two atoms is called the **bond order**. The single, double, and triple bonds described above have bond orders of 1, 2, and 3, respectively. It was just shown that bond strength increases and bond length decreases with increasing numbers of electrons in the bonds connecting two atom nuclei. Therefore, increased bond order is associated with increased bond energy and decreased bond length. These relationships were illustrated above for C–C, C = C, and C ≡ C bonds.

The absorption of infrared radiation by bonds in molecules provides the chemist with a powerful probe to determine bond orders, lengths, and energies. Infrared radiation was mentioned as a form of electromagnetic radiation in Chapter 3. When absorbed by molecules, infrared energy causes atoms to vibrate and (when more than two atoms are present) bend relative to each other. The chemical bond can be compared to a spring connecting two atomic nuclei. A short, strong bond acts like a short, strong spring, so that the atoms connected by the bond vibrate rapidly relative to each other. The infrared radiation

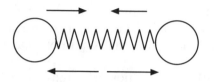

Figure 4.14. Representation of vibration of a chemical bond as a spring connecting two atom nuclei (circles).

absorbed by the bond causes it to vibrate. For a particular mode of bond vibration, the infrared radiation absorbed must be of a specific energy, which results in specific frequencies (v) and wavelengths (λ) for the infrared radiation absorbed (see Section 3.11 in Chapter 3 for the relationships among energies, frequencies, and wavelengths of electromagnetic radiation).

Electronegativity and Covalent Bonding

   Electronegativity refers to the ability of a bonded atom to attract electrons to itself. Whereas the previously-discussed ionization energy and electron affinity deal with the attraction for electrons of isolated gas-phase atoms, electronegativity applies to bonded atoms. Tabulated values of electronegativity range from 0.79 for cesium to 4.0 for fluorine. Unlike ionization energies and electron affinities, which are directly measurable intrinsic properties of atoms, electronegativities are numerical estimates calculated from several elemental properties. Electronegativity values for the first 20 elements are shown in Figure 4.15.

   Examination of Figure 4.15 shows that electronegativity values increase from left to right across a period of the periodic table (there is little discernible trend

1							2
H							He
2.2							-
3	4	5	6	7	8	9	10
Li	Be	B	C	N	O	F	Ne
1.0	1.6	1.8	2.5	3.0	3.4	4.0	-
11	12	13	14	15	16	17	18
Na	Mg	Al	Si	P	S	Cl	Ar
0.93	1.3	1.6	1.9	2.2	2.6	3.2	-
19	20						
K	Ca						
0.82	1.0						

Figure 4.15. Electronegativities of the first 20 elements in the periodic table.

for the transition elements). Within a group of elements, electronegativity decreases with increasing atomic number. For the elements in the abbreviated version of the periodic table shown, this is most obvious for the active metals in the first two groups.

As will be seen in the next section, electrons are often shared unequally between atoms in chemical bonds. The electronegativity concept is useful in dealing with this important phenomenon in chemical bonding.

Sharing Electrons — Unequally

The Lewis formula for hydrogen chloride,

$$H \!\cdot\! \overset{\displaystyle\cdot\cdot}{\underset{\displaystyle\cdot\cdot}{Cl}} :$$

Shared electron pair
in covalent bond

might leave the impression that the two electrons in the covalent bond between H and Cl are shared equally between the two atoms. However, the numbers in Figure 4.15 show that the electronegativity of Cl, 3.2, is significantly higher than that of H, 2.2. This indicates that the Cl atom has a greater attraction for electrons than does the H atom. Therefore, the 2 electrons shared in the bond spend a greater fraction of time in the vicinity of the Cl atom than they do around the H atom. The result of this unequal sharing of electrons is that the chloride end of the molecule acquires a partial negative charge and the hydrogen end a partial positive charge. Although the overall charge of the molecule is zero, it is distributed unequally over the molecule. A body with an unequal electrical charge distribution is said to be **polar.** A covalent bond in which the electrons are not shared equally is called a **polar covalent bond.** A covalent bond in which the sharing of electrons is exactly equal is a **nonpolar covalent bond.**

$$H \!\cdot\! H \qquad\qquad : N \!:\!: N :$$

Electrons shared between two identical atoms are
shared equally and the covalent bond is polar.

The ultimate in unequal sharing of electrons is the ionic bond, in which there is a complete transfer of electrons.

The two common ways of showing a polar bond are illustrated below. The δ's represent partial positive and partial negative charge; the point of the arrow is toward the more electronegative atom that attracts electrons more strongly than the other atom.

$$\overset{\delta+\quad\delta-}{H-Cl} \qquad\qquad \overset{\longmapsto}{H-Cl}$$

Coordinate Covalent Bonds

The covalent bonds examined so far have consisted of electrons contributed equally from both of the atoms involved in the bond. It is possible to have covalent bonds in which only one of the two atoms contributes the two electrons in the bond. This is called a **coordinate covalent bond** or a **dative bond**. Such a bond forms when ammonia gas and hydrogen chloride gas react. This sometimes happens accidentally in the laboratory when these gases are evolved from beakers of concentrated ammonia solution and concentrated hydrochloric acid (a solution of HCl gas) that are left uncovered. When these two gases meet, a white chemical fog is formed. It is ammonium chloride, NH_4Cl, produced by the reaction,

Site of coordinate covalent bond formation (after the
bond has formed, the four N-H bonds are identical)

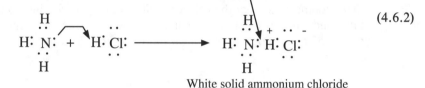

(4.6.2)

White solid ammonium chloride

in which the unshared pair of electrons from NH_3 forms a coordinate covalent bond with an H^+ ion released from HCl. The product is an ion of NH_4. After the coordinate covalent bond has formed, the ammonium ion has four equivalent N-H bonds; the one formed by coordinate covalent bonding is indistinguishable from the other three bonds.

Compounds That Do Not Conform to the Octet Rule

By now a number of examples of bonding have been shown that are explained very well by the octet rule. There are numerous exceptions to the octet rule, however. These fall into the three following major categories:

1. Molecules with an uneven number of valence electrons. A typical example is nitrogen oxide, NO. Recall that the nitrogen atom has 5 valence electrons and the oxygen atom 6, so that NO must have 11, an uneven number. The Lewis structure of NO may be represented as one of the two following forms (these are resonance structures, which are discussed later in this section):

        :N::O.   N::O.

Nitrogen atom has only 7   **Oxygen atom has only 7**
outer-shell electrons.   **outer-shell electrons**

2. Molecules in which an atom capable of forming an octet has fewer than eight outer electrons. A typical example is highly reactive, toxic boron trichloride, BCl_3.

The dashed circle around the central B atom shows that it has only 6 outer shell electrons. However, it can accept two more from another compound such as NH_3 to fill a vacant orbital and complete an octet.

3. Molecules in which an atom has more than eight outer electrons. This can occur with elements in the third and higher periods of the periodic table because such elements have underlying d orbitals capable of accepting electron pairs, so that the valence shell is no longer confined to one s orbital and three p (orbitals that make up an octet of electrons). An example of a compound with more than 8 outer electrons is chlorine trifluoride,

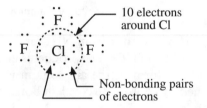

composed of 4 halide atoms (Cl and F), each of which has 7 valence electrons for a total of 28. These are accommodated by placing 5 electron pairs – two of them non-bonding – around the central Cl atom.

Resonance Structures

For some compounds and multi-atom ions it is possible to draw two or more equivalent arrangements of electrons. As an example consider the following for sulfur dioxide:

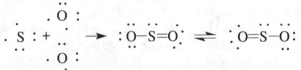

An atom of sulfur and two atoms of oxygen compose a molecule of sulfur dioxide, SO_2, for which two equivalent electron structures may be drawn.

It is seen that in order to have an octet of electrons around each atom, it is necessary to have a double bond between S and one of the Os. However, either O atom may be chosen, so that there are two equivalent structures, which are called **resonance structures**. Resonance structures are conventionally shown with a double arrow between them, as in the example above. They differ only in the locations of bonds and unshared electrons, not in the positions of the atoms. Although the name and structures might lead one to believe that the molecule shifts back and forth between the structures, this is not the case, and the molecule is a hybrid between the two.

Another compound with different resonance forms is nitrogen dioxide, NO_2:

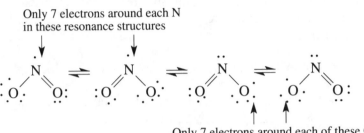

Only 7 electrons around each N
in these resonance structures

Only 7 electrons around each of these Os

Four resonance structures of NO_2.

This compound is produced in smog-forming atmospheres from pollutant NO emitted by automobile engines and is a major air pollutant. It is seen above that there are two aspects of the electron structure of NO_2 relating to material that is under discussion. First, there are four resonance structures. Second, in each of these resonance structures, there is one atom with an incomplete octet of outer shell electrons. This is because NO_2 has an uneven number of electrons. This type of compound is, therefore, an exception to the octet rule. The molecule has an "extra" electron for which there is no other electron to form an electron pair. Compounds consisting of molecules with such an electron are attracted by a magnetic field and are said to be **paramagnetic**.

4.7. CHEMICAL FORMULAS OF COMPOUNDS

What a Chemical Formula States

Chemical formulas were introduced in Chapter 1, and a number of them have been given so far, such as H_2 for water and CO_2 for carbon dioxide. In this and later sections chemical formulas and their meaning are discussed in more detail. A chemical formula contains a lot of information. It is one of the most important tools for communicating in chemical language. Therefore, it is important to understand what a chemical formula means, and how to use it properly.

In order to best illustrate all of the information contained in a chemical formula, consider the moderately complicated example of calcium phosphate, $Ca_3(PO_4)_2$. This compound is composed of Ca^{2+} ions, each with a $+2$ charge, and phosphate ions, PO_4^{3-}, each of which has a -3 charge. The four oxygen atoms and the phosphorus atom in the phosphate ion are held together with covalent bonds. The calcium and phosphate ions are bonded together ionically in a lattice composed of these two kinds of ions. Therefore, both ionic and covalent bonds are involved in calcium phosphate. Figure 4.16 summarizes the information contained in the calcium phosphate formula.

Exercise: In the formula for aluminum sulfate, $Al_2(SO_4)_3$.
(a) The elements present are _____, _____, and _____.
(b) In each formula unit there are _____ Al^{3+} ions and _____ SO_4^{2-} ions.
(c) Each sulfate ion contains _____ sulfur atoms and _____ oxygen atoms.
(d) The total number of each kind of atom in a formula unit of $Al_2(SO_4)_3$ is _____ Al, _____ S, and _____ O.

Answers: (a) aluminum, sulfur, and oxygen; (b) 2, 3; (c) 1, 4; (d) 2, 3, 12.

Percentage Composition from Chemical Formulas

The percentage elemental composition of a chemical compound is readily calculated from either its empirical or molecular formula. One way to calculate the percentage composition of a compound is to consider a mole of the compound and figure out the number of grams of each element in 1 mole of the compound, then calculate the percentages. This can be shown with glucose (blood sugar), molecular formula $C_6H_{12}O_6$. The subscripts indicate that each mole of glucose contains 6 mol C, 12 mol H, and 6 mol O. The mass of each

The elements present
in calcium phosphate are: calcium   phosphorus   oxygen

In each formula
unit there are: 3 calcium ions   and 2 phosphate ions

In each phosphate ion there is 1
phosphorus atom and 4 oxygen atoms

Figure 4.16. Summary of information contained in the chemical formula of calcium phosphate.

element in a mole of glucose and the mass of a mole of glucose are given by the following:

For C: $6 \text{ mol} \times \dfrac{12.01 \text{ g}}{1 \text{ mol}} = 72.06 \text{ g}$

For H: $12 \text{ mol} \times \dfrac{1.01 \text{ g}}{1 \text{ mol}} = 12.12 \text{ g}$

For O: $6 \text{ mol} \times \dfrac{16.00 \text{ g}}{1 \text{ mol}} = 96.00 \text{ g}$

Molar mass of glucose $= 180.18 \text{ g}$

The percentage of each element in glucose is given as follows:

Percent C: $\dfrac{72.06 \text{ g}}{180.18 \text{ g}} \times 100 = 40.0\%$

Percent H: $\dfrac{12.12 \text{ g}}{180.18 \text{ g}} \times 100 = 6.7\%$

Percent O: $\dfrac{96.0 \text{ g}}{180.18 \text{ g}} \times 100 = 53.3\%$

Example: What is the percentage elemental composition of calcium oxalate, CaC_2O_4.

Answer: Exactly 1 mole of CaC_2O_4 contains 1 mol Ca, 2 mol C and 4 mol O. Therefore,

For Ca: $1 \text{ mol} \times \dfrac{40.08 \text{ g}}{1 \text{ mol}} = 40.08 \text{ g}$ Percent Ca $= \dfrac{40.08 \text{ g}}{128.10 \text{ g}} \times 100 = 31.29\%$

For C: $2 \text{ mol} \times \dfrac{12.01 \text{ g}}{1 \text{ mol}} = 24.02 \text{ g}$ Percent Ca $= \dfrac{24.02 \text{ g}}{128.10 \text{ g}} \times 100 = 18.75\%$

For O: $4 \text{ mol} \times \dfrac{16.00 \text{ g}}{1 \text{ mol}} = 64.00 \text{ g}$ Percent Ca $= \dfrac{64.00 \text{ g}}{128.10 \text{ g}} \times 100 = 49.96\%$

Molar mass $CaC_2O_4 = 128.10 \text{ g}$ (per mol)   Total $= 100.0\%$

Calculation of Chemical Formulas

In earlier sections of this chapter it was shown how the chemical formulas of some compounds follow logically from the sharing or exchange of electrons in accordance with the octet rule. However, long before the nature of atoms or the existence of electrons were known, accurate chemical formulas were written for many common compounds. In this section it is shown how chemical formulas are calculated from laboratory data.

First consider the calculation of the **empirical formula** of a chemical compound. The empirical formula gives the lowest whole number values of atoms in the formula. This may not be the same as the **molecular formulas** of substances that exist as molecules, which show the total number of each kind of atom in a molecule. For example, the empirical formula of benzene, one of the hydrocarbons with good engine performance properties in gasoline, is CH. However, each molecule of benzene contains 6 C atoms and 6 H atoms, so the molecular formula is C_6H_6, illustrating the true number of each of the two kinds of atoms in the molecule.

Before calculating some empirical formulas, it is important to know what is meant by a mole of substance. The mole concept was discussed in detail in Section 2.3 and is summarized below:

1. The mole is based upon the number of particles or entities in a substance. These may be molecules of an element or compound, atoms of an element, atoms of a specified element in a particular mass of a compound, or ions in a compound.
2. In expressing moles, it is essential to state clearly the entity being described, such as atoms of Si, molecules of F_2, molecules of H_2O, formula units (see Section 2.3) of ionic Na_2SO_4, or ions of Na^+.
3. A mole of an entity contains Avogadro's number, 6.02×10^{23}, of that entity.
4. The mass of a mole (molar mass) of an element, compound, or portion of a compound (such as the SO_4^{2-} ion in Na_2SO_4) is the mass in grams equal numerically to the atomic mass, molecular mass, or formula mass. For example, given 23.0, 32.1, and 16.0 for the atomic masses of Na, S, and O, respectively, the following can be stated: (a) A mole of Na atoms has a mass of 23.0 g, (b) a mole of Na_2SO_4 has a mass of $2 \times 23.0 + 32.1 + 4 \times 16.0 = 142.1$ g, and (c) a mole of SO_4^{2-} ions has a mass of $32.1 + 4 \times 16.0 = 96.1$ g.
5. Number of moles of substance =

$$\frac{\text{mass of substance in grams}}{\text{molar mass of substance in grams per mole}} \qquad (4.7.1)$$

Empirical Formula from Percentage Composition

Suppose that the only two elements in a compound are phosphorus and oxygen and that it is 43.64% P and 56.36% O by mass. The following steps can be used to find the empirical formula of the compound:

1. From the percentages it follows that for each 100 grams of compound there are 43.64 g of P and 56.36 g of O. The number of moles of each of these elements in exactly 100 g of the compound is calculated from the number of grams of each per mole of compound as follows:

$$43.64 \text{ g P} \times \frac{1 \text{ mol P}}{30.97 \text{ g P}} = 1.409 \text{ mol P} \qquad (4.7.2)$$

↑	↑	↑
In 100.0 g of the compound there are 43.64 g of P	In 1 mol of P there are 30.97 g of P (from the atomic mass of P)	Therefore, 100.0 g of the compound contains 1.409 mol of P

$$56.36 \text{ g O} \times \frac{1 \text{ mol O}}{16.00 \text{ g O}} = 3.523 \text{ mol O} \qquad (4.7.3)$$

↑	↑	↑
In 100.0 g of the compound there are 56.36 g of O	In 1 mol of O there are 16.00 g of O (from the atomic mass of O)	Therefore, 100.0 g of the compound contains 3.523 mol of O

2. Calculate the relative number of moles of O per mole of P.

$$\frac{3.523 \text{ mol O}}{1.409 \text{ mol P}} = \frac{2.500 \text{ mol O}}{1.000 \text{ mol P}} \qquad (4.7.4)$$

3. Calculate the ratio of moles of O to moles of P in terms of whole numbers. The denominator of the ratio, 2.500 mol O/1.000 mol P is already a whole number, and the numerator can be converted to a whole number (5) if it is multiplied by 2. Therefore, both the numerator and the denominator are multiplied by 2 to get the whole number ratio,

$$\frac{2.500 \text{ mol O}}{1.000 \text{ mol P}} \times \frac{2}{2} = \frac{5.000 \text{ mol O}}{2.000 \text{ mol P}} \qquad (4.7.5)$$

4. Knowing the ratio of moles, write the empirical formula of the compound.

$$P_2O_5$$

The empirical formula does not necessarily give the actual number of atoms per molecule; it gives the smallest whole number ratio for the number of each kind of atom in the molecule. The molecular formula of this compound is P_4O_{10}, indicating that a molecule of it contains 2 times the number of each kind of atom indicated by the empirical formula. The correct name of the compound is tetraphosphorus decoxide. It is made commercially by burning elemental phosphorus in dry air. It reacts violently with water to produce commercially important orthophosphoric acid, H_3PO_4. The strong affinity of P_4O_{10} for water makes P_4O_{10} very useful as a drying and dehydrating agent.

Empirical formulas are actually determined in the laboratory by chemical analyses for each element in the compound. There is always some error involved in such an analysis, so the numbers obtained for the ratios of one element to another may not be as nicely rounded as in the example of the calculation of the empirical formula P_2O_5. As an example of this, consider that a compound used as an inert filler in laundry detergent was found by chemical analysis to be 32.59% Na, 22.74% S, and 44.67% O. What is the empirical formula of the compound?

1. According to the percentages given, there are 32.59 g of Na, 22.74 g of S and 44.67 g of O in 100.0 g of the compound. Using the appropriate atomic masses gives the following number of moles of each element in 100 g of the compound:

$$32.59 \text{ g Na} \times \frac{1 \text{ mol Na}}{22.99 \text{ g Na}} = 1.418 \text{ mol Na} \qquad (4.7.6)$$

$$22.74 \text{ g S} \times \frac{1 \text{ mol S}}{32.06 \text{ g S}} = 0.7093 \text{ mol S} \qquad (4.7.7)$$

$$44.67 \text{ g O} \times \frac{1 \text{ mol O}}{16.00 \text{ g O}} = 2.792 \text{ mol O} \qquad (4.7.8)$$

2. Of the 3 elements present, the element with the least number of moles is sulfur. Therefore, the ratios of the number of moles of each of the other elements relative to a mole of S is calculated.

$$\frac{1.418 \text{ mol Na}}{0.7093 \text{ mol S}} \times = \frac{1.999 \text{ mol Na}}{1.000 \text{ mol S}} \qquad (4.7.9)$$

$$\frac{2.792 \text{ mol O}}{0.7093 \text{ mol S}} \times \frac{3.936 \text{ mol O}}{1.000 \text{ mol S}} \qquad (4.7.10)$$

3. Examination of the results above shows that, rounding 1.999 to 2 and 3.936 to 4, there are 2 moles of Na and 4 moles of O per mole of S, leading to the empirical formula,

$$Na_2SO_4$$

This compound is called sodium sulfate.

Example: Analysis of a compound shows it to be 26.49% potassium, 35.58% Cr, and 37.93% O. What is the empirical formula of the compound?

Answer: In 100.0 g of the compound there are 26.49 g K, 35.58 g Cr, and 37.93 g O, and the following number of moles of each element:

$$26.49 \text{ g K} \times \frac{1 \text{ mol K}}{39.10 \text{ g K}} = 0.6775 \text{ mol K} \qquad (4.7.11)$$

$$35.58 \text{ g Cr} \times \frac{1 \text{ mol Cr}}{52.00 \text{ g Cr}} = 0.6842 \text{ mol Cr} \qquad (4.7.12)$$

$$37.93 \text{ g O} \times \frac{1 \text{ mol O}}{16.00 \text{ g })} = 2.371 \text{ mol O} \qquad (4.7.13)$$

Examination of the above figures reveals that there is essentially the same number of moles of K and Cr in a gram of the compound (rounding gives 0.68 mole of each per gram of compound), whereas there is a higher number of moles of O. Calculating the number of moles of K and O relative to the number of moles of Cr gives the following:

$$\frac{0.6775 \text{ mol K}}{0.6842 \text{ mol Cr}} \times 100.0 \text{ g compound} = 0.9902 \text{ mol K} \qquad (4.7.14)$$

$$\frac{2.371 \text{ mol O}}{0.6842 \text{ mol Cr}} \times 100.0 \text{ g compound} = 3.465 \text{ mol O} \qquad (4.7.15)$$

Rounding these numbers to the nearest 0.1 mol gives 1.0 mol of K and 3.5 mol of O per mol of Cr. One can write the empirical formula as,

$$KCrO_{3.5}$$

but this does not give a whole number of O atoms in the formula unit. Multiplying the subscripts by 2 (in the formula above there is understood to be a subscript 1 after K and after Cr) yields an empirical formula with whole numbers of atoms:

$$K_2Cr_2O_7$$

This is the formula of the compound potassium dichromate, each formula unit of which consists of 2 K^+ ions and a dichromate ion, $Cr_2O_7^{2-}$.

4.8.   THE NAMES OF CHEMICAL COMPOUNDS

One of the more important aspects of chemical language is the correct use of names for chemical compounds — chemical nomenclature. So far in the text the names of a number of compounds have been given. This section presents a systematic approach to the naming of inorganic compounds. These are all of the compounds that do not contain carbon plus a few — such as CO_2 and Na_2CO_3 — that do contain carbon. Most carbon-containing compounds are organic compounds; organic compound nomenclature will not be considered here. Inorganic nomenclature will be divided into several categories so that it may be approached in a systematic manner.

Binary Molecular Compounds

In this section the rules for naming binary molecular compounds — those composed of only two elements bonded together as molecules — are examined. Most such compounds are composed of nonmetallic elements. In a binary molecular compound, one of the elements is usually regarded as being somewhat more positive than the other. This is a result of unequal sharing of electrons in covalent bonds, a characteristic of bonds considered briefly in Section 4.6. The symbol of the element with a more positive nature is given first in the chemical formula of a binary compound. For example, in HCl hydrogen is regarded as having a somewhat more positive, and Cl a somewhat more negative character.

In naming binary molecular compounds, the first part of the name is simply that of the first element in the chemical formula, and the second part of the name is that of the second element with an -*ide* ending. Therefore, the name of HCl is hydrogen chloride. In most cases, however, it is necessary to have prefixes to designate the relative number of atoms in the molecular formula. These prefixes are the following:

1-mono	3–tri	5-penta	7-hepta	9-nona
2-di	4-tetra	6-hexa	8-octa	10-deca

The uses of these prefixes are illustrated by the names of several oxygen compounds, or oxides in Table 4.2. Other examples of compounds in which prefixes are used for naming are $SiCl_4$, silicon tetrachloride; Si_2F_6, disilicon hexafluoride; PCl_5, phosphorus pentachloride; and SCl_2, sulfur dichloride.

It is seen from the examples given that the prefix is omitted before the name of the first element in the chemical formula when there is only 1 atom of that element per molecule. In cases where the name of the second element in a binary molecular compound formula begins with a vowel, the "a" or "o" at the end of the prefix may be deleted. Thus, CO is called carbon monoxide, rather than carbon monooxide and Cl_2O_7 is called dichlorine heptoxide, not dichlorine heptooxide.

A number of compounds, including those other than binary molecular compounds, have been known for so long and have been so widely used that they have acquired common names that do not describe their chemical formulas. The best example of these is H_2O, which is always known as water, not dihydrogen

Table 4.2. Some Oxides and Their Names

Compound Formula	Compound Name	Compound Formula	Compound Name
CO	Carbon monoxide	SO_2	Sulfur dioxide
CO_2	Carbon dioxide	SO_3	Sulfur trioxide
N_2O_5	Dinitrogen pentoxide	SiO_2	Silicon dioxide
N_2O_4	Dinitrogen tetroxide	P_2O_4	Diphosphorus tetroxide
NO_2	Nitrogen dioxide	P_4O_7	Tetraphosphorus heptoxide
NO	Nitrogen monoxide	P_4O_{10}	Tetraphosphorus decoxide
N_2O	Dinitrogen oxide (nitrous oxide)	Cl_2O_7	Dichlorine heptoxide

monoxide. Other examples are H_2O_2, hydrogen peroxide; NH_3, ammonia; and N_2O, nitrous oxide.

Names of Ionic Compounds

Recall that ionic compounds consist of aggregates of positively charged cations and negatively charged anions held together by the mutual attraction of their opposite electrical charges. The easiest of these compounds to visualize are those that are composed of only two elements, the cation formed by the loss of all its outer shell electrons and the anion formed by the gain of enough outer shell electrons to give it a stable octet of outer electrons (as shown for sodium and chlorine in Equation 4.3.1). The formation of such ions is readily shown by consideration of some of the elements in the abbreviated version of the periodic table in Figure 4.2 as illustrated by the following half-reactions (literally "half a reaction" in which electrons are reactants or products):

Half-reactions for cation formation

$$Li: \cdot \longrightarrow Li:^+ + e^- \qquad Mg: \longrightarrow :Mg:^{2+} + 2e^-$$

$$Na \cdot \longrightarrow :Na:^+ + e^- \qquad Ca: \longrightarrow :Ca:^{2+} + 2e^-$$

$$K \cdot \longrightarrow :K:^+ + e^- \qquad Al: \longrightarrow :Al:^{3+} + 3e^-$$

Half-reactions for anion formation

$$\cdot \overset{\cdot}{\underset{\cdot}{N}} \colon + 3e^- \longrightarrow \colon \overset{\cdot \cdot}{\underset{\cdot \cdot}{N}} \colon^{3-} \qquad H \cdot + e^- \longrightarrow H \colon^-$$

$$\cdot \overset{\cdot \cdot}{\underset{\cdot}{O}} \colon + 2e^- \longrightarrow \colon \overset{\cdot \cdot}{\underset{\cdot \cdot}{O}} \colon^{2-} \qquad \colon \overset{\cdot}{\underset{\cdot \cdot}{F}} \colon + e^- \longrightarrow \colon \overset{\cdot \cdot}{\underset{\cdot \cdot}{F}} \colon^-$$

$$\cdot \overset{\cdot \cdot}{\underset{\cdot}{S}} \colon + 2e^- \longrightarrow \colon \overset{\cdot \cdot}{\underset{\cdot \cdot}{S}} \colon^{2-} \qquad \colon \overset{\cdot}{\underset{\cdot \cdot}{Cl}} \colon + e^- \longrightarrow \colon \overset{\cdot \cdot}{\underset{\cdot \cdot}{Cl}} \colon^-$$

The cations formed as shown by the half-reactions above are simply given the names of the metals that produced them, such as sodium for Na^+ and calcium for Ca^{2+}. Since they consist of only one element, the name for each anion has an -ide ending, that is, N^{3-}, nitride; O^{2-}, oxide; S^{2-}, sulfide; H^-, hydride; F^-, fluoride; and Cl^-, chloride. An advantage in the nomenclature of ionic compounds is that it is not usually necessary to use prefixes to specify the numbers of each kind of ion in a formula unit. This is because the charges on the ions determine the relative numbers of each as shown by the examples in Table 4.3.

Exercise: For the compounds listed below, where x and y are unspecified subscripts, give the compound formula and name.

(a) K_xCl_y	(d) Li_xCl_y	(g) K_xN_y	(j) Mg_xO_y
(b) K_xS_y	(e) Li_xF_y	(h) Ca_xN_y	(k) Ca_xO_y
(c) Al_xF_y	(f) Li_xO_y	(i) Mg_xN_y	(l) Ca_xS_y

Answers: (a) KCl, potassium chloride; (b) K_2S, potassium sulfide; (c) AlF_3, aluminum fluoride; (d) LiCl, lithium chloride; (e) LiF, lithium fluoride; (f) Li_2O, lithium oxide, (g) K_3N, potassium nitride; (h) Ca_3N_2, calcium nitride; (i) Mg_3N_2, magnesium nitride; (j) MgO, magnesium oxide; (k) CaO, calcium oxide; (l) CaS, calcium sulfide.

Table 4.3. Formulas and Names of Ionic Compounds Formed from Two Elements

Ionic Compound Formula	Compound Name	Neutral Compound Formed from the Ions Below
NaF	Sodium fluoride	One +1 ion and one –1 ion
K_2O	Potassium oxide	Two +1 ions and one –2 ion
Na_3N	Sodium nitride	Three +1 ions and one –3 ion
$MgCl_2$	Magnesium chloride	One +2 ion and two –1 ions
CaH_2	Calcium hydride	One +2 ion and two –1 ions
$AlCl_3$	Aluminum chloride	One +3 ion and three –1 ions
Al_2O_3	Aluminum oxide	Two +3 ions and three –2 ions

4.9. ACIDS, BASES, AND SALTS

Most inorganic chemical compounds belong to one of three classes — acids, bases, or salts. These three categories of compounds are addressed briefly in this section and in more detail in Chapter 6.

Acids

For the present an acid may be defined as a substance that dissolves in water to produce hydrogen ion, $H^+(aq)$. (Recall that aq in parentheses after a species formula shows that it is dissolved in water.) The most obviously acidic compounds are those that contain ionizable H in their formulas. Typically, hydrogen chloride gas $HCl(g)$ dissolves in water,

$$HCl(g) \xrightarrow{\text{H}_2\text{O}} H^+(aq) + Cl^-(aq) \tag{4.9.1}$$

and completely dissociates (falls apart) to give an aqueous solution containing H^+ cation and Cl^- anion. Note that the H_2O above the arrow in this chemical equation shows the reaction occurs in water. The presence of a substantial concentration of H^+ ions in water provides an acidic solution, in this case a solution of hydrochloric acid.

Some acids produce more than one hydrogen ion per molecule of acid. For example, sulfuric acid dissociates in two steps,

$$\underset{\text{sulfuric acid}}{H_2SO_4(aq)} \rightarrow H^+(aq) + \underset{\text{hydrogen sulfate ion}}{HSO_4^-(aq)} \tag{4.9.2}$$

$$HSO_4^-(aq) \rightarrow H^+(aq) + SO_4^{2-}(aq) \tag{4.9.3}$$

to yield $2\,H^+$ per H_2SO_4 molecule; the second of the two hydrogen ions comes off much less readily than the first. Some acids have hydrogen atoms that do not produce H^+ ion in water. For example, each molecule of acetic acid produces only one H^+ ion when it dissociates in water; the other 3 H atoms stay covalently bonded to the acetate ion:

$$\begin{array}{c} \text{H O} \\ | \ \| \\ \text{H–C–C–OH} \\ | \\ \text{H} \end{array} \longrightarrow \begin{array}{c} \text{H O} \\ | \ \| \\ \text{H–C–C–O}^- \\ | \\ \text{H} \end{array} + \; H^+ \tag{4.9.4}$$

Some of the most widely produced industrial chemicals are acids. Sulfuric acid ranks first among all chemicals produced in the United States with production of almost 40 million metric tons per year. Its greatest single use is in the production of phosphate fertilizers, and it has applications in many other areas including petroleum refining, alcohol synthesis, iron and steel pickling (corrosion removal), and storage battery manufacture. Nitric acid ranks about 10th among U.S. chemicals with annual production of 7–8 million metric tons, and hydrochloric acid is about 25th at about 3 million metric tons (annual production and rank vary from year to year).

The naming of acids is addressed in more detail in Chapter 6. Briefly, acids that contain only H and another atom are "hydro-ic" acids, such as hydrochloric acid, HCl. For acids that contain two different amounts of oxygen in the anion part, the one with more oxygen is an "-ic" acid and the one with less is an "-ous" acid. This is illustrated by nitric acid, HNO_3, and nitrous acid, HNO_2. An even greater amount of oxygen is denoted by a "per-" prefix and less by the "hypo-ous" name. The guidelines discussed above are illustrated for acids formed by chlorine ranging from 0 to 4 oxygen atoms per acid molecule as follows: HCl, hydrochloric acid; HClO, hypochlorous acid; $HClO_2$, chlorous acid; $HClO_3$, chloric acid; $HClO_4$, perchloric acid.

Bases

A **base** is a substance that contains hydroxide ion OH^-, or produces it when dissolved in water. Most of the best known inorganic bases have a formula unit composed of a metal cation and one or more hydroxide ions. Typical of these are sodium hydroxide, NaOH, and calcium hydroxide, $Ca(OH)_2$, which dissolve in water, to yield hydroxide ion and their respective metal ions. Other bases such as ammonia, NH_3, do not contain hydroxide ion, but react with water,

$$NH_3 + H_2O \rightarrow NH_4^+ + OH^- \qquad (4.9.5)$$

to produce hydroxide ion (this reaction proceeds only to a limited extent; most of the NH_3 is in solution as the NH_3 molecule).

Bases are named for the cation in them plus "hydroxide." Therefore, KOH is potassium hydroxide.

Salts

A **salt** is an ionic compound consisting of a cation other than H^+ and an anion other than OH^-. A salt is produced by a chemical reaction between an acid and a base. The other product of such a reaction is always water. Typical salt-producing reactions are given below:

$$KOH + HCl \rightarrow KCl + H_2O \qquad (4.9.6)$$

base   acid   a salt, potassium chloride   water

$$Ca(OH)_2 + H_2SO_4 \quad \rightarrow \quad CaSO_4 \quad + \quad 2H_2O \qquad (4.9.7)$$

$$\underset{\text{base}}{} \quad \underset{\text{acid}}{} \qquad \underset{\text{a salt, calcium sulfate}}{} \quad \underset{\text{water}}{}$$

Except for those compounds in which the cation is H^+ (acids) or the anion is OH^- (bases), the compounds that consist of a cation and an anion are salts. Therefore, the rules of nomenclature discussed in Section 4.8 are those of salts. The salt product of Reaction 4.9.6, above, consists of K^+ cation and Cl^- anion, so the salt is called potassium chloride. The salt product of Reaction 4.9.7, above, is made up of Ca^{2+} cation and SO_4^{2-} anion and is called calcium sulfate. The reaction product of LiOH base with H_2SO_4 acid is composed of Li^+ ions and SO_4^{2-} ions. It takes 2 singly charged Li^+ ions to compensate for the 2^- charge of the SO_4^{2-} anion, so the formula of the salt is Li_2SO_4. It is called simply lithium sulfate. It is not necessary to call it dilithium sulfate because the charges on the ions denote the relative numbers of ions in the formula.

CHAPTER SUMMARY

The chapter summary below is presented in a programmed format to review the main points covered in this chapter. It is used most effectively by filling in the blanks, referring back to the chapter as necessary. The correct answers are given at the end of the summary.

Chemical bonds are normally formed by the transfer or sharing of (1)_____

_____, which are those in the (2)_____.

An especially stable group of electrons attained by many atoms in chemical

compounds is an (3)_____. An ion consists of

(4)_____

_____.

A cation has (5)_____ and an anion has

(6)_____. An ionic compound is one that contains

(7)_____ and is held together by (8)_____.

Both F^- and Mg^{2+} have the electron configuration (9)_____

identical to that of the neutral atom (10)_____.

In visualizing a neutral metal atom reacting with a neutral nonmetal atom to

produce an ionic compound, the three major energy factors are (11)_____

_____.

The energy required to separate all of the ions in a crystalline ionic compound and remove them a sufficient distance from each other so that there is no interaction between them is called the (12)_____.

The ion formed from Ca is (13)_____, the ion formed from Cl is (14)_____, and the formula of the compound formed from these ions is (15)_____.

A covalent bond may be described as (16)_____

_____

_____

In the figure,

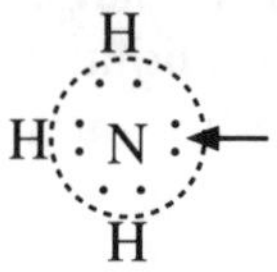

$$\text{H} \cdot\!: \text{N} :\!\!\leftarrow$$

each pair of dots between N and H represents (17)_____

_____ the arrow points to

(18)_____

and the circle outlines (19)_____

_____. A central atom refers to (20)_____

_____

_____. A triple bond

consists of (21)_____

and is represented in a structural chemical formula as (22)_____

_____. With increasing bond order

(single < double < triple), bond length (23)_____ and bond strength (24)_____.

Electronegativity refers to (25)_____

_____. A polar covalent bond is one in which

(26)_____

_____. A coordinate covalent bond is

(27)_____

_____. Three major exceptions to the octet rule are (28)_____

_____

_____

Resonance structures are those for which (29)_____.

Chemical formulas consist of (30)_____

that tell the following three things about a compound (31)_____.

The percentage elemental composition of a chemical compound is calculated by (32)_____.

The empirical formula of a chemical compound is calculated by computing the masses of each constituent element in (33)_____,

dividing each of the resulting values by (34)_____,

dividing each value by (35)_____, and rounding to (36)_____.

The prefixes for numbers 1–10 used to denote relative numbers of atoms of each kind of atom in a chemical formula are (37)_____.

The compound N_2O_5 is called dinitrogen pentoxide.

The ionic compound $AlCl_3$ is called (38)_____, which does not contain a "tri-" because (39)_____.

Ammonium ion, NH_4^+, is classified as a (40)_____ meaning that it consists of (41)_____.

An acid is (42)_____

_____.

A base is (43)_____

_____. A salt is (44)_____

_____ and is produced

by (45)_____. The other product of such a

reaction is always (46)_____.

Answers

1. valence electrons
2. outermost shell of the atom
3. octet of outer electrons
4. an atom or group of atoms having an unequal number of electrons and protons and, therefore, a net electrical charge
5. a positive charge
6. a negative charge
7. cations and anions
8. ionic bonds
9. $1s^2 2s^2 2p^6$
10. neon
11. ionization energy, electron affinity, and lattice energy
12. lattice energy
13. Ca^{2+}
14. Cl^-
15. $CaCl_2$
16. one that joins two atoms through the sharing of 1 or more pairs of electrons between them.
17. a pair of electrons shared in a covalent bond;
18. an unshared pair of electrons,
19. a stable octet of electrons around the N atom
20. an atom to which several other atoms are bonded
21. 3 pair (total of 6) electrons shared in a covalent bond
22.
23. decreases
24. increases
25. the ability of a bonded atom to attract electrons to itself
26. the electrons involved are not shared equally
27. one in which only one of the two atoms contributes the two electrons in the bond

28. molecules with an uneven number of valence electrons, molecules in which an atom capable of forming an octet has fewer than eight outer electrons, and molecules in which an atom has more than eight outer electrons
29. it is possible to draw two or more equivalent arrangements of electrons
30. atomic symbols, subscripts, and sometimes parentheses and charges
31. elements in it, relative numbers of each kind of atom, and charge (if an ion)
32. dividing the mass of a mole of a compound into the mass of each of the constituent elements in a mole
33. 100 g of the element
34. the atomic mass of the element
35. the smallest value
36. the smallest whole number for each element
37. 1-mono, 2-di, 3-tri, 4-tetra, 5-penta, 6-hexa, 7-hepta, 8-octa, 9-nona, 10-deca
38. aluminium chloride
39. the charges on ions are used to deduce chemical formulas
40. polyatomic ion
41. two or more atoms per ion
42. a substance that dissolves in water to produce hydrogen ion, H^+ (aq)
43. a substance that contains hydroxide ion OH^-, or produces it when dissolved in water
44. an ionic compound consisting of a cation other than H^+ and an anion other than OH^-
45. a chemical reaction between an acid and a base
46. water

QUESTIONS AND PROBLEMS

1. Atoms can attain a stable octet of outer shell electrons by losing electrons to become (a)_____, gaining electrons to become (b)_____, or by sharing electrons with other atoms in (c)_____.

2. Chlorofluorocarbons (Freons) are composed of molecules in which Cl and F atoms are bonded to 1 or 2 C atoms. These compounds do not break down well in the atmosphere until they drift high into the stratosphere where very short wavelength ultraviolet electromagnetic radiation from the sun is present, leading to the production of free Cl atoms that react to deplete the stratospheric ozone layer. Recalling what has been covered so far about the energy of electromagnetic radiation as a function of wavelength, what does this say about the strength of C–Cl and C–F bonds?

3. Illustrate the octet rule with examples of (a) a cation, (b) an anion, (c) a diatomic elemental gas, and (d) a covalently bound chemical compound.

4. When elements with atomic numbers 6 through 9 are covalently or ionically bound, or when Na, Mg, or Al have formed ions, which single element do their outer electron configurations most closely resemble? Explain your answer in terms of filled orbitals.

5. Ionic bonds exist because of (a)_____ between (b) ___
_____.

6. In the crystal structure of NaCl, what are the number and type of ions that are nearest neighbors to each Na^+ ion?

7. What major aspect of ions in crystals is not shown in Figure 4.5?

8. What are five energy factors that should be considered in the formation of ionic NaCl from solid Na and gaseous Cl_2?

9. What are two major factors that increase the lattice energy of ions in an ionic compound?

10. Is energy released or is it absorbed when gaseous ions come together to form an ionic crystal?

11. What is incomplete about the statement that "the energy change from lattice energy for NaCl is 785 kilojoules of energy released?"

12. How do the sizes of anions and cations compare with their parent atoms?

13. How do the sizes of monatomic (one-atom) cations and anions compare in the same period?

14. Define covalent bond.

15. Explain the energy minimum in the diagram illustrating the H-H covalent bond in Figure 4.9.

16. What may be said about the likelihood of H atoms being involved in double covalent bonds?

17. What is represented by a dashed line, –, in a chemical formula?

18. What is represented by the two dots in the formula of phosphine, PH_3, below:

19. What are multiple bonds? Which three elements are most likely to form multiple bonds?

20. What can be said about the nature of covalent bonds between (a) two atoms with almost identical electronegativity values and (b) two atoms with substantially different electronegativity values?

21. What symbols are used to show bond polarity?

22. What is the "ultimate" in polar bonds?

23. A molecule of NH_3 will combine with one of BF_3. Describe the kind of bond formed in the resulting compound.

24. What is required for an atom to form a compound with more than 8 electrons in the central atom's outer shell?

25. What are resonance structures?

26. How many total valence electrons are in the nitrate ion, NO_3^-? What are the resonance structures of this ion?

27. Summarize the steps involved in calculating the percentage composition of a compound from its formula.

28. Phosgene, $COCl_2$, is a poisonous gas that was used for warfare in World War II. What is its percentage composition?

29. The molecular formula of acetylsalicylic acid (aspirin) is $C_9H_8O_4$. What is its percentage composition?

30. Hydrates are compounds in which each formula unit is associated with a definite number of water molecules. A typical hydrate is copper (II) sulfate pentahydrate, $CuSO_4 \cdot 5H_2O$, which represents a pair of Cu^{2+} and SO_4^{2-} ions associated with 5 H_2O molecules. The water of hydration can be driven off by heating, leaving the anhydrous compound. Answer the following pertaining to $CuSO_4 \cdot 5H_2O$: (a) mass of 1 mole of the compound, (b) mass of H_2O in 1 mole of $CuSO_4 \cdot 5H_2O$, (c) percentage of H_2O in $CuSO_4 \cdot 5H_2O$.

31. What are the percentages of oxygen in (a) perchloric acid, (b) chloric acid, (c) chlorous acid, and (d) hypochlorous acid (these acids were discussed in this chapter).

32. Answer the following pertaining to sodium oxalate, $Na_2C_2O_4$: (a) What is the simplest (empirical) formula?

33. What is the formula of dichlorine heptoxide? What is its percentage composition?

34. Chlorine dioxide, ClO_2, is used as a substitute for chlorine gas in the disinfection of drinking water. What is the percentage composition of ClO_2?

35. Summarize in steps the calculation of empirical formula from the percentage composition of a compound.

36. If the empirical formula and the actual formula mass (molecular mass) of a compound are known, how is the true formula calculated?

37. A refrigerant Freon gas is 9.93% C, 31.43% F, and 58.64% Cl. What is its empirical formula?

38. A 100.0 g portion of the Freon gas (preceding question) was found to occupy 25.3 L at 100°C and 1.000 atm pressure. Assuming that the gas behaved ideally and using the ideal gas equation, how many moles of the gas are in 100.0 g of Freon? What is its molar mass? What is its molecular formula? (To answer this question it may be necessary to refer back to a discussion of the gas laws in Section 2.6).

39. A compound is 5.88% H and 94.12% O. What is its empirical formula?

40. The compound from the preceding problem has a molecular mass of 34.0. What is its molecular formula?

41. A pure liquid compound with an overpowering vinegar odor is 40.0% C, 6.67% H, and 53.3% O. Its molecular mass is 60.0. What is its empirical formula? What is its molecular formula?

42. A compound is 29.1% Na, 40.5% S, and 30.4% O. It has a formula mass of 158.0. Fill out the table below pertaining to the compound, give its true formula, and give its actual formula. It is an ionic compound. What is the anion?

Element	Grams of Element in 100 g Compound	Mol Element in 100 g Compound	$\dfrac{\text{Mol Element}}{\text{mol Element with Least Moles}}$
Na	(a)_____	(d)_____	(g)_____
S	(b)_____	(e)_____	(h)_____
O	(c)_____	(f)_____	(i)_____

43. An ionic compound is 41.7% Mg, 54.9% O, and 3.4% H. What is its empirical formula? Considering the ions in Table 4.1 and elsewhere in this chapter, what is the actual formula?

44. Ethylenediamine is 40.0% C, 13.4% H, and 46.6% N; its formula mass is 60.1. What are its empirical and molecular formulas?

45. The empirical formula of butane is C_2H_5 and its molecular mass is 58.14. What is the molecular formula?

5 CHEMICAL REACTIONS, EQUATIONS, AND STOICHIOMETRY

5.1. THE SENTENCES OF CHEMISTRY

As noted earlier, chemistry is a language. Success in the study of chemistry depends upon how well chemical language is learned. This chapter presents the last of the most basic parts of the chemical language. When it has been learned, the reader will have the essential tools needed to speak and write chemistry and to apply it in environmental and other areas.

Recall that the discussion of chemical language began by learning about the *elements*, the *atoms* composing the elements, and the *symbols* used to designate these elements and their atoms. Atoms of the elements bond together in various combinations to produce *chemical compounds*. These are designated by *chemical formulas* consisting of symbols for the kinds of atoms in the compound and subscripts indicating the relative numbers of atoms of each kind in the compound. In chemical language the symbols of the elements are the letters of the chemical alphabet and the formulas are the words of chemistry.

Chemical Reactions and Equations: The Sentences of the Chemical Language

The formation of chemical compounds, their decomposition, and their interactions with one another fall under the category of **chemical reactions**. Chemical reactions are involved in the annual production of millions of kilograms of industrial chemicals, bacterially mediated degradation of water pollutants, the chemical analysis of the kinds and quantities of components of a sample, and practically any other operation involving chemicals. To a very large extent, chemistry is the study of chemical reactions expressed on paper as chemical equations. A **chemical equation** is a sentence of chemistry, made up of words consisting of chemical formulas. A sentence should be put together according to rules understood by all those literate in the language. The rules of the chemical language are particularly rigorous. Although a grammatically sloppy sentence in a spoken language can still convey a meaningful message, a chemical equation with even a small error is misleading and often meaningless.

Quantitative Calculations from Chemical Equations

Chemistry is a quantitative science, and it is important to know how to do some of the basic chemical calculations early in a beginning chemistry course. Among the most important of these are the calculations of the quantities of substances consumed or produced in a chemical reaction. Such calculations are classified as **stoichiometry**. Heat is normally evolved or taken up in the course of a chemical reaction. The calculation of the quantity of heat involved in a reaction falls in the branch of chemistry called **thermochemistry**.

5.2.   THE INFORMATION IN A CHEMICAL EQUATION

Chemical Reactions

A chemical reaction is a process involving the breaking and/or formation of chemical bonds and a change in the chemical composition of the materials participating in the reaction. A chemical reaction might involve the combination of two elements to form a compound. An example of this is the reaction of elemental hydrogen and oxygen to produce the compound water. Passage of an electrical current through water can cause the compound to break down and produce elemental hydrogen and oxygen. When wood burns, cellulose, a compound in the wood, reacts with elemental oxygen in air to produce two compounds, carbon dioxide and water. If the carbon dioxide produced is bubbled through a solution of the compound calcium hydroxide dissolved in water it produces the compounds calcium carbonate (a form of limestone) and water. Energy is involved in chemical reactions; some reactions produce energy, others require it in order for them to occur.

Expressing a Chemical Reaction as a Chemical Equation

A **chemical equation** is a means of expressing what happens when a chemical reaction occurs. It tells what reacts, what is produced, and the relative quantities of each. The information provided can best be understood by examining a typical chemical equation. For example, consider the burning of propane, a gas extracted from petroleum that is widely used for heating, cooking, grain drying, and other applications in which a clean-burning fuel is needed in areas where piped natural gas is not available. When propane burns in a camp stove, it reacts with oxygen in the air. The chemical equation for this reaction and the information in it are the following:

$$C_3H_8 + 5O_2 \rightarrow 3CO_2 + 4H_2O \qquad (5.2.1)$$

- Propane reacts with oxygen to give carbon dioxide and water.
- There are two **reactants** on the left side of the equation – propane, chemical formula C_3H_8, and oxygen, chemical formula O_2.
- There are two **products** on the right side of the equation – carbon dioxide, chemical formula CO_2, and water, chemical formula H_2O.
- For the smallest possible unit of this reaction, 1 propane molecule reacts with 5 oxygen molecules to produce 3 carbon dioxide molecules, and 4 water molecules as shown by the respective numbers preceding the chemical formulas (there is understood to be a 1 in front of the C_3H_8).
- There are 3 C atoms altogether on the left side of the equation, all contained in the C_3H_8 molecule, and 3 C atoms on the right side contained in 3 molecules of CO_2.
- There are 8 H atoms among the reactants, all in the C_3H_8 molecule, and 8 H atoms among the 4 molecules of H_2O in the products.
- There are 10 O atoms in the 5 O_2 molecules on the left side of the equation and 10 O atoms on the right side. The 10 O atoms in the products are present in 3 CO_2 molecules and 4 H_2O molecules.

Like mathematical equations, the left side of a chemical equation must be equivalent to the right side. Chemical equations are **balanced** in terms of atoms:

A correctly written chemical equation has equal numbers of each kind of atom on both sides of the equation.

As was just seen, the chemical equation being discussed has 3 carbon atoms, 8 hydrogen atoms and 10 oxygen atoms on both the left and right sides of the equation. Therefore, the equation is balanced. Balancing a chemical equation is a very important operation that is discussed in the next section.

Symbols Used in Chemical Equations

Several symbols are used in chemical equations. The two sides of the equation may be separated by an arrow, $\rightarrow$; an equals sign, $=$; or a double arrow, $\rightleftarrows$. The double arrow denotes a reversible reaction; that is, one that can go in either direction. The physical state (see Section 2.5) of a reaction component is indicated by letters in parentheses immediately following the formula. Therefore, (s) stands for a solid, (l) for a liquid, (g) for a gas, and (aq) for a substance dissolved in aqueous (water) solution. An arrow pointing up, $\uparrow$, immediately after the formula of a product indicates that the product is evolved as a gas, whereas $\downarrow$ shows that it is a precipitate (solid forming from a reaction in solution and settling to the bottom of the container). These two symbols are not used extensively in this book, but they are encountered in some of the older chemical literature. The symbol, Δ, over the arrow dividing products from reactants shows that heat is applied to the reaction. As an example of the uses of some of the symbols just defined above, consider the following reaction:

$$CaCO_3(s) + H_2SO_4(aq) \xrightarrow{\Delta} CaSO_4(s) + CO_2(g) + H_2O(l) \qquad (5.2.2)$$

This reaction shows that solid calcium carbonate reacts with a heated solution of sulfuric acid dissolved in water to form solid calcium sulfate, carbon dioxide gas, and liquid water.

5.3. BALANCING CHEMICAL EQUATIONS

As indicated in the preceding section, a correctly written chemical equation has equal numbers of atoms of each element on both sides of the equation. Balancing a chemical equation is accomplished by placing the correct number in front of each formula in the chemical equation. However, the following must be remembered:

Only the numbers in front of the chemical formulas may be changed to balance a chemical equation. The chemical formulas, themselves, (subscript numbers), may not be changed in balancing the equation.

Balancing an equation is best accomplished by considering one element at a time, balancing it by changing the numbers preceding the formulas in which it is contained, then successively balancing other elements in the formulas contained in the equation.

Balancing the Equation for the Reaction of Hydrogen Sulfide with Sulfur Dioxide

As an example of how to balance a chemical equation, consider the reaction of hydrogen sulfide gas (H_2S) with sulfur dioxide (SO_2) to yield elemental sulfur (S) and water (H_2O). This reaction is the basis of the Claus process by which commercially valuable elemental sulfur is recovered from pollutant sulfur dioxide and from toxic hydrogen sulfide in "sour" natural gas. The steps used in balancing the equation are the following:

1. Write the correct formulas of the reactants and products on either side of the equation. *These must remain the same throughout the balancing process.*

$$SO_2 \quad + \quad H_2S \quad \rightarrow \quad S \; + H_2O$$

sulfur dioxide   hydrogen sulfide   sulfur   water

Reactants   Products

2. Choose an element to balance initially, preferably one that is contained in only one reactant and one product. In this case oxygen may be chosen. The 2 oxygen atoms in the SO_2 molecule on the left may be balanced by placing a 2 in front of the H_2O product.

$$SO_2 + H_2S \rightarrow S + 2H_2O$$

3. Choose another element in one of the formulas involved in the preceding operation and balance it on both sides of the equation. In this case, the H in H_2O may be balanced by placing a 2 in front of H_2S.

$$SO_2 + 2H_2S \rightarrow S + 2H_2O$$

4. Proceed to the remaining element. So far, sulfur has not yet been considered. There are 3 sulfur atoms on the left, contained in 1 SO_2 molecule and 2 H_2S molecules. We have already considered these molecules in preceding operations and should avoid changing the numbers of either one. However, sulfur can be balanced by placing a 3 in front of the S product.

$$SO_2 + 2H_2S \rightarrow 3S + 2H_2O$$

5. Add up the numbers of each kind of element on both sides of the equation to see if they balance. In this case it is seen that there are 3 S atoms, 4 H atoms, and 2 O atoms on both sides, so that the equation is in fact balanced.

Some Other Examples of Balancing Equations

Two other examples of balancing equations are considered here. The first of these is for the combustion in a moist atmosphere of aluminum phosphide, AlP, to give aluminum oxide and phosphoric acid, H_3PO_4. The unbalanced equation for this reaction is

$$AlP + O_2 + H_2O \rightarrow Al_2O_3 + H_3PO_4$$

The steps in balancing this equation are the following, starting with Al:

$$2AlP + O_2 + H_2O \rightarrow Al_2O_3 + H_3PO_4$$

Balance P:

$$2AlP + O_2 + H_2O \rightarrow Al_2O_3 + 2H_3PO_4$$

Balance H:

$$2AlP + O_2 + 3H_2O \rightarrow Al_2O_3 + 2H_3PO_4$$

Balance O:

$$2AlP + 4O_2 + 3H_2O \rightarrow Al_2O_3 + 2H_3PO_4 \qquad (5.3.1)$$

Check each element for balance:

Reactants	Products
2 Al in 2 AlP	2 Al in 1 Al_2O_3
11 O in $4O_2$ and $3H_2O$	11 O in 1 Al_2O_3 and $2H_3PO_4$
6 H in 3 H_2O	6 H in 2 H_3PO_4

The second example of balancing equations is illustrated for trimethylchloro-silane, a flammable liquid used to produce high-purity silicon for semiconductor applications. Transportation accidents have resulted in spillage of this chemical and fires that produce carbon dioxide and a fog of silicon dioxide and hydrogen chloride dissolved in water droplets. The unbalanced equation for this reaction is

$$(CH_3)_3SiCl + O_2 \rightarrow CO_2 + H_2O + SiO_2 + HCl$$

Si and Cl are balanced as the equation stands, so balance H:

$$(CH_3)_3SiCl + O_2 \rightarrow CO_2 + 4H_2O + SiO_2 + HCl$$

Balance C:

$$(CH_3)_3SiCl + O_2 \rightarrow 3CO_2 + 4H_2O + SiO_2 + HCl$$

Balance O:

$$(CH_3)_3SiCl + 6O_2 \rightarrow 3CO_2 + 4H_2O + SiO_2 + HCl \qquad (5.3.2)$$

Checking the quantities of each of the elements in the products and reactants shows that the equation is balanced.

In some cases the presence of a diatomic species such as O_2 necessitates doubling quantities of everything else. As an example, consider the combustion of methane (natural gas, CH_4) in an oxygen-deficient atmosphere such that toxic carbon monoxide is produced. The unbalanced equation is

$$CH_4 + O_2 \rightarrow CO + H_2O$$

Balancing C and H gives

$$CH_4 + O_2 \rightarrow CO + 2H_2O$$

The presence of 3 oxygen atoms to the reactants side would balance oxygen. This can be done by taking $^3/_2O_2$ to give

$$CH_4 + ^3/_2O_2 \rightarrow CO + 2H_2O$$

Ordinarily, however, integer numbers should be used for the coefficients. The fraction can be avoided by multiplying everything by 2 to give the balanced equation:

$$2CH_4 + 3O_2 \rightarrow 2CO + 4H_2O \tag{5.3.3}$$

Exercise: Balance the following:

1. $Fe_2O_3 + CO \rightarrow Fe + CO_2$
2. $FeSO_4 + O_2 + H_2O \rightarrow Fe(OH)_3 + H_2SO_4$
3. $C_2H_2 + O_2 \rightarrow CO_2 + H_2O$
4. $Mg_3N_2 + H_2O \rightarrow Mg(OH)_2 + NH_3$
5. $NaAlH_4 + H_2O \rightarrow H_2 + NaOH + Al(OH)_3$
6. $Zn(C_2H_5)_2 + O_2 \rightarrow ZnO + CO_2 + H_2O$

Answers: (1) $Fe_2O_3 + 3CO \rightarrow 2Fe + 3CO_2$, (2) $4FeSO_4 + O_2 + 10H_2O \rightarrow 4Fe(OH)_3 + 4H_2SO_4$, (3) $2C_2H_2 + 5O_2 \rightarrow 4CO_2 + 2H_2O$, (4) $Mg_3N_2 + 6H_2O \rightarrow 3Mg(OH)_2 + 2NH_3$ (5) $NaAlH_4 + 4H_2O \rightarrow 4H_2 + NaOH + Al(OH)_3$, (6) $Zn(C_2H_5)_2 + 7O_2 \rightarrow ZnO + 4CO_2 + 5H_2O$

Summary of Steps in Balancing an Equation

Below is a summary of steps that can be followed to balance a chemical equation. Keep in mind that there is a limit to the usefulness of following a set of rules for this procedure. Ultimately, it is a matter of experience and good judgment. In general, the best sequence of steps to follow is the following:

1. Express the equation in words representing the compounds, elements, and (where present) the ions participating in the reaction.
2. Write down the correct formulas of all the reactants and all the products.
3. Examine the unbalanced equation for groups of atoms, such as those in

the SO_4 ion, that go through the reaction intact. Balancing is simplified by considering these atoms as a group.

4. Examine the unbalanced equation for diatomic molecules, such as O_2, whose presence may require doubling the numbers in front of the other reaction participants.

5. Choose an element, preferably one that is found in only one reactant and one product, and balance that element by placing the appropriate numbers in front of both the reactants and products involved.

6. Balance another element that appears in one of the species balanced in the preceding step.

7. Continue the balancing process, one element at a time, until all the elements have been balanced.

8. Check to make sure that the same number of atoms of each kind of element appear on both sides of the equation and that the charges from charged species (ions, whose appearance in chemical equations will be considered later) in the equation also balance on the left and right.

Exercise: The reaction of liquid hydrazine, N_2H_4, with liquid dinitrogen tetroxide, N_2O_4, to produce nitrogen gas and water is used in some rocket engines for propulsion. Balance the equation for this reaction by going through the following steps:

(a) What is the unbalanced equation for the reaction?
(b) What is the equation after balancing O?
(c) What is the equation after balancing H?
(d) What is the equation after balancing N?
(e) How many atoms of each element are on both sides of the equation after going through these steps:

Answers: (a) $N_2O_4 + N_2H_4 \rightarrow H_2O + N_2$, (b) $N_2O_4 + N_2H_4 \rightarrow 4H_2O + N_2$, (c) $N_2O_4 + 2N_2H_4 \rightarrow 4H_2O + N_2$, (d) $N_2O_4 + 2N_2H_4 \rightarrow 4H_2O + 3N_2$, (e) 6 N, 4 O, 8 H.

5.4.   WILL A REACTION OCCUR?

It is possible to write chemical equations for reactions that do not occur, or which occur only to a limited extent. This may be illustrated with a couple of examples. Consider the laboratory problem faced by a technician doing studies of plant nutrient metal ions leached from soil by water. The technician was using atomic absorption analysis, a sensitive instrumental technique for the determination of metal ions in solution. While determining the concentration of zinc ion, Zn^{2+}, dissolved in the soil leachate, the technician ran out of standard zinc solution used to provide known concentrations of zinc to calibrate the instrument, so that its readings would give known values from the sample solutions. Each liter of the standard solution contained exactly 1 mg of zinc in the form of dissolved zinc chloride, $ZnCl_2$. The technician reasoned that such a solution could be prepared by weighing out 100 mg of pure zinc metal, dissolving it in a solution of hydrochloric acid (HC1), diluting the solution to a volume of 1000 milliliters (mL), and in turn diluting 10 mL of that solution to 1000 mL to give the desired solution containing 1 mg of zinc per L. After thinking a bit, the technician came up with the equation,

$$Zn(s) + 2HCl(aq) \rightarrow H_2(g) + ZnCl_2(aq) \qquad (5.4.1)$$

to describe the chemical reaction. When the 100-mg piece of zinc metal was added to some hydrochloric acid in a flask, bubbles of hydrogen gas were evolved, the zinc dissolved as zinc chloride, and the standard solution containing the desired concentration of dissolved zinc was prepared according to the plan.

Later in the investigation, the technician ran out of standard copper solution containing 1 mg/L of copper in the form of dissolved copper (II) chloride, $CuCl_2$. The same procedure that was used to prepare the standard zinc solution was tried, with a 100-mg piece of copper wire substituted for the zinc metal. However, nothing happened to the copper metal when it was placed in hydro-chloric acid. No amount of heating, stirring or waiting could persuade the copper wire to dissolve. The technician wrote the chemical equation,

$$Cu(s) + 2HCl(aq) \rightarrow H_2(g) + CuCl_2(aq) \qquad (5.4.2)$$

for the reaction analogous to that of zinc, but copper metal and hydrochloric acid simply do not react. Even though a plausible chemical equation can be written for a reaction, it does not tell whether or not the chemical reaction will, in fact, occur. Consideration of whether or not particular reactions take place is considered in the realms of chemical thermodynamics and chemical equilibrium covered in later chapters.

Figure 5.1.   A piece of zinc metal in contact with hydrochoric acid solution reacts rapidly, giving off hydrogen gas and going into solution as $ZnCl_2$. A piece of copper metal (wire) placed in hydrochloric acid solution does not react.

5.5   HOW FAST DOES A REACTION GO?

Consider a spoonful of table sugar, sucrose, exposed to air. Will sucrose, chemical formula $C_{12}H_{22}O_{11}$, react with oxygen in the air? The following equation can be written for the chemical reaction that might occur as follows:

$$C_{12}H_{22}O_{11}(s) + 12O_2(g) \rightarrow 12CO_2(g) + 11H_2O(l) \qquad (5.5.1)$$

One may have an intuitive feeling that this reaction should occur from having seen sugar burn in a fire, or from knowing that the human body "burns" sugar to obtain energy. Furthermore, it is true that from the standpoint of energy, the atoms shown in the above equation are more stable when present as 12 molecules of CO_2 and 11 molecules of H_2O, rather than as one molecule of $C_{12}H_{22}O_{11}$ and 12 molecules of O_2. But anyone knows from experience that a spoonful of sugar can be exposed to dry air for a very long time without the occurrence of any visible change.

The answer to the question raised above lies in the **rate of reaction**. Sucrose does, indeed, tend to react with O_2 as shown in the chemical equation above. But at room temperature, the reaction is just too slow to be significant. Of course, if the sugar were thrown into a roaring fire in a fireplace, it would burn rapidly. Special proteins known as enzymes inside of living cells can bring about the reaction of sugar and oxygen at body temperature of about 37°C, enabling the body to use the energy from the reaction. The enzymes, themselves, are not used up in the reaction, though they speed it up greatly; a substance that acts in such a manner is called a **catalyst**.

An important distinction must be made between reactions, such as that between copper and hydrochloric acid that will not occur under any circumstances, and others that "want to occur," but which are just too slow to be perceptible at moderate temperatures or in the absence of a catalyst. The latter type of reaction often does take place under the proper conditions or with a catalyst. Rates of reactions are quite important in chemistry.

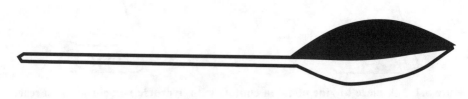

Figure 5.2.   Sugar exposed to air at room temperature does not react with the oxygen in the air at a detectable rate.

5.6. CLASSIFICATION OF CHEMICAL REACTIONS

Chemical reactions may involve several kinds of processes. Reactions may consist of putting elements together to make compounds, or taking compounds apart to produce the component elements. Reactions may occur between compounds, between compounds and elements, or between ions. Many reactions involve the transfer of electrons, whereas others do not. Given these possibilities, plus others, it is helpful to categorize chemical reactions in several major classes. These are defined below.

A **combination reaction** is one in which two reactants bond together to form a single product. An example of such a reaction is provided by the burning of elemental phosphorus as one of the steps in the manufacture of phosphoric acid, a widely used industrial chemical and fertilizer ingredient. The reaction is,

$$P_4 + 5O_2(g) \rightarrow P_4O_{10} \qquad\qquad (5.6.1)$$

Elemental phosphorus,   Tetraphosphorus decoxide
which occurs as the mole-
cule with 4 P atoms.

This is one example of the many combination reactions in which two elements combine to form a compound. The general classification, however, may be applied to combinations of two compounds or of a compound and an element to form a compound. For example, the P_4O_{10} produced in the preceding reaction is combined with water to yield phosphoric acid:

$$6H_2O + P_4O_{10} \rightarrow 4H_3PO_4 \qquad\qquad (5.6.2)$$

Phosphoric acid

A **decomposition reaction** is the opposite of a combination reaction. An example of a decomposition reaction in which a compound decomposes to form the elements in it is provided by the manufacture of carbon black. This material is a finely divided form of pure carbon, C, and is used as a filler in rubber tire manufacture and as an ingredient in the paste used to fill electrical dry cells. It is made by heating methane (natural gas) to temperatures in the range of 1260–1425°C in a special furnace causing the following reaction to occur:

$$CH_4(g) \overset{\Delta}{\rightarrow} C(s) + 2H_2(g) \qquad\qquad (5.6.3)$$

The finely divided carbon black product is collected in a special device called a cyclone collector, shown in Figure 5.3, and the hydrogen gas by-product is recycled as a fuel to the furnace that heats the methane. As illustrated by the preceding reaction, a reaction in which a compound is broken down into its component elements is a decomposition reaction.

Decomposition reactions may also involve the breakdown of a compound to

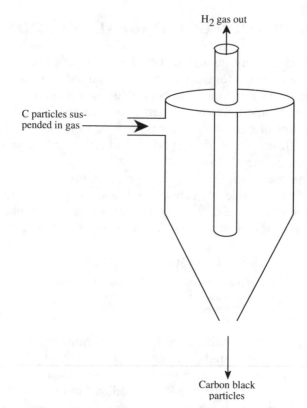

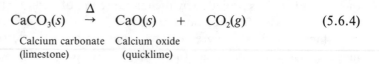

Figure 5.3.   A cyclone collector is used to collect carbon black particles from a gas stream.

another compound and an element, or to two or more compounds. As an example of the latter consider the reaction for the manufacture of calcium oxide, CaO, commonly called quicklime:

$$CaCO_3(s) \xrightarrow{\Delta} CaO(s) + CO_2(g) \tag{5.6.4}$$

Calcium carbonate   Calcium oxide
(limestone)   (quicklime)

In this reaction, high temperatures are used to decompose limestone ($CaCO_3$) to CaO and carbon dioxide gas. This is an important reaction because quicklime, CaO, ranks second only to sulfuric acid in annual chemical production. It is used in the manufacture of cement, for water treatment, and in many other applications.

   Substitution or **replacement** reactions occur when a component of a chemical compound is replaced by something else. For example, zinc displaces hydrogen in Reaction 5.4.1. A **double replacement** or **metathesis** reaction occurs when there is a two-way exchange of ions between compounds. This happens, for example, when a solution of sulfuric acid reacts with a solution of barium hydroxide,

$$H_2SO_4(aq) + Ba(OH)_2(aq) \rightarrow BaSO_4(s) + 2H_2O(l) \qquad (5.6.5)$$

to yield solid barium sulfate and water. This reaction also falls into two other categories. Because it involves the combination of H^+ ions (from H_2SO_4) and OH^- ions [from $Ba(OH)_2$] to produce water, it is a **neutralization** reaction. This reaction is

> *The reaction of H^+ from any acid with OH^- to produce water is a neutralization reaction.*

also a **precipitation** reaction because of the formation of a solid, $BaSO_4$. Such a solid formed by the reaction of two dissolved chemicals is called a precipitate.

Evolution of a gas may also be used as a basis for classifying reactions. An example is provided by the treatment of industrial wastewater containing dissolved ammonium chloride, NH_4Cl. This compound is composed of the ammonium ion, NH_4^+, and the chloride ion, Cl^-. Commercially valuable by-product ammonia gas may be recovered from such water by the addition of calcium hydroxide,

$$Ca(OH)_2(aq) + 2NH_4Cl(aq) \rightarrow 2NH_3(g) + 2H_2O(l) + CaCl_2(aq) \qquad (5.6.6)$$

resulting in the evolution of ammonia gas, NH_3, which can be recovered.

Exercise: Classify each of the following reactions as combination, decomposition, substitution, metathesis, neutralization, precipitation, or evolution of a gas. In some cases a reaction will fit into more than one category.

(a) $2Mg(s) + O_2(g) \rightarrow 2MgO(s)$

(b) $2KClO_3(s) \xrightarrow{\Delta} 2KCl(s) + 3O_2(g)$

(c) $SO_2(g) + H_2O(l) \rightarrow H_2SO_3(aq)$

(d) $CaCO_3(s) + 2HCl(aq) \rightarrow CaCl_2(aq) + H_2O(l) + CO_2(g)$

(e) $Fe(s) + CuCl_2(aq) \rightarrow Cu(s) + FeCl_2(aq)$

(f) $NaOH(aq) + HCl(aq) \rightarrow NaCl(aq) + H_2O(l)$

(g) $MgCl_2(aq) + 2NaOH(aq) \rightarrow Mg(OH)_2(s) + 2NaCl(aq)$

> *Answers:* (a) Combination, (b) decomposition, evolution of a gas, (c) combination, (d) metathesis, evolution of a gas, (e) substitution, (f) neutralization, metathesis, (g) precipitation, metathesis.

5.7.   QUANTITATIVE INFORMATION FROM CHEMICAL REACTIONS

Review of Quantitative Chemical Terms

So far, chemical equations have been described largely in terms of individual atoms and molecules. Chemistry deals with much larger quantities, of course. On an industrial level, kilograms, tons, or even thousands of tons are commonly used. It is easy to scale up to such large quantities, because the relative quantities of materials involved remain the same, whether one is dealing with just a few atoms and molecules, or train carloads of material. Before proceeding with the discussion of quantitative calculations with chemical equations, it will be helpful to consider some terms that have been defined previously:

Formula mass: The sum of the atomic masses of all the atoms in a formula unit of a compound. Although the average masses of atoms and molecules may be expressed in atomic mass units (amu or u), formula mass is generally viewed as being relative and without units.

Molar mass: Where X is the formula mass, the molar mass is X grams of an element or compound; that is, the mass in grams of 1 mole of the element or compound.

Mole: The fundamental unit for quantity of material. Each mole contains Avogadro's number (6.022×10^{23}) of formula units of the element or compound.

Formula	Formula Mass	Molar Mass	Number of Formula Units mole
H	1.01	1.01 g/mol	6.022×10^{23} H atoms/mol
N	14.01	14.01 g/mol	6.022×10^{23} N atoms/mol
N_2	2×14.01 = 28.02	28.02 g/mol	6.022×10^{23} N_2 molecules/mol
NH_3	14.01 + $3 \times 1.01 = 17.04$	17.04 g/mol	6.022×10^{23} NH_3 molecules/ mol
CaO	40.08 + 16.00 = 56.08	56.08 g/mol	6.022×10^{23} $\dfrac{\text{formula units CaO}^a}{\text{mol}}$

*Since CaO consists of Ca^{2+} and O^{2-} ions, there are not really individual CaO molecules, so it is more correct to refer to a formula unit of CaO consisting of 1 Ca^{2+} ion and 1 O^{2-} ions.

Calcination of Limestone

To illustrate some of the quantitative information that may be obtained from chemical equations, consider the calcination of limestone to make quicklime for water treatment:

$$CaCO_3(s) \xrightarrow{\Delta} CaO(s) + CO_2(g) \tag{5.7.1}$$

Limestone, calcium   Quicklime, calcium
carbonate   oxide

The quantitative information contained in this equation may be summarized as follows:

$$CaCO_3(s) \xrightarrow{\Delta} CaO(s) + CO_2(g)$$

At the formula unit (molecular) level

1 formula unit $CaCO_3$	1 formula unit CaO	1 molecule CO_2
1 Ca atom (atomic mass 40.1)	1 Ca atom	1 C atom
1 C atom (atomic mass 12.0)	1 O atom	2 O atoms
3 O atoms (atomic mass 16.0)		
100. 1 u	56.1 u	44.0 u

At the mole level

1 mole	1 mole CaO	1 mole CO_2
100. 1 g	56.1 g CaO	44.0 g CO_2

From the quantities above, it is seen that the equation may be viewed in terms as small as the smallest number of molecules and formula units. In this case that involves simply 1 formula unit of $CaCO_3(s)$, 1 formula unit of $CaO(s)$, and 1 molecule of CO_2. This would involve a total of 1 Ca atom, 1 C atom, and 3 O atoms. From such a small scale it is possible to expand to moles by scaling up by 6.022×10^{23} (Avogadro's number), giving 100.1 g of $CaCO_3$, 56.1 g of CaO, and 44.0 g of CO_2. Actually, these quantitative relationships are applicable to any amount of matter and they enable the calculation of the amounts of material

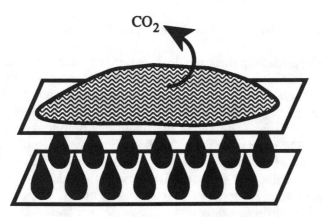

Figure 5.4. Heating calcium carbonate to a high temperature results in the production of quicklime, CaO, and the evolution of carbon dioxide. The reaction is a decomposition reaction and the process is called calcination.

reacting and produced in a chemical reaction. Next it is shown how these kinds of calculations are performed.

5.8. WHAT IS STOICHIOMETRY AND WHY IS IT IMPORTANT?

Stoichiometry is the calculation of the quantities of reactants or products involved in a chemical reaction. The importance of stoichiometry can be appreciated by visualizing industrial operations that process hundreds or thousands of tons of chemicals per day. The economics of many chemical manufacturing processes are such that an unnecessary excess of only a percent or so of a reacting chemical can lead to waste that can make the operation unprofitable. Obtaining accurate values in chemical analysis, which may need to be known to within about a part per thousand, often involves highly exacting stoichiometric calculations.

Stoichiometric calculations are based upon **the law of conservation of mass**, which states that:

The total mass of reactants in a chemical reaction equals the total mass of products; Matter is neither created nor destroyed in chemical reactions.

The key to doing stoichiometric calculations correctly is the following:

The relative masses (or number of moles, number of atoms, or molecules) of the participants in a designated chemical reaction remain the same, regardless of the overall quantities of reaction participants.

The Mole Ratio Method of Stoichiometric Calculations

In a chemical reaction there is a definite ratio between the number of moles of a particular reactant or product and the number of moles of any other reactant or product. These ratios are readily seen by simply examining the coefficients in front of the reaction species in the chemical equation. Normally, a stoichiometric calculation is made to relate the quantities of only two of the reaction participants. The objective may be to figure out how much of one reactant will react with a given quantity of another reactant. Or, a particular quantity of a product may be desired, so that it is necessary to calculate the quantity of a specific reactant needed to give the amount of product. To perform stoichiometric calculations involving only two reaction participants, it is necessary only to know the relative number of moles of each and their molar masses. The most straightforward type of stoichiometric calculation is the **mole ratio method** defined below:

*The **mole ratio method** is a means of performing stoichiometric calculations based upon the constant ratios of the numbers of moles of various reactants and products regardless of the overall quantity of reaction taking place.*

The mole ratio method greatly simplifies stoichiometric calculations. It can even be used to relate relative quantities of reaction participants in a series of reactions. For example, if a particular quantity of reactant is involved in a reaction followed by one or more additional reactions, the amount of a product in the final reaction is readily calculated by the mole ratio method.

To illustrate the mole ratio method, consider a typical reaction, that of hydrogen gas and carbon monoxide gas to produce methane:

$$3H_2 \quad + \quad CO \quad \rightarrow \quad CH_4 \quad + \quad H_2O \qquad\qquad (5.8.1)$$

3 moles	1 mole	1 mole	1 mole
6.0 g	28.0 g	16.0 g	18.0 g

This reaction is called the **methanation reaction**, and is used in the petroleum and synthetic fuels industry for the manufacture of nonpolluting synthetic natural gas (CH_4). From examination of the chemical equation it is easy to get the ratio of the number of moles of any reaction participant to the number of moles of any other reaction participant as shown in the following table:

These ratios enable calculation of the number of moles of any reaction participant, if we know the number of moles of any other participant. For example, if it is known that 1.00 mole of H_2 reacts, the calculation of the number of moles of CH_4 produced is simply the following:

Table 5.1.   Mole Ratios Used in Calculations with the Methanation Reaction

Equality of Number of Moles	Mole Ratios	
3 mol H_2 = 1 mol CO	$\dfrac{3 \text{ mol } H_2}{1 \text{ mol CO}}$	$\dfrac{1 \text{ mol CO}}{3 \text{ mol } H_2}$
3 mol H_2 = 1 mol CH_4	$\dfrac{3 \text{ mol } H_2}{1 \text{ mol } CH_4}$	$\dfrac{1 \text{ mol } CH_4}{3 \text{ mol } H_2}$
3 mol H_2 = 1 mol H_2O	$\dfrac{3 \text{ mol } H_2}{1 \text{ mol } H_2O}$	$\dfrac{1 \text{ mol } H_2O}{3 \text{ mol } H_2}$
1 mol CO = 1 mol CH_4	$\dfrac{1 \text{ mol CO}}{1 \text{ mol } CH_4}$	$\dfrac{1 \text{ mol } CH_4}{1 \text{ mol CO}}$
1 mol CO = 1 mol H_2O	$\dfrac{1 \text{ mol CO}}{1 \text{ mol } H_2O}$	$\dfrac{1 \text{ mol } H_2O}{1 \text{ mol CO}}$
1 mol CH_4 = 1 mol H_2O	$\dfrac{1 \text{ mol } CH_4}{1 \text{ mol } H_2O}$	$\dfrac{1 \text{ mol } H_2O}{1 \text{ mol } CH_4}$

$$1.00 \text{ mol } H_2 \times \frac{1 \text{ mol } CH_4}{3 \text{ mol } H_2} = 0.333 \text{ mol } CH_4 \qquad (5.8.2)$$

Calculation of the mass of a substance requires conversion between moles and mass. Suppose that one needs to know the mass of H_2 required to produce 4.00 g of CH_4. The first step is to convert the mass of CH_4 to moles:

$$4.00 \text{ g } CH_4 \times \frac{1 \text{ mol } CH_4}{16.0 \text{ g } CH_4} = 0.250 \text{ mol } CH_4 \qquad (5.8.3)$$

The molar mass of
CH_4 is 16.0 g/mol

$$0.250 \text{ mol } CH_4 \times \frac{3 \text{ mol } H_2}{1 \text{ mol } CH_4} \times \frac{2.00 \text{ g } H_2}{1 \text{ mol } H_2} = 1.50 \text{ g } H_2 \qquad (5.8.4)$$

This operation gives the number of   Multiplying by this
moles of H_2.   gives the mass of H_2.

Several problems will be shown that illustrate the mole ratio method. First, however, it will be helpful to learn the following steps used in solving a stoichiometric problem by this method:

1. Write the balanced chemical equation for the reaction involved.
2. Identify the reactant or product whose quantity is known (known substance) and the one whose quantity is being calculated (desired substance).
3. Express the number of moles of the known substance, which usually must be calculated from its mass.
4. Multiply the number of moles of known substance times the mole ratio of desired substance to obtain the number of moles of desired substance.

   Moles desired = moles of known × mole ratio of desired substance
   substance   substance   to known substance

5. Calculate the number of grams of desired substance by multiplying its molar mass times the number of moles.

   Mass in grams of = molar mass of × moles of desired
   desired substance   desired substance   substance

To illustrate these steps, consider the preparation of ammonia, NH_3, from hydrogen gas and atmospheric nitrogen. The reaction in words is,

Hydrogen plus nitrogen yields ammonia

Insertion of the correct formulas gives,

$$H_2 + N_2 \rightarrow NH_3 \qquad\qquad (5.8.5)$$

and the equation is balanced by placing the correct coefficients in front of each formula:

$$3H_2 + N_2 \rightarrow 2NH_3 \qquad\qquad (5.8.6)$$
$$\text{3 mol} \quad \text{1 mol} \quad \text{2 mol}$$

As an example, calculate the number of grams of H_2 required for the synthesis of 4.25 g of NH_3 using the following steps:

1. Calculate the number of moles of NH_3 (molar mass 17.0 g/mole).

$$\text{Mol } NH_3 = 4.25 \text{ g } NH_3 \times \frac{1 \text{ mol } NH_3}{17.0 \text{ g } NH_3} = 0.250 \text{ mol } NH_3$$

2. Express the mole ratio of H_2 to NH_3 from examination of Equation 5.8.6.

$$\frac{3 \text{ mol } H_2}{2 \text{ mol } NH_3}$$

3. Calculate the number of moles of H_2.

$$\text{Mol } H_2 = 0.250 \text{ mol } NH_3 \times \frac{3 \text{ mol } H_2}{2 \text{ mol } NH_3} = 0.375 \text{ mol } H_2$$

4. Calculate the mass of H_2.

$$0.375 \text{ mol } H_2 \times \frac{2.00 \text{ g } H_2}{1 \text{ mol } H_2} = 0.750 \text{ g } H_2$$

Once the individual steps involved are understood, it is easy to combine them all into a single calculation as follows:

$$4.25 \text{ g } NH_3 \times \frac{1 \text{ mol } NH_3}{17.0 \text{ g } NH_3} \times \frac{3 \text{ mol } H_2}{2 \text{ mol } NH_3} \times \frac{2 \text{ g } H_2}{1 \text{ mol } H_2} \qquad (5.8.7)$$

As a second example of a stoichiometric calculation by the mole ratio method, consider the reaction of iron(III) sulfate, $Fe(SO_4)_3$, with calcium hydroxide, $Ca(OH)_2$. This reaction is used in water treatment processes for the preparation of gelatinous iron(III) hydroxide, $Fe(OH)_3$, which settles in the water carrying solid particles with it. The iron(III) hydroxide acts to remove suspended matter (turbidity) from water The $Ca(OH)_2$ (slaked lime) is added as a base (source of OH^- ion) to react with iron(III) sulfate. The reaction is,

$$Fe_2(SO_4)_3(aq) + 3Ca(OH)_2(aq) \rightarrow 2Fe(OH)_3(s) + 3CaSO_4(s) \quad (5.8.8)$$

Suppose that a mass of 1000 g of iron(III) sulfate is to be used to treat a tankful of water. What mass of calcium hydroxide is required to react with the iron(III) sulfate? The steps required to solve this problem are the following:

1. $\dfrac{\text{Formula mass}}{Fe_2(SO_4)_3} = \dfrac{2 \times 55.8}{2 \text{ Fe atoms}} + \dfrac{3 \times 32.0}{3 \text{ S atoms}} + \dfrac{12 \times 16.0}{12 \text{ O atoms}} = 400$

   $\dfrac{\text{Formula mass}}{Ca(OH)_2} = \dfrac{1 \times 40.1}{1 \text{ Ca atom}} + \dfrac{2 \times 16.0}{2 \text{ O atoms}} + \dfrac{2 \times 1.0}{2 \text{ H atoms}} = 74.1$

2. $\text{Mol } Fe_2(SO_4)_3 = 1000 \text{ g } Fe_2(SO_4)_3 \times \dfrac{1 \text{ mol } Fe_2(SO_4)_3}{400 \text{ g } Fe_2(SO_4)_3}$

3. $\text{Mol ratio} = \dfrac{3 \text{ mol } Ca(OH)_2}{1 \text{ mol } Fe_2(SO_4)_3}$

4. $\dfrac{\text{Mass of}}{Ca(OH)_2} = 1000 \text{ g } Fe_2(SO_4)_3 \times \dfrac{1 \text{ mol } Fe_2(SO_4)_3}{400 \text{ g } Fe_2(SO_4)_3} \times \dfrac{3 \text{ mol } Ca(OH)_2}{1 \text{ mol } Fe_2(SO_4)_3} \times$

$$\dfrac{74.1 \text{ g } Ca(OH)_2}{1 \text{ mol } Ca(OH)_2} = 556 \text{ g } Ca(OH)_2$$

Exercise: Gelatinous aluminum hydroxide for the treatment of water can be generated by the following reaction with sodium bicarbonate:

$$Al_2(SO_4)_3(aq) + 6NaHCO_3(aq) \rightarrow 2Al(OH)_3(s) + 3Na_2SO_4(aq) + 6CO_2(g)$$

Calculate the mass in g of $NaHCO_3$ required to react with and precipitate 80.0 g of $Al_2(SO_4)_3$.

Answer: The molar mass of $NaHCO_3$ is 72.0 and that of $Al_2(SO_4)_3$ is 342.3. From the above reaction it is seen that 6 mol of $NaHCO_3$ are required per mol of $Al_2(SO_4)_3$. Therefore, the calculation is

$$\dfrac{\text{Mass of}}{NaHCO_3} = 80.0 \text{ g } Al_2(SO_4)_3 \times \dfrac{1 \text{ mol } Al_2(SO_4)_3}{342.3 \text{ g } Al_2(SO_4)_3} \times \dfrac{6 \text{ mol } NaHCO_3}{1 \text{ mol } Al_2(SO_4)_3} \times$$

$$\dfrac{72.0 \text{ g } NaHCO_3}{1 \text{ mol } NaHCO_3} = 101.0 \text{ g } NaHCO_3$$

CHAPTER SUMMARY

The chapter summary below is presented in a programmed format to review the main points covered in this chapter. It is used most effectively by filling in the blanks, referring back to the chapter as necessary. The correct answers are given at the end of the summary.

The formation of chemical compounds, their decomposition, and their interactions with one another fall under the category of (1)_____. A chemical equation may be viewed as a (2)_____in chemical language. Calculations of the quantities of substances consumed or produced in a chemical reaction are classified as (3)_____. The calculation of the quantity of heat involved in a reaction falls in the branch of chemistry called (4)_____.

Substances on the left side of a chemical equation are called (5)_____ _____ whereas those on the right are (6)_____. A correctly written chemical equation has equal numbers of (7)_____ on both sides of the equation. When used in a chemical equation, the symbols →, ⇌, (s), (l), (g), (aq), ↑, ↓, and Δ stand for (8)_____

_____

_____

_____

_____, respectively.

Chemical formulas (9)_____ _____ in balancing chemical equations.

After expressing an equation in words, the next step in balancing it is to (10)_____.

Balancing an equation may be simplified by considering together groups of atoms that (11)_____ _____. The presence of (12)_____

may require doubling the numbers in front of the other reaction participants. The final step in balancing an equation is (13)_____
_____
_____
_____.

Simply because it is possible to write a chemical equation for a reaction does not necessarily mean that (14)_____
_____.

Rates of reactions (15)_____ but may be increased by a (16)_____.

The reaction $C + O_2 \rightarrow CO_2$ is an example of a (17)_____ reaction. The reaction $2H_2O \rightarrow 2H_2 + O_2$ is an example of a (18)_____ _____ reaction. The reaction $H_2SO_4 + Fe \rightarrow FeSO_4(s) + H_2$ is a (19)_____ reaction. A neutralization reaction is (20)_____ _____ and a precipitation reaction is (21)_____.

Formula mass is (22)_____
_____. Molar mass is (23)_____
_____. Each mole of an element or compound contains (24)_____
_____.

The molar mass of NH_3 is (25)_____ a mass that contains (26)_____ individual H atoms. For the smallest whole numbers of moles, the equation

$$CaCO_3(s) \xrightarrow{\Delta} CaO(s) + CO_2(g)$$

states that (27)_____ g of $CaCO_3$ are heated to produce (28)_____ g of CaO and (29)_____ g of CO_2.

Stoichiometry is (30)_____
_____. Stoichiometric

calculations are based upon the law of conservation of mass, which states that

(31)_____

_____

_____.

The "key to doing stoichiometric calculations correctly" is the fact that (32)__

_____

_____

_____

_____.

In a chemical reaction there is a (33)_____ between the

number of moles of a particular reactant or product and the number of moles of

any other reactant or product. Using the mole ratio method for stoichiometric

calculations, the steps involved after writing the balanced chemical equation for

the reaction are (34)_____

_____

_____

_____

_____

_____

_____

_____

_____

_____

Answers

1. chemical reactions
2. sentence
3. stoichiometry
4. thermochemistry
5. reactants
6. products
7. each kind of atom

8. division between reactants and products, reversible reaction, solid, liquid, gas, substance dissolved in water, evolution of a gas, formation of a precipitate, and application of heat
9. may not be changed
10. write down the correct formulas of all the reactants and all the products
11. go through the reaction intact
12. diatomic molecules, such as O_2
13. to check to make sure that the same number of atoms of each kind of element appear on both sides of the equation and that the charges from ions in the equation also balance on the left and right
14. the reaction will occur
15. vary greatly
16. catalyst
17. combination
18. decomposition
19. displacement
20. the reaction of H^+ from any acid with OH^- to produce water
21. one in which a solid forms from a reaction in solution
22. the sum of the atomic masses of all the atoms in a formula unit of a compound
23. the mass in grams of 1 mole of an element or compound
24. Avogadro's number (6.022×10^{23}) of formula units of the element or compound
25. 17.0 g/mol
26. $3 \times 6.022 \times 10^{23}$
27. 100.1
28. 56.1
29. 44.0
30. the calculation of the quantities of reactants or products involved in a chemical reaction
31. the total mass of reactants in a chemical reaction equals the total mass of products; that is, matter is neither created nor destroyed in chemical reactions
32. the relative masses (or number of moles, number of atoms, or molecules) of the participants in a designated chemical reaction remain the same, regardless of the overall quantities of reaction participants.
33. definite ratio
34. (1) identify the reactant or product whose quantity is known and the one whose quantity is being calculated, (2) express the number of moles of the known substance, (3) multiply the number of moles of known substance times the mole ratio of desired substance to known substance to obtain the number of moles of desired substance, (4) calculate the number of grams of desired substance by multiplying its molar mass times its number of moles.

QUESTIONS AND PROBLEMS

1. Describe chemical reactions and chemical equations and distinguish between them.

2. Briefly summarize the information in the chemical equation
$$CS_2 + 3O_2 \rightarrow CO_2 + 2SO_2$$

3. Summarize the information in the following chemical equation, including the meanings of the terms in italics.
$$H_2O(l) + Na(s) \rightarrow NaOH(aq) + H_2(g)$$

4. What are the meanings of the following symbols in a chemical equation, $\rightarrow$, $=$, $\rightleftharpoons$, $\uparrow$, $\downarrow$, Δ, (s), (l), (g), (aq),

5. A typical word statement of a chemical reaction is, "Ammonia reacts with sulfuric acid to yield ammonium sulfate." The compounds or the ions involved in them have been given earlier in this book. Using correct chemical formulas, show the unbalanced and balanced chemical equations.

6. What is wrong with balancing $Ca + O_2 \rightarrow CaO$ as $Ca + O_2 \rightarrow CaO_2$?

7. Iron (II) sulfate dissolved in acid mine water, a pollutant from coal mines, reacts with oxygen from air according to the unbalanced equation,

   (a) $FeSO_4 + H_2SO_4 + O_2 \rightarrow Fe_2(SO_4)_3 + H_2O$

   Examination of this equation reveals two things about groups of atoms that simplify the balancing of the equation. What are these two factors, and how may awareness of them help to balance the equation?

8. Balance the equation from Question 6.

9. Balance each of the following: (a) $C_2H_6 + O_2 \rightarrow CO_2 + H_2O$, (b) $KClO_3 \rightarrow KCl + O_2$, (c) $Ag_2SO_4 + BaI_2 \rightarrow AgI + BaSO_4$, (d) $KClO_4 + C_6H_{12}O_6 \rightarrow KCl + CO_2 + H_2O$, (e) $Fe + O_2 \rightarrow Fe_2O_3$, (f) $P + Cl_2 \rightarrow PCl_3$

10. Having studied Section 5.4, and knowing something about silver metal and its uses, suggest what would happen if a small item of silver jewelry were placed in a solution of hydrochloric acid.

11. From the information given about Reactions 5.4.1 and 5.4.2, suggest a reaction that might occur if a piece of zinc metal were placed in a solution of $CuCl_2$. Explain. If a chemical reaction does occur, write an equation describing the reaction.

12. Steel wool heated in a flame and thrust into a bottle of oxygen gas burns vigorously, implying that a reaction such as $4Fe + 3O_2 \rightarrow 2Fe_2O_3$ does in fact occur. Why does a steel girder, such as one used for bridge construction, not burn when exposed to air?

13. What does a catalyst do?

14. Classify each of the following reactions according to the categories that are given in Section 5.6:

(a) $2NaCl(l, \text{ melted liquid}) \xrightarrow[\text{current}]{\text{Electrical}} 2Na(l) + Cl_2(g)$

(b) $Ca(OH)_2(aq) + CO_2(g) \rightarrow CaCO_3(s) + H_2O(l)$

(c) $Zn(s) + CuCl_2(aq) \rightarrow Cu(s) + ZnCl_2(aq)$

(d) $2NaOH(aq) + H_3PO_4(aq) \rightarrow Na_2HPO_4(aq) + 2H_2O(l)$

(e) $BaCl_2(aq) + Ag_2SO_4(aq) \rightarrow BaSO_4(s) + 2AgCl(s)$

(f) $(NH_4)_2SO_4(s) + Ca(OH)_2(s) \xrightarrow{\Delta} 2NH_3(g) + 2H_2O(g) + CaSO_4(s)$

15. Define (a) molecular mass (formula mass when the formula unit is a molecule), (b) mole, (c) molar mass.

16. What are the three main entities that can compose a formula unit of a substance?

17. A total of 7.52 g of $AlCl_3$ contains (a) _____ moles and (b) _____ formula units of the compound.

18. A total of 336 g of methane, CH_4, contains (a) _____ moles and (b) _____ molecules of the compound.

19. Ammonium nitrate, NH_4NO_3, can be made by a reaction between NH_3 and HNO_3. A "recipe" for the manufacture of ammonium nitrate specifies mixing 270 kg of NH_3 with 1000 kg of HNO_3. What is the significance of the relative quantities of these two ingredients?

20. What are some reasons that stoichiometric ratios of reactants are used for many industrial processes? In what cases are stoichiometric ratios not used?

21. What is the relationship constituting the basis of stoichiometry between the masses of the reactants and products in a chemical reaction?

22. In addition to the law of conservation of mass, another important stoichiometric relationship involves the proportions in mass of the reaction participants. What is this relationship?

23. Briefly define and explain the mole ratio method of stoichiometric calculations.

24. Carbon disulfide, CS_2, burns rapidly in air because of the reaction with oxygen. What is the mole ratio of O_2 to CS_2 in this reaction?

25. What is the first step required to solve a problem by the mole ratio method?

26. What is the mass (g) of NO_2 produced by the reaction of 10.44 g of O_2 with NO, yielding NO_2?

27. Plants utilize light energy in the photosynthesis process to synthesize glucose, $C_6H_{12}O_6$, from CO_2 and H_2O by way of the reaction $6CO_2 + 6H_2O \rightarrow C_6H_{12}O_6 + 6O_2$. How many grams of CO_2 are consumed in the production of 90 g of glucose?

28. Bacteria in water utilize organic material as a food source, consuming oxygen in a process called respiration. If glucose sugar is the energy source, the reaction is $C_6H_{12}O_6 + 6O_2 \rightarrow 6CO_2 + 6H_2O$ (chemically, the reverse of the photosynthesis reaction in the preceding problem). At 25°C the maximum amount of oxygen in 1.00 liter of water due to dissolved air is 8.32 mg. What is the mass of glucose in mg that will cause all the oxygen in 1.50 L of water to be used up by bacterial respiration?

29. In a blast furnace, the overall reaction by which carbon in coke is used to produce iron metal from iron ore is $2Fe_2O_3 + 3C \rightarrow 4Fe + 3CO_2$. How many tons of C are required to produce 100 tons of iron metal?

30. The gravimetric chemical analysis of NaC1 may be carried out by precipitating dissolved chloride from solution by the following reaction with silver nitrate solution: $AgNO_3(aq) + NaCl(aq) \rightarrow AgCl(s) + NaNO_3(aq)$. If this reaction produced 1.225 g of AgC1, what was the mass of NaC1?

31. For the reaction of 100.0 g of NaOH with Cl_2, $2NaOH + Cl_2 \rightarrow NaClO + NaCl + H_2O$, give the masses of each of the products.

32. Hydrochloric acid (HCl gas dissolved in water) reacts with calcium carbonate in a piece of limestone as follows: $CaCO_3 + 2HCl \rightarrow CaCl_2 + CO_2 + H_2O$. If 14.6 g of CO_2 are produced in this reaction, what is the total mass of reactants and the total mass of the products?

33. Given the reaction in the preceding problem, how many moles of HC1 are required to react with 0.618 moles of $CaCO_3$?

34. Given the reaction in Problem 32, how many moles of $CaCO_3$ are required to produce 100.0 g of $CaCl_2$?

35. Silicon tetrachloride, $SiCl_4$, is used to make organosilicon compounds (silicones) and produces an excellent smokescreen for military operations. In the latter application the $SiCl_4$ reacts with atmospheric moisture (water), $SiCl_4(g) + 2H_2O(g) \rightarrow SiO_2(s) + 4HCl(g)$, to form particles of silicon dioxide and hydrogen chloride gas. The HC1 extracts additional moisture from the atmosphere to produce droplets of hydrochloric acid which, along with small particles of SiO_2, constitute the "smoke" in the smokescreen. Air in a smokescreen was sampled by drawing 100 m³ of the air through water to collect HCl and to cause any unreacted $SiCl_4$ to react according to the above reaction. After the sampling was completed, the water was found to contain 1.85 g of HC1. What was the original concentration of $SiCl_4$ in the atmosphere (before any of the above reaction occurred) in units of milligrams of $SiCl_4$ per cubic meter?

6 ACIDS, BASES, AND SALTS

6.1. THE IMPORTANCE OF ACIDS, BASES, AND SALTS

Almost all inorganic compounds and many organic compounds can be classified as acids, bases, or salts. Some of these types of compounds were mentioned in earlier chapters and are discussed in greater detail in this chapter.

Acids, bases, and salts are vitally involved with life processes, agriculture, industry, and the environment. The most widely produced chemical is an acid, sulfuric acid. The second-ranking chemical, lime, is a base. Another base, ammonia, ranks fourth in annual chemical production. Among salts, sodium chloride is widely produced as an industrial chemical, potassium chloride is a source of essential potassium fertilizer, and sodium carbonate is used in huge quantities for glass and paper manufacture, and for water treatment.

The salt content and the acid-base balance of blood must stay within very narrow limits to keep a person healthy, or even alive. Soil with too much acid or excessive base will not support good crop growth. Too much salt in irrigation water may prevent crops from growing. This is a major agricultural problem in arid regions of the world such as the mid-East and California's Imperial Valley. The high salt content of irrigation water discharged to the Rio Grande River has been a source of dispute between the U.S. and Mexico that has been resolved to a degree by installation of a large desalination (salt removal) plant by the U.S.

From the above discussion it is seen that acids, bases and salts are important to human health and welfare. This chapter discusses their preparation, properties, and naming.

6.2. THE NATURE OF ACIDS, BASES, AND SALTS

Hydrogen Ion and Hydroxide Ion

Recall that an ion is an atom or group of atoms having an electrical charge. In discussing acids and bases two very important ions are involved. One of these is the **hydrogen ion, H^+**. It is always produced by acids. The other is the **hydroxide ion, OH^-**. It is always produced by bases. These two ions react together,

$$H^+ + OH^- \rightarrow H_2O \qquad (6.2.1)$$

to produce water. As noted in Section 5.6, this is called a **neutralization reaction**. It is one of the most important of all chemical reactions.

Acids

An **acid** is a substance which produces hydrogen ions. For example, HCl in water is entirely in the form of H^+ ions and Cl^- ions. These two ions in water form hydrochloric acid. Acetic acid, which is present in vinegar, also produces hydrogen ions in water:

$$\underset{\substack{|\\H}}{H\!-\!\overset{\substack{H\\|}}{C}\!-\!\overset{\substack{O\\||}}{C}\!-\!OH} \underset{\longleftarrow}{\overset{H_2O}{\longrightarrow}} \underset{\substack{|\\H}}{H\!-\!\overset{\substack{H\\|}}{C}\!-\!\overset{\substack{O\\||}}{C}\!-\!O^-} + H^+ \qquad (6.2.2)$$

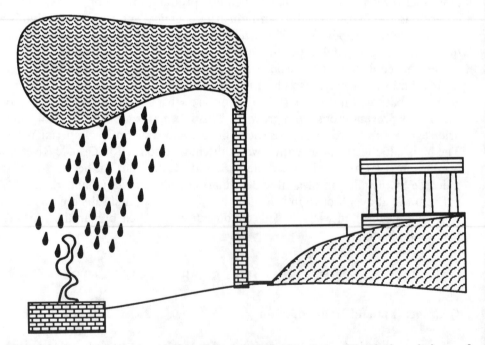

Figure 6.1.   Acid rain resulting from the introduction of sulfuric, nitric, and hydrochloric acids into the atmosphere by the burning of fossil fuels damages buildings, statues, crops, and electrical equipment in some areas of the world, including parts of the northeastern U.S.

Acetic acid demonstrates two important characteristics of acids. First, many acids contain H which is not released by the acid molecule to form H^+. Of the 4 hydrogens in CH_3CO_2H, only the one bonded to oxygen is ionizable to form H^+. The second important point about acetic acid has to do with how much of it is ionized to form H^+ and acetate ion, CH_3CO^{2-}. Most of the acetic acid remains as molecules of CH_3CO_2H in solution. In a 1 molar solution of acetic acid (containing 1 mol of acetic acid per liter of solution) only about 0.5% of the acid is ionized to produce an acetate ion and a hydrogen ion. Of a thousand molecules of acetic acid, 995 remain as un-ionized CH_3CO_2H. Therefore, acetic acid is said to be a weak acid. This term will be discussed later in the chapter.

A hydrogen ion in water is strongly attracted to water molecules. Hydrogen ions react with water,

$$\text{H} \qquad\qquad \text{H}$$
$$\text{H}\!:\!\ddot{\text{O}}\!:\ +\ \text{H}^+\ \longrightarrow\ \text{H}\!:\!\ddot{\text{O}}\!:\!\text{H}^+ \qquad\qquad (6.2.3)$$

$$\text{Hydronium ion, } H_3O^+$$

to form H_3O^+ or clusters with even more water molecules such as $H_5O_2{}^+$ or $H_7O_3{}^+$. The hydrogen ion in water is frequently shown as H_3O^+. In this book, however, it is simply indicated as H^+.

Bases

A **base** is a substance that produces hydroxide ion and/or accepts H^+. Many bases consist of metal ions and hydroxide ions. For example, solid sodium hydroxide dissolves in water,

$$NaOH(s) \rightarrow Na^+(aq) + OH^-(aq) \qquad\qquad (6.2.4)$$

to yield a solution containing OH^- ions. When ammonia gas is bubbled into water, a few of the NH_3 molecules remove hydrogen ion from water and produce ammonium ion, $NH_4{}^+$, and hydroxide ion as shown by the following reaction:

$$NH_3 + H_2O \rightarrow NH_4{}^+ + OH^- \qquad\qquad (6.2.5)$$

Only about 0.5% of the ammonia in a 1M solution goes to $NH_4{}^+$ and OH^-. Therefore, as discussed later in the chapter, NH_3 is called a **weak base**.

Salts

Whenever an acid and a base are brought together, water is always a product. But a negative ion from the acid and a positive ion from the base are always left over as shown in the following reaction:

$$H^+ + Cl^- + Na^+ + OH^- \rightarrow Na^+ + Cl^- + H_2O \qquad (6.2.6)$$

hydrochloric
acid   sodium hydroxide   sodium chloride   water

Sodium chloride dissolved in water is a solution of a **salt**. A salt is made up of a positively charged ion called a *cation* and a negatively charged ion called an *anion*. If the water were evaporated, the solid salt made up of cations and anions would remain as crystals. A salt is a chemical compound made up of a cation (other than H^+) and an anion (other than OH^-).

Amphoteric Substances

Some substances called amphoteric substances can act both as an acid and a base. The simplest example is water. Water can split apart to form a hydrogen ion and a hydroxide ion.

$$H_2O \rightarrow H^+ + OH^- \qquad (6.2.7)$$

Since it produces a hydrogen ion, water is an acid. However, the fact that it produces a hydroxide ion also makes it a base. This reaction occurs only to a very small extent. In pure water only one out of 10 million molecules of water is in the form of H^+ and OH^-. Except for this very low concentration of these two ions that can exist together, H^+ and OH^- react strongly with each other to form water.

Another important substance that can be either an acid or base is glycine. Glycine is one of the amino acids that is an essential component of the body's protein. It can give off a hydrogen ion

$$
\begin{array}{c}
\text{H} \quad \text{H} \quad \text{O} \\
\mid \quad \mid \quad \parallel \\
\text{H--N}^{\pm}\text{C--C--O}^-
\\
\mid \quad \mid \\
\text{H} \quad \text{H}
\end{array}
\underset{}{\overset{H_2O}{\rightleftharpoons}}
\begin{array}{c}
\text{H}_{\diagdown} \quad \text{H} \quad \text{O} \\
\quad \mid \quad \parallel \\
\quad \text{N--C--C--O}^- + \text{H}^+ \\
\text{H}^{\diagup} \quad \mid \\
\quad \text{H}
\end{array}
\qquad (6.2.8)
$$

or it can react with water to release a hydroxide ion from the water:

$$
\begin{array}{c}
\text{H} \quad \text{H} \quad \text{O} \\
\mid \quad \mid \quad \parallel \\
\text{H--N}^{\pm}\text{C--C--O}^- + \text{H}_2\text{O}
\\
\mid \quad \mid \\
\text{H} \quad \text{H}
\end{array}
\rightleftharpoons
\begin{array}{c}
\text{H} \quad \text{H} \quad \text{O} \\
\mid \quad \mid \quad \parallel \\
\text{H--N}^{\pm}\text{C--C--OH} + \text{OH}^- \\
\mid \quad \mid \\
\text{H} \quad \text{H}
\end{array}
\quad (6.2.9)
$$

Substances which can be either acids or bases are termed **amphoteric**.

Metal Ions as Acids

Some metal ions are acids. As an example consider iron(III) ion, Fe^{3+}. This ion used to be commonly called ferric ion. When iron(III) chloride, $FeCl_3$, is dissolved in water,

$$FeCl_3 + 6H_2O \rightarrow Fe(H_2O)_6^{3+} + 3Cl^- \qquad (6.2.10)$$

it produces chloride ions and triply-charged iron(III) ions. Each iron(III) ion is bonded to 6 water molecules. The iron(III) ion surrounded by water is called a **hydrated ion**. This hydrated ferric ion can lose hydrogen ions and form a slimy brown precipitate of iron(III) hydroxide, $Fe(OH)_3$:

$$Fe(H_2O)_6^{3+} \rightarrow Fe(OH)_3 + 3H_2O + 3\ H^+ \qquad (6.2.11)$$

It is this reaction that is partly responsible for the acid in iron-rich acid mine water. It is also used to purify drinking water. The gelatinous $Fe(OH)_3$ settles out, carrying the impurities to the bottom of the container, and the water clears up.

Salts That Act as Bases

Some salts that do not contain hydroxide ion produce this ion in solution. The most widely used of these is sodium carbonate, Na_2CO_3, which is commonly known as soda ash. Millions of pounds of soda ash are produced each year for the removal of hardness from boiler water, for the treatment of waste acid, and for many other industrial processes. Sodium carbonate reacts in water,

$$2Na^+ + CO_3^{2-} + H_2O \rightarrow Na^+ + HCO_3^- + Na^+ + OH^- \qquad (6.2.12)$$

sodium carbonate	sodium bicarbonate	sodium hydroxide

to produce hydroxide ion. If H^+, such as from hydrochloric acid, is already present in the water, sodium carbonate reacts with it as follows:

$$2Na^+ + CO_3^{2+} + H^+ + Cl^- \rightarrow Na^+ + Cl^- + Na^+ + HCO_3^- \qquad (6.2.13)$$

sodium carbonate	hydrochloric acid	sodium chloride	sodium bicarbonate

Salts That Act as Acids

Some salts act as acids. Salts that act as acids react with hydroxide ions. Ammonium chloride, NH_4Cl, is such a salt. This salt is also called "sal ammoniac." As a "flux" added to solder used to solder copper plumbing or automobile radiators, ammonium chloride dissolves coatings of corrosion on the metal surfaces so that the solder can stick. In the presence of a base, NH_4Cl reacts with the hydroxide ion,

$$NH_4^+ + Cl^- + Na^+ + OH^- \longrightarrow NH_3 + H_2O + Na^+ + Cl^- \qquad (6.2.14)$$

ammonium chloride   sodium hydroxide   sodium chloride

to produce ammonia gas and water.

6.3. CONDUCTANCE OF ELECTRICITY BY ACIDS, BASES, AND SALTS IN SOLUTION

In Section 6.2 it was seen that when acids, bases, or salts are dissolved in water, charged ions are formed. When HCl gas is dissolved in water,

$$HCl(g) \xrightarrow{\text{Water}} H^+(aq) + Cl^-(aq) \qquad (6.3.1)$$

all of it goes to H^+ and Cl^- ions. Acetic acid in water also forms a few ions,

$$CH_3CO_2H \xrightarrow{\text{Water}} H^+ + CH_3CO_2^- \qquad (6.3.2)$$

but most of it stays as CH_3CO_2H. Sodium hydroxide in water is all in the form of Na^+ and OH^- ions. The salt, NaCl, is all present as Na^+ and Cl^- ions in water.

One of the most important properties of ions is that they conduct electricity in water. Water containing ions from an acid, base, or salt will conduct electricity much like a metal wire. Consider what would happen if very pure distilled water were made part of an electrical circuit as shown in Figure 6.2. The light bulb would not glow at all. This is because pure water does not conduct electricity. However, if a solution of salt water, such as oil well brine, is substituted for the distilled water, the bulb will glow brightly as shown in Figure 6.2. Salty water conducts electricity because of the ions that it contains. Even tap water has some ions dissolved in it, which is why one may experience a painful, even fatal, electric shock by touching an electrical fixture while bathing.

Electrolytes

Materials which conduct electricity in water are called **electrolytes**. These materials form ions in water. The charged ions allow the electrical current to flow through the water. Materials, such as sugar, that do not form ions in water are called nonelectrolytes. Solutions of nonelectrolytes in water do not conduct electricity. A solution of brine conducts electricity very well because it contains dissolved NaCl. All of the NaCl in the water is in the form of Na^+ and Cl^-. The NaCl is completely ionized, and it is a strong electrolyte. An ammonia water

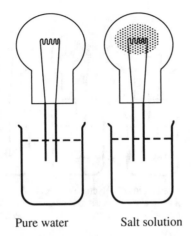

Pure water   Salt solution

Figure 6.2. Pure water does not conduct electricity whereas water containing dissolved salt conducts very well.

solution (used for washing windows) does not conduct electricity very well. That is because only a small fraction of the NH_3 molecules react,

$$NH_3 + H_2O \rightarrow NH_4^+ + OH^- \qquad (6.3.3)$$

to form the ions which let electricity pass through the water. Ammonia is a weak electrolyte. (Recall that it is also a weak base.) Nitric acid, HNO_3 is a strong electrolyte because it is completely ionized to H^+ and NO_3^- ions. Acetic acid is a weak electrolyte, as well as a weak acid. The base, sodium hydroxide, is a strong electrolyte. All salts are strong electrolytes because they are always completely ionized in water. Acids and bases may be weak or strong electrolytes.

In the laboratory the strength of an electrolyte can be measured by how well it conducts electricity in solution, as shown in Figure 6.3. The ability of a solution to conduct electrical current is called its **conductivity**.

When electricity is passed through solutions of acids, bases, or salts, chemical reactions occur. One such reaction is the breakdown of water to hydrogen and oxygen. Electricity passing through a solution is widely used to separate and purify various substances.

6.4. DISSOCIATION OF ACIDS AND BASES IN WATER

It has already been seen that acids and bases come apart in water to form ions. When acetic acid splits up in water,

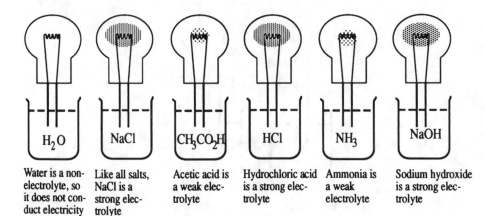

Figure 6.3.   The electrical conductivity of a solution can be determined by placing the solution in an electrical circuit and observing how well electricity is conducted by the solution. *Strong electrolytes* **conduct electricity well;** *weak electrolytes* **conduct it poorly. This principle is used in water analysis to determine the total salt concentrations in water.**

$$CH_3CO_2H \rightarrow CH_3CO_2^- + H^+ \tag{6.4.1}$$

it forms hydrogen ions and acetate ions. The process of forming ions is called **ionization**. Another term is commonly employed. When the acetic acid molecule comes apart, it is said to **dissociate**. The process is called **dissociation**.

There is a great difference in how much various acids and bases dissociate. Some, like HCl or NaOH, are completely dissociated in water. Because of this, hydrochloric acid is called a **strong acid**. Sodium hydroxide is a **strong base**. Some acids such as acetic acid are only partly dissociated in water. They are called weak acids. Ammonia, NH_3, reacts only a little bit in water to form an ammonium ion (NH_4^+) and a hydroxide ion (OH^-). It is a weak base.

The extent of dissociation is a very important property of an acid or base. The 3% or so acetic acid solution used to make up oil and vinegar salad dressing lends a pleasant taste to the lettuce and tomatoes. There is not much of the H^+ ion in the acetic acid. If 3% HCl had been used instead, nobody could eat the salad. All of the H in HCl is in the form of H^+, and a 3% solution of hydrochloric acid is very sour, indeed. Similarly, a several percent solution of NH_3 in water makes a good window wash, helping to dissolve the grease and grime. If a similar concentration of sodium hydroxide were used to clean windows, they would soon become permanently fogged because the OH^- in the strong base eventually reacts with glass and etches it. However, sodium hydroxide solutions are used to clean ovens where a very strong base is required to break down the charred, baked-on grease.

Table 6.1 shows some acids and the degree to which they are dissociated. It allows comparison of the strengths of these acids.

The percentage of acid molecules which are dissociated depends upon the concentration of the acid. The lower the concentration, the higher the percentage of dissociated molecules. This may be understood by looking again at the reaction,

$$CH_3CO_2H \rightarrow CH_3CO_2^- + H^+ \qquad (6.4.1)$$

for the dissociation of acetic acid. At high concentrations there will be more crowding together of H^+ and $CH_3CO_2^-$ ions. This forces them back together to form CH_3CO_2H again. At low concentrations there are fewer H^+ and $CH_3CO_2^-$ ions. They are more free to roam around the solution alone, and there is less pressure for them to form CH_3CO_2H. It is somewhat like the seating which occurs on a bus. If there are few passengers, they will spread around the cabin and not sit next to each other; that is, they will be dissociated. If there are many passengers, they will, of course, have to occupy adjacent seats.

An idea of the effect of concentration upon the dissociation of a weak acid can be obtained from the percentage of acid molecules that have dissociated to ions at several different concentrations. This is shown for acetic acid in Table 6.2.

Table 6.2 shows that in a 1 M solution less than 1% of acetic acid is dissociated. In a one-thousandth M (0.001 M) solution 12 out of 100 molecules of acetic acid are in the form of H^+ and acetate ions. In a one-millionth M (0.000001 M) solution only 5 out of 100 acetic acid molecules are present as CH_3CO_2H.

It is important to know the difference between the strength of an acid or base in solution and the concentration of the solution. A strong acid is one which is all in the form of H^+ ions and anions. It may be very concentrated or very dilute. A weak acid does not give off much H^+ to water solution. It may also range in concentration from a very dilute solution to a very concentrated one. Similar reasoning applies to bases.

6.5.   THE HYDROGEN ION CONCENTRATION AND BUFFERS

It is important to make the distinction between the concentration of H^+ and the concentration of an acid. To show this difference, compare 1 M solutions of acetic acid and hydrochloric acid. The concentration of H^+ in a 1 M solution of CH_3CO_2H is only 0.0042 mole/liter. The concentration of H^+ in a 1 M solution of HCl is 1 mole/liter. A liter of a 1 M solution of HCl contains 240 times as many H^+ ions as a liter of a 1 M solution of acetic acid.

Table 6.1.   Dissociation of Acids

Acid Formula	Acid Name	Common Uses	Percent Dissociated in 1 M Solution	Strength
H_2SO_4	Sulfuric	Industrial chemical	100	Strong
HNO_3	Nitric	Industrial chemical	100	Strong
H_3PO_4	Phosphoric	Fertilizer, food additive	8	Moderately weak
$H_3C_6H_5O_7$	Citric	Fruit drinks	3	Weak
CH_3CO_2H	Acetic	Foods, industry	0.4	Weak
HClO	Hypochlorous	Disinfectant	0.02	Weak
HCN	Hydrocyanic	Very poisonous industrial chemical, electroplating waste	0.002	Very weak
H_3BO_4	Boric acid	Antiseptic, ceramics	0.002	Very weak

Table 6.2. **Percent Dissociation of Acetic Acid at Various Concentrations**

Total Acetic Acid Concentration	Percent Dissociated to H^+ and $CH_3CO_2^-$
1 mol/liter	0.4
0.1 mol/liter	1.3
0.01 mol/liter	4.1
1×10^{-3} mol/liter	12
1×10^{-4} mol/liter	34
1×10^{-5} mol/liter	71
1×10^{-6} mol/liter	95
1×10^{-7} mol/liter	99

Consider, however, the amount of NaOH which will react with 1.00 liter of 1.00 M acetic acid. The reaction is

$$CH_3CO_2H + Na^+ + OH^- \rightarrow Na^+ + CH_3CO_2^- + H_2O \qquad (6.5.1)$$

Acetic acid   Sodium hydroxide   Sodium acetate

Exactly 1.00 mole of NaOH reacts with the 1.00 mole of acetic acid contained in a liter of a 1.00 M solution of this acid. Exactly the same amount of NaOH reacts with the HCl in 1.00 liter of 1.00 M HCl.

$$H^+ + Cl^- + Na^+ + OH^- \rightarrow Na^+ + Cl^- + H_2O \qquad (6.5.2)$$

Hydrochloric acid   Sodium hydroxide   Sodium chloride

Therefore, even though acetic acid is a weaker acid than hydrochloric acid, equal volumes of each, with the same molar concentration, will react with the same number of moles of base.

In many systems the concentration of H^+ is very important. For a person to remain healthy, the H^+ concentration in blood must stay within a very narrow range. If the H^+ concentration is too high in a boiler system, the pipes may be corroded through in a short time. If the H^+ concentration becomes too high or too low in a lake, plant and animal life cannot thrive in it.

Fortunately, there are mixtures of chemicals which keep the H^+ concentration of a solution relatively constant. Reasonable quantities of acid or base added to such solutions do not cause large changes in H^+ concentration. Solutions that resist changes in H^+ concentration are called **buffers**.

To understand how a buffer works, consider a typical buffer system. A solution containing both acetic acid and sodium acetate is a good buffer. The acetic acid in the solution is present as undissociated CH_3CO_2H. The H^+, which is in solution, is there because a very small amount of the CH_3CO_2H has dissociated to H^+ and $CH_3CO_2^-$ ions. The sodium acetate is present as Na^+ ion and $CH_3CO_2^-$ ion. If some base, such as NaOH, is added, some of the acetic acid reacts.

$$CH_3CO_2H + Na^+ + OH^- \rightarrow Na^+ + CH_3CO_2^- + H_2O \qquad (6.5.3)$$
$$\text{Sodium acetate}$$

This reaction changes some of the acetic acid to sodium acetate, but it does not change the hydrogen ion concentration much. If a small amount of hydrochloric acid is added to the buffer mixture of acetic acid and sodium acetate, some of the sodium acetate is changed to acetic acid.

$$Na^+ + CH_3CO_2^- + H^+ + Cl^- \rightarrow CH_3CO_2H + Na^+ + Cl^- \qquad (6.5.4)$$

The acetate ion acts like a sponge for H^+ and prevents the concentration of the added hydrogen ion from becoming too high.

Buffers can also be made from a mixture of a weak base and a salt of the base. A mixture of NH_3 and NH_4Cl is such a buffer. Mixtures of two salts can be buffers. A mixture of NaH_2PO_4 and Na_2HPO_4 is a buffer made from salts. It is one of the very common phosphate buffers, such as those that occur in body fluids.

6.6.   pH AND THE RELATIONSHIP BETWEEN HYDROGEN ION AND HYDROXIDE ION CONCENTRATIONS

Because of the fact that water, itself, produces both hydrogen ion and hydroxide ion,

$$H_2O \rightarrow H^+ + OH^- \qquad (6.6.1)$$

there is always some H^+ and some OH^- in any solution. Of course, in an acid solution the concentration of OH^- must be very low. In a solution of base the concentration of H^+ is very low. There is a definite relationship between the concentration of H^+ and the concentration of OH^-. It varies a little with temperature. At 25°C (about room temperature) the following relationship applies:

$$[H^+][OH^-] = 1.00 \times 10^{-14} = K_w \text{ (at 25°C)} \qquad (6.6.2)$$

If the value of either $[H^+]$ or $[OH^-]$ is known, the value of the other can be calculated. For example, in a solution of 0.100 M HCl in which $[H^+] = 0.100$ M,

$$[OH^-] = \frac{K_w}{[H^+]} = \frac{1.00 \times 10^{-14}}{0.100} = 1.00 \times 10^{-13} \text{ M} \qquad (6.6.3)$$

Acids, such as HCl and H_2SO_4, produce H^+ ion, whereas bases, such as sodium hydroxide and calcium hydroxide [NaOH and $Ca(OH)_2$, respectively], produce hydroxide ion, OH^-. Molar concentrations of hydrogen ion, $[H^+]$, range over many orders of magnitude and are conveniently expressed by pH defined as

$$pH = -log[H^+] \qquad (6.6.4)$$

In absolutely pure water the value of $[H^+]$ is exactly 1×10^{-7} mole/L, the pH is 7.00, and the solution is **neutral** (neither acidic nor basic). **Acidic** solutions have pH values of less than 7 and **basic** solutions have pH values of greater than 7. Table 6.3 gives some example hydrogen ion concentrations and the corresponding pH values.

As seen in Table 6.3, when the H^+ ion concentration is 1 times 10 to a power (the superscript number, such as –2, –7, etc.) the pH is simply the negative value of that power. Thus, when $[H^+]$ is 1×10^{-3}, the pH is 3; when $[H^+]$ is 1×10^{-4}, the pH is 4. That is because the log of 1×10^{-3} is –3 and that of 1×10^{-4} is –4. Therefore, the negative logs are 3 and 4, respectively, because the sign is reversed. What about the pH of a solution with a hydrogen ion concentration between 1×10^{-4} and 1×10^{-3}, such as 3.16×10^{-4}? The pH is obviously going to be between 3 and 4. The pH is calculated very easily on an electronic calculator by entering 3.16×10^{-4} on the keyboard and pressing the "log" button. The log of the number is –3.50, and the pH is 3.50.

Acid-Base Equilibria

Many of the phenomena in aquatic chemistry and geochemistry involve **solution equilibrium**. In a general sense solution equilibrium deals with the extent to which **reversible** acid-base, solubilization (precipitation), complexation, or

Table 6.3. Values of [H^+] and Corresponding pH Values

[H^+], mol/L	pH
1.00	0
0.100	1.00
1.00×10^{-3}	3.00
2.25×10^{-6} ($10^{-5.65}$)	5.65
1.00×10^{-7}	7.00
1.00×10^{-9}	9.00
5.17×10^{-9} ($10^{-8.29}$)	8.29
1.00×10^{-13}	13.00
1.00×10^{-14}	14.00
1.00×10^{-2}	2.00

oxidation-reduction reactions proceed in a forward or backward direction. This is expressed for a generalized equilibrium reaction,

$$aA + bB \rightleftharpoons cC + dD \qquad (6.6.5)$$

There are several major kinds of equilibria in aqueous solution. The one under consideration here is acid-base equilibrium as exemplified by the ionization of acetic acid, HAc,

$$HAc \rightleftharpoons H^+ + Ac^- \qquad (6.6.6)$$

$$Ac^- \text{ represents } H-\overset{\overset{\displaystyle H}{|}}{\underset{\underset{\displaystyle H}{|}}{C}}-\overset{\overset{\displaystyle O}{\|}}{C}-O^-$$

for which the acid dissociation constant is

$$\frac{[H^+][Ac^-]}{[HAc]} = K = 1.75 \times 10^{-5} \text{ (at 25°C)} \qquad (6.6.7)$$

As an example of an acid-base equilibrium problem consider water in equilibrium with atmospheric carbon dioxide. The value of $[CO_2(aq)]$ in water at 25°C in equilibrium with air that is 350 parts per million CO_2 (the current concentration of this gas in the atmosphere) is 1.146×10^{-5} moles/liter (M). The carbon dioxide dissociates partially in water to produce equal concentrations of H^+ and HCO_3^-:

$$CO_2 + H_2O \rightleftharpoons HCO_3^- + H^+ \qquad (6.6.8)$$

so that:

$$[H^+] = [HCO_3^-] \qquad (6.6.9)$$

The concentrations of H^+ and HCO_3^- are calculated from K_{a1}:

$$K_{a1} = \frac{[H^+][HCO_3^-]}{[CO_2]} = \frac{[H^+]^2}{1.146 \times 10^{-5}} = 4.45 \times 10^{-7} \qquad (6.6.10)$$

$$[H^+][HCO_3^-] = [CO_2]K_{a1} \qquad (6.6.11)$$

Since $[H^+] = [HCO_3^-]$, this relationship simplifies to

$$[H^+] = [HCO_3^-] = (1.146 \times 10^{-5} \times 4.45 \times 10^{-7})^{1/2} = 2.25 \times 10^{-6} \quad (6.6.12)$$

$$pH = 5.65$$

This calculation explains why pure water that has equilibrated with the unpolluted atmosphere is slightly acidic with a pH somewhat less than 7.

6.7.   PREPARATION OF ACIDS

Acids can be prepared in several ways. In discussing the preparation of acids, it is important to keep in mind that acids usually contain nonmetals. All acids either contain ionizable hydrogen or produce H^+ ion when dissolved in water. Finally, more often than not, acids contain oxygen.

A simple way to make an acid is to react hydrogen with a nonmetal. Hydrochloric acid can be made by reacting hydrogen and chlorine

$$H_2 + Cl_2 \rightarrow 2HCl \quad (6.7.1)$$

and adding the hydrogen chloride product to water. Other acids which consist of hydrogen combined with a nonmetal are HF, HBr, HI, and H_2S. Hydrocyanic acid, HCN, is an "honorary member" of this family of acids, even though it contains three elements.

Sometimes a nonmetal reacts directly with water to produce acids. The best example of this is the reaction of chlorine with water,

$$Cl_2 + H_2O \rightarrow HCl + HClO \quad (6.7.2)$$

to produce hydrochloric acid and hypochlorous acid.

Many very important acids are produced when nonmetal oxides react with water. One of the best examples is the reaction of sulfur trioxide with water

$$SO_3 + H_2O \rightarrow H_2SO_4 \quad (6.7.3)$$

to produce sulfuric acid. Other examples are shown in Table 6.4.

Volatile acids—those which evaporate easily—can be made from salts and nonvolatile acids. The most common nonvolatile acid so used is sulfuric acid, H_2SO_4. When solid NaCl is heated in contact with concentrated sulfuric acid,

$$2NaCl(s) + H_2SO_4(l) \rightarrow 2HCl(g) + Na_2SO_4(s) \quad (6.7.4)$$

Table 6.4.　Important Acids Produced When Nonmetal Oxides React with Water

Oxide Reacted	Acid Formula	Acid Name	Use and Significance of Acid
SO_3	H_2SO_4	Sulfuric	Major industrial chemical, constituent of acid rain
SO_2	H_2SO_3	Sulfurous	Paper-making, scrubbed from stack gas containing SO_2
N_2O_5	HNO_3	Nitric	Synthesis of chemicals, constituent of acid rain
N_2O_3	HNO_2	Nitrous	Unstable, toxic to ingest, few uses
P_4O_{10}	H_3PO_4	Phosphoric	Fertilizer, chemical synthesis

HCl gas is given off. This gas can be collected in water to make hydrochloric acid. Similarly, when calcium sulfite is heated with sulfuric acid,

$$CaSO_3(s) + H_2SO_4(l) \rightarrow CaSO_4(s) + SO_2(g) + H_2O \qquad (6.7.5)$$

sulfur dioxide is given off as a gas. It can be collected in water to produce sulfurous acid, H_2SO_3.

Organic acids, such as acetic acid, CH_3CO_2H, have the group,

$$\underset{\text{(–CO}_2\text{H)}}{-\overset{\overset{\textstyle O}{\|}}{C}-OH}$$

attached to a hydrocarbon group. These **carboxylic acids** are discussed in Chapter 9.

6.8.　PREPARATION OF BASES

Bases can be prepared in several ways. Many bases contain metals and some metals react directly with water to produce a solution of base. Lithium, sodium, and potassium react very vigorously with water to produce their hydroxides:

$$2K + 2H_2O \rightarrow 2K^+ + 2OH^- + H_2(g) \qquad (6.8.1)$$

Potassium hydroxide
(strong base)

Many metal oxides form bases when they are dissolved in water. When waste liquor (a concentrated solution of salts and materials extracted from wood) from the sulfite paper-making process is burned to produce energy and reclaim magnesium hydroxide, the magnesium in the ash is recovered as MgO. This is added to water

$$MgO + H_2O \rightarrow Mg(OH)_2 \qquad (6.8.2)$$

to produce the magnesium hydroxide used with other chemicals to break down the wood and produce paper fibers. Other important bases and the metal oxides from which they are prepared are given in Table 6.5.

Many important bases cannot be isolated as the hydroxides but produce OH^- ion in water. A very good example is ammonia, NH_3. Ammonium hydroxide, NH_4OH, cannot be obtained in a pure form. Even when ammonia is dissolved in water, very little NH_4OH is present in the solution. However, ammonia does react with water,

$$NH_3 + H_2O \rightarrow NH_4^+ + OH^- \qquad (6.8.3)$$

to give an ammonium ion and a hydroxide ion. Since only a small percentage of the ammonia molecules react this way, ammonia is a weak base.

Many salts that do not themselves contain hydroxide ion act as bases by reacting with water to produce OH^-. Sodium carbonate, Na_2CO_3, is the most widely used of these salts. When sodium carbonate is placed in water, the carbonate ion reacts with water,

$$CO_3^{2-} + H_2O \rightarrow HCO_3^- + OH^- \qquad (6.8.4)$$

to form a hydroxide ion and a bicarbonate ion, HCO_3^-. Commercial grade sodium carbonate, soda ash, is used very widely for neutralizing acid in water treatment and other applications. It is used in phosphate-free detergents. It is a much easier base to handle and use than sodium hydroxide. Whereas sodium

Table 6.5.   Important Bases Produced When Metal Oxides React with Water

Oxide Reacted	Base Formula	Base Name	Use and Significance of Base
Li_2O	LiOH	Lithium hydroxide	Constituent of some lubricating greases
Na_2O	NaOH	Sodium hydroxide	Soap making, many industrial uses, removal of H_2S from petroleum
K_2O	KOH	Potassium hydroxide	Alkaline battery manufacture
MgO	$Mg(OH)_2$	Magnesium hydroxide	Paper-making, medicinal uses
CaO	$Ca(OH)_2$	Calcium hydroxide	Water purification, soil treatment to neutralize excessive acidity

hydroxide rapidly absorbs enough water from the atmosphere to dissolve itself to make little puddles of highly concentrated NaOH solution that are very harmful to the skin, sodium carbonate does not absorb water nearly so readily. It is not as dangerous to the skin.

Trisodium phosphate, Na_3PO_4, is an even stronger base than sodium carbonate. The phosphate ion reacts with water

$$PO_4^{3-} + H_2O \rightarrow HPO_4^{2-} + OH^- \tag{6.8.5}$$

to yield a high concentration of hydroxide ions. This kind of reaction with water is called a **hydrolysis reaction**.

Many organic compounds are bases. Most of these contain nitrogen. One of these is trimethylamine, $(CH_3)_3N$. This compound is one of several that give dead fish their foul smell. It reacts with water

$$(CH_3)_3N + H_2O \rightarrow (CH_3)_3NH^+ + OH^- \tag{6.8.6}$$

to produce hydroxide ion. Like most organic bases it is a weak base.

6.9. PREPARATION OF SALTS

Many salts are important industrial chemicals. Others are used in food preparation or medicine. A huge quantity of Na_2CO_3 is used each year, largely to treat water and to neutralize acid. Over 1.5 million tons of Na_2SO_4 are used in applications such as inert filler in powdered detergents. Approximately 30,000 tons of sodium thiosulfate, $Na_2S_2O_3$, are consumed each year in developing photographic film and in other applications. Canadian mines produce more than 10 million tons of KCl each year for use as fertilizer. Lithium carbonate, Li_2CO_3, is employed as a medicine to treat some kinds of manic-depressive illness. Many other examples of the importance of salts could be given.

Whenever possible, salts are obtained by simply mining them. Many kinds of salts can be obtained by evaporating water from a few salt-rich inland sea waters or from brines pumped from beneath the ground. However, most salts cannot be obtained so directly and must be made by chemical processes. Some of these processes will be discussed.

One way of making salts has already been discussed in this chapter. That method is to react an acid and a base to produce a salt and water. Calcium propionate, which is used to preserve bread could be made by reacting calcium hydroxide and propionic acid, $HC_3H_5O_2$:

$$Ca(OH)_2 + 2HC_3H_5O_2 \rightarrow Ca(C_3H_5O_2)_2 + 2H_2O \tag{6.9.1}$$

Calcium propionate

Almost any salt can be made by the reaction of the appropriate acid and base.

In some cases a metal and a nonmetal will react directly to make a salt. If a strip of magnesium burns (explodes would be a better description) in an atmosphere of chlorine gas,

$$Mg + Cl_2 \rightarrow MgCl_2 \tag{6.9.2}$$

magnesium chloride salt is produced.

Metals react with acids to produce a salt and hydrogen gas. Calcium placed in sulfuric acid will yield calcium sulfate.

$$Ca + H_2SO_4 \rightarrow H_2(g) + CaSO_4(s) \tag{6.9.3}$$

Some metals react with strong bases to produce salts. Aluminum metal reacts with sodium hydroxide to yield sodium aluminate, Na_3AlO_3.

$$2Al + 6NaOH \rightarrow 2Na_3AlO_3 + 3H_2(g) \tag{6.9.4}$$

In cases where a metal forms an insoluble hydroxide, addition of a base to a salt of that metal can result in the formation of a new salt. If potassium hydroxide is added to a solution of magnesium sulfate

$$2KOH + MgSO_4 \rightarrow Mg(OH)_2(s) + K_2SO_4(aq) \tag{6.9.5}$$

the insoluble magnesium hydroxide precipitates out of the solution leaving potassium sulfate salt in solution.

If the anion in a salt can form a volatile acid, a new salt can be formed by adding a nonvolatile acid. If nonvolatile sulfuric acid is heated with NaCl,

$$H_2SO_4 + 2NaCl \rightarrow 2HCl(g) + Na_2SO_4 \tag{6.9.6}$$

HCl gas is given off and sodium sulfate remains behind.

Some metals will displace other metals from a salt. Advantage is taken of this for the removal of toxic heavy metals from water solutions of the metals' salts by reaction with a more active metal, a process called **cementation**. For example, metallic iron can be reacted with wastewater containing dissolved toxic cadmium sulfate,

$$Fe(s) + CdSO_4(aq) \rightarrow Cd(s) + FeSO_4(aq) \tag{6.9.7}$$

to isolate solid cadmium metal, forming a new soluble salt, iron(II) sulfate.

Finally, there are many special commercial processes for making specific salts. One such example is the widely used Solvay Process for making sodium bicarbonate and sodium carbonate. In this process a sodium chloride solution is saturated with ammonia gas, then saturated with carbon dioxide and finally cooled. The reaction which occurs is

$$NaCl + NH_3 + CO_2 + H_2O \rightarrow NaHCO_3(s) + NH_4Cl \qquad (6.9.8)$$

and sodium bicarbonate (baking soda) precipitates from the cooled solution. When the sodium bicarbonate is heated, it is converted to sodium carbonate:

$$2NaHCO_3 + heat \rightarrow Na_2CO_3 + H_2O(g) + CO_2(g) \qquad (6.9.9)$$

6.10.   ACID SALTS AND BASIC SALTS

Acid Salts

Some compounds are crosses between acids and salts. Other salts are really crosses between bases and salts. The acid salts contain hydrogen ion. This hydrogen ion can react with bases. One example of this is sodium hydrogen sulfate, $NaHSO_4$, which reacts with sodium hydroxide,

$$NaHSO_4 + NaOH \rightarrow Na_2SO_4 + H_2O \qquad (6.10.1)$$

to give sodium sulfate and water. Some other examples of acid salts are shown in Table 6.6.

Some salts contain hydroxide ions. These are known as **basic salts**. One of the most important of these is calcium hydroxyapatite, $Ca_5OH(PO_4)_3$. Commonly known as *hydroxyapatite*, this salt occurs in the mineral, rock phosphate, which is the source of essential phosphate fertilizer. The heavy metals in particular have a tendency to form basic salts. Many rock-forming minerals are basic salts.

Table 6.6.   Some Important Salts

Acid Salt Formula	Acid Salt Name	Typical Use
$NaHCO_3$	Sodium hydrogen carbonate	Food preparation (baking soda)
NaH_2PO_4	Sodium dihydrogen phosphate	To prepare buffers
Na_2HPO_4	Disodium hydrogen phosphate	To prepare buffers
$KHC_4H_4O_6$	Potassium hydrogen tartrate	Dry acid in baking powder[a]

[a]Cream of tartar baking powder consists of a mixture of potassium hydrogen tartrate and sodium hydrogen carbonate. When this mixture contacts water in a batch of dough, the reaction

$$KHC_4H_4O_6 + NaHCO_3 \rightarrow KNaC_4H_4O_6 + H_2O + CO_2(g)$$

occurs, and the small bubbles of carbon dioxide it produces cause the dough to rise.

6.11.   WATER OF HYDRATION

Water is frequently bound to other chemical compounds. Water bound to a salt in a definite proportion is called **water of hydration**. An important example is sodium carbonate decahydrate:

$$Na_2CO_3 \cdot 10\ H_2O$$

Sodium   decahydrate
carbonate

This salt is used in detergents, as a household cleaner, and for water softening. When dissolved in water, it yields a basic solution by reacting with the water to produce hydroxide ion, OH^-. Such a solution tends to dissolve grease, so it can be used as a household cleaner. Sodium carbonate decahydrate is used in detergents, because they work best in basic solutions. The carbonate ion from this salt reacts with calcium ion (which causes water hardness),

$$Ca^{2+} + CO_3^{2-} \rightarrow CaCO_3(s) \tag{6.11.1}$$

This removes hardness from the water by producing solid calcium carbonate, $CaCO_3$. Because of its ability to remove water hardness, sodium carbonate decahydrate is a good water softener, another reason that it is used in detergents. Sodium carbonate without water, Na_2CO_3, is said to be **anhydrous**. It should not be used in consumer products such as powdered detergents because of its strong attraction for water. If anhydrous Na_2CO_3 were present in a detergent that was accidentally ingested, it would draw water from the tissue in the mouth and throat, greatly increasing the harm done.

6.12.   NAMES OF ACIDS, BASES, AND SALTS

Acids

Acids consisting of hydrogen and one other element are named for the element with a *hydro-* prefix and an ending of *-ic*. These acids include HF, hydrofluoric; HCl, hydrochloric; HBr, hydrobromic; and H_2S, hydrosulfuric acid. Another acid named in this manner is HCN, hydrocyanic acid.

The names of many common acids end with the suffix *-ic*, as is the case with acetic acid, nitric acid (HNO_3), and sulfuric acid. In some cases where the anion of the acid contains oxygen, there are related acids with different numbers of oxygen atoms in the anion. When this occurs, the acid with one less oxygen than the "-ic" acid has a name ending with *-ous*. For example, H_2SO_4 is sulfuric acid and H_2SO_3 is sulfurous acid. Similarly, HNO_3 is nitric acid and HNO_2 is nitrous acid. One more oxygen in the anion than the "-ic" acid is indicated by a *per-*

Table 6.7. Names of the Oxyacids of Chlorine

Acid Formula	Acid Name[a]	Anion Name[a,b]
$HClO_4$	*Per*chlor*ic* acid	*Per*chlor*ate*
$HClO_3$	Chlor*ic* acid	Chlor*ate*
$HClO_2$	Chlor*ous* acid	Chlor*ite*
$HClO$	*Hypo*chlor*ous* acid	*Hypo*chlor*ite*

[a]Italicized letters are used with the names only to emphasize the prefixes and suffixes.
[b]Names of anions, such as ClO_4^-, formed by removal of H^+ ion from the acid.

prefix and an *-ic* suffix on the acid name. One less oxygen in the anion than in the "-ous" acid gives the acid a name with a *hypo-* prefix and an *-ous* suffix. These rules are shown for the oxyacids (acids containing oxygen in the anion) of chlorine in Table 6.7.

Bases

Bases that contain hydroxide ion are named very simply by the rules of nomenclature for ionic compounds. The name consists of the name of the metal followed by hydroxide. As examples, LiOH is lithium hydroxide, KOH is potassium hydroxide, and $Mg(OH)_2$ is magnesium hydroxide.

Salts

Salts are named according to the name of the cation followed by the name of the anion. The names of the more important ions are listed in Table 6.8. One important cation, ammonium ion, NH_4^+, and a number of anions are polyatomic ions, meaning that they consist of two or more atoms per ion. Table 6.8 includes some polyatomic ions. Note that most polyatomic ions contain oxygen as one of the elements.

As illustrated in Table 6.8, the names of anions are based on the names of the acids from which they are formed by removal of H^+ ions. A "*hydro -ic*" acid yields an "*-ide*" anion and, therefore, an "*-ide*" salt. For example, hydrochloric acid reacts with a base to give a chloride salt. An "*-ic*" acid yields an "*-ate*" salt; for example, calcium sulf*ate* is the salt that results from a reaction of sulfur*ic* acid with calcium hydroxide. The anion contained in an "*-ous*" acid is designated by "*-ite*" in a salt; sulfur*ous* acid, H_2SO_3 reacts NaOH to give the salt Na_2SO_3 called sodium sulfite. A "*per -ic*" acid, such as perchloric acid, reacts with a base, such as NaOH, to give a "*per -ate*" salt, for example, sodium perchlorate, $NaClO_4$. A "*hypo -ous*" acid, such as hypochlorous acid, reacts with a base, KOH, for example, to give a "*hypo -ite*" salt, such as potassium hypochlorite, KClO. Some additional examples are illustrated by the reactions below:

Table 6.8.   Some Important Ions

+1 Charge	+2 Charge	+3 Charge	−1 Charge	−2 Charge	−3 Charge
H^+, hydrogen	Mg^{2+}, magnesium	Al^{3+}, aluminum	H^-, hydride	O^{2-}, oxide	N^{3-} nitride
Li^+, lithium	Ca^{2+}, calcium	Fe^{3+}, iron(III)[b]	F^-, fluoride	S^{2-}, sulfide	PO_4^{3-} phosphate
Na^+, sodium	Ba^{2+}, barium	Cr^{3+}, chromium(III)[b]	Cl^-, chloride	SO_4^{2-}, sulfate	
K^+, potassium	Fe^{2+}, iron (II)[a] Cu^{2+}, copper (II)[b]		$Br-$, bromide	SO_3^{2-}, sulfite	
Cu^+, copper(I)[a]	Cr^{2+}, chromium(II)[a] Zn^{2+}, zinc		I^-, iodide	CO_3^{2-}, carbonate	
Ag^+, silver	Pb^{2+}, lead (II)[a] Mn^{2+}, manganese(II)[a]		$C_2H_3O_2^-$, acetate	CrO_4^{2-}, chromate	
NH_4^+, ammonium	Hg^{2+}, mercury (II)[b] Sn^{2+}, tin(II)[c]		CN^-, cyanide	$Cr_2O_7^{2-}$, dichromate	
			NO_3^-, nitrate	O_2^{2-}, peroxide	
			OH^-, hydroxide	HPO_4^{2-}, monohydrogen phosphate	
			$H_2PO_4^-$, dihydrogen phosphate		
			HCO_3^-, hydrogen carbonate[c]		
			ClO_3^-, chlorate		
			HSO_4^-, hydrogen sulfate[c]		
			ClO_4^-, perchlorate		
			MnO_4^-, permanganate		

[a]These metals can also exist as ions with a higher charge and may also be designated by their Latin names with an -*ous* ending as follows: Cu^+, cuprous; Fe^{2+}, ferrous; Cr^{2+}, chromous; Pb^{2+}, plumbous; Mn^{2+}, manganous; Sn^{2+}, stannous.

[b]These metals can also exist as ions with a lower charge and may also be designated by their Latin names with an -*ic* ending as follows: Cu^{2+}, cupric; Hg^{2+}, mercuric; Fe^{3+}, ferric; Cr^{3+}, chromic.

[c]The ions HCO_3^- and HSO_4^- are known as bicarbonate and bisulfate, respectively.

$$2HClO_4 + Zn(OH)_2 \rightarrow Zn(ClO_4)_2 + 2H_2O \qquad (6.12.1)$$

perchloric acid   zinc perchlorate

$$2HNO_3 + Cu(OH)_2 \rightarrow Cu(NO_3)_2 + 2H_2O \qquad (6.12.2)$$

nitric acid   copper(II) nitrate

$$HNO_2 + KOH \quad \rightarrow \quad KNO_2 + H_2O \qquad (6.12.3)$$

nitrous acid   potassium nitrite

$$2HClO + Ca(OH)_2 \rightarrow Ca(ClO)_2 + 2H_2O \qquad (6.12.4)$$

hypochlorous acid   calcium hypochlorite

Using the information given in Table 6.8, it is possible to figure out the formulas and give the names of a very large number of ionic compounds. To do that, simply observe the following steps:

1. Choose the cation and the anion of the compound. The name of the compound is simply the name of the cation followed by the name of the anion. For example, when Fe^{3+} is the cation and SO_4^{2-} is the anion, the name of the ionic compound is iron(III) sulfate.
2. Choose subscripts to place after the cation and anion in the chemical formula of the compound such that multiplying the subscript of the cation times the charge of the cation gives a number equal in magnitude and opposite in sign from that of the product of the anion's subscript times the anion's charge. In the example of iron(III) sulfate, a subscript of 2 for Fe^{3+} gives $2 \times (3+) = 6+$, and a subscript of 3 for SO_4^{2-} gives $3 \times (2-) = 6-$, thereby meeting the condition for a neutral compound.
3. Write the compound formula. If the subscript after any polyatomic ion is greater than 1, put the formula of the ion in parentheses to show that the subscript applies to all the atoms in the ion. Omit the charges on the ions because they make the compound formula too cluttered. In the example under consideration, the formula of iron(III) sulfate is $Fe_2(SO_4)_3$.

Exercise: Match each cation in the left column below with each anion in the right column and give the formulas and names of each of the resulting ionic compounds.

(a) Na^+	(1) Br^-
(b) Ca^{2+}	(2) CO_3^{2-}
(c) Al^{3+}	(3) PO_4^{3-}

Answers: a-1, NaBr, sodium bromide; a-2, Na_2CO_3, sodium carbonate; a-3, Na_3PO_4, sodium phosphate; b-1, $CaBr_2$, calcium bromide; b-2, $CaCO_3$, calcium carbonate; b-3, $Ca_3(PO_4)_2$, calcium phosphate; c-1, $AlBr_3$, aluminum bromide; c-2, $Al_2(CO_3)_3$, aluminum carbonate; c-3, $AlPO_4$, aluminum phosphate.

The names and formulas of ionic compounds that contain hydrogen in the name and formula of the anion are handled just like any other ionic compound. Therefore, $NaHCO_3$ is sodium hydrogen carbonate, $Ca(H_2PO_4)_2$ is calcium dihydrogen phosphate, and K_2HPO_4 is potassium monohydrogen phosphate. The acetate ion, $C_2H_3O_2^-$ also contains hydrogen, but its hydrogen is covalently bonded to a C atom and cannot form H^+ ions, whereas the anions listed with hydrogen in their names can produce H^+ ion when dissolved in water.

CHAPTER SUMMARY

The chapter summary below is presented in a programmed format to review the main points covered in this chapter. It is used most effectively by filling in the blanks, referring back to the chapter as necessary. The correct answers are given at the end of the summary.

(1)_____ ion is produced by acids and (2)_____ by bases. A neutralization

reaction is (3)_____. Hydrogen ion, H^+, in water

is bonded to (4)_____ and is often represented

as (5)_____.

A **base** is a substance that accepts H^+ and produces hydroxide ion. Although

NH_3 does not contain hydroxide ions, it undergoes the reaction (6)_____

to produce OH^- water.

The two products produced whenever an acid and a base react together are

(7)_____. A salt is made up of (8)_____

_____. An amphoteric substance is one that

(9)_____. In

water, a metal ion is bonded to (10)_____ in

a form known as a (11)_____. In terms of acid-base behavior,

some metal ions act as (12)_____. In water solution sodium carbonate

acts as a (13)_____ and undergoes the reaction (14)_____

_____. Salts that act as acids react

with (15)_____.

Pure water conducts electricity (16)_____, a solution of

acetic acid conducts (17)_____, and a solution of HCl (18)_____

_____. These differences are due to differences in concentrations of

(19)_____ in the water. Materials which conduct electricity in water are called

(20)_____. Materials that do not form ions in water are called (21)_____.

The reaction

$$CH_3CO_2H \rightarrow CH_3CO_2^- + H^+$$

may be classified as (22)_____ or (23)_____ and when the acetic acid molecule comes apart, it is said to (24)_____. A base that is completely dissociated in water is called a (25)_____ and an acid that is only slightly dissociated is called a (26)_____. At high concentrations the percentage of dissociation of a weak acid is (27)_____ _____ than at lower concentrations.

Buffers are (28)_____.
A buffer can be made from a mixture of a weak base and (29)_____ _____.

The reaction that results in the production of very low concentrations of ions in even pure water is (30)_____ and the relationship between the concentrations of these ions in water is (31)_____. In absolutely pure water the value of $[H^+]$ is exactly (32)_____, the pH is (33)_____, so that the solution is said to be (34)_____ _____. Acidic solutions have pH values of (35)_____ and basic solutions have pH values of (36)_____.

In a general sense solution equilibrium deals with the extent to which reversible acid-base, solubilization (precipitation), complexation, or oxidation-reduction reactions (37)_____ _____. As an example of acid-base equilibrium the reaction for the ionization of acetic acid, HAc, is (38)_____ for which the acid dissociation constant is (39)_____.

Some ways to prepare acids are (40)_____ _____ _____

_____

_____

_____

_____

_____.

Some ways to prepare bases are (41)_____

_____

_____

_____

_____.

The reaction of an ion with water such as

$$PO_4^{3-} + H_2O \rightarrow HPO_4^{2-} + OH^-$$

is an example of a (42)_____.

   The most obvious way to prepare a salt is by (43)_____

_____. Active metals react

with acids to produce (44)_____

_____.

Other than reacting with acids, some metals react with (45)_____.

If the anion in a salt can form a volatile acid, a new salt can be formed by

(46)_____. Some metals will displace

other metals from a salt. If magnesium, a highly reactive metal is added to a

solution of copper sulfate, the reaction that occurs is (47)_____.

   $NaHSO_4$ which has an ionizable hydrogen is an example of (48)_____

_____, whereas $Ca_5OH(PO_4)_3$ is an example of (49)_____.

The water in $CuSO_4 \cdot 5\,H_2O$ is called (50)_____.

The names of $HClO_4$, $HClO_3$, $HClO_2$, and $HClO$ are, respectively, (51)_____

_____.

The names of $NaClO_4$, $NaClO_3$, $NaClO_2$, and $NaClO$ are, respectively, (52)___

_____.

The name of a base containing a metal consists of (53)_____

_____. The name of a salt is (54)_____

_____. The

names of the ions Ca^{2+}, Fe^{3+}, H^-, SO_3^{2-}, and $C_2H_3O_2^-$ are, respectively,

55)_____.

Answers

1. H^+
2. OH^-
3. $H^+ + OH^- \rightarrow H_2O$
4. water molecules
5. hydronium ion, H_3O^+
6. $NH_3 + H_2O \rightarrow NH_4^+ + OH^-$
7. water and a salt
8. a cation (other than H^+) and an anion (other than OH^-)
9. can act as either an acid or a base
10. water molecules
11. hydrated ion
12. acids
13. base
14. $2Na^+ + CO_3^{2+} + H_2O \rightarrow Na^+ + HCO_3^- + Na^+ + OH^-$
15. hydroxide ions
16. not at all
17. poorly
18. very well
19. ions
20. electrolytes
21. nonelectrolytes
22. ionization
23. dissociation
24. dissociate or ionize
25. strong base
26. weak acid
27. lower
28. solutions that resist changes in H^+ concentration
29. a salt of the base
30. $H_2O \rightarrow H^+ + OH^-$
31. $[H^+][OH^-] = 1.00 \times 10^{-14} = K_w$
32. 1×10^{-7} mole/L

33. 7.00
34. neutral
35. less than 7
36. greater than 7
37. proceed in a forward or backward direction
38. $HAc \rightleftharpoons H^+ + Ac^-$
39. $\dfrac{[H^+][Ac^-]}{[HAc]} = K = 1.75 \times 10^{-5}$
40. reaction of hydrogen with a nonmetal, reaction of a nonmetal directly with water, reaction of a nonmetal oxide with water, production of volatile acids by reaction of salts of the acids with nonvolatile acids
41. reaction of active metals directly with water, reaction of metal oxides with water, reaction of a basic compound that does not itself contain hydroxide with water
42. hydrolysis
43. reaction of an acid with a base
44. a salt and hydrogen gas
45. strong bases
46. adding a nonvolatile acid
47. $Mg(s) + CuSO_4(aq) \rightarrow MgSO_4(aq)$
48. an acid salt
49. a basic salt
50. water of hydration
51. perchloric acid, chloric acid, chlorous acid, and hypochlorous acid
52. sodium perchlorate, sodium chlorate, sodium chlorite, and sodium hypochlorite
53. the name of the metal followed by hydroxide
54. the name of the cation followed by the name of the anion
55. calcium, iron(III), hydride, sulfite, and acetate

QUESTIONS AND PROBLEMS

1. Give the neutralization reaction for each of the acids in the left column reacting with each of the bases in the right column, below:

   (1) Hydrocyanic acid   (a) Ammonia
   (2) Acetic acid   (b) Sodium hydroxide
   (3) Phosphoric acid   (c) Calcium hydroxide

2. Methylamine is an organic amine, formula $H_3C\text{-}NH_2$, that acts as a weak base in water. By analogy with ammonia, suggest how it might act as a base.

3. In a 1 molar solution of acetic acid (containing 1 mol of acetic acid per liter of solution) only about 0.5% of the acid is ionized to produce an acetate ion and a hydrogen ion. Calculate the number of moles of H^+ in a liter of such a solution.

4. What does H_3O^+ represent in water?

5. When exactly 1 mole of NaOH reacts with exactly 1 mole of H_2SO_4, the product is an acid salt. Show the production of the acid salt with a chemical reaction.

6. An amphoteric substance can be viewed as one that may either accept or produce an ion of H^+. Using that definition, explain how H_2O is amphoteric.

7. Explain how Fe^{3+} ion dissolved in water can be viewed as an acidic hydrated ion.

8. Cyanide ion, CN^-, has a strong attraction for H^+. Show how this explains why NaCN acts as a base.

9. Separate solutions containing 1 mole per liter of NH_3 and 1 mole per liter of acetic acid conduct electricity poorly, whereas when such solutions are mixed, the resulting solution conducts well. Explain.

10. What characteristic of solutions of electrolytes enables them to conduct electricity well?

11. A solution containing 6 moles of NH_3 dissolved in a liter of solution would be relatively highly concentrated. Explain why it would not be correct, however, to describe such a solution as a "strong base" solution.

12. The dissociation of acetic acid may be represented by

$$CH_3CO_2H \rightleftharpoons CH_3CO_2^- + H^+$$

Explain why this reaction may be characterized as an ionization of acetic acid. Explain on the basis of the "crowding" concept why the percentage of acetic acid molecules dissociated is less in relatively concentrated solutions of the acid.

13. A solution containing 0.1 mole of HCl per liter of solution has a low pH of 1, whereas a solution containing 0.1 mole of acetic acid per liter of solution has a significantly higher pH. Explain.

14. Explain why a solution containing both NH_3 and NH_4Cl acts as a buffer. In so doing, consider reactions of NH_3, NH_4^+ ion, H^+ ion, OH^- ion, and H_2O.

15. NaH_2PO_4 and Na_2HPO_4 dissolved in water produce $H_2PO_4^-$ and HPO_4^{2-} ions, respectively. Show by reactions of these ions with H^+ and OH^- ions why a solution consisting of a mixture of both NaH_2PO_4 and Na_2HPO_4 dissolved in water acts as a buffer.

16. What is the expression and value for K_w? What is the reaction upon which this expression is based?

17. On the basis of pH, distinguish among acidic, basic, and neutral solutions.

18. Give the pH values corresponding to each of the following values of $[H^+]$: (a) 1.00×10^{-4} mol/L, (b) 1.00×10^{-8} mol/L, (c) 5.63×10^{-9} mol/L, (d) 3.67×10^{-6} mol/L.

19. Why does solution equilibrium deal only with reversible reactions?

20. Write an equilibrium constant expression for the reaction

$$CO_3^{2-} + H_2O \rightleftharpoons HCO_3^- + OH^-$$

21. Calculate $[H^+]$ in a solution of carbon dioxide in which $[CO_2(aq)]$ is 3.25×10^{-4} moles/liter.

22. Cl_2 and F_2 are both halogens. Suggest acids that might be formed from the reaction of F_2 with H_2 and with H_2O.

23. Suggest the acid that might be formed by reacting S with H_2.

24. Suggest the acid or base that might be formed by the reaction of each of the following oxides with water: (a) N_2O_3, (b) CO_2, (c) SO_3, (d) Na_2O, (e) CaO, (f) Cl_2O.

25. Knowing that H_2SO_4 is a nonvolatile acid, suggest the acid that might be formed by the reaction of sodium acetate, $Na_2C_2H_3O_2$, with water.

26. Acetic acid is a carboxylic acid. Formic acid is the lowest carboxylic acid, and it contains only 1 C atom per molecule. What is its formula?

27. A base can be prepared by the reaction of calcium metal with hot water. Give the reaction and the name of the base product.

28. A base can be prepared by the reaction of sodium oxide with water. Give the reaction and the name of the base product.

29. Give the reactions by which the following act as bases in water: (a) NH_3, (b) Na_2CO_3, (c) Na_3PO_4, and dimethylamine, $(CH_3)_2NH$.

30. Choosing from the reagents H_2SO_4, HCl, Mg(OH), and LiOH, give reactions that illustrate "the most straightforward" means of preparing salts.

31. Choosing from the reagents NaOH, HCl, CaO, Mg, F_2 and Al, give reactions that illustrate the preparation of salts by (a) reaction of a metal and a nonmetal directly to make a salt, (b) reaction of a metal with acid, (c) reaction of a metal with strong base, (d) reaction of a salt with a nonvolatile acid, (e) cementation.

32. Describe what is meant by an acid salt.

33. Describe what is meant by a basic salt.

34. Explain how sodium carbonate decahydrate illustrates water of hydration. Why is it less hazardous to skin than is anhydrous sodium carbonate? Illustrate with a chemical reaction why it is also acts as a base.

35. Give the names of each of the following acids: (a) HBr, (b) HCN, (c) HClO, (d) $HClO_2$, (e) $HClO_3$, (f) $HClO_4$, (g) HNO_2.

36. Give the names of (a) LiOH, (b) $Ca(OH)_2$, and (c) $Al(OH)_3$.

37. Give the names of (a) $MgSO_3$, (b) NaClO, (c) $Ca(ClO_4)_2$, (d) KNO_3, (e) $Ca(NO_2)_2$

38. Match each cation in the left column below with each anion in the right column and give the formulas and names of each of the resulting ionic compounds.

(a) Li^+	(1) CN^-
(b) Ca^{2+}	(2) SO_3^{2-}
(c) Fe^{3+}	(3) NO_3^-

7 SOLUTIONS

7.1. WHAT ARE SOLUTIONS? WHY ARE THEY IMPORTANT?

To help understand solutions, consider a simple experiment. Run some water from a faucet into a glass and examine it. Depending on the condition of the local water supply, it will probably appear clear. If tasted, it will probably not have any particularly strong flavors. Now, add a teaspoon of sugar and stir the contents of the glass; you will notice the sugar begin to disappear, and the water around the sugar will start to appear cloudy or streaked. With continued stirring, the sugar will seem to vanish completely, and the water will look as clear as it did when it came out of the tap. However, it will have a sweet taste; some of its properties have therefore been changed by the addition of sugar.

This experiment illustrates several important characteristics of chemicals and how they are used. Sugar **dissolves** in water; when this happens, a **solution** is formed. The sugar molecules were originally contained in hard, rigid sugar crystals. In water, the molecules break away from the crystals and spread throughout the liquid. The sugar is still there, but it is dissolved in the water, which is called the **solvent**. The dissolved sugar is called the **solute**; it can no longer be seen, but it can be tasted. The water appears unchanged, but it is different. It has a different taste. Some of its other properties have changed, too; for example, it now boils at a higher temperature and freezes at a lower temperature than pure water. In some cases there is readily visible evidence that a solute is present in a solvent. The solution may have a strong color, as is the case for intensely purple solutions of potassium permanganate, $KMnO_4$. It may have a strong odor, such as that of ammonia, NH_3, dissolved in water.

As illustrated in Figure 7.1, tap water, itself, is a solution containing many things. It has some dissolved oxygen and carbon dioxide in it. It almost certainly has some dissolved calcium compounds, making it "hard." There may be a small amount of iron present, which causes the water to stain clothing. It contains some chlorine, added to kill bacteria. In unfavorable cases it may even contain some toxic lead and cadmium dissolved from plumbing and the solder used to

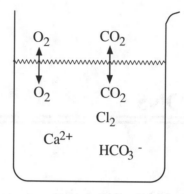
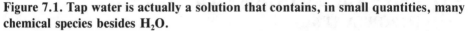

Figure 7.1. Tap water is actually a solution that contains, in small quantities, many chemical species besides H₂O.

connect copper pipes. Thus, it is clear that many of the liquids that people come in contact with are actually solutions.

There are many examples of important and useful solutions. Sugar, for instance, must be dissolved before it can be used by the body for food. Candy, which is sugar with added flavoring, dissolves in the mouth to form a solution of sugar with saliva. The coolant in an automobile's cooling system is a solution of antifreeze in water. This solution freezes at a much lower temperature than pure water, thus preventing the liquid from freezing and cracking the engine block. A solution containing the simple sugar glucose and some other substances may be injected directly into the veins of an ill or injured person who cannot take food through the mouth. Chemists use many different kinds of solutions that undergo chemical reactions with other kinds of chemicals. By measuring how much of a solution is required to complete a reaction in a procedure called *titration*, the chemist can tell how much of a particular kind of chemical is in a solution.

Reactions in Solution

One of the most important properties of solutions is their ability to allow chemical species to come into close contact so that they can react. For example, if perfectly dry crystals of calcium chloride, $CaCl_2$, were mixed with dry crystals of sodium fluoride, NaF, a chemical reaction would not occur. However, if each is dissolved in separate solutions which are then mixed, a precipitation reaction occurs,

$$CaCl_2(aq) \; + \; 2 \, NaF(aq) \; \rightarrow \; CaF_2(s) \; + \; 2 \, NaCl(aq) \qquad (7.1.1.)$$

in which calcium chloride and sodium fluoride in aqueous solution (*aq*) react to produce calcium fluoride solid (*s*) and a solution of sodium chloride. This reaction takes place because in solution the Ca,$^{2+}$ ions (from dissolved $CaCl_2$) and the F$^-$ ions (from dissolved NaF) move around and easily come together to form CaF_2. The calcium fluoride product does not stay in solution but forms a precipitate; it is **insoluble**.

In other cases, solutions enable chemical reactions to occur that result in materials being dissolved. Some of these reactions are important in geology. Consider limestone, which is made of calcium carbonate, $CaCO_3$. Limestone does not react with dry CO_2 gas, nor is it soluble in pure water. However, when water containing dissolved CO_2 contacts limestone, a chemical reaction occurs:

$$CaCO_3(s) + CO_2(g) + H_2O \rightarrow Ca^{2+}(aq) + 2\ HCO_3^-(aq) \qquad (7.1.2)$$

The calcium ion and the bicarbonate ion, HCO_3^-, remain dissolved in water; this solution dissolves the limestone, leaving a cave or hole in the limestone formation. In some regions, such as parts of southern Missouri, this has occurred to such an extent that the whole area is underlain by limestone caves and potholes; these are called *karst* regions.

Solutions in Living Systems

For living things the most important function of solutions is to carry molecules and ions to and from cells. Body fluids consist of complex solutions. Digestion is largely a process of breaking down complex, insoluble food molecules to simple, soluble molecules that may be carried by the blood to the body cells, which need them for energy and production of more cell material. On the return trip the blood carries waste products in solution, such as carbon dioxide, which are eliminated from the body.

Solutions in the Environment

Solutions are of utmost importance in the environment. Solutions transport environmental chemical species in the aquatic environment and are crucial participants in geochemical processes. Dissolution in rainwater is the most common process by which atmospheric pollutants are removed from air. Acid rain is a solution of strong mineral acids in water. Many important environmental chemical processes occur in solution and at the interface of solutions with solids and gases. Pollutant pesticides and hazardous waste chemicals are transported in solution as surface water or groundwater. Many hazardous waste chemicals are dissolved in solution; often the large amount of water in which they are dissolved makes their treatment relatively more difficult and expensive.

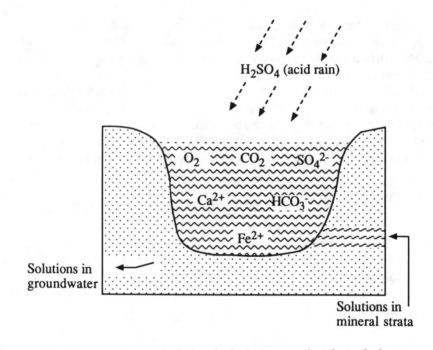

Figure 7.2. Many environmental chemical phenomena involve solutions.

Industrial Uses of Solutions

Solutions are used throughout industry. Many chemical reactions that are part of the manufacture of important industrial chemicals occur in solutions. *Natural brines* are solutions that contain a lot of dissolved materials and that occur underground and in some lakes (saline lakes). Commercially valuable chemicals are recovered from brines, including borax, a compound containing boron and oxygen, which is used as an antiseptic, in making ceramics, and in some cleaning formulations. Solutions of detergents are used for cleaning; some dyes are applied as solutions. Ammonia fertilizer may be added to the soil as a solution of NH_3 in water.

Many organic materials do not dissolve well in water. Such substances may dissolve in organic solvents, such as those discussed in Section 7.2. Solutions of organic compounds in organic solvents have a number of uses, such as for extractions and degreasing.

7.2.   SOLVENTS

Water is the solvent for most of the solutions discussed in this chapter. However, it should be mentioned that many other liquids are also used as solvents. Other than water, most solvents are organic (carbon-containing) liquids. Some of the more important organic solvents are shown in Table 7.1.

There are many uses for solvents. One of the most important of these is their role as media in which chemical reactions may occur. In the chemical industry, solvents are employed for purification, separation, and physical processing. Solvents are also used for cleaners; one important example is the use of organic solvents to dissolve grease and oil from metal parts. The chemicals that make up synthetic fibers, such as rayon, are dissolved in solvents, then forced under very high pressure through small holes in a special die to make individual filaments. One of the most important uses for solvents is in coatings, which include paint, printing inks, lacquers, and antirust formulations. In order to apply these coatings, it is necessary to dissolve them in a solvent (**vehicle**) so that they may be spread around on the surface to be coated. The vehicle is a **volatile** liquid, one that evaporates quickly to form a vapor; when it evaporates, it leaves the coating behind as a thin, protective layer.

Fire and toxicity are major hazards associated with the use of many solvents. Some organic solvents, such as benzene, are even more of a fire hazard than gasoline. Interestingly, carbon tetrachloride does not burn and was even used in fire extinguishers in the past. Both benzene and carbon tetrachloride are toxic and can damage the body in cases of excess exposure. Benzene is suspected of

Table 7.1.   Important Organic Solvents

Solvent	Solvent Use (may have many other *non*solvent uses)	Approximate Annual U.S. Production, millions of kilograms
Benzene	Dissolves grease and other organic compounds.	5,000
Carbon tetrachloride	Formerly used in dry cleaning, largely discontinued because of toxicity. Used for industrial degreasing.	450
Perchloroethylene	Best solvent for dry cleaning, also used for degreasing metals and extraction of fats.	300
Acetone	Solvent for spinning cellulose acetate fibers and for spreading paints and other protective coatings.	1,000

causing leukemia, and worker exposure to this solvent is now carefully regulated. The toxicity hazard of solvents arises from absorption through the skin and inhalation through the lungs. One solvent, dimethyl sulfoxide, is relatively harmless by itself but has the property of carrying toxic solutes through the skin and into the body. Exposure to solvent vapor is limited by occupational health regulations which include a threshold limiting value (TLV). This is the measure of solvent vapor concentration in the atmosphere considered safe for exposure to healthy humans over a normal 40-hour work week.

7.3.   WATER—A UNIQUE SOLVENT

The remainder of this chapter deals with water as a solvent. Water is such an important compound that all of Chapter 11 is spent discussing it in detail. Here, just those properties of water that relate directly to its characteristics as a solvent are summarized.

At room temperature H_2O is a colorless, tasteless, odorless liquid. It boils at 100°C (212°F) and freezes at 0°C (32°F). Water by itself is a very stable compound; it is very difficult to break up by heating. However, as explained in Section 8.6, when electrically conducting ions are present in water, a current may be passed through the water, causing it to break up into hydrogen gas and oxygen gas.

Water is an excellent solvent for a variety of materials; these include many ionic compounds (acids, bases, salts). Some gases dissolve well in water, particularly those that react with it chemically. Sugars and many other biologically important compounds are also soluble in water. Greases and oils generally are not soluble in water but dissolve in organic solvents instead.

Some of water's solvent properties can best be understood by considering the structure and bonding of the water molecule:

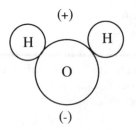

The water molecule is made up of two hydrogen atoms bonded to an oxygen atom. The three atoms are not in a straight line; instead, as shown above, they form an angle of 105°.

Because of water's bent structure and the fact that the oxygen atom attracts the negative electrons more strongly than do the hydrogen atoms, the water molecule behaves like a body having opposite electrical charges at either end or pole. Such a body is called a *dipole*. Due to the fact that it has opposite charges at opposite ends, the water dipole may be attracted to either positively or negatively charged ions. Recall that NaCl dissolves in water to form positive Na^+ ions and negative Cl^- ions in solution. The positive sodium ions are surrounded by water molecules with their negative ends pointed at the ions, and the chloride ions are surrounded by water molecules with their positive ends pointing at the negative ions, as shown in Figure 7.3. This kind of attraction for ions is the reason why water dissolves many ionic compounds and salts that do not dissolve in other liquids. Some noteworthy examples are sodium chloride in the ocean; waste salts in urine; calcium bicarbonate, which is very important in lakes and in geological processes; and widely used industrial acids (such as HNO_3, HCl, and H_2SO_4).

In addition to being a polar molecule, the water molecule has another important property which gives it many of its special characteristics: the ability to form **hydrogen bonds**. Hydrogen bonds are a special type of bond that can form between the hydrogen in one water molecule and the oxygen in another water molecule. This bonding takes place because the oxygen has a partly negative charge and the hydrogen, a partly positive charge. Hydrogen bonds, shown in Figure 7.4 as dashed lines, hold the water molecules together in large groups.

Hydrogen bonds also help to hold some solute molecules or ions in solution. This happens when hydrogen bonds form between the water molecules and hydrogen or oxygen atoms on the solute molecule (see Figure 7.4). Hydrogen bonding is one of the main reasons that some proteins can be put in water solution or held suspended in water as extremely small particles called *colloidal particles* (see Section 7.9).

7.4.   THE SOLUTION PROCESS AND SOLUBILITY

Very little happens to simple molecules, such as N_2 and O_2, when they dissolve in water. They mingle with the water molecules and occupy spaces that open up between water molecules to accommodate the N_2 and O_2 molecules. If the water

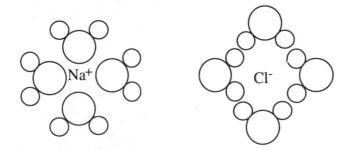

Figure 7.3. Polar water molecules surrounding Na⁺ ion (left) and Cl⁻ ion (right).

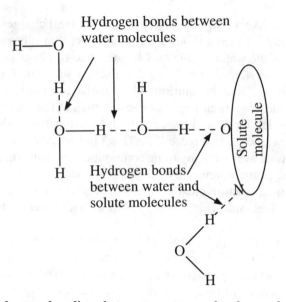

Figure 7.4. Hydrogen bonding between water molecules and between water molecules and a solute molecule in solution.

is heated, some of the gases are driven out of solution. This may be observed as the small bubbles that appear in heated water just before it boils. A fish can extract some of the oxygen in water by "breathing" through its gills; just 6 or 7 parts of oxygen in a million parts of water is all that fish require. Water saturated with air at 25°C contains about 8 parts per million oxygen. Chapter 12 discusses how only a small amount of an oxygen-consuming substance can use up this tiny portion of oxygen in water and cause the fish to suffocate and die.

Although N_2 and O_2 dissolve in water in the simple form of their molecules, the situation is much different when hydrogen chloride gas, HCl, dissolves in water. The hydrogen chloride molecule consists of a hydrogen atom bonded to a chlorine atom with a covalent bond. (Recall that covalent bonds are formed by *sharing* electrons between atoms.) Water can absorb large amounts of hydrogen chloride: 100 grams of water at 0°C will dissolve 82 g of this gas. When HCl dissolves in water (Figure 7.5), the solution is not simply hydrogen chloride

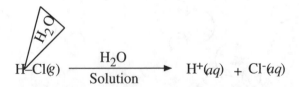

Figure 7.5. HCl dissolving in water. Water breaks apart a hydrogen chloride molecule to form a hydrogen ion, H⁺, and a chloride ion, Cl⁻.

molecules mixed with water molecules. The water has a strong effect upon the HCl molecule and breaks it into two parts, with the 2 electrons in the chemical bond staying with the chlorine. This forms a positively charged hydrogen ion, H^+, and a negatively charged chlorine ion, Cl^-.

In water solution, the chloride ion is surrounded by the positive ends of the water molecules, which are attracted to the negatively charged Cl^- ion. This kind of attraction of water molecules for a negative ion has already been shown in Figure 7.3. The H^+ ion from the HCl molecule does not remain in water as an isolated ion; it attaches to an unshared electron pair on a water molecule. This water molecule, with its extra hydrogen ion and extra positive charge, becomes a different ion with a formula of H_3O^+; it is called a **hydronium ion**. Although a hydrogen ion in solution is indicated as H^+ for simplicity, it is really present as part of a hydronium ion or larger ion aggregates ($H_5O_2^+$, $H_7O_3^+$).

The solution of hydrogen chloride in water illustrates a case in which a neutral molecule dissolves and forms electrically charged ions in water. While this happens with other substances dissolved in water, the hydrogen ion resulting when substances like HCl dissolve in water is particularly important because it results in the formation of a solution of acid. So rather than calling this a solution of hydrogen chloride, it is called a **hydrochloric acid** solution.

7.5. SOLUTION CONCENTRATIONS

In describing a solution it is necessary to do so *qualitatively*, that is, to specify what the solvent is and what the solutes are. For example, it was just seen that HCl gas is the solute placed in water to form hydrochloric acid. It is also necessary to know what happens to the material when it dissolves. For instance, one needs to know that hydrogen chloride molecules dissolved in water form H^+ ions and Cl^- ions. In many cases it is necessary to have *quantitative* information about a solution, its **concentration**. The concentration of a solution is the *amount* of solute material dissolved in a particular amount of solution, or by a particular amount of solvent. Solution concentration may be expressed in a

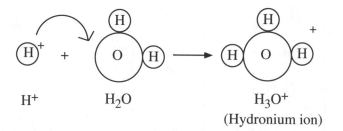

$$H^+ \qquad\qquad H_2O \qquad\qquad\qquad H_3O^+$$
$$\text{(Hydronium ion)}$$

Figure 7.6. A hydrogen ion, H^+, bonds to a water molecule, H_2O, to produce a hydronium ion, H_3O^+. Bonding to additional water molecules may form larger aggregates, $H_5O_2^+$, $H_7O_3^+$.

number of different ways. Solutions used in technical applications, such as for cleaning, are often made up of a specified number of grams of solute per 100 mL (milliliters) of solvent added. On the other hand, a person involved in crop spraying may mix the required solution by adding several pounds of pesticide to a specified number of barrels of water.

The concentrations of water pollutants frequently are given in units of milligrams per liter (mg/L). Most of the chemicals that commonly pollute water are harmful at such low levels that milligrams per liter of water is the most convenient way of expressing their concentrations. For example, water containing more than about one-third of a milligram of iron per liter of water can stain clothing and bathroom fixtures. To get an idea of how small this quantity is, consider that a liter of water (strictly speaking, at 4°C) weighs 1 million milligrams. Water containing one-third part per million of iron contains only 1 mg of iron in 3 million milligrams (3 liters) of water. Because 1 liter of water weighs 1 million milligrams, 1 mg of a solute dissolved in a liter is a **part per million**, abbreviated as **ppm**. The terms *part per million* and *milligrams per liter* are both frequently used in reference to levels of pollutants in water.

Some pollutants are so poisonous that their concentrations are given in micrograms per liter (μg/Liter). A particle weighing a microgram is so small that it cannot be seen with the naked eye. Since a microgram is a millionth of a gram, a liter of water weighs 1 billion micrograms. So, 1 microgram per liter is **1 part per billion (ppb)**. Sometimes it is necessary to think in terms of concentrations that are this low; a good example of this involves the long-banned pesticide *Endrin*. It is so toxic that, at a concentration of only two-thirds of a microgram per liter, it can kill half of the fingerlings (young fish) in a body of water over a four-day period.

At the other end of the scale, it may be necessary to consider very high concentration; these are often given as *percent by weight*. As indicated in Figure 7.7, the concentrations of commercial acids and bases are often expressed in this way. For example, a solution of concentrated ammonia purchased for labora-

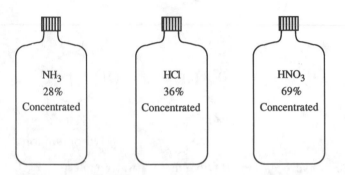

Figure 7.7. Percentages of solutes in commercial solutions of ammonia, hydrochloric acid, and nitric acid.

tory use is *28% by mass* NH_3; this means that out of 100 g of the ammonia solution, 28 g are NH_3 and 72 g are water. Commercial hydrochloric acid is about 36% by mass hydrogen chloride, HCl: of 100 g of concentrated hydrochloric acid, 36 g are HCl and 64 g are water.

Molar Concentration

The *mass* of a solute in solution often does not provide full information about its effect. For example, a solution of either NaCl or sodium iodide (NaI) will remove silver from a solution of silver nitrate ($AgNO_3$) when the two solutions are mixed. The two possible chemical reactions are,

$$AgNO_3(aq) + NaCl(aq) \rightarrow NaNO_3(aq) + AgCl(s) \qquad (7.5.1)$$

$$AgNO_3(aq) + NaI(aq) \rightarrow NaNO_3(aq) + AgI(s) \qquad (7.5.2)$$

in which solid precipitates of AgCl and AgI, respectively, come out of solution. The molecular mass of NaCl is 58.5 and that of NaI is 149.9. By comparing the removal of silver from solution by solutions of NaCl and NaI, both containing the same mass of solid in a specific volume of solution, it can be seen that a given volume of NaCl solution removes more silver. In chemical reactions such as these, it is the number of ions or molecules that is important, not their masses in solution. It would be convenient to have a way of expressing concentrations in quantities that would be directly related to the number of molecules or ions in solution. Instead of expressing such huge numbers, however, the chemist normally works with *moles*. (Recall from Section 2.3 that a mole is the number of grams of a substance that is equal numerically to the mass in u of the smallest unit of the substance. For example, since the molecular mass of NaCl is 58.5, a mole of NaCl weighs 58.5 g.) The **molar concentration** of a solution is *the number of moles of solute dissolved in a liter of solution*. Molar concentration is abbreviated with the letter M. A 1 M solution has 1 mole of solute dissolved in a liter of solution. To understand this concept better, consider the following:

- The molecular mass of HCl is 36.5; a 1 M solution of HCl has 1 mole of HCl (36.5 g) dissolved in a liter of solution.
- The molecular mass of NH_3 is 17; a 1 M solution of NH_3 has one mole of NH_3 (17 g) in a liter of solution.
- The molecular mass of glucose, $C_6H_{12}O_6$, is 180; a 1 M solution of glucose has 1 mole of glucose (180 g) in a liter of solution.

Of course, solutions are not always exactly 1 M in concentration. However, it is easy to perform calculations involving molar concentration using the relationships expressed in the following equation:

$$M = \frac{\text{Moles solute}}{\text{Volume solution, L}} = \frac{\text{Mass solute, g}}{\text{Molar mass solute} \times \text{Volume solution, L}} \quad (7.5.3)$$

To see how this equation can be used, consider the following examples:

Example: What is the molar concentration of a solution that contains 2.00 moles of HCl in 0.500 L of solution?

Answer:

$$M = \frac{2.00 \text{ mol}}{0.500 \text{ L}} = 4.00 \text{ mol/L}$$

Example: What is the mass of HCl, molar mass 36.5 g/mol, in 2.75 L of a 0.800 M solution?

Answer:

$$0.800 \text{ mol/L} = \frac{\text{Mass solute}}{36.5 \text{ g/mol} \times 2.75 \text{ L}}$$

$$\text{Mass solute} = 0.800 \text{ mol/L} \times 36.5 \text{ g/mol} \times 2.75 \text{ L} = 80.3 \text{ g}$$

Example: What is the molar concentration of a solution containing 87.6 g of HCl in a total volume of 3.81 L of solution?

Answer:

$$M = \frac{87.6 \text{ g}}{36.5 \text{ g/mol} \times 3.81 \text{L}} = 0.630 \text{ mol/L}$$

To summarize the steps involved in making up a certain quantity of a solution with a specified concentration, consider the task of making up 4.60 liters of 0.750M NaCl, which would involve the following steps:

Step 1.   Calculate the mass of NaCl, molar mass 58.5 g/mol, required.

$$0.750 \text{ mol/L} = \frac{\text{Mass solute}}{58.5 \text{ g/mol} \times 4.60 \text{ L}}$$

$$\text{Mass solute} = 0.750 \text{ mol/L} \times 58.5 \text{ g/mol} \times 4.60 \text{ L} = 202 \text{ g}$$

Step 2. Weigh out 202 g NaCl

Step 3. Add NaCl to a container with a mark at 4.60 L

Step 4. Add water and mix to a final volume of 4.60 L

Diluting Solutions

Often it is necessary to make a less concentrated solution from a more concentrated solution; this process is called **dilution**. This situation usually occurs in the laboratory because it is more convenient to store more concentrated solutions in order to save shelf space. Also, laboratory acid solutions, such as those of hydrochloric acid, sulfuric acid, and phosphoric acid, are almost always purchased as highly concentrated solutions; these are more economical because there is more of the active ingredient per bottle. For standard solutions to use in chemical analysis it is more accurate to weigh out a relatively large quantity of solute to make a relatively concentrated solution, then dilute the solution quantitatively to prepare a more dilute standard solution.

For example, the concentration of commercial concentrated hydrochloric acid is 12 M. A laboratory technician needs 1 liter of 1 M hydrochloric acid. How much of the concentrated acid is required? The key to this problem is to realize that when a volume of the concentrated acid is diluted with water, the total *amount of solute acid* in the solution remains the same. The problem can then be solved by considering the equation,

$$M = \frac{\text{Mol solute}}{\text{Volume solution}} \qquad (7.5.3)$$

using subscripts "1" and "2" to indicate values before and after dilution, respectively,

$$M_1 = \frac{(\text{Mol solute})_1}{(\text{Volume solution})_1} \qquad (7.5.4)$$

$$M_2 = \frac{(\text{Mol solute})_2}{(\text{Volume solution})_2} \qquad (7.5.5)$$

and setting "Moles solute" before and after dilution equal to each other:

$$M_1 \times (\text{Volume solution})_1 = M_2 \times (\text{Volume solution})_2 \qquad (7.5.6)$$

In the example cited above, the volume of HCl before dilution is to be calculated:

$$(\text{Volume solution})_1 = \frac{M_2 \times (\text{Volume solution})_2}{M_1} \qquad (7.5.7)$$

$$\text{(Volume solution)}_1 = \frac{1 \text{ Mol/L} \times 1 \text{ L}}{12 \text{ mol/L}} = 0.083 \text{ L} \qquad (7.5.8)$$

The result of this calculation shows that 0.083 L, or 83 mL of 12 M HCl must be taken to make 1.00 L of 1 M solution.

The same general approach used in solving these dilution problems may be used when concentrations are expressed in units other than molar concentration. Concentrations of metals dissolved in water are frequently measured by *atomic absorption spectroscopy*. To measure the concentration of metal in an unknown solution, it is necessary to have a standard solution (see Section 7.6) of known concentration. As purchased, these standard solutions contain 1000 mg of the desired metal dissolved in 1 liter of solution. Suppose one had a standard solution of $CaCl_2$ containing 1000 mg of Ca^{2+} per liter of solution. How would this solution be diluted to make 5 liters of a solution containing 20 mg of calcium per liter.

Step 1. Consider how many milligrams of calcium are in the desired 5 L of solution containing 20 mg of calcium per liter.

$$5 \text{ L} \times 20 \text{ mg calcium/L} = 100 \text{ mg calcium} \qquad (7.5.9)$$

Step 2. Find how much standard solution containing 1000 mg of calcium per liter contains 100 mg of calcium.

$$\frac{100 \text{ mg of calcium}}{1000 \text{ mg calcium/L}} = 0.100 \text{ L of standard solution} \qquad (7.5.10)$$

Step 3. Dilute 0.100 L (100 mL) of the standard calcium solution to 5 L to obtain the desired solution containing 20 mg of calcium per liter.

Molar Concentration of H⁺ Ion and pH

Concentrations are important in expressing the degree to which solutions are acidic or basic. Recall that **acids**, such as HCl and H_2SO_4, produce H^+ ion, whereas **bases**, such as sodium hydroxide and calcium hydroxide [NaOH and $Ca(OH)_2$, respectively], produce hydroxide ion, OH^-. Molar concentrations of hydrogen ion, $[H^+]$, range over many orders of magnitude and are conveniently expressed by pH defined as

$$pH = -\log[H^+] \qquad (7.5.11)$$

In absolutely pure water, the value of $[H^+]$ is exactly 1×10^{-7} mole/L, the pH is 7.00, and the solution is **neutral** (neither acidic nor basic). **Acidic** solutions have pH values of less than 7 and **basic** solutions have pH values of greater than 7.

Solubility

If a chemist were to attempt to make a solution containing calcium ion by stirring calcium carbonate, $CaCO_3$, with water, not much of anything would happen. The chemist would not observe any calcium carbonate dissolving because its **solubility** in water is very low. Only about 5 mg (just a small white "speck") of calcium carbonate will dissolve in a liter of water.

This illustrates an important point that people working in laboratories should keep in mind. When preparing a solution, it is a good idea to look up the solubility of the compound being dissolved. Make sure that the solubility is high enough to give the concentration that is desired. Many futile hours have been spent shaking bottles, trying to get something to dissolve whose solubility is just too low to make the desired solution.

A solution that has dissolved as much of a solute as possible is said to be **saturated** with regard to that solute. The *concentration* of the substance in the *saturated* solution is the **solubility**. Solubilities of different substances in water vary enormously. At the low end of the scale, even glass will dissolve a little bit in water, although the concentration of glass in solution is very, very low. Even this very small solubility can cause difficulties in the analysis of silicon (one of the elements in glass). This is because the dissolved glass increases the silicon content of the water in the solution, which causes errors in the results of the analysis. Antifreeze (composed of an organic compound called ethylene glycol) is an example of a substance that is completely soluble in water. One could pour a cup of water into a barrel and keep adding antifreeze without ever getting a saturated solution of antifreeze in water, even if the barrel were filled to the point of overflowing.

Supersaturated Solutions

Consider what happens when a bottle of carbonated beverage at an elevated temperature is opened. As the cap comes off, there is a loud pop, and the contents pour forth in a geyser of foam. This is because the solution of CO_2 in the bottle is a **supersaturated** solution. The carbon dioxide is added to the cold beverage under pressure. When the contents of the bottle become warm the solubility of the CO_2 is significantly lowered. If pressure is released by removing the bottle cap, the solution suddenly contains more CO_2 than it can hold in an unpressurized solution.

Factors Affecting Solubility

Solubility is affected by several factors. The preceding example has shown two of the most important factors affecting *gas* solubility: temperature and pressure. The solubilities of gases such as CO_2 *decrease* with *increasing* temperature. A gas is much less soluble in water just about hot enough to boil than it is in water just about cold enough to freeze. However, once the solution freezes, the gas is not at all soluble in the ice formed. This can be illustrated vividly with soft drinks, which also contain carbon dioxide dissolved under pressure. A can of soda placed in the freezer to cool quickly and then forgotten may burst from the pressure of the carbon dioxide coming out of solution as the liquid freezes, causing havoc in its immediate surroundings.

The solubilities of most solids *increase* with *rising* temperature, although not in all cases. Sugar, for example, is much easier to dissolve in a cup of hot tea than in a glass of iced tea.

One of the most important things affecting solubility is the presence of other chemicals in the solution that react with the solute. For example, CO_2 is really not very soluble in water. But, if the water already contains some NaOH (sodium hydroxide), the carbon dioxide is very soluble because a chemical reaction occurs between NaOH and CO_2 to form highly soluble sodium bicarbonate, $NaHCO_3$.

$$NaOH + CO_2 \text{ (not very soluble)} \rightarrow NaHCO_3 \text{ (highly soluble)} \qquad (7.5.12)$$

In general, the solubility of solids is relatively lower at lower temperatures, behavior opposite to that of gases. Therefore, supersaturated solutions of some solids may be prepared by preparing a solution at an elevated temperature, then cooling it. The solid may be caused to come out of solution by adding a small crystal of the solid or by scratching the side of the container.

7.6. STANDARD SOLUTIONS AND TITRATIONS

Standard solutions are those of known concentration that are widely used in chemical analysis. Basically, a standard solution is one against which a solution of unknown concentration can be compared to determine the concentration of the latter. One of the most common and straightforward means of comparison is by way of **titration**, which consists of measuring the amount of a standard solution that reacts with a sample using the apparatus illustrated in Figure 7.8. As an example consider a hospital incinerator with a water scrubbing system to clean exhaust gases. Suppose it is observed that the scrub water contains hydrochloric acid from the burning of chlorine-containing plastics and that this acid is corroding iron drain pipes, a problem that can be eliminated by adding sodium hydroxide to neutralize the acid. The amount of base needed may be determined by titrating a measured volume of the hydrochloric acid-containing scrub solution by allowing it to react with a standard solution of NaOH,

**Figure 7.8. A long glass tube with marks on it and a stopcock on the end called a
buret is used to measure the volume of standard solution added to a sample during
titration.**

$$HCl(aq) \; + \; NaOH(aq) \; \rightarrow \; NaCl(aq) \; + H_2O \qquad (7.6.1)$$

until just enough base has been added to react with all the HCl in the measured
volume of sample. The point at which this occurs is called the **end point** and is
shown by an abrupt change in color of a dissolved dye called an **indicator**.
Phenolphthalein, a dye that is red in base and colorless in acid is commonly used
as an indicator.

Suppose that 50.0 mL (0.0500 L) of waste incinerator scrubber water required
40.0 mL (0.0400 L) of 0.0100 M standard NaOH for neutralization of HCl in the
wastewater. The molar concentration of the HCl in the sample may be calculated
from an equation derived from the following:

$$M \; = \; \frac{\text{Moles solute}}{\text{Volume solution}} \qquad (7.5.3)$$

In this case, as shown by Reaction 7.6.1, the number of moles of HCl in the
sample exactly equals the number of moles of NaOH at the end point, which
leads to the relationship,

$$M_{HCl} \times \text{Volume HCl} \; = \; M_{NaOH} \times \text{Volume NaOH} \qquad (7.6.2)$$

where the subscript formulas denote HCl and NaOH solutions. This equation
can be rearranged and values substituted into it to give the following:

$$M_{HCl} = \frac{M_{NaOH} \times \text{Volume NaOH}}{\text{Volume HCl}} \qquad (7.6.3)$$

$$M_{HCl} = \frac{0.100 \text{ mol/L} \times 0.0400 \text{ L}}{0.0500 \text{ L}} = 0.0800 \text{ mol/L} \qquad (7.6.4)$$

From this concentration, determined by titration, the quantity of base that must be added to neutralize the waste acid can be calculated.

7.7. PHYSICAL PROPERTIES OF SOLUTIONS

The presence of solutes in water may have profound effects upon the properties of the solvent. These effects include lowering the freezing point, elevating the boiling point, and osmosis, all **colligative properties**, which depend upon the concentration of solute, rather than its particular identity. The effects of solutes and solution concentrations on colligative properties are addressed briefly here.

Freezing Point Depression

One of the most practical uses of solutions depends upon the effect that materials dissolved in water have upon the temperature at which the water, or the solution, freezes. Solutions freeze at lower temperatures than does water. This phenomenon is applied in the cooling systems of automobiles. Most automobile engines are water-cooled, that is, water circulates through the engine picking up heat and then goes to a radiator where the excess heat is given off. If pure water is left in an engine at freezing temperatures, it will, of course, freeze. When water freezes, it expands, so that the ice which is formed has a larger volume than the original liquid. This increase in volume produces enormous forces which can crack the stoutest engine block as if it were an egg shell.

To prevent an engine block from damage by freezing and expansion of the coolant, antifreeze is mixed with water in the cooling system. The chemical name of antifreeze is ethylene glycol, and it has the chemical formula $C_2H_6O_2$. A solution containing 40% ethylene glycol and 60% water by mass freezes at $-8°$ Fahrenheit, or $-22.2°$ Celsius. (Recall that pure water freezes at $32°F$, or $0°C$; it boils at $212°F$, or $100°C$.) A solution containing exactly half ethylene glycol and half water by mass freezes at $-29°F$.

Boiling Point Elevation

Solutions can also keep water from boiling. For instance, mixing antifreeze or other materials that do not boil easily with water raises the boiling temperature. Because of the higher boiling temperature, a solution of antifreeze in water makes a good "summer coolant." In fact, antifreeze is now marketed as "antifreeze-antiboil" and is virtually required in an engine to keep the cooling system from boiling in summer's heat.

Osmosis

Human blood consists chiefly of red blood cells suspended in a fairly concentrated solution or **plasma**. The dissolved material making up the solution is mostly sodium chloride at around 0.15 molar concentration. The red blood cells also contain a solution much like the plasma in which they float. If one were to take some blood, force the red cells to settle out by spinning test tubes full of blood samples in a centrifuge and then place the red cells in pure water, a strange effect would be observed. When viewed under a microscope, the cells would be seen to swell and finally burst. This is an example of **osmosis**, the process whereby a substance passes through a membrane, from an area of higher concentration of the substance to an area of lower concentration of the substance. Water can go through the membrane that holds the red blood cell together. When the cell is placed in pure water, the water tends to pass through the membrane to the more concentrated salt solution inside the cell. One way of looking at it is to say that the solution inside the cell has relatively less water, so that water from the outside has a tendency to enter the cell. Finally, so much water gets in that it breaks the cell apart (Figure 7.9). However, when a blood cell is placed in a very strong solution of sodium chloride, the opposite effect is observed: water passes from inside the cell to the outside, where there is less water, relative to sodium chloride, and the cell shrivels up (Figure 7.10).

Many biological membranes, such as the "wall" surrounding a red blood cell, allow material to pass through. Water in the ground tends to move through cell membranes in root cells and finally to plant leaves, where it evaporates to form water vapor in the atmosphere (a process called *transpiration*). Osmosis is a major part of the driving force behind these processes.

The concentration of a solution relative to another solution determines which

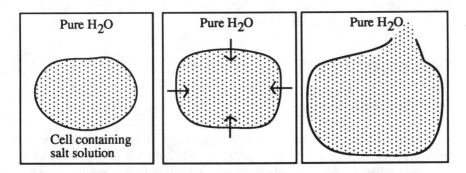

Figure 7.9. Osmosis causes pure water to be absorbed by a red blood cell, increasing the pressure in the cell and causing it to burst.

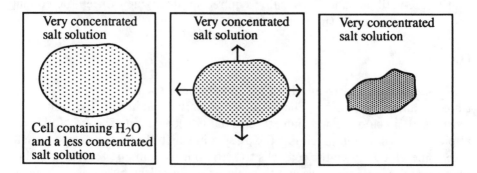

Figure 7.10. Osmosis causes water inside a red blood cell to pass out into a concentrated salt solution, causing the cell to shrivel.

solution water will have a tendency to enter. *Water* tends to enter a solution of higher concentration (of some other substance), since it contains *less water*, relatively speaking. This can be particularly important in blood. Solutions injected into the veins for feeding, or to replace body fluids lost due to illness, must be prepared so that they will not cause water to move in or out of the cells by osmosis. Such a solution, which prevents osmosis from taking place, is called an **isotonic** solution. It is sometimes referred to as a *physiological* solution. Its use prevents damage to blood cells by either shrinking or swelling.

Osmosis may result in the buildup of a very high pressure called **osmotic pressure**. It is this pressure which can get so high in a red blood cell suspended in water that the cell bursts. Osmotic pressure can be several times atmospheric pressure.

Pure water tends to move by osmosis through a membrane to a more "contaminated" solution. Application of sufficient opposite pressure can "squeeze" water out of a contaminated solution, producing pure water. This process is called **reverse osmosis**, and is commonly used to purify small quantities of water. Some very large reverse osmosis installations are now being used to remove salt from seawater, a process called **desalination**.

7.8.   SOLUTION EQUILIBRIA

Solution Equilibria

Many phenomena in aquatic chemistry and geochemistry involve **solution equilibrium**. In a general sense solution equilibrium deals with the extent to which **reversible** acid-base, solubilization (precipitation), complexation, or oxidation-reduction reactions proceed in a forward or backward direction. This is expressed for a generalized equilibrium reaction,

$$aA + bB \rightleftarrows cC + dD \qquad (7.8.1)$$

by the following **equilibrium constant expression:**

$$\frac{[C]^c[D]^d}{[A]^a[B]^b} = K \qquad (7.8.2)$$

where K is the **equilibrium constant.**

A reversible reaction may approach equilibrium from either direction. In the example above, if A were mixed with B, or C were mixed with D, the reaction would proceed in a forward or reverse direction such that the concentrations of species—[A], [B], [C], and [D]—substituted into the equilibrium expression gave a value equal to K.

As expressed by **Le Châtelier's principle**, a stress placed upon a system in equilibrium will shift the equilibrium to relieve the stress. For example, adding product "D" to a system in equilibrium will cause Reaction 7.8.1 to shift to the left, consuming "C" and producing "A" and "B," until the equilibrium constant expression is again satisfied. This is called the **mass action effect**, and it is the driving force behind many environmental chemical phenomena.

There are several major kinds of equilibria in aqueous solution. One of these is **acid-base equilibrium** as exemplified by the ionization of acetic acid, HAc,

$$HAc \rightleftarrows H^+ + Ac^- \qquad (7.8.3)$$

for which the acid dissociation constant is

$$\frac{[H^+][Ac^-]}{[HAc]} = K = 1.75 \times 10^{-5} \text{ (at } 25°C) \qquad (7.8.4)$$

Very similar expressions are obtained for the formation and dissociation of metal **complexes** or **complex ions,** formed by the reaction of a metal ion in solution with a **complexing agent** or **ligand,** both of which are capable of independent existence in solution. This can be shown by the reaction of iron(III) ion and thiocyanate ligand

$$Fe^{3+} + SCN^- \rightleftarrows FeSCN_2^+ \qquad (7.8.5)$$

for which the **formation constant expression** is:

$$\frac{[FeSCN^{2+}]}{[Fe^{3+}][SCN^-]} = K_f = 1.07 \times 10^3 \text{ (at } 25°C) \qquad (7.8.6)$$

The bright red color of the $FeSCN^{2+}$ complex formed could be used to test for the presence of iron(III) in acid mine water.

An example of an **oxidation-reduction reaction**, those that involve the transfer of electrons between species, is

$$MnO_4^- + 5Fe^{2+} + 8H^+ \rightleftharpoons Mn^{2+} + 5Fe^{3+} + 4H_2O \qquad (7.8.7)$$

for which the equilibrium expression is:

$$\frac{[Mn^{2+}][Fe^{3+}]^5}{[MnO_4^-][Fe^{2+}]^5[H^+]^8} = K = 3 \times 10^{62} \text{ (at 25°C)} \qquad (7.8.8)$$

(The concentration of water, $[H_2O]$ is omitted from the K expression because it is a constant in solvent water and is incorporated with the value of K.)

The value of K is calculated from the Nernst equation (Chapter 8).

Distribution Between Phases

Many important environmental chemical phenomena involve distribution of species between phases. This most commonly involves the equilibria between species in solution and in a solid phase. **Solubility equilibria** deal with reactions such as,

$$AgCl(s) \rightleftharpoons Ag^+ + Cl^- \qquad (7.8.9)$$

in which one of the participants is a slightly soluble (virtually insoluble) salt and for which the equilibrium constant is a **solubility product, K_{sp}**:

$$[Ag+][Cl^-] = K_{sp} = 1.82 \times 10^{-10} \text{ (at 25°C)} \qquad (7.8.10)$$

Note that in the equilibrium constant expression there is not a value given for the solid AgCl. This is because the activity of a solid (in this case the tendency of Ag^+ and Cl^- ions to break away from solid AgCl) is constant at a specific temperature and is contained in the value of K_{sp}.

An important example of distribution between phases is that of a hazardous waste species partitioned between water and a body of immiscible organic liquid in a hazardous waste site. The equilibrium for such a reaction,

$$X(aq) \rightleftharpoons X(org) \qquad (7.8.11)$$

is described by the **distribution law** expressed by a **distribution coefficient** or **partition coefficient** in the following form:

$$\frac{[X(org)]}{[X(aq)]} = K_d \qquad (7.8.12)$$

Solubilities of Gases

The solubilities of gases in water are described by Henry's law which states that *at constant temperature the solubility of a gas in a liquid is proportional to the partial pressure of the gas in contact with the liquid.* For a gas, "X," this law applies to equilibria of the type

$$X(g) \rightleftharpoons X(aq) \tag{7.8.13}$$

and does not account for additional reactions of the gas species in water such as,

$$NH_3 + H_2O \rightleftharpoons NH_4^+ + OH^- \tag{7.8.14}$$

$$SO_2 + HCO_3^- \text{ (From water alkalinity)} \rightleftharpoons CO_2 + HSO_3^- \tag{7.8.15}$$

which may result in much higher solubilities than predicted by Henry's law alone.

Mathematically, Henry's law is expressed as

$$[X(aq)] = K P_X \tag{7.8.16}$$

where $[X(aq)]$ is the aqueous concentration of the gas, P_X is the partial pressure of the gas, and K is the Henry's law constant applicable to a particular gas at a specified temperature. For gas concentrations in units of moles per liter and gas pressures in atmospheres, the units of K are $mol \times L^{-1} \times atm^{-1}$. Some values of K for dissolved gases that are significant in water are given in Table 7.1.

In calculating the solubility of a gas in water, a correction must be made for the partial pressure of water by subtracting it from the total pressure of the gas. At 25°C the partial pressure of water is 0.0313 atm; values at other temperatures are readily obtained from standard handbooks. The concentration of oxygen in water saturated with air at 1.00 atm and 25°C may be calculated as an example of a simple gas solubility calculation. Considering that dry air is 20.95% by

Table 7.2. Henry's Law Constants for Some Gases in Water at 25°C.

Gas	$K, mol \times L^{-1} \times atm^{-1}$
O_2	1.28×10^{-3}
CO_2	3.38×10^{-2}
H_2	7.90×10^{-4}
CH_4	1.34×10^{-3}
N_2	6.48×10^{-4}
NO	2.0×10^{-4}

volume oxygen and factoring in the partial pressure of water gives the following:

$$P_{O_2} = (1.0000 \text{ atm} - 0.0313 \text{ atm}) \times 0.2095 = 0.2029 \text{ atm} \quad (7.8.17)$$

$$[O_2(aq)] = K \times P_{O_2} = 1.28 \times 10^{-3} \text{ mol} \times L^{-1} \times \text{atm}^{-1} \times 0.2029 \text{ atm}$$
$$= 2.60 \times 10^{-4} \text{ mol} \times L^{-1} \quad (7.8.18)$$

Since the molecular mass of oxygen is 32, the concentration of dissolved oxygen in water in equilibrium with air under the conditions given above is 8.32 mg/L, or 8.32 parts per million (ppm).

The solubilities of gases decrease with increasing temperature. Account is taken of this factor with the **Clausius-Clapeyron** equation,

$$\log \frac{C_2}{C_1} = \frac{\Delta H}{2.303R} \left[\frac{1}{T_1} - \frac{1}{T_2} \right] \quad (7.8.19)$$

where C_1 and C_2 denote the gas concentration in water at absolute temperatures of T_1 and T_2, respectively, ΔH is the heat of solution, and R is the gas constant. The value of R is 1.987 cal $\times$ deg^{-1} $\times$ mol^{-1}, which gives ΔH in units of cal/mol. Detailed calculations involving the Clausius-Clapeyron equation are beyond the scope of this chapter.

7.9.   COLLOIDAL SUSPENSIONS

If some fine sand is poured into a bottle of water and shaken vigorously, the sand will float around in the water for a very brief time, then rapidly settle to the bottom. Particles like sand, which float briefly in water and then rapidly settle out, are called **suspensions**. It has already been seen that molecules or ions that dissolve in water and mingle individually with the water molecules make up *solutions*. Between these two extremes are particles which are much smaller than can be seen with the naked eye, but much larger than individual molecules. Such particles are called **colloidal particles**. They play important roles in living systems, transformations of minerals, and a number of industrial processes. Colloidal particles in water form **colloidal suspensions**. Unlike sand stirred in a jar, or soil granules carried by a vigorously running stream, *colloidal particles do not settle out of colloidal suspension by gravity alone*. Colloidal particles range in diameter from about 0.001 micrometer (μm) to about 1 μm, have some characteristics of both species in solution and larger particles in suspension, and in general exhibit unique properties and behavior. An important characteristic of colloidal particles is their ability to scatter light. Such light scattering is called the **Tyndall effect** and is observed as a light blue hue at right angles to incident white light. This phenomenon results from colloidal particles being the same order of size as the wavelength of light.

Individual cells of bacteria are colloidal particles and can form colloidal suspensions in water. Many of the green algae in water are present as colloidal suspensions of individual cells of the algae. Milk, mayonnaise, and paint are all colloidal suspensions. Colloids are used in making rubber, glue, plastics, and grease. The formation of a colloidal suspension of butterfat in milk is the process by which milk is *homogenized*, so that the cream does not rise to the top.

Kinds of Colloidal Particles

Colloids may be classified as *hydrophilic colloids, hydrophobic colloids*, or *association colloids*. These three classes are briefly summarized below.

Hydrophilic colloids generally consist of macromolecules, such as proteins and synthetic polymers, that are characterized by strong interaction with water resulting in spontaneous formation of colloids when they are placed in water. In a sense, hydrophilic colloids are solutions of very large molecules or ions. Suspensions of hydrophilic colloids are less affected by the addition of salts to water than are suspensions of hydrophobic colloids.

Hydrophobic colloids interact to a lesser extent with water and are stable because of their positive or negative electrical charges as shown in Figure 7.11. The charged surface of the colloidal particle and the **counter-ions** that surround it compose an **electrical double layer**, which causes the particles to repel each other. Hydrophobic colloids are usually caused to settle from suspension by the

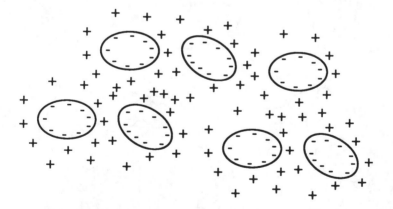

Figure 7.11. Representation of negatively charged hydrophobic colloidal particles surrounded in solution by positively charged counter-ions, forming an electrical double layer. (Colloidal particles suspended in water may have either a negative or positive charge.)

addition of salts. Examples of hydrophobic colloids are clay particles, petroleum droplets, and very small gold particles.

Association colloids consist of special aggregates of ions and molecules called **micelles**. To understand the nature of micelles, consider sodium stearate, a typical soap with the structural formula shown below:

$$H-\overset{\overset{\displaystyle H}{|}}{\underset{\underset{\displaystyle H}{|}}{C}}-\overset{\overset{\displaystyle H}{|}}{\underset{\underset{\displaystyle H}{|}}{C}}-\overset{\overset{\displaystyle H}{|}}{\underset{\underset{\displaystyle H}{|}}{C}}-\overset{\overset{\displaystyle H}{|}}{\underset{\underset{\displaystyle H}{|}}{C}}-\overset{\overset{\displaystyle H}{|}}{\underset{\underset{\displaystyle H}{|}}{C}}-\overset{\overset{\displaystyle H}{|}}{\underset{\underset{\displaystyle H}{|}}{C}}-\overset{\overset{\displaystyle H}{|}}{\underset{\underset{\displaystyle H}{|}}{C}}-\overset{\overset{\displaystyle H}{|}}{\underset{\underset{\displaystyle H}{|}}{C}}-\overset{\overset{\displaystyle H}{|}}{\underset{\underset{\displaystyle H}{|}}{C}}-\overset{\overset{\displaystyle H}{|}}{\underset{\underset{\displaystyle H}{|}}{C}}-\overset{\overset{\displaystyle H}{|}}{\underset{\underset{\displaystyle H}{|}}{C}}-\overset{\overset{\displaystyle H}{|}}{\underset{\underset{\displaystyle H}{|}}{C}}-\overset{\overset{\displaystyle H}{|}}{\underset{\underset{\displaystyle H}{|}}{C}}-\overset{\overset{\displaystyle H}{|}}{\underset{\underset{\displaystyle H}{|}}{C}}-\overset{\overset{\displaystyle H}{|}}{\underset{\underset{\displaystyle H}{|}}{C}}-\overset{\overset{\displaystyle H}{|}}{\underset{\underset{\displaystyle H}{|}}{C}}-\overset{\overset{\displaystyle O}{\|}}{C}-O^- \ Na^+$$

Represented as ⟋⟍⟋⟍⟋⟍⟋⟍⟋⟍⟋⟍⟍◯_(−)

The stearate ion has both a hydrophilic $-CO_2^-$ head and a long organophilic tail $CH_3(CH_2)_{16}$. As a result, stearate anions in water tend to form clusters consisting of as many as 100 anions clustered together with their hydrocarbon "tails" on the inside of a spherical colloidal particle and their ionic "heads" on the surface in contact with water and with Na^+ counter-ions. This results in the formation of **micelles** as illustrated in Figure 7.12.

Water-insoluble organic matter may be entrained in the micelle.

Figure 7.12. Representation of colloidal soap micelle particles.

Colloid Stability

The stability of colloids is a prime consideration in determining their behavior. It is involved in important aquatic chemical phenomena including the formation of sediments, dispersion and agglomeration of bacterial cells, and dispersion and removal of pollutants (such as crude oil from an oil spill).

Discussed above, the two main phenomena contributing to the stabilization of colloids are **hydration** and **surface charge**. The layer of water on the surface of hydrated colloidal particles prevents contact, which would result in the formation of larger units. A surface charge on colloidal particles may prevent aggregation, since like-charged particles repel each other. The surface charge is frequently pH-dependent; around pH 7 most colloidal particles in natural waters are negatively charged. Negatively charged aquatic colloids include algal cells, bacterial cells, proteins, and colloidal petroleum droplets.

One of the three major ways in which a particle may acquire a surface charge is by **chemical reaction at the particle surface**. This phenomenon, which frequently involves hydrogen ion and is pH-dependent, is typical of hydroxides and oxides. As an illustration of pH-dependent charge on colloidal particle surfaces consider the effects of pH on the surface charge of hydrated manganese oxide, $MnO_2(H_2O)(s)$. In a relatively acidic medium, the reaction

$$MnO_2(H_2O)(s) + H^+ \rightarrow MnO_2(H_3O)^+(s) \qquad (7.9.1)$$

may occur on the surface giving the particle a net positive charge. In a more basic medium, hydrogen ion may be lost from the hydrated oxide surface to yield negatively charged particles:

$$MnO_2(H_2O)(s) \rightarrow MnO_2(OH)^-(s) + H^+ \qquad (7.9.2)$$

Ion absorption is a second way in which colloidal particles become charged. This phenomenon involves attachment of ions onto the colloidal particle surface by means other than conventional covalent bonding, including hydrogen bonding and London (van der Waals) interactions.

Ion replacement is a third way in which a colloidal particle may gain a net charge. For example, replacement of some of the Si(IV) with Al(III) in the basic SiO_2 chemical unit in the crystalline lattice of some clay minerals,

$$[SiO_2] + Al(III) \rightarrow [AlO_2^-] + Si(IV) \qquad (7.9.3)$$

yields sites with a net negative charge. Similarly, replacement of Al(III) by a divalent metal ion such as Mg(II) in the clay crystalline lattice produces a net negative charge.

Coagulation and Flocculation of Colloidal Particles

The **aggregation** of colloidal particles is very important in a number of processes, including industrial processes and aquatic chemical phenomena. The ways in which colloidal particles of minerals come together and settle out of water are crucial in the formation of mineral deposits. The bacteria in colloidal suspension that devour sewage in the water at a sewage treatment plant must eventually be brought out of colloidal suspension to obtain purified water. Crude oil frequently comes out of the ground as a colloidal suspension in water; the colloidal particles of hydrocarbon must be separated from the water before the crude oil can be refined.

It is possible to force electrically-charged colloidal particles to clump together and settle. This is done by adding charged ions, such as Na^+ and Cl^- ions from sodium chloride. These charged ions tend to neutralize the electrical charges on the colloidal particles and allow them to come together; this process is called **coagulation**. In some cases colloidal particles are caused to settle by virtue of *bridging groups* called *flocculants* that join the particles together. This phenomenon is called **flocculation**.

CHAPTER SUMMARY

The chapter summary below is presented in a programmed format to review the main points covered in this chapter. It is used most effectively by filling in the blanks, referring back to the chapter as necessary. The correct answers are given at the end of the summary.

A solid that seems to disappear when stirred with water is said to (1)_____, forming a (2)_____. The substance that dissolves is called the (3)_____, and the liquid in which it dissolves is the (4)_____. A substance that does not dissolve appreciably in a solvent is said to be (5)_____. Four typical organic solvents are (6)_____, _____, _____, and

_____.

Two characteristics of water molecules that make water an excellent solvent for many substances are (7)_____. When O_2 dissolves in water, the oxygen molecules are in solution as (8)_____. When NaCl dissolves in water it is present as (9)_____, and when HCl

dissolves in water it forms (10)_____. The hydrogen ion in water
is bonded to a water molecule to form a (11)_____.

The amount of solute in a particular amount of solution or solvent is the
(12)_____. Low concentrations of pollutants in water are often
expressed as parts per (13)_____, abbreviated (14)_____, or,
at even lower concentrations, as (15)_____, abbreviated (16)_____.
The molar concentration of a solution is the number of (17)_____
of solute per (18)_____. The molecular mass of NH_3 is (19)_____,
so a mole of ammonia weighs (20)_____. The molar concentration of a
solution prepared by dissolving 8.5 g of NH_3 gas in 2 liters of solution is
(21)_____ M. The relationships among the volume of a solution taken for
dilution (V_1), its molar concentration (M_1), the volume of the diluted solution
(V_2), and the concentration of the diluted solution (M_2) is (22)_____.
The volume of 2 M HCl that must be diluted with water to make 5 liters of 1.5 M
HCl is (23)_____ liters.

The equation that defines pH is (24)_____. A pH less
than 7 indicates a solution that is (25)_____.

The maximum amount of solute that can dissolve in a particular volume of
solution is called the solute's (26)_____. A solution containing the
maximum amount of solute that it can normally hold is said to be
(27)_____.

A solution of known concentration used in chemical analyses is called a
(28)_____. An operation in which the quantity of such
a solution required to react with a substance in another solution is measured is
known as (29)_____. The reaction is finished at the (30)_____, which
is shown by a change in color of an (31)_____.

Effects including lowering of freezing point, elevation of boiling point, and
osmosis, which depend upon the concentration of solute, rather than its particular identity are called (32)_____. Solutes (33)_____

the freezing temperature of water. A solution of antifreeze in water freezes at a (34)_____ temperature and boils at a (35)_____ temperature than pure water.

Solution equilibrium deals with the extent to which (36)_____ reactions proceed in a forward or backward direction. Four major classes of these reactions are (37)_____

_____.

An expression such as

$$\frac{[C]^c[D]^d}{[A]^a[B]^b} = K$$

that describes an equilibrium reaction represented by

$$aA + bB \rightleftharpoons cC + dD$$

is called (38)_____. A statement of the law governing gas solubilities is (39)_____

_____

_____,

which is called (40)_____. For a gas, "X," a mathematical expression of this law is (41)_____.

Extremely small particles suspended uniformly in water form a (42)_____ suspension. Colloidal particles are stabilized in water by electrical (43)_____or attraction to (44)_____. Two foods that are colloidal suspensions are (45)_____. Coagulation is a term given to the process by which colloidal particles (46)_____.

Water and solutes pass through certain membranes by a process called (47)_____. Because of this phenomenon, red blood cells placed in pure water (48)_____, and those placed in a solution containing much more salt than blood plasma contains will (49)_____.

Answers

1. dissolve
2. solution
3. solute
4. solvent
5. insoluble
6. benzene, carbon tetrachloride, perchloroethylene, and acetone
7. hydrogen bonding and the fact that water molecules are dipoles
8. O_2 molecules
9. Na^+ and Cl^- ions
10. H^+ and Cl^- ions
11. hydronium ion, H_3O^+
12. concentration
13. million
14. ppm
15. parts per billion
16. ppb
17. moles
18. liter of solution
19. 17
20. 17 g
21. 0.25 mol/L
22. $M_1 \times V_1 = M_2 \times V_2$
23. 3.75
24. $pH = -\log[H^+]$
25. acidic
26. solubility
27. saturated
28. standard solution
29. titration
30. end point
31. indicator
32. colligative properties
33. lower
34. lower
35. higher
36. reversible
37. acid-base, solubilization (precipitation), complexation, oxidation-reduction
38. an equilibrium constant expression
39. at constant temperature the solubility of a gas in a liquid is proportional to the partial pressure of the gas in contact with the liquid
40. Henry's law

41. $[X(aq)] = K P_X$
42. colloidal
43. charge
44. water
45. milk and mayonnaise
46. aggregate or come together
47. osmosis
48. swell
49. shrink

QUESTIONS AND PROBLEMS

1. Of the following, the **untrue** statement is: (a) Sugar dissolves in water. (b) When this happens, the sugar becomes the solute. (c) A solution is formed. (d) In the solution, the sugar is present as aggregates of molecules in the form of very small microcrystals. (e) The water is the solvent.

2. Is tap water a solution? If so, what does it contain besides water?

3. What does a reaction such as,

   $$Ca^{2+}(aq) + CO_3^{2-}(aq) \rightarrow CaCO_3(s)$$

   illustrate about the role of solutions in chemical reactions?

4. Summarize the industrial uses of solutions.

5. What is the significance of solutions in the environment?

6. Why are volatile solvents especially useful in industrial applications?

7. What are some of the health considerations involved with the use of industrial solvents?

8. Sketch the structure of the water molecule and explain why it is particularly significant in respect to water's solvent properties. Why is the water molecule called a dipole?

9. What are hydrogen bonds? Why are they significant in regard to water's solvent properties?

10. Distinguish among the nature of the solution process and the species in solution for the dissolution of (a) O_2, (b) NaCl, and (c) HCl in water.

11. What is represented by the following species:

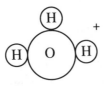

   What is its formula, and what are the formulas of related species? How is it formed? What is it called?

12. Why is a solution formed by dissolving $HCl(g)$ in water not called a "hydrogen chloride" solution?

13. The molecular mass of ammonia, NH_3, is 17. How many grams of ammonia are in 1 liter of 2 M solution?

14. How many grams of ammonia are in 3 liters of 1 M solution?

15. How many grams of ammonia are in 3 liters of 2 M solution?

16. If one wished to prepare 2 liters of 1.2 M NaCl, how many liters of 2 M NaCl would be required?

17. How much of a standard solution of a sodium compound containing 1000 mg of Na^+ per liter should be taken to prepare 5 liters of a standard sodium solution containing 10 mg of sodium per liter?

18. Exactly 1.50 liters of 2.00 M HCl was added to a 2.00-liter volumetric flask, and the volume brought to 2.00 liters with water. What was the molar concentration of HCl in the resulting solution?

19. Give an equation that relates molar concentration, volume of solution, mass of solute, and molar mass of solute.

20. What is the process called when a measured volume of a solution is diluted with water to a specific volume? What remains the same during this process?

21. What are the relationships among the terms *neutral, acidic, basic,* and *pH*?

22. What does a *saturated* solution have to do with the *solubility* of a solute? What is meant by *supersaturated*?

23. What are the common effects of different temperatures upon the solubilities of (a) solids and (b) gases?

24. Explain how a *standard solution, buret,* and *indicator* are used in *titrations.*

25. Exactly 1.00 liter of 1.00 M HCl was added to a 2.00-liter volumetric flask. Next, a total of 0.25 mole of solid NaOH was added. After the chemical reaction between the HCl and the NaOH was complete, water was added to bring the total volume to 2.00 liters. What chemical reaction occurred when the HCl was added? How many moles of HCl were left after this reaction occurred? What was the molar concentration of HCl after the solution was diluted to 2.00 liters?

26. A total of 2.50 liters of 0.100 M NaCl was mixed with 3.00 liters 0.0800 M NaCl. Assume that the volumes of the solutions add together to give 5.50 liters (volumes may not be additive with more concentrated solutions). What was the molar concentration of NaCl in the final solution?

27. What is the purpose of an equation such as

$$M_{HCl} \times \text{Volume HCl} = M_{NaOH} \times \text{Volume NaOH} \tag{7.6.2}$$

in titration calculations?

28. How are *colligative properties* defined and what are three major colligative properties?

29. Why is it desirable to have ethylene glycol mixed with water in an automobile's cooling system even during the summer?

30. Placed in a solution that is about 0.10 molar in sodium chloride, blood cells would be somewhat turgid, or "swollen." What property of solutions does this observation illustrate? Explain.

31. What is an *isotonic solution*?

32. What are the nature and uses of *reverse osmosis*?

33. Consider the following generalized reaction where all species are in solution:

$$aA + bB \leftrightarrows cC + dD$$

What is the significance of the **double** arrows? Explain *solution equilibrium* in the context of this reaction. Be sure to include a discussion of the *equilibrium constant expression.*

34. Match each reaction on the left with the type of equilibrium that it represents on the right, below:

1. $Fe^{3+} + SCN^- \rightleftharpoons FeSCN^{2+}$ (a) Acid-base
2. $AgCl(s) \rightleftharpoons Ag^+ + Cl^-$ (b) Solubility
3. $HAc \rightleftharpoons H^+ + Ac^-$ (c) Oxidation-reduction
4. $MnO_4^- + 5Fe^{2+} + 8H^+ \rightleftharpoons$ (d) Complex ion formation
      $Mn^{2+} + 5Fe^{3+} + 4H_2O$

35. Name, state, and give the mathematical expression for the law that describes the following kind of equilibrium:

$$X(g) \rightleftharpoons X(aq)$$

36. How do reactions such as the ones below affect gas solubility?

$$NH_3 + H_2O \rightleftharpoons NH_4^+ + OH$$

$$SO_2 + HCO_3^- \text{ (From water alkalinity)} \rightleftharpoons CO_2 + HSO_3^-$$

37. What does the equation,

$$P_{O_2} = (1.0000 \text{ atm} - 0.0313 \text{ atm}) \times 0.2095 = 0.2029 \text{ atm}$$

express? What is the "0.0313 atm?"

38. Describe the nature and characteristics of *colloidal particles* that are in colloidal suspension in water. What is there about colloidal particles that causes them to exhibit the *Tyndall effect*?

39. Distinguish among *hydrophilic colloids, hydrophobic colloids*, and *association colloids*. Which consists of *micelles*? What are micelles?

8   CHEMISTRY AND ELECTRICITY

8.1.   CHEMISTRY AND ELECTRICITY

Electricity can have a strong effect on chemical systems. For example, passing electricity through a solution of sodium sulfate in water breaks the water down into H_2 gas and O_2 gas. In Section 6.3 it was discussed how solutions of completely ionized strong electrolytes conduct electricity well, weak electrolytes conduct it poorly, and nonelectrolytes not at all. Electrical current basically involves the activity of electrons, as well as their flow and exchange between chemical species. Since electrons are so important in determining the chemical bonding and behavior of atoms, it is not surprising that electricity is strongly involved with many chemical processes.

The exchange of electrons between chemical species is part of the more general phenomenon of oxidation-reduction, which is defined in more detail in Section 8.2. Oxidation-reduction processes are of particular importance in environmental chemistry. Organic pollutants are degraded and nutrient organic matter utilized for energy and as a carbon source for biomass by oxidation-reduction reactions involving fungi in water and soil. The nature of inorganic species in water and soil depends upon whether the medium is oxidizing (oxygen present, low electron activity) or reducing (O_2 absent, high electron activity). Strong oxidants, such as ozone (O_3), and organic peroxides formed by photochemical processes in polluted atmospheres, are noxious pollutants present in photochemical smog.

The flow of electricity through a chemical system can cause chemical reactions to occur. Similarly, chemical reactions may be used to produce electricity. Such phenomena are called **electrochemical phenomena** and are covered under the category of **electrochemistry**. Examples of electrochemistry abound (Figure 8.1). The transmission of nerve impulses in animals, and even human thought processes, are essentially electrochemical. A dry cell produces electricity from chemical reactions. An automobile storage battery stores electrical energy as chemical energy, and reverses the process when it is discharged. An electrochemical process called electrodialysis can be used for the purification of water. The analytical chemist can use an electrical potential developed at a probe called a glass electrode to measure the pH of water. Obviously, the electrical aspects of chemistry are important; they are covered in this chapter.

8.2.   OXIDATION AND REDUCTION

It has been seen that many chemical reactions can be regarded as the transfer of electrically charged electrons from one atom to another. Such a transfer can be caused to occur through a wire connected to an electrode in contact with atoms that either gain or lose electrons. In that way, electricity can be used to bring about chemical reactions, or chemical reactions can be used to generate electricity. First, however, consider the transfer of electrons between atoms that are in contact with each other.

As shown in Figure 8.2, an oxygen atom, which has a strong appetite for electrons, accepts two valence (outer shell) electrons from a calcium atom, to form a calcium ion, Ca^{2+}, and an oxide ion, O^{2-}. The loss of electrons by the

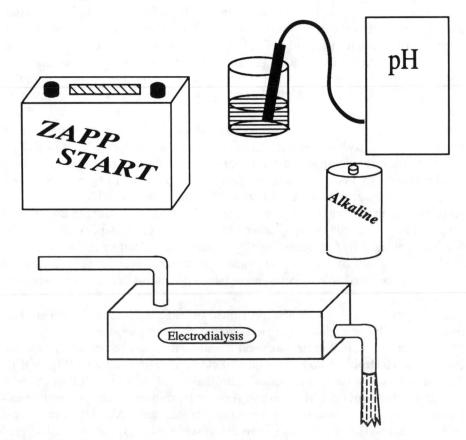

Figure 8.1   The interaction of electricity and chemistry is an important phenomenon in many areas.

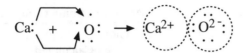

Figure 8.2.   Oxygen atoms have a strong tendency to accept electrons, and an atom which has lost electrons is said to be oxidized.

calcium atom is called oxidation, and it is said to have been oxidized. These terms are derived from the name of oxygen because of its tendency to take electrons from other atoms, such as calcium. The gain of electrons by the oxygen is called **reduction,** and the oxygen atom has been **reduced**; atoms that gain electrons in a transfer such as this are always reduced. The overall reaction is called an **oxidation-reduction reaction,** sometimes abbreviated as **redox**. Oxidation and reduction always occur together. Whenever something is oxidized, something else is always reduced. In this case, Ca is oxidized when O is reduced.

It is easy to see what is meant by oxidation and reduction when there is a transfer of electrons to form ionic bonds (see Section 4.3). However, the concept can be extended to compounds that are bonded together by covalent bonds. For example, when H combines with O to form H_2O,

$$H\cdot   H$$

$$+ \; \ddot{\underset{\cdot\cdot}{O}}\!\!: \; \rightarrow \; H\!:\!\ddot{\underset{\cdot\cdot}{O}}\!: (8.2.1)$$

$$H\cdot$$

electrons are shared between H and O atoms in the covalent bonds joining them together. However, the sharing is not equal because the O atom has a much stronger tendency to grab onto electrons. Therefore, the O atom is regarded as "taking" an electron from each of the H atoms. If O really did take both of these electrons completely away from H, it would have a -2 charge. Although the oxygen does not have a charge as such, the oxygen atom in H_2O is assigned an oxidation number of -2 because of its tendency to attract the 2 electrons from the 2 H atoms. The oxidation number is the hypothetical charge that an atom would have if it gained or lost a particular number of electrons when it formed a chemical compound. In virtually all cases encountered in this book, the oxidation number of chemically combined O is -2. In H_2O, each H atom can be visualized as losing its electron to the O atom because of the unequal sharing. This gives H an oxidation number of +1. In practically all the compounds encountered in this book, the oxidation number of chemically combined H is +1.

The oxidation number of any element is zero in the elemental form. Therefore, the oxidation number of O in O_2 is zero, and the oxidation number of H in H_2 is zero.

Table 8.1.   Oxidation Numbers of Elements in Some Compounds and Ions

Compound or Ion	Element Whose Oxidation Number is to be Calculated	Oxidation Number of Element
SO_2	S	+4
SO_3	S	+6
NH_3	N	-3
CH_4	C	-4
HNO_3	N	+5
Na_3PO_4	P	+5
CO_3^{2-}	C	+4
NO_2^-	N	+3

Figure 8.3 gives the oxidation numbers of the chemically combined forms of the first 20 elements in the periodic table. From this figure, observe that the chemically combined forms of the elements in the far left column (H plus the alkali metals, Li, Na, and K) have an oxidation number of +1. Those in the next column over (the alkaline earths, Be, Mg, and Ca) have an oxidation number of +2 and those in the following column (B and Al) have an oxidation number of +3. It has already been mentioned that chemically combined O almost always has an oxidation number of –2. That of F is always –1. The oxidation number of chemically combined Cl is generally –1. As shown in Figure 8.3, the oxidation numbers of the other elements are variable.

Even though the oxidation numbers of some elements vary, they can usually be figured out by applying the following rules:

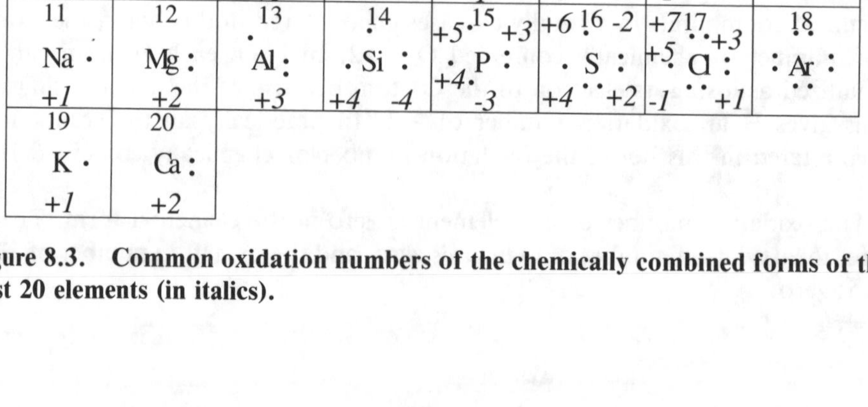

Figure 8.3.   Common oxidation numbers of the chemically combined forms of the first 20 elements (in italics).

- The sum of the oxidation numbers of the elements in a compound equals 0.
- The sum of the oxidation numbers of the elements in an ion equals the charge on the ion.

The application of these rules can be seen in Table 8.1. For each compound in this table, the oxidation number of each element, except for one, is definitely known. For example, in SO_2, each O has an oxidation number of –2. There are 2 O's for a total of –4. The –4 must be balanced with a $+4$ for S. In SO_3, the total of –6 for 6 O's, each with an oxidation number of –2, must be balanced with a $+6$ for S. In Na_3PO_4, each Na has an oxidation number of $+1$, and each O has –2. Therefore, the oxidation number of P is calculated as follows:

$$3 \text{ Na times } +1 \text{ per Na } = +3$$
$$4 \text{ O times } -2 \text{ per O } = -8$$

That leaves a sum of –5 which must be balanced with a $+5$

$$(+3 \text{ for 3 Na}) + (x \text{ for 1 P}) + (-8 \text{ for 4 O}) = 0$$
$$x = \text{oxidation number of P} = 0 -3 + 8 = 5 \qquad (8.2.2)$$

In the NO_2^- ion, each of the 2 O's has an oxidation number of –2, for a total of –4. The whole ion has a net charge of –1, so that the oxidation number of N is $+3$.

8.3. OXIDATION-REDUCTION IN SOLUTION

Many important oxidation-reduction reactions occur involving species dissolved in solution. Many years ago, a struggling college student worked during the summer in a petroleum refinery to earn money for the coming academic year. One of the duties assigned to the labor gang at that time involved carrying a solution of copper sulfate, $CuSO_4$, dissolved in water from one place to another. Castoff steel buckets were used for that purpose. After about the fifth or sixth trip, the bottom would drop out of the bucket, spilling the copper sulfate solution on the ground. This unfortunate phenomenon can be explained by an oxidation-reduction reaction. $CuSO_4$ dissolved in water is present as Cu^{2+} and SO_4^{2-} ions. The Cu^{2+} ion acts as an oxidizing agent and reacts with the iron (Fe) in the bucket as follows:

$$Cu^{2+} + SO_4^{2-} + Fe \rightarrow Fe^{2+} + SO_4^{2-} + Cu \qquad (8.3.1)$$

copper(II) sulfate $\qquad$ iron(II) sulfate

This results in the formation of a solution of iron (II) sulfate and leaves little pieces of copper metal in the bottom of the bucket. A hole is eventually eaten through the side or bottom of the bucket as the iron goes into solution (Figure 8.4).

The kind of reaction just described is used to purify some industrial wastewaters that contain dissolved metal ions by a process called **cementation**. The water is allowed to flow over iron scraps and so-called heavy metal ions, such as Cu^{2+}, Cd^{2+}, and Pb^{2+} precipitate from solution as they are replaced by Fe^{2+}. Cementation results in the replacement of poisonous heavy metal ions in solution by relatively harmless iron.

The reaction of iron and copper sulfate involves the transfer of electrons. These electrons can be forced to go through a wire as electricity and do useful work. To do that, an **electrochemical cell** would be set up, as shown in Figure 8.5. The diagram of this cell shows that Cu and a $CuSO_4$ solution are kept in a container that is separated from a bar of Fe dipping into a solution of $FeSO_4$. These two containers are called **half-cells**. On the left side, Cu^{2+} is reduced to Cu by the **reduction half-reaction**,

$$Cu^{2+} + 2e^- \rightarrow Cu \tag{8.3.2}$$

The electrons required for the reduction of Cu^{2+} are picked up from the bar of copper. On the right side, Fe is oxidized by the **oxidation half-reaction**,

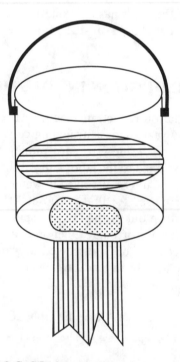

Figure 8.4.   A solution of $CuSO_4$ in a steel bucket soon eats its way through the steel, allowing a leak to occur.

$$Fe \rightarrow Fe^{2+} + 2e^- \qquad (8.3.3)$$

In this case the electrons are left behind on the iron bar, and they go through a wire to the copper bar. In so doing, they may be forced through a light bulb or electric motor and made to do useful work. This is an example of the conversion of chemical energy to electrical energy. The salt bridge shown in Figure 8.5 is just a tube filled with a solution of a salt, such as Na_2SO_4. It completes the circuit by allowing charged ions to move between the two half-cells so that neither has an excess of positive or negative charge. The two half-reactions may be added together,

$$
\begin{array}{l}
Cu^{2+} + 2e^- \rightarrow Cu \\
\underline{Fe \rightarrow Fe^{2+} + 2e^-} \\
Cu^{2+} + Fe \rightarrow Fe^{2+} + Cu
\end{array} \qquad (8.3.4)
$$

to give the total reaction that occurs in the electrochemical cell. Recall that it is the same reaction that occurs when $CuSO_4$ solution contacts iron in a steel bucket (when that reaction was written, SO_4^{2-} was shown as a spectator ion — one that does not take part in the reaction).

In order for an electrical current to flow and light up the light bulb shown in Figure 8.5, there must be a **voltage difference** (difference in electrical potential) between the iron and copper bar. These bars are used to transfer electrons between a wire and solution, and in this case they actually participate in the oxidation-reduction reaction that occurs. The metal bars are called **electrodes**. Because this cell is used to generate a voltage and to extract electricity from a chemical reaction, it is called a **voltaic cell**.

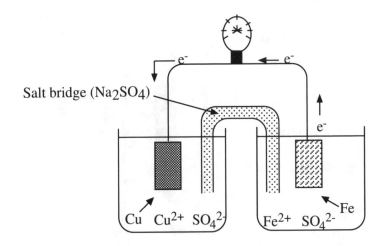

Figure 8.5.   An electrochemical cell

8.4. THE DRY CELL

Dry cells are used as sources of portable electrical energy for flashlights, radios, portable instruments, and many other devices. As shown in Figure 8.6, a dry cell consists of a graphite (carbon) rod in the center of a zinc cylinder, which is filled with a moist paste of manganese dioxide, ammonium chloride, and carbon black. The carbon rod and the zinc cylinder make up the two electrodes in the dry cell. When the two terminals on the electrodes are connected, such as through an electrical circuit powering a portable radio, half-reactions occur at both electrodes. At the graphite electrode the half-reaction is,

$$MnO_2(s) + NH_4^+(aq) + e^- \rightarrow MnO(OH)(s) + NH_3(aq) \qquad (8.4.1)$$

where abbreviations have been used to show which of the materials are solids (s) and which are dissolved in water (aq). This half-reaction takes negatively charged electrons away from the graphite electrode, leaving it with a + charge. It is a reduction half-reaction. The electrode at which reduction occurs is always called the **cathode**. Therefore, the graphite rod is the cathode in a dry cell. The half-reaction that occurs at the zinc electrode is,

$$Zn(s) \rightarrow Zn^{2+}(aq) + 2e^- \qquad (8.4.2)$$

This reaction leaves a surplus of negatively charged electrons on the zinc electrode so that it has a $-$ charge. It is an oxidation reaction. The electrode at which oxidation occurs in an electrochemical cell is always called the **anode**. If the two electrodes are connected, electrons will flow from the zinc anode to the carbon cathode. Such a flow of electrons is an electrical current from which useful work may be extracted. Every time 2 MnO_2 molecules are reduced, as shown in the

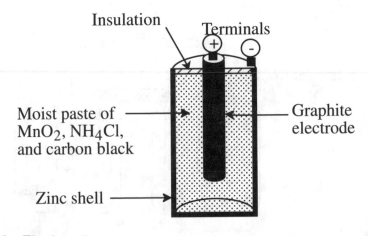

Figure 8.6. The dry cell.

first half-reaction, one Zn is oxidized. Therefore, the overall oxidation-reduction reaction is obtained by multiplying everything in the reduction half-reaction at the cathode by 2 and adding it to the oxidation half-reaction that occurs at the anode. The 2 e⁻ cancel on both sides of the equation yielding the following overall oxidation-reduction reaction:

$$2 MnO_2(s) + 2 NH_4^+(aq) + Zn(s) \rightarrow$$
$$Zn^{2+}(aq) + 2MnO(OH)(s) + 2NH_3(aq) \qquad (8.4.3)$$

8.5.   STORAGE BATTERIES

One of society's generally faithful technological servants is the lead storage battery used in automobiles. This kind of battery provides a steady source of electricity for starting, lights, and other electrical devices on the automobile. Unlike the electrochemical cells already discussed, it can both store electrical energy as chemical energy and convert chemical energy to electrical energy. In the former case it is being charged, and in the latter case it is being discharged.

A 12-volt lead storage battery actually consists of 6 electrochemical cells, each delivering 2 volts. The simplest unit of a lead storage battery that can be visualized consists, in the charged state, of two lead grids, one of which is covered by a layer of lead dioxide, PbO_2. These two electrodes are immersed in a sulfuric acid solution, as shown in Figure 8.7. When the battery is discharged, the pure lead electrode acts as an anode where the oxidation reaction occurs. This reaction is,

$$Pb + SO_4^{2-} \rightarrow PbSO_4 + 2e^- \qquad (8.5.1)$$

in which a layer of solid lead sulfate, $PbSO_4$, is plated onto the lead electrode. The PbO_2 on the other electrode is reduced as part of the cathode reaction:

$$PbO_2 + 4H^+ + SO_4^{2-} + 2e^- \rightarrow PbSO_4 + 2H_2O \qquad (8.5.2)$$

The electrons needed by this half-reaction are those produced at the anode. These electrons are forced to go through the battery leads and through the automobile's electrical system to do useful work. Adding these two half-reactions together gives the overall reaction that occurs as the battery is discharged:

$$Pb + PbO_2 + 4H^+ + 2SO_4^{2-} \rightarrow 2PbSO_4 + 2H_2O \qquad (8.5.3)$$

As the battery is discharged, solid $PbSO_4$ is deposited on both electrodes.

As the battery is charged, everything is reversed. The half-reactions and overall reaction are,

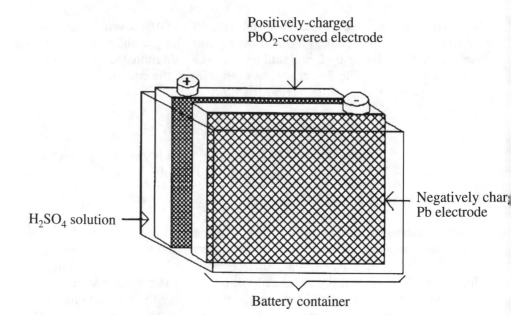

Figure 8.7.   Two electrodes from a lead storage battery.

Cathode: $PbSO_4 + 2e^- \rightarrow Pb + SO_4^{2-}$

Anode: $PbSO_4 + 2H_2O \rightarrow PbO_2 + 4H^+ + SO_4^{2-} + 2e^-$

Overall: $2PbSO_4 + 2H_2O \rightarrow Pb + PbO_2 + 4H^+ + 2SO_4^{2-}$ (8.5.4)

Another kind of battery that is frequently used is the rechargeable nickel-cadmium battery employed in electronic calculators, electronic camera flash attachments, and other applications. During discharge, the reactions that occur in this battery are,

Cathode: $NiO_2 + 2H_2O + 2e^- \rightarrow Ni(OH)_2 + 2OH^-$

Anode: $Cd + 2OH^- \rightarrow Cd(OH)_2 + 2e^-$

Overall: $Cd + NiO_2 + 2H_2O \rightarrow Cd(OH)_2 + Ni(OH)_2$ (8.5.5)

As the battery is being charged, both half-reactions and the overall reaction are simply reversed. Larger versions of the "Nicad" battery are used in some automobiles and motorcycles.

High capacity, readily charged storage batteries are very important to the success of newly developing electrically powered vehicles. The successful development and widespread use of such vehicles would be very helpful in alleviating atmospheric pollution problems, such as carbon monoxide emissions and photochemical smog.

8.6. USING ELECTRICITY TO MAKE CHEMICAL REACTIONS OCCUR

Electricity is commonly used as an energy source to make chemical reactions occur that do not occur by themselves. Such a reaction is called an **electrolytic reaction**. The cell in which it occurs is an **electrolytic cell**. A direct electrical current passing through a solution of a salt in water causes the water to break up and form H_2 and O_2. The cell is shown in Figure 8.8. The process may now be examined in a little more detail. As always, oxidation occurs at the anode.

In this case the external electrical power supply withdraws electrons from the anode, which in turn takes electrons away from water, as shown for the following oxidation half-reaction:

$$2H_2O \rightarrow 4H^+ + O_2(g) + 4e^- \qquad (8.6.1)$$

As a result, oxygen gas, O_2, is given off at the anode. The power supply forces electrons onto the cathode where the following reduction half-reaction occurs

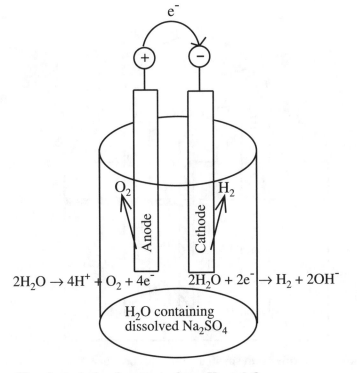

Figure 8.8. The electrolysis of water to form H_2 and O_2.

$$2H_2O + 2e^- \rightarrow H_2(g) + 2OH^- \qquad (8.6.2)$$

As a result, hydrogen gas is produced at the cathode. The H^+ ions produced at the anode and the OH^- ions produced at the cathode combine according to the neutralization reaction,

$$H^+ + OH^- \rightarrow H_2O \qquad (8.6.3)$$

The overall electrolysis reaction is,

$$2H_2O + \text{electrical energy} \rightarrow 2H_2(g) + O_2(g) \qquad (8.6.4)$$

Electrolytic processes are widely used to manufacture chemicals. One of the simplest of these is the Downs' process for manufacturing liquid sodium. This chemically active metal is used to synthesize organic compounds, such as tetra-ethyllead once widely used as a gasoline additive, and in many other applications. It is made by passing an electrical current through melted sodium chloride as shown in Figure 8.9.

Electrons forced onto the iron cathode by the external source of electricity bring about the reduction half-reaction,

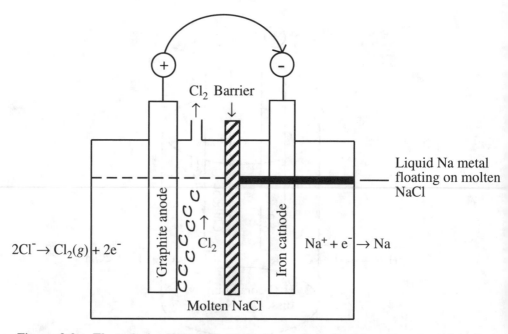

Figure 8.9.   Electrolysis of melted NaCl to produce liquid sodium metal and chlorine gas.

$$Na^+ + e^- \rightarrow Na \qquad\qquad (8.6.5)$$

Withdrawal of electrons from the graphite anode by the external electrical circuit applies the driving force for the oxidation half-reaction:

$$2Cl^- \rightarrow Cl_2(g) + 2e^- \qquad\qquad (8.6.6)$$

The overall oxidation-reduction reaction is,

$$2Na^+ + 2Cl^- \rightarrow 2Na + Cl_2(g) \qquad\qquad (8.6.7)$$

In industrial practice a Downs cell is especially designed to prevent contact of the Na and Cl_2, which would of course react violently with each other.

The Cl_2 that is produced during the manufacture of liquid Na is a valuable by-product that is used to manufacture chlorinated solvents, pesticides, and in other applications. Far more Cl_2 is needed than sodium. More economical processes involving the electrolysis of NaCl solutions in water are used to manufacture Cl_2 along with NaOH.

8.7. ELECTROPLATING

One of the most common uses of electrochemistry is the plating of a thin layer of an expensive or attractive metal onto a base of cheaper metal. The shiny layer of chromium metal that used to be very common on automobile bumpers and other trim before it became too expensive for this application is plated onto steel with an electrochemical process. A thin layer of silver is commonly plated onto "silverware" to make eating utensils look like the "real thing" using an electrochemical process. The use of electrochemistry for plating metals onto surfaces is called **electroplating**.

Many electroplating processes are actually quite complicated. However, an idea of how these processes work can be gained by considering the electroplating of silver metal onto a copper object as shown in Figure 8.10. The copper object and a piece of silver metal make up the two electrodes. They are dipped into a solution containing dissolved silver nitrate, $AgNO_3$, and other additives that result in a smooth deposit of silver metal. Silver ion in solution is reduced at the negatively charged cathode,

$$Ag^+ + e^- \rightarrow Ag \qquad\qquad (8.7.1)$$

leaving a thin layer of silver metal. The silver ion is replaced in solution by oxidation of silver metal at the silver metal anode:

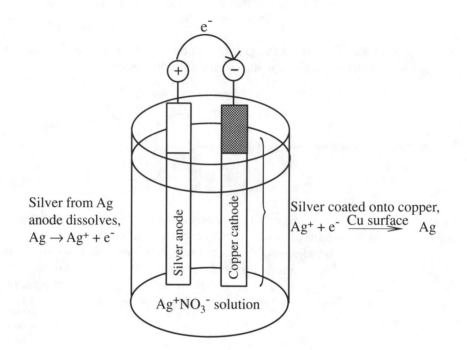

Silver from Ag anode dissolves, $Ag \rightarrow Ag^+ + e^-$

Silver coated onto copper, $Ag^+ + e^- \xrightarrow{Cu\ surface} Ag$

Silver anode

Copper cathode

$Ag^+NO_3^-$ solution

Figure 8.10.   Electrochemical cell for electroplating silver onto copper.

$$Ag \rightarrow Ag^+ + e^- \qquad\qquad (8.7.2)$$

Electroplating is commonly used to prevent corrosion (rusting) of metal. Zinc metal electroplated onto steel prevents rust. The corrosion of "tin cans" is inhibited by a very thin layer of tin plated onto rolled steel.

8.8.   FUEL CELLS

Current methods for the conversion of chemical energy to electrical energy are rather cumbersome and wasteful. Typically, coal is burned in a boiler to generate steam, the steam goes through a turbine, the turbine drives a generator, and the generator produces electricity. This whole process wastes about 60% of the energy originally in the coal. More energy is wasted in transmitting electricity through power lines to users. It is easy to see the desirability of converting chemical energy directly to electricity. This can be done in fuel cells.

A fuel cell that uses the chemical combination of H_2 and O_2 to produce electricity directly from a chemical reaction is shown in Figure 8.11. This device consists of two porous graphite (carbon) electrodes dipping into a solution of potassium hydroxide, KOH. At the anode, H_2 is oxidized, giving up electrons,

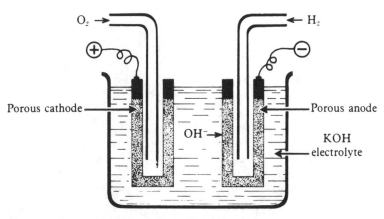

H₂ going through anode reacts to give off electrons: $2\,H_2 + 4\,OH^- \rightarrow 4\,H_2O + 4\,e^-$
O₂ going through cathode reacts to take up electrons: $O_2 + 2\,H_2O + 4\,e^- \rightarrow 4\,OH^-$
Overall reaction: $2\,H_2 + O_2 \rightarrow 2\,H_2O$

Figure 8.11. Fuel cell for direct production of electricity by oxidation of hydrogen.

and at the cathode, O_2 is reduced, taking up electrons. The two half-reactions and the overall chemical reaction are,

$$\text{Anode: } 2H_2 + 4OH^- \rightarrow 4H_2O + 4e^-$$

$$\text{Cathode: } O_2 + 2H_2O + 4e^- \rightarrow 4OH^-$$

$$\text{Overall: } 2H_2 + O_2 \rightarrow 2H_2O \tag{8.8.1}$$

This reaction obviously provides a nonpolluting source of energy. Unfortunately, both hydrogen and oxygen are rather expensive to generate.

8.9. SOLAR CELLS

The direct conversion of light energy to electrical energy may be accomplished in **photovoltaic cells**, also called **solar cells**. These consist basically of two layers of silicon, Si, one **doped** with about one atom per million of arsenic, As, and the other doped with about one atom per million of boron, B. To understand what happens, consider the Lewis symbols of the three elements involved:

$$\cdot \, \overset{\displaystyle \cdot}{\underset{}{Si}} : \qquad \cdot \, \overset{\displaystyle \cdot}{\underset{\displaystyle \cdot}{As}} : \qquad \overset{\displaystyle \cdot}{B} :$$

Each silicon atom has 4 valence electrons. In a crystal of Si, each of the silicon atoms is covalently bonded to 4 other Si atoms, as shown in Figure 8.12. If a B

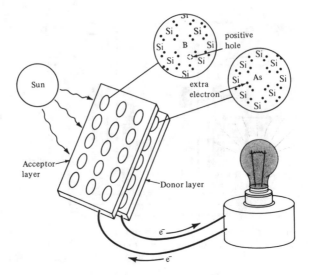

Figure 8.12.   A photovoltaic cell. Note the positive hole, *, at the site of the B atom and the extra electron around the As atom.

atom replaces one of the Si atoms, the 3 valence electrons in B result in a shortage of 1 electron, leaving what is called a **positive hole**. Each As atom has 5 valence electrons, so that replacement of one of the Si atoms by As leaves a surplus of 1 electron, which can move about in the Si crystal. A layer of Si atoms doped with As makes up a **donor** layer and Si doped with B makes up an **acceptor** layer. When two such layers are placed in contact, the positive holes near the surface of the acceptor layer become filled with excess electrons from the donor layer. Nothing else happens unless light falls on the solar cell. This light provides the energy required to push electrons back across the boundary. These electrons can be withdrawn by a wire from the donor layer and returned by a wire to the acceptor layer resulting in a flow of useable electrical current.

8.10.   REACTION TENDENCY

It has already been seen that whole oxidation-reduction reactions can be constructed from half-reactions. The direction in which a reaction goes is a function of the relative tendencies of its constituent half-reactions to go to the right or left. These tendencies, in turn, depend upon the concentrations of the half-reaction reactants and products and their relative tendencies to gain or lose electrons. The latter is expressed by a **standard electrode potential, E^0.** The tendency of the whole reaction to proceed to the right as written is calculated from the **Nernst equation**, which contains both E^0 and the concentrations of the reaction participants. These concepts are explained further in this section and the following section.

Measurement of E^0

To visualize the measurement of E^0, consider the electrochemical cell shown in Figure 8.13. When the two electrodes are connected by an electrical conductor, the reaction,

$$2Ag^+ + H_2 \rightleftharpoons 2H^+ + 2Ag \qquad (8.10.1)$$

occurs in which hydrogen gas would reduce silver ion to silver metal. This reaction is composed of the two following half-reactions:

$$
\begin{array}{lll}
2Ag^+ + 2e^- \rightleftharpoons 2Ag & E^0 = 0.799 \text{ volt} & (8.10.2) \\
-(2H^+ + 2e^- \rightleftharpoons H_2 & E^0 = 0.00 \text{ volt}) & (8.10.3) \\
\hline
2Ag^+ + H_2 \rightleftharpoons 2H^+ + 2Ag & E^0 = 0.799 \text{ volt} & (8.10.1)
\end{array}
$$

If, in the cell shown in Figure 8.13, the activities of both Ag^+ and H^+ were exactly 1 (approximated by concentrations of 1 mol/L) and the pressure of H_2 exactly 1 atm, the potential registered between the two electrodes by a voltmeter, "E," would be 0.799 volt. The platinum electrode, which serves as a conducting surface to exchange electrons, would be negative because of the prevalent tendency for H_2 molecules to leave negatively-charged electrons behind on it as they go into solution as H^+ ions. The silver electrode would be positive because of the

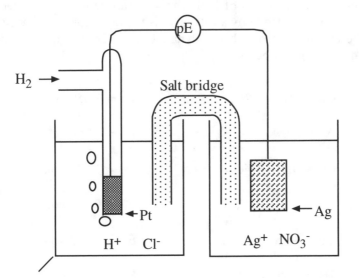

Figure 8.13.   Electrochemical cell in which the reaction $2Ag^+ + H_2 \rightarrow 2Ag + 2H^+$ can be carried out in two half-cells.

prevalent tendency for Ag^+ ions to pick up electrons from it and to be deposited as Ag atoms.

The left electrode shown in Figure 8.13 is of particular importance because it is the standard electrode against which all other electrode potentials are compared. It is called the **standard hydrogen electrode**, SHE, and has been assigned a value of exactly 0 volts by convention; its half-reaction is written as the following:

$$2H^+ + 2e^- \rightleftharpoons H_2 \qquad\qquad E^0 = +0.00 \text{ volts} \qquad (8.10.3)$$

The measured potential of the right-hand electrode in Figure 8.13 versus the standard hydrogen electrode is called the **electrode potential, E.** If Ag^+ and other ions in solution are at unit activity (approximated by a concentration of 1 mole/liter), the potential is the **standard electrode potential, E^0.** The standard electrode potential for the Ag^+/Ag couple is 0.799 volt expressed conventionally as follows:

$$Ag^+ + e^- \rightleftharpoons Ag \qquad\qquad E^0 = 0.799 \text{ volt} \qquad (8.10.2)$$

E^0 Values and Reaction Tendency

Recall that half-reactions can be combined to give whole reactions and that half-reactions can occur in separate half-cells of an electrochemical cell. The inherent tendency of a half-reaction to occur is expressed by a characteristic E^0 value. In favorable cases, E^0s of half-reactions can be measured directly vs. a standard hydrogen electrode by a cell such as the one shown in Figure 8.13, or they can be calculated from thermodynamic data. Several half-reactions and their E^0 values are given below:

$$Cl_2 + 2e^- \rightleftharpoons 2Cl^- \qquad\qquad E^0 = 1.359 \text{ volt} \qquad (8.10.4)$$

$$Ag^+ + e^- \rightleftharpoons Ag \qquad\qquad E^0 = 0.799 \text{ volt} \qquad (8.10.2)$$

$$O_2 + 4H^+ + 4e^- \rightleftharpoons 2H_2O \qquad\qquad E^0 = 1.229 \text{ volt} \qquad (8.10.5)$$

$$Fe^{3+} + e^- \rightleftharpoons Fe^{2+} \qquad\qquad E^0 = 0.771 \text{ volt} \qquad (8.10.6)$$

$$Cu^{2+} + 2e^- \rightleftharpoons Cu \qquad\qquad E^0 = 0.337 \text{ volt} \qquad (8.10.7)$$

$$2H^+ + 2e^- \rightleftharpoons H_2 \qquad\qquad E^0 = 0.00 \text{ volt} \qquad (8.10.3)$$

$$Pb^{2+} + 2e^- \rightleftharpoons Pb \qquad\qquad E^0 = -0.126 \text{ volt} \qquad (8.10.8)$$

$$Zn^{2+} + 2e^- \rightleftarrows Zn \qquad\qquad E^0 = -0.763 \text{ volt} \qquad (8.10.9)$$

Basically, the E^0 values of these half–reactions express the tendency for the reduction half-reaction to occur when all reactants and products are present at unit activity; the more positive the value of E^0, the greater the tendency of the reduction half-reaction to proceed. (In a simplified sense, the activity of a substance is 1 when its concentration in aqueous solution is 1 mole/liter, its pressure as a gas is 1 atm, or it is present as a solid.) With these points in mind, examination of the E^0 values above show the following:

- The highest value of E^0 shown above is for the reduction of Cl_2 gas to Cl^- ion. This is consistent with the strong oxidizing tendency of Cl_2; chlorine much "prefers" to exist as chloride ion rather than highly reactive chlorine gas.
- The comparatively high E^0 value of 0.7994 volt for the reduction of Ag^+ ion to Ag metal indicates that silver is relatively stable as a metal, which is consistent with its uses in jewelry and other applications where resistance to oxidation is important.
- The value of exactly $E^0 = 0.000$ volt is assigned by convention for the half-reaction in which H^+ ion is reduced to H_2 gas; all other E^0s are relative to this value.
- The lowest (most negative) E^0 value shown above is –0.763 volt for the half-reaction $Zn^{2+} + 2e^- \rightleftarrows Zn$. This reflects the strong tendency for zinc to leave the metallic state and become zinc ion; that is, the half-reaction tends to lie strongly to the left. Since zinc metal gives up electrons when it is oxidized to Zn^{2+} ion, zinc metal is a good reducing agent.

Half-reactions and their E^0 values can be used to explain observations such as the following: A solution of Cu^{2+} flows through a lead pipe and the lead acquires a layer of copper metal through the reaction

$$Cu^{2+} + Pb \rightarrow Cu + Pb^{2+} \qquad (8.10.10)$$

This reaction occurs because the copper(II) ion has a greater tendency to acquire electrons than the lead ion has to retain them. This reaction can be obtained by subtracting the lead half-reaction, Equation 8.10.8, from the copper half-reaction, Equation 8.10.7:

$$
\begin{array}{ll}
Cu^{2+} + 2e^- \rightleftarrows Cu & E^0 = 0.337 \text{ volt} \\
-(Pb^{2+} + 2e^- \rightleftarrows Pb & E^0 = -0.126 \text{ volt}) \\
\hline
Cu^{2+} + Pb \rightleftarrows Cu + Pb^{2+} & E^0 = 0.463 \text{ volt}
\end{array}
\qquad (8.10.10)
$$

The appropriate mathematical manipulation of the E^0s of the half-reactions enables calculation of an E^0 for the overall reaction, the positive value of which

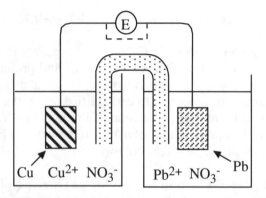

Figure 8.14.   Electrochemical cell in which the tendency for the reaction Cu^{2+} + Pb $\rightleftarrows$ Cu + Pb^{2+} may be measured. In this configuration "E" has a very high resistance and current cannot flow.

indicates that Reaction 8.10.10 tends to go to the right as written. This is in fact what occurs when lead metal directly contacts a solution of copper(II) ion. Therefore, if a waste solution containing cupric ion, a relatively innocuous pollutant, comes into contact with lead in plumbing, toxic lead may go into solution.

Based on the above calculation, if the electrochemical cell shown in Figure 8.14 were set up with activities of both Cu^{2+} and Pb^{2+} at exactly 1 (approximated by concentrations of 1 mole/liter), the potential registered by "E," a meter that measures voltage but does not allow any current to flow, would be 0.463 volts. The lead electrode would be negative because the reaction tendency is for Pb metal to give up negative electrons to the external circuit and go into solution as Pb^{2+} ion, whereas Cu^{2+} ions tend to remove electrons from the copper electrode, giving it a + charge and coming out of solution as Cu metal. The effects of different concentrations on the potential that would be measured in such a cell are discussed in the following section.

8.11.   EFFECT OF CONCENTRATION: NERNST EQUATION

The **Nernst equation** is used to account for the effect of different activities upon electrode potential. Referring to Figure 8.14, if the Cu^{2+} ion concentration is increased with everything else remaining constant, it is readily visualized that the potential of the left electrode will become more positive because the higher concentration of electron-deficient Cu^{2+} ions clustered around it tends to draw electrons from the electrode. Decreased Cu^{2+} ion concentration has the opposite effect. If Pb^{2+} ion concentration in the right electrode is increased, it is "harder" for Pb atoms to leave the Pb electrode as positively charged ions; therefore, there is less of a tendency for electrons to be left behind on the Pb electrode, and its potential tends to be more positive. At a lower value of $[Pb^{2+}]$ in the right half-cell, the opposite is true. Such concentration effects upon E are expressed by the **Nernst equation**. As applied to the reaction in question,

$$Cu^{2+} + Pb \rightleftarrows Cu + Pb^{2+} \qquad E^0 = 0.463 \text{ volt} \qquad (8.11.1)$$

the potential of the cell, E, is given by the Nernst equation,

$$E = E^0 + \frac{2.303RT}{nF} \log \frac{[Cu^{2+}]}{[Pb^{2+}]} = 0.463 + \frac{0.0591}{2} \log \frac{[Cu^{2+}]}{[Pb^{2+}]} \qquad (8.11.2)$$

$$\uparrow$$
$$\text{(at } 25°C)$$

where R is the molar gas constant, T is the absolute temperature, F is the Faraday constant, n is the number of electrons involved in the half-reaction (2 in this case), and the activities are approximated by concentrations. The value of $2.303RT/F$ is 0.0591 at 25°C.

As an example of the application of the Nernst equation, suppose that $[Cu^{2+}]$ = 3.33×10^{-4} mol/L and $[Pb^{2+}]$ 0.0137 mol/L. Substituting into the Nernst equation above gives the following:

$$E = 0.463 \text{ volt} + \frac{0.0591}{2} \log \frac{3.33 \times 10^{-4}}{0.0137} = 0.415 \text{ volt} \qquad (8.11.3)$$

The value of E is still positive and Reaction 8.11.1 still proceeds to the right as written.

8.12. POTENTIOMETRY

Most of this discussion of electrochemistry has dealt with effects involving a flow of electrical current. It has been seen that chemical reactions can be used to produce an electrical current. It has also been seen that the flow of an electrical current through an electrochemical cell can be used to make a chemical reaction occur. Another characteristic of electricity is its voltage, or electrical potential, which was discussed as E and E^0 values above as a kind of the "driving force" behind oxidation-reduction reactions.

As noted above, voltage is developed between two electrodes in an electrochemical cell. The voltage depends upon the kinds and concentrations of dissolved chemicals in the solutions contacted by the electrodes. In some cases this voltage can be used to measure concentrations of some substances in solution. This gives rise to the branch of analytical chemistry known as **potentiometry**. Potentiometry uses **ion-selective electrodes** or **measuring electrodes** whose potentials relative to a **reference electrode** vary with the concentrations of particular ions in solution. The reference electrode that serves as the ultimate standard for potentiometry is the standard hydrogen electrode shown in Figure 8.13. In practice, other electrodes, such as the silver/silver chloride or calomel (mercury metal in contact with Hg_2Cl_2) are used. A reference electrode is hooked to the reference terminal input of a voltmeter and a measuring electrode is hooked to the measuring terminal. To measure a concentration, both electrodes are immersed in the solution being analyzed, the potential of the measuring electrode is read vs. the reference electrode, and this potential is used to calculate an ion concentration from the Nernst equation.

Two very useful ion-selective electrodes are shown in Figure 8.15. The easier of these to understand is the fluoride ion-selective electrode consisting of a disc of lanthanum fluoride, LaF_3, molded into the end of a plastic body and connected to a wire. When this electrode is placed in solution, its potential relative to a reference electrode (an electrode whose potential does not vary with the composition of the solution) shifts more negative with increasing concentrations of F^- ion in solution. By comparing the potential of the fluoride electrode in a solution of unknown F^- concentration with its potential in a solution of known F^- concentration, it is possible to calculate the value of the unknown concentration.

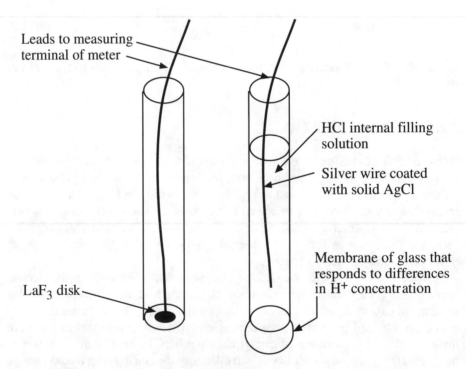

Leads to measuring terminal of meter

HCl internal filling solution

Silver wire coated with solid AgCl

Membrane of glass that responds to differences in H^+ concentration

LaF_3 disk

Figure 8.15. A fluoride ion-selective electrode and a glass electrode used to measure pH.

The potential of the fluoride measuring electrode at 25°C responds according to the Nernst equation in the form,

$$E = E_a - 0.0591 \log [F^-] \tag{8.12.1}$$

where E is the measured potential of the fluoride electrode vs. a reference electrode. (Strictly speaking, E is a function of *activity* of fluoride ion, which is approximated very well by $[F^-]$ at relatively lower concentrations of F^-.) E_a is not a true E^0, but is treated just like E^0 in the Nernst expression. Equation 8.12.1 shows that for every 10-fold increase in $[F^-]$, the potential of the fluoride electrode shifts in a negative direction by 0.0591 volt (59.1 millivolt). Increasing the concentration of a *negatively charged* ion shifts the potential to more *negative* values.

The other electrode shown in Figure 8.15 is the glass electrode used to measure H^+ concentration. Its potential varies with the concentration of H^+ ion in solution. The electrode has a special glass membrane on its end to which H^+ ions tend to adsorb. The more H^+ ions in the solution in which the electrode is dipped, the more that are absorbed to the membrane. This adsorption of *positively charged* ions shifts the potential *positive* with increasing H^+. This shift in potential is detected by a special voltmeter with a high resistance called a **pH meter**. The pH meter makes contact with the glass membrane by way of a silver chloride (AgCl) coated silver wire dipping into a solution of HCl contained inside the electrode. The potential is, of course, measured relative to a reference electrode, also dipping into the solution. Recall from Section 7.5 that pH is equal to the negative logarithm of the hydrogen ion concentration, so that as the potential of the glass electrode becomes more positive, a lower pH is indicated. The Nernst equation that applies to the glass electrode is,

$$E = E_a + 0.0591 \log [H^+] \tag{8.12.2}$$

and, since $pH = -\log [H^+]$, the following applies:

$$E = E_a - 0.0591 \, pH \tag{8.12.3}$$

To measure pH, the pH meter must first be calibrated. This is done by placing the electrodes in a **standard buffer solution** (a buffer solution is one that resists changes in pH with added acid, base, or water, and a standard buffer solution is one made up to an accurately known pH) of known pH. After calibration, the pH meter is then adjusted to read that pH. Next the electrodes are removed from the buffer solution, rinsed to remove any buffer, and placed in the solution of unknown pH. This pH is then read directly from the pH meter.

8.13. CORROSION

Corrosion is defined as *the destructive alteration of metal through interactions with its surroundings*—in short, rust. It is a redox phenomenon resulting from the fact that most metals are unstable in relation to their surroundings. Thus the steel in cars really prefers to revert back to the iron oxide ore that it came from by reacting with oxygen in the air. The corrosion process is accelerated by exposure to water, which may contain corrosive salt placed on road ice or acid from acid rain. It is only by the application of anticorrosive coatings and careful maintenance that the process is slowed down.

Corrosion occurs when an electrochemical cell is set up on a metal surface as shown in Figure 8.16. A layer of moisture serves to dissolve ions and act as a salt bridge between the anode and cathode. The area in which the metal is oxidized is the anode. Typically, when iron is corroded, the anode reaction is,

$$Fe(s) \rightarrow Fe^{2+}(aq) + 2e^- \qquad (8.13.1)$$

so that the solid iron goes into solution as Fe^{2+} ion. Several cathode reactions are possible. Usually oxygen is involved. A typical cathode reaction is,

$$O_2(g) + 4H^+(aq) + 4e^- \rightarrow 2H_2O \qquad (8.13.2)$$

The overall corrosion process is normally very complicated and involves a number of different reactions. Very commonly bacteria are involved in corrosion. The bacterial cells derive energy by acting as catalysts in the corrosion reactions.

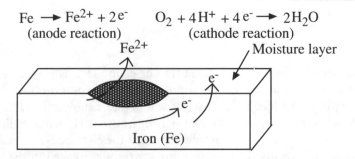

Figure 8.16.   Electrochemical cell on a metal surface on which corrosion occurs.

CHAPTER SUMMARY

The chapter summary below is presented in a programmed format to review the main points covered in this chapter. It is used most effectively by filling in the blanks, referring back to the chapter as necessary. The correct answers are given at the end of the summary.

Passing electricity through a solution of sodium sulfate in water causes the water to (1)_____.
To a large extent, the nature of inorganic species in water depends upon whether the water is (2)_____ or (3)_____.

The study of the interaction of chemistry and electricity is called (4)_____.
The loss of electrons by an atom is called (5)_____, and the atom is said to have been (6)_____. A gain of electrons by an atom is called (7)_____, and the atom is said to have been (8)_____.
The hypothetical charge that an atom would have if it gained or lost a particular number of electrons in forming a chemical compound is called its (9)_____. For chemically combined O the oxidation number is almost always (10)_____ and for chemically combined H it is almost always (11)_____. The oxidation number of any element in its elemental form is (12)_____. The sum of the oxidation numbers of the elements in a compound equals (13)_____, and the sum of the oxidation numbers of the elements in an ion equals (14)_____.

In chemical compounds, Na, Ca, and F have oxidation numbers of (15)_____, _____, and _____, respectively.

When a solution containing Cu^{2+} ions comes into contact with iron metal, the products are (16)_____. As a result of this process Fe metal is (17)_____ and Cu^{2+} ion is (18)_____. A device designed to carry out a reduction half-reaction and an oxidation half-reaction in physically separate locations such that electrical energy may be extracted is called (19)_____. Metal or graphite bars which transfer electrons between wires and the solution in an electrochemical cell are called (20)_____. The one of these at which reduction occurs is called the (21)_____, and the one at which oxidation occurs is the (22)_____. The function of a salt bridge is to (23)_____. In a dry cell the cathode consists of (24)_____ and the anode is (25)_____. The overall reaction in a dry cell is (26)_____

_____.

A reaction that is made to occur by the passage of electrical current through a solution is called (27)_____ and it takes place in a device called (28)_____. When an electrical current is passed through a solution of melted NaCl, (29)_____ is produced at the graphite anode and (30)_____ is produced at the iron cathode. When an automobile battery is discharged, (31)_____

_____ is oxidized at the anode and (32)_____ is reduced at the cathode. The device in which the overall reaction, $Cd + NiO_2 + 2 H_2O \rightarrow Cd(OH)_2 + Ni(OH)_2$ occurs is the (33)_____. A device in which H_2 and O_2 are used for the direct production of electrical energy from a chemical reaction is a (34)_____. The direct conversion of light energy to electrical energy is accomplished in a (35)_____. Such a device consists of two layers known as the (36)_____ and the (37)_____. The process by which electricity is used to plate a layer of metal onto an object made from another metal is called (38)_____.

In a fuel cell powered with H_2 and O_2, electrons are produced at the anode by oxidation of (39)_____ and electrons are taken up at the cathode by reduction of (40)_____. In a solar cell, doping silicon with boron atoms produces (41)_____ and doping with arsenic atom produces (42)_____.

The tendency of a half-reaction to go to the right or to the left as written is expressed by a (43)_____. Account is taken of the influence of differences in concentration on this tendency by the (44)_____.

The standard electrode against which all other electrode potentials are compared is called the (45)_____ for which the assigned E^0 value is (46)_____ and the half-reaction is (47)_____ _____. The measured potential of an electrode *vs.* the standard hydrogen electrode when the activities of all the reaction constituents are exactly 1 is called the (48)_____. The fact that the reaction $Cu^{2+} + Pb \rightleftharpoons Cu + Pb^{2+}$ goes to the right as written indicates that E^0 for the half-reaction $Cu^{2+} + 2e^- \rightleftharpoons Cu$ has a (49)_____ value than E^0 for the half-reaction $Pb^{2+} + 2e^- \rightleftharpoons Pb$.

The Nernst equation applied to the reaction $Cu^{2+} + Pb \rightleftharpoons Cu + Pb^{2+}$ ($E^0 = 0.463$ volt) at $25°C$ is (50)_____.

The branch of analytical chemistry in which the voltage developed by an electrode is used to measure the concentration of an ion in solution is called (51)_____. The electrodes used for this purpose are called by the general name of (52)_____ and they must always be used with a (53)_____. The voltage developed at a glass membrane is used to measure (54)_____ or (55)_____. Before measurement of an unknown pH, a pH meter and its electrode system must be calibrated using a known (56)_____.

The destructive alteration of metal through interactions with its surroundings is known as (57)_____. In the electrochemical cell typically involved with this phenomenon, the metal is oxidized at the (58)_____.

Answers

1. break down into H_2 and O_2.
2. oxidizing
3. reducing
4. Electrochemistry
5. oxidation
6. oxidized
7. reduction

8. reduced
9. oxidation number
10. –2
11. +1
12. 0
13. 0
14. the charge on the ion
15. +1, +2, –1
16. Cu metal and Fe^{2+} ion in solution
17. oxidized
18. reduced
19. an electrochemical cell
20. electrodes
21. cathode
22. anode
23. allow for transfer of ions between half-cells
24. a graphite rod
25. a zinc cylinder
26. $2MnO_2(s) + 2NH_4^+(aq) + Zn(s) \rightarrow Zn^{2+}(aq) + 2MnO(OH)(s) + 2NH_3(aq)$
27. an electrolytic reaction
28. an electrolytic cell
29. chlorine gas
30. liquid sodium
31. Pb metal
32. PbO_2
33. nickel-cadmium storage battery
34. fuel cell
35. solar cell
36. donor layer
37. acceptor layer
38. electroplating
39. H_2
40. O_2
41. holes
42. a donor layer
43. standard electrode potential, E^0
44. Nernst equation
45. standard hydrogen electrode, SHE
46. 0.000 volts
47. $2H^+ + 2e^- \rightleftharpoons H_2$
48. standard electrode potential, E^0
49. higher

50. $E = 0.463 + \dfrac{0.0591}{2} \log \dfrac{[Cu^{2+}]}{[Pb^{2+}]}$

51. potentiometry
52. ion-selective electrodes
53. reference electrode
54. H^+ ion concentration
55. pH
56. buffer solution
57. corrosion
58. anode

QUESTIONS AND PROBLEMS

1. Which element has such a "strong appetite" for electrons that its name is the basis for the term used to describe loss of electrons?

2. Give the oxidation number of each of the elements marked with an asterisk in the following species: Pb^*O_2, $N^*H_4{}^+$, $Cl^*{}_2$, CaS^*O_4, $Cd^*(OH)_2$

3. For what purpose is a Downs cell used?

4. How is cementation used to purify water?

5. Small mercury batteries are often used in cameras, wristwatches and similar applications. The chemical species involved in a mercury battery are Zn, ZnO, Hg, and HgO, where Zn is the chemical symbol for zinc and Hg is that of mercury. If one were to coat a moist paste of HgO onto Zn, a reaction would occur in which Hg and Zn are produced. From this information write the cathode, anode, and overall reactions involved in the discharge of a mercury battery. You may use the oxide ion, O^{2-}, in the half-reactions.

6. What are the major ingredients of the paste used to fill a dry cell?

7. Pure water does not conduct electricity. However, by passing an electrical current through water, pure H_2 and O_2 may be prepared. What is done to the water to make this possible?

8. In this chapter it was mentioned that Cl_2 gas could be prepared by electrolysis of a solution of NaCl in water. Elemental Na reacts with water to give NaOH and H_2 gas. Recall that the electrolysis of melted NaCl gives Na metal and Cl_2 gas. From this information give the cathode, anode, and overall reactions for the electrolysis of a solution of NaCl in water.

9. What is the definition of cathode? What is the definition of anode?

10. Give the cathode, anode, and overall reactions when a nickel-cadmium battery is charged.

11. From a knowledge of automobile exhausts and coal-fired power plants, justify the statement that "a fuel-cell is a nonpolluting source of energy."

12. What is meant by a positive hole in the acceptor layer of Si in a solar cell? What distinguishes the donor layer?

13. From the E^0 values given for half-reactions in Section 8.10, give the E^0 values of the following whole reactions:

(A) $2Fe^{3+} + Pb \rightleftharpoons 2Fe^{2+} + Pb^{2+}$
(B) $2Cl_2 + 2H_2O \rightleftharpoons 4Cl^- + O_2 + 4H^+$
(C) $2Ag^+ + Zn \rightleftharpoons 2Ag + Zn^{2+}$
(D) $2Fe^{3+} + Pb \rightleftharpoons 2Fe^{2+} + Pb^{2+}$

14. How may it be deduced that metallic zinc and metallic lead displace H from strong acid whereas metallic copper and silver do not?

15. Given $Fe^{3+} + e^- \rightleftharpoons Fe^{2+}$, $E^0 = 0.771$ volt, and $Ce^{4+} + e^- \rightleftharpoons Ce^{3+}$, $E^0 = 0.771$ volt, calculate E^0 for the reaction $Ce^{4+} + Fe^{2+} \rightleftharpoons Ce^{3+} + Fe^{3+}$. Using the Nernst equation calculate E for this reaction when $[Ce^{4+}] = 9.00 \times 10^{-3}$ mol/L, $[Ce^{3+}] = 1.25 \times 10^{-3}$ mol/L, $[Fe^{3+}] = 8.60 \times 10^{-4}$ mol/L, and $[Fe^{2+}] = 2.00 \times 10^{-2}$ mol/L.

16. In millivolts (mv) the Nernst equation (Equation 8.12.1) applied to the fluoride ion-selective electrode is $E = E_a - 59.1 \log [F^-]$. Suppose that the potential of a fluoride electrode vs. a reference electrode is -88.3 mv in a solution that is 1.37×10^{-4} mol/L in F^-. What is the potential of this same electrode system in a solution for which $[F^-] = 4.68 \times 10^{-2}$ mol/L?

17. Suppose that the potential of a fluoride electrode vs. a reference electrode is -61.2 mv in a solution that is 5.00×10^{-5} mol/L in F^-. What is the concentration of fluoride in a solution in which the fluoride electrode registers a potential of -18.2 mv?

18. A glass electrode has a potential of 33.3 mv vs. a reference electrode in a medium for which $[H^+] = 1.13 \times 10^{-6}$ mol/L. What is the value $[H^+]$ in a solution in which the potential of the electrode registers 197.0 mv?

19. A glass electrode has a potential of 127 mv vs. a reference electrode in a pH 9.03 buffer. What is the pH of a solution in which the electrode registers a potential of 195 mv?

20. Consider an electrochemical cell in which a standard hydrogen electrode is connected by way of a salt bridge to a lead electrode in contact with a Pb^{2+} solution. How does the potential of the lead electrode change when the concentration of Pb^{2+} ion increases? Explain.

9   ORGANIC CHEMISTRY

9.1. ORGANIC CHEMISTRY

Most carbon-containing compounds are **organic chemicals** and are addressed by the subject of **organic chemistry**. Organic chemistry is a vast, diverse, discipline because of the enormous number of organic compounds that exist as a consequence of the versatile bonding capabilities of carbon.[1] Such diversity is due to the ability of carbon atoms to bond to each other through single (2 shared electrons) bonds, double (4 shared electrons) bonds, and triple (6 shared electrons) bonds, in a limitless variety of straight chains, branched chains, and rings.

Among organic chemicals are included the majority of important industrial compounds, synthetic polymers, agricultural chemicals, biological materials, and most substances that are of concern because of their toxicities and other hazards. Pollution of the water, air, and soil environments by organic chemicals is an area of significant concern.

Chemically, most organic compounds can be divided among hydrocarbons, oxygen-containing compounds, nitrogen-containing compounds, sulfur-containing compounds, organohalides, phosphorus-containing compounds, or combinations of these kinds of compounds. Each of these classes of organic compounds is discussed briefly here.

All organic compounds of course contain carbon. Virtually all also contain hydrogen and have at least one C-H bond. The simplest organic compounds, and those easiest to understand, are those that contain only hydrogen and carbon. These compounds are called **hydrocarbons** and are addressed first among the organic compounds discussed in this chapter. Hydrocarbons are used here to illustrate some of the most fundamental points of organic chemistry, including organic formulas, structures, and names.

Molecular Geometry in Organic Chemistry

The three-dimensional shape of a molecule; that is, its molecular geometry, is particularly important in organic chemistry. This is because its molecular geometry determines in part the properties of an organic molecule, particularly its interactions with biological systems. Shapes of molecules are represented in drawings by lines of normal, uniform thickness for bonds in the plane of the paper, broken lines for bonds extending away from the viewer, and heavy lines for bonds extending toward the viewer. These conventions are shown by the example of dichloromethane, CH_2Cl_2, an important organochloride solvent and extractant, illustrated in Figure 9.1.

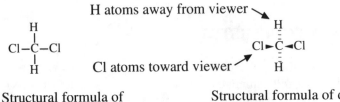

Structural formula of
dichloromethane in
two dimensions

Structural formula of dichloromethane represented in three
dimensions

Figure 9.1.   Structural formulas of dichloromethane, CH_2Cl_2; the formula on the right provides a three-dimensional representation.

9.2.   HYDROCARBONS

As noted above, hydrocarbon compounds contain only carbon and hydrogen. The major types of hydrocarbons are alkanes, alkenes, alkynes, and aryl compounds. Examples of each are shown in Figure 9.2.

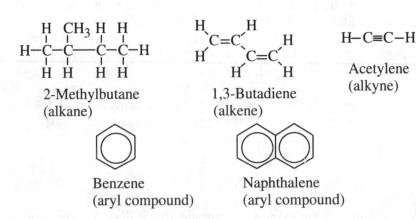

2-Methylbutane
(alkane)

1,3-Butadiene
(alkene)

Acetylene
(alkyne)

Benzene
(aryl compound)

Naphthalene
(aryl compound)

Figure 9.2.   Examples of major types of hydrocarbons.

Alkanes

Alkanes, also called **paraffins** or **aliphatic hydrocarbons**, are hydrocarbons in which the C atoms are joined by single covalent bonds (sigma bonds) consisting of two shared electrons (see Section 1.3). Some examples of alkanes are shown in Figure 9.2. As with other organic compounds, the carbon atoms in alkanes may form straight chains or branched chains. These three kinds of alkanes are, respectively, **straight-chain alkanes, branched-chain alkanes,** and **cycloalkanes**. As shown in Figure 9.2, a typical branched chain alkane is 2-methylbutane, a volatile, highly flammable liquid. It is a component of gasoline, which may explain why it is commonly found as an air pollutant in urban air. The general molecular formula for straight- and branched-chain alkanes is C_nH_{2n+2}, and that of cyclic alkanes is C_nH_{2n}. The four hydrocarbon molecules in Figure 9.3 contain 8 carbon atoms each. In one of the molecules, all of the carbon atoms are in a straight chain and in two they are in branched chains, whereas in a fourth, 6 of the carbon atoms are in a ring.

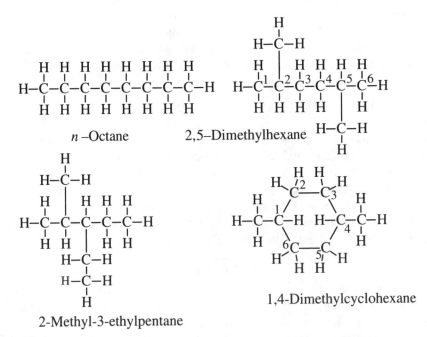

Figure 9.3. Structural formulas of four hydrocarbons, each containing 8 carbon atoms, that illustrate the structural diversity possible with organic compounds. Numbers used to denote locations of atoms for purposes of naming are shown on two of the compounds.

Formulas of Alkanes

Formulas of organic compounds present information at several different levels of sophistication. **Molecular formulas,** such as that of octane (C_8H_{18}), give the number of each kind of atom in a molecule of a compound. As shown in Figure 9.3, however, the molecular formula of C_8H_{18} may apply to several alkanes, each one of which has unique chemical, physical, and toxicological properties. These different compounds are designated by **structural formulas** showing the order in which the atoms in a molecule are arranged. Compounds that have the same molecular, but different structural formulas are called **structural isomers.** Of the compounds shown in Figure 9.3, *n*-octane, 2,5-dimethylhexane, and 2-methyl-3–ethylpentane are structural isomers, all having the formula C_8H_{18}, whereas 1,4-dimethylcyclohexane is not a structural isomer of the other three compounds because its molecular formula is C_8H_{16}.

Alkanes and Alkyl Groups

Most organic compounds can be derived from alkanes. In addition, many important parts of organic molecules contain one or more alkane groups minus a hydrogen atom bonded as substituents onto the basic organic molecule. As a consequence of these factors the names of many organic compounds are based upon alkanes and it is useful to know the names of some of the more common alkanes and substituent groups derived from them as shown in Table 9.1.

Names of Alkanes and Organic Nomenclature

Systematic names, from which the structures of organic molecules can be deduced, have been assigned to all known organic compounds. The more common organic compounds, including many toxic and hazardous organic substances, likewise have **common names** that have no structural implications. Although it is not possible to cover organic nomenclature in any detail in this chapter, the basic approach to nomenclature (naming) is presented in the chapter along with some pertinent examples. The simplest approach is to begin with names of alkanes.

Consider the alkanes shown in Figure 9.3. The fact that *n*-octane has no side chains is denoted by "*n*", that it has 8 carbon atoms is denoted by "oct," and that it is an alkane is indicated by "ane." The names of compounds with branched chains or atoms other than H or C attached make use of numbers that stand for positions on the longest continuous chain of carbon atoms in the molecule. This convention is illustrated by the second compound in Figure 9.3. It gets the hexane part of the name from the fact that it is an alkane with 6 carbon atoms in its longest continuous chain ("hex" stands for 6). However, it has a methyl group (CH_3) attached on the second carbon atom of the chain and another on the fifth.

Table 9.1.   Some Alkanes and Substituent Groups Derived from Them

Alkane	Substituent groups derived from alkane

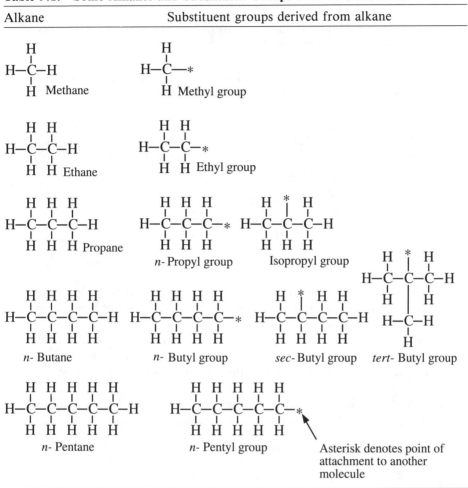

Asterisk denotes point of attachment to another molecule

Hence the full systematic name of the compound is 2,5-dimethylhexane, where "di" indicates two methyl groups. In the case of 2-methyl-3–ethylpentane, the longest continuous chain of carbon atoms contains 5 carbon atoms, denoted by *pen*tane, a methyl group is attached to the second carbon atom, and an ethyl group, C_2H_5, on the third carbon atom. The last compound shown in the figure has 6 carbon atoms in a ring, indicated by the prefix "cyclo," so it is a cyclo*hex*-ane compound. Furthermore, the carbon in the ring to which one of the methyl groups is attached is designated by "1" and another methyl group is attached to the fourth carbon atom around the ring. Therefore, the full name of the compound is 1,4-dimethylcyclohexane.

Summary of Organic Nomenclature as Applied to Alkanes

Naming relatively simple alkanes is a straightforward process. The basic rules to be followed are the following:

1. The name of the compound is based upon the longest continuous chain of carbon atoms. (The structural formula may be drawn so that this chain is not immediately obvious.)
2. The carbon atoms in the longest continuous chain are numbered sequentially from one end. The end of the chain from which the numbering is started is chosen to give the lower numbers for substituent groups in the final name. For example, the compound,

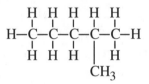

   could be named 4-methylpentane (numbering the 5-carbon chain from the left), but should be named 2-methylpentane (numbering the 5-carbon chain from the right).
3. All groups attached to the longest continuous chain are designated by the number of the carbon atom to which they are attached and by the name of the substituent group ("2-methyl" in the example cited in Step 2, above).
4. A prefix is used to denote multiple substitutions by the same kind of group. This is illustrated by 2,2,3–trimethylpentane

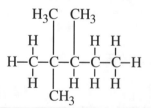

   in which the prefix *tri* is used to show that *three* methyl groups are attached to the pentane chain.
5. The complete name is assigned so that it denotes the longest continuous chain of carbon atoms and the name and location on this chain of each substituent group.

Reactions of Alkanes

Alkanes contain only C-C and C-H bonds, both of which are relatively strong. For that reason they have little tendency to undergo many kinds of reactions common to some other organic chemicals, such as acid-base reactions or low-temperature oxidation-reduction reactions. However, at elevated temperatures alkanes readily undergo oxidation, more specifically combustion, with molecular oxygen in air as shown by the following reaction of propane:

$$C_3H_8 + 5O_2 \rightarrow 3CO_2 + 4H_2O + heat \tag{9.2.1}$$

Common alkanes are highly flammable and the more volatile lower molecular mass alkanes form explosive mixtures with air. Furthermore, combustion of alkanes in an oxygen-deficient atmosphere or in an automobile engine produces significant quantities of carbon monoxide, CO, the toxic properties of which are discussed in Section 21.8.

In addition to combustion, alkanes undergo **substitution reactions** in which one or more H atoms on an alkane are replaced by atoms of another element. The most common such reaction is the replacement of H by chlorine, to yield **organochlorine** compounds. For example, methane reacts with chlorine to give chloromethane. This reaction begins with the dissociation of molecular chlorine, usually initiated by ultraviolet electromagnetic radiation:

$$Cl_2 + UV \text{ energy} \rightarrow Cl\cdot + Cl\cdot \tag{9.2.2}$$

The Cl· product is a **free radical** species in which the chlorine atom has only 7 outer shell electrons as shown by the Lewis symbol,

instead of the favored octet of 8 outer-shell electrons. In gaining the octet required for chemical stability, the chlorine atom is very reactive. It abstracts a hydrogen from methane,

$$Cl\cdot + CH_4 \rightarrow HCl + CH_3\cdot \tag{9.2.3}$$

to yield HCl gas and another reactive species with an unpaired electron, $CH_3\cdot$, called methyl radical. The methyl radical attacks molecular chlorine,

$$CH_3\cdot + Cl_2 \rightarrow CH_3Cl + Cl\cdot \tag{9.2.4}$$

to give the chloromethane (CH_3Cl) product and regenerate Cl·, which can attack additional methane as shown in Reaction 9.2.3. The reactive Cl· and $CH_3\cdot$ species continue to cycle through the two preceding reactions.

The reaction sequence shown above illustrates three important aspects of chemistry that will be shown to be very important in the discussion of atmospheric chemistry in Section 16.4. The first of these is that a reaction may be initiated by a **photochemical process** in which a photon of "light" (electromagnetic radiation) energy produces a reactive species, in this case the Cl· atom. The second point illustrated is the high chemical reactivity of **free radical species** with unpaired electrons and incomplete octets of valence electrons. The third point illustrated is that of **chain reactions**, which can multiply many-fold the effects of a single reaction-initiating event, such as the photochemical dissociation of Cl_2.

Alkenes and Alkynes

Alkenes or **olefins** are hydrocarbons that have double bonds consisting of 4 shared electrons. The simplest and most widely manufactured alkene is ethylene,

$$
\begin{array}{ccc}
H & & H \\
\diagdown & & \diagup \\
& C = C & \\
\diagup & & \diagdown \\
H & & H
\end{array}
\qquad \text{Ethylene (ethene)}
$$

used for the production of polyethylene polymer. Another example of an important alkene is 1,3–butadiene (Figure 9.2), widely used in the manufacture of polymers, particularly synthetic rubber. The lighter alkenes, including ethylene and 1,3–butadiene, are highly flammable and form explosive mixtures with air. This was illustrated tragically by a massive explosion and fire involving leaking ethylene that destroyed a 20-billion pound per year Phillips Petroleum Company plastics plant in Pasadena, Texas, on October 23, 1989, killing more than 20 workers.

Acetylene (Figure 9.2) is an **alkyne**, a class of hydrocarbons characterized by carbon-carbon triple bonds consisting of 6 shared electrons. Highly flammable acetylene is used in large quantities as a chemical raw material and fuel for oxyacetylene torches. It forms dangerously explosive mixtures with air.

Addition Reactions

The double and triple bonds in alkenes and alkynes have "extra" electrons capable of forming additional bonds. Therefore, the carbon atoms attached to these bonds can add atoms without losing any atoms already bonded to them and the multiple bonds are said to be **unsaturated**. Therefore, alkenes and alkynes both undergo **addition reactions** in which pairs of atoms are added across unsaturated bonds as shown in the reaction of ethylene with hydrogen to give ethane:

$$\underset{H}{\overset{H}{\diagdown}}C=C\underset{H}{\overset{H}{\diagup}} \; + \; H-H \; \longrightarrow \; H-\underset{\underset{H}{|}}{\overset{\overset{H}{|}}{C}}-\underset{\underset{H}{|}}{\overset{\overset{H}{|}}{C}}-H \qquad (9.2.5)$$

This is an example of a **hydrogenation reaction**, a very common reaction in organic synthesis, food processing (manufacture of hydrogenated oils), and petroleum refining. Another example of an addition reaction is that of HCl gas with acetylene to give vinyl chloride:

$$\underset{H}{\overset{H}{\diagdown}}C=C\underset{H}{\overset{H}{\diagup}} \; + \; H-Cl \; \longrightarrow \; H-\underset{\underset{H}{|}}{\overset{\overset{H}{|}}{C}}-\underset{\underset{H}{|}}{\overset{\overset{H}{|}}{C}}-Cl \qquad (9.2.6)$$

This kind of reaction, which is not possible with alkanes, adds to the chemical and metabolic versatility of compounds containing unsaturated bonds and is a factor contributing to their generally higher toxicities. It makes unsaturated compounds much more chemically reactive, more hazardous to handle in industrial processes, and more active in atmospheric chemical processes, such as smog formation (see Chapter 18).

Alkenes and *cis-trans* Isomerism

As shown by the two simple compounds in Figure 9.4, the two carbon atoms connected by a double bond in alkenes cannot rotate relative to each other. For this reason, another kind of isomerism, called ***cis-trans*** isomerism, is possible for alkenes. *Cis-trans* isomers have different parts of the molecule oriented differently in space, although these parts occur in the same order. Both alkenes illustrated in Figure 9.4 have a molecular formula of C_4H_8. In the case of *cis*-2-butene, the two CH_3 (methyl) groups attached to the $C=C$ carbon atoms are on the same side of the molecule, whereas in *trans*-2-butene they are on opposite sides.

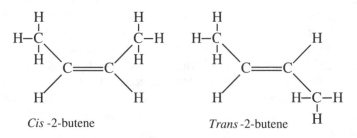

Cis -2-butene   Trans -2-butene

Figure 9.4.   *Cis* and *trans* isomers of the alkene, 2-butene.

Condensed Structural Formulas

To save space, structural formulas are conveniently abbreviated as **condensed structural formulas**. The condensed structural formula of 2-methyl-3–ethylpentane is $CH_3CH(CH_3)CH(C_2H_5)CH_2CH_3$ where the CH_3 (methyl) and C_2H_5 (ethyl) groups are placed in parentheses to show that they are branches attached to the longest continuous chain of carbon atoms, which contains 5 carbon atoms. It is understood that each of the methyl and ethyl groups is attached to the carbon immediately preceding it in the condensed structural formula (methyl attached to the second carbon atom, ethyl to the third).

As illustrated by the examples in Figure 9.5, the structural formulas of organic molecules may be represented in a very compact form by lines and by figures such as hexagons. The ends and intersections of straight line segments in these formulas indicate the locations of carbon atoms. Carbon atoms at the terminal ends of lines are understood to have three H atoms attached, C atoms at the intersections of two lines are understood to have *two* H atoms attached to each, *one* H atom is attached to a carbon represented by the intersection of three lines, and *no* hydrogen atoms are bonded to C atoms where four lines intersect. Other atoms or groups of atoms, such as the Cl atom or OH group, that are substituted for H atoms are shown by their symbols attached to a C atom with a line.

Aromatic Hydrocarbons

Benzene (Figure 9.6) is the simplest of a large class of **aromatic** or **aryl** hydrocarbons. Many related compounds have substituent groups containing atoms of elements other than hydrogen and carbon and are called **aryl compounds** or **aromatic compounds**. Most aromatic compounds discussed in this book contain 6-carbon-atom benzene rings as shown for benzene, C_6H_6, in Figure 9.6. Aromatic compounds have ring structures and are held together in part by particularly stable bonds that contain delocalized clouds of so-called π (pi, pronounced "pie") electrons. In an oversimplified sense the structure of benzene can be visualized as resonating between the two equivalent structures shown on the left in Figure 9.6 by the shifting of electrons in chemical bonds. This structure can be shown more simply and accurately by a hexagon with a circle in it.

Aryl compounds, also called aromatic compounds or arenes, have special characteristics of **aromaticity**, which include a low hydrogen:carbon atomic ratio; $C-C$ bonds that are quite strong and of intermediate length between such bonds in alkanes and those in alkenes; tendency to undergo substitution reactions rather than the addition reactions characteristic of alkenes; and delocalization of π electrons over several carbon atoms. The last phenomenon adds substantial stability to aromatic compounds and is known as **resonance stabilization**.

Many toxic substances, environmental pollutants, and hazardous waste com-

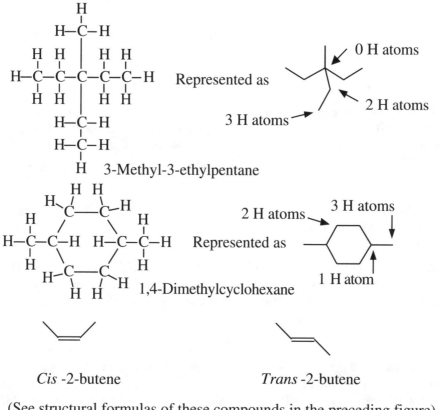

3-Methyl-3-ethylpentane

1,4-Dimethylcyclohexane

Cis -2-butene   *Trans* -2-butene

(See structural formulas of these compounds in the preceding figure)

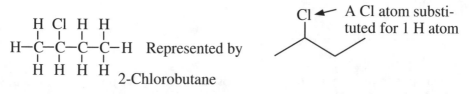

2-Chlorobutane

Figure 9.5.   Representation of structural formulas with lines. A carbon atom is understood to be at each corner and at the end of each line. The numbers of hydrogen atoms attached to carbons at several specific locations are shown with arrows.

pounds, such as benzene, toluene, naphthalene, and chlorinated phenols, are aryl compounds (see Figure 9.7). As shown in Figure 9.7, some aryl compounds, such as naphthalene and the polycyclic aromatic compound, benzo(a)pyrene, contain fused rings.

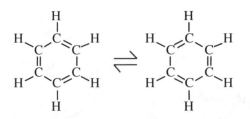

Figure 9.6. Representation of the aromatic benzene molecule with two resonance structures (left) and, more accurately, as a hexagon with a circle in it (right). Unless shown by symbols of other atoms, it is understood that a C atom is at each corner and that one H atom is bonded to each C atom.

Benzene and Naphthalene

Benzene is a volatile, colorless, highly flammable liquid that is consumed as a raw material for the manufacture of phenolic and polyester resins, polystyrene plastics, alkylbenzene surfactants, chlorobenzenes, insecticides, and dyes. It is hazardous both for its ignitability and toxicity (exposure to benzene causes blood abnormalities that may develop into leukemia). Naphthalene is the simplest member of a large number of multicyclic aromatic hydrocarbons having two or more fused rings. It is a volatile white crystalline solid with a characteristic odor and has been used to make mothballs. The most important of the many chemical derivatives made from naphthalene is phthalic anhydride, from which phthalate ester plasticizers are synthesized.

Polycyclic Aromatic Hydrocarbons

Benzo(a)pyrene (Figure 9.7) is the most studied of the polycyclic aromatic hydrocarbons (PAHs), which are characterized by condensed ring systems ("chicken wire" structures). These compounds are formed by the incomplete combustion of other hydrocarbons, a process that consumes hydrogen in preference to carbon. The carbon residue is left in the thermodynamically favored condensed aromatic ring system of the PAH compounds.

Because there are so many partial combustion and pyrolysis processes that favor production of PAHs, these compounds are encountered abundantly in the atmosphere, soil, and elsewhere in the environment from sources that include engine exhausts, wood stove smoke, cigarette smoke, and charbroiled food. Coal tars and petroleum residues such as road and roofing asphalt have high levels of PAHs. Some PAH compounds, including benzo(a)pyrene, are of toxicological concern because they are precursors to cancer-causing metabolites.

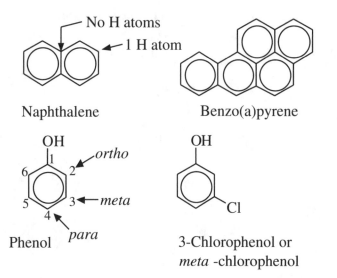

Figure 9.7. Aromatic compounds containing fused rings (top) and showing the numbering of carbon atoms for purposes of nomenclature.

9.3. ORGANIC FUNCTIONAL GROUPS AND CLASSES OF ORGANIC COMPOUNDS

The discussion of organic chemistry so far in this chapter has emphasized hydrocarbon compounds, those that contain only hydrogen and carbon. It has been shown that hydrocarbons may exist as alkanes, alkenes, and aryl compounds (arenes), depending upon the kinds of bonds between carbon atoms. The presence of elements other than hydrogen and carbon in organic molecules greatly increases the diversity of their chemical behavior. **Functional groups** consist of specific bonding configurations of atoms in organic molecules. Most functional groups contain at least one element other than carbon or hydrogen, although two carbon atoms joined by a double bond (alkenes) or triple bond (alkynes) are likewise considered to be functional groups. Table 9.2 shows some of the major functional groups that determine the nature of organic compounds.

Organooxygen Compounds

The most common types of compounds with oxygen-containing functional groups are epoxides, alcohols, phenols, ethers, aldehydes, ketones, and carboxylic acids. The functional groups characteristic of these compounds are illustrated by the examples of oxygen-containing compounds shown in Figure 9.8.

Ethylene oxide is a moderately to highly toxic sweet-smelling, colorless, flammable, explosive gas used as a chemical intermediate, sterilant, and fumigant. It

Table 9.2.   Examples of Some Important Functional Groups

Type of Functional Group	Example Compound	Structural Formula of Functional Group[a]
Alkene (olefin)	Propene (propylene)	
Alkyne	Acetylene	
Alcohol (-OH attached to alkyl group)	2-Propanol	
Phenol (-OH attached to aryl group)	Phenol	
Ketone	Acetone	
Amine	Methylamine	
Nitro compounds	Nitromethane	
Sulfonic acids	Benzenesulfonic acid	
Organohalides	1,1–Dichloro-ethane	

(When $-\overset{\overset{\displaystyle O}{\|}}{C}-H$ group is on end carbon, compound is an aldehyde)

[a]Functional group outlined by dashed line.

is a mutagen and a carcinogen to experimental animals. It is classified as hazardous for both its toxicity and ignitability. **Methanol** is a clear, volatile, flammable liquid alcohol used for chemical synthesis, as a solvent, and as a fuel. It is being advocated strongly in some quarters as an alternative to gasoline that would result in significantly less photochemical smog formation than currently used gasoline formulations. Ingestion of methanol can be fatal and blindness can

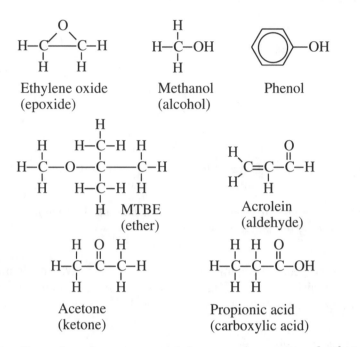

Ethylene oxide (epoxide)   Methanol (alcohol)   Phenol

MTBE (ether)   Acrolein (aldehyde)

Acetone (ketone)   Propionic acid (carboxylic acid)

Figure 9.8. Examples of oxygen-containing organic compounds that may be significant as wastes, toxic substances, or environmental pollutants.

result from sublethal doses. Phenol is a dangerously toxic aryl alcohol widely used for chemical synthesis and polymer manufacture. **Methyltertiarybutyl ether**, MTBE, is an ether that has become the octane booster of choice to replace tetraethyllead in gasoline. **Acrolein** is an alkenic aldehyde and a volatile, flammable, highly reactive chemical. It forms explosive peroxides upon prolonged contact with O_2. An extreme lachrimator and strong irritant, acrolein is quite toxic by all routes of exposure. **Acetone** is the lightest of the ketones. Like all ketones, acetone has a carbonyl ($C=O$) group that is bonded to *two* carbon atoms (that is, it is somewhere in the middle of a carbon atom chain). Acetone is a good solvent and is chemically less reactive than the aldehydes which all have the functional group,

in which binding of the $C=O$ to H makes the molecule significantly more reactive. **Propionic acid** is a typical organic carboxylic acid. The $-CO_2H$ group characteristic of carboxylic acids may be viewed as the most oxidized functional

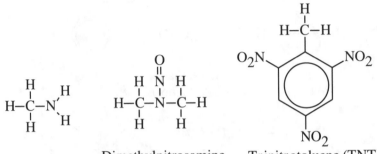

Methylamine   Dimethylnitrosamine   Trinitrotoluene (TNT)
(N-nitrosodimethylamine)

Figure 9.9.   Examples of organonitrogen that may be significant as wastes, toxic substances, or environmental pollutants.

group on an oxygenated organic compound, and carboxylic acids may be synthesized by oxidizing aldehydes and alcohols that have an –OH group or $C=O$ group on an end carbon atom.

Organonitrogen Compounds

Figure 9.9 shows examples of three classes of the many kinds of compounds that contain N (amines, nitrosamines, and nitro compounds). Nitrogen occurs in many functional groups in organic compounds, some of which contain nitrogen in ring structures, or along with oxygen.

Methylamine is a colorless, highly flammable gas with a strong odor. It is a severe irritant affecting eyes, skin, and mucous membranes. Methylamine is the simplest of the **amine** compounds, which have the general formula,

where the R's are hydrogen or hydrocarbon groups, at least one of which is the latter.

Dimethylnitrosamine is an N-nitroso compound, all of which contain the $N-N=O$ functional group. It was once widely used as an industrial solvent, but was observed to cause liver damage and jaundice in exposed workers. Subsequently, numerous other N-nitroso compounds, many produced as by-products of industrial operations and food and alcoholic beverage processing, were found to be carcinogenic.

Solid **trinitrotoluene** (TNT) has been widely used as a military explosive. TNT is moderately to very toxic and has caused toxic hepatitis or aplastic anemia in exposed individuals, a few of whom have died from its toxic effects. It belongs to the general class of nitro compounds characterized by the presence of $-NO_2$ groups bonded to a hydrocarbon structure.

Some organonitrogen compounds are chelating agents that bind strongly to metal ions and play a role in the solubilization and transport of heavy metal wastes. Prominent among these are salts of the aminocarboxylic acids which, in the acid form, have $-CH_2CO_2H$ groups bonded to nitrogen atoms. An important example of such a compound is the monohydrate of trisodium nitrilotriacetate (NTA):

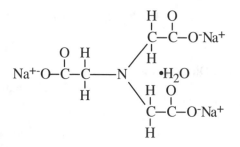

This compound is widely used in Canada as a substitute for detergent phosphates to bind to calcium ion and make the detergent solution basic. NTA is used in metal plating formulations. It is highly water soluble and quickly eliminated with urine when ingested. It has a low acute toxicity and no chronic effects have been shown for plausible doses. However, concern does exist over its interaction with heavy metals in waste treatment processes and in the environment.

Organohalide Compounds

Organohalides (Figure 9.10) exhibit a wide range of physical and chemical properties. These compounds consist of halogen-substituted hydrocarbon molecules, each of which contains at least one atom of F, Cl, Br, or I. They may be saturated (**alkyl halides**), unsaturated (**alkenyl halides**), or aromatic (**aryl halides**). The most widely manufactured organohalide compounds are chlorinated hydrocarbons, many of which are regarded as environmental pollutants or as hazardous wastes.

Alkyl Halides

Substitution of halogen atoms for one or more hydrogen atoms on alkanes gives **alkyl halides**, example structural formulas of which are given in Figure 9.10. Most of the commercially important alkyl halides are derivatives of alkanes of low molecular mass. A brief discussion of the uses of the compounds listed in Figure 9.10 is given here to provide an idea of the versatility of the alkyl halides.

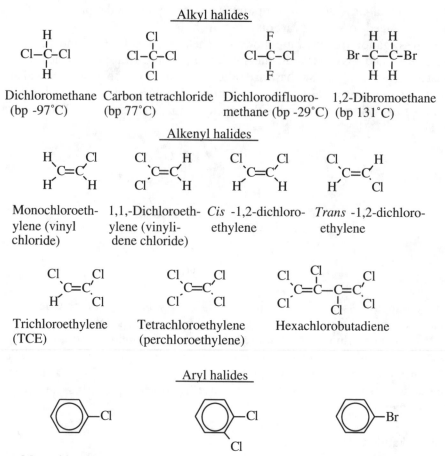

Figure 9.10.   Some example organohalide compounds.

Dichloromethane is a volatile liquid with excellent solvent properties for non-polar organic solutes. It has been used as a solvent for the decaffeination of coffee, in paint strippers, as a blowing agent in urethane polymer manufacture, and to depress vapor pressure in aerosol formulations. Once commonly sold as a solvent and stain remover, highly toxic **carbon tetrachloride** is now largely restricted to uses as a chemical intermediate under controlled conditions, primarily to manufacture chlorofluorocarbon refrigerant fluid compounds, which are also discussed in this section. Insecticidal **1,2-dibromoethane** has been consumed in large quantities as a lead scavenger in leaded gasoline and to fumigate soil, grain, and fruit (fumigation with this compound has been discontinued because of toxicological concerns). An effective solvent for resins, gums, and waxes, it serves as a chemical intermediate in the syntheses of some pharmaceutical compounds and dyes.

Alkenyl Halides

Viewed as hydrocarbon-substituted derivatives of alkenes, the **alkenyl** or **olefinic organohalides** contain at least one halogen atom and at least one carbon-carbon double bond. The most significant of these are the lighter chlorinated compounds, such as those illustrated in Figure 9.10.

Vinyl chloride is consumed in large quantities as a raw material to manufacture pipe, hose, wrapping, and other products fabricated from polyvinylchloride plastic. This highly flammable, volatile, sweet-smelling gas is a known human carcinogen.

As shown in Figure 9.10, there are three possible dichloroethylene compounds, all clear, colorless liquids. Vinylidene chloride forms a copolymer with vinyl chloride used in some kinds of coating materials. The geometrically isomeric 1,2-dichloroethylenes are used as organic synthesis intermediates and as solvents. **Trichloroethylene** is a clear, colorless, nonflammable, volatile liquid. It is an excellent degreasing and drycleaning solvent and has been used as a household solvent and for food extraction (for example, in decaffeination of coffee). Colorless, nonflammable liquid **tetrachloroethylene** has properties and uses similar to those of trichloroethylene. **Hexachlorobutadiene**, a colorless liquid with an odor somewhat like that of turpentine, is used as a solvent for higher hydrocarbons and elastomers, as a hydraulic fluid, in transformers, and for heat transfer.

Aryl Halides

Aryl halide derivatives of benzene and toluene have many uses in chemical synthesis, as pesticides and raw materials for pesticides manufacture, as solvents, and a diverse variety of other applications. These widespread uses over many decades have resulted in substantial human exposure and environmental contamination. Three example aryl halides are shown in Figure 9.10. Monochlorobenzene is a flammable liquid boiling at 132°C. It is used as a solvent, heat transfer fluid, and synthetic reagent. Used as a solvent, 1,2-dichlorobenzene is employed for degreasing hides and wool. It also serves as a synthetic reagent for dye manufacture. Bromobenzene is a liquid boiling at 156°C that is used as a solvent, motor oil additive, and intermediate for organic synthesis.

Halogenated Naphthalene and Biphenyl

Two major classes of halogenated aryl compounds containing two benzene rings are made by the chlorination of naphthalene and biphenyl and have been sold as mixtures with varying degrees of chlorine content. Examples of chlorinated naphthalenes, and polychlorinated biphenyls (PCBs discussed later), are shown in Figure 9.11. The less highly chlorinated of these compounds are liquids and those with higher chlorine contents are solids. Because of their physical and chemical stabilities and other desirable qualities, these compounds have had many uses, including heat transfer fluids, hydraulic fluids, and dielectrics. Poly-brominated biphenyls (PBBs) have served as flame retardants. However, because chlorinated naphthalenes, PCBs, and PBBs are environmentally extremely persistent, their uses have been severely curtailed.

Chlorofluorocarbons, Halons, and Hydrogen-Containing Chlorofluorocarbons

Chlorofluorocarbons (CFCs) are volatile 1- and 2-carbon compounds that contain Cl and F bonded to carbon. These compounds are notably stable and nontoxic. They have been widely used in recent decades in the fabrication of flexible and rigid foams and as fluids for refrigeration and air conditioning. The most widely manufactured of these compounds are CCl_3F (CFC-11), CCl_2F_2 (CFC-12), $C_2Cl_3F_3$ (CFC-113), $C_2Cl_2F_4$ (CFC-114), and C_2ClF_5 (CFC-115).

Halons are related compounds that contain bromine and are used in fire extinguisher systems. The most commonly produced commercial halons are $CBrClF_2$ (Halon-1211), $CBrF_3$ (Halon-1301), and $C_2Br_2F_4$ (Halon-2402), where the sequence of numbers denotes the number of carbon, fluorine, chlorine, and bromine atoms, respectively, per molecule. Halons are particularly effective fire extinguishing agents because of the way in which they stop combustion. Some fire suppressants, such as carbon dioxide, act by depriving the flame of oxygen by a smothering effect, whereas water cools a burning substance to a tempera-ture below which combustion is supported. Halons act by chain reactions (dis-

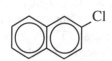

2-Chloronaphthalene

Polychlorinated
naphthalenes

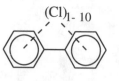

Polychlorinated
biphenyls (PCBs)

Polybrominated
biphenyls (PBBs)

Figure 9.11.   Halogenated naphthalenes and biphenyls.

cussed previously in this section) that destroy hydrogen atoms which sustain combustion. The basic sequence of reactions involved is outlined in the following reactions:

$$CBrClF_2 \ + \ H\cdot \ \longrightarrow \ CClF_2\cdot \ + \ HBr \tag{9.3.1}$$

$$HBr \ + \ H\cdot \ \longrightarrow \ Br\cdot \ + \ H_2 \tag{9.3.2}$$

Chain
reaction

$$Br\cdot \ + \ H\cdot \ \longrightarrow \ HBr \tag{9.3.3}$$

Halons are used in automatic fire extinguishing systems, such as those located in flammable solvent storage areas, and in specialty fire extinguishers, such as those on aircraft. It has been difficult to find substitutes for halons that have their same excellent performance characteristics.

All of the chlorofluorocarbons and halons discussed above have been implicated in the halogen atom-catalyzed destruction of atmospheric ozone. As a result of U. S. Environmental Protection Agency regulations imposed in accordance with the 1986 Montreal Protocol on Substances that Deplete the Ozone Layer, production of CFCs and halocarbons in the U. S. was curtailed starting in 1989. The most likely substitutes for these halocarbons are hydrogen-containing chlorofluorocarbons (HCFCs) and hydrogen-containing fluorocarbons (HFCs). The substitute compounds to be produced commercially first are CH_2FCF_3 (HFC-134a, a substitute for CFC-12 in automobile air conditioners and refrigeration equipment), $CHCl_2CF_3$ (HCFC-123, substitute for CFC-11 in plastic foam-blowing), CH_3CCl_2F (HCFC-141b, substitute for CFC-11 in plastic foam-blowing), $CHClF_2$ (HCFC-22, air conditioners and manufacture of plastic foam food containers). Because of the more readily broken H-C bonds that they contain, these compounds are more easily destroyed by atmospheric chemical reactions (particularly with hydroxyl radical, see Section 16.4 and Chapter 18) before they reach the stratosphere. Relative to a value of 1.0 for CFC-11, the ozone-depletion potentials of these substitutes are HFC-134a, 0; HCFC-123, 0.016; HCFC-141b, 0.081; and HCFC-22, 0.053. Concern has been expressed over toxicities and fire hazards of HCFCs, which are more reactive both chemically and biochemically than the CFCs and halons that they are designed to replace. The four leading substitutes, like the CFCs that they are designed to replace, have shown no evidence of causing skin or eye irritation, birth defects, or other short-term toxic effects.[2]

The Du Pont Company, which introduced chlorofluorocarbons in the 1930s and is the largest manufacturer of them, has announced that it intends to cease production shortly after the year 2000. The company began to manufacture HFC substitutes in 1991.

Chlorinated Phenols

The chlorinated phenols, particularly **pentachlorophenol,**

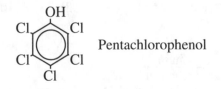

Pentachlorophenol

and the trichlorophenol isomers are significant hazardous wastes. These compounds are biocides that are used to treat wood to prevent rot by fungi and to prevent termite infestation. They are toxic, causing liver malfunction and dermatitis; contaminant polychlorinated dibenzodioxins may be responsible for some of the observed effects. Wood preservative chemicals such as pentachlorophenol are encountered at many hazardous waste sites in wastewaters and sludges.

Organosulfur Compounds

Sulfur is chemically similar to, but more diverse than oxygen. Whereas, with the exception of peroxides, most chemically combined organic oxygen is in the -2 oxidation state, sulfur occurs in the -2, $+4$, and $+6$ oxidation states. Many organosulfur compounds are noted for their foul, "rotten egg" or garlic odors. A number of example organosulfur compounds are shown in Figure 9.12.

Thiols and Thioethers

Substitution of alkyl or aryl hydrocarbon groups such as phenyl and methyl for H on hydrogen sulfide, H_2S, leads to a number of different organosulfur thiols (mercaptans, R–SH) and **sulfides,** also called thioethers (R–S–R). Structural formulas of examples of these compounds are shown in Figure 9.12.

Methanethiol and other lighter alkyl thiols are fairly common air pollutants that have "ultragarlic" odors; both 1- and 2-butanethiol are associated with skunk odor. Gaseous methanethiol is used as an odorant leak-detecting additive for natural gas, propane, and butane; it is also employed as an intermediate in pesticide synthesis. A toxic, irritating volatile liquid with a strong garlic odor, 2-propene-1-thiol (allyl mercaptan) is a typical alkenyl mercaptan. Benzenethiol (phenyl mercaptan) is the simplest of the aryl thiols. It is a toxic liquid with a severely "repulsive" odor.

Alkyl sulfides or thioethers contain the C-S-C functional group. The lightest of these compounds is dimethyl sulfide, a volatile liquid (bp 38°C) that is moderately toxic by ingestion. Cyclic sulfides contain the C-S-C group in a ring struc-

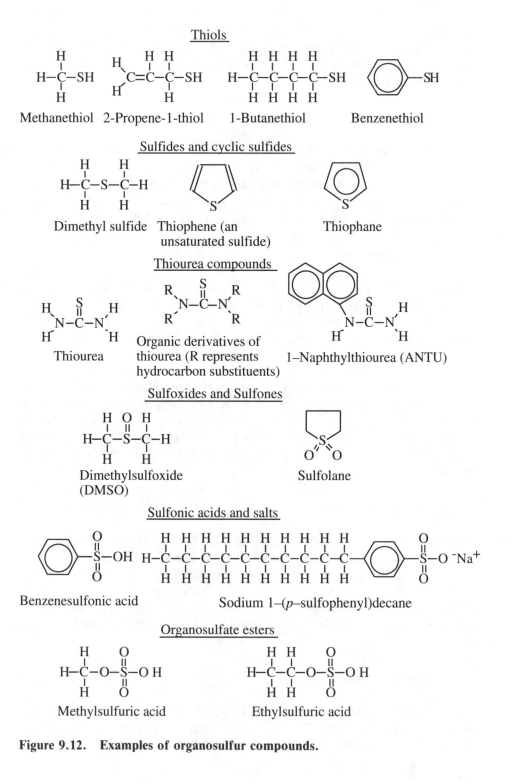

Figure 9.12.   Examples of organosulfur compounds.

ture. The most common of these compounds is thiophene, a heat-stable liquid (bp 84°C) with a solvent action much like that of benzene, which is used in the manufacture of pharmaceuticals, dyes, and resins. Its saturated analog is tetra-hydrothiophene, or thiophane.

Nitrogen-Containing Organosulfur Compounds

Many important organosulfur compounds also contain nitrogen. One such compound is **thiourea**, the sulfur analog of urea. Its structural formula is shown in Figure 9.12. Thiourea and **phenylthiourea** have been used as rodenticides. Commonly called ANTU, **1-naphthylthiourea** is an excellent rodenticide that is virtually tasteless and has a very high rodent:human toxicity ratio.

Sulfoxides and Sulfones

Sulfoxides and **sulfones** (Figure 9.12) contain both sulfur and oxygen. **Dimethylsulfoxide** (DMSO) is a liquid with numerous uses and some very interesting properties. It is used to remove paint and varnish, as a hydraulic fluid, mixed with water as an antifreeze solution, and in pharmaceutical applications as an anti-inflammatory and bacteriostatic agent. A polar aprotic (no ionizable H) solvent with a relatively high dielectric constant, **sulfolane** dissolves both organic and inorganic solutes. It is the most widely produced sulfone because of its use in an industrial process called BTX processing in which it selectively extracts benzene, toluene, and xylene from aliphatic hydrocarbons; as the solvent in the Sulfinol process by which thiols and acidic compounds are removed from natural gas; as a solvent for polymerization reactions; and as a polymer plasticizer.

Sulfonic Acids, Salts, and Esters

Sulfonic acids and sulfonate salts contain the $-SO_3H$ and $-SO_3^-$ groups, respectively, attached to a hydrocarbon moiety. The structural formula of benzenesulfonic acids and of sodium 1-(p-sulfophenyl)decane, a biodegradable detergent surfactant, are shown in Figure 9.12. The common sulfonic acids are water-soluble strong acids that lose virtually all ionizable H^+ in aqueous solution. They are used commercially to hydrolyze fat and oil esters to fatty acids and glycerol.

Organic Esters of Sulfuric Acid

Replacement of 1 H on sulfuric acid, H_2SO_4, yields an acid ester and replacement of both yields an ester. Examples of these esters are shown in Figure 9.12. Sulfuric acid esters are used as alkylating agents, which act to attach alkyl groups (such as methyl) to organic molecules, in the manufacture of agricultural chemicals, dyes, and drugs. **Methylsulfuric acid** and **ethylsulfuric acid** are oily water-soluble liquids that are strong irritants to skin, eyes, and mucous tissue.

Organophosphorus Compounds

Alkyl and Aryl Phosphines

The first two examples in Figure 9.13 illustrate that the structural formulas of alkyl and aryl phosphine compounds may be derived by substituting organic groups for the H atoms in phosphine (PH_3), the hydride of phosphorus, a toxic inorganic compound. **Methylphosphine** is a colorless, reactive gas. Crystalline, solid **triphenylphosphine** has a low reactivity and moderate toxicity when inhaled or ingested.

As shown by the reaction,

$$4C_3H_9P + 26O_2 \rightarrow 12CO_2 + 18H_2O + P_4O_{10} \qquad (9.3.4)$$

combustion of aryl and alkyl phosphines produces P_4O_{10}, a corrosive irritant toxic substance that reacts with moisture in the air to produce droplets of corrosive orthophosphoric acid, H_3PO_4.

Organophosphate Esters

The structural formulas of three esters of orthophosphoric acid (H_3PO_4) and an ester of pyrophosphoric acid ($H_4P_2O_6$) are shown in Figure 9.13. Although **trimethylphosphate** is considered to be only moderately toxic, **tri-o-cresylphosphate, TOCP,** has a notorious record of poisonings. **Tetraethylpyrophosphate, TEPP,** was developed in Germany during World War II as a substitute for insecticidal nicotine. Although it is a very effective insecticide, its use in that application was of very short duration because it kills almost everything else, too.

Phosphorothionate Esters

Parathion, shown in Figure 9.13, is an example of **phosphorothionate** esters. These compounds are used as insecticidal acetylcholinesterase inhibitors. They contain the $P = S$ (thiono) group, which increases their insect:mammal toxicity ratios. Since the first organophosphate insecticides were developed in Germany during the 1930s and 1940s, many insecticidal organophosphate compounds have been synthesized. One of the earliest and most successful of these is **parathion,** O,O-diethyl-O-p-nitrophenylphosphorothionate (banned from use in the U.S. in 1991 because of its acute toxicity to humans). From a long-term

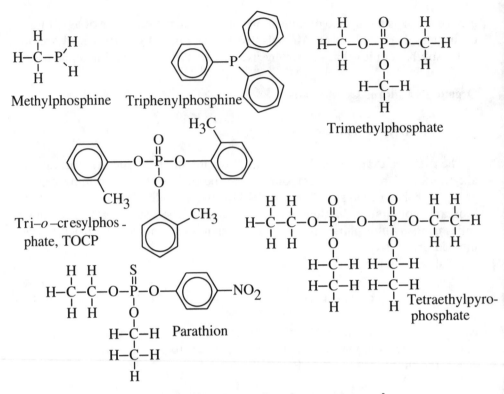

Figure 9.13.   Some representative organophosphorus compounds.

environmental standpoint organophosphate insecticides are superior to the organohalide insecticides that they largely displaced because the organophosphates readily undergo biodegradation and do not bioaccumulate.

9.4. SYNTHETIC POLYMERS

A large fraction of the chemical industry worldwide is devoted to polymer manufacture, which is very important in the area of hazardous wastes, as a source of environmental pollutants, and in the manufacture of materials used to alleviate environmental and waste problems. Synthetic **polymers** are produced when small molecules called **monomers** bond together to form a much smaller number of very large molecules. Many natural products are polymers; for example, cellulose in wood, paper, and many other materials is a polymer of the sugar glucose. Synthetic polymers form the basis of many industries, such as rubber, plastics, and textiles manufacture.

An important example of a polymer is that of polyvinylchloride, shown in Figure 9.14. This polymer is synthesized in large quantities for the manufacture of water and sewer pipe, water-repellant liners, and other plastic materials.

Other major polymers include polyethylene (plastic bags, milk cartons), polypropylene, (impact-resistant plastics, indoor-outdoor carpeting), polyacrylonitrile (Orlon, carpets), polystyrene (foam insulation), and polytetrafluoroethylene (Teflon coatings, bearings); the monomers from which these substances are made are shown in Figure 9.15.

Many of the hazards from the polymer industry arise from the monomers used as raw materials. Many monomers are reactive and flammable, with a tendency to form explosive vapor mixtures with air. All have a certain degree of toxicity; vinyl chloride is a known human carcinogen. The combustion of many polymers may result in the evolution of toxic gases, such as hydrogen cyanide (HCN) from polyacrylonitrile or hydrogen chloride (HCl) from polyvinylchloride. Another hazard presented by plastics results from the presence of **plasticizers** added to provide essential properties, such as flexibility. The most widely used plasticizers are phthalates, which are environmentally persistent, resistant to treatment processes, and prone to undergo bioaccumulation.

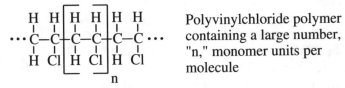

"n" vinyl chloride monomers

Polyvinylchloride polymer containing a large number, "n," monomer units per molecule

Figure 9.14. Polyvinylchloride polymer.

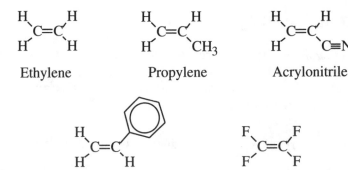

Ethylene　　　　Propylene　　　　Acrylonitrile

Styrene　　　　Tetrafluoroethylene

Figure 9.15.　Monomers from which commonly used polymers are synthesized.

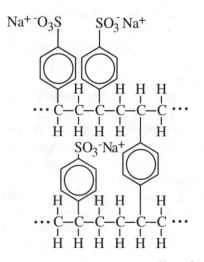

Figure 9.16.   Polymeric cation exchanger in the sodium form.

Polymers have a number of applications in waste treatment and disposal. Waste disposal landfill liners are made from synthetic polymers as are the fiber filters which remove particulate pollutants from flue gas in baghouses. Membranes used for ultrafiltration and reverse osmosis treatment of water are composed of very thin sheets of synthetic polymers. Organic solutes can be removed from water by sorption onto hydrophobic (water-repelling) organophilic beads of Amberlite XAD resin. Heavy metal pollutants are removed from wastewater by cation exchange resins made of polymers with anionic functional groups. Typically, these resins exchange harmless sodium ion, Na^+, on the solid resin for toxic heavy metal ions in water. Figure 9.16 shows a segment of the polymeric structure of a cation exchange resin in the sodium form.

CHAPTER SUMMARY

The chapter summary below is presented in a programmed format to review the main points covered in this chapter. It is used most effectively by filling in the blanks, referring back to the chapter as necessary. The correct answers are given at the end of the summary.

The diversity of organic chemistry is due to the ability of carbon atoms to

(1)_____ in a large variety of (2)_____

_____. All organic compounds

contain the element (3)_____ and virtually all contain the element

(4)_____ . Organic compounds that contain only these two elements are called (5)_____ . The molecular shape of a molecule is its (6)_____ .

The major types of hydrocarbons are (7)_____
_____. Alkanes are hydrocarbons in which the C atoms are (8)_____. Based upon their chain and ring structures, the three kinds of alkanes are (9)_____
_____. Formulas that only give the number of each kind of atom in a molecule of a compound are called (10)_____. Compounds that have the same molecular, but different structural, formulas are called (11)_____
_____. CH_4 is called (12)_____ and the CH_3 group is the (13)_____ group, whereas C_2H_6 is called (14)_____ and C_2H_5 is called the (15)_____ group. Using standard rules for naming organic compounds, the systematic name of the compound

$$
\begin{array}{ccccccc}
\text{H} & \text{H} & \text{H} & \text{H} & \text{H} & \text{H} & \text{H} \\
| & | & | & | & | & | & | \\
\text{H}-\text{C}- & \text{C}- & \text{C}- & \text{C}- & \text{C}- & \text{C}- & \text{C}-\text{H} \\
| & | & | & | & | & | & | \\
\text{H} & \text{CH}_3 & \text{H} & \text{H} & \text{H} & \text{H} \\
& \text{C}_2\text{H}_5 & & & &
\end{array}
$$

is (16)_____.

The major reaction that alkanes undergo at high temperature is an (17)_____
_____. The only possible reaction of alkanes with Cl_2 is a (18)_____ in which (19)_____ substitutes for H. Highly reactive species with unpaired single electrons, such as CH_3·, are known as (20)_____.

Alkenes are hydrocarbons that have (21)_____ bonds consisting of (22)_____. Alkynes are a class of hydrocarbons characterized by (23)_____ bonds. Because they contain unsaturated bonds alkenes and alkynes both undergo (24)_____ reactions, which are not possible with alkanes. The compound 2-butene can exist in the two forms of (25)_____.

Benzene is the simplest member of a large class of hydrocarbons known as (26)_____ . Delocalization of π electrons over several carbon atoms in aromatic compounds results in a stability effect called (27)_____ . Exposure to benzene causes abnormalities of (28)_____. Benzo(a)pyrene is a common example of a class of compounds known as (29)_____.

Groups of atoms in specific bonding configurations in organic molecules constitute (30)_____. Of these, the C=O group on a hydrocarbon that is not at the end of a hydrocarbon chain characterizes (31)_____, whereas the–CO_2H group is characteristic of (32)_____.

Compounds with the general formula,

$$R-N\begin{smallmatrix} R' \\ \\ R'' \end{smallmatrix}$$

where the R's are hydrogen or hydrocarbon groups, at least one of which is the latter are called (33)_____. All N-nitroso compounds contain the (34)_____. Nitro compounds are characterized by the presence of (35)_____ groups bonded to a hydrocarbon structure.

Organohalide compounds consist of (36)_____-substituted hydro-carbon molecules. The compound of this type in which all the H atoms on methane, CH_4, are substituted by Cl is called (37)_____. Vinyl chloride is an unsaturated organohalide compound used to make (38)_____ _____ and noted toxicologically because it is a (39)_____ _____. An environmentally-significant class of organohalide compounds containing two benzene rings joined at a single C atom and having various numbers of Cl atoms substituted onto the rings are

the (40)_____. Chlorofluorocarbons (CFCs) are
notably (41)_____.

Organosulfur compounds containing the –SH group are called (42)_____.
They are noted for their (43)_____.
Sulfonic acids contain the (44)_____ group attached to a hydrocarbon moiety.

Tetraethylpyrophosphate, TEPP, is an example of a class of compounds
called (45)_____ and is noted for its extreme
(46)_____. The compound O,O-diethyl-O-p-nitrophenylphosphoro-
thionate is an organophosphate ester commonly called (47)_____
and once widely used as (48)_____.

Synthetic polymers are produced when small molecules called (49)_____
bond together to form (50)_____
_____. Some polymers contain plasticizers, of which the
most widely used are the (51)_____.

Answers

1. bond to each other
2. straight chains, branched chains, and rings
3. carbon
4. hydrogen
5. hydrocarbons
6. molecular structure
7. alkanes, alkenes, alkynes, and aryl compounds
8. joined by single covalent bonds
9. straight-chain alkanes, branched-chain alkanes, and cycloalkanes
10. molecular formulas
11. structural isomers
12. methane
13. methyl
14. ethane
15. ethyl
16. 2-methyl,3-ethylheptane
17. oxidation
18. substitution
19. Cl
20. free radicals

21. double
22. 4 shared electrons
23. carbon-carbon triple
24. addition
25. *cis* and *trans* isomers
26. aromatic or aryl hydrocarbons
27. resonance stabilization
28. blood
29. polycyclic aromatic hydrocarbons
30. functional groups
31. ketones
32. carboxylic acids
33. amines
34. $>N-N=O$
35. $-NO_2$
36. halogen
37. carbon tetrachloride
38. polyvinylchloride plastic
39. known human carcinogen
40. polychlorinated biphenyls, PCBs
41. stable and nontoxic
42. thiols
43. foul, "rotten egg" or garlic odors
44. $-SO_3H$
45. organophosphate esters
46. toxicity
47. parathion
48. an insecticide
49. monomers
50. a much smaller number of very large molecules
51. phthalates

QUESTIONS AND PROBLEMS

1. Explain the bonding properties of carbon that makes organic chemistry so diverse.

2. Distinguish among alkanes, alkenes, alkynes, and aryl compounds. To which general class of organic compounds do all belong?

3. In what sense are alkanes saturated? Why are alkenes more reactive than alkanes?

4. Name the compound below:

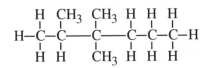

5. What is indicated by "*n*" in a hydrocarbon name?

6. Discuss the chemical reactivity of alkanes. Why are they chemically reactive or unreactive?

7. Discuss the chemical reactivity of alkenes. Why are they chemically reactive or unreactive?

8. What are the characteristics of aromaticity? What are the chemical reactivity characteristics of aromatic compounds?

9. Describe chain reactions, discussing what is meant by free radicals, and photochemical processes.

10. Define, with examples, what is meant by isomerism.

11. Describe how the two forms of 1,2-dichloroethylene can be used to illustrate *cis-trans* isomerism.

12. Give the structural formula corresponding to the condensed structural formula of $CH_3CH(C_2H_5)CH(C_2H_5)CH_2CH_3$.

13. Discuss how organic functional groups are used to define classes of organic compounds.

14. Give the functional groups corresponding to (a) alcohols, (b) aldehydes, (c) carboxylic acids, (d) ketones, (e) amines, (f) thiol compounds, and (g) nitro compounds.

15. Give an example compound of each of the following: epoxides, alcohols, phenols, ethers, aldehydes, ketones, and carboxylic acids.

16. Which functional group is characteristic of N-nitroso compounds, and why are these compounds toxicologically significant?

17. Give an example of each of the following: Alkyl halides, alkenyl halides, aryl halides.

18. Give an example compound of a chlorinated naphthalene and of a PCB.

19. What explains the tremendous chemical stability of CFCs? What kinds of compounds are replacing CFCs? Why?

20. How does a thio differ from a thioether?

21. How does a sulfoxide differ from a sulfone?

22. Which inorganic compound is regarded as the parent compound of alkyl and aryl phosphines? Give an example of each of these.

23. What are organophosphate esters and what is their toxicological significance?

24. Define what is meant by a polymer and give an example of one.

LITERATURE CITED

1. "Organic Chemistry," Chapter 26 in *Chemistry, The Central Science,* T. L. Brown, H. E. LeMay, Jr., and B. E. Bursten, (Englewood Cliffs, NJ: Prentice Hall. 1991) pp. 938–969.
2. Zurer, P. S., "CFC Substitutes: Candidates Pass Early Toxicity Tests," *Chemical and Engineering News*, October 9, 1989, p. 4.

10   BIOLOGICAL CHEMISTRY

10.1. BIOCHEMISTRY

Most people have had the experience of looking through a microscope at a single cell. It may have been an amoeba, alive and oozing about like a blob of jelly on the microscope slide, or a cell of bacteria, stained with a dye to make it show up more plainly. Or, it may have been a beautiful cell of algae with its bright green chlorophyll. Even the simplest of these cells is capable of carrying out a thousand or more chemical reactions. These life processes fall under the heading of **biochemistry**, that branch of chemistry that deals with the chemical properties, composition, and biologically-mediated processes of complex substances in living systems.

Biochemical phenomena that occur in living organisms are extremely sophisticated. In the human body complex metabolic processes break down a variety of food materials to simpler chemicals, yielding energy and the raw materials to build body constituents, such as muscle, blood, and brain tissue. Impressive as this may be, consider a humble microscopic cell of blue-green algae only about a micrometer in size, which requires only a few simple inorganic chemicals and sunlight for its existence. This cell uses sunlight energy to convert carbon from CO_2, hydrogen and oxygen from H_2O, nitrogen from NO_3^-, sulfur from SO_4^{2-}, and phosphorus from inorganic phosphate into all the proteins, nucleic acids, carbohydrates, and other materials that it requires to exist and reproduce. Such a simple cell accomplishes what could not even be done by human endeavors in a vast chemical factory costing billions of dollars.

Ultimately, most environmental pollutants and hazardous substances are of concern because of their effects upon living organisms. The study of the adverse effects of substances on life processes requires some basic knowledge of biochemistry. Biochemistry is discussed in this chapter, with emphasis upon aspects that are especially pertinent to environmentally hazardous and toxic substances, including cell membranes, DNA, and enzymes.

Biochemical processes not only are profoundly influenced by chemical species in the environment, they largely determine the nature of these species, their degradation, and even their syntheses, particularly in the aquatic and soil environments. The study of such phenomena forms the basis of **environmental biochemistry**.[1]

Biomolecules

The biomolecules that constitute matter in living organisms are often polymers with molecular masses of the order of a million or even larger. As discussed later in this chapter, these biomolecules may be divided into the categories of carbohydrates, proteins, lipids, and nucleic acids. Proteins and nucleic acids consist of macromolecules, lipids are usually relatively small molecules, carbohydrates range from relatively small sugar molecules to high molecular mass macromolecules such as those in cellulose.

The behavior of a substance in a biological system depends to a large extent upon whether the substance is hydrophilic ("water-loving") or hydrophobic ("water-hating"). Some important toxic substances are hydrophobic, a characteristic that enables them to traverse cell membranes readily. Part of the detoxification process carried on by living organisms is to render such molecules hydrophilic, therefore water-soluble and readily eliminated from the body.

10.2.   BIOCHEMISTRY AND THE CELL

The focal point of biochemistry and biochemical aspects of toxicants is the cell, the basic building block of living systems where most life processes are carried out.[2] Bacteria, yeasts, and some algae consist of single cells. However, most living things are made up of many cells. In a more complicated organism the cells have different functions. Liver cells, muscle cells, brain cells, and skin cells in the human body are quite different from each other and do different things. Cells are divided into two major categories depending upon whether or not they have a nucleus: **eukaryotic** cells have a nucleus and **prokaryotic** cells do not. Prokaryotic cells are found predominately in single-celled organisms such as bacteria. Eukaryotic cells occur in multicelled plants and animals — higher life forms.

Major Cell Features

Figure 10.1 shows the major features of the **eukaryotic cell**, which is the basic structure in which biochemical processes occur in multicelled organisms. These features are the following:

- **Cell membrane**, which encloses the cell and regulates the passage of ions, nutrients, lipid-soluble ("fat-soluble") substances, metabolic products, toxicants, and toxicant metabolites into and out of the cell interior

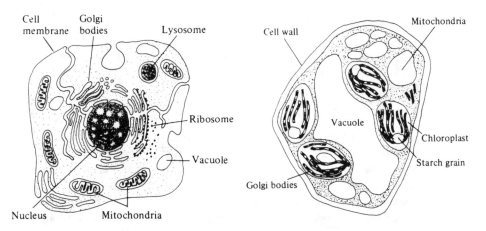

Figure 10.1. Some major features of the eukaryotic cell in animals (left) and plants (right).

because of its varying **permeability** for different substances. The cell membrane protects the contents of the cell from undesirable outside influences. Cell membranes are composed in part of phospholipids that are arranged with their hydrophilic ("water-seeking") heads on the cell membrane surfaces and their hydrophobic ("water-repelling") tails inside the membrane. Cell membranes contain bodies of proteins that are involved in the transport of some substances through the membrane. One reason the cell membrane is very important in toxicology and environmental biochemistry is because it regulates the passage of toxicants and their products into and out of the cell interior. Furthermore, when its membrane is damaged by toxic substances, a cell may not function properly and the organism may be harmed.

- **Cell nucleus**, which acts as a sort of "control center" of the cell. It contains the genetic directions the cell needs to reproduce itself. The key substance in the nucleus is **deoxyribonucleic** acid (**DNA**). **Chromosomes** in the cell nucleus are made up of combinations of DNA and proteins. Each chromosome stores a separate quantity of genetic information. Human cells contain 46 chromosomes. When DNA in the nucleus is damaged by foreign substances, various toxic effects, including mutations, cancer, birth defects, and defective immune system function may occur.
- **Cytoplasm**, which fills the interior of the cell not occupied by the nucleus. Cytoplasm is further divided into a water-soluble proteinaceous filler called **cytosol**, in which are suspended bodies called **cellular organelles**, such as mitochondria or, in photosynthetic organisms, chloroplasts.

- **Mitochondria,** "powerhouses" which mediate energy conversion and utilization in the cell. Mitochondria are sites in which food materials — carbohydrates, proteins, and fats — are broken down to yield carbon dioxide, water, and energy, which is then used by the cell. The best example of this is the oxidation of the sugar glucose, $C_6H_{12}O_6$:

$$C_6H_{12}O_6 + 6O_2 \rightarrow 6CO_2 + 6H_2O + energy$$

This kind of process is called cellular respiration.
- **Ribosomes,** which participate in protein synthesis.
- **Endoplasmic reticulum,** which is involved in the metabolism of some toxicants by enzymatic processes.
- **Lysosome,** a type of organelle that contains potent substances capable of digesting liquid food material. Such material enters the cell through a "dent" in the cell wall, which eventually becomes surrounded by cell material. This surrounded material is called a **food vacuole**. The vacuole merges with a lysosome, and the substances in the lysosome bring about digestion of the food material. The digestion process consists largely of **hydrolysis reactions** in which large, complicated food molecules are broken down into smaller units by the addition of water.
- **Golgi bodies,** that occur in some types of cells. These are flattened bodies of material that serve to hold and release substances produced by the cells.
- **Cell walls** of plant cells. These are strong structures that provide stiffness and strength. Cell walls are composed mostly of cellulose, which will be discussed later in this chapter.
- **Vacuoles** inside plant cells that often contain materials dissolved in water.
- **Chloroplasts** in plant cells that are involved in photosynthesis (the chemical process which uses energy from sunlight to convert carbon dioxide and water to organic matter). Photosynthesis occurs in these bodies. Food produced by photosynthesis is stored in the chloroplasts in the form of **starch grains**.

10.3. PROTEINS

Proteins are nitrogen-containing organic compounds which are the basic units of living systems. Cytoplasm, the jelly-like liquid filling the interior of cells is made up largely of protein. Enzymes, which act as catalysts of life reactions, are made of proteins; they are discussed later in the chapter. Proteins are made up of **amino acids** joined together in huge chains. Amino acids are organic compounds which contain the carboxylic acid group, $-CO_2H$, and the amino group, $-NH_2$. They are sort of a hybrid of carboxylic acids and amines (Section 9.3). Proteins are polymers or **macromolecules** of amino acids containing from approximately forty to several thousand amino acid groups joined by peptide linkages. Smaller molecule amino acid polymers, containing only about ten to about forty amino acids per molecule, are called **polypeptides**. A portion of the amino acid left after the elimination of H_2O during polymerization is called a **residue**. The amino acid sequence of these residues is designated by a series of three-letter abbreviations for the amino acid.

Natural amino acids all have the following chemical group:

In this structure the $-NH_2$ group is always bonded to the carbon next to the $-CO_2H$ group. This is called the "alpha" location, so natural amino acids are alpha-amino acids. Other groups, designated as "R," are attached to the basic alpha-amino acid structure. The R groups may be as simple as an atom of H found in glycine,

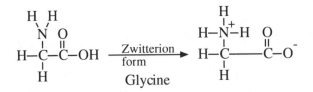

or, they may be as complicated as the structure,

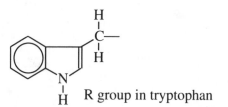

found in tryptophan. As shown in Table 10.1, there are 20 common amino acids in proteins. These are shown with uncharged $-NH_2$ and $-CO_2H$ groups. Actually, these functional groups exist in the charged **zwitterion** form as shown for glycine, above.

Amino acids in proteins are joined together in a specific way. These bonds are called the **peptide linkage**. The formation of peptide linkages is a condensation

Table 10.1. Amino Acids That Occur in Protein

Name	Structural Formula
Glycine (gly)	$H-\overset{\overset{\displaystyle NH_2}{\mid}}{CH}-CO_2H$
Alanine (ala)	$H_3C-\overset{\overset{\displaystyle NH_2}{\mid}}{CH}-CO_2H$
Valine (val)*	$\begin{array}{c} H_3C \\ \diagdown \\ CH-\overset{\overset{\displaystyle NH_2}{\mid}}{CH}-CO_2H \\ \diagup \\ H_3C \end{array}$
Phenylalanine (phe)*	$C_6H_5-CH_2-\overset{\overset{\displaystyle NH_2}{\mid}}{CH}-CO_2H$
Serine (ser)	$HO-CH_2-\overset{\overset{\displaystyle NH_2}{\mid}}{CH}-CO_2H$
Threonine (thr)*	$H_3C-\overset{\overset{\displaystyle OH}{\mid}}{CH}-\overset{\overset{\displaystyle NH_2}{\mid}}{CH}-CO_2H$
Asparagine (asn)	$H_2N-\overset{\overset{\displaystyle O}{\|\|}}{C}-CH_2-\overset{\overset{\displaystyle NH_2}{\mid}}{CH}-CO_2H$
Leucine (leu)*	$\begin{array}{c} H_3C \\ \diagdown \\ CH-CH_2-\overset{\overset{\displaystyle NH_2}{\mid}}{CH}-CO_2H \\ \diagup \\ H_3C \end{array}$
Isoleucine (ile)*	$\begin{array}{c} H_3C \\ \diagdown \\ CH-\overset{\overset{\displaystyle NH_2}{\mid}}{CH}-CO_2H \\ \diagup \\ H_3C-CH_2 \end{array}$
Proline (pro)	$\begin{array}{c} H_2C\!-\!-\!-CH_2 \\ \mid \mid \\ H_2C CH-CO_2H \\ \diagdown_{N}\diagup \\ \mid \\ H \end{array}$
Methionine (met)*	$H_3C-S-CH_2-CH_2-\overset{\overset{\displaystyle NH_2}{\mid}}{CH}-CO_2H$
Cysteine (cys)	$HS-CH_2-\overset{\overset{\displaystyle NH_2}{\mid}}{CH}-CO_2H$
Tyrosine (tyr)	$HO-C_6H_4-CH_2-\overset{\overset{\displaystyle NH_2}{\mid}}{CH}-CO_2H$

Table 10.1, continued

Name	Structural Formula
Glutamine (gin)	$$\underset{\text{H}_2\text{N}}{}\overset{\displaystyle\text{O}}{\underset{\|}{\text{C}}}-\text{CH}_2-\text{CH}_2-\overset{\displaystyle\text{NH}_2}{\underset{\|}{\text{CH}}}-\text{CO}_2\text{H}$$
Tryptophan (try)*	indole ring$-\text{CH}_2-\overset{\displaystyle\text{NH}_2}{\underset{\|}{\text{CH}}}-\text{CO}_2\text{H}$
Aspartic acid (asp)	$$\text{HO}-\overset{\displaystyle\text{O}}{\underset{\|}{\text{C}}}-\text{CH}_2-\overset{\displaystyle\text{NH}_2}{\underset{\|}{\text{CH}}}-\text{CO}_2\text{H}$$
Histidine (his)	imidazole ring$-\text{CH}_2-\overset{\displaystyle\text{NH}_2}{\underset{\|}{\text{CH}}}-\text{CO}_2\text{H}$
Glutamic acid (glu)	$$\text{HO}-\overset{\displaystyle\text{O}}{\underset{\|}{\text{C}}}-\text{CH}_2-\text{CH}_2-\overset{\displaystyle\text{NH}_2}{\underset{\|}{\text{CH}}}-\text{CO}_2\text{H}$$
Lysine (lys)*	$$\overset{+}{\text{H}_3\text{N}}-\text{CH}_2-\text{CH}_2-\text{CH}_2-\text{CH}_2-\overset{\displaystyle\text{NH}_2}{\underset{\|}{\text{CH}}}-\text{CO}_2\text{H}$$
Arginine (arg)	$$\text{H}_2\text{N}-\overset{\displaystyle\overset{+}{\text{NH}_2}}{\underset{\|}{\text{C}}}-\text{NH}-\text{CH}_2-\text{CH}_2-\text{CH}_2-\overset{\displaystyle\text{NH}_2}{\underset{\|}{\text{CH}}}-\text{CO}_2\text{H}$$

* Amino acids designated with an asterisk are the essential amino acids for humans. Unlike the other amino acids, these cannot be synthesized by the human body, and must be present in the diet. The average human requires about 70 grams of "first-class" protein (meat, fish, cheese, or egg) to supply the essential amino acids.

process involving the loss of water. Consider as an example the condensation of alanine, leucine, and tyrosine shown in Figure 10.2. When these three amino acids join together, two water molecules are eliminated. The product is a *tri*peptide since there are three amino acids involved. The amino acids in proteins are linked as shown for this tripeptide, except that many more monomeric amino acid groups are involved.

Proteins may be divided into several major types that have widely varying functions. These are given in Table 10.2.

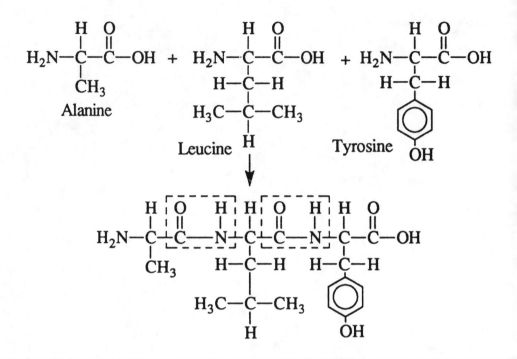

Figure 10.2. Condensation of alanine, leucine, and tyrosine to form a tripeptide consisting of three amino acids joined by peptide linkages (outlined by dashed lines).

Protein Structure

The order of amino acids in protein molecules, and the resulting three-dimensional structures that form, provide an enormous variety of possibilities for **protein structure**. This is what makes life so diverse. Proteins have primary, secondary, tertiary, and quaternary structures. The structures of protein molecules determine the behavior of proteins in crucial areas such as the processes by which the body's immune system recognizes substances that are foreign to the body. Proteinaceous enzymes depend upon their structures for the very specific functions of the enzymes.

The order of amino acids in the protein molecule determines its primary structure. **Secondary protein structures** result from the folding of polypeptide protein chains to produce a maximum number of hydrogen bonds between peptide linkages:

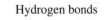

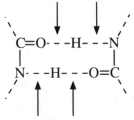

Illustration of hydrogen bonds between N and O atoms in peptide linkages, which constitute protein secondary structures

Table 10.2. Major Types of Proteins

Type of Protein	Example	Function and Characteristics
Nutrient	Casein (milk protein)	Food source. People must have an adequate supply of nutrient protein with the right balance of amino acids for adequate nutrition.
Storage	Ferritin	Storage of iron in animal tissues
Structural	Collagen (tendons) keratin (hair)	Structural and protective components in organisms
Contractile	Actin, myosin in muscle tissue	Strong, fibrous proteins that can contract and cause movement to occur
Transport	Hemoglobin	Transport inorganic and organic species across cell membranes, in blood, between organs
Defense	– –	Antibodies against foreign agents such as viruses produced by the immune system
Regulatory	Insulin, human growth hormone	Regulate biochemical processes such as sugar metabolism or growth by binding to sites inside cells or on cell membranes
Enzymes	Acetylcholinesterase	Catalysts of biochemical reactions (see Section 10.6.)

Further folding of the protein molecules held in place by attractive forces between amino acid side chains give proteins **a secondary structure**. The nature of the R groups on the amino acids determines the secondary structure. Small R groups enable protein molecules to be hydrogen-bonded together in a parallel arrangement. With larger R groups the molecules tend to take a spiral form. Such a spiral is known as an **alpha-helix.**

Tertiary structures are formed by the twisting of alpha-helices into specific shapes. They are produced and held in place by the interactions of amino side chains on the amino acid residues constituting the protein macromolecules. Tertiary protein structure is very important in the processes by which enzymes identify specific proteins and other molecules upon which they act. It is also involved with the action of antibodies in blood which recognize foreign proteins by their shape and react to them. This is basically what happens in the case of immunity to a disease where antibodies in blood recognize specific proteins from viruses or bacteria and reject them.

Two or more protein molecules consisting of separate polypeptide chains may be further attracted to each other to produce a **quaternary structure**.

Some proteins are **fibrous proteins**, which occur in skin, hair, wool, feathers, silk, and tendons. The molecules in these proteins are long and threadlike and are laid out parallel in bundles. Fibrous proteins are quite tough and they do not dissolve in water.

An interesting fibrous protein is keratin, which is found in hair. The cross-linking bonds between protein molecules in keratin are $-S-S-$ bonds formed from two $HS-$ groups in two molecules of the amino acid, cysteine. These bonds largely hold hair in place, thus keeping it curly or straight. A "permanent" consists of breaking the bonds chemically, setting the hair as desired, then reforming the cross-links to hold the desired shape.

Aside from fibrous protein, the other major type of protein form is the **globular protein**. These proteins are in the shape of balls and oblongs. Globular proteins are relatively soluble in water. A typical globular protein is hemoglobin, the oxygen-carrying protein in red blood cells. Enzymes are generally globular proteins.

Denaturation of Proteins

Secondary, tertiary, and quaternary protein structures are easily changed by a process called **denaturation**. These changes can be quite damaging. Heating, exposure to acids or bases, and even violent physical action can cause denaturation to occur. The albumin protein in egg white is denatured by heating so that it forms a semisolid mass. Almost the same thing is accomplished by the violent physical action of an egg beater in the preparation of meringue. Heavy metal poisons such as lead and cadmium change the structures of proteins by binding to functional groups on the protein surface.

10.4. CARBOHYDRATES

Carbohydrates have the approximate simple formula CH_2O and include a diverse range of substances composed of simple sugars such as glucose:

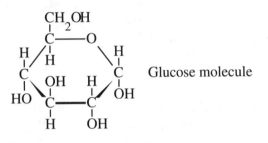

Glucose molecule

High-molecular-mass **polysaccharides**, such as starch and glycogen ("animal starch"), are biopolymers of simple sugars.

When photosynthesis occurs in a plant cell, the energy from sunlight is converted to chemical energy in a carbohydrate, $C_6H_{12}O_6$. This carbohydrate may be transferred to some other part of the plant for use as an energy source. It may be converted to a water-insoluble carbohydrate for storage until it is needed for energy. Or it may be transformed to cell wall material and become part of the structure of the plant. If the plant is eaten by an animal, the carbohydrate is used for energy by the animal.

The simplest carbohydrates are the **monosaccharides**. These are also called **simple sugars**. Because they have 6 carbon atoms, simple sugars are sometimes called *hex*oses. Glucose (formula shown above) is the most common simple sugar involved in cell processes. Other simple sugars with the same formula but somewhat different structures are fructose, mannose, and galactose. These must be changed to glucose before they can be used in a cell. Because of its use for energy in body processes, glucose is found in the blood. Normal levels are from 65 to 110 mg glucose per 100 mL of blood. Higher levels may indicate diabetes.

Units of two monosaccharides make up several very important sugars known as **disaccharides**. When two molecules of monosaccharides join together to form a disaccharide,

$$C_6H_{12}O_6 + C_6H_{12}O_6 \rightarrow C_{12}H_{22}O_{11} + H_2O \qquad (10.4.1)$$

a molecule of water is lost. Recall that proteins are also formed from smaller amino acid molecules by condensation reactions involving the loss of water molecules. Disaccharides include sucrose (cane sugar used as a sweetener), lactose (milk sugar), and maltose (a product of the breakdown of starch).

Polysaccharides consist of many simple sugar units hooked together. One of the most important polysaccharides is **starch**, which is produced by plants for food storage. Animals produce a related material called **glycogen**. The chemical formula of starch is $(C_6H_{10}O_5)_n$, where n may represent a number as high as several hundreds. What this means is that the very large starch molecule consists of many units of $C_6H_{10}O_5$ joined together. For example, if n is 100, there are 6 times 100 carbon atoms, 10 times 100 hydrogen atoms, and 5 times 100 oxygen atoms in the molecule. Its chemical formula is $C_{600}H_{1000}O_{500}$. The atoms in a starch molecule are actually present as linked rings represented by the structure shown in Figure 10.3. Starch occurs in many foods, such as bread and cereals. It is readily digested by animals, including humans.

Cellulose is a polysaccharide which is also made up of $C_6H_{10}O_5$ units. Molecules of cellulose are huge, with molecular weights of around 400,000. The cellulose structure (Figure 10.4) is similar to that of starch. Cellulose is produced by plants and forms the structural material of plant cell walls. Wood is about 60% cellulose, and cotton contains over 90% of this material. Fibers of cellulose are extracted from wood and pressed together to make paper.

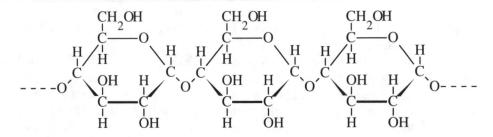

Figure 10.3. Part of a starch molecule showing units of $C_6H_{10}O_5$ condensed together.

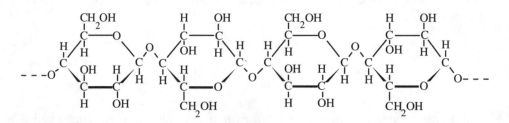

Figure 10.4. Part of the structure of cellulose.

Humans and most other animals cannot digest cellulose. Ruminant animals (cattle, sheep, goats, moose) have bacteria in their stomachs that break down cellulose into products which can be used by the animal. Chemical processes are available to convert cellulose to simple sugars by the reaction

$$(C_6H_{10}O_5)_n + nH_2O \rightarrow nC_6H_{12}O_6 \qquad (10.4.2)$$

$$\text{cellulose} \qquad\qquad\qquad \text{glucose}$$

where n may be 2000–3000. This involves breaking the linkages between units of $C_6H_{10}O_5$ by adding a molecule of H_2O at each linkage, a hydrolysis reaction. Large amounts of cellulose from wood, sugar cane, and agricultural products go to waste each year. The hydrolysis of cellulose enables these products to be converted to sugars, which can be fed to animals.

Carbohydrate groups are attached to protein molecules in a special class of materials called **glycoproteins**. Collagen is a crucial glycoprotein that provides structural integrity to body parts. It is a major constituent of skin, bones, tendons, and cartilage.

10.5.   LIPIDS

Lipids are substances that can be extracted from plant or animal matter by organic solvents such as chloroform, diethyl ether, or toluene (Figure 10.5). Whereas carbohydrates and proteins are characterized predominately by the monomers (monosaccharides and amino acids) from which they are composed, lipids are defined essentially by their physical characteristic of organophilicity. The most common lipids are fats and oils composed of **triglycerides** formed from the alcohol glycerol, $CH_2(OH)CH(OH)CH_2(OH)$ and a long-chain fatty acid such as stearic acid, $CH_3(CH_2)_{16}C(O)OH$ (Figure 10.6). Numerous other biological materials, including waxes, cholesterol, and some vitamins and hormones, are classified as lipids. Common foods, such as butter and salad oils are lipids. The longer chain fatty acids, such as stearic acid, are also organic-soluble and are classified as lipids.

Lipids are toxicologically important for several reasons. Some toxic substances interfere with lipid metabolism, leading to detrimental accumulation of lipids. Many toxic organic compounds are poorly soluble in water, but are lipid-soluble, so that bodies of lipids in organisms serve to dissolve and store toxicants.

An important class of lipids consists of **phosphoglycerides** (glycerophosphatides). These compounds may be regarded as triglycerides in which one of the acids bonded to glycerol is orthophosphoric acid (see example structure below). These lipids are especially important because they are essential constituents of cell membranes. These membranes consist of bilayers in

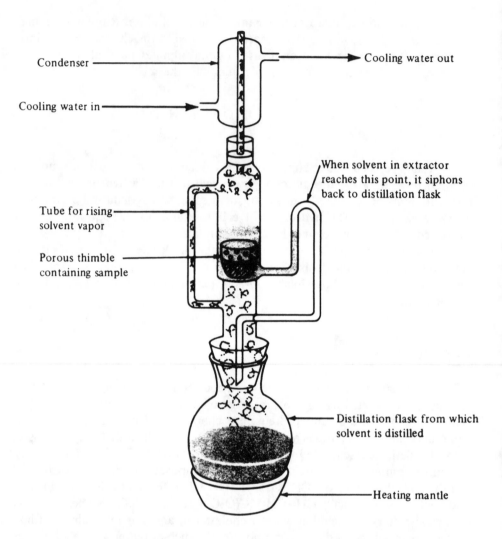

Figure 10.5. Lipids are extracted from some biological materials with a Soxhlet extractor (above). The solvent is vaporized in the distillation flask by the heating mantle, rises through one of the exterior tubes to the condenser, and is cooled to form a liquid. The liquid drops onto the porous thimble containing the sample. Siphon action periodically drains the solvent back into the distillation flask. The extracted lipid collects as a solution in the solvent in the flask.

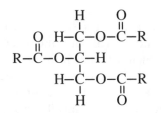

Figure 10.6. General formula of triglycerides, which make up fats and oils. The R group is from a fatty acid and is a hydrocarbon chain, such as $-(CH_2)_{16}CH_3$.

which the hydrophilic phosphate ends of the molecules are on the outside of the membrane and the hydrophobic "tails" of the molecules are on the inside.

Waxes are also esters of fatty acids. However, the alcohol in a wax is not glycerol. The alcohol in waxes is often a very long chain alcohol. For example, one of the main compounds in beeswax is myricyl palmitate.

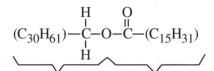

Alcohol portion   Fatty acid portion
of ester   of ester

Myricyl palmitate

in which the alcohol portion of the ester has a very large hydrocarbon chain. Waxes are produced by both plants and animals, largely as protective coatings. Waxes are found in a number of common products. Lanolin is one of these. It is the "grease" in sheep's wool. When mixed with oils and water, it forms stable colloidal emulsions consisting of extremely small oil droplets suspended in water. This makes lanolin useful for skin creams and pharmaceutical ointments. Carnauba wax occurs as a coating on the leaves of some Brazilian palm trees. Spermaceti wax is composed largely of cetyl palmitate,

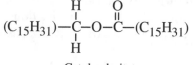

Cetyl palmitate

extracted from the blubber of the sperm whale. It is very useful in some cosmetics and pharmaceutical preparations.

Steroids are lipids found in living systems which all have the ring system shown in Figure 10.7 for cholesterol. Steroids occur in bile salts, which are produced by the liver and then secreted into the intestines. Their breakdown products give feces its characteristic color. Bile salts act upon fats in the intestine. They suspend very tiny fat droplets in the form of colloidal emulsions. This enables the fats to be broken down chemically and digested.

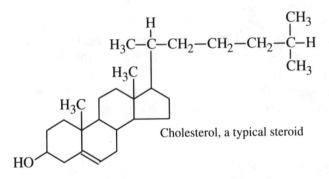

Cholesterol, a typical steroid

Figure 10.7. Steroids are characterized by the ring structure shown above for cholesterol.

Some steroids are **hormones**. Hormones act as "messengers" from one part of the body to another. As such, they start and stop a number of body functions. Male and female sex hormones are examples of steroid hormones. Hormones are given off by glands in the body called **endocrine glands.** The locations of the important endocrine glands are shown in Figure 10.8.

10.6.   ENZYMES

Catalysts are substances that speed up a chemical reaction without themselves being consumed in the reaction. The most sophisticated catalysts of all are those found in living systems. They bring about reactions that could not be performed at all, or only with great difficulty, outside a living organism. These catalysts are called **enzymes**. In addition to speeding up reactions by as much as ten- to a hundred million-fold, enzymes are extremely selective in the reactions which they promote.

Enzymes are proteinaceous substances with highly specific structures that interact with particular substances or classes of substances called **substrates**. Enzymes act as catalysts to enable biochemical reactions to occur, after which they are regenerated intact to take part in additional reactions. The extremely high specificity with which enzymes interact with substrates results from their

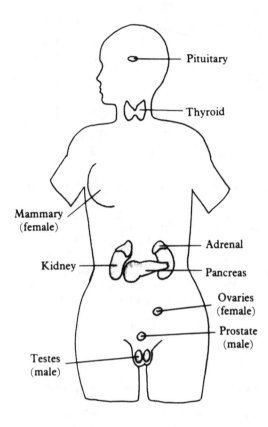

Figure 10.8. Locations of important endocrine glands.

"lock and key" action based upon the unique shapes of enzymes as illustrated in Figure 10.9.

This illustration shows that an enzyme "recognizes" a particular substrate by its molecular structure and binds to it to produce an **enzyme-substrate complex**. This complex then breaks apart to form one or more products different from the original enzyme, regenerating the unchanged enzyme, which is then available to catalyze additional reactions. The basic process for an enzyme reaction is, therefore,

$$\text{enzyme} + \text{substrate} \rightleftharpoons \text{enzyme-substrate complex} \rightleftharpoons$$
$$\text{enzyme} + \text{product} \qquad (10.6.1)$$

Several important things should be noted about this reaction. As shown in Figure 10.9, an enzyme acts on a specific substrate to form an enzyme-substrate complex because of the fit between their structures. As a result,

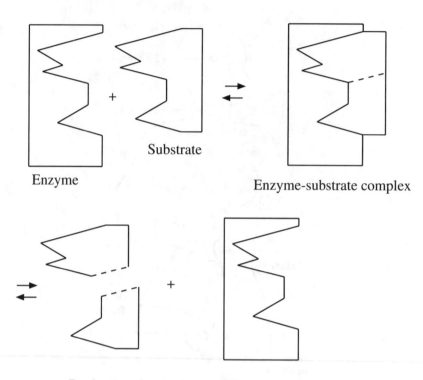

Figure 10.9. Representation of the "lock-and-key" mode of enzyme action which enables the very high specificity of enzyme-catalyzed reactions.

something happens to the substrate molecule. For example, it might be split in two at a particular location. Then the enzyme-substrate complex comes apart, yielding the enzyme and products. The enzyme is not changed in the reaction and is now free to react again. Note that the arrows in the formula for enzyme reaction point both ways. This means that the reaction is **reversible**. An enzyme-substrate complex can simply go back to the enzyme and the substrate. The products of an enzymatic reaction can react with the enzyme to form the enzyme-substrate complex again. It, in turn, may again form the enzyme and the substrate. Therefore, the same enzyme may act to cause a reaction to go either way.

Some enzymes cannot function by themselves. In order to work, they must first be attached to **coenzymes**. Coenzymes normally are not protein materials. Some of the vitamins are important coenzymes.

Enzymes are named for what they do. For example, the enzyme given off by the stomach, which splits proteins as part of the digestion process, is called *gastric proteinase*. The "gastric" part of the name refers to the enzyme's origin in the stomach. The "proteinase" denotes that it splits up protein molecules. The common name for this enzyme is pepsin. Similarly, the enzyme produced by the pancreas that breaks down fats (lipids) is called *pancreatic lipase*. Its common name is steapsin. In general, lipase enzymes cause lipid triglycerides to dissociate and form glycerol and fatty acids.

The enzymes mentioned above are **hydrolyzing enzymes**, which bring about the breakdown of high molecular weight biological compounds by the addition of water. This is one of the most important types of the reactions involved in digestion. The three main classes of energy-yielding foods that animals eat are carbohydrates, proteins, and fats. Recall that the higher carbohydrates humans eat are largely disaccharides (sucrose, or table sugar) and polysaccharides (starch). These are formed by the joining together of units of simple sugars, $C_6H_{12}O_6$, with the elimination of an H_2O molecule at the linkage where they join. Proteins are formed by the condensation of amino acids, again with the elimination of a water molecule at each linkage. Fats are esters which are produced when glycerol and fatty acids link together. A water molecule is lost for each of these linkages when a protein, fat, or carbohydrate is synthesized. In order for these substances to be used as a food source, the reverse process must occur to break down large, complicated molecules of protein, fat, or carbohydrate to simple, soluble substances which can penetrate a cell membrane and take part in chemical processes in the cell. This reverse process is accomplished by hydrolyzing enzymes.

Biological compounds with long chains of carbon atoms are broken down into molecules with shorter chains by the breaking of carbon-carbon bonds. This commonly occurs by the elimination of $-CO_2H$ groups from carboxylic acids. For example, *pyruvic decarboxylase* enzyme acts upon pyruvic acid,

$$
\underset{\text{Pyruvic acid}}{H-\overset{\overset{\displaystyle H}{|}}{\underset{\underset{\displaystyle H}{|}}{C}}-\overset{\overset{\displaystyle O}{\|}}{C}-\overset{\overset{\displaystyle O}{\|}}{C}-OH} \xrightarrow[\text{decarboxylase}]{\text{Pyruvate}} \underset{\text{Acetaldehyde}}{H-\overset{\overset{\displaystyle H}{|}}{\underset{\underset{\displaystyle H}{|}}{C}}-\overset{\overset{\displaystyle O}{\|}}{C}-H} + CO_2 \qquad (10.6.2)
$$

to split off CO_2 and produce a compound with one less carbon. It is by such carbon-by-carbon breakdown reactions that long chain compounds are eventually degraded to CO_2 in the body, or that long-chain hydrocarbons undergo biodegradation by the action of microorganisms in the water and soil environments.

Oxidation and reduction are the major reactions for the exchange of energy in living systems. Cellular respiration mentioned in Section 10.2 is an oxidation reaction in which a carbohydrate, $C_6H_{12}O_6$, is broken down to carbon dioxide and water with the release of energy.

$$C_6H_{12}O_6 + 6O_2 \rightarrow 6CO_2 + 6H_2O + energy \qquad (10.6.3)$$

Actually, such an overall reaction occurs in living systems by a complicated series of individual steps. Some of these steps involve oxidation. The enzymes that bring about oxidation in the presence of free O_2 are called **oxidases**. In general, biological oxidation-reduction reactions are catalyzed by **oxidoreductase enzymes**.

In addition to the types of enzymes discussed above, there are many enzymes that perform miscellaneous duties in living systems. Typical of these are **isomerases**, which form isomers of particular compounds. For example, there are several simple sugars with the formula $C_6H_{12}O_6$. However, only glucose can be used directly for cell processes. The other isomers are converted to glucose by the action of isomerases. **Transferase enzymes** move chemical groups from one molecule to another, **lyase enzymes** remove chemical groups without hydrolysis and participate in the formation of $C = C$ bonds or addition of species to such bonds, and **ligase enzymes** work in conjunction with ATP (adenosine triphosphate, a high-energy molecule that plays a crucial role in energy-yielding, glucose-oxidizing metabolic processes) to link molecules together with the formation of bonds such as carbon-carbon or carbon-sulfur bonds.

Enzyme action may be affected by many different things. Enzymes require a certain hydrogen ion concentration to function best. For example, gastric proteinase requires the acid environment of the stomach to work well. When it passes into the much less acidic intestines, it stops working. This prevents damage to the intestine walls, which would occur if the enzyme tried to digest them. Temperature is critical. Not surprisingly, the enzymes in the human body work best at around 98.6°F (37°C), which is the normal body temperature. Heating these enzymes to around 140°F permanently destroys them. Some bacteria that thrive in hot springs have enzymes which work best at relatively high temperatures. Other "cold-seeking" bacteria have enzymes adapted to near the freezing point of water.

One of the greatest concerns regarding the effects of surroundings upon enzymes is the influence of toxic substances. A major mechanism of toxicity is the alteration or destruction of enzymes by agents such as cyanide, heavy metals, or organic compounds, such as insecticidal parathion. An enzyme that has been destroyed obviously cannot perform its designated function, whereas one that has been altered may either not function at all or may act improperly. Toxicants can affect enzymes in several ways. Parathion, for example, bonds covalently to the nerve enzyme acetylcholinesterase, which can then no longer serve to stop nerve impulses. Heavy metals tend to bind to sulfur atoms in enzymes (such as sulfur from the amino acid cysteine shown in Table 10.1) thereby altering the shape and function of the enzyme. Enzymes are denatured by some poisons causing them to "unravel" so that the enzyme no longer has its crucial specific shape.

10.7. NUCLEIC ACIDS

The "essence of life" is contained in **deoxyribonucleic acid (DNA,** which stays in the cell nucleus) and **ribonucleic acid (RNA,** which functions in the cell cytoplasm). These substances, which are known collectively as **nucleic acids,** store and pass on essential genetic information that controls reproduction and protein synthesis.

The structural formulas of the monomeric constituents of nucleic acids are given in Figure 10.10. These are pyrimidine or purine nitrogen-containing bases,

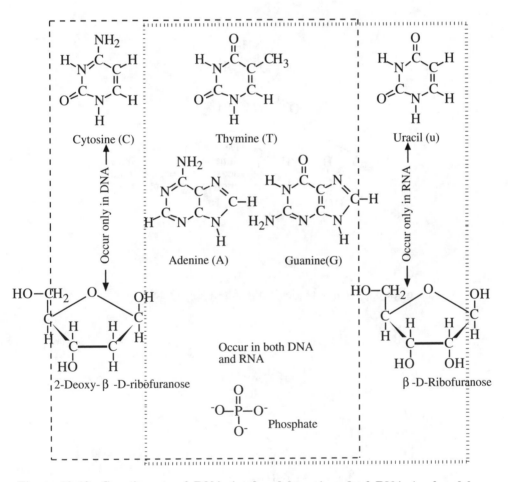

Figure 10.10. Constituents of DNA (enclosed by --) and of RNA (enclosed by -----).

two sugars, and phosphate. DNA molecules are made up of the nitrogen-containing bases adenine, guanine, cytosine, and thymine; phosphoric acid (H_3PO_4), and the simple sugar 2-deoxy-β-D-ribofuranose (commonly called deoxyribose). RNA molecules are composed of the nitrogen-containing bases adenine, guanine, uracil, and thymine; phosphoric acid (H_3PO_4), and the simple sugar β-D-ribofuranose (commonly called ribose).

The formation of nucleic acid polymers from their monomeric constituents may be viewed as the following steps:

- Monosaccharide (simple sugar) + cyclic nitrogenous base yields **nucleoside**:

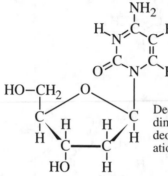

Deoxyctidine formed by the dimerization of cytosine and deoxyribose with the elimination of a molecule of H_2O.

- Nucleoside + phosphate yields **phosphate ester nucleotide**.

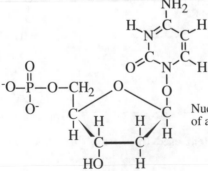

Nucleotide formed by the bonding of a phosphate group to deoxyctidine

- Polymerized nucleotide yields **nucleic acid**. In the nucleic acid the phosphate negative charges are neutralized by metal cations (such as Mg^{2+}) or positively-charged proteins (histones).

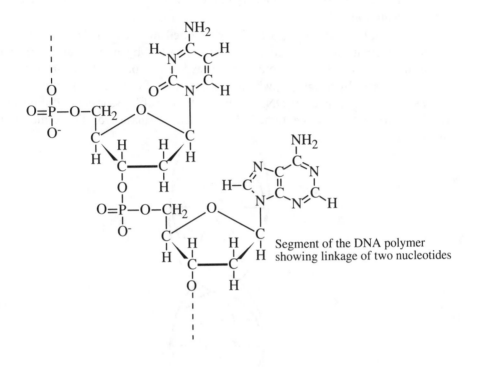

Segment of the DNA polymer showing linkage of two nucleotides

Molecules of DNA are huge. Their molecular weights are greater than one billion. Molecules of RNA are also quite large. The structure of DNA is that of the famed "double helix." It was figured out in 1953 by an American scientist, James D. Watson, and Francis Crick, a British scientist. They received the Nobel prize for this scientific milestone in 1962. This model visualizes DNA as a so-called double α-helix structure of oppositely-wound polymeric strands held together by hydrogen bonds between opposing pyrimidine and purine groups. As a result, DNA has both a primary and a secondary structure; the former is due to the sequence of nucleotides in the individual strands of DNA and the latter results from the α-helix interaction of the two strands. In the secondary structure of DNA, only cytosine can be opposite guanine and only thymine can be opposite adenine and *vice versa*. Basically, the structure of DNA is that of two spiral ribbons "counter-wound" around each other (Figure 10.11). The two strands of DNA are **complementary**. This means that a particular portion of one strand fits like a key in a lock with the corresponding portion of another strand. If the two strands are pulled apart, each manufactures a new complementary strand, so that two copies of the original double helix result. This occurs during cell reproduction.

The molecule of DNA is sort of like a coded message. This "message," the genetic information contained in, and transmitted by nucleic acids, depends upon the sequence of bases from which they are composed. It is somewhat like the message sent by telegraph, which consists only of dots, dashes, and spaces in between. The key aspect of DNA structure that enables storage and replication of this information is the famed double helix structure of DNA mentioned above.

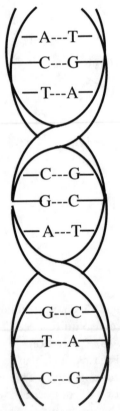

Figure 10.11. Representation of the double helix structure of DNA showing the allowed base pairs held together by hydrogen bonding between the phosphate/sugar polymer "backbones" of the two strands of DNA. The letters stand for adenine (A), cytosine (C), guanine (G), and thymine (T). The dashed lines, ---, represent hydrogen bonds.

Portions of the DNA double helix may unravel, and one of the strands of DNA may produce a strand of RNA. This substance then goes from the cell nucleus out into the cell and regulates the synthesis of new protein. In this way, DNA regulates the function of the cell and acts to control life processes.

Nucleic Acids in Protein Synthesis

Whenever a new cell is formed, the DNA in its nucleus must be accurately reproduced from the parent cell. Life processes are absolutely dependent upon accurate protein synthesis as regulated by cell DNA. The DNA in a single cell must be capable of directing the synthesis of up to 3000 or even more different proteins. The directions for the synthesis of a single protein are contained in a segment of DNA called a **gene**. The process of transmitting information from DNA to a newly-synthesized protein involves the following steps:

- The DNA undergoes **replication**. This process involves separation of a segment of the double helix into separate single strands which then replicate such that guanine is opposite cytosine (and *vice versa*) and adenine is opposite thymine (and *vice versa*). This process continues until a complete copy of the DNA molecule has been produced.
- The newly replicated DNA produces **messenger RNA (m-RNA)**, a complement of the single strand of DNA, by a process called **transcription**.
- A new protein is synthesized using m-RNA as a template to determine the order of amino acids in a process called **translation**.

Modified DNA

DNA molecules may be modified by the unintentional addition or deletion of nucleotides or by substituting one nucleotide for another. The result is a **mutation** that is transmittable to offspring. Mutations can be induced by chemical substances. This is a major concern from a toxicological viewpoint because of the detrimental effects of many mutations and because substances that cause mutations often cause cancer as well. DNA disruption results in birth defects. The failure to control cell reproduction results in cancer. Radiation from X-rays and radioactivity may also alter DNA and may cause mutation.

10.8. RECOMBINANT DNA AND GENETIC ENGINEERING

As noted above, segments of DNA contain information for the specific syntheses of particular proteins. Within the last two decades it has become possible to transfer this information between organisms by means of **recombinant DNA technology**, which has resulted in a new industry based on **genetic engineering**. Most often the recipient organisms are bacteria, which can be reproduced (cloned) over many orders of magnitude from a cell that has acquired the desired qualities. Therefore, to synthesize a particular substance, such as human insulin or growth hormone, the required genetic information can be transferred from a human source to bacterial cells, which then produce the substance as part of their metabolic processes.

The first step in recombinant DNA gene manipulation is to lyze, or "open up" a cell that has the genetic material needed and to remove this material from the cell. Through enzyme action the sought-after genes are cut from the donor DNA chain. These are next spliced into small DNA molecules. These molecules, called **cloning vehicles**, are capable of penetrating the host cell and becoming incorporated into its genetic material. The modified host cell is then reproduced many times and carries out the desired biosynthesis.

Early concerns about the potential of genetic engineering to produce "monster organisms" or new and horrible diseases have been largely allayed, although caution is still required with this technology. In the environmental area genetic engineering offers some hope for the production of bacteria engineered to safely destroy troublesome wastes and to produce biological substitutes for environmentally damaging synthetic pesticides.

10.9. METABOLIC PROCESSES

Biochemical processes that involve the alteration of biomolecules fall under the category of **metabolism**. Metabolic processes may be divided into the two major categories of **anabolism** (synthesis) and **catabolism** (degradation of substances). An organism may use metabolic processes to yield energy or to modify the constituents of biomolecules.

Energy-Yielding Processes

Organisms can gain energy by the following three processes:

- **Respiration** in which organic compounds undergo catabolism that requires molecular oxygen (**aerobic respiration**) or that occurs in the absence of molecular oxygen (**anaerobic respiration**). Aerobic respiration uses the **Krebs cycle** to obtain energy from the following reaction:

$$C_6H_{12}O_6 + 6O_2 \rightarrow 6CO_2 + 6H_2O + \text{energy}$$

About half of the energy released is converted to short-term stored chemical energy, particularly through the synthesis of **adenosine triphosphate**

(ATP) nucleoside. For longer-term energy storage, glycogen or starch polysaccharides are synthesized, and for still longer-term energy storage lipids (fats) are generated and retained by the organism.

- **Fermentation**, which differs from respiration in not having an electron transport chain. Yeasts produce ethanol from sugars by respiration:

$$C_6H_{12}O_6 \rightarrow 2CO_2 + 2C_2H_5OH$$

- **Photosynthesis** in which light energy captured by plant and algal chloroplasts is used to synthesize sugars from carbon dioxide and water:

$$6CO_2 + 6H_2O + h\nu \rightarrow C_6H_{12}O_6 + 6O_2$$

Plants cannot always get the energy that they need from sunlight. During the dark they must use stored food. Plant cells, like animal cells, contain mitochondria in which stored food is converted to energy by cellular respiration.

Plant cells, which use sunlight as a source of energy and CO_2 as a source of carbon, are said to be **autotrophic**. In contrast, animal cells must depend upon organic material manufactured by plants for their food. These are called **heterotrophic** cells. They mediate the chemical reaction between oxygen and food material and use the energy from the reaction to carry out their life processes.

CHAPTER SUMMARY

The chapter summary below is presented in a programmed format to review the main points covered in this chapter. It is used most effectively by filling in the blanks, referring back to the chapter as necessary. The correct answers are given at the end of the summary.

The biomolecules that constitute matter in living organisms are often high-molecular-mass (1)_____ divided into the categories of

(2)_____. In respect to affinity for water vs. lipids, the behavior of a substance in a biological system depends to a large extent upon whether the substance is (3)_____

or (4)_____.

The two major kinds of cells are (5)_____ cells that

have a defined nucleus and (6)_____ cells that do not. Several major features of the eukaryotic cell are (7)_____ _____. Three other features characteristic of plant cells are (8)_____ _____.

Proteins are macromolecular biomolecules composed of (9)_____. A generalized structure for amino acids is (10)_____, where "R" represents groups characteristic of each amino acid. Amino acids are bonded together in proteins by a specific arrangement of atoms and bonds called a (11)_____. Casein is a protein that functions as a (12)_____, hemoglobin is a (13)_____ protein, and collagen is a (14)_____ protein. Proteins exhibit (15)_____ different orders of structure, the first of which is determined by the (16)_____. Protein denaturation consists of disruption of (17)_____.

Carbohydrates have the approximate simple formula (18)_____. The molecular formula of glucose is (19)_____. Polysaccharides are carbohydrates that consist of many (20)_____. An important polysaccharide produced by plants for food storage is (21)_____. Cellulose is a polysaccharide produced by (22)_____ that forms (23)_____ _____.

Substances that can be extracted from plant or animal matter by organic solvents, such as chloroform are called (24)_____. The most common lipids are triglycerides formed from (25)_____ _____. Two reasons that lipids are toxicologically important are that (26)_____ _____ _____. A special class of lipids with a characteristic ring system, such as that in cholesterol consists of (27)_____. Some steroids are (28)_____ that act as "messengers" from one part of the body to another.

Enzymes are proteinaceous substances with highly specific structures that function as biochemical (29)_____ and interact with particular substances or classes of substances called (30)_____. Enzymes are very specific in the kinds of substance that they interact with because of their unique (31)_____ _____. In order to work, some enzymes must first be attached to (32)_____. Enzymes are named for (33)_____. For example, *gastric proteinase* comes from the (34)_____ and acts to (35)_____. It is an example of a general class of enzymes called (36)_____. As another example, biological oxidation-reduction reactions are catalyzed by (37)_____ _____.

Deoxyribonucleic acid (DNA) and ribonucleic acid (RNA) are the two major kinds of (38)_____. The monomers from which they are composed are (39)_____ _____. Cell DNA regulates synthesis of (40)_____.

Biochemical processes that involve the alteration of biomolecules are termed (41)_____. The three processes by which organisms gain energy are (42)_____.

Answers

1. polymers
2. carbohydrates, proteins, lipids, and nucleic acids
3. hydrophilic
4. hydrophobic
5. eukaryotic
6. prokaryotic
7. cell membrane, cell nucleus, cytoplasm, mitochondria, ribosomes, endoplasmic reticulum, lysosome, Golgi bodies
8. cell walls, vacuoles, and chloroplasts

9. amino acids

10.

$$R \overset{\overset{\displaystyle H}{\underset{\displaystyle H}{|}}}{\underset{\displaystyle \underset{H}{|}}{C}} \overset{\overset{\displaystyle O}{||}}{C} - OH$$

11. peptide linkage
12. nutrient
13. transport
14. structural
15. four
16. order of amino acids
17. secondary, tertiary, and quaternary protein structures
18. CH_2O
19. $C_6H_{12}O_6$
20. simple sugar units hooked together
21. starch
22. plants
23. the structural material of plant cell walls
24. lipids
25. glycerol and long-chain fatty acids
26. some toxic substances interfere with lipid metabolism, and many toxic organic compounds are lipid-soluble
27. steroids
28. hormones
29. catalysts
30. substrates
31. shapes or structures
32. coenzymes
33. what they do
34. stomach
35. split apart protein molecules
36. hydrolyzing enzymes
37. oxidoreductase enzymes
38. nucleic acids
39. nitrogen-containing bases, phosphoric acid, and simple sugars
40. proteins
41. metabolism
42. respiration, fermentation, and photosynthesis

QUESTIONS AND PROBLEMS

1. What is the toxicological importance of lipids? How do lipids relate to hydrophobic ("water-disliking") pollutants and toxicants?

2. What is the function of a hydrolase enzyme?

3. Match the cell structure on the left with its function on the right, below:

   1. Mitochondria
   2. Endoplasmic reticulum
   3. Cell membrane
   4. Cytoplasm
   5. Cell nucleus

   (a) Toxicant metabolism
   (b) Fills the cell
   (c) Contains deoxyribonucleic acid
   (d) Mediate energy conversion and utilization
   (e) Encloses the cell and regulates the passage of materials into and out of the cell interior

4. The formula of simple sugars is $C_6H_{12}O_6$. The simple formula of higher carbohydrates is $C_6H_{10}O_5$. Of course, many of these units are required to make a molecule of starch or cellulose. If higher carbohydrates are formed by joining together molecules of simple sugars, why is there a difference in the ratios of C, H, and O atoms in the higher carbohydrates as compared to the simple sugars?

5. Why does wood contain so much cellulose?

6. What would be the chemical formula of a *tri*saccharide made by the bonding together of three simple sugar molecules?

7. The general formula of cellulose may be represented as $(C_6H_{10}O_5)_x$. If the molecular weight of a molecule of cellulose is 400,000, what is the estimated value of x?

8. During one month a factory for the production of simple sugars, $C_6H_{12}O_6$, by the hydrolysis of cellulose processes one million pounds of cellulose. The percentage of cellulose that undergoes the hydrolysis reaction is 40%. How many pounds of water are consumed in the hydrolysis of cellulose each month?

9. What is the structure of the largest group of atoms common to all amino acid molecules?

10. Glycine and phenylalanine can join together to form two different dipeptides. What are the structures of these two dipeptides?

11. One of the ways in which two parallel protein chains are joined together, or cross linked, is by way of an $-S-S-$ link. What amino acid do you think might be most likely to be involved in such a link? Explain your choice.

12. Fungi, which break down wood, straw, and other plant material, have what are called "exoenzymes." Fungi have no teeth and cannot break up plant material physically by force. Knowing this, what do you suppose an exoenzyme is? Explain how you think it might operate in the process by which fungi break down something as tough as wood.

13. Many fatty acids of lower molecular weight have a bad odor. Speculate as to the reasons that rancid butter has a bad odor. What chemical compound is produced that has a bad odor? What sort of chemical reaction is involved in its production?

14. The long-chain alcohol with 10 carbons is called decanol. What do you think would be the formula of decyl stearate? To what class of compounds would it belong?

15. Write an equation for the chemical reaction between sodium hydroxide and cetyl stearate. What are the products?

16. What are two endocrine glands that are found only in females? What are two of these glands found only in males?

17. The action of bile salts is a little like that of soap. What function do bile salts perform in the intestine? Look up the action of soaps in Sections 7.9 and 12.9, and explain how you think bile salts may function somewhat like soap.

18. If the structure of an enzyme is illustrated as,

how should the structure of its substrate be represented?

19. Look up the structures of ribose and deoxyribose. Explain where the "deoxy" came from in the name, deoxyribose.

20. In what respect are an enzyme and its substrate like two opposite strands of DNA?

21. For what discovery are Watson and Crick noted?

22. Why does an enzyme no longer work if it is denatured?

LITERATURE CITED

1. Manahan, S. E., *Toxicological Chemistry*, 2nd ed., (Chelsea, MI: Lewis Publishers, Inc., 1992).
2. Feigl, D. M., J. W. Hill, and E. Boschman, *Foundations of Life: An Introduction to General, Organic, and Biological Chemistry*, 3rd ed. (New York: Macmillan Publishing Co., 1991).

11 ENVIRONMENTAL CHEMISTRY OF WATER

11.1. INTRODUCTION

Water has been mentioned prominently throughout this book. It was the first example of a chemical compound cited in Chapter 1. Its bonding and molecular structure were considered in Chapter 4, and it appeared often among the reactants and products of chemical reactions discussed in Chapter 5. Water is essential in the formation and reactions of acids, bases, and salts discussed in Chapter 6, it was the solvent used for solutions discussed in Chapter 7, the medium for electrochemical processes that were the subject of Chapter 8, and the medium for essential life processes covered in Chapter 10. Water composes one of the four "spheres" in which environmental chemistry is discussed. This chapter introduces the environmental chemistry of water. Chapter 12 covers water pollution, and Chapter 13 discusses water treatment.

11.2. THE PROPERTIES OF WATER, A UNIQUE SUBSTANCE

Water has a number of unique properties that are essential to life and that determine its environmental chemical behavior. Many of these properties are due to water's polar molecular structure and its ability to form hydrogen bonds, both discussed in Section 7.3. The more important special characteristics of water are summarized in Table 11.1.

11.3. SOURCES AND USES OF WATER: THE HYDROLOGIC CYCLE

The world's water supply is found in the five parts of the hydrologic cycle (Figure 11.1). A large portion of the water is found in the oceans.

Table 11.1. Important Properties of Water

Property	Effects and Significance
Excellent solvent	Transport of nutrients and waste products, making biological processes possible in an aqueous medium
Highest dielectric constant of any common liquid	High solubility of ionic substances and their ionization in solution
Higher surface tension than any other liquid	Controlling factor in physiology; governs drop and surface phenomena
Transparent to visible and longer-wavelength fraction of ultraviolet light	Colorless, allowing light required for photosynthesis to reach considerable depths in bodies of water
Maximum density as a liquid at 4°C	Ice floats; vertical circulation restricted in stratified bodies of water
Higher heat of evaporation than any other material	Determines transfer of heat and water molecules between the atmosphere and bodies of water
Higher latent heat of fusion than any other liquid except ammonia	Temperature stabilized at the freezing point of water
Higher heat capacity than any other liquid except ammonia	Stabilization of temperatures of organisms and geographical regions

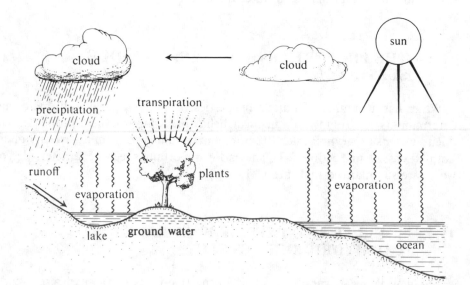

Figure 11.1.   The hydrologic cycle.

Another fraction is present as water vapor in the atmosphere (clouds). Some water is contained in the solid state as ice and snow in snowpacks, glaciers, and the polar ice caps. Surface water is found in lakes, streams, and reservoirs. Groundwater is located in aquifers underground.

There is a strong connection between the *hydrosphere*, where water is found, and the *lithosphere*, or land; human activities affect both. For example, disturbance of land by conversion of grasslands or forests to agricultural land or intensification of agricultural production may reduce vegetation cover, decreasing **transpiration** (loss of water vapor by plants) and affecting the microclimate. The result is increased rain runoff, erosion, and accumulation of silt in bodies of water. The nutrient cycles may be accelerated, leading to nutrient enrichment of surface waters. This, in turn, can profoundly affect the chemical and biological characteristics of bodies of water.

The water that humans use is primarily fresh surface water and groundwater. In arid regions, a small fraction of the water supply comes from the ocean, a source that is becoming more important as the world's supply of fresh water dwindles relative to demand and as massive water desalination plants are installed in arid regions of the world. Saline or brackish groundwaters may also be utilized in some areas.

Groundwater and surface water have appreciably different characteristics. Many substances either dissolve in surface water or become suspended in it on its way to the ocean. Surface water in a lake or reservoir that contains essential mineral nutrients may support a heavy growth of algae. Surface water with a high level of biodegradable organic material, used as food by bacteria, normally contains a large population of bacteria. All these factors have a profound effect upon the quality of surface water.

Groundwater may dissolve minerals from the formations through which it passes. Most microorganisms originally present in groundwater are gradually filtered out as it seeps through mineral formations. Occasionally, the content of undesirable salts may become excessively high in groundwater, although it is generally superior to surface water as a domestic water source.

At present in the continental United States, water usage is about 1.6×10^{12} liters per day, or roughly 10 percent of the average annual precipitation. Agricultural and industrial use each account for approximately 46% of total consumption. Municipal use consumes the remaining 8%.

A major problem with global water supply is its nonuniform distribution with location and time. As illustrated for the continental U.S. by Figure 11.2, precipitation falls unevenly. This is a problem because people in areas with low precipitation, such as the Southwestern continental U.S., often consume more water than people in regions with more rainfall.

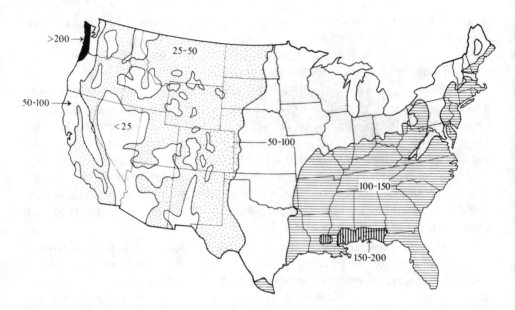

Figure 11.2.   Distribution of precipitation in the continental U.S., showing average annual rainfall in centimeters.

11.4.   THE CHARACTERISTICS OF BODIES OF WATER

The physical condition of a body of water strongly influences the chemical and biological processes that occur in water. **Surface water** is found primarily in streams, lakes, and reservoirs. **Wetlands** are productive flooded areas in which the water is shallow enough to enable growth of bottom-rooted plants. **Estuaries** constitute another type of body of water, consisting of arms of the ocean into which streams flow. The mixing of fresh and salt water gives estuaries unique chemical and biological properties. Estuaries are the breeding grounds of much marine life, which makes their preservation very important.

Water's unique temperature-density relationship results in the formation of distinct layers within nonflowing bodies of water, as shown in Figure 11.3. During the summer a surface layer (**epilimnion**) is heated by solar radiation and, because of its lower density, floats upon the bottom layer, or **hypolimnion**. This phenomenon is called **thermal stratification**. When an appreciable temperature difference exists between the two layers, they do not mix but behave independently and have very different chemical and biological properties. The epilimnion, which is exposed to light, may have a heavy growth of algae. As a result of exposure to the atmosphere and (during daylight hours) because of the photosynthetic activity of algae, the epilimnion contains relatively higher levels of dissolved oxygen and generally is aerobic. Because of the presence of oxygen, oxidized species predominate in the epilimnion. In the hypolimnion, consump-

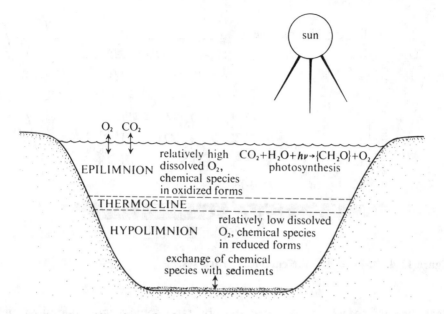

Figure 11.3.   Stratification of a lake.

tion of O_2 by bacterial action on biodegradable organic material may cause the water to become anaerobic. As a consequence, chemical species in a relatively reduced form tend to predominate in the hypolimnion.

The chemistry and biology of the Earth's vast oceans are unique because of the ocean's high salt content, great depth, and other factors. Oceanographic chemistry is a discipline in its own right. The environmental problems of the oceans have increased greatly in recent years because of ocean dumping of pollutants, oil spills, and increased utilization of natural resources from the oceans.

11.5.   AQUATIC CHEMISTRY

Figure 11.4 summarizes the more important aspects of **aquatic chemistry** as it applies to environmental chemistry. As shown in this figure, a number of chemical phenomena occur in water. Many aquatic chemical processes are influenced by the action of algae and bacteria in water. For example, Figure 11.4 shows that algal photosynthesis fixes inorganic carbon from HCO_3^- ion in the form of biomass (represented as CH_2O), in a process that also produces carbonate ion, CO_3^{2-}. Carbonate undergoes an acid-base reaction to produce OH^- ion and raise the pH, or it reacts with Ca^{2+} ion to precipitate solid $CaCO_3$. Most of the many oxidation-reduction reactions that occur in water are mediated (catalyzed) by bacteria. For example, bacteria convert inorganic nitrogen largely to ammonium ion, NH_4^+, in the oxygen-deficient (anaerobic) lower layers of a body of water.

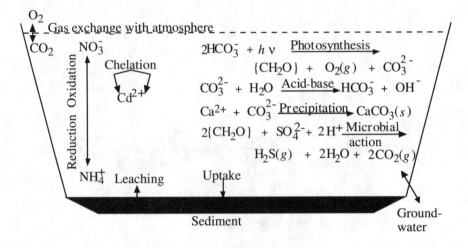

Figure 11.4.   Major aquatic chemical processes.

Near the surface, where O_2 is available, bacteria convert inorganic nitrogen to nitrate ion, NO_3^-. Metals in water may be bound to organic chelating agents, such as pollutant nitrilotriacetic acid (NTA) or naturally-occurring fulvic acids. Gases are exchanged with the atmosphere, and various solutes are exchanged between water and sediments in bodies of water.

Several important characteristics of unpolluted water should be noted. One of these is **gas solubility**. Since it is required to support aquatic life and maintain water quality, oxygen is the most important dissolved gas in water. Water in equilibrium with air at 25°C contains 8.3 milligrams per liter (mg/L) of dissolved O_2. Water **alkalinity** (see Section 11.6) is defined as the ability of solutes in water to neutralize added strong acid. Water **hardness** is due to the presence of calcium ion, Ca^{2+}, and, to a lesser extent, magnesium ion, Mg^{2+}.

11.6.   ALKALINITY AND ACIDITY

Alkalinity

The capacity of water to accept H^+ ions (protons) is called **alkalinity**. Alkalinity is important in water treatment and in the chemistry and biology of natural waters. Frequently, the alkalinity of water must be known to calculate the quantities of chemicals to be added in treating the water. Highly alkaline water often has a high pH and generally contains elevated levels of dissolved solids. These characteristics may be detrimental for water to be used in boilers, food processing, and municipal water systems. Alkalinity serves as a pH buffer and reservoir for inorganic carbon, thus helping to determine the ability of water to support algal growth and other aquatic life. It is used by biologists as a measure of water fertility. Generally, the basic species responsible for alkalinity in water are bicarbonate ion, carbonate ion, and hydroxide ion:

$$HCO_3^- + H^+ \rightarrow CO_2 + H_2O \qquad (11.6.1)$$

$$CO_3^{2-} + H^+ \rightarrow HCO_3^- \qquad (11.6.2)$$

$$OH^- + H^+ \rightarrow H_2O \qquad (11.6.3)$$

Other, usually minor, contributors to alkalinity are ammonia and the conjugate bases of phosphoric, silicic, boric, and organic acids.

It is important to distinguish between high *basicity*, manifested by an elevated pH, and high *alkalinity*, the capacity to accept H^+. Whereas pH is an *intensity* factor, alkalinity is a *capacity* factor. This may be illustrated by comparing a solution of 1.00×10^{-3} M NaOH with a solution of 0.100 M HCO_3^-. The sodium hydroxide solution is quite basic, with a pH of 11, but a liter of it will neutralize only 1.00×10^{-3} mole of acid. The pH of the sodium bicarbonate solution is 8.34, much lower than that of the NaOH. However, a liter of the sodium bicarbonate solution will neutralize 0.100 mole of acid; therefore, its alkalinity is 100 times that of the more basic NaOH solution.

As an example of a water-treatment process in which water alkalinity is important, consider the use of *filter alum*, $Al_2(SO_4)_3 \cdot 18H_2O$ as a coagulant. The hydrated aluminum ion is acidic, and when added to water it reacts with base to form gelatinous aluminum hydroxide,

$$Al(H_2O)_6^{3+} + 3OH^- \rightarrow Al(OH)_3(s) + 6H_2O \qquad (11.6.4)$$

which settles and carries suspended matter with it. This reaction removes alkalinity from the water. Sometimes the addition of more alkalinity is required to prevent the water from becoming too acidic.

In engineering terms, alkalinity frequently is expressed in units of mg/L of $CaCO_3$, based upon the following acid-neutralizing reaction:

$$CaCO_3 + 2H^+ \rightarrow Ca^{2+} + CO_2 + H_2O \qquad (11.6.5)$$

The equivalent weight of calcium carbonate is one-half its formula weight because only one-half of a $CaCO_3$ molecule is required to neutralize one H^+. Expressing alkalinity in terms of mg/L of $CaCO_3$ can, however, lead to confusion, and equivalents/L is a preferable notation for the chemist.

Acidity

Acidity as applied to natural water systems is the capacity of the water to neutralize OH^-. Acidic water is not frequently encountered, except in cases of severe pollution. Acidity generally results from the presence of weak acids such as $H_2PO_4^-$, CO_2, H_2S, proteins, fatty acids, and acidic metal ions, particularly Fe^{3+}. Acidity is more difficult to determine than is alkalinity. One reason for the difficulty in determining acidity is that two of the major contributors are CO_2 and H_2S, both volatile solutes which are readily lost from the sample. The acquisition and preservation of representative samples of water to be analyzed for these gases is difficult.

The term *free mineral acid* is applied to strong acids such as H_2SO_4 and HCl in water. Pollutant acid mine water contains an appreciable concentration of free mineral acid. Whereas total acidity is determined by titration with base to the phenolphthalein endpoint (pH 8.2, where both strong and weak acids are neutralized), free mineral acid is determined by titration with base to the methyl orange endpoint (pH 4.3, where only strong acids are neutralized).

The acidic character of some hydrated metal ions may contribute to acidity as shown by the following example:

$$Al(H_2O)_6^{3+} + H_2O \rightleftharpoons Al(H_2O)_5OH^{2+} + H_3O^+ \qquad (11.6.6)$$

For brevity in this book, the hydronium ion, H_3O^+, is abbreviated simply as H^+ and H^+-accepting water is omitted so that the above equation becomes

$$Al(H_2O)_6^{3+} \rightleftharpoons Al(H_2O)_5OH^{2+} + H^+ \qquad (11.6.7)$$

Some industrial wastes, for example, pickling liquor used to remove corrosion from steel, contain acidic metal ions and often some excess strong acid. For such wastes the determination of acidity is important in calculating the amount of lime, or other chemicals, that must be added to neutralize the acid.

11.7.   METAL IONS AND CALCIUM IN WATER

Metal ions in water, commonly denoted M^{n+}, exist in numerous forms. Despite what the formula implies, a bare metal ion, Mg^{2+}, for example, cannot exist as a separate entity in water. In order to secure the highest stability of their outer electron shells, metal ions in water are bonded, or *coordinated*, to water molecules in forms such as the hydrated metal cation $M(H_2O)_x^{n+}$, or other stronger bases (electron-donor partners) that might be present. Metal ions in aqueous solution seek to reach a state of maximum stability through chemical reactions, including acid-base,

$$Fe(H_2O)_6^{3+} \rightleftharpoons FeOH(H_2O)_5^{2+} + H^+ \qquad (11.7.1)$$

precipitation,

$$Fe(H_2O)_6^{3+} \rightarrow Fe(OH)_3(s) + 3H_2O + 3H^+ \qquad (11.7.2)$$

and oxidation-reduction reactions:

$$Fe(H_2O)_6^{2+} \rightleftharpoons Fe(OH)_3(s) + 3H_2O + e^- + 3H^+ \qquad (11.7.3)$$

These all provide means through which metal ions in water are transformed to more stable forms. Because of reactions such as these and the formation of dimeric species, such as,

$$(H_2O)_4Fe \overset{\displaystyle \overset{H}{O}}{\underset{\displaystyle \underset{H}{O}}{<>}} Fe(H_2O)_4^{4+}$$

the concentration of simple hydrated $Fe(H_2O)_6^{3+}$ ion in water is vanishingly small; the same holds true for many other ionic species dissolved in water.

The properties of metals dissolved in water depend largely upon the nature of metal species dissolved in the water. Therefore, **speciation** of metals plays a crucial role in their environmental chemistry in natural waters and wastewaters. In addition to the hydrated metal ions, for example, $Fe(H_2O)_6^{3+}$ and hydroxo species such as $FeOH(H_2O)_5^{2+}$ discussed above, metals may exist in water reversibly bound to inorganic anions or to organic compounds as **metal complexes**, or they may be present as **organometallic** compounds containing carbon-to-metal bonds. The solubilities, transport properties, and biological effects of such species are often vastly different from those of the metal ions themselves. Subsequent sections of this chapter consider metal species with an emphasis upon metal complexes. Special attention is given to *chelation*, in which particularly strong metal complexes are formed.

Hydrated Metal Ions as Acids

Hydrated metal ions, particularly those with a charge of $+3$ or more, are Brönsted acids, that is, they are species that tend to lose H^+ in aqueous solution. The acidity of a metal ion increases with charge and decreases with increasing radius. Hydrated iron(III) ion is a relatively strong acid, ionizing as follows:

$$Fe(H_2O)_6^{3+} \rightleftharpoons Fe(H_2O)_5OH^{2+} + H^+ \qquad (11.7.1)$$

Hydrated trivalent metal ions, such as iron(III), generally are minus at least one hydrogen ion at neutral pH values or above. Generally, divalent metal ions do not lose a hydrogen ion at pH values below 6, whereas monovalent metal ions such as Na^+ do not act as acids and exist in water solution as simple hydrated ions.

The tendency of hydrated metal ions to behave as acids may have a profound effect upon the aquatic environment. A good example is *acid mine water* (see Sections 11.13 and 12.7), which derives part of its acidic character from the tendency of hydrated iron(III) to lose H^+:

$$Fe(H_2O)_6^{3+} \rightleftharpoons Fe(OH)_3(s) + 3H^+ + 3H_2O \qquad (11.7.4)$$

Hydroxide, OH^-, bonded to a metal ion, may function as a bridging group to join two or more metals together as shown below for iron(III) that has lost H^+:

$$2Fe(H_2O)_5OH^{2+} \rightarrow (H_2O)_4Fe \begin{matrix} H \\ O \\ \diagup \diagdown \\ \diagdown \diagup \\ O \\ H \end{matrix} Fe(H_2O)_4^{4+} + 2H_2O \qquad (11.7.5)$$

The process may continue with formation of higher hydroxy polymers terminating with precipitation of solid metal hydroxide.

Calcium and Hardness

Of the cations found in most freshwater systems, calcium generally has the highest concentration and often has the most influence on aquatic chemistry and water uses and treatment. The chemistry of calcium, although complicated enough, is simpler than that of the transition metal ions found in water. Calcium is a key element in many geochemical processes, and minerals constitute the primary sources of calcium ion in waters. Among the primary contributing minerals are gypsum, $CaSO_4 \cdot 2H_2O$; anhydrite, $CaSO_4$; dolomite, $CaMg(CO_3)_2$; and calcite and aragonite, which are different mineral forms of $CaCO_3$.

Calcium is present in water as a consequence of equilibria between calcium and magnesium carbonate minerals and CO_2 dissolved in water, which it enters from the atmosphere and from decay of organic matter in sediments. These relationships are depicted in Figure 11.5. Water containing a high level of carbon dioxide readily dissolves calcium from its carbonate minerals:

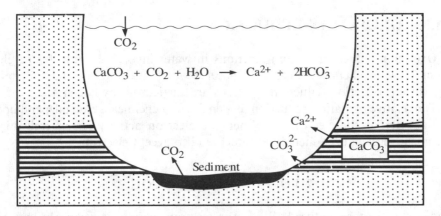

Figure 11.5.   Carbon dioxide-calcium carbonate equilibria

$$CaCO_3(s) + CO_2(aq) + H_2O \rightleftharpoons Ca^{2+} + 2HCO_3^- \qquad (11.7.6)$$

When the above equation is reversed and CO_2 is lost from the water, calcium carbonate deposits are formed. The concentration of CO_2 in water determines the extent of dissolution of calcium carbonate. The carbon dioxide that water may gain by equilibration with the atmosphere is not sufficient to account for the levels of calcium dissolved in natural waters, especially groundwaters. Rather, the respiration of microorganisms degrading organic matter in water, sediments, and soil accounts for the high levels of CO_2 required to dissolve $CaCO_3$ in water. This is an extremely important factor in aquatic chemical processes and geochemical transformations.

Calcium ion, along with magnesium and sometimes iron(II) ion, accounts for **water hardness**. The most common manifestation of water hardness is the curdy precipitate formed by the reaction of soap, a soluble sodium salt of a long-chain fatty acid, with calcium ion in hard water:

$$2C_{17}H_{35}COO^-Na^+ + Ca^{2+} \rightarrow Ca(C_{17}H_{35}CO_2)_2(s) + 2Na^+ \qquad (11.7.7)$$

Temporary hardness is due to the presence of calcium and bicarbonate ions in water and may be eliminated by boiling the water, thus causing the reversal of Equation 11.7.6:

$$Ca^{2+} + 2HCO_3^- \rightleftharpoons CaCO_3(s) + CO_2(g) + H_2O \qquad (11.7.8)$$

Increased temperature may force this reaction to the right by evolving CO_2 gas and a white precipitate of calcium carbonate may form in boiling water having temporary hardness.

11.8.   OXIDATION-REDUCTION

Oxidation-reduction (redox) reactions in water involve the transfer of electrons between chemical species. In natural water, wastewater, and soil, most significant oxidation-reduction reactions are carried out by bacteria.

The relative oxidation-reduction tendencies of a chemical system depend upon the activity of the electron e^-. When the electron activity is relatively high, chemical species (even including water) tend to accept electrons,

$$2H_2O + 2e^- \rightleftarrows H_2(g) + 2OH^- \tag{11.8.1}$$

and are said to be reduced. When the electron activity is relatively low, the medium is **oxidizing**, and chemical species such as H_2O may be **oxidized** by the loss of electrons:

$$2H_2O \rightleftarrows O_2(g) + 4H^+ + 4e^- \tag{11.8.2}$$

The relative tendency toward oxidation or reduction is based upon the electrode potential, E, which is relatively more positive in an oxidizing medium and negative in a reducing medium (see Sections 8.10 and 8.11). It is defined in terms of the half-reaction,

$$2H^+ + 2e^- \rightleftarrows H_2 \tag{11.8.3}$$

for which E is defined as exactly zero when the activity of H^+ is exactly 1 (concentration approximately 1 mole per liter) and the pressure of H_2 gas is exactly 1 atmosphere. Because electron activity in water varies over many orders of magnitude, environmental chemists find it convenient to discuss oxidizing and reducing tendencies in terms of pE, a parameter analogous to pH (pH = $-\log a_{H^+}$) and defined conceptually as the negative log of the electron activity:

$$pE = -\log a_{e^-} \tag{11.8.4}$$

The value of pE is calculated from E by the relationship,

$$pE = \frac{E}{\dfrac{2.303RT}{F}} \tag{11.8.5}$$

At 25°C for E in volts, pE = E/0.0591

where R is the gas constant, T is the absolute temperature, and F is the Faraday.

pE-pH Diagram

The nature of chemical species in water is usually a function of both pE and pH. A good example of this is shown by a simplified pE-pH diagram for iron in water, assuming that iron is in one of the four forms of Fe^{2+} ion, Fe^{3+} ion, solid $Fe(OH)_3$, or solid $Fe(OH)_2$, as shown in Figure 11.6. Water in which the pE is higher than that shown by the upper dashed line is thermodynamically unstable toward oxidation (Reaction 11.8.2), and below the lower dashed line water is thermodynamically unstable toward reduction (Reaction 11.8.3). It is seen that Fe^{3+} ion is stable only in a very oxidizing, acidic medium such as that encountered in acid mine water, whereas Fe^{2+} ion is stable over a relatively large region as reflected by the common occurrence of soluble iron(II) in oxygen-deficient groundwaters. Highly insoluble $Fe(OH)_3$ is the predominant iron species over a very wide pE-pH range.

11.9. COMPLEXATION AND CHELATION

As noted in Section 11.7, metal ions in water are always bonded to water molecules in the form of hydrated ions represented by the general formula, $M(H_2O)_x^{n+}$, from which the H_2O is often omitted for simplicity. Other species may be present that bond to the metal ion more strongly than does water. Specifically, a metal ion in water may combine with an ion or compound that contributes electron pairs to the metal ion. Such a substance is an electron donor, or Lewis base. Called a **ligand,** it bonds to a metal ion to form a **complex** or **coordination compound** (or ion). Thus, cadmium ion in water combines with a cyanide ion ligand to form a complex ion as shown below:

$$Cd^{2+} + CN^- \rightleftharpoons CdCN^+ \qquad (11.9.1)$$

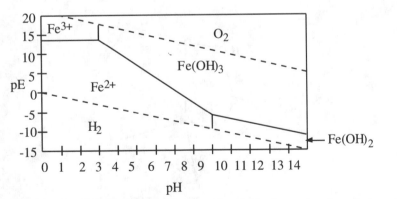

Figure 11.6.   A simplified pE-pH diagram for iron in water (maximum total soluble iron concentration 1.0×10^{-5} M).

Additional cyanide ligands may be added to form the progressively weaker (more easily dissociated) complexes $Cd(CN)_2$, $Cd(CN)_3^-$, and $Cd(CN)_4^{2-}$.

In this example, the cyanide ion is a **unidentate ligand**, which means that it possesses only one site that bonds to the cadmium metal ion. Complexes of unidentate ligands are of relatively little importance in solution in natural waters. Of considerably more importance are complexes with **chelating agents**. A chelating agent has more than one atom that may be bonded to a central metal ion at one time to form a ring structure. One such chelating agent is the nitrilotriacetate (NTA) ligand, which has the following formula:

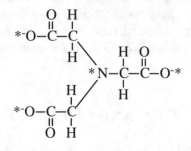

This ion has four binding sites, each marked with an asterisk in the preceding illustration, which may simultaneously bond to a metal ion, forming a structure with three rings. Such a species is known as a **chelate**, and NTA is a chelating agent. In general, since a chelating agent may bond to a metal ion in more than one place simultaneously, chelates are more stable than complexes involving unidentate ligands. Stability tends to increase with the number of chelating sites available on the ligand.

Structures of metal chelates take a number of different forms, all characterized by rings in various configurations. The structure of a tetrahedrally coordinated chelate of nitrilotriacetate ion is shown in Figure 11.7.

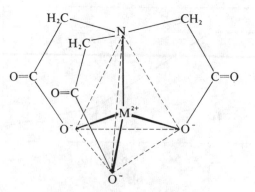

Figure 11.7.   Nitrilotriacetate chelate of a divalent metal ion in a tetrahedral configuration.

The ligands found in natural waters and wastewaters contain a variety of functional groups which can donate the electrons required to bond the ligand to a metal ion. Among the most common of these groups are:

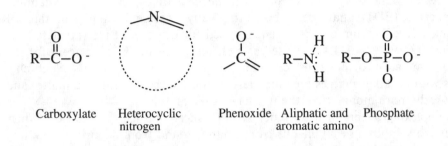

| Carboxylate | Heterocyclic nitrogen | Phenoxide | Aliphatic and aromatic amino | Phosphate |

These ligands complex most metal ions found in unpolluted waters and biological systems (Mg^{2+}, Ca^{2+}, Mn^{2+}, Fe^{2+}, Fe^{3+}, Cu^{2+}, Zn^{2+}, VO^{2+}). They also bind to contaminant metal ions such as Co^{2+}, Ni^{2+}, Sr^{2+}, Cd^{2+}, and Ba^{2+}.

Complexation may have a number of effects, including reactions of both ligands and metals. Among the ligand reactions are oxidation-reduction, decarboxylation, and hydrolysis. Complexation may cause changes in the oxidation state of the metal and may result in a metal becoming solubilized from an insoluble compound. The formation of insoluble complex compounds removes metal ions from solution. For example, complexation with negatively charged ligands may convert a soluble metal species from a cation to an anion, such as $Ni(CN)_4^{2-}$. Whereas cationic species are readily bound and immobilized by ion exchange processes in soil, anionic species are not strongly held by soil. Thus, codisposal of metal salts and chelating agents in wastes can result in increased hazards from heavy metals. On the other hand, some chelating agents are used for the treatment of heavy metal poisoning and insoluble chelating agents, such as chelating resins, can be used to remove metals from waste streams. Metal ions chelated by hazardous waste chelating agents, such as NTA from metal plating bath solutions, may be especially mobile in water.

Complex compounds and chelates of metals such as iron (in hemoglobin) and magnesium (in chlorophyll) are vital to life processes. Naturally occurring chelating agents, such as humic substances and amino acids, are found in water and soil. The high concentration of chloride ion in seawater results in the formation of some chloro complexes. Synthetic chelating agents such as sodium tripolyphosphate, sodium ethylenediaminetetraacetate (EDTA), sodium nitrilotriacetate (NTA), and sodium citrate are produced in large quantities for use in metal-plating baths, industrial water treatment, detergent formulations, and food preparation. Small quantities of these compounds enter aquatic systems through waste discharges.

Occurrence and Importance of Chelating Agents in Water

Chelating agents are common potential water pollutants. These substances can occur in sewage effluent and industrial wastewater such as metal plating wastewater. Chelates formed by the strong chelating agent ethylenediamine-tetraacetate (EDTA) have been shown to greatly increase the migration rates of radioactive ^{60}Co from pits and trenches used for disposal of intermediate-level radioactive waste. EDTA was used as a cleaning and solubilizing agent for the decontamination of hot cells, equipment, and reactor components. Such chelates with negative charges are much less strongly sorbed by mineral matter and are vastly more mobile than the unchelated metal ions.

Complexing agents in wastewater are of concern primarily because of their ability to solubilize heavy metals from plumbing and from deposits containing heavy metals. Complexation may increase the leaching of heavy metals from waste disposal sites and reduce the efficiency with which heavy metals are removed with sludge in conventional biological waste treatment. Removal of chelated iron is difficult with conventional municipal water treatment processes. Iron(III) and perhaps several other essential micronutrient metal ions are kept in solution by chelation in algal cultures. The yellow-brown color of some natural waters is due to naturally-occurring chelates of iron.

Complexation by Humic Substances

The most important class of complexing agents that occur naturally are the **humic substances**. These are degradation-resistant materials formed during the decomposition of vegetation that occurs as deposits in soil, marsh sediments, peat, coal, lignite, or in almost any location where large quantities of vegetation have decayed. They are commonly classified on the basis of solubility. If a material containing humic substances is extracted with strong base, and the resulting solution is acidified, the products are (a) a nonextractable plant residue called **humin**; (b) a material that precipitates from the acidified extract, called **humic acid**; and (c) an organic material that remains in the acidified solution, called **fulvic acid**. Because of their acid-base, sorptive, and complexing properties, both the soluble and insoluble humic substances have a strong effect upon the properties of water. In general, fulvic acid dissolves in water and exerts its effects as the soluble species. Humin and humic acid remain insoluble and affect water quality through exchange of species, such as cations or organic materials, with water.

Humic substances are high-molecular-weight, polyelectrolytic macromolecules. Molecular weights range from a few hundred for fulvic acid to tens of thousands for the humic acid and humin fractions.

Some feeling for the nature of humic substances may be obtained by considering the structure of a hypothetical molecule of fulvic acid below:

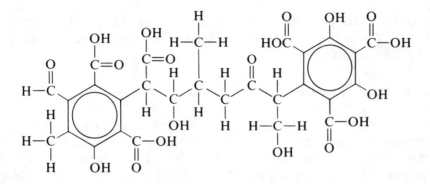

The binding of metal ions by humic substances is one of the most important environmental qualities of humic substances. This binding can occur as chelation between a carboxyl group and a phenolic hydroxyl group, as chelation between two carboxyl groups, or as complexation with a carboxyl group (see Figure 11.8).

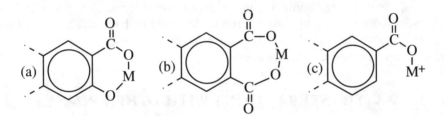

Figure 11.8.   Binding of a metal ion, M^{2+}, by humic substances (a) by chelation between carboxyl and phenolic hydroxyl, (b) by chelation between two carboxyl groups, and (c) by complexation with a carboxyl group.

Soluble fulvic acid complexes of metals may be important in natural waters. They probably keep some of the biologically important transition-metal ions in solution and are particularly involved in iron solubilization and transport. Fulvic acid-type compounds are associated with color in water. These yellow materials, called **Gelbstoffe**, frequently are encountered along with soluble iron.

Insoluble humic substances, the humins and humic acids, effectively exchange cations with water and may accumulate large quantities of metals. Lignite coal, which is largely a humic-acid material, tends to remove some metal ions from water.

Special attention has been given to humic substances since about 1970, following the discovery of **trihalomethanes** (THMs, such as chloroform and dibromochloromethane) in water supplies. It is now generally believed that these suspected carcinogens can be formed in the presence of humic substances during the disinfection of raw municipal drinking water by chlorination (see Chapter 13).

The humic substances produce THMs by reaction with chlorine. The formation of THMs can be reduced by removing as much of the humic material as possible prior to chlorination.

Metals Bound as Organometallic Compounds

Another major type of metal species important in hazardous wastes consists of **organometallic compounds,** which differ from complexes and chelates in that the organic portion is bonded to the metal by a carbon-metal bond and the organic ligand is frequently not capable of existing as a stable separate species. Typical examples of organometallic compound species are monomethylmercury ion and dimethylmercury:

$$Hg^{2+} \qquad\qquad HgCH_3{}^+ \qquad\qquad Hg(CH_3)_2$$
Mercury(II) ion   Monomethylmercury ion   Dimethylmercury

Organometallic compounds may enter the environment directly as pollutant industrial chemicals and some, including organometallic mercury, tin, selenium, and arsenic compounds, are synthesized biologically by bacteria. Some of these compounds are particularly toxic because of their mobilities in living systems and abilities to cross cell membranes.

11.10.   WATER INTERACTIONS WITH OTHER PHASES

Most of the important chemical phenomena associated with water do not occur in solution, but rather through interaction of solutes in water with other phases. For example, the oxidation-reduction reactions catalyzed by bacteria occur in bacterial cells. Many organic hazardous wastes are carried through water as emulsions of very small particles suspended in water. Some hazardous wastes are deposited in sediments in bodies of water, from which they may later enter the water through chemical or physical processes and cause severe pollution effects.

Figure 11.9 summarizes some of the most significant types of interactions between water and other phases, including solids, immiscible liquids, and gases. Films of organic compounds, such as hydrocarbon liquids, may be present on the surface of water. Exposed to sunlight, these compounds are subject to photochemical reactions (see Section 16.4). Gases such as O_2, CO_2, CH_4, and H_2S are exchanged with the atmosphere. Photosynthesis occurs in suspended cells of algae, and other biological processes, such as biodegradation of organic wastes, occur in bacterial cells. Particles contributing to the turbidity of water may be introduced by physical processes, including the erosion of streams or sloughing of water impoundment banks. Chemical processes, such as the formation of solid $CaCO_3$, illustrated in Figure 11.9, may also form particles in water.

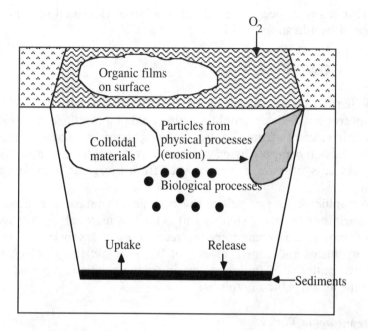

Figure 11.9.   Aquatic chemical processes at interfaces between water and gases, solids, or other liquids.

Sediments

Sediments are repositories of a wide variety of chemical species and the site of many chemical and biochemical processes. Anaerobic fermentation of organic matter by bacteria produces methane gas that is evolved from sediments. Similar bacteria produce mobile $HgCH_3^+$ and $Hg(CH_3)_2$ from insoluble, relatively harmless inorganic mercury compounds. Sediments are sinks for many hazardous organic compounds and heavy metal salts that have gotten into water.

Colloids

The most important interactions between species dissolved in aqueous solution and species in other phases occur with colloidal particles. Colloidal particles are very small particles ranging from 0.001 micrometer (μm) to 1 μm in diameter. Colloids have a strong influence on aquatic chemistry. Because of their extremely small size, these particles have a very high surface-to-volume ratio. They are small enough to remain suspended in water, enabling maximum exposure to the water and solutes dissolved in it. Toxic substances in colloidal form are much more available to organisms in water than are such substances in bulk form. Special measures are required to remove colloidal particles from water. Usually, chemical treatment measures are applied to cause colloidal particles to aggregate together (processes called **coagulation** or **flocculation**), and the solids are removed by filtration.

11.11.   AQUATIC LIFE

The living organisms (**biota**) in an aquatic ecosystem may be classified as either autotrophic or heterotrophic. **Autotrophic** biota utilize solar or chemical energy to fix elements from simple, nonliving inorganic material into complex life molecules that compose living organisms. Autotrophic organisms that utilize solar energy to synthesize organic matter from inorganic materials are called **producers.**

Heterotrophic organisms utilize the organic substances produced by auto-trophic organisms as energy sources and as the raw materials for the synthesis of their own biomass. **Decomposers** (or **reducers**) are a subclass of the hetero-trophic organisms and consist chiefly of bacteria and fungi, which ultimately break down material of biological origin to the simple compounds originally fixed by the autotrophic organisms.

Microorganisms in Water

Microorganisms comprise a diverse group of organisms generally capable of existing as single cells that can be seen only under a microscope. Microscopically small single-celled microorganisms, consisting of bacteria, fungi, and algae are of the utmost importance in water for a number of reasons as listed below:

- Through their ability to fix inorganic carbon, algae and photosynthetic bacteria are the predominant producers of the biomass that supports the rest of the food chain in bodies of water.
- As catalysts of aquatic chemical reactions, bacteria mediate most of the significant oxidation-reduction processes that occur in water.
- By breaking down biomass and mineralizing essential elements, especially nitrogen and phosphorus, aquatic microorganisms play a key role in nutrient cycling.
- Aquatic microorganisms are essential for the major biogeochemical cycles.
- Aquatic bacteria are responsible for the breakdown and detoxication of many xenobiotic pollutants that get into the hydrosphere.

From the viewpoint of environmental chemistry, the small size of microorganisms is particularly significant because it gives them a very high surface/volume ratio enabling very rapid exchange of nutrients and metabolic products with their surroundings, and resulting in exceptionally high rates of metabolic reactions. This, combined with the spectacularly fast geometric increase in population of single-celled microorganisms by fission during the log phase of growth (see Figure 11.11), enables microganisms to multiply very rapidly on environmental chemical substrates, such as biodegradable organic matter.

Microorganisms function as living catalysts that enable a vast number of chemical processes to occur in water and soil. A majority of the important chemical reactions that take place in water and soil, particularly those involving organic matter and oxidation-reduction processes, occur through bacterial intermediaries. Algae are the primary producers of biological organic matter (biomass) in water. Microorganisms are responsible for the formation of many sediment and mineral deposits; they also play the dominant role in secondary wastewater treatment. Pathogenic microorganisms must be eliminated from water purified for domestic use.

Algae

For the purposes of discussion here, **algae** may be considered as generally microscopic organisms that subsist on inorganic nutrients and produce organic matter from carbon dioxide by photosynthesis. In a highly simplified form, the production of organic matter by algal photosynthesis is described by the reaction

$$CO_2 + H_2O \xrightarrow{h\nu} \{CH_2O\} + O_2(g) \qquad (11.11.1)$$

where $\{CH_2O\}$ represents a unit of carbohydrate and $h\nu$ stands for the energy of a quantum of light.

Fungi

Fungi are nonphotosynthetic organisms. The morphology (structure) of fungi covers a wide range, and is frequently manifested by filamentous structures. Discovered in 1992, the largest and oldest organism known is a huge fungus covering about 40 acres of Northern Michigan forest land. Fungi are aerobic (oxygen-requiring) organisms and generally tolerate more acidic media and higher concentrations of heavy metal ions than bacteria.

Although fungi do not grow well in water, they play an important role in determining the composition of natural waters and wastewaters because of the large amount of their decomposition products that enter water from the breakdown of cellulose in wood and other plant materials. To accomplish this, fungal cells secrete an extracellular enzyme (exoenzyme), *cellulase*. An environmentally important by-product of fungal decomposition of plant matter is humic material (Section 11.9), which interacts with hydrogen ions and metals.

11.12.   BACTERIA

Bacteria are single-celled prokaryotic microorganisms shaped as rods (**bacillus**), spheres (**coccus**), or spirals (**vibrios, spirilla, spirochetes**) that are uniquely important in environmental chemistry. Characteristics of most bacteria include a semirigid cell wall, motility with flagella for those capable of movement, unicellular nature (although clusters of cloned bacterial cells are common), and multiplication by binary fission in which each of two daughter cells is genetically identical to the parent cell.

Bacteria obtain the energy and raw materials needed for their metabolic processes and reproduction by mediating chemical reactions. Nature provides a large number of such reactions, and bacterial species have evolved that utilize many of these. As a consequence of their participation in such reactions, bacteria are involved in many biogeochemical processes and elemental transitions and cycles in water and soil.

The metabolic activity of bacteria is greatly influenced by their small size, which is of the order of a micrometer in magnitude. Their surface-to-volume ratio is extremely large, so that the inside of a bacterial cell is highly accessible to a chemical substance in the surrounding medium. Thus, for the same reason that a finely divided catalyst is more efficient than a more coarsely divided one, bacteria may bring about very rapid chemical reactions compared to those mediated by larger organisms.

An example of **autotrophic bacteria** is *Gallionella*, which, like all autotrophic bacteria, employs inorganic carbon as a carbon source and derives its energy from mediating a chemical reaction:

$$4FeS(s) + 9O_2 + 10H_2O \leftrightarrow 4Fe(OH)_3(s) + 4SO_4^{2-} + 8H^+ \qquad (11.12.1)$$

Because of their consumption and production of a wide range of minerals, autotrophic bacteria are involved in many geochemical transformations.

Heterotrophic bacteria are much more common in occurrence than autotrophic bacteria. They are the microorganisms primarily responsible for the breakdown of pollutant organic matter in waters and of organic wastes in biological waste treatment processes.

Aerobic bacteria require molecular oxygen as an electron receptor:

$$O_2 + 4H^+ + 4e^- \rightarrow 2H_2O \qquad (11.12.2)$$

Anaerobic bacteria function only in the complete absence of molecular oxygen, using substances such as nitrate ion and sulfate ion as substitutes for O_2. Frequently, molecular oxygen is quite toxic to anaerobic bacteria. A third class of bacteria, **facultative bacteria**, utilize free oxygen when it is available and use other substances as electron receptors (oxidants) when molecular oxygen is not available.

The Prokaryotic Bacterial Cell

Unicellular bacteria and cyanobacteria consist of prokaryotic cells, which differ in several major respects from the eukaryotic cells of higher organisms. Illustrated in Figure 11.10, prokaryotic bacterial cells are enclosed in a **cell wall,** which in many bacteria is frequently surrounded by a **slime layer** (capsule). The thin **cell membrane** or **cytoplasmic membrane** on the inner surface of the cell wall encloses the cellular cytoplasm, controls the nature and quantity of materials transported into and out of the cell, and is susceptible to damage from some toxic substances. Hairlike **pili** on the surface of a bacterial cell enable the cell to stick to surfaces. **Flagella** are moveable appendages that enable motile bacterial cells to move by their whipping action. Bacterial cells are filled with **cytoplasm,** an aqueous solution and suspension containing proteins, lipids, carbohydrates, nucleic acids, ions, and other materials constituting the medium in which the cell's metabolic processes are carried out. The major constituents of cytoplasm are a **nuclear body** composed of a single DNA macromolecule that controls metabolic processes and reproduction; **inclusions** of reserve food material, usually consisting of fats and carbohydrates; and **ribosomes,** which function as sites of protein synthesis.

Rate of Bacterial Growth

The population size of bacteria and unicellular algae as a function of time in a growth culture is illustrated by Figure 11.11, which shows a **population curve** for a bacterial culture. Such a culture is started by inoculating a rich nutrient

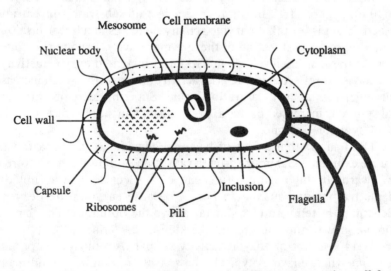

Figure 11.10.   Generic prokaryotic bacterial cell illustrating major cell features.

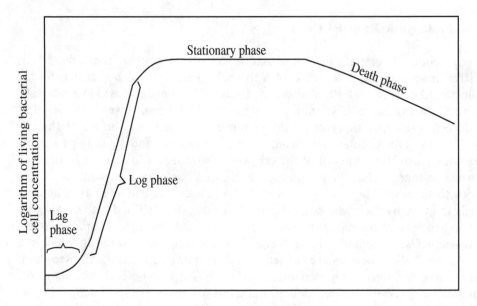

Figure 11.11.   Population curve for a bacterial culture.

medium with a small number of bacterial cells. Bacteria multiply exponentially in the log phase.

Bacterial Metabolism

Metabolic reactions by which bacteria break down substances and extract energy, as well as synthesizing biological materials, are mediated by enzymes (see Section 10.6). Figure 11.12 illustrates the effect of **substrate concentration** on enzyme activity. It is seen that enzyme activity increases in a linear fashion up to a value that represents saturation of the enzyme activity. Beyond this concentration, increasing substrate levels do not result in increased enzyme activity. This kind of behavior is reflected in bacterial activity and growth, which increase with available nutrients up to a saturation value. Superimposed on this plot in a bacterial system is increased bacterial population, which, in effect, increases the amount of available enzyme.

Figure 11.13 shows that a plot of bacterial enzyme activity as a function of temperature exhibits a maximum growth rate at an optimum temperature that is skewed toward the high temperature end of the curve. The abrupt dropoff beyond the temperature maximum occurs because enzymes are destroyed by being denatured at temperatures not far above the optimum. The temperature for optimum growth rate varies with the kind of bacteria.

Figure 11.14 is a plot of enzyme activity vs. pH. Although the optimum pH will vary somewhat, enzymes typically have a pH optimum around neutrality. Enzymes tend to become denatured at pH extremes. This behavior likewise is

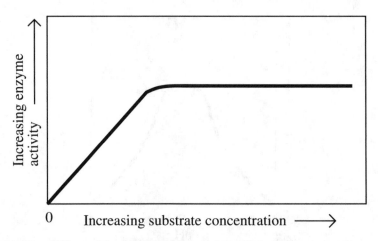

Figure 11.12.   Effect of increasing substrate concentration on enzyme activity.

reflected in plots of bacterial metabolism as a function of pH. For some bacteria that thrive in acid, such as those that generate sulfuric acid by the oxidation of sulfide or organic acids by fermentation processes, the optimum pH may be quite low.

11.13.   MICROBIALLY MEDIATED ELEMENTAL TRANSITIONS AND CYCLES

Microbially mediated transitions between elemental species and the elemental cycles resulting therefrom constitute one of the most significant aspects of geo-

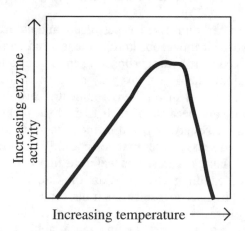

Figure 11.13.   Enzyme activity as a function of temperature. A plot of bacterial growth vs. temperature has the same shape.

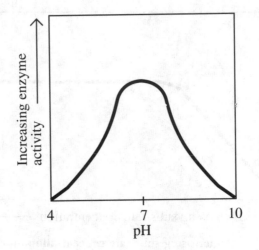

Figure 11.14. Enzyme activity as a function of pH.

chemistry as influenced by biological processes. This section addresses such cycles for several important elements.

Microbial Transformations of Carbon

The carbon cycle is represented in Figure 11.15. A relatively small, but highly significant, portion of global carbon is in the atmosphere as CO_2. A very large amount of carbon is present as minerals, particularly calcium and magnesium carbonates.

Another fraction of carbon is fixed as petroleum and natural gas, with a much larger amount as hydrocarbonaceous kerogen, coal, and lignite. Manufacturing processes are used to convert hydrocarbons to xenobiotic compounds with functional groups containing halogens, oxygen, nitrogen, phosphorus, or sulfur. Though a very small amount of total environmental carbon, these compounds are particularly significant because of their toxicological chemical effects.

Microorganisms are strongly involved in the carbon cycle, mediating crucial biochemical reactions discussed later in this section. Photosynthetic algae are the predominant carbon-fixing organisms in water; as they consume CO_2, the pH of the water is raised, enabling precipitation of $CaCO_3$ and $CaCO_3 \cdot MgCO_3$. Organic carbon fixed by microorganisms is transformed by biogeochemical processes to fossil petroleum, kerogen, coal, and lignite. Microorganisms degrade organic carbon from biomass, petroleum, and xenobiotic sources, ultimately returning it to the atmosphere as CO_2. The following summarize the prominent ways in which microorganisms are involved in the carbon cycle:

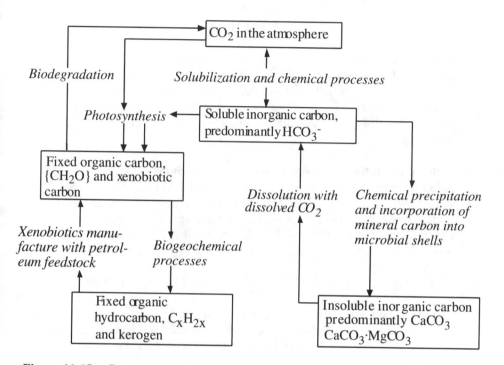

Figure 11.15.   Important aspects of the biogeochemical carbon cycle.

- **Photosynthesis** in which algae, higher plants, and photosynthetic bacteria use light energy to fix inorganic carbon in a high-energy organic form:

$$CO_2 + H_2O \xrightarrow{h\nu} [CH_2O] + O_2(g)$$

- **Respiration** in which organic matter is oxidized in the presence of molecular O_2 **(aerobic respiration)**

$$\{CH_2O\} + O_2(g) \rightarrow CO_2 + H_2O$$

or **anaerobic respiration**, which uses oxidants other than O_2, such as NO_3^- or SO_4^{2-}.

- **Degradation of biomass** by bacteria and fungi. Biodegradation of dead organic matter consisting predominantly of plant residues prevents accumulation of excess waste residue and converts organic carbon, nitrogen, sulfur, and phosphorus to simple organic forms that can be utilized by plants. It is a key part of the biogeochemical cycles of these elements and also leaves a humus residue that is required for optimum physical condition of soil.

- **Methane production** by methane-forming bacteria, such as *Methanobacterium*, in anoxic (oxygen-less) sediments,

$$2\{CH_2O\} \rightarrow CH_4 + CO_2$$

plays a key role in local and global carbon cycles as the final step in the anaerobic decomposition of organic matter. It is the source of about 80% of the methane entering the atmosphere. Microbial methane production is a **fermentation reaction**, defined as an oxidation-reduction process in which both the oxidizing agent and reducing agent are organic substances.

- **Bacterial utilization and degradation of hydrocarbons**. The oxidation of higher hydrocarbons under aerobic conditions by *Micrococcus, Pseudomonas, Mycobacterium*, and *Nocardia* is an important environmental process by which petroleum wastes are eliminated from water and soil. The initial step in the microbial oxidation of alkanes is conversion of a terminal $-CH_3$ group to a $-CO_2$ group followed by β-oxidation,

$$CH_3CH_2CH_2CH_2CO_2H + 3O_2 \rightarrow CH_3CH_2CO_2H + 2CO_2 + 2H_2O$$

wherein carbon atoms are removed in two-carbon fragments. The overall process leading to ring cleavage in aromatic hydrocarbons is the following in which cleavage is preceded by addition of OH to adjacent carbon atoms:

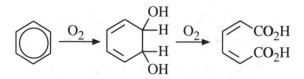

- **Biodegradation of organic matter**, such as occurs in treatment of municipal wastewater by reactions represented in a general sense by,

$$\{CH_2O\} + O_2(g) \rightarrow CO_2 + H_2O + biomass$$

Microbial Transformations of Nitrogen

Some of the most important microorganism-mediated chemical reactions in aquatic and soil environments are those involving nitrogen compounds. They are key constituents of the **nitrogen cycle** shown in Figure 11.16, which describes the dynamic processes through which nitrogen is interchanged among the atmosphere, organic matter, and inorganic compounds. The key microbially-mediated processes in the nitrogen cycle are the following:

- **Nitrogen fixation,** the binding of atmospheric nitrogen in a chemically combined form:

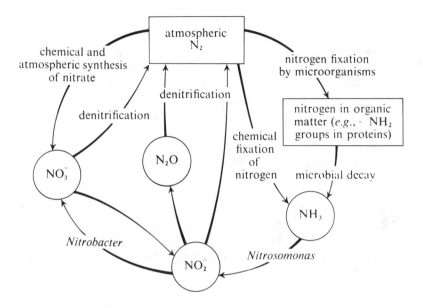

Figure 11.16. Aspects of the nitrogen cycle.

$$3\{CH_2O\} + 2N_2 + 3H_2O + 4H^+ \rightarrow 3CO_2 + 4NH_4^+$$

Biological nitrogen fixation is a key biochemical process in the environment and is essential for plant growth in the absence of synthetic fertilizers.

- **Nitrification**, the conversion of N(-III) to N(V) catalyzed by *Nitrosomonas* and *Nitrobacter*.

$$2O_2 + NH_4^+ \rightarrow NO_3^- + 2H^+ + H_2O$$

Nitrification is especially important in nature because nitrogen is absorbed by plants primarily as nitrate. When fertilizers are applied in the form of ammonium salts or anhydrous ammonia, the ammonia is microbially oxidized to nitrate, which can be assimilated by plants.

- **Nitrate reduction** by which nitrogen in chemical compounds is reduced by microbial action to lower oxidation states in the absence of free oxygen:

$$2NO_3^- + \{CH_2O\} \rightarrow 2NO_2^- + H_2O + CO_2$$

$$2NO_2^- + 3\{CH_2O\} + 4H^+ \rightarrow 2NH_4^+ + 3CO_2 + H_2O$$

- **Denitrification**, which produces N_2 gas from chemically fixed nitrogen:

$$4NO_3^- + 5\{CH_2O\} + 4H^+ \rightarrow 2N_2 + 5CO_2 + 7H_2O$$

Denitrification is the mechanism by which fixed nitrogen is returned to the atmosphere and is useful in advanced water treatment for the removal of nutrient nitrogen. Loss of nitrogen to the atmosphere may also occur through the formation of N_2O and NO by bacterial action on nitrate and nitrite catalyzed by several types of bacteria.

Microbial Transformations of Sulfur

The sulfur cycle involves interconversions among a number of sulfur species, including inorganic soluble sulfates, insoluble sulfates, soluble sulfide, gaseous hydrogen sulfide, and insoluble sulfides; biologically bound sulfur; and sulfur in synthetic organic compounds. The major microbially mediated processes in this cycle are the following:

- **Sulfate reduction** to sulfide by bacteria such as *Desulfovibrio,* which utilizes sulfate as an electron acceptor in the oxidation of organic matter:

$$SO_4^{2-} + 2\{CH_2O\} + 2H^+ \rightarrow H_2S + 2CO_2 + 2H_2O$$

The odiferous and toxic H_2S product may cause serious problems with water quality.

- **Sulfide oxidation** by bacteria such as *Thiobacillus*:

$$2H_2S + 4O_2 \rightarrow 4H^+ + 2SO_4^{2-}$$

Oxidation of sulfur in a low oxidation state to sulfate ion produces sulfuric acid, a strong acid. Some of the bacteria that mediate this reaction, such as *Thiobacillus thiooxidans*, are remarkably acid-tolerant. Acid-tolerant sulfur-oxidizing bacteria produce, and thrive in, environmentally-damaging acidic waters, such as acid mine water.

- **Degradation of organic sulfur compounds** by bacterially-mediated processes that can result in production of strong-smelling, noxious, volatile organic sulfur compounds, such as methyl thiol, CH_3SH, and dimethyl disulfide, CH_3SSCH_3. The formation of these compounds, in addition to that of H_2S, accounts for much of the odor associated with the biodegradation of sulfur-containing organic compounds. Hydrogen sulfide is formed from a large variety of organic compounds through the action of a number of different kinds of microorganisms.

Microbial Transformations of Phosphorus

The phosphorus cycle involves natural and pollutant sources of phosphorus including biological, organic, and inorganic phosphorus. Biological phosphorus is a key constituent of cellular DNA. Organic phosphorus occurs in organophosphate insecticides. The major inorganic phosphorus species are soluble $H_2PO_4^-$ and HPO_4^{2-} and insoluble $Ca_5(OH)(PO_4)_3$. Soil and aquatic microbial processes are very important in the phosphorus cycle. Of particular importance is the fact that phosphorus is the most common limiting nutrient in water, particularly for the growth of algae. Bacteria are even more effective than algae in taking up phosphate from water, accumulating it as excess cellular phosphorus that can be released to support additional bacterial growth, if the supply of phosphorus becomes limiting. Microorganisms that die release phosphorus that can support additional organisms.

Biodegradation of phosphorus compounds is important in the environment for two reasons. The first of these is that it is a *mineralization* process that releases inorganic phosphorus from the organic form, thereby providing a source of algal nutrient orthophosphate. Secondly, biodegradation deactivates highly toxic organophosphate compounds, such as the phosphate ester insecticides.

Microbial Transformations of Halogens and Organohalides

Among the more important microbial processes that operate on pollutant xenobiotic compounds in soil and water are those involving the degradation of organohalide compounds. Such compounds, particularly the organochloride compounds, are among the more abundant air and water pollutants and hazardous waste constituents. Some are relatively toxic and even carcinogenic, and they tend to accumulate in lipid tissues. The key step in biodegradation of organohalide compounds is **dehalogenation**, which involves the replacement of a halogen atom:

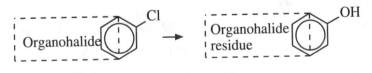

Microbial Transformations of Iron

Some bacteria, including *Ferrobacillus, Gallionella,* and some forms of *Sphaerotilus*, utilize iron compounds in obtaining energy for their metabolic needs by oxidizing iron(II) to iron(III) with molecular oxygen:

$$4Fe(II) + 4H^+ + O_2 \rightarrow 4Fe(III) + 2H_2O \qquad (11.13.1)$$

The carbon source for some of these bacteria is CO_2. Since they do not require organic matter for carbon, and because they derive energy from the oxidation of inorganic matter, these bacteria may thrive in environments where organic matter is absent. The iron(III) product is amassed as large quantities of solid $Fe(OH)_3$, so that large deposits of hydrated iron(III) oxide form in areas where iron-oxidizing bacteria thrive. Some of the iron bacteria, notably *Gallionella*, secrete large quantities of hydrated iron(III) oxide in the form of intricately branched structures that grow at the end of a twisted stalk of the iron oxide.

Acid Mine Waters

One consequence of bacterial action on metal compounds is acid mine drainage, a common and damaging water pollution problem. Acid mine water results from the presence of sulfuric acid produced by the oxidation of pyrite, FeS_2. Microorganisms are closely involved in the overall process, which consists of several reactions. Acid mine water is discussed further in Section 12.6.

CHAPTER SUMMARY

The chapter summary below is presented in a programmed format to review the main points covered in this chapter. It is used most effectively by filling in the blanks, referring back to the chapter as necessary. The correct answers are given at the end of the summary.

The most important unique properties of water that largely determine its environmental chemical behavior are (1)_____

Thermal stratification of bodies of water results from water's (2)_____

_____. Most of the many oxidation-reduction reactions that occur in water are mediated by (3)_____. The ability of solutes in water to neutralize added strong acid is called (4)_____

and water hardness is due mostly to the presence of (5)_____. In water near neutral pH the major contributor to alkalinity is (6)_____.

A major pollutant contributor to acidity is free mineral acid manifested by the presence of (7)_____. The hydronium ion, H_3O^+, indicates (8)_____, and can be abbreviated simply as (9)_____.

A bare metal ion cannot exist as a separate entity in water, but is present instead as (10)_____. Because they tend to lose H^+ in aqueous solution, hydrated metal ions with a charge of $+3$ or more are (11)_____. Calcium is present in water as a consequence of (12)_____.

The reaction $2C_{17}H_{35}COO^-Na^+ + Ca^{2+} \rightarrow Ca(C_{17}H_{35}CO_2)_2(s) + 2Na^+$ is a manifestation of (13)_____.

Oxidation-reduction reactions in water involve the transfer of (14)_____ _____, and in natural water, wastewater, and soil are carried out by (15)_____. The relative oxidation-reduction tendencies of a chemical system depend upon (16)_____. The parameter pE is defined conceptually as (17)_____. In the pE-pH diagram for iron in water, the species that predominates at low pE and low pH is (18)_____, whereas at high pE and higher pH it is (19)_____. In the pE-pH diagram, the upper and lower dashed lines show (20)_____ _____, respectively.

A ligand bonds to a metal ion to form a (21)_____. The NTA anion as a ligand has (22)_____ binding sites, and it forms chelates with three rings. Some of the more significant environmental effects of complexation and chelation are (23)_____ _____. Complexing agents in wastewater are of concern primarily because of (24)_____. The most important class of complexing agents that occur naturally are (25)_____, which are (26)_____. These substances are divided into the three classes of (27)_____ and are commonly divided on the basis of (28)_____. They are most significant in drinking water supplies because of (29)_____ _____. Organometallic compounds differ from complexes and chelates in that (30)_____

_____.

Most of the important chemical phenomena associated with water do not occur in solution, but rather through (31)_____,

which may include (32)_____.

Sediments are (33)_____

_____ and the site of (34)_____.

Very small particles ranging from 0.001 micrometer (μm) to 1 μm in diameter have a very high surface-to-volume ratio and are called (35)_____.

Living organisms that utilize solar or chemical energy to fix elements from simple, nonliving inorganic material into complex life molecules that compose living organisms are called (36)_____, whereas

heterotrophic organisms (37)_____

_____. Four

reasons that microorganisms are especially important in water are (38)_____

_____

_____.

The small size of microorganisms is particularly significant because (39)_____.

In a simplified sense the basic reaction by which algae produce biomass is (40)_____. Fungi are important in the

environmental chemistry of water because (41)_____

_____.

Three classifications of bacteria based upon their requirement for oxygen are (42)_____.

The kinds of cells that bacteria have are (43)_____.

The major phases in a bacterial growth curve are (44)_____

_____.

A plot of bacterial enzyme activity as a function of temperature is shaped such that it (45)_____

_____

because (46)_____

_____.

The prominent ways in which microorganisms are involved in the carbon cycle
are (47)_____

_____.

The steps in the nitrogen cycle that involve bacteria are (48)_____

_____. Microbial transformations of sulfur

involve (49)_____

_____. The two reasons for

which biodegradation of phosphorus compounds is important in the environ-

ment are (50)_____

_____. The key step in biodegradation

of organohalide compounds is (51)_____. The

bacterially-catalyzed oxidation of iron(II) to iron(III) by bacteria can result in

the formation of large deposits of (52)_____. Acid mine water

results from the presence of (53)_____ produced by the oxidation of

(54)_____.

Answers

1. Properties of water as listed in Table 11.1.
2. unique temperature-density relationship
3. bacteria
4. alkalinity
5. Ca^{2+} ion
6. HCO_3^- ion
7. strong acids such as H_2SO_4 and HCl in water
8. H^+ ion bound to H_2O
9. H^+
10. the hydrated metal cation $M(H_2O)_x^{n+}$
11. Brönsted acids
12. equilibria between calcium and magnesium carbonate minerals and CO_2 dissolved in water
13. water hardness
14. electrons between chemical species
15. bacteria

16. the activity of the electron e⁻
17. the negative log of the electron activity
18. Fe^{2+}
19. solid $Fe(OH)_3$
20. the oxidizing and reducing limits of water stability, respectively
21. complex or coordination compound
22. four
23. reactions of both ligands and metals, changes in oxidation state of the metal, and effects on life processes
24. their ability to solubilize heavy metals
25. humic substances
26. degradation-resistant materials formed during the decomposition of vegetation
27. fulvic acid, humic acid, and humin
28. solubility, particularly in acid and base
29. their potential to form trihalomethanes
30. the organic portion is bonded to the metal by a carbon-metal bond and the organic ligand is frequently not capable of existing as a stable separate species
31. interaction of solutes in water with other phases
32. solids, immiscible liquids, and gases
33. repositories of a wide variety of chemical species
34. many chemical and biochemical processes
35. colloids
36. autotrophic organisms
37. utilize the organic substances produced by autotrophic organisms as energy sources and as the raw materials for the synthesis of their own biomass
38. They are the predominant producers of the biomass in water, they mediate most of the significant oxidation-reduction processes that occur in water, they play a key role in nutrient cycling, they are essential for the major biogeochemical cycles, and they are responsible for the breakdown and detoxication of many xenobiotic pollutants that get into the hydrosphere.
39. it gives them a very high surface/volume ratio
40. $$CO_2 + H_2O \xrightarrow{h\nu} \{CH_2O\} + O_2(g)$$
41. they play an important role in determining the composition of natural waters and wastewaters due to the large amount of their decomposition products that enter water from the breakdown of cellulose in wood and other plant materials
42. aerobic, anaerobic bacteria, and facultative
43. prokaryotic cells
44. lag phase, log phase, stationary phase, and death phase
45. exhibits a maximum growth rate at an optimum temperature that is skewed toward the high temperature end of the curve

46. enzymes are destroyed by being denatured at temperatures not far above the optimum
47. photosynthesis, respiration, degradation of biomass, methane production, bacterial utilization and degradation of hydrocarbons, and biodegradation of organic matter
48. nitrogen fixation, nitrification, nitrate reduction, and denitrification
49. sulfate reduction, sulfide oxidation, and degradation of organic sulfur compounds
50. mineralization, which releases inorganic phosphorus from the organic form, and deactivation of highly toxic organophosphate compounds
51. dehalogenation
52. iron(III) hydroxide
53. sulfuric acid
54. pyrite, FeS_2

QUESTIONS AND PROBLEMS

1. A sample of groundwater heavily contaminated with soluble inorganic iron is brought to the surface and the alkalinity is determined without exposing the sample to the atmosphere. Why does a portion of such a sample exposed to the atmosphere for some time exhibit a decreased alkalinity?

2. What indirect role is played by bacteria in the formation of limestone caves?

3. Over the long term, irrigation must be carried out so that there is an appreciable amount of runoff, although a much smaller quantity of water would be sufficient to wet the ground. Why must there be some runoff?

4. Alkalinity is **not** (a) a measure of the degree to which water can support algal growth, (b) the capacity of water to neutralize acid, (c) a measure of the capacity of water to resist a decrease in pH, (d) a measure of pH, (e) important in considerations of water treatment.

5. Suggest why oxygen levels may become rather low at night in water supporting a heavy growth of algae

6. What is the molar concentration of O_2 in water in equilibrium with atmospheric air at 25°C?

7. Explain why a solution of $Fe_2(SO_4)_3$ in water is acidic.

8. How does the ionic radius of ions such as Ca^{2+}, Fe^{2+}, and Mn^{2+} correlate with their relative tendencies to be acidic?

9. An individual measured the pH of a water sample as 11.2 and reported it to be "highly alkaline." Is that statement necessarily true?

10. Match the following pertaining to properties of water (left) and their effects and significance (right):

    (a) Highest dielectric constant of any common liquid

    (b) Higher surface tension than any other liquid

    (c) Maximum density as a liquid

    (d) Higher heat capacity than any liquid other than ammonia

    1. Stabilization of temperatures of organisms and geographical regions

    2. Controlling factor in physiology; governs drop and surface phenomena

    3. High solubility of ionic substances and their ionization in aqueous solution

    4. Ice floats

11. How does the temperature-density relationship of water influence the presence of oxidized and reduced species in a body of water?

12. What is a particularly unique aspect of oceanographic chemistry?

13. Explain with appropriate chemical reactions how the fixation of inorganic carbon by algal photosynthesis in water can result in the eventual precipitation of $CaCO_3$.

14. The chemical species that is the predominant contributor to water alkalinity is _____ and the predominant contributor to water hardness is _____.

15. In what sense is pH an *intensity* factor, and alkalinity a *capacity* factor? How do the comparative pHs and alkalinities of 1.00×10^{-3} M NaOH and 0.100 M HCO_3^- illustrate this point?

16. Illustrate with a chemical reaction how the addition of $Al(H_2O)_6^{3+}$ might reduce alkalinity in water.

17. Explain what the chemical formula Mg^{2+} used to represent magnesium ion dissolved in water really means.

18. Explain what the following chemical formulas have to do with speciation of metals dissolved in water: $Fe(H_2O)_6^{3+}$, $Fe(CN)_6^{3-}$, $Pb(C_2H_5)_4$, where C_2H_5 is an ethyl group bound to lead through a carbon atom.

19. What does the reaction below show about water chemistry, hardness, and alkalinity? How is microbial degradation of organic matter tied in with the phenomenon illustrated?

$$CaCO_3(s) + CO_2(aq) + H_2O \rightleftharpoons Ca^{2+} + 2HCO_3^-$$

20. Explain how the reaction below illustrates a detrimental aspect of water hardness.

$$2C_{17}H_{35}COO^-Na^+ + Ca^{2+} \rightarrow Ca(C_{17}H_{35}CO_2)_2(s) + 2Na^+$$

21. Explain what the two reactions below show about oxidation-reduction phenomena in water. How are they tied in with pE in water? How is pE conceptually defined?

$$2H_2O + 2e^- \rightleftharpoons H_2(g) + 2OH^-$$
$$2H_2O \rightleftharpoons O_2(g) + 4H^+ + 4e^-$$

22. Define ligand, complex, and chelating agent. Why is a chelate often particularly stable? What is a unidentate ligand?

23. Explain in what sense the species $Fe(H_2O)_6^{2+}$ is a complex ion.

24. Explain some of the major environmental implications of complexation and chelation. What is the particular environmental harm that might come from codisposal of a toxic heavy metal salt and a chelating compound?

25. In what sense does an organometallic compound differ from a complex compound?

26. What are the three classes of humic substances? What is the source of humic substances? What are their chelating properties?

27. What are some of the unique characteristics of colloidal particles that make them so important in water? What is the approximate size of colloidal particles?

28. How are sediments formed? Discuss the significance of sediments in aquatic chemistry.

29. Define the following terms as applied to aquatic organisms and give an example of each: Autotrophic organisms, heterotrophic organisms, producers, decomposers, aerobic bacteria, anaerobic bacteria.

30. List and discuss the reasons that microorganisms are of particular importance in water.

31. In the pE-pH diagram below, suppose that the upper dashed line represents equilibrium of the system shown with atmospheric oxygen and that the lower dashed line represents equilibrium with an anaerobic system, such as in the sediment of a body of water. That being the case, use arrows and reactions to explain the following in a system in which some iron is available: (1) Acid mine water typically contains dissolved iron in equilibrium with air. Which region on the diagram might represent acid mine water? (2) What is observed as the pH of iron-containing acid mine water is raised by dilution or exposure to base? (3) Explain the presence of soluble iron in anaerobic groundwater at pH 7. (4) Explain what might happen, and why a precipitate might be observed, as acid mine water is raised to the surface and exposed to air.

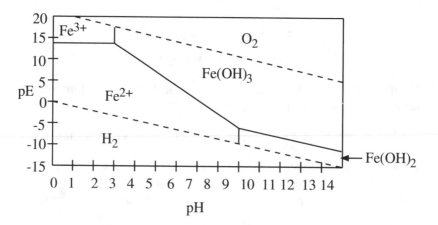

32. What are the important implications for environmental chemistry of the fact that aquatic bacteria are extremely small and may reproduce very rapidly?

33. Define algae and fungi. Give a simplified biochemical reaction characteristic of algae and one characteristic of fungi. In what sense are fungi and algae "opposites"? Why is *cellulase* particularly important in environmental chemical processes?

34. Define each of the following applied to aquatic bacteria: Cell wall, slime layer, cell membrane, flagella, cytoplasm, nuclear body, inclusions.

35. Offer plausible explanations for the four regions shown in a bacterial growth curve. What are the special implications of the log region?

36. Explain the plot showing the effect of increasing substrate concentration on enzyme activity.

37. Explain the plot showing the effect of temperature on bacterial enzyme activity and growth. Why is it unsymmetrical?

38. Explain the significance of each of the following in the carbon cycle: Photosynthesis, the biochemical reaction $\{CH_2O\} + O_2(g)$ $CO_2 + H_2O$, methane production, β-oxidation.

39. Give the name and significance of each of the following microbially-mediated reactions involving nitrogen: (1) $3\{CH_2O\} + 2N_2 + 3H_2O + 4H^+$ $\rightarrow$ $3CO_2 + 4NH_4^+$, (2) $2O_2 + NH_4^+ \rightarrow NO_3^- + 2H^+ + H_2O$, (3) $2NO_3^- + \{CH_2O\} \rightarrow 2NO_2^- + H_2O + CO_2$, (4) $2NO_2^- + 3\{CH_2O\} + 4H^+ \rightarrow 2NH_4^+ + 3CO_2 + H_2O$, (5) $4NO_3^- + 5\{CH_2O\} + 4H^+ \rightarrow 2N_2 + 5CO_2 + 7H_2O$

40. Give the name and significance of each of the following compounds or microbially-mediated reactions involving sulfur: (1) $SO_4^{2-} + 2\{CH_2O\} + 2H^+ \rightarrow H_2S + 2CO_2 + 2H_2O$, (2) $2H_2S + 4O_2 \rightarrow 4H^+ + 2SO_4^{2-}$, (3) CH_3SH

SUPPLEMENTARY REFERENCE

1. Stumm, W., and J.J. Morgan, *Aquatic Chemistry*, 2nd ed. (New York, NY: Wiley-Interscience, 1981).

12   WATER POLLUTION

12.1. NATURE AND TYPES OF WATER POLLUTANTS

Throughout history, the quality of drinking water has been a factor in determining human welfare. Waterborne diseases in drinking water have decimated the populations of whole cities. Unwholesome water polluted by natural sources has caused great hardship for people forced to drink it or use it for irrigation. Although there are still occasional epidemics of bacterial and viral diseases caused by infectious agents carried in drinking water, such as a cholera epidemic in South America in 1991–92, waterborne diseases have in general been well controlled. Currently, toxic chemicals pose the greatest threat to the safety of water supplies in industrialized nations. There are many possible sources of chemical contamination in water. These include chlorinated hydrocarbon wastes from industrial chemical production, heavy metal contamination from metal plating operations, pesticides, and salinity in the runoff from agricultural lands.

Since World War II there has been a tremendous growth in the manufacture and use of synthetic chemicals. Many of the chemicals have contaminated water supplies. Two examples are insecticide and herbicide runoff from agricultural land, and industrial discharge into surface waters. Most serious, though, is the threat to groundwater from waste chemical dumps and landfills, storage lagoons, treating ponds, and other facilities.

It is clear that water pollution should be a concern of every citizen. Understanding the sources, interactions, and effects of water pollutants is essential for controlling pollutants in an environmentally safe and economically acceptable manner. Above all, an understanding of water pollution and its control depends upon a basic knowledge of aquatic environmental chemistry. That is why this text covers the basics of aquatic chemistry (Chapter 11) prior to discussing pollution. Water pollution may be studied much more effectively with a sound background in the basic properties of water, aquatic microbial reactions, sediment-water interactions, and other factors involved with the reactions, transport, and effects of these pollutants.

In considering water pollution, it is useful to keep in mind an overall picture

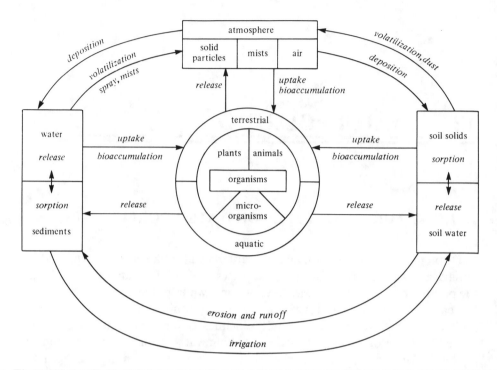

Figure 12.1.   Pollutant interchange among the biotic, terrestrial, atmospheric, and aquatic environments.

of possible pollutant cycles and interchange in the environment as shown in Figure 12.1. Water pollutants can be divided among some general classifications, as summarized in Table 12.1. Most of these categories of pollutants, and several subcategories, are discussed in this chapter. An enormous amount of material is published on this subject each year, and it is impossible to cover it all in one chapter. In order to be up to date on this subject the reader may want to survey journals and books dealing with water pollution, such as those listed in the Supplementary References section at the end of this chapter.

12.2.   ELEMENTAL POLLUTANTS

Elements that occur at very low levels of a few parts per million or less in the environment are commonly called **trace elements**. The term **trace substance** is a more general one applied to both elements and chemical compounds. Table 12.2 summarizes the more important trace elements encountered in natural waters. Some of these are recognized as nutrients required for animal and plant life. Of these, many are essential at low levels but toxic at higher levels.

Some of the **heavy metals** are among the most harmful of the elemental pollutants. These elements, which are in general the metals in the lower right-

Pollutants

Table 12.1. General Types of Water Pollutants

Class of Pollutant	Significance
Trace Elements	Health, aquatic biota
Metals bound by organic species	Metal transport
Inorganic pollutants	Toxicity, aquatic biota
Asbestos	Human health
Algal nutrients	Eutrophication
Radionuclides	Toxicity
Acidity, alkalinity, salinity (in excess)	Water quality, aquatic life
Sewage	Water quality, oxygen levels
Biochemical oxygen demand	Water quality, oxygen levels
Trace organic pollutants	Toxicity
Pesticides	Toxicity, aquatic biota, wildlife
Polychlorinated biphenyls	Possible biological effects
Chemical carcinogens	Incidence of cancer
Petroleum wastes	Effect on wildlife, esthetics
Pathogens	Health effects
Detergents	Eutrophication, wildlife, esthetics
Sediments	Water quality, aquatic biota, wildlife
Taste, odor, and color	Esthetics

hand corner of the periodic table, include essential elements like iron as well as toxic metals like lead, cadmium, and mercury. Most of them have a tremendous affinity for sulfur and attack sulfur bonds in enzymes, thus immobilizing the enzymes. Protein carboxylic acid ($-CO_2H$) and amino ($-NH_2$) groups are also chemically bound by heavy metals. Cadmium, copper, lead and mercury ions bind to cell membranes, hindering transport processes through the cell wall. Heavy metals may also precipitate phosphate biocompounds or catalyze their decomposition. Some biochemical effects of metals are discussed in Chapter 21.

Some of the **metalloids**, elements on the borderline between metals and non-metals, are significant water pollutants. Arsenic, selenium, and antimony are of particular interest.

Heavy Metals

Cadmium

Pollutant **cadmium** in water may arise from industrial discharges and mining wastes. Cadmium is widely used in metal plating. Chemically, cadmium is very similar to zinc, and these two metals frequently undergo geochemical processes together. Both metals are found in water in the +2 oxidation state.

The effects of acute cadmium poisoning in humans are very serious. Among

Table 12.2. Some Important Trace Elements in Natural Waters

Element	Sources	Effects and Significance
Arsenic	Mining byproduct, pesticides, chemical waste	Toxic, possibly carcinogenic
Beryllium	Coal, industrial	Toxicity, possibly carcinogenic
Cadmium	Industrial, mining, metal	Replaces zinc biochemically, toxic
Chromium	Metal plating, other	Possibly carcinogenic as Cr(VI)
Copper	Metal plating, industrial waste, mining	Not very toxic to animals, toxic to algae and plants
Fluorine (F⁻ ion)	Natural, industrial, water additive	Prevents tooth decay at around 1 mg/L, causes tooth and bone damage at around 5 mg/L
Iodine (I⁻ ion)	Industrial, natural brines, seawater intrusion	Prevents goiter
Iron	Low pE water in contact with iron sources, acid mine water	Not very toxic, damages bathroom fixtures and clothing because of deposits of iron oxides
Lead	Industrial, mining, fuels, plumbing	Toxicity, destruction of wildlife
Manganese	Low pE water in contact with Mn sources, industrial waste	Not very toxic, damages bathroom fixtures and clothing because of deposits of Mn oxides
Mercury	Industrial wastes, mining, coal	Acute and chronic toxicity
Molybdenum	Industrial waste, natural sources	Essential to plants, some toxicity to animals
Selenium	Natural geological sources, coal, sulfur	Essential micronutrient at lower levels, toxic at higher levels
Silver	Natural geological sources, electroplating photographic processes	Causes blue-grey discoloration of skin, mucous membranes, eyes
Zinc	Industrial wastes, metal plating, plumbing	Essential element, phytotoxic at higher levels

them are high blood pressure, kidney damage, destruction of testicular tissue, and destruction of red blood cells. It is believed that much of the physiological action of cadmium arises from its chemical similarity to zinc. Specifically, cadmium may replace zinc in some enzymes, thereby altering the stereostructure of

impaired catalytic activity

the enzyme and impairing its catalytic activity. Disease symptoms ultimately result.

Lead

Lead occurs in water in the $+II$ oxidation state and arises from a number of industrial and mining sources. Lead from leaded gasoline used to be a major source of atmospheric and terrestrial lead, much of which eventually entered natural water systems. In addition to pollutant sources, lead-bearing limestone and galena (PbS) contribute lead to natural waters in some locations.

Despite greatly increased total use of lead by industry, evidence from hair samples and other sources indicates that body burdens of this toxic metal have decreased during recent decades. This may be the result of less lead use in plumbing and other products that come in contact with food or drink.

Acute lead poisoning in humans causes severe dysfunction in the kidneys, reproductive system, liver, and the brain and central nervous system. Sickness or death results. Lead poisoning from environmental exposure is thought to have caused mental retardation in many children. Mild lead poisoning causes anemia. The victim may have headaches and sore muscles and may feel generally fatigued and irritable.

Lead is probably not a major problem in drinking water, although the potential exists in cases where old lead pipe is still in use. Lead used to be a constituent of solder and some pipe-joint formulations, so that household water does have some contact with lead. Water that has stood in household plumbing for some time may accumulate spectacular levels of lead (along with zinc, cadmium, and copper) and should be let run for a while before use.

Mercury

Mercury generates the most concern of any of the heavy-metal pollutants. Mercury is found as a trace component of many minerals. Mercury enters the environment from a large number of miscellaneous sources related to human use of the element. These include discarded laboratory chemicals, batteries, broken thermometers, lawn fungicides, amalgam tooth fillings, and pharmaceutical products. Taken individually, each of these sources may not contribute much of the toxic metal, but the total effect can be substantial. Sewage effluent sometimes contains up to 10 times the level of mercury found in typical natural waters.

Among the more severe toxicological effects of mercury are neurological damage, including irritability, paralysis, blindness, or insanity; chromosome breakage; and birth defects. Milder symptoms of mercury poisoning, such as depression and irritability, have a psychopathological character, and may escape detection. The toxicity of mercury was tragically illustrated in the Minamata Bay area of Japan during the period 1953–1960. A total of 111 cases of mercury poisoning and 43 deaths were reported among people who had consumed seafood from the bay that had been contaminated with mercury waste from a chemical plant that drained into Minamata Bay. Congenital defects were observed in 19 babies whose mothers had consumed seafood contaminated with mercury. The level of metal in the contaminated seafood was 5–20 parts per million.

Unexpectedly high concentrations of mercury found in water and in fish tissues in the U.S. and Canada around 1970 were later attributed to the formation of soluble monomethylmercury ion, CH_3Hg^+, and volatile dimethylmercury, $(CH_3)_2Hg$, by anaerobic bacteria in sediments. Mercury from these compounds becomes concentrated in fish lipid (fat) tissue and the concentration factor from water to fish may exceed 10^3. The methylating agent by which inorganic mercury is converted to methylmercury compounds is methylcobalamin, a vitamin B_{12} analog:

$$HgCl_2 \xrightarrow{\text{Methylcobalamin}} CH_3HgCl + Cl^- \qquad (12.2.1)$$

It is believed that the bacteria that synthesize methane produce methylcobalamin as an intermediate in the synthesis. Thus, waters and sediments in which anaerobic decay is occurring provide the conditions under which methylmercury production takes place. In neutral or alkaline waters, the formation of dimethyl mercury, $(CH_3)_2Hg$, is favored. This volatile compound can escape to the atmosphere.

Metalloids

The most significant water pollutant metalloid element is arsenic, a toxic element that has been the chemical villain of more than a few murder plots. Acute arsenic poisoning can result from the ingestion of more than about 100 mg of the element. Chronic poisoning occurs with the ingestion of small amounts of arsenic over a long period of time. There is some evidence that this element is also carcinogenic.

Arsenic occurs in the Earth's crust at an average level of 2–5 ppm. The combustion of fossil fuels, particularly coal, introduces large quantities of arsenic into the environment, much of it reaching natural waters. Arsenic occurs with phosphate minerals and enters into the environment along with some phosphorus compounds. Some formerly-used pesticides, particularly those from before World War II, contain highly toxic arsenic compounds. The most common of these are lead arsenate, $Pb_3(AsO_4)_2$; sodium arsenite, Na_3AsO_3; and Paris Green, $Cu_3(AsO_3)_2$. Another major source of arsenic is mine tailings. Arsenic produced as a by-product of copper, gold, and lead refining greatly exceeds the commercial demand for arsenic, and it accumulates as waste material.

Like mercury, arsenic may be converted to more mobile and toxic methyl derivatives by bacteria, according to the following reactions:

$$H_3AsO_4 + 2H^+ + 2e^- \rightarrow H_3AsO_3 + H_2O \qquad (12.2.2)$$

$$H_3AsO_3 \xrightarrow{\text{Methylcobalamin}} \underset{\substack{\text{Methylarsinic} \\ \text{acid}}}{CH_3AsO(OH)_2} \qquad (12.2.3)$$

$$CH_3AsO(OH)_2 \xrightarrow{\text{Methylcobalamin}} \underset{\text{(Dimethylarsinic acid)}}{(CH_3)_2AsO(OH)} \qquad (12.2.4)$$

$$(CH_3)_2AsO(OH) + 4H^+ + 4e^- \rightarrow (CH_3)_2AsH + 2H_2O \qquad (12.2.5)$$

12.3.   ORGANICALLY BOUND METALS AND METALLOIDS

There are two major types of metal-organic interactions to be considered in an aquatic system. The first of these is complexation, usually chelation when organic ligands are involved. Normally, complexation by organics in natural water and wastewater systems occurs such that a species is present that reversibly dissociates to a metal ion and an organic complexing species as a function of hydrogen ion concentration:

$$ML + 2H^+ \rightarrow M^{2+} + H_2L \qquad (12.3.1)$$

In this equation, M^{2+} is a metal ion and H_2L is the acidic form of a complexing — frequently chelating — ligand, L^{2-}.

Organometallic compounds (Section 11.9), on the other hand, contain metals bound to organic entities by way of a carbon atom and do not dissociate reversibly at lower pH or greater dilution. Furthermore, the organic component, and sometimes the particular oxidation state of the metal involved, may not be stable apart from the organometallic compound. The more important kinds of organometallic compounds are those in which the organic group is an alkyl group such as ethyl in tetraethyllead, $Pb(C_2H_5)_4$; carbonyls, having carbon monoxide CO bonded to metals; those in which the organic group is a π electron donor, such as ethylene; or combinations of these kinds of compounds.

The interaction of trace metals with organic compounds in natural waters is too vast an area to cover in detail in this chapter; however, it may be noted that metal-organic interactions may involve organic species of both pollutant (such as EDTA) and natural (such as fulvic acids) origin. These interactions are influenced by, and sometimes play a role in, redox equilibria; formation and dissolution of precipitates; colloid formation and stability; acid-base reactions; and microorganism-mediated reactions in water. Metal-organic interactions may increase or decrease the toxicity of metals in aquatic ecosystems, and they have a strong influence on the growth of algae in water.

Organotin Compounds

Of all the metals, tin has the greatest number of organometallic compounds in commercial use, with global production on the order of 40,000 metric tons per year. In addition to synthetic organotin compounds, methylated tin species can be produced biologically in the environment. A typical organotin compound is tetra-*n*-butyltin, below:

$$C_4H_9-\underset{\underset{C_4H_9}{|}}{\overset{\overset{C_4H_9}{|}}{Sn}}-C_4H_9$$

Bactericidal, fungicidal, and insecticidal tributyl tin chloride and related tributyl tin (TBT) compounds are of particular environmental significance because of growing use as industrial biocides. A major use of TBT is in boat and ship hull coatings to prevent the growth of fouling organisms. Other applications include preservation of wood, leather, paper, and textiles. Because of their anti-fungal activity, TBT compounds are used as slimicides in cooling tower water. In addition to tributyl tin chloride, other tributyl tin compounds used as biocides include the hydroxide, the naphthenate, bis(tributyltin) oxide, and tris(tributyl-stannyl) phosphate.

12.4.   INORGANIC SPECIES

In addition to inorganic pollutants that contribute acidity, alkalinity, or salinity to water; algal nutrients; and the important inorganic water pollutants mentioned in Section 12.2, as part of the discussion of pollutant trace elements, there are some other important inorganic pollutant species, of which cyanide ion, CN^-, is probably the most important. Still others are ammonia, carbon dioxide, hydrogen sulfide, nitrite, and sulfite discussed briefly in this chapter.

Cyanide

Cyanide, a deadly poisonous substance, exists in water as HCN, a very weak acid. The cyanide ion has a strong affinity for many metal ions, forming relatively less-toxic ferrocyanide, $Fe(CN)_6^{4-}$, with iron(II), for example. Volatile HCN is very toxic and has been used in gas chamber executions in the U.S. Large quantities of cyanide are used in industry, especially for metal cleaning and electroplating. It is also one of the main gas and coke scrubber effluent pollutants from gas works and coke ovens. Cyanide is widely used in certain mineral-processing operations. The presence of cyanide in water is indicative of a serious pollution problem.

Ammonia and Other Inorganic Pollutants

Excessive levels of ammoniacal nitrogen cause water-quality problems. **Ammonia** is the initial product of the decay of nitrogenous organic wastes, and its presence frequently is indicative of such wastes. It is a normal constituent of low-pE (reducing) groundwaters and is sometimes added to drinking water, where it reacts with chlorine to provide residual disinfectant chlorine. Since the pK_a of ammonium ion is 9.26, most ammonia in water is present as NH_4^+ rather than as NH_3.

Hydrogen sulfide, H_2S, is a product of the anaerobic decay of organic matter containing sulfur. It is also produced in the anaerobic reduction of sulfate by microorganisms (see Chapter 11) and is evolved as a gaseous pollutant from geothermal waters. Wastes from chemical plants, paper mills, textile mills, and tanneries may also contain H_2S. Its presence is easily detected by its characteristic rotten-egg odor. In water, H_2S is a weak diprotic acid with pK_{a1} of 6.99 and pK_{a2} of 12.92; S^{2-} is not present in normal natural waters. The sulfide ion has tremendous affinity for many heavy metals, and precipitation of metallic sulfides often accompanies production of H_2S.

Free **carbon dioxide**, CO_2, is frequently present in water at high levels due to decay of organic matter. It is also added to softened water during water treatment as part of a recarbonation process, which prevents $CaCO_3$ from precipitating in water distribution pipes. Excessive carbon dioxide levels may make water more corrosive and may be harmful to aquatic life.

Nitrite ion, NO_2^-, occurs in water as an intermediate oxidation state of nitrogen. Its pE range of stability is relatively narrow. Nitrite is added to some industrial process water to inhibit corrosion; it is rarely found in drinking water at levels over 0.1 mg/L.

Sulfite ion, SO_3^{2-}, is found in some industrial wastewaters. Sodium sulfite is commonly added to boiler feedwaters as an oxygen scavenger:

$$2SO_3^{2-} + O_2 \rightarrow 2SO_4^{2-} \tag{12.4.1}$$

Since pK_{a1} of sulfurous acid is 1.76 and pK_{a2} is 7.20, sulfite exists as either HSO_3^- or SO_3^{2-} in natural waters, depending upon pH. It may be noted that hydrazine, N_2H_4, also functions as an oxygen scavenger:

$$N_2H_4 + O_2 \rightarrow 2H_2O + N_2(g) \tag{12.4.2}$$

Asbestos in Water

It is not known for sure whether asbestos, a substance that causes lung cancer when inhaled, is toxic in drinking water. This has been a matter of considerable concern because of the dumping of taconite (iron ore tailings) containing asbestos-like fibers into Lake Superior. The fibers have been found in drinking waters of cities around the lake. After having dumped the tailings into Lake Superior since 1952, the Reserve Mining Company at Silver Bay on Lake Superior solved the problem in 1980 by constructing a 6-square-mile containment basin inland from the lake. This $370-million facility keeps the taconite tailings covered with a 3–meter layer of water to prevent escape of fiber dust.

12.5. ALGAL NUTRIENTS AND EUTROPHICATION

The term **eutrophication**, derived from the Greek word meaning "well-nourished," describes a condition of lakes or reservoirs involving excess algal growth, which may eventually lead to severe deterioration of the body of water. The first step in eutrophication of a body of water is an input of plant nutrients (Table 12.3) from watershed runoff or sewage. The nutrient-rich body of water then produces a great deal of plant biomass by photosynthesis, along with a smaller amount of animal biomass. Dead biomass accumulates in the bottom of the lake, where it partially decays, recycling nutrient carbon dioxide, phosphorus, nitrogen, and potassium, which is beneficial, but also consuming dis-

Table 12.3. Essential Plant Nutrients: Sources and Functions

Nutrient	Source	Function
Macronutrients		
Carbon (CO_2)	Atmosphere, decay	Biomass constituent
Hydrogen	Water	Biomass constituent
Oxygen	Water	Biomass constituent
Nitrogen (NO_3^-)	Decay, atmosphere (from nitrogen-fixing organisms), pollutants	Protein constituent
Phosphorus (phosphate)	Decay, minerals, pollutants	DNA/RNA constituent
Potassium	Minerals, pollutants	Metabolic function
Sulfur (sulfate)	Minerals	Proteins, enzymes
Magnesium	Minerals	Metabolic function
Calcium	Minerals	Metabolic function
Micronutrients		
B, Cl, Co, Cu, Fe, Mo, Mn, Na, Si V, Zn	Minerals, pollutants	Metabolic function and/or constituent of enzymes

solved oxygen, which is not. If the lake is not too deep, bottom-rooted plants begin to grow, accelerating the accumulation of solid material in the basin. Eventually a marsh is formed, which finally fills in to produce a meadow or forest.

Human activity can greatly accelerate the eutrophication process. To understand why this is so, refer to Table 12.3, which shows the chemical elements needed for plant growth. Most of these are present at a level more than sufficient to support plant life in the average lake or reservoir. Hydrogen and oxygen come from the water itself. Carbon is provided by CO_2 from the atmosphere or from decaying vegetation. Sulfate, magnesium, and calcium are normally present in abundance from mineral strata in contact with the water. The micronutrients are required at only very low levels (for example, approximately 40 ppb for copper). Therefore, the nutrients most likely to be limiting are the "fertilizer" elements: nitrogen, phosphorus, and potassium. These are all present in sewage and are, of course, found in runoff from heavily fertilized fields. They are also constituents of various kinds of industrial wastes. Each of these elements can also come from natural sources—phosphorus and potassium from mineral formations, and nitrogen fixed by bacteria, blue-green algae, or discharge of lightning in the atmosphere.

The single plant nutrient most likely to be limiting is phosphorus, and it is generally named as the culprit in excessive eutrophication. Household detergents once were a common source of phosphate in wastewater, and eutrophication control has concentrated upon eliminating phosphates from detergents, removing phosphate at the sewage-treatment plant, and preventing phosphate-laden sewage effluents (treated or untreated) from entering bodies of water.

12.6.   ACIDITY, ALKALINITY, AND SALINITY

Aquatic biota are sensitive to extremes of pH. Largely because of osmotic effects, they cannot live in a medium having a salinity to which they are not adapted. Thus, a freshwater fish soon succumbs in the ocean, and sea fish normally cannot live in fresh water. Excess salinity soon kills plants not adapted to it. There are, of course, ranges in salinity and pH in which organisms live. These ranges frequently may be represented by a reasonably symmetrical curve, along the fringes of which an organism may live without really thriving.

The most common source of **pollutant acid** in water is acid mine drainage. The sulfuric acid in such drainage arises from the microbial oxidation of pyrite (FeS_2) or other sulfide minerals. The values of pH encountered in acid-polluted water may fall below 3, a condition deadly to most forms of aquatic life except the culprit bacteria mediating the pyrite and iron(II) oxidation. Industrial wastes frequently contribute strong acid to water. Sulfuric acid produced by the air oxidation of pollutant sulfur dioxide enters natural waters as acidic rainfall. In cases where the water does not have contact with a basic mineral, such as limestone, the water pH may become dangerously low. This condition occurs in some Canadian lakes, for example.

Acid Mine Waters

One consequence of bacterial action on metal compounds is acid mine drainage, a common and damaging water pollution problem. Acid mine water results from the presence of sulfuric acid produced by the oxidation of pyrite, FeS_2. Microorganisms are closely involved in the overall process, which consists of several reactions. The first of these reactions is the oxidation of pyrite:

$$2FeS_2(s) + 2H_2O + 7O_2 \rightarrow 4H^+ + 4SO_4^{2-} + 2Fe^{2+} \qquad (12.6.1)$$

The next step is the oxidation of iron(II) ion to iron(III) ion,

$$4Fe^{2+} + O_2 + 4H^+ \rightarrow 4Fe^{3+} + 2H_2O \qquad (12.6.2)$$

a process that occurs very slowly as a purely chemical reaction at the low pH values found in acid mine waters. However, below pH 3.5, the iron oxidation is catalyzed by the iron bacterium *Thiobacillus ferrooxidans*, and in the pH range 3.5–4.5 it may be catalyzed by a variety of *Metallogenium*, a filamentous iron bacterium. Other bacteria that may be involved in acid mine water formation are *Thiobacillus thiooxidans* and *Ferrobacillus ferrooxidans*. The iron(III) ion further dissolves pyrite by chemical interaction,

$$FeS_2(s) + 14Fe^{3+} + 8H_2O \rightarrow 15Fe^{2+} + 2SO_4^{2-} + 16H^+ \qquad (12.6.3)$$

which in conjunction with Reaction 12.6.2 constitutes a cycle for the dissolution of pyrite. The $Fe(H_2O)_6^{3+}$ is acidic, and at pH values much above 3, the iron(III) precipitates as unsightly amorphous, semigelatinous hydrated iron(III) oxide:

$$Fe^{3+} + 3H_2O \rightleftharpoons Fe(OH)_3(s) + 3H^+ \qquad (12.6.4)$$

Alkalinity

Excess **alkalinity**, and frequently accompanying high pH, generally are not introduced directly into water by human activity. However, in many geographic areas, the soil and mineral strata are alkaline and impart a high alkalinity to water. Human activity can aggravate the situation; for example, by exposure of alkaline overburden from strip mining to surface water or groundwater. Excess alkalinity in water is manifested by a characteristic fringe of white salts at the edges of a body of water or on the banks of a stream.

Water **salinity** may be increased by a number of human activities. Water passing through a municipal water system inevitably picks up salt from a number of processes; for example, recharging water softeners with sodium chloride. Salts can leach from spoil piles. One of the major environmental constraints on the production of shale oil, for example, is the high percentage of leachable sodium sulfate in piles of spent shale. Irrigation adds a great deal of salt to water, a phenomenon responsible for the Salton Sea in California, and is a source of conflict between the United States and Mexico over saline contamination of the Rio Grande and Colorado rivers. Irrigation and intensive agricultural production have caused saline seeps in some of the Western states. These occur when water seeps into a slight depression in tilled, sometimes irrigated, fertilized land, carrying salts (particularly sodium, magnesium, and calcium sulfates) along with it. The water evaporates in the dry summer heat, leaving a salt-laden area behind which no longer supports much plant growth. With time, these areas spread, removing once fertile crop land from production.

12.7. OXYGEN, OXIDANTS, AND REDUCTANTS

Oxygen is a vitally important species in water. In water, oxygen is consumed rapidly by the oxidation of organic matter, CH_2O:

$$\{CH_2O\} + O_2 \xrightarrow{\text{Microorganisms}} CO_2 + H_2O \qquad (12.7.1)$$

Unless the water is reaerated efficiently, as by turbulent flow in a shallow stream, it rapidly becomes depleted in oxygen and will not support higher forms of aquatic life.

In addition to the microorganism-mediated oxidation of organic matter, oxygen in water may be consumed by the biooxidation of nitrogenous material,

$$NH_4^+ + 2O_2 \rightarrow 2H^+ + NO_3^- + H_2O \qquad (12.7.2)$$

and by the chemical or biochemical oxidation of chemical reducing agents:

$$4Fe^{2+} + O_2 + 10H_2O \rightarrow 4Fe(OH)_3(s) + 8H^+ \qquad (12.7.3)$$

$$2SO_3^{2-} + O_2 \rightarrow 2SO_4^{2-} \qquad (12.7.4)$$

All these processes contribute to the deoxygenation of water.

The degree of oxygen consumption by microbially-mediated oxidation of contaminants in water is called the **biochemical oxygen demand** (or biological oxygen demand), **BOD**. This parameter is commonly measured by determining the quantity of oxygen utilized by suitable aquatic microorganisms during a five-day period. Though the choice of a five-day period is somewhat arbitrary, a five-day BOD test remains a respectable measure of the short-term oxygen demand exerted by a pollutant.

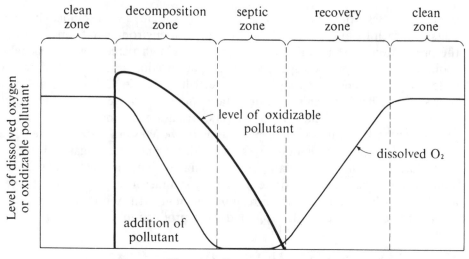

Figure 12.2.   Oxygen sag curve resulting from the addition of oxidizable pollutant material to a stream.

The addition of oxidizable pollutants to streams produces a typical **oxygen sag curve** as shown in Figure 12.2. Initially, a well-aerated, unpolluted stream is relatively free of oxidizable material; the oxygen level is high; and the bacterial population is relatively low. With the addition of oxidizable pollutant, the oxygen level drops because reaeration cannot keep up with oxygen consumption. In the decomposition zone, the bacterial population rises. The septic zone is characterized by a high bacterial population and very low oxygen levels. The septic zone terminates when the oxidizable pollutant is exhausted, and then the recovery zone begins. In the recovery zone, the bacterial population decreases and the dissolved oxygen level increases until the water regains its original condition.

12.8.   ORGANIC POLLUTANTS

Sewage

As shown in Table 12.4, sewage from domestic, commercial, food-processing, and industrial sources contains a wide variety of pollutants, including organic pollutants. Some of these pollutants, particularly oxygen-demanding substances (see Section 12.7), oil, grease, and solids, are removed by primary and secondary sewage-treatment processes. Others, such as salts, heavy metals, and refractory (degradation-resistant) organics, are not efficiently removed.

Disposal of inadequately treated sewage can cause severe problems. For example, offshore disposal of sewage results in the formation of beds of sewage

Table 12.4. Some of the Primary Constituents of Municipal Sewage

Constituent	Potential Sources	Effects in Water
Oxygen-demanding substances	Mostly organic materials, particularly human feces	Consume dissolved oxygen
Refractory organics	Industrial wastes, household products	Toxic to aquatic life
Viruses	Human wastes	Cause disease (possibly cancer); major deterrent to sewage recycle through water systems
Detergents	Household detergents	Esthetics, prevent grease and oil removal, toxic to aquatic life
Phosphates	Detergents	Algal nutrients
Grease and oil	Cooking, food processing, industrial wastes	Esthetics, harmful to some aquatic life
Salts	Human wastes, water softeners, industrial wastes	Increase water salinity
Heavy metals	Industrial wastes, chemical laboratories	Toxicity
Chelating agents	Some detergents, industrial wastes	Heavy metal ion solubilization and transport
Solids	All sources	Esthetics, harmful to aquatic life

residues. Now that most municipal sewage is treated in sewage treatment plants, a significant problem can be the sludge produced as a product of the sewage treatment processes employed (Section 13.4). This sludge contains organic material which continues to degrade slowly; refractory organics; and heavy metals. The amounts of sludge produced are truly staggering, ranging up to a million tons of sludge or more each year for a major metropolitan area. A major consideration in the safe disposal of such amounts of sludge is the presence of potentially dangerous components such as heavy metals.

Better control of sewage sources is needed to minimize sewage pollution problems. Particularly, heavy metals and refractory organic compounds need to be controlled at the source to enable use of sewage, or treated sewage effluents, for irrigation, recycle to the water system, or groundwater recharge.

Soaps, Detergents, and Detergent Builders

Soaps, detergents, and associated chemicals are potential sources of organic pollutants. These pollutants are discussed briefly here.

Soaps are salts of higher fatty acids, such as sodium stearate, $C_{17}H_{35}COO^-Na^+$. The cleaning action of soap results largely from its emulsifying power caused by the dual nature of the soap anion. A soap ion consists of an ionic carboxyl "head" and a long hydrocarbon "tail":

In the presence of oils, fats, and other water-insoluble organic materials, the "tail" of the anion tends to dissolve in the organic matter, whereas the "head" remains in aquatic solution. Thus, the soap emulsifies, or suspends, organic material in water. In the process, the anions form colloidal soap micelles, as shown in Figure 7.12. Soap lowers the surface tension of water, thus making the water "wetter."

The primary disadvantage of soap as a cleaning agent comes from its reaction with divalent cations to form insoluble salts of fatty acids:

$$2C_{17}H_{35}COO^-Na^+ + Ca^{2+} \rightarrow Ca(C_{17}H_{35}CO_2)_2(s) + 2Na^+ \quad (12.8.1)$$

These insoluble products, usually salts of magnesium or calcium, are not at all effective as cleaning agents. In addition, the insoluble "curds" form unsightly deposits on clothing and in washing machines. If sufficient soap is used, all of the divalent cations may be removed by their reaction with soap and the water containing excess soap will have good cleaning qualities. This is the approach commonly used when soap is employed in the bathtub or wash basin, where the insoluble calcium and magnesium salts can be tolerated. However, in applications such as washing clothing, the water must be softened by the removal of calcium and magnesium or their complexation by substances such as polyphosphates.

Although the formation of insoluble calcium and magnesium salts has resulted in the essential elimination of soap as a cleaning agent for clothing, dishes, and most other materials, it has distinct advantages from the environmental standpoint. As soon as soap gets into sewage or an aquatic system, it generally precipitates as calcium and magnesium salts. Hence, any effects that soap might have in solution are eliminated. As it eventually biodegrades, the soap is completely eliminated from the environment. Therefore, aside from the occasional formation of unsightly scum, soap does not cause any substantial pollution problems.

Synthetic **detergents** have good cleaning properties and do not form insoluble salts with "hardness ions" such as calcium and magnesium. Such synthetic detergents have the additional advantage of being the salts of relatively strong acids, and, therefore, they do not precipitate out of acidic waters as insoluble acids, an undesirable characteristic of soaps.

Synthetic detergents usually have a surface-active agent, or surfactant, added to them that lowers the surface tension of water to which the detergent is added, making the water "wetter." Until the early 1960s, the most common surfactant used was an alkyl benzene sulfonate, ABS, a sulfonation product of an alkyl derivative of benzene:

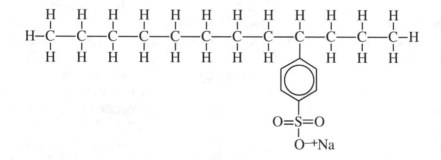

ABS suffered the distinct disadvantage of being only very slowly biodegradable because of its branched-chain structure. The most objectionable manifestation of the nonbiodegradable detergents, insofar as the average citizen was concerned, was the "head" of foam that began to appear in glasses of drinking water in areas where sewage was recycled through the domestic water supply. Sewage-plant operators were disturbed by spectacular beds of foam which appeared near sewage outflows and in sewage treatment plants. Occasionally, the entire aeration tank of an activated sludge plant would be smothered by a blanket of foam. Among the other undesirable effects of persistent detergents upon waste-treatment processes were lowered surface tension of water; deflocculation of colloids; flotation of solids; emulsification of grease and oil; and destruction of useful bacteria. Consequently, ABS was replaced by biodegradable LAS surfactant.

LAS, α-benzenesulfonate, has the general structure:

where the benzene ring may be attached at any point on the alkyl chain except at the ends. LAS is more biodegradable than ABS because the alkyl portion of LAS is not branched and does not contain the tertiary carbon which is so detrimental to biodegradability. Since LAS has replaced ABS in detergents, problems arising from the surface-active agent in the detergents (such as toxicity to fish fingerlings) have greatly diminished and the levels of surface-active agents found in water have decreased markedly.

Most of the environmental problems most recently associated with detergents have arisen from builders rather than surface-active agents. Builders added to detergent formulations bind to hardness ions, making the detergent solution alkaline and greatly improving the action of the detergent surfactant. A commercial solid detergent contains only 10–30% surfactant. In addition, some detergents still contain polyphosphates added to complex calcium and to function as a builder. Other ingredients include anticorrosive sodium silicates, amide foam stabilizers, soil-suspending carboxymethylcellulose, diluent sodium sulfate, and water absorbed by other components. Of these materials, the polyphosphates have caused the most concern as environmental pollutants, although these problems have largely been resolved.

Biorefractory Organic Pollutants

Millions of tons of organic compounds are manufactured globally each year. Significant quantities of several thousand such compounds appear as water pollutants. Most of these compounds, particularly the less biodegradable ones, are substances to which living organisms have not been exposed until recent years. Frequently, their effects upon organisms are not known, particularly for long-term exposures at very low levels. The potential of synthetic organics for causing genetic damage, cancer, or other ill effects is uncomfortably high. On the positive side, organic pesticides enable a level of agricultural productivity without which millions would starve.

Biorefractory organics are the organic compounds of most concern in wastewater, particularly when they are found in sources of drinking water. These are nonbiodegradable, low-molecular-weight compounds of low volatility. Generally, biorefractory compounds are aromatic or chlorinated hydrocarbons, or both. Included in the list of refractory organic industrial wastes are acetone, benzene, bornyl alcohol, bromobenzene, bromochlorobenzene, butylbenzene, camphor chloroethyl ether, chloroform, chloromethylethyl ether, chloronitrobenzene, chloropyridine, dibromobenzene, dichlorobenzene, dichloroethyl ether, dinitrotoluene, ethylbenzene, ethylene dichloride, 2-ethylhexanol, isocyanic acid, isopropylbenzene, methylbiphenyl, methyl chloride, nitrobenzene, styrene, tetrachloroethylene, trichloroethane, toluene, and 1,2-dimethoxy benzene. Many of these compounds have been found in drinking water, and some are known to cause taste and odor problems in water. Biorefractory compounds are not completely removed by biological treatment, and water contaminated with these compounds must be treated by physical and chemical means, including air stripping, solvent extraction, ozonation, and carbon absorption.

Pesticides in Water

The introduction of DDT during World War II marked the beginning of a period of very rapid growth in pesticide use. Pesticides are employed for many different purposes. Chemicals used in the control of invertebrates include insecticides, molluscicides for the control of snails and slugs, and nematicides for the control of microscopic roundworms. Vertebrates are controlled by rodenticides which kill rodents, avicides used to repel birds, and piscicides used in fish control. Herbicides are used to kill plants. Plant growth regulators, defoliants, and plant desiccants are used for various purposes in the cultivation of plants. Fungicides are used against fungi, bactericides against bacteria, and algicides against algae. Annual U.S. pesticide production is several hundred million kilograms of active ingredients. Although insecticide production has remained about level during the last two or three decades, herbicide production has increased greatly as chemicals have increasingly replaced cultivation of land in the control of weeds. Large quantities of pesticides enter water either directly, in applications such as mosquito control, or indirectly, primarily from drainage of agricultural lands. These generally belong to the classes of chlorinated hydrocarbons, organic phosphates, and carbamates, which are derived from carbamic acid,

$$\underset{\text{H}-\text{N}-\text{C}-\text{OH}}{\overset{\text{H} \quad \text{O}}{}}$$

The toxicities of pesticides vary widely. Parathion,

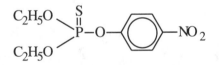

banned for most uses in 1991, is so toxic that as little as 120 mg of it has been known to kill an adult human and a dose of 2 mg has killed a child. Most accidental poisonings have occurred by absorption through the skin. Since its use began, several hundred people have been killed by parathion.

In contrast, **malathion** shows how differences in structural formula can cause pronounced differences in the properties of organophosphate pesticides. Malathion has two carboxyester linkages which are hydrolyzable by carboxylase enzymes to relatively nontoxic products as shown by the following reaction:

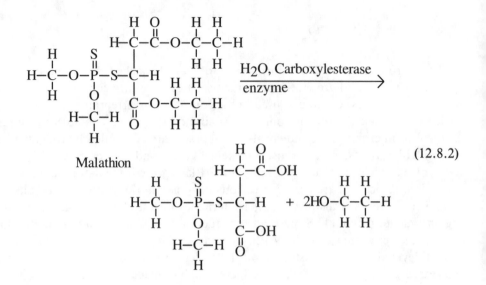

Malathion

$$(12.8.2)$$

The enzymes that accomplish malathion hydrolysis are possessed by mammals, but not by insects, so that mammals can detoxify malathion, whereas insects cannot. The result is that malathion has selective insecticidal activity. For example, although malathion is a very effective insecticide, its LD_{50} (dose required to kill 50% of test subjects) for adult male rats is about 100 times that of parathion, reflecting the much lower mammalian toxicity of malathion compared to some of the more toxic organophosphate insecticides, such as parathion.

The chlorinated hydrocarbon DDT (dichlorodiphenyltrichloroethane or 1,1,1-trichloro-2,2-di-(4-chlorophenyl)-ethane),

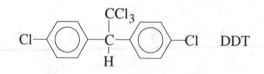

was used in massive quantities following World War II. Although its toxicity is generally low, its persistence and accumulation in food chains have led to a ban on DDT use in the United States; it is still employed in some countries.

A number of water pollution and health problems have been associated with the manufacture of pesticides. For example, degradation-resistant hexachlorobenzene,

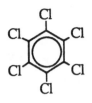

is used as a raw material for the synthesis of other pesticides and has often been found in water.

The most notorious by-products of pesticide manufacture are **polychlorinated dibenzodioxins**, which have the same basic structure as that of TCDD (2,3,7,8 tetrachlorodibenzo-*p*-dioxin shown in Figure 12.3), but different numbers and arrangements of chlorine atoms on the ring structure. Commonly referred to as "dioxins," these species have a high environmental and toxicological significance.[1]

From 1 to 8 Cl atoms may be substituted for H atoms on dibenzo-*p*-dioxin, giving a total of 75 possible chlorinated derivatives. Of these, the most notable pollutant and hazardous waste compound is **2,3,7,8-tetrachlorodibenzo-*p*-dioxin (TCDD)**, often referred to simply as "**dioxin**." This compound, which is one of the most toxic of all synthetic substances to some animals, was produced as a low-level contaminant in the manufacture of some aromatic, oxygen-containing organohalide compounds such as chlorophenoxy herbicides and hex-achlorophene (Figure 12.4) manufactured by processes used until the 1960s.

The chlorophenoxy herbicides, including 2,4,5-trichlorophenoxyacetic acid (2,4,5-T) were manufactured on a large scale for weed and brush control and as military defoliants. Fungicidal and bactericidal hexachlorophene was once widely applied to crops in the production of vegetables and cotton and was used as an antibacterial agent in personal care products, an application that has been discontinued because of toxic effects and possible TCDD contamination.

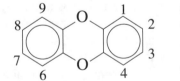
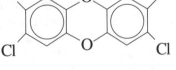

Dibenzo-*p*-dioxin   2,3,7,8-Tetrachlorodibenzo-
                                  p-dioxin

Figure 12.3. Dibenzo-*p*-dioxin and 2,3,7,8-tetrachlorodibenzo-*p*-dioxin (TCDD), often called simply "dioxin." In the structure of dibenzo-*p*-dioxin each number refers to a numbered carbon atom to which an H atom is bound and the names of derivatives are based upon the carbon atoms where another group has been substituted for the H atoms, as is seen by the structure and name of 2,3,7,8-tetrachlorodibenzo-*p*-dioxin.

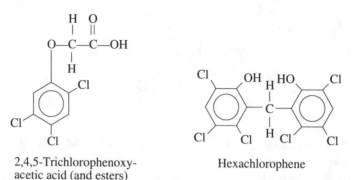

2,4,5-Trichlorophenoxy-
acetic acid (and esters)

Hexachlorophene

Figure 12.4.   Two chemicals whose manufacture resulted in the production of by-product TCDD contaminant.

TCDD has a very low vapor pressure of only 1.7×10^{-6} mm Hg at 25°C, a high melting point of 305°C, and a water solubility of only 0.2 μg/L. It is stable thermally up to about 700°C, has a high degree of chemical stability, and is poorly biodegradable. It is very toxic to some animals, with an LD_{50} of only about 0.6 μg/kg body mass in male guinea pigs. (The type and degree of its toxicity to humans is largely unknown; it is known to cause a severe skin condition called chloracne). Because of its properties, TCDD is a stable, persistent environmental pollutant and hazardous waste constituent of considerable concern. It has been identified in some municipal incineration emissions, and has been a widespread environmental pollutant from improper waste disposal.

The most notable case of TCDD contamination resulted from the spraying of waste oil mixed with TCDD on roads and horse arenas in Missouri in the early 1970s. The oil was used to try to keep dust down in these areas. The extent of contamination was revealed by studies conducted in late 1982 and early 1983. As a result, the U.S. EPA bought out the entire TCDD-contaminated town of Times Beach, Missouri, in March, 1983, at a cost of $33 million. On December 31, 1990 final permission was granted by the courts to incinerate the soil at Times Beach, as well as TCDD-contaminated soil from other areas at a total cost of $80 million. TCDD has been released in a number of industrial accidents, the most massive of which exposed several tens of thousands of people to a cloud of chemical emissions spread over an approximately 3-square-mile area at the Givaudan-La Roche Icmesa manufacturing plant near Seveso, Italy, in 1976. On an encouraging note from a toxicological perspective, no abnormal occurrences of major malformations were found in a study of 15,291 children born in the area within 6 years after the release.[2]

One of the greater environmental disasters ever to result from pesticide manufacture involved the production of Kepone, structural formula

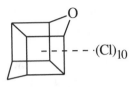

This pesticide has been used for the control of banana-root borer, tobacco wireworm, ants, and cockroaches. Kepone exhibits acute, delayed, and cumulative toxicity in birds, rodents, and humans, and it causes cancer in rodents. It was manufactured in Hopewell, Virginia, during the mid-1970s. During this time, workers were exposed to Kepone and are alleged to have suffered health problems as a result. The plant was connected to the Hopewell sewage system, and frequent infiltration of Kepone wastes caused the Hopewell sewage treatment plant to become inoperative at times. As much as 53,000 kg of Kepone may have been dumped into the sewage system during the years that the plant was operated. The sewage effluent was discharged to the James River, resulting in extensive environmental dispersion and toxicity to aquatic organisms. Decontamination of the river would have required dredging and detoxification of 135 million cubic meters of river sediment at a prohibitively high cost of several billion dollars.

Polychlorinated Biphenyls

First discovered as environmental pollutants in 1966, polychlorinated biphenyls (PCB compounds) have been found throughout the world in water, sediments, bird tissue, and fish tissue. These compounds constitute an important class of special wastes. They are made by substituting from 1 to 10 Cl atoms onto the biphenyl aromatic structure as shown on the left in Figure 12.5. This substitution can produce 209 different compounds (congeners), of which one example is shown on the right in Figure 12.5.

Polychlorinated biphenyls have very high chemical, thermal, and biological stability; low vapor pressure; and high dielectric constants. These properties have led to the use of PCBs as coolant-insulation fluids in transformers and

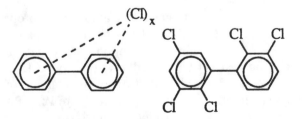

Figure 12.5.   General formula of polychlorinated biphenyls (left, where X may range from 1 to 10) and a specific 5-chlorine congener (right).

capacitors; for the impregnation of cotton and asbestos; as plasticizers; and as additives to some epoxy paints. The same properties that made extraordinarily stable PCBs so useful also contributed to their widespread dispersion and accumulation in the environment. Under regulations issued in the U.S. under the authority of the Toxic Substances Control Act passed in 1976, the manufacture of PCBs was discontinued in the U.S. and their uses and disposal were strictly controlled.

Several chemical formulations have been developed to substitute for PCBs in electrical applications. Disposal of PCBs from discarded electrical equipment and other sources remains a problem, particularly since PCBs can survive ordinary incineration by escaping as vapors through the smokestack. However, they can be destroyed by special incineration processes.

PCBs are especially prominent pollutants in the sediments of the Hudson River as a result of waste discharges from two capacitor manufacturing plants, which operated about 60 km upstream from the southernmost dam on the river from 1950 to 1976. The river sediments downstream from the plants exhibit PCB levels of about 10 ppm, 1–2 orders of magnitude higher than levels commonly encountered in river and estuary sediments. Recent evidence suggests that significant amounts of the PCBs in sediments have undergone biodegradation and there is some hope that the process may be hastened by addition of oxygen and essential nutrients.

Askarel

Askarel is the generic name of PCB-containing dielectric fluids in transformers. These fluids are 50–70 percent PCBs and may contain 30–50 percent trichlorobenzenes (TCBs). As of 1989 an estimated 100,000 askarel-containing transformers were still in use in the U.S. Although the dielectric fluid may be replaced in these transformers enabling their continued use, PCBs tend to leach into the replacement fluid from the transformer core and other parts of the transformer over a several-month period.

12.9. RADIONUCLIDES IN THE AQUATIC ENVIRONMENT

The massive production of **radionuclides** (radioactive isotopes) by weapons and nuclear reactors since World War II has been accompanied by increasing concern about the effects of radioactivity upon health and the environment. Radionuclides are produced as fission products of heavy nuclei of such elements as uranium or plutonium. They also result from the reaction of neutrons with stable nuclei. These phenomena are illustrated in Figure 12.6. Radionuclides are formed in large quantities as waste products in nuclear power generation. Their ultimate disposal is a problem that has caused much controversy regarding the widespread use of nuclear power. Artificially produced radionuclides are also

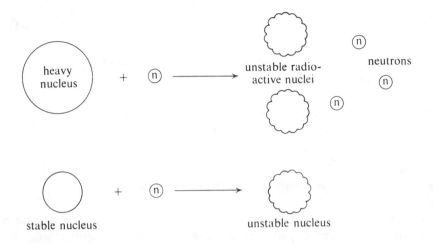

Figure 12.6.   A heavy nucleus, such as that of ^{235}U, may absorb a neutron and break up (undergo fission), yielding lighter radioactive nuclei. A stable nucleus may absorb a neutron to produce a radioactive nucleus.

widely used in industrial and medical applications, particularly as "tracers." With so many possible sources of radionuclides, it is impossible to entirely eliminate radioactive contamination of aquatic systems. Furthermore, radionuclides may enter aquatic systems from natural sources. Therefore, the transport, reactions, and biological concentration of radionuclides in aquatic ecosystems are of great importance to the environmental chemist.

Radionuclides differ from other nuclei in that they emit **ionizing radiation** — alpha particles, beta particles, and gamma rays. The most massive of these emissions is the **alpha particle**, a helium nucleus of atomic mass 4, consisting of two neutrons and two protons and denoted $^{4}_{2}\alpha$. An example of alpha production is found in the radioactive decay of uranium-238:

$$^{238}_{92}U \rightarrow {}^{234}_{90}Th + {}^{4}_{2}\alpha + energy \qquad (12.9.1)$$

This transformation consists of a uranium nucleus, atomic number 92 and atomic mass 238, losing an alpha particle, atomic number 2 and atomic mass 4 to yield a thorium nucleus, atomic number 90 and atomic mass 234.

Beta radiation consists of either highly energetic, negative electrons, which are designated $^{0}_{-1}\beta$, or positive electrons, called positrons and designated $^{0}_{1}\beta$. A typical beta emitter, chlorine-38, may be produced by irradiating chlorine with neutrons. The chlorine-37 nucleus, natural abundance 24.5%, absorbs a neutron to produce chlorine-38 and gamma radiation:

$$\ce{^{37}_{17}Cl} + \ce{^{1}_{0}n} \rightarrow \ce{^{38}_{17}Cl} + \gamma \tag{12.9.2}$$

The chlorine-38 nucleus is radioactive and loses a negative beta particle to become an argon-38 nucleus:

$$\ce{^{38}_{17}Cl} \rightarrow \ce{^{38}_{18}Ar} + \ce{^{0}_{-1}\beta} \tag{12.9.3}$$

Since the negative beta particle has essentially no mass and a –1 charge, the stable product isotope, argon-38, has the same mass and a charge 1 greater than chlorine-38.

Gamma rays are electromagnetic radiation similar to X-rays, though more energetic. Since the energy of gamma radiation is often a well-defined property of the emitting nucleus, it may be used in some cases for the qualitative and quantitative analysis of radionuclides.

The primary effect of alpha particles, beta particles, and gamma rays upon materials is the production of ions; therefore, they are called **ionizing radiation**. Due to their large size, alpha particles do not penetrate matter deeply, but cause an enormous amount of ionization along their short path of penetration. Therefore, alpha particles present little hazard outside the body, but are very dangerous when ingested. Although beta particles are more penetrating than alpha particles, they produce much less ionization per unit path length. Gamma rays are much more penetrating than particulate radiation. Their degree of penetration is proportional to their energy.

The **decay** of a specific radionuclide follows first-order kinetics; that is, the number of nuclei disintegrating in a short time interval is directly proportional to the number of radioactive nuclei present. This means that each radionuclide has a characteristic **half-life**. As the term implies, a half-life is the period of time during which half of a given number of atoms of a specific kind of radionuclide decay. Ten half-lives are required for the loss of 99.9% of the activity of a radionuclide.

Radiation damages living organisms by initiating harmful chemical reactions in tissues. For example, bonds are broken in the macromolecules that carry out life processes. In cases of acute radiation poisoning, bone marrow which produces red blood cells is destroyed and the concentration of red blood cells is diminished. Radiation-induced genetic damage is of great concern. Such damage may not become apparent until many years after exposure. As humans have learned more about the effects of ionizing radiation, the dosage level considered to be safe has steadily diminished. For example, the United States Atomic Energy Commission has dropped the maximum permissible concentration of some radioisotopes to levels of less than one ten-thousandth of those considered safe in the early 1950s. Although it is possible that even the slightest exposure to ionizing radiation entails some damage, some radiation is unavoidably received

Table 12.5. Radionuclides in Water

Radionuclide	Half-life	Nuclear Reaction, Description, Source
Naturally occurring and from cosmic reactions		
Carbon-14	5730 years	$^{14}N(n,p)$ ^{14}C,[a] thermal neutrons from cosmic or nuclear-weapon sources reacting with N_2
Silicon-32	~300 years	^{40}Ar (p,x) ^{32}Si, nuclear spallation (splitting of the nucleus) of atmospheric argon by cosmic-ray protons
Potassium-40	~1.4×10^9 years	0.0119% of natural potassium
Naturally occurring from ^{238}U series		
Radium-226	1620 years	Diffusion from sediments, atmosphere
Lead-210	21 years	$^{226}Ra \rightarrow 6$ steps $\rightarrow ^{210}Pb$
Thorium-230	75,200 years	$^{238}U \rightarrow 3$ steps $\rightarrow ^{230}Th$ produced in situ
Thorium-234	24 days	$^{238}U \rightarrow ^{234}Th$ produced in situ
From reactor and weapons fission		
Strontium-90	28 years	These are the fission-product radioisotopes of greatest significance because of their high yields and biological activity
Iodine-131	8 days	
Cesium-137	30 years	
Barium-140	13 days	The isotopes from barium-140 through krypton-85 are listed in generally decreasing order of fission yield.
Zirconium-95	65 days	
Cerium-141	33 days	
Strontium-89	51 days	
Ruthenium-103	40 days	
Krypton-85	10.3 years	
Cobalt-60	5.25 years	From nonfission neutron reactions in reactors
Manganese-54	310 years	From nonfission neutron reactions in reactors
Iron-55	2.7 years	$^{56}Fe(n,2n)$ ^{55}Fe, from high-energy neutrons acting on iron in weapon hardware
Plutonium-239	24,300 years	^{238}U (n,γ) ^{239}Pu, neutron capture by uranium

[a]This notation denotes the isotope nitrogen-14 reacting with a neutron, n, giving off a proton, p, and forming the isotope carbon-14; other nuclear reactions may be similarly deduced from the notation shown. (Note that x represents nuclear fragments from the spallation reaction.)

from natural sources. For the majority of the population, exposure to natural radiation exceeds that from artificial sources.

Table 12.5 summarizes the major natural and artificial radionuclides likely to be encountered in water. The study of the ecological and health effects of radionuclides involves consideration of many factors. Among these are the type and energy of the radiation emitter and the half-life of the source. In addition, the degree to which the particular element is absorbed by living species and the chemical interactions and transport of the element in aquatic ecosystems are important factors. Radionuclides having very short half-lives may be hazardous when produced but decay too rapidly to affect the environment into which they are introduced. Radionuclides with very long half-lives may be quite persistent in the environment but of such low activity that little environmental damage is caused. Therefore, in general, radionuclides with intermediate half-lives are the most dangerous. They persist long enough to enter living systems while still retaining a high activity. Because they may be incorporated within living tissue, radionuclides of "life elements" are particularly dangerous. Much concern has been expressed over strontium-90, a common waste product of nuclear testing. This element is interchangeable with calcium in bone. Strontium-90 fallout drops onto pasture and crop land and is ingested by cattle. Eventually, it enters the bodies of infants and children by way of cows' milk.

Some radionuclides found in water, primarily radium and potassium-40, originate from natural sources, particularly leaching from minerals. Others come from pollutant sources, primarily nuclear power plants and testing of nuclear weapons. The levels of radionuclides found in water typically are measured in units of picoCuries/liter, where a Curie is 3.7×10^{10} disintegrations per second, and a picoCurie is 1×10^{-12} that amount, or 3.7×10^{-2} disintegrations per second. (2.2 disintegrations per minute).

The radionuclide of most concern in drinking water is **radium**, Ra. Areas in the United States where significant radium contamination of water has been observed include the uranium-producing regions of the western U.S., Iowa, Illinois, Wisconsin, Missouri, Minnesota, Florida, North Carolina, Virginia, and the New England states.

CHAPTER SUMMARY

The chapter summary below is presented in a programmed format to review the main points covered in this chapter. It is used most effectively by filling in the blanks, referring back to the chapter as necessary. The correct answers are given at the end of the summary.

The general types of water pollutants most likely to be involved with metal transport are (1)_____, and those most likely to be involved with eutrophication are (2)_____.
Lead, cadmium and mercury are all examples of water pollutants called (3)_____, whereas arsenic is an example of (4)_____.
One of these pollutants that is chemically very similar to zinc is (5)_____, one that potentially can get into water from plumbing made of it is (6)_____, and one that is converted to methylated organometallic forms by bacteria is

(7)_____. A metalloid pollutant that occurs with phosphate minerals and enters into the environment along with some phosphorus compounds is (8)_____.

The two major types of metal-organic interactions to be considered in an aquatic system are (9)_____. Of all the metals, (10)_____ has the greatest number of organometallic compounds in commercial use. Bactericidal, fungicidal, and insecticidal tributyl tin chloride and related tributyl tin (TBT) compounds are of particular environmental significance because of their growing use as (11)_____.

Some classes and specific examples of inorganic water pollutants are (12)_____

_____.

A highly toxic water pollutant that is widely used in certain mineral-processing operations is (13)_____. A water pollutant that is the initial product of the decay of nitrogenous organic wastes is (14)_____ and one produced by bacteria acting on sulfate is (15)_____. A substance that causes lung cancer when inhaled, but which may not be particularly dangerous in drinking water is (16)_____.

Eutrophication describes a condition of lakes or reservoirs involving (17)_____. Generally, the single plant nutrient that is generally named as the culprit in excessive eutrophication is (18)_____.

The most common source of pollutant acid in water is (19)_____ formed by the microbial oxidation of (20)_____. Another water quality parameter discussed along with salinity that is, however, generally not introduced directly into water by human activity is (21)_____. The recharging of water softeners can introduce excessive amounts of (22)_____ into water.

The overall biologically mediated reaction by which oxygen is consumed in water is (23)_____. Biochemical oxygen demand is a measurement of (24)_____

_____. Primary and secondary sewage treatment processes remove (25)_____ _____ from water. Substances that are especially troublesome in improperly disposed sewage sludge are (26)_____ _____.

A soap anion exhibits a dual nature because it consists of (27)_____ _____ that likes water and a (28)_____ _____. Soap's primary disadvantage as a cleaning agent is (29)_____ _____. The surface-active agent, or surfactant, in detergents acts by (30)_____ _____.

The advantage of ABS surface active agent over LAS is that the latter is (31)_____. Most of the environmental problems that have been more recently associated with the use of detergents arise from the (32)_____ in detergents.

Nonbiodegradable, low-molecular-weight organic compounds of low volatility are commonly called (33)_____. Some general classes of pesticides are (34)_____ _____. Chemically, both parathion and malathion are classified as (35)_____, but the latter is safer because (36)_____. Although the toxicity of DDT is low, it is environmentally damaging because of its (37)_____. The most notorious by-products of pesticide manufacture are (38)_____ of which the most notorious example is (39)_____.

A general and specific example of PCB structures are (40)_____. A transformer coolant fluid containing 50–70 percent PCBs is known as (41)_____. Two ways of producing radionuclides are (42)_____ _____.

Radionuclides produce (43)_____ radiation in the three major

forms of (44)_____. A time

interval characteristic of radionuclides is called the (45)_____ defined

as (46)_____

_____.

Radiation damages living organisms by (47)_____

_____. In

general, radionuclides with intermediate half-lives are the most dangerous

because (48)_____.

The radionuclide of most concern in drinking water is (49)_____.

Answers

1. metal-binding organic species
2. algal nutrients and detergents
3. heavy metals
4. metalloids
5. cadmium
6. lead
7. mercury
8. arsenic
9. complexation and organometallic compound formation
10. tin
11. industrial biocides
12. those that contribute acidity, alkalinity, or salinity to water; algal nutrients; cyanide ion; ammonia; carbon dioxide; hydrogen sulfide; nitrite; and sulfite
13. cyanide
14. ammonia (ammonium ion)
15. hydrogen sulfide
16. asbestos
17. excess algal growth
18. phosphate
19. acid mine drainage
20. pyrite
21. alkalinity
22. salinity
23. $\{CH_2O\} + O_2 \xrightarrow{\text{Microorganisms}} CO_2 + H_2O$

24. the degree of oxygen consumption by microbially-mediated oxidation of contaminants in water
25. oxygen-demanding substances, oil, grease, and solids
26. organic material which continues to degrade slowly, refractory organics, and heavy metals
27. an ionic carboxyl "head"
28. a long hydrocarbon "tail"
29. its reaction with divalent cations to form insoluble salts of fatty acids
30. lowering the surface tension of water to which the detergent is added, making the water "wetter"
31. biodegradable
32. builders
33. biorefractory
34. insecticides, molluscicides, nematicides, rodenticides, herbicides, plant growth regulators, defoliants, plant desiccants, fungicides, bactericides, and algicides.
35. organophosphates
36. mammals can detoxify it
37. persistence and accumulation in food chains
38. polychlorinated dibenzodioxins
39. TCDD
40. structures shown in Figure 12.5
41. askarel
42. as fission products of heavy nuclei and by the reaction of neutrons with stable nuclei
43. ionizing
44. alpha particles, beta particles, and gamma rays
45. half-life
46. time during which half of a given number of atoms of a specific kind of radionuclide decay
47. initiating harmful chemical reactions in tissues
48. they persist long enough to enter living systems while still retaining a high activity
49. radium

QUESTIONS AND PROBLEMS

1. What do mercury and arsenic have in common in regard to their interactions with bacteria in sediments?

2. What are some characteristics of radionuclides that make them especially hazardous to humans?

3. To what class do pesticides containing the following group belong?

4. Consider the following compound:

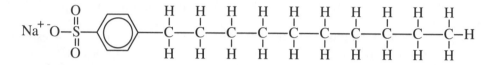

Which of the following characteristics is not possessed by the compound: (a) one end of the molecule is hydrophilic and the other end is hydrophobic, (b) surface-active qualities, (c) the ability to lower surface tension of water, (d) good biodegradability, (e) tendency to persist in water.

5. A certain pesticide is fatal to fish fingerlings at a level of 0.50 parts per million in water. A leaking metal can containing 5.00 kg of the pesticide was dumped into a stream with a flow of 10.0 liters per second moving at 1 kilometer per hour. The container leaks pesticide at a constant rate of 5 mg/sec. For what distance (in km) downstream is the water contaminated by fatal levels of the pesticide by the time the container is empty?

6. A pesticide sprayer got stuck while trying to ford a stream flowing at a rate of 136 liters per second. Pesticide leaked into the stream for exactly 1 hour and at a rate that contaminated the stream at a uniform 0.25 ppm of methoxychlor. How much pesticide was lost from the sprayer during this time?

7. A sample of water contaminated by the accidental discharge of a radionuclide used for medicinal purposes showed an activity of 23,956 counts per second at the time of sampling and 2,993 cps exactly 46 days later. Estimate the half-life of the radionuclide.

8. What are the two reasons that soap is environmentally less harmful than ABS surfactant used in detergents?

9. Why can one not give an exact chemical formula of the specific compound designated as PCB?

10. A radioisotope has a nuclear half-life of 24 hours and a biological half-life of 16 hours (half of the element is eliminated from the body in 16 hours). A person accidentally swallowed sufficient quantities of this isotope to give an

initial "whole body" count rate of 1000 counts per minute. What was the count rate after 16 hours?

11. What is the primary detrimental effect upon organisms of salinity in water arising from dissolved NaCl and Na_2SO_4?

12. Give a specific example of each of the following general classes of water pollutants: (a) trace elements, (b) organically bound metal, (c) pesticides

13. Match each compound in the left column with the description corresponding to it in the right column.

   (a) CdS

     (1) Pollutant released to a U.S. stream by a poorly controlled manufacturing process.

   (b) $(CH_3)_2AsH$

     (2) Insoluble form of a toxic trace element likely to be found in anaerobic sediments.

   (c)

     (3) Common environmental pollutant formerly used as a transformer coolant.

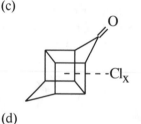

   (d)

     (4) Chemical species thought to be produced by bacterial action.

LITERATURE CITED

1. Esposito, M. P., "Dioxin Wastes," Section 4.3 in *Standard Handbook of Hazardous Waste Treatment and Disposal*, H. M. Freeman, Ed., (New York, NY: McGraw Hill Book Company, 1989), pp. 4.25–4.34.
2. "Dioxin is Found Not to Increase Birth Defects," *New York Times*, March 18, 1988, p. 12.

SUPPLEMENTARY REFERENCES

1. Brown, L. R., *State of the World 1990*, Worldwatch Institute, Washington, DC, 1989.
2. Hassell, K. A., *The Biochemistry and Uses of Pesticides*, 2nd ed., (New York: NY: VCH Publishers, Inc., 1991).

3. Hites, R. A. and S. J. Eisenreich, Eds., *Sources and Fates of Aquatic Pollutants*, Advances in Chemistry Series 216. American Chemical Society, Washington, DC, 1987.
4. Howard, P. H., *Handbook of Environmental Fate and Exposure Data for Organic Chemicals* (Three Volumes), (Chelsea, MI: Lewis Publishers, 1989).
5. Mance, G., *Pollution Threat of Heavy Metals in Aquatic Environments*. (Essex, England: Elsevier Applied Science Publishers, 1987).
6. Paddock, T., *Dioxins and Furans: Questions and Answers*, Academy of Natural Sciences, Philadelphia, PA, 1989.
7. Rump, H. H., and H. Krist, *Laboratory Manual for the Examination of Water, Waste Water, and Soil*. (New York, NY: VCH Publishers, 1989).
8. Worobec, M. D., and G. Ordway, *Toxic Substances Controls Guide*, BNA Books Distribution Center, Edison, NJ, 1989.

13 WATER TREATMENT

13.1. WATER TREATMENT AND WATER USE

The treatment of water may be divided into three major categories:

- Purification for domestic use
- Treatment for specialized industrial applications
- Treatment of wastewater to make it acceptable for release or reuse.

The type and degree of treatment are strongly dependent upon the source and intended use of the water. Water for domestic use must be thoroughly disinfected to eliminate disease-causing microorganisms, but may contain appreciable levels of dissolved calcium and magnesium (hardness). Water to be used in boilers may contain bacteria but must be quite soft to prevent scale formation. Wastewater being discharged into a large river may require less rigorous treatment than water to be reused in an arid region. As world demand for limited water resources grows, more sophisticated and extensive means will have to be employed to treat water.

Most physical and chemical processes used to treat water involve similar phenomena, regardless of their application to the three main categories of water treatment listed above. Therefore, after introducing water treatment for municipal use, industrial use, and disposal, this chapter discusses each major kind of treatment process as it applies to all of these applications.

13.2. MUNICIPAL WATER TREATMENT

The function of a modern water treatment plant is to produce clear, safe, even tasteful drinking water from raw water that may consist of a murky liquid pumped from a polluted river laden with mud and swarming with bacteria; well water, much too hard for domestic use and containing high levels of stain-producing dissolved iron and manganese; or other questionable sources of raw water. A schematic diagram of a typical municipal water treatment plant is

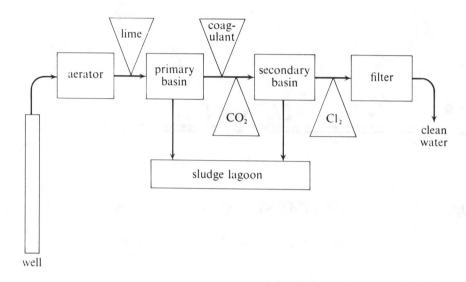

Figure 13.1.   Schematic of a municipal water treatment plant.

shown in Figure 13.1. This particular facility treats water containing excessive hardness and a high level of iron. The raw water taken from wells first goes to an aerator. Contact of the water with air removes volatile solutes such as hydrogen sulfide, carbon dioxide, methane, and volatile odorous substances such as thiomethane (CH_3SH) and bacterial metabolites. Contact with oxygen also aids iron removal by oxidizing soluble iron(II) to insoluble iron(III). The addition of lime as CaO or $Ca(OH)_2$ after aeration raises the pH and results in the formation of precipitates containing the hardness ions Ca^{2+} and Mg^{2+}. These precipitates settle from the water in a primary basin. Much of the solid material remains in suspension and requires the addition of coagulants [such as iron(III) and aluminum sulfates, which form gelatinous metal hydroxides] to settle the colloidal particles. Activated silica or synthetic polyelectrolytes may also be added to stimulate coagulation or flocculation. The settling occurs in a secondary basin after the addition of carbon dioxide to lower the pH. Sludge from both the primary and secondary basins is pumped to a sludge lagoon. The water is finally chlorinated, filtered, and pumped to the city water mains.

13.3.   TREATMENT OF WATER FOR INDUSTRIAL USE

Water is widely used in various process applications in industry. Other major industrial uses are boiler feedwater and cooling water. The kind and degree of treatment of water in these applications depends upon the end use. As examples, cooling water may require only minimal treatment; removal of corrosive substances and scale-forming solutes is essential for boiler feedwater; and water used in food processing must be free of pathogens and toxic substances. Improper treatment of water for industrial use can cause problems, such as corrosion, scale formation, reduced heat transfer in heat exchangers, reduced water flow, and product contamination. These effects may cause reduced equipment performance or equipment failure, increased energy costs due to inefficient heat utilization or cooling, increased costs for pumping water, and product deterioration. Obviously, the effective treatment of water at minimum cost for industrial use is a very important area of water treatment.

Numerous factors must be taken into consideration in designing and operating an industrial water treatment facility. These include the following:

- Water requirement
- Quantity and quality of available water sources
- Sequential use of water (successive uses for applications requiring progressively lower water quality)
- Water recycle
- Discharge standards

The various specific processes employed to treat water for industrial use are discussed in later sections of this chapter. **External treatment**, usually applied to the plant's entire water supply, uses processes such as aeration, filtration, and clarification to remove material from water that may cause problems. Such substances include suspended or dissolved solids, hardness, and dissolved gases. Following this basic treatment, the water may be divided into different streams, some to be used without further treatment and the rest to be treated for specific applications.

Internal treatment is designed to modify the properties of water for specific applications. Examples of internal treatment include the following:

- Removal of dissolved oxygen by reaction with hydrazine or sulfite
- Addition of chelating agents to react with dissolved Ca^{2+} and prevent formation of calcium deposits
- Addition of precipitants, such as phosphate used for calcium removal
- Treatment with dispersants to inhibit scale
- Addition of inhibitors to prevent corrosion
- Adjustment of pH
- Disinfection for food processing uses or to prevent bacterial growth in cooling water

13.4. SEWAGE TREATMENT

Typical municipal sewage contains oxygen-demanding materials, sediments, grease, oil, scum, pathogenic bacteria, viruses, salts, algal nutrients, pesticides, refractory organic compounds, heavy metals, and an astonishing variety of flotsam ranging from children's socks to sponges. It is the job of the waste treatment plant to remove as much of this material as possible.

Several characteristics are used to describe sewage. These include turbidity (international turbidity units); suspended solids (ppm); total dissolved solids (ppm); acidity (H^+ ion concentration or pH); and dissolved oxygen (in ppm O_2). Biochemical oxygen demand (Section 12.7) is used as a measure of oxygen-demanding substances.

Current processes for the treatment of wastewater may be divided into three main categories of primary treatment, secondary treatment, and tertiary treatment, each of which is discussed separately. Also discussed are total wastewater treatment systems, based largely upon physical and chemical processes.

Waste from a municipal water system is normally treated in a **publicly owned treatment works, POTW.** In the United States these systems are allowed to discharge only effluents that have attained a certain level of treatment, as mandated by federal law.

Primary Waste Treatment

Primary treatment of wastewater consists of the removal of insoluble matter such as grit, grease, and scum from water. The first step in primary treatment normally is screening. Screening removes or reduces the size of trash and large solids that get into the sewage system. These solids are collected on screens and scraped off for subsequent disposal. Most screens are cleaned with power rakes. Comminuting devices shred and grind solids in the sewage. Particle size may be reduced to the extent that the particles can be returned to the sewage flow.

Grit in wastewater consists of such materials as sand and coffee grounds which do not biodegrade well and generally have a high settling velocity. **Grit removal** is practiced to prevent its accumulation in other parts of the treatment system, to reduce clogging of pipes and other parts, and to protect moving parts from abrasion and wear. Grit normally is allowed to settle in a tank under conditions of low flow velocity, and it is then scraped mechanically from the bottom of the tank.

Primary sedimentation removes both settleable and floatable solids. During primary sedimentation there is a tendency for flocculent particles to aggregate for better settling, a process that may be aided by the addition of chemicals. The material that floats in the primary settling basin is known collectively as grease. In addition to fatty substances, the grease consists of oils, waxes, free fatty acids, and insoluble soaps containing calcium and magnesium. Normally, some of the grease settles with the sludge and some floats to the surface, where it may be removed by a skimming device.

Secondary Waste Treatment by Biological Processes

The most obvious harmful effect of biodegradable organic matter in waste-water is BOD, consisting of a biochemical oxygen demand for dissolved oxygen by microorganism-mediated degradation of the organic matter. **Secondary wastewater treatment** is designed to remove BOD, usually by taking advantage of the same kind of biological processes that would otherwise consume oxygen in water receiving the wastewater. Secondary treatment by biological processes takes many forms but consists basically of the following: Microorganisms provided with added oxygen are allowed to degrade organic material in solution or in suspension until the BOD of the waste has been reduced to acceptable levels. The waste is oxidized biologically under conditions controlled for optimum bacterial growth and at a site where this growth does not influence the environment.

One of the simplest biological waste treatment processes is the **trickling filter** (Figure 13.2) in which wastewater is sprayed over rocks or other solid support material covered with microorganisms. The structure of the trickling filter is such that contact of the wastewater with air is allowed and degradation of organic matter occurs by the action of the microorganisms.

Rotating biological reactors, another type of treatment system, consist of groups of large plastic discs mounted close together on a rotating shaft. The device is positioned so that at any particular instant half of each disc is immersed in wastewater and half exposed to air. The shaft rotates constantly, so that the submerged portion of the discs is always changing. The discs, usually made of high-density polyethylene or polystyrene, accumulate thin layers of attached biomass, which degrades organic matter in the sewage. Oxygen is absorbed by the biomass and by the layer of wastewater adhering to it during the time that the biomass is exposed to air.

Both trickling filters and rotating biological reactors are examples of fixed-film biological (FFB) processes. The greatest advantage of these processes is their low energy consumption. The energy consumption is minimal because it is not necessary to pump air or oxygen into the water, as is the case with the popular activated sludge process described below. The trickling filter has long been a standard means of wastewater treatment; for example, a 31-acre trickling

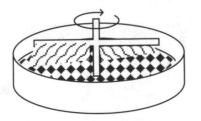

Figure 13.2.   Trickling filter for secondary waste treatment.

filter installation in Baltimore which became operational in 1907 was used throughout most of the century.

The **activated sludge process**, Figure 13.3, is probably the most versatile and effective of all waste treatment processes. Microorganisms in the aeration tank convert organic material in wastewater to microbial biomass and CO_2. Organic nitrogen is converted to ammonium ion or nitrate. Organic phosphorus is converted to orthophosphate. The microbial cell matter formed as part of the waste degradation processes is normally kept in the aeration tank until the microorganisms are past the log phase of growth (Figure 11.11), at which point the cells flocculate relatively well to form settleable solids. These solids settle out in a settler and a fraction of them is discarded. Part of the solids, the return sludge, is recycled to the head of the aeration tank and comes into contact with fresh sewage. The combination of a high concentration of "hungry" cells in the return sludge and a rich food source in the influent sewage provides optimum conditions for the rapid degradation of organic matter.

The degradation of organic matter that occurs in an activated sludge facility also occurs in streams and other aquatic environments. However, in general, when a degradable waste is put into a stream, it encounters only a relatively small population of microorganisms capable of carrying out the degradation process. Thus, several days may be required for the buildup of a sufficient population of organisms to degrade the waste. In the activated sludge process, continual recycling of active organisms provides the optimum conditions for waste degradation, and a waste may be degraded within the very few hours that it is present in the aeration tank.

The activated sludge process provides two pathways for the removal of BOD: (1) oxidation of organic matter to provide energy for the metabolic processes of the microorganisms, and (2) synthesis, incorporation of the organic matter into

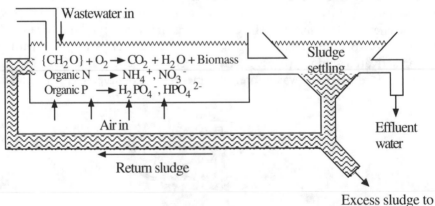

Figure 13.3.   Activated sludge process.

cell mass. In the first pathway, carbon is removed in the gaseous form as CO_2. The second pathway provides for removal of carbon as a solid in biomass. That portion of the carbon converted to CO_2 is vented to the atmosphere and does not present a disposal problem. The disposal of waste sludge, however, is a problem, primarily because it is only about 1% solids and contains many undesirable components. Normally, partial water removal is accomplished by drying on sand filters, vacuum filtration, or centrifugation. The dewatered sludge may be incinerated or used as landfill, both of which are facing increased regulatory and public opposition. To a certain extent, sewage sludge may be digested in the absence of oxygen by methane-producing anaerobic bacteria to produce methane and carbon dioxide,

$$2\{CH_2O\} \rightarrow CH_4 + CO_2 \tag{13.4.1}$$

a process that reduces both the volatile-matter content and the volume of the sludge by about 60%. A carefully designed plant may produce enough methane to provide for all of its power needs.

One of the most desirable means of sludge disposal is to use it to fertilize and condition soil. However, care has to be taken that excessive levels of heavy metals are not applied to the soil as sludge contaminants. Problems with various kinds of sludges resulting from water treatment are discussed further in Section 13.10.

Nitrification (the microbially mediated conversion of ammonium nitrogen to nitrate; see Section 11.13), is a significant process that occurs during biological waste treatment. Ammonium ion is normally the first inorganic nitrogen species produced in the biodegradation of nitrogenous organic compounds. It is oxidized, under the appropriate conditions, first to nitrite by *Nitrosomonas* bacteria,

$$2NH_4^+ + 3O_2 \rightarrow 4H^+ + 2NO_2^- + 2H_2O \tag{13.4.2}$$

then to nitrate by *Nitrobacter*:

$$2NO_2^- + O_2 \rightarrow 2NO_3^- \tag{13.4.3}$$

These reactions occur in the aeration tank of the activated sludge plant and are favored in general by long retention times, low organic loadings, large amounts of suspended solids, and high temperatures. Nitrification can reduce sludge settling efficiency because the denitrification reaction

$$4NO_3^- + 5\{CH_2O\} + 4H^+ \rightarrow 2N_2(g) + 5CO_2(g) + 7H_2O \tag{13.4.4}$$

occurring in the oxygen-deficient settler causes bubbles to form on the sludge floc (aggregated sludge particles), making it so buoyant that it floats to the top.

This prevents settling of the sludge and increases the organic load in the receiving waters. Under the appropriate conditions, however, advantage can be taken of this phenomenon to remove nutrient nitrogen from water (see Section 13.9).

Tertiary Waste Treatment

Unpleasant as the thought may be, many people drink used water — water that has been discharged from a municipal sewage treatment plant or from some industrial process. This raises serious questions about the presence of pathogenic organisms or toxic substances in such water. Because of high population density and rapid industrial development, the problem is especially acute in Europe. About one-third of the drinking water supplies in England and Wales contain "used" water; in Paris and surrounding areas, the extent of reused water ranges from 50 to 70 percent; and in West Germany, the Ruhr River has at times consisted of as much as 40 percent treated wastewater. Obviously, there is a great need to treat wastewater in a manner that makes it amenable to reuse. This requires treatment beyond the secondary processes.

Tertiary waste treatment (sometimes called **advanced waste treatment**) is a term used to describe a variety of processes performed on the effluent from secondary waste treatment. The contaminants removed by tertiary waste treatment fall into the general categories of (1) suspended solids; (2) dissolved organic compounds; and (3) dissolved inorganic materials, including the important class of algal nutrients. Each of these categories presents its own problems with regard to water quality. Suspended solids are primarily responsible for residual biological oxygen demand in secondary sewage effluent waters. The dissolved organics are the most hazardous from the standpoint of potential toxicity. The major problem with dissolved inorganic materials is that presented by algal nutrients, primarily nitrates and phosphates. In addition, potentially hazardous toxic metals may be found among the dissolved inorganics.

In addition to these chemical contaminants, secondary sewage effluent often contains a number of disease-causing microorganisms, requiring disinfection in cases where humans may later come into contact with the water. Among the bacteria that may be found in secondary sewage effluent are organisms causing tuberculosis, dysenteric bacteria (*Bacillus dysenteriae, Shigella dysenteriae, Shigella paradysenteriae, Proteus vulgaris*), cholera bacteria (*Vibrio cholerae*), bacteria causing mud fever (*Leptospira icterohemorrhagiae*), and bacteria causing typhoid fever (*Salmonella typhosa, Salmonella paratyphi*). In addition, viruses causing diarrhea, eye infections, infectious hepatitis, and polio may be encountered. Ingestion of sewage still causes disease, even in more developed nations.

Physical-Chemical Treatment of Municipal Wastewater

Complete physical-chemical wastewater treatment systems offer both advantages and disadvantages relative to biological treatment systems. The capital costs of these facilities can be less than those of biological treatment facilities, and they usually require less land. They are better able to cope with toxic materials and overloads. However, they require careful operator control and consume relatively large amounts of energy.

Basically, a physical-chemical treatment process involves:

- Removal of scum and solid objects.
- Clarification, generally with addition of a coagulant, and frequently with the addition of other chemicals (such as lime for phosphorus removal).
- Filtration to remove filterable solids.
- Activated carbon adsorption.
- Disinfection.

The basic steps of a complete physical-chemical wastewater treatment facility are shown in Figure 13.4.

During the early 1970s, it appeared likely that physical-chemical treatment would largely replace biological treatment. However, dramatically higher chemical and energy costs since then have significantly slowed the development of physical/chemical waste treatment processes.

13.5.   INDUSTRIAL WASTEWATER TREATMENT

Wastewater to be treated must be characterized fully, particularly with a thorough chemical analysis of possible waste constituents and their chemical and metabolic products. The biodegradability of wastewater constituents should also be determined. The options available for the treatment of wastewater are summarized briefly in this section and discussed in greater detail in later sections.

One of two major ways of removing organic wastes is biological treatment by an activated sludge, or related process (see Section 13.4 and Figure 13.3). It may be necessary to acclimate microorganisms to the degradation of constituents that are not normally biodegradable. Consideration needs to be given to possible hazards of biotreatment sludges, such as those containing excessive levels of heavy metal ions. The other major process for the removal of organics from wastewater is sorption by activated carbon (see Section 13.8), usually in columns of granular activated carbon. Activated carbon and biological treatment can be combined with the use of powdered activated carbon in the activated sludge process. The powdered activated carbon sorbs some constituents that may be toxic to microorganisms and is collected with the sludge. A major consideration with the use of activated carbon on wastewater is the hazard that spent activated carbon may present from the wastes it retains. These hazards may include those of toxicity or reactivity, such as those posed by explosives manufacture wastes sorbed to activated carbon. Regeneration of the carbon is expensive and can be hazardous in some cases.

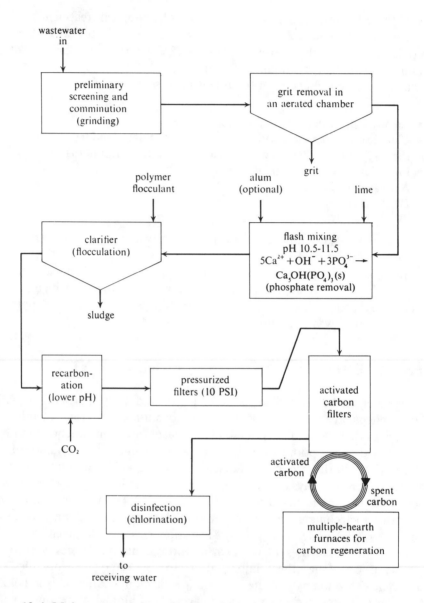

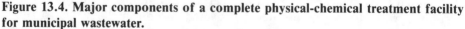

Figure 13.4. Major components of a complete physical-chemical treatment facility for municipal wastewater.

Wastewater can be treated by a variety of chemical processes, including acid/base neutralization, precipitation, and oxidation/reduction. In some cases these treatment steps must precede biological treatment; for example, wastewater exhibiting extremes of pH must be neutralized in order for microorganisms to thrive in it. Cyanide in the wastewater may be oxidized with chlorine and organics with ozone, hydrogen peroxide promoted with ultraviolet radiation, or dissolved oxygen at high temperatures and pressures. Heavy metals may be precipitated with base, carbonate, or sulfide.

Wastewater can be treated by several physical processes. In some cases, simple density separation and sedimentation can be used to remove water-immiscible liquids and solids. Filtration is frequently required and flotation by gas bubbles generated on particle surfaces may be useful. Wastewater solutes can be concentrated by evaporation, distillation, and membrane processes, including reverse osmosis, hyperfiltration, and ultrafiltration. Organic constituents can be removed by solvent extraction, air stripping, or steam stripping.

Synthetic resins are useful for removing some pollutant solutes from wastewater. Organophilic resins have proved useful for the removal of alcohols; aldehydes; ketones; hydrocarbons; chlorinated alkanes, alkenes, and aryl compounds; esters, including phthalate esters; and pesticides. Cation exchange resins are effective for the removal of heavy metals.

13.6. REMOVAL OF SOLIDS

Relatively large solid particles are removed from water by simple **settling** and **filtration**. A special type of filtration procedure known as **microstraining** is especially effective in the removal of the very small particles. These filters are woven from stainless steel wire so fine that it is barely visible. This enables preparation of filters with openings only 60–70 μm across. These openings may be reduced to 5–15 μm by partial clogging with small particles, such as bacterial cells. The cost of this treatment is likely to be substantially lower than the costs of competing processes. High flow rates at low back pressures are normally achieved.

The removal of colloidal solids from water usually requires coagulation. Salts of aluminum and iron are the coagulants most often used in water treatment. Of these, alum or filter alum is most commonly used. This substance is a hydrated aluminum sulfate, $Al_2(SO_4)_3 \cdot 18H_2O$. When added to water, the aluminum ion in this salt hydrolyzes by reactions that consume alkalinity in the water, such as:

$$Al(H_2O)_6^{3+} + 3HCO_3^- \rightarrow Al(OH)_3(s) + 3CO_2 + 6H_2O \qquad (13.6.1)$$

The gelatinous hydroxide thus formed carries suspended material with it as it settles. In addition, however, it is likely that positively charged hydroxyl-bridged dimers such as

$$(H_2O)_4Al \underset{O}{\overset{O}{\underset{H}{\overset{H}{<}}}} Al(H_2O)_4^{4+}$$

and higher polymers are formed which interact specifically with colloidal particles, bringing about coagulation. Metal ions in coagulants also react with virus proteins and destroy up to 99% of the virus in water.

Anhydrous iron(III) sulfate added to water forms ferric hydroxide in a reaction analogous to Equation 13.6.1. An advantage of iron(III) sulfate is that it works over a wide pH range of approximately 4–11. Hydrated iron(II) sulfate, $FeSO_4 \cdot 7H_2O$, or copperas, is also commonly used as a coagulant. It forms a gelatinous precipitate of hydrated iron(III) oxide; in order to function, it must be oxidized to iron(III) by dissolved oxygen in the water at an elevated pH, or by chlorine, which can oxidize iron(II) at lower pH values. Sodium silicate partially neutralized by acid aids coagulation, particularly when used with alum.

Natural and synthetic polyelectrolytes are used in flocculating particles. Among the natural compounds so used are starch and cellulose derivatives, proteinaceous materials, and gums composed of polysaccharides. Synthetic polymers that are effective flocculants are now widely used. Neutral polymers and both anionic and cationic polyelectrolytes have been employed successfully as flocculants in various applications.

Coagulation-filtration is a much more effective procedure than filtration alone for the removal of suspended material from water. As the term implies, the process consists of the addition of coagulants that aggregate the particles into larger size particles, followed by filtration. Either alum or lime, often with added polyelectrolytes, is most commonly employed for coagulation.

The filtration step of coagulation-filtration is usually performed on a medium such as sand or anthracite coal. Often, to reduce clogging, several media with progressively smaller interstitial spaces are used. One example is the **rapid sand filter,** which consists of a layer of sand supported by layers of gravel particles, the particles becoming progressively larger with increasing depth. The substance that actually filters the water is coagulated material that collects in the sand. As more material is removed, the buildup of coagulated material eventually clogs the filter and must be removed by back-flushing.

An important class of solids that must be removed from wastewater consists of suspended solids in secondary sewage effluent that arise primarily from sludge that was not removed in the settling process. These solids account for a large part of the BOD in the effluent and may interfere with other aspects of tertiary waste treatment. For example, these solids may clog membranes in reverse osmosis water treatment processes. The quantity of material involved may be rather high. Processes designed to remove suspended solids often will remove 10–20 mg/L of organic material from secondary sewage effluent. In addition, a small amount of the inorganic material is removed as well.

13.7.   REMOVAL OF CALCIUM AND OTHER METALS

Calcium and magnesium salts, which generally are present in water as bicarbonates or sulfates, cause water hardness. One of the most common manifestations of water hardness is the insoluble "curd" formed by the reaction of soap with calcium or magnesium ions. The formation of these insoluble soap salts is discussed in Section 12.8. Although ions that cause water hardness do not form insoluble products with detergents, they do adversely affect detergent performance. Therefore, calcium and magnesium must be complexed or removed from water for detergents to function properly.

Another problem caused by hard water is the formation of mineral deposits. For example, when water containing calcium and bicarbonate ions is heated, insoluble calcium carbonate is formed:

$$Ca^{2+} + 2HCO_3^- \rightarrow CaCO_3(s) + CO_2(g) + H_2O \qquad (13.7.1)$$

This product coats the surfaces of hot water systems, clogging pipes and reducing heating efficiency. Dissolved salts such as calcium and magnesium bicarbonates and sulfates can be especially damaging in boiler feedwater. Clearly, the removal of water hardness is essential for many uses of water.

Several processes are used for softening water. On a large scale in a centralized water treatment plant the lime-soda process utilizing lime, $Ca(OH)_2$, and soda ash, Na_2CO_3 is employed. Calcium is precipitated as $CaCO_3$ and magnesium as $Mg(OH)_2$. When the calcium is present primarily as "bicarbonate hardness," it can be removed by the addition of $Ca(OH)_2$ alone:

$$Ca^{2+} + 2HCO_3^- + Ca(OH)_2 \rightarrow 2CaCO_3(s) + 2H_2O \qquad (13.7.2)$$

When bicarbonate ion is not present at substantial levels, a source of CO_3^{2-} must be provided at a high enough pH to prevent conversion of most of the carbonate to bicarbonate. These conditions are obtained by the addition of Na_2CO_3. For example, calcium present as the chloride can be removed from water by the addition of soda ash:

$$Ca^{2+} + 2Cl^- + 2Na^+ + CO_3^{2-} \rightarrow CaCO_3(s) + 2Cl^- + 2Na^+ \qquad (13.7.3)$$

Note that the removal of bicarbonate hardness results in a net removal of soluble salts from solution, whereas removal of nonbicarbonate hardness involves the addition of at least as many equivalents of ionic material as are removed.

The precipitation of magnesium as the hydroxide requires a higher pH than the precipitation of calcium as the carbonate:

$$Mg^{2+} + 2OH^- \rightarrow Mg(OH)_2(s) \qquad (13.7.4)$$

The high pH required may be provided by the basic carbonate ion from soda ash:

$$CO_3^{2-} + H_2O \rightarrow HCO_3^- + OH^- \qquad (13.7.5)$$

Some large-scale, lime-soda softening plants make use of the precipitated calcium carbonate product as a source of additional lime. The calcium carbonate is first heated to at least 825°C to produce quicklime, CaO:

$$CaCO_3 + heat \rightarrow CaO + CO_2(g) \qquad (13.7.6)$$

The quicklime is then slaked with water to produce calcium hydroxide:

$$CaO + H_2O \rightarrow Ca(OH)_2 \qquad (13.7.7)$$

The water softened by lime-soda softening plants usually suffers from two defects. First, because of supersaturation effects, some $CaCO_3$ and $Mg(OH)_2$ usually remain in solution. If not removed, these compounds will precipitate at a later time and cause harmful deposits or undesirable cloudiness in water. The second problem results from the use of highly basic sodium carbonate, which gives the product water an excessively high pH, up to pH 11. To overcome these problems, the water is recarbonated by bubbling CO_2 into it. The carbon dioxide converts the slightly soluble calcium carbonate and magnesium hydroxide to their soluble bicarbonate forms:

$$CaCO_3(s) + CO_2 + H_2O \rightarrow Ca^{2+} + 2HCO_3^- \qquad (13.7.8)$$

$$Mg(OH)_2(s) + 2CO_2 \rightarrow Mg^{2+} + 2HCO_3^- \qquad (13.7.9)$$

The CO_2 also neutralizes excess hydroxide ion:

$$OH^- + CO_2 \rightarrow HCO_3^- \qquad (13.7.10)$$

The pH generally is brought within the range 7.5 to 8.5 by recarbonation. The source of CO_2 used in the recarbonation process may be from the combustion of carbonaceous fuel. Scrubbed stack gas from a power plant frequently is utilized. Water adjusted to a pH, alkalinity, and Ca^{2+} concentration very close to $CaCO_3$ saturation is labeled *chemically stabilized*. It neither precipitates $CaCO_3$ in water mains, which can clog the pipes, nor dissolves protective $CaCO_3$ coatings from the pipe surfaces. Water with a Ca^{2+} concentration much below $CaCO_3$ saturation is called an *aggressive* water.

Calcium may be removed from water very efficiently by the addition of orthophosphate:

$$5Ca^{2+} + 3PO_4^{3-} + OH^- \rightarrow Ca_5OH(PO_4)_3(s) \qquad (13.7.11)$$

It should be pointed out that the chemical formation of a slightly soluble product for the removal of undesired solutes such as hardness ions, phosphate, iron, and manganese must be followed by sedimentation in a suitable apparatus. Frequently, coagulants must be added, and filtration employed for complete removal of these sediments.

Water may be purified by ion exchange, the reversible transfer of ions between aquatic solution and a solid material capable of bonding ions. The removal of NaCl from solution by two ion exchange reactions is a good illustration of this process. First the water is passed over a solid cation exchanger in the hydrogen form, represented by $H^{+-}\{Cat(s)\}$:

$$H^{+-}\{Cat(s)\} + Na^+ + Cl^- \rightarrow Na^{+-}\{Cat(s)\} + H^+ + Cl^- \qquad (13.7.12)$$

Next the water is passed over an anion exchanger in the hydroxide ion form, represented by $OH^{-+}\{An(s)\}$:

$$OH^{-+}\{An(s)\} + H^+ + Cl^- \rightarrow Cl^{-+}\{An(s)\} + H_2O \qquad (13.7.13)$$

Thus, the cations in solution are replaced by hydrogen ion and the anions by hydroxide ion, yielding water as the product.

The softening of water by ion exchange does not require the removal of all ionic solutes, just those cations responsible for water hardness. Generally, therefore, only a cation exchanger is necessary. Furthermore, the sodium rather than the hydrogen form of the cation exchanger is used, and the divalent cations are replaced by sodium ion. Sodium ion at low concentrations is harmless in water to be used for most purposes, and sodium chloride is a cheap and convenient substance with which to recharge the cation exchangers (though it does add a salinity burden to the water).

A number of materials have ion-exchanging properties. Among the minerals especially noted for their ion exchange properties are the aluminum silicate minerals, or **zeolites**. An example of a zeolite which has been used commercially in water softening is glauconite, $K_2(MgFe)_2Al_6(Si_4O_{10})_3(OH)_{12}$. Synthetic zeolites have been prepared by drying and crushing the white gel produced by mixing solutions of sodium silicate and sodium aluminate.

The discovery in the mid-1930s of synthetic ion exchange resins composed of organic polymers with attached functional groups marked the beginning of modern ion exchange technology. Structures of typical synthetic ion exchangers are shown in Figures 13.5 and 13.6. The cation exchanger shown in Figure 13.5 is called a **strongly acidic cation exchanger** because the parent $-SO_3^-H^+$ group is

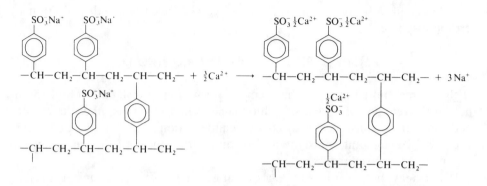

Figure 13.5. Strongly acidic cation exchanger. Sodium exchange for calcium in water is shown.

a strong acid. When the functional group binding the cation is the $-CO_2^-$ group, the exchange resin is called a **weakly acidic cation exchanger** because the $-CO_2H$ group is a weak acid. Figure 13.6 shows a **strongly basic anion exchanger** in which the functional group is a quaternary ammonium group, $-N^+(CH_3)_3$. In the hydroxide form, $-N^+(CH_3)_3OH^-$, the hydroxide ion is readily released; hence the exchanger is classified as **strongly basic**.

The water-softening capability of a cation exchanger is shown in Figure 13.5, where sodium ion on the exchanger is exchanged for calcium ion in solution. The same reaction occurs with magnesium ion. Water softening by cation exchange is now a widely used, effective, and economical process. In many areas having a low water flow, however, it is not likely that home water softening by ion exchange may be used universally without some deterioration of water quality arising from the contamination of wastewater by sodium chloride. Such contamination results from the periodic need to regenerate a water softener with sodium chloride, in order to displace calcium and magnesium ions from the resin and replace these hardness ions with sodium ions:

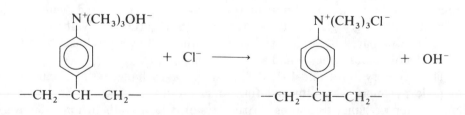

Figure 13.6. Strongly basic anion exchanger. Chloride exchange for hydroxide ion is shown.

$$\text{Ca}^{2+}\text{-}\{\text{Cat}(s)\}_2 + 2\text{Na}^+ + 2\text{Cl}^- \rightarrow 2\text{Na}^+\text{-}\{\text{Cat}(s)\} + \text{Ca}^{2+} + 2\text{Cl}^- \quad (13.7.14)$$

During the regeneration process, a large excess of sodium chloride must be used — several pounds for a home water softener. Appreciable amounts of dissolved sodium chloride can be introduced into sewage by this route.

Strongly acidic cation exchangers are used for the removal of water hardness. Weakly acidic cation exchangers having $-\text{CO}_2\text{H}$ as the functional group are useful for removing alkalinity along with hardness. Alkalinity generally is manifested by bicarbonate ion. This species is a sufficiently strong base to neutralize the acid of a weak acid cation exchanger:

$$2\text{R-CO}_2\text{H} + \text{Ca}^{2+} + 2\text{HCO}_3^- \rightarrow \{\text{R-CO}_2^-\}_2\text{Ca}^{2+} + 2\text{H}_2\text{O} + 2\text{CO}_2 \quad (13.7.15)$$

However, weak bases such as sulfate ion or chloride ion are not strong enough to remove hydrogen ion from the carboxylic acid exchanger. An additional advantage of these exchangers is that they may be regenerated almost stoichiometrically with dilute strong acids, thus avoiding the potential pollution problem caused by the use of excess sodium chloride to regenerate strongly acidic cation changers.

Chelation or, as it is sometimes known, *sequestration*, is an effective method of softening water without actually having to remove calcium and magnesium from solution. A complexing agent is added which greatly reduces the concentrations of free hydrated cations. For example, chelating calcium ion with excess EDTA anion (Y^{4-}),

$$\text{Ca}^{2+} + Y^{4-} \rightarrow \text{Ca}Y^{2-} \quad (13.7.16)$$

reduces the concentration of hydrated calcium ion, preventing the precipitation of calcium carbonate:

$$\text{Ca}^{2+} + \text{CO}_3^{2-} \rightarrow \text{CaCO}_3(s) \quad (13.7.17)$$

Polyphosphate salts, EDTA, and NTA (see Section 11.9) are chelating agents commonly used for water softening. Polysilicates are used to complex iron.

Removal of Iron, Manganese, and Heavy Metals

Soluble iron and manganese are found in many groundwaters because of reducing conditions which favor the soluble $+2$ oxidation state of these metals. Iron is the more commonly encountered of the two metals. In groundwater, the level of iron seldom exceeds 10 mg/L, and that of manganese is rarely higher than 2 mg/L. The basic method for removing both of these metals depends upon oxidation to higher insoluble oxidation states. The oxidation is generally accomplished by aeration. The rate of oxidation is pH-dependent in both cases, with a high pH favoring more rapid oxidation. The oxidation of soluble Mn(II) to insoluble MnO_2 is a complicated process. It appears to be catalyzed by solid MnO_2, which is known to adsorb Mn(II). This adsorbed Mn(II) is slowly oxidized on the MnO_2 surface.

Chlorine and potassium permanganate are sometimes employed as oxidizing agents for iron and manganese. There is some evidence that organic chelating agents with reducing properties hold iron(II) in a soluble form in water. In such cases, chlorine is effective because it destroys the organic compounds and enables the oxidation of iron(II).

In water with a high level of carbonate, $FeCO_3$ and $MnCO_3$ may be precipitated directly by raising the pH above 8.5 by the addition of sodium carbonate or lime. This approach is less popular than oxidation, however.

Relatively high levels of insoluble iron(III) and manganese(IV) frequently are found in water as colloidal material which is difficult to remove. These metals may be associated with humic colloids or "peptizing" organic material that binds to colloidal metal oxides, stabilizing the colloid.

Heavy metals such as copper, cadmium, mercury, and lead are found in wastewaters from a number of industrial processes. Because of the toxicity of many heavy metals, their concentrations must be reduced to very low levels prior to release of the wastewater. A number of approaches are used in heavy metals removal.

Lime treatment (Section 13.7) removes heavy metals as insoluble hydroxides, basic salts, or coprecipitated with calcium carbonate or ferric hydroxide. This process does not completely remove mercury, cadmium, or lead, so their removal is aided by addition of sulfide (most heavy metals are sulfide-seekers):

$$Cd^{2+} + S^{2-} \rightarrow CaS(s) \qquad (13.7.18)$$

Heavy chlorination is frequently necessary to break down metal-solubilizing ligands (see Chapter 11). Lime precipitation does not normally permit recovery of metals and is sometimes undesirable from the economic viewpoint.

Electrodeposition (reduction of metal ions to metal by electrons at an electrode), *reverse osmosis* (see Section 13.9), and *ion exchange* are frequently employed for metal removal. Solvent extraction using organic-soluble chelating substances is also effective in removing many metals. **Cementation**, a process by which a metal deposits by reaction of its ion with a more readily oxidized metal, may be employed:

$$Cu^{2+} + Fe \text{ (iron scrap)} \rightarrow Fe^{2+} + Cu \qquad (13.7.19)$$

Activated carbon adsorption effectively removes some metals from water at the part per million level. Sometimes a chelating agent is sorbed to the charcoal to increase metal removal.

Even when not specifically designed for the removal of heavy metals, most waste treatment processes remove appreciable quantities of the more trouble-some heavy metals encountered in wastewater. These metals accumulate in the sludge from biological treatment, so sludge disposal must be given careful con-sideration (see Section 13.10). Various physical-chemical treatment processes effectively remove heavy metals from wastewaters. One such treatment is lime precipitation followed by activated-carbon filtration. Activated-carbon filtra-tion may also be preceded by treatment with ferric chloride to form a ferric hydroxide floc, which is an effective heavy metals scavenger. Similarly, alum, which forms aluminum hydroxide, may be added prior to activated-carbon filtration.

It should be noted that the form of the heavy metal has a strong effect upon the efficiency of metal removal. For instance, chromium(VI) is normally more difficult to remove than chromium(III). Chelation may prevent metal removal by solubilizing metals (see Section 11.9).

In the past, removal of heavy metals has been largely a fringe benefit of wastewater treatment processes. Currently, however, more consideration is being given to design and operating parameters that specifically enhance heavy-metals removal as part of wastewater treatment.

13.8. REMOVAL OF DISSOLVED ORGANICS

Very low levels of exotic organic compounds in drinking water are suspected of contributing to cancer and other maladies. Some of these are chlorinated organic compounds produced by chlorination of organics in water, especially humic substances. Removal of organics to very low levels prior to chlorination has been found to be effective in preventing trihalomethane formation. In addi-tion, many organic compounds survive, or are produced by, secondary waste-water treatment. Almost half of these are humic substances (see Section 11.9) with a molecular-weight range of 1000–5000. Among the remainder are found ether-extractable materials, carbohydrates, proteins, detergents, tannins, and lignins. The humic compounds, because of their high molecular weight and anionic character, influence some of the physical and chemical aspects of waste treatment. The ether-extractables contain many of the compounds that are resistant to biodegradation and are of particular concern regarding potential toxicity, carcinogenicity, and mutagenicity. In the ether extract are found many fatty acids, hydrocarbons of the *n*-alkane class, naphthalene, diphenylmethane, diphenyl, methylnaphthalene, isopropyl benzene, dodecyl benzene, phenol, dioctylphthalate, and triethylphosphate.

The standard method for the removal of dissolved organic material is adsorp-tion on activated carbon, a product that is produced from a variety of carbona-ceous materials, including wood, pulp-mill char, peat, and lignite. The carbon is produced by charring the raw material anaerobically below 600°C, followed by an activation step consisting of partial oxidation. Carbon dioxide may be employed as an oxidizing agent at 600–700°C.

$$CO_2 + C \rightarrow 2CO \qquad (13.8.1)$$

or the carbon may be oxidized by water at 800–900°C:

$$H_2O + C \rightarrow H_2 + CO \qquad (13.8.2)$$

These processes develop porosity, increase the surface area, and leave the C atoms in arrangements that have affinities for organic compounds.

Activated carbon comes in two general types: granulated activated carbon, consisting of particles 0.1 to 1 mm in diameter, and powdered activated carbon, in which most of the particles are 50–100 μm in diameter.

The exact mechanism by which activated carbon holds organic materials is not known. However, one reason for the effectiveness of this material as an adsorbent is its tremendous surface area. A solid cubic foot of carbon particles may have a combined pore and surface area of approximately 10 square miles!

Although interest is increasing in the use of powdered activated carbon for water treatment, currently granular carbon is more widely used. It may be employed in a fixed bed, through which water flows downward. Accumulation of particulate matter requires periodic backwashing. An expanded bed in which particles are kept slightly separated by water flowing upward may be used with less chance of clogging.

Economics require regeneration of the carbon, which is accomplished by heating it to 950°C in a steam-air atmosphere. This process oxidizes adsorbed organics and regenerates the carbon surface, with an approximately 10% loss of carbon.

Removal of organics may also be accomplished by adsorbent synthetic polymers. Such polymers as Amberlite XAD-4 have hydrophobic surfaces and strongly attract relatively insoluble organic compounds, such as chlorinated pesticides. The porosity of these polymers is up to 50% by volume, and the surface area may be as high as 850 m²/g. They are readily regenerated by solvents such as isopropanol and acetone. Under appropriate operating conditions, these polymers remove virtually all nonionic organic solutes; for example, phenol at 250 mg/L is reduced to less than 0.1 mg/L by appropriate treatment with Amberlite XAD-4.

Oxidation of dissolved organics holds some promise for their removal. Ozone, hydrogen peroxide, molecular oxygen (with or without catalysts), chlorine and its derivatives, or permanganate can be used. Electrochemical oxidation may be possible in some cases.

13.9.   REMOVAL OF DISSOLVED INORGANICS

In order for complete water recycling to be feasible, inorganic solute removal is essential. The effluent from secondary waste treatment generally contains 300–400 mg/L more dissolved inorganic material than does the municipal water supply. It is obvious, therefore, that 100% water recycle without removal of inorganics would cause the accumulation of an intolerable level of dissolved material. Even when water is not destined for immediate reuse, the removal of the inorganic nutrients phosphorus and nitrogen is highly desirable to reduce eutrophication downstream. In some cases, the removal of toxic trace metals is needed.

One of the most obvious methods for removing inorganics from water is distillation. Unfortunately, the energy required for distillation is generally too high for the process to be economically feasible. Furthermore, volatile materials such as ammonia and odorous compounds are carried over to a large extent in the distillation process, unless special preventative measures are taken. Freezing produces a very pure water, but is considered uneconomical with present technology. Membrane processes considered most promising for bulk removal of inorganics from water are electrodialysis, ion exchange, and reverse osmosis. Other membrane processes used in water purification are nanofiltration, ultrafiltration, microfiltration, and dialysis.

Ion Exchange

The ion exchange method for softening water is described in detail in Section 13.7. The ion exchange process used for removal of inorganics consists of passing the water successively over a solid cation exchanger and a solid anion exchanger, which replace cations and anions by hydrogen ion and hydroxide ion, respectively. The net result is that each equivalent of salt is replaced by a mole of water. For the hypothetical ionic salt MX, the reactions are

$$H^{+-}\{Cat(s)\} + M^+ + X^- \rightarrow M^{+-}\{Cat(s)\} + H^+ + X^- \quad (13.9.1)$$

$$OH^{-+}\{An(s)\} + H^+ + X^- \rightarrow X^{-+}\{An(s)\} + H_2O \quad (13.9.2)$$

where $^-\{Cat(s)\}$ represents the solid cation exchanger and $^+\{An(s)\}$ represents the solid anion exchanger. The cation exchanger is regenerated with strong acid and the anion exchanger with strong base.

Demineralization by ion exchange generally produces water of a very high quality. Unfortunately, some organic compounds in wastewater foul ion exchangers, and microbial growth on the exchangers can diminish their efficiency. In addition, regeneration of the resins is expensive, and the concentrated wastes from regeneration require disposal in a manner that will not damage the environment.

Reverse Osmosis

Reverse osmosis is a very useful technique for the purification of water. Basically, reverse osmosis consists of forcing pure water through a semipermeable membrane that allows the passage of water but not of other material. This process depends on the preferential sorption of water on the surface of the membrane, which is composed of porous cellulose acetate or polyamide. Pure water from the sorbed layer is forced through pores in the membrane under pressure. If the thickness of the sorbed water layer is d, the pore diameter for optimum separation should be 2d. The optimum pore diameter depends upon the thickness of the sorbed pure water layer and may be several times the diameters of the solute and solvent molecules. Therefore, reverse osmosis is not a simple sieve separation or ultrafiltration process. The principle of reverse osmosis is illustrated in Figure 13.7.

Phosphorus Removal

Advanced waste treatment normally requires removal of phosphorus to reduce algal growth. Algae may grow at PO_4^{3-} levels as low as 0.05 mg/L. Growth inhibition requires levels well below 0.5 mg/L. Since municipal wastes typically contain approximately 25 mg/L of phosphate (as orthophosphates, polyphosphates, and insoluble phosphates), the efficiency of phosphate removal must be quite high to prevent algal growth. This removal may occur in the sewage treatment process (1) in the primary settler; (2) in the aeration chamber of the activated sludge unit; or (3) after secondary waste treatment.

Normally, the activated sludge process removes about 20% of the phosphorus from sewage. Thus, an appreciable fraction of largely biological phosphorus is removed with the sludge. Detergents and other sources contribute significant amounts of phosphorus to domestic sewage and considerable phosphate ion

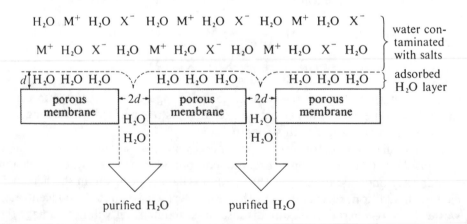

Figure 13.7. Solute removal from water by reverse osmosis.

remains in the effluent. However, some wastes, such as carbohydrate wastes from sugar refineries, are so deficient in phosphorus that supplementation of the waste with inorganic phosphorus is required for proper growth of the microorganisms degrading the wastes.

Under some sewage plant operating conditions, much greater than normal phosphorus removal has been observed. In such plants, characterized by high dissolved oxygen and high pH levels in the aeration tank, removal of 60–90% of the phosphorus has been attained, yielding two or three times the normal level of phosphorus in the sludge. In a conventionally operated aeration tank of an activated sludge plant, the CO_2 level is relatively high because the aeration rate is insufficient to sweep out sufficient dissolved carbon dioxide to bring its concentration down to low levels. Thus, the pH generally is low enough that phosphate is maintained primarily in the form of the $H_2PO_4^-$ ion. However, at a higher rate of aeration in a relatively hard water, the CO_2 is swept out, the pH rises, and reactions such as the following occur:

$$5Ca^{2+} + 3HPO_4^{2-} + H_2O \rightarrow Ca_5OH(PO_4)_3(s) + 4H^+ \qquad (13.9.3)$$

The precipitated hydroxyapatite or other form of calcium phosphate is incorporated in the sludge floc.

Chemically, phosphate is most commonly removed by precipitation. Some common precipitants and their products are shown in Table 13.1. Precipitation processes are capable of at least 90–95% phosphorus removal at reasonable cost.

Lime, $Ca(OH)_2$, is the chemical most commonly used for phosphorus removal:

$$5Ca(OH)_2 + 3HPO_4^{2-} \rightarrow Ca_5OH(PO_4)_3(s) + 3H_2O + 6OH^- \quad (13.9.4)$$

Lime has the advantages of low cost and ease of regeneration. The efficiency with which phosphorus is removed by lime is not as high as would be predicted by the low solubility of hydroxyapatite, $Ca_5OH(PO_4)_3$. Some of the possible reasons for this are slow precipitation of $Ca_5OH(PO_4)_3$; formation of nonsettling colloids; precipitation of calcium as $CaCO_3$ in certain pH ranges; and the fact

Table 13.1. Chemical Precipitants for Phosphate and Their Products

Precipitant(s)	Products
$Ca(OH)_2$	$Ca_5OH(PO_4)_3$ (hydroxyapatite)
$Ca(OH)_2 + NaF$	$Ca_5F(PO_4)_3$ (fluorapatite)
$Al_2(SO_4)_3$	$AlPO_4$
$FeCl_3$	$FePO_4$
$MgSO_4$	$MgNH_4PO_4$

that phosphate may be present as condensed phosphates (polyphosphates) which form soluble complexes with calcium ion.

Phosphate can be removed from solution by adsorption on some solids, particularly activated alumina, Al_2O_3. Removals of up to 99.9% of orthophosphate have been achieved with this method.

Nitrogen Removal

Next to phosphorus, nitrogen is the algal nutrient most commonly removed as part of advanced wastewater treatment. Nitrogen in municipal wastewater generally is present as organic nitrogen or ammonia. Ammonia is the primary nitrogen product produced by most biological waste treatment processes. This is because it is expensive to aerate sewage sufficiently to oxidize the ammonia to nitrate through the action of nitrifying bacteria. If the activated sludge process is operated under conditions such that the nitrogen is maintained in the form of ammonia, the latter may be stripped in the form of NH_3 gas from the water by air. For ammonia stripping to work, the ammoniacal nitrogen must be converted to volatile NH_3 gas, which requires a pH substantially higher than the pK_a of the NH_4^+ ion. In practice, the pH is raised to approximately 11.5 by the addition of lime (which also serves to remove phosphate). The ammonia is stripped from the water by air.

Nitrification followed by denitrification is a promising technique for the removal of nitrogen from wastewater. The first step is an essentially complete conversion of ammonia and organic nitrogen to nitrate under strongly aerobic conditions, achieved by more extensive than normal aeration of the sewage:

$$NH_4^+ + 2O_2 \text{ (Nitrifying bacteria)} \rightarrow NO_3^- + 2H^+ + H_2O \quad (13.9.5)$$

The second step is the reduction of nitrate to nitrogen gas. This reaction is also bacterially catalyzed and requires a carbon source and a reducing agent such as methanol, CH_3OH.

$$6NO_3^- + 5CH_3OH + 6H^+ \text{ (Denitrifying bacteria)} \rightarrow$$
$$3N_2(g) + 5CO_2 + 13H_2O \quad (13.9.6)$$

The denitrification process may be carried out either in a tank or on a carbon column. In pilot plant operation, conversions of 95% of the ammonia to nitrate and 86% of the nitrate to nitrogen have been achieved.

13.10. SLUDGE

Perhaps the most pressing water treatment problem at this time has to do with sludge collected or produced during water treatment. Finding a safe place to put the sludge or a use for it has proved troublesome, and the problem is aggravated by the growing numbers of water treatment systems.

Improper disposal of wastes continues to be a subject of public and governmental concern. One of the more recent problems to be addressed by legislative action in the U.S. is ocean dumping of sewage sludge. For many years sludge from New York and New Jersey has been disposed in the Atlantic Ocean's 106-Mile Deepwater Municipal Sewage Sludge Disposal Site. In response to this kind of activity, the U.S. Congress passed the Ocean Dumping Ban Act of 1988. Among the act's requirements are an immediate prohibition against new ocean dumping of sewage sludge and stiff penalties for ocean disposal by communities already engaged in the practice starting January 1, 1992.

Some sludge is present in wastewater and may be collected from it. Such sludge includes human wastes, garbage grindings, organic wastes and inorganic silt and grit from storm water runoff, and organic and inorganic wastes from commercial and industrial sources. There are two major kinds of sludge generated in a waste treatment plant. The first of these is organic sludge from activated sludge, trickling filter, or rotating biological reactors. The second is inorganic sludge from the addition of chemicals, such as in phosphorus removal (see Section 13.9).

Most commonly, sewage sludge is subjected to anaerobic digestion in a digester designed to allow bacterial action to occur in the absence of air. This reduces the mass and volume of sludge and ideally results in the formation of a stabilized humus. Disease agents are also destroyed in the process.

Following digestion, sludge is generally conditioned and thickened to concentrate and stabilize it and make it more dewaterable. Relatively inexpensive processes, such as gravity thickening, may be employed to get the moisture content down to about 95%. Sludge may be further conditioned chemically by the addition of iron or aluminum salts, lime, or polymers.

Sludge dewatering is employed to convert the sludge from an essentially liquid material to a damp solid containing not more than about 85% water. This may be accomplished on sludge drying beds consisting of layers of sand and gravel. Mechanical devices may also be employed, including vacuum filtration, centrifugation, and filter presses. Heat may be used to aid the drying process.

Some of the alternatives for the ultimate disposal of sludge include land spreading, ocean dumping, and incineration. Each of these choices has disadvantages, such as the presence of toxic substances in sludge spread on land, or the high fuel cost of incineration.

Sewage sludge may contain a number of undesirable constituents. Included among these are biorefractory organic compounds (see Section 12.8), such as

PCBs; heavy metals, such as cadmium, lead, and mercury; and pathogenic microorganisms, including viruses, salmonella, and fecal coliforms.

Rich in nutrients, waste sewage sludge contains around 5% N, 3% P, and 0.5% K on a dry-weight basis and can be used to fertilize and condition soil. The humic material in the sludge improves the physical properties and cation-exchange capacity of the soil. Among the factors limiting this application of sludge are excess nitrogen pollution of runoff water and groundwater, survival of pathogens, and the presence of heavy metals in the sludge.

Possible accumulation of heavy metals is of the greatest concern insofar as the use of sludge on cropland is concerned. Sewage sludge is an efficient heavy metals scavenger. On a dry basis, sludge samples from industrial cities have shown levels of up to 9,000 ppm zinc, 6,000 ppm copper, 600 ppm nickel, and up to 800 ppm cadmium! These and other metals tend to remain immobilized in soil by chelation with organic matter, adsorption on clay minerals, and precipitation as insoluble compounds, such as oxides or carbonates. However, increased application of sludge on cropland has caused distinctly elevated levels of zinc and cadmium in both leaves and grain of corn. Therefore, caution has been advised in heavy or prolonged application of sewage sludge to soil. The problem of heavy metals in sewage sludge is one of the many reasons for not allowing mixture of wastes to occur prior to treatment. Sludge does, however, contain nutrients which should not be wasted, given the possibility of eventual fertilizer shortages. Prior control of heavy metal contamination from industrial sources should greatly reduce the heavy metal content of sludge and enable it to be used more extensively on soil.

Sludge sidestreams consisting of water removed from sludge by various treatment processes may cause significant problems. Sewage treatment processes can be divided into **mainstream treatment processes** (primary clarification, trickling filter, activated sludge, and rotating biological reactor) and **sidestream processes**. During sidestream treatment sludge is dewatered, degraded, and disinfected by a variety of processes, including gravity thickening, dissolved air flotation, anaerobic digestion, aerobic digestion, vacuum filtration, centrifugation, belt-filter press filtration, sand-drying-bed treatment, sludge-lagoon settling, wet air oxidation, pressure filtration, and Purifax treatment. Each of these produces a liquid by-product sidestream which is circulated back to the mainstream. These add to the biochemical oxygen demand and suspended solids of the mainstream.

A variety of chemical sludges are produced by various water treatment and industrial processes. Among the most abundant of such sludges is alum sludge produced by the hydrolysis of Al(III) salts used in the treatment of water, which creates gelatinous aluminum hydroxide:

$$Al^{3+} + 3OH^-(s) \rightarrow Al(OH)_3(aq) \tag{13.10.1}$$

Alum sludges normally are 98% or more water and are very difficult to dewater.

Both iron(II) and iron(III) compounds are used for the precipitation of impurities from wastewater via the precipitation of $Fe(OH)_3$. The sludge contains $Fe(OH)_3$ in the form of soft, fluffy precipitates that are difficult to dewater beyond 10 or 12% solids.

The addition of either lime, $Ca(OH)_2$, or quicklime, CaO, to water is used to raise the pH to about 11.5 and cause the precipitation of $CaCO_3$, along with metal hydroxides and phosphates. Calcium carbonate is readily recovered from lime sludges and can be recalcined to produce CaO, which can be recycled through the system.

Metal hydroxide sludges are produced in the removal of metals such as lead, chromium, nickel, and zinc from wastewater by raising the pH to such a level that the corresponding hydroxides or hydrated metal oxides are precipitated. The disposal of these sludges is a substantial problem because of their toxic heavy metal content. Reclamation of the metals is a possibility for some of these sludges.

Pathogenic (disease-causing) microorganisms may persist in the sludge left from the treatment of sewage. Many of these organisms present potential health hazards, and there is risk of public exposure when the sludge is applied to soil. Therefore, it is necessary both to be aware of pathogenic microorganisms in municipal wastewater treatment sludge and to find a means of reducing the hazards caused by their presence.

The most significant organisms in municipal sewage sludge include the following: (1) indicators, including fecal and total coliform: (2) pathogenic bacteria, including *Salmonellae* and *Shigellae*; (3) enteric (intestinal) viruses, including enterovirus and poliovirus; and (4) parasites, such as *Entamoeba histolytica* and *Ascaris lumbricoides*.

Several ways are recommended to significantly reduce levels of pathogens in sewage sludge. Aerobic digestion involves aerobic agitation of the sludge for periods of 40 to 60 days (longer times are employed with low sludge temperatures). Air drying involves draining and/or drying of the liquid sludge for at least three months in a layer 20–25 cm thick. This operation may be performed on underdrained sand beds or in basins. Anaerobic digestion involves maintenance of the sludge in an anaerobic state for periods of time ranging from 60 days at 20°C to 15 days at temperatures exceeding 35°C. Composting involves mixing dewatered sludge cake with bulking agents subject to decay, such as wood chips or shredded municipal refuse, and allowing the action of bacteria to promote decay at temperatures ranging up to 45–65°C. The higher temperatures tend to kill pathogenic bacteria. Finally, pathogenic organisms may be destroyed by lime stabilization in which sufficient lime is added to raise the pH of the sludge to 12 or higher.

13.11.   WATER DISINFECTION

Chlorine is the most commonly used disinfectant employed for killing bacteria in water. When chlorine is added to water, it rapidly hydrolyzes according to the reaction

$$Cl_2 + H_2O \rightarrow H^+ + Cl^- + HOCl \qquad (13.11.1)$$

which has the following equilibrium constant:

$$K = \frac{[H^+][Cl^-][HOCl]}{[Cl_2]} = 4.5 \times 10^{-4} \qquad (13.11.2)$$

Hypochlorous acid, HOCl, is a weak acid that dissociates according to the reaction,

$$HOCl \rightleftarrows H^+ + OCl^- \qquad (13.11.3)$$

with an ionization constant of 2.7×10^{-8}. From the above it can be calculated that the concentration of elemental Cl_2 is negligible at equilibrium above pH 3 when chlorine is added to water at levels below 1.0 g/L.

Sometimes, hypochlorite salts are substituted for chlorine gas as a disinfectant. Calcium hypochlorite, $Ca(OCl)_2$, is commonly used. The hypochlorites are safer to handle than gaseous chlorine.

The two chemical species formed by chlorine in water, HOCl and OCl^-, are known as **free available chlorine**. Free available chlorine is very effective in killing bacteria. In the presence of ammonia, monochloramine, dichloramine, and trichloramine are formed:

$$NH_4^+ + HOCl \rightarrow NH_2Cl \text{ (monochloramine)} + H_2O + H^+ \qquad (13.11.4)$$

$$NH_2Cl + HOCl \rightarrow NHCl_2 \text{ (dichloramine)} + H_2O \qquad (13.11.5)$$

$$NHCl_2 + HOCl \rightarrow NCl_3 \text{ (trichloramine)} + H_2O \qquad (13.11.6)$$

The chloramines are called **combined available chlorine**. Chlorination practice frequently provides for formation of combined available chlorine which, although a weaker disinfectant than free available chlorine, is more readily retained as a disinfectant throughout the water distribution system. Too much ammonia in water is considered undesirable because it exerts excess demand for chlorine.

At sufficiently high Cl:N molar ratios in water containing ammonia, some HOCl and OCl^- remain unreacted in solution, and a small quantity of NCl_3 is formed. The ratio at which this occurs is called the **breakpoint**. Chlorination beyond the breakpoint ensures disinfection. It has the additional advantage of destroying the more common materials that cause odor and taste in water.

At moderate levels of NH_3-N (approximately 20 mg/L), when the pH is between 5.0 and 8.0, chlorination with a minimum 8:1 weight ratio of Cl to NH_3-nitrogen produces efficient denitrification:

$$NH_4^+ + HOCl \rightarrow NH_2Cl + H_2O + H^+ \qquad (13.11.4)$$

$$2NH_2Cl + HOCl \rightarrow N_2(g) + 3H^+ + 3Cl^- + H_2O \qquad (13.11.7)$$

This reaction is used to remove pollutant ammonia from wastewater. However, problems can arise from chlorination of organic wastes. Typical of such by-products is chloroform, produced by the chlorination of humic substances in water.

Chlorine is used to treat water other than drinking water. It is employed to disinfect effluent from sewage treatment plants, as an additive to the water in electric power plant cooling towers, and to control microorganisms in food processing.

Chlorine Dioxide

Chlorine dioxide, ClO_2, is an effective water disinfectant that is of particular interest because, in the absence of impurity Cl_2, it does not produce toxic contaminant trihalomethanes in water treatment. In acidic and neutral water, respectively, the two half-reactions for ClO_2 acting as an oxidant are the following:

$$ClO_2 + 4H^+ + 5e^- \rightleftharpoons Cl^- + 2H_2O \qquad (13.11.8)$$

$$ClO_2 + e^- \rightleftharpoons ClO_2^- \qquad (13.11.9)$$

In the neutral pH range, chlorine dioxide in water remains largely as molecular ClO_2 until it contacts a reducing agent with which to react. Chlorine dioxide is a gas that is violently reactive with organic matter and explosive when exposed to light. For these reasons, ClO_2 is not shipped, but is generated on-site by processes such as the reaction of chlorine gas with solid sodium hypochlorite:

$$2NaClO_2(s) + Cl_2(g) \rightleftharpoons 2ClO_2(g) + 2NaCl(s) \qquad (13.11.10)$$

A high content of elemental chlorine in the product may require its purification to prevent unwanted side reactions from Cl_2.

As a water disinfectant, chlorine dioxide does not chlorinate or oxidize ammonia or other nitrogen-containing compounds. Some concern has been raised over possible health effects of its main degradation by-products, ClO_2^- and ClO_3^-.

Ozone

Ozone is sometimes used as a disinfectant in place of chlorine, particularly in Europe. Figure 13.8 shows the basic components of an ozone water treatment system. Basically, air is filtered, cooled, dried, and pressurized, then subjected to an electrical discharge of approximately 20,000 volts. The ozone produced is then pumped into a contact chamber where water contacts ozone for 10–15 minutes. Concern over possible production of toxic organochlorine compounds by water chlorination processes has increased interest in ozonation. Furthermore, ozone is more destructive to viruses than is chlorine. Unfortunately, the disinfective power of ozone is limited by its relatively low solubility in water.

13.12. NATURAL WATER PURIFICATION PROCESSES

Virtually all of the materials that waste treatment processes are designed to eliminate may be absorbed by soil or degraded in soil. In fact, most of these materials are essential for soil fertility. Wastewater may provide the water that is essential to plant growth, in addition to the nutrients—phosphorus, nitrogen and potassium—usually provided by fertilizers. Wastewater also contains essential trace elements and vitamins. Stretching the point a bit, the degradation of organic wastes provides the CO_2 essential for photosynthetic production of plant biomass.

Soil may be viewed as a natural filter for wastes. Most organic matter is readily degraded in soil and, in principle, soil constitutes an excellent treatment system—primary, secondary and tertiary—for water. Soil has physical, chemical, and biological characteristics that can enable wastewater detoxification, biodegradation, chemical decomposition, and physical and chemical fixation.

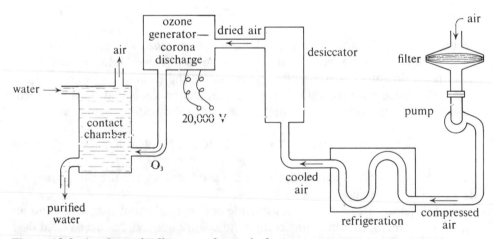

Figure 13.8. A schematic diagram of a typical ozone water-treatment system.

A number of soil characteristics are important in determining its use for land treatment of wastes. These characteristics include physical form, ability to retain water, aeration, organic content, acid-base characteristics, and oxidation-reduction behavior. Soil is a natural medium for a number of living organisms that may have an effect upon biodegradation of wastewaters, including those that contain industrial wastes. Of these, the most important are bacteria, including those from the genera *Agrobacterium, Arthrobacteri, Bacillus, Flavobacterium*, and *Pseudomonas*. Actinomycetes and fungi are important in decay of vegetable matter and may be involved in biodegradation of wastes. Other unicellular organisms that may be present in or on soil are protozoa and algae. Soil animals, earthworms for example, affect soil parameters such as soil texture. The growth of plants in soil may have an influence on its waste treatment potential in such aspects as uptake of soluble wastes and erosion control.

Some early civilizations used human organic wastes to increase soil fertility, and the practice continues in some countries today. The ability of soil to purify water was noted well over a century ago. In 1850 and 1852, J. Thomas Way, a consulting chemist to the Royal Agricultural Society in England, presented two papers to the Society entitled "Power of Soils to Absorb Manure." Mr. Way's experiments showed that soil is an ion exchanger. Much practical and theoretical information on the ion exchange process resulted from his work.

Experiments involving the direct application of wastewater to soil have shown appreciable increases in soil productivity. A process called *overland flow* has been used in which wastewater containing pulverized solids is allowed to trickle over sloping soil. Suspended solids, BOD, and nutrients are largely removed. This method is most applicable to rural communities in relatively warm climates. Approximately one acre of land is required to handle the sewage from 200 persons.

If such systems are not properly designed and operated, odor can become an overpowering problem. The author of this book is reminded of driving into a small town, recalled from some years before as a very pleasant place, and being assaulted with a virtually intolerable odor. The disgruntled residents pointed to a large spray irrigation system on a field in the distance — unfortunately upwind — spraying liquified pig manure as part of an experimental feedlot waste treatment operation. The experiment was not deemed a success and was discontinued by the investigators, presumably before they met with violence from the local residents.

Industrial Wastewater Treatment by Soil

Wastes that are amenable to land treatment are biodegradable organic substances, particularly those contained in municipal sewage and in wastewater from some industrial operations, such as food processing. However, through acclimation over a long period of time, soil bacterial cultures may develop that are effective in degrading normally recalcitrant compounds that occur in indus-trial wastewater. Acclimated microorganisms most commonly occur at contami-nated sites, such as those where soil has been exposed to crude oil for many years.

A variety of enzyme activities are exhibited by microorganisms in soil that enable them to degrade synthetic substances. Even sterile soil may show enzyme activity due to extracellular enzymes secreted by microorganisms in soil. Some of these enzymes are hydrolase enzymes (see Section 10.6), such as those that catalyze the hydrolysis of organophosphate compounds as shown by the reaction,

$$
\begin{array}{c}
\overset{\displaystyle X}{\underset{\displaystyle R}{\overset{\|}{\underset{\displaystyle |}{R-O-\overset{\displaystyle |}{\underset{\displaystyle O}{P}}-O-Ar}}}} + H_2O \xrightarrow{\text{Phosphatase enzyme}} \\
\\
R-O-\overset{\displaystyle X}{\underset{\displaystyle R}{\overset{\|}{\underset{\displaystyle |}{\underset{\displaystyle O}{P}}}}}-OH + HOAr
\end{array}
\qquad (13.12.1)
$$

where R is an alkyl group, Ar is a substituent group that is frequently aromatic, and X is either S or O. Another example of a reaction catalyzed by soil enzymes is the oxidation of phenolic compounds by diphenol oxidase:

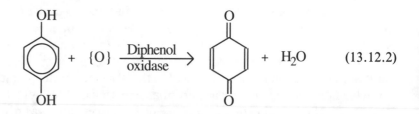

$$(13.12.2)$$

Land treatment is most used for petroleum refining wastes and is applicable to the treatment of fuels and wastes from leaking underground storage tanks. It can also be applied to biodegradable organic chemical wastes, including some organohalide compounds. Land treatment is not suitable for the treatment of wastes containing acids, bases, toxic inorganic compounds, salts, heavy metals, and organic compounds that are excessively soluble, volatile, or flammable.

CHAPTER SUMMARY

The chapter summary below is presented in a programmed format to review the main points covered in this chapter. It is used most effectively by filling in the blanks, referring back to the chapter as necessary. The correct answers are given at the end of the summary.

The major categories for water treatment are (1)_____

_____. External treatment of water is usually applied to (2)_____

_____whereas internal treatment

is designed to (3)_____

_____.

Current processes for the treatment of wastewater may be divided into the

three main categories of (4)_____

_____. Primary treatment of water is designed to remove

(5)_____. The major

constituent removed from water by secondary wastewater treatment is (6)_____

_____. The reaction by which sewage sludge

may be digested in the absence of oxygen by methane-producing anaerobic

bacteria is (7)_____ and it reduces (8)_____

_____. Tertiary

waste treatment normally is applied to (9)_____

_____, and the three

general kinds of contaminants that it removes are (10)_____

_____

_____. Five major operations

in a physical-chemical wastewater treatment process are (11)_____

_____

_____.

Two factors that have prevented development of physical-chemical wastewater

treatment are (12)_____. Some
physical processes used in industrial wastewater treatment are (13)_____
_____
_____.

Before colloidal solids can be removed by filtration, they usually must be subjected to (14)_____.

A reaction by which water becomes softened when heated is (15)_____
_____. The reaction by which water containing "bicarbonate
hardness" can be treated by addition of lime is (16)_____
_____. When bicarbonate ion is not present at substantial
levels, softening water by $CaCO_3$ removal requires (17)_____
_____. Two reactions involving a solid
ion exchanger by which dissolved NaCl may be removed from water are
(18)_____
_____.

The basic method for removing both soluble iron and manganese from water
is (19)_____.
The removal of mercury, cadmium, or lead by lime treatment is aided by addition of (20)_____.

The standard method of removing organic compounds from water is
(21)_____. Some methods for removing dissolved inorganic
material from water are (22)_____
_____.

A water purification process that consists of forcing pure water through a semipermeable membrane that allows the passage of water but not of other material
is (23)_____. Phosphorus is removed in advanced
wastewater treatment to reduce (24)_____.
Phosphorus removal is usually accomplished by addition of (25)_____ ,
for which the reaction is (26)_____.

At a very high aeration rate in an activated sludge treatment process, phosphate is commonly removed because (27)_____

_____.

The two overall biological reactions by which nitrogen originally present as NH_4^+ can be removed from water are (28)_____ _____ and (29)_____.

Anaerobic digestion of sewage sludge in a digester serves to (30)_____

_____.

Major plant nutrients contained in sewage sludge are (31)_____ _____. Sludge

created by the water treatment reaction $Al^{3+} + 3OH^-(s) \rightarrow Al(OH)_3(aq)$ is known as alum sludge and causes problems because it is (32)_____

_____.

The most commonly used water disinfectant is (33)_____ which reacts with water according to the reaction (34)_____.

Free available chlorine consists of (35)_____ in water, and combined available chlorine consists of (36)_____.

Chlorine dioxide is of particular interest for water disinfection because it does not produce (37)_____. An oxidizing disinfectant that does not contain chlorine is (38)_____ produced by (39)_____

_____.

Soil has physical, chemical, and biological characteristics that can enable (40)_____ of

impurities in wastewater. The soil characteristics that are important in determining its use for land treatment of wastes are (41)_____

_____. Acclimated microorganisms

adapted to degradation of organic compounds are found most commonly at (42)_____

_____.

Answers

1. purification for domestic use, treatment for specialized industrial applications, treatment of wastewater to make it acceptable for release or reuse
2. the plant's entire water supply
3. modify the properties of water for specific applications
4. primary treatment, secondary treatment, and tertiary treatment
5. insoluble matter such as grit, grease, and scum from water
6. biochemical oxygen demand
7. $2\{CH_2O\} \rightarrow CH_4 + CO_2$
8. both the volatile-matter content and the volume of the sludge by about 60%
9. a variety of processes performed on the effluent from secondary waste treatment
10. suspended solids, dissolved organic compounds, and dissolved inorganic materials
11. removal of scum and solid objects, clarification, generally with addition of a coagulant, and frequently with the addition of other chemicals (such as lime for phosphorus removal), filtration to remove filterable solids, activated carbon adsorption, disinfection
12. high costs of chemicals and energy
13. density separation, filtration, flotation, evaporation, distillation, reverse osmosis, hyperfiltration, ultrafiltration, solvent extraction, air stripping, or steam stripping
14. coagulation
15. $Ca^{2+} + 2HCO_3^- \rightarrow CaCO_3(s) + CO_2(g) + H_2O$
16. $Ca^{2+} + 2HCO_3^- + Ca(OH)_2 \rightarrow 2CaCO_3(s) + 2H_2O$
17. a source of CO_3^{2-} at a relatively high pH
18. $H^{+-}\{Cat(s)\} + Na^+ + Cl^- \rightarrow Na^{+-}\{Cat)s)\} + H^+ + Cl^-$ and
    $OH^{-+}\{An(s)\} + H^+ + Cl^- \rightarrow Cl^{-+}\{An(s)\} + H_2O$
19. oxidation to higher insoluble oxidation states
20. sulfide
21. activated carbon sorption
22. distillation, electrodialysis, ion exchange, reverse osmosis, nanofiltration, ultrafiltration, microfiltration, and dialysis
23. reverse osmosis
24. algal growth
25. lime
26. $5Ca(OH)_2 + 3HPO_4^{2-} \rightarrow Ca_5OH(PO_4)_3(s) + 3H_2O + 6OH^-$
27. the CO_2 is swept out, the pH rises, and reactions occur such as $5Ca^{2+} + 3HPO_4^{2-} + H_2O \rightarrow Ca_5OH(PO_4)_3(s) + 4H^+$
28. $NH_4^+ + 2O_2$ (Nitrifying bacteria) $\rightarrow NO_3^- + 2H^+ + H_2O$
29. $6NO_3^- + 5CH_3OH + 6H^+$ (Denitrifying bacteria) $\rightarrow 3N_2(g) + 5CO_2 + 13H_2O$

30. reduce the mass and volume of sludge and destroy disease agent
31. 5% N, 3% P, and 0.5% K on a dry-weight basis
32. very difficult to dewater
33. Cl_2
34. $Cl_2 + H_2O \rightarrow H^+ + Cl^- + HOCl$
35. HOCl and OCl⁻
36. the chloramines
37. trihalomethanes
38. ozone
39. an electrical discharge through dry air
40. detoxification, biodegradation, chemical decomposition, and physical and chemical fixation
41. physical form, ability to retain water, aeration, organic content, acid-base characteristics, and oxidation-reduction behavior
42. sites contaminated with the kinds of wastes degraded

QUESTIONS AND PROBLEMS

1. What is the purpose of the return sludge step in the activated sludge process?

2. What are the two processes by which the activated sludge process removes soluble carbonaceous material from sewage?

3. Why might hard water be desirable as a medium if phosphorus is to be removed by an activated sludge plant operated under conditions of high aeration?

4. How does reverse osmosis differ from a simple sieve separation or ultra-filtration process?

5. How many liters of methanol would be required daily to remove the nitro-gen from a 200,000-L/day sewage treatment plant producing an effluent containing 50 mg/L of nitrogen? Assume that the nitrogen has been con-verted to NO_3^- in the plant. The denitrifying reaction is Reaction 13.9.6.

6. Discuss some of the advantages of physical-chemical treatment of sewage as opposed to biological wastewater treatment. What are some disadvantages?

7. Why is recarbonation necessary when water is softened by the lime-soda process?

8. Assume that a waste contains 300 mg/L of biodegradable $\{CH_2O\}$ and is processed through a 200,000-L/day sewage-treatment plant which converts

40% of the waste to CO_2 and H_2O. Calculate the volume of air (at 25°, 1 atm) required for this conversion. Assume that the O_2 is transferred to the water with 20% efficiency.

9. If all of the $\{CH_2O\}$ in the plant described in Question 8 could be converted to methane by anaerobic digestion, how many liters of methane (STP) could be produced daily?

10. Assuming that aeration of water does not result in the precipitation of calcium carbonate, which of the following would not be removed by aeration: hydrogen sulfide; carbon dioxide; volatile odorous bacterial metabolites; alkalinity; iron.

11. In which of the following water supplies would moderately high water hardness be most detrimental: municipal water; irrigation water; boiler feedwater; drinking water (in regard to potential toxicity).

12. A wastewater containing dissolved Cu^{2+} ion is to be treated to remove copper. Which of the following processes would *not* remove copper in an insoluble form; lime precipitation; cementation; treatment with NTA; ion exchange; reaction with metallic Fe.

13. Match each water contaminant in the left column with its preferred method of removal in the right column.

(a) Mn^{2+} (1) Activated carbon
(b) Ca^{2+} and HCO_3^- (2) Raise pH by addition of Na_2CO_3
(c) Trihalomethane compounds   (3) Addition of lime
(d) Mg^{2+} (4) Oxidation

14. A cementation reaction employs iron to remove Cd^{2+} present at a level of 350 mg/L from a wastewater stream. Given that the atomic weight of Cd is 112.4 and that of Fe is 55.8, how many kg of Fe are consumed in removing all the Cd from 4.50×10^6 liters of water?

15. Consider municipal drinking water from two different kinds of sources, one a flowing, well-aerated stream with a heavy load of particulate matter and the other an anaerobic groundwater. Describe possible differences in the water treatment strategies for these two sources of water.

16. In treating water for industrial use, consideration is often given to "sequential use of the water." What might be meant by this term? Give some plausible examples of sequential use of water.

17. Active biomass is used in the secondary treatment of municipal wastewater. Describe three ways of supporting a growth of the biomass, contacting it with wastewater, and exposing it to air.

18. Using appropriate chemical reactions for illustration, show how calcium present as the dissolved HCO_3^- salt in water is easier to remove than other forms of hardness, such as dissolved $CaCl_2$.

19. Label each of the following as primary treatment (pr), secondary treatment (sec), or tertiary treatment (tert): screening, comminuting, ( ) grit removal, ( ) BOD removal, ( ) activated carbon filtration removal of dissolved organic compounds, ( ) removal of dissolved inorganic materials

20. Both activated sludge waste treatment and natural processes in streams and bodies of water remove degradable material by biodegradation. Explain why activated sludge treatment is so much more effective.

21. Of the following, the one that does not belong with the rest is ( ) removal of scum and solid objects, ( ) clarification, ( ) filtration, ( ) degradation with activated sludge, ( ) activated carbon adsorption, ( ) disinfection.

22. Explain why complete physical-chemical wastewater treatment systems are better than biological systems in dealing with toxic substances and overloads.

23. What are the two major ways in which dissolved carbon (organic compounds) is removed from water in industrial wastewater treatment. How do these two approaches differ fundamentally?

24. What is the reaction for the hydrolysis of aluminum ion in water? How is this reaction used for water treatment?

25. Explain why coagulation is used with filtration.

26. What are two major problems that arise from the use of excessively hard water?

27. Show with chemical reactions how the removal of bicarbonate hardness with lime results in a net removal of ions from solution, whereas removal of nonbicarbonate hardness does not.

28. Why is cation exchange normally used without anion exchange for softening water?

29. Show with chemical reactions how oxidation is used to remove soluble iron and manganese from water.

30. Show with chemical reactions how lime treatment, sulfide treatment, and cementation are used to remove heavy metals from water.

31. How is activated carbon prepared? What are the chemical reactions involved? What is remarkable about the surface area of activated carbon?

32. How is the surface of the membrane employed involved in the process of reverse osmosis?

33. Describe with a chemical reaction how lime is used to remove phosphate from water. What are some other chemicals that may be used for phosphate removal?

34. Why is nitrification required as a preliminary step in removal of nitrogen from water by biological denitrification?

35. What are some possible beneficial uses for sewage sludge? What are some of its characteristics that may make such uses feasible?

36. Distinguish between free available chlorine and combined available chlorine in water disinfection.

37. What is one major advantage and one major disadvantage of using ozone for water disinfection?

38. What is one major advantage and one major disadvantage of using ozone dioxide for water disinfection?

39. Discuss how soil may be viewed as a natural filter for wastes. How does soil aid waste treatment? How may waste treatment be of benefit to soil in some cases?

40. Suggest a source of microorganisms to use in a waste treatment process. Where should an investigator look for microorganisms to use in such an application? What are some kinds of wastes for which soil is particularly unsuitable as a treatment medium?

SUPPLEMENTARY REFERENCES

Sussman, S., "Industrial Water Treatment," in *Kirk-Othmer Concise Encyclopedia of Chemical Technology*, D. Eckroth, Ed. (New York, NY: John Wiley and Sons, 1985).

Hammer, D. A., Ed., *Constructed Wetlands for Wastewater Treatment*, (Chelsea, MI: Lewis Publishers, Inc., 1989).

Perkins, R. J., *Onsite Wastewater Disposal*, (Chelsea, MI: Lewis Publishers, 1989).

Freeman, H. M., Ed., *Standard Handbook of Hazardous Waste Treatment and Disposal*, (New York, NY: McGraw-Hill, 1988).

Scholze, R. J., Ed., *Biotechnology for Degradation of Toxic Chemicals in Hazardous Wastes*, (Park Ridge, NJ: Noyes Publications, 1988).

Forster, C. F., *Biotechnology and Wastewater Treatment*, (New York, NY: Cambridge University Press, 1985).

Sposito, G., *The Chemistry of Soils*, (New York, NY: Oxford University Press, 1989).

14 THE GEOSPHERE AND GEOCHEMISTRY

14.1. INTRODUCTION

The **geosphere**, or solid Earth, is that part of the Earth upon which humans live and from which they extract most of their food, minerals, and fuels. Once thought to have an almost unlimited buffering capacity against the perturbations of humankind, the geosphere is now known to be rather fragile and subject to harm by human activities. For example, some billions of tons of Earth material are mined or otherwise disturbed each year in the extraction of minerals and coal. Two atmospheric pollutant phenomena — excess carbon dioxide and acid rain (see Chapter 17) — have the potential to cause major changes in the geosphere. Too much carbon dioxide in the atmosphere may cause global heating ("greenhouse effect"), which could significantly alter rainfall patterns and turn currently productive areas of the Earth into desert regions. The low pH characteristic of acid rain can bring about drastic changes in the solubilities and oxidation-reduction rates of minerals. Erosion caused by intensive cultivation of land is washing away vast quantities of topsoil from fertile farmlands each year. In some areas of industrialized countries, the geosphere has been the dumping ground for toxic chemicals (see the discussion of hazardous wastes in Chapters 19-20). Ultimately, the geosphere must provide disposal sites for the nuclear wastes of the approximately 400 nuclear reactors now operating worldwide, as well as those yet to be completed. It may be readily seen that the preservation of the geosphere in a form suitable for human habitation is one of the greatest challenges facing humankind.

Human activities on the Earth's surface may have an effect upon climate. The most direct such effect is through the change of surface albedo, defined as the percentage of incident solar radiation reflected by a land or water surface. For example, if the sun radiates 100 units of energy per minute to the outer limits of the atmosphere, and the Earth's surface receives 60 units per minute of the total, then reflects 30 units upward, the albedo is 50 percent. Some typical albedo values for different areas on the Earth's surface are: evergreen forests, 7-15%; dry, plowed fields, 10-15%; deserts, 25-35%; fresh snow, 85-90%; asphalt,

8%. In some heavily developed areas, anthropogenic (human-produced) heat release is comparable to the solar input. The anthropogenic energy release over the 60 square kilometers of Manhattan Island averages about 4 times the solar energy falling on the area; over the 3,500 km² of Los Angeles the anthropogenic energy release is about 13% of the solar flux.

One of the greater impacts of humans upon the geosphere is the creation of desert areas through abuse of land with marginal amounts of rainfall. This process, called **desertification**, is manifested by declining groundwater tables, salinization of topsoil and water, reduction of surface waters, unnaturally high soil erosion, and desolation of native vegetation. The problem is severe in some parts of the world, particularly Africa's Sahel (southern rim of the Sahara), where the Sahara advanced southward at a particularly rapid rate during the period 1968–73, contributing to widespread starvation in Africa during the 1980s. Large, arid areas of the western U.S. are experiencing at least some of the symptoms of desertification as the result of human activities and a severe drought that lasted for several years during the late 1980s and early 1990s. As the populations of the Western states increase, one of the greatest challenges facing the residents is to prevent additional conversion of land to desert.

The most important part of the geosphere for life on earth is soil. It is the medium upon which plants grow and virtually all terrestrial organisms depend upon it for their existence. The productivity of soil is strongly affected by environmental conditions and pollutants. Because of the importance of soil, all of Chapter 15 is devoted to its environmental chemistry.

With increasing population and industrialization, one of the more important aspects of human use of the geosphere has to do with the protection of water sources. Mining, agricultural, chemical, and radioactive wastes all have the potential for contaminating both surface water and groundwater. Sewage sludge spread on land may contaminate water by release of nitrate and heavy metals. Landfills may likewise be sources of contamination. Leachates from unlined pits and lagoons containing hazardous liquids or sludges may pollute drinking water.

It should be noted, however, that many soils have the ability to assimilate and neutralize pollutants. Various chemical and biochemical phenomena in soils operate to reduce the harmful nature of pollutants. These phenomena include oxidation-reduction processes, hydrolysis, acid-base reactions, precipitation, sorption, and biochemical degradation. Some hazardous organic chemicals may be degraded to harmless products on soil, and heavy metals may be sorbed by it. In general, however, extreme care should be exercised in disposing of chemicals, sludges, and other potentially hazardous materials on soil, particularly where the possibility of water contamination exists.

14.2. THE NATURE OF SOLIDS IN THE GEOSPHERE

The earth is divided into layers, including the solid iron-rich inner core, molten outer core, mantle, and crust. Environmental chemistry is most concerned with the **lithosphere**, which consists of the outer mantle and the **crust**. The latter is the earth's outer skin that is accessible to humans. It is extremely thin compared to the diameter of the earth, ranging from 5 to 40 km thick.

Most of the solid earth crust consists of rocks. Rocks are composed of minerals, where a **mineral** is a naturally-occurring inorganic solid with a definite internal structure and chemical composition. A **rock** is a mass of pure mineral or an aggregate of two or more minerals.

Structure and Properties of Minerals

Many minerals have very well-defined crystalline structures which result from the ways in which the atoms and ions of different kinds and sizes are chemically bonded together in the mineral. Physical properties of minerals can be used to classify them. The characteristic external appearance of a pure crystalline mineral is its **crystal form**. Because of space constrictions on the ways that minerals grow, the pure crystal form of a mineral is often not expressed. **Color** is an obvious characteristic of minerals, but can vary widely due to the presence of impurities. The appearance of a mineral surface in reflected light describes its **luster**. Minerals may have a metallic luster or appear partially metallic (or submetallic), vitreous (like glass), dull or earthy, resinous, or pearly. The color of a mineral in its powdered form as observed when the mineral is rubbed across an unglazed porcelain plate is known as **streak**. **Hardness** is expressed on Mohs scale, which ranges from 1 to 10 and is based upon 10 minerals that vary from talc, hardness 1, to diamond, hardness 10. **Cleavage** denotes the manner in which minerals break along planes and the angles in which these planes intersect. For example, mica cleaves to form thin sheets. Most minerals **fracture** irregularly, although some fracture along smooth curved surfaces or into fibers or splinters. **Specific gravity** — density relative to that of water — is another important physical characteristic of minerals.

Kinds of Minerals

Although over two thousand minerals are known, only about 25 **rock-forming minerals** make up most of the earth's crust. The nature of these minerals may be better understood with a knowledge of the elemental composition of the crust. Oxygen and silicon make up 46.6% and 27.7% by mass of the earth's crust, respectively. Therefore, most minerals are **silicates** such as quartz, SiO_2, or orthoclase, $KAlSi_3O_8$. In descending order of abundance, the other elements in the earth's crust are aluminum (8.1%), iron (5.0%), calcium (3.6%), sodium

Table 14.1. Major Mineral Groups in the Earth's Crust

Mineral Group	Examples	Formula
Silicates	Quartz	SiO_2
	Olivine	$(Mg,Fe)_2SiO_4$
	Potassium feldspar	$KAlSi_3O_8$
Oxides	Corundum	Al_2O_3
	Magnetite	Fe_3O_4
Carbonates	Calcite	$CaCO_3$
	Dolomite	$CaCO_3 \cdot MgCO_3$
Sulfides	Pyrite	FeS_2
	Galena	PbS
Sulfates	Gypsum	$CaSO_4 \cdot 2H_2O$
Halides	Halite	$NaCl$
	Fluorite	CaF_2
Native elements	Copper	Cu
	Sulfur	S

(2.8%), potassium (2.6%), magnesium (2.1%), and other (1.5%). Table 14.1 summarizes the major kinds of minerals in the earth's crust.

Secondary minerals are formed by alteration of parent mineral matter. **Clays** are silicate minerals, usually containing aluminum, that constitute one of the most significant classes of secondary minerals. Olivine, augite, hornblende, and feldspars all form clays. Clays are discussed in detail in Section 14.4.

Evaporites

Evaporites are soluble salts that precipitate from solution under special arid conditions, commonly as the result of the evaporation of seawater. The most common evaporite is **halite**, $NaCl$. Other simple evaporite minerals are sylvite (KCl), thenardite (Na_2SO_4), and anhydrite ($CaSO_4$). Many evaporites are hydrates, including bischofite ($MgCl_2 \cdot 6H_2O$), gypsum ($CaSO_4 \cdot 2H_2O$), keiserite ($MgSO_4 \cdot H_2O$), and epsomite ($MgSO_4 \cdot 7H_2O$). Double salts, such as kainite with a formula of $KMgClSO_4 \cdot 11/4H_2O$, glaserite ($K_3Na(SO_4)_2$), polyhalite ($K_2MgCa_2(SO_4)_4 \cdot 2H_2O$), carnallite ($KMgCl_3 \cdot 6H_2O$), and loeweite ($Na_{12}Mg_7(SO_4)_{13} \cdot 15H_2O$), are commonly found in evaporites.

The precipitation of evaporites from marine and brine sources depends upon a number of factors. Prominent among these are the concentrations of the evaporite ions in the water and the solubility products of the evaporite salts. The presence of a common ion decreases solubility; for example, $CaSO_4$ precipitates more readily from a brine that contains Na_2SO_4 than it does from a solution that contains no other source of sulfate. The presence of other salts that do not have a common ion increases solubility because it decreases activity coefficients. Differences in temperature result in significant differences in solubility.

The nitrate deposits that occur in the hot and extraordinarily dry regions of northern Chile are chemically unique because of the stability of highly oxidized nitrate salts. The dominant salt, which has been mined for its nitrate content for use in explosives and fertilizers, is Chile saltpeter, $NaNO_3$. Traces of highly oxidized $CaCrO_4$ and $Ca(ClO_4)_2$ are also encountered in these deposits, and some regions contain enough $Ca(IO_3)_2$ to serve as a commercial source of iodine.

Volcanic Sublimates

A number of mineral substances are gaseous at the magmatic temperatures of volcanoes and are mobilized with volcanic gases. These kinds of substances condense near the mouths of volcanic fumaroles and are called sublimates. Elemental sulfur is a common **sublimate**. Some oxides, particularly of iron and silicon, are deposited as sublimates. Most other sublimates consist of chloride and sulfate salts. The cations most commonly involved are monovalent cations of ammonium ion, sodium, and potassium; magnesium; calcium; aluminum; and iron. Fluoride and chloride sublimates are sources of gaseous HF and HCl formed by their reactions at high temperatures with water, such as the following:

$$2H_2O + SiF_4 \rightarrow 4HF + SiO_2 \qquad (14.2.1)$$

Igneous and Sedimentary Rock

The solidification of molten rock, called magma, produces **igneous rock**. Common igneous rocks are granite, basalt, quartz (SiO_2), pyroxene [$(Mg,Fe)SiO_3$], feldspar [$(Ca,Na,K)AlSi_3O_8$], olivine [$(Mg,Fe)_2SiO_4$], and magnetite (Fe_3O_4). Igneous rocks are formed under water-deficient, chemically reducing conditions of high temperature and high pressure. Exposed igenous rocks, therefore, are not in chemical equilibrium with their surroundings and disintegrate by a process called **weathering**. Weathering tends to be slow because igneous rocks are often hard, nonporous, and of low reactivity. Erosion from wind, water, or glaciers picks up materials from weathering rocks and converts it to **sedimentary rock** and **soil**, which in contrast to the parent igneous rocks are porous, soft, and chemically reactive. Heat and pressure convert sedimentary rock to **metamorphic rock**.

Sedimentary rocks may be **detrital rocks** consisting of solid particles eroded from igneous rocks as a consequence of weathering; quartz is the most likely to survive weathering and transport from its original location chemically intact. A second kind of sedimentary rocks consists of **chemical sedimentary rocks** produced by the precipitation or coagulation of dissolved or colloidal weathering products. **Organic sedimentary rocks** contain residues of plant and animal remains. Carbonate minerals of calcium and magnesium—**limestone** or **dolomite**—are especially abundant in sedimentary rocks. Important examples of sedimentary rocks are the following:

- Sandstone produced from sand-sized particles of minerals such as quartz
- Conglomerates made up of relatively larger particles of variable size
- Shale formed from very fine particles of silt or clay
- Limestone, $CaCO_3$, produced by the chemical or biochemical precipitation of calcium carbonate:

$$Ca^{2+} + CO_3^{2-} \rightarrow CaCO_3(s)$$

$$Ca^{2+} + 2HCO_3^- + h\nu\text{(algal photosynthesis)} \rightarrow$$
$$\{CH_2O\}\text{(biomass)} + CaCO_3(s) + O_2(g)$$

- Chert consisting of microcrystalline, SiO_2

Stages of Weathering

Weathering can be classified into **early**, **intermediate**, and **advanced stages**. The stage of weathering to which a mineral is exposed depends upon time; chemical conditions, including exposure to air, carbon dioxide, and water; and physical conditions, such as temperature and mixing with water and air.

Reactive and soluble minerals such as carbonates, gypsum, olivine, feldspars, and iron(II)-rich substances can survive only early weathering. This stage is characterized by dry conditions, low leaching, absence of organic matter, reducing conditions, and limited time of exposure. Quartz, vermiculite, and smectites can survive the intermediate stage of weathering manifested by retention of silica, sodium, potassium, magnesium, calcium, and iron(II) not present in iron(II) oxides. These substances are mobilized in advanced-stage weathering, other characteristics of which are intense leaching by fresh water, low pH, oxidizing conditions [iron(II) $\rightarrow$ iron(III)], presence of hydoxy polymers of aluminum, and dispersion of silica.

14.3.   SEDIMENTS

Vast areas of land, as well as lake and stream sediments, are formed from sedimentary rocks. The properties of these masses of material depend strongly upon their origins and transport. Water is the main vehicle of sediment transport, although wind can also be significant. Enormous amounts of sediment are carried by major rivers.

The action of flowing water in streams cuts away stream banks and carries sedimentary materials for great distances. Sedimentary materials may be carried by flowing water in streams as the following:

- **Dissolved load** from sediment-forming minerals in solution.
- **Suspended load** from solid sedimentary materials carried along in suspension
- **Bed load** dragged along the bottom of the stream channel

The transport of calcium carbonate as dissolved calcium bicarbonate provides a straightforward example of dissolved load. Water with a high dissolved carbon dioxide content (usually present as the result of bacterial action) in contact with calcium carbonate formations contains Ca^{2+} and HCO_3^- ions. Flowing water containing calcium as such *temporary hardness* may become more basic by loss of CO_2 to the atmosphere, consumption of CO_2 during algal photosynthesis, or contact with dissolved base, resulting in the deposition of insoluble $CaCO_3$:

$$Ca^{2+} + 2HCO_3^- \rightarrow CaCO_3(s) + CO_2(g) + H_2O \qquad (14.3.1)$$

Most flowing water that contains dissolved load originates underground, where it dissolves minerals from the rock strata that it flows through.

Most sediments are transported by streams as suspended load, obvious in the observation of "mud" in the flowing water of rivers draining agricultural areas or finely divided rock in Alpine streams fed by melting glaciers. Under normal conditions, finely divided silt, clay, or sand make up most of the suspended load, although larger particles are transported in rapidly flowing water. The degree and rate of movement of suspended sedimentary material in streams are functions of the velocity of water flow and the settling velocity of the particles in suspension.

Bed load is moved along the bottom of a stream by the action of water "pushing" particles along. Particles carried as bed load do not move continuously. The grinding action of such particles is an important factor in stream erosion.

Typically, about $2/3$ of the sediment carried by a stream is transported in suspension, about $1/4$ in solution, and the remaining relatively small fraction as bed load. The ability of a stream to carry sediment increases with both the overall rate of flow of the water (mass per unit time) and the velocity of the water. Both of these are higher under flood conditions, so floods are particularly important in the transport of sediments.

Streams mobilize sedimentary materials through **erosion, transport** materials along with stream flow, and release them in a solid form during **deposition**. Deposits of stream-borne sediments are called **alluvium**. As conditions such as lowered stream velocity begin to favor deposition, larger, more settleable particles are released first. This results in **sorting** such that particles of a similar size and type tend to occur together in alluvium deposits. Much sediment is deposited in flood plains where streams overflow their banks.

14.4.   CLAYS

Clays are extremely common and important in mineralogy. Furthermore, in general (see Section 14.5 and Chapter 15), clays predominate in the inorganic components of most soils and are very important in holding water and in plant nutrient cation exchange. All clays contain silicate and most contain aluminum and water. Physically, clays consist of very fine grains having sheet-like structures. For purposes of discussion here, **clay** is defined as a group of microcrystalline secondary minerals consisting of hydrous aluminum silicates that have sheet-like structures. Clay minerals are distinguished from each other by general chemical formula, structure, and chemical and physical properties. The three major groups of clay minerals are **illite** [$K_{0-2}Al_4(Si_{8-6}Al_{0-2})O_{20}(OH)_4$], **montmorillonite** [$Al_2(OH)_2Si_4O_{10}$], and **kaolinite** [$Al_2Si_2O_5(OH)_4$]. Many clays contain large amounts of sodium, potassium, magnesium, calcium, and iron, as well as trace quantities of other metals. Clays bind cations such as Ca^{2+}, Mg^{2+}, K^+, Na^+, and NH_4^+, which protects the cations from leaching by water but keeps them available in soil as plant nutrients. Since many clays are readily suspended in water as colloidal particles, they may be leached from soil or carried to lower soil layers.

Olivine, augite, hornblende, and feldspars are all parent minerals that form clays. An example is the formation of kaolinite [$Al_2Si_2O_5(OH)_4$] from potassium feldspar rock ($KAlSi_3O_8$):

$$2KAlSi_3O_8(s) + 2H^+ + 9H_2O \rightarrow Al_2Si_2O_5(OH)_4(s)$$
$$+ 2K^+(aq) + 4H_4SiO_4(aq) \qquad (14.4.1)$$

The layered structures of clays consist of sheets of silicon oxide alternating with sheets of aluminum oxide. The silicon oxide sheets are made up of tetrahedra in which each silicon atom is surrounded by four oxygen atoms. Of the four oxygen atoms in each tetrahedron, three are shared with other silicon atoms that are components of other tetrahedra. This sheet is called the **tetrahedral sheet**. The aluminum oxide is contained in an **octahedral sheet**, so named because each aluminum atom is surrounded by six oxygen atoms in an octahedral configuration. The structure is such that some of the oxygen atoms are shared between aluminum atoms and some are shared with the tetrahedral sheet.

Structurally, clays may be classified as either **two-layer clays** in which oxygen atoms are shared between a tetrahedral sheet and an adjacent octahedral sheet, and **three-layer clays**, in which an octahedral sheet shares oxygen atoms with tetrahedral sheets on either side. These layers composed of either two or three sheets are called **unit layers**. A unit layer of a two-layer clay typically is around 0.7 nanometers (nm) thick, whereas that of a three-layer clay exceeds 0.9 nm in thickness. The structure of the two-layer clay kaolinite is represented in Figure 14.1. Some clays, particularly the montmorillonites, may absorb large quantities of water between unit layers, a process accompanied by swelling of the clay.

Clay minerals may attain a net negative charge by **ion replacement**, in which Si(IV) and Al(III) ions are replaced by metal ions of similar size but lesser charge. Compensation must be made for this negative charge by association of cations with the clay layer surfaces. Since these cations need not fit specific sites in the crystalline lattice of the clay, they may be relatively large ions, such as K^+, Na^+, or NH_4^+. They are called **exchangeable cations** because they are exchangeable for other cations in water. The amount of exchangeable cations, expressed as milliequivalents (millimoles of monovalent cations) per 100 g of dry clay, is

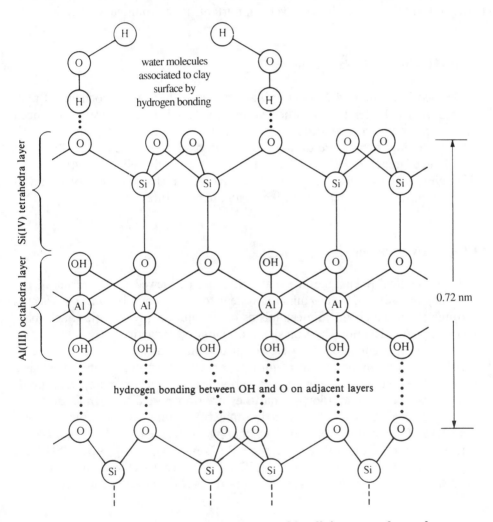

Figure 14.1.   Representation of the structure of kaolinite, a two-layer clay.

called the **cation-exchange capacity, CEC**, of the clay and is a very important characteristic of colloids and sediments that have cation-exchange capabilities.

14.5. GEOCHEMISTRY

Geochemistry deals with chemical species, reactions, and processes in the lithosphere and their interactions with the atmosphere and hydrosphere. The branch of geochemistry that explores the complex interactions among the rock/water/air/life systems that determine the chemical characteristics of the surface environment is **environmental geochemistry**. Obviously, geochemistry and its environmental subdiscipline are very important in environmental chemistry.

Physical Aspects of Weathering

Defined in Section 14.2, *weathering* is discussed here as a geochemical phenomenon. Rocks tend to weather more rapidly when there are pronounced differences in physical conditions—alternate freezing and thawing and wet periods alternating with severe drying. Other mechanical aspects are swelling and shrinking of minerals with hydration and dehydration as well as growth of roots through cracks in rocks. Temperature is involved in that the rates of chemical weathering (below) increase with increasing temperature.

Chemical Weathering

As a chemical phenomenon, weathering can be viewed as the result of the tendency of the rock/water/mineral system to attain equilibrium. This occurs through the usual chemical mechanisms of dissolution/precipitation, acid-base reactions, complexation, hydrolysis, and oxidation-reduction.

Weathering occurs extremely slowly in dry air. Water increases the rate of weathering by many orders of magnitude for several reasons. Water, itself, is a chemically active substance in the weathering process. Furthermore, water holds weathering agents in solution so that they are transported to chemically active sites on rock minerals and contact the mineral surfaces at the molecular and ionic level. Prominent among such weathering agents are CO_2, O_2, organic acids (including humic and fulvic acids, see Section 11.9), sulfur acids [$SO_2(aq)$, H_2SO_4], and nitrogen acids (HNO_3, HNO_2). Water provides the source of H^+ ion needed for acid-forming gases to act as acids as shown by the following:

$$CO_2 + H_2O \rightarrow H^+ + HCO_3^- \qquad (14.5.1)$$

$$SO_2 + H_2O \rightarrow H^+ + HSO_3^- \tag{14.5.2}$$

Rainwater is essentially free of mineral solutes. It is usually slightly acidic due to the presence of dissolved carbon dioxide or more highly acidic because of acid-rain forming constitutents. Therefore, rainwater is *chemically aggressive* (see Section 13.7) toward some kinds of mineral matter, which it breaks down by a process called **chemical weathering**. Because of this process, river water has a higher concentration of dissolved inorganic solids than does rainwater.

Chemical weathering may be divided into the following major categories:

- Hydration/dehydration, for example:

$$CaSO_4(s) + 2H_2O \rightarrow CaSO_4 \cdot 2H_2O(s)$$

$$2Fe(OH)_3 \cdot xH_2O(s) \rightarrow Fe_2O_3(s) + (3 + 2x)H_2O$$

- Dissolution, for example:

$$CaSO_4 \cdot 2H_2O(s) \text{ (water)} \rightarrow Ca^{2+}(aq) + SO_4^{2-}(aq) + 2H_2O$$

- Oxidation, such as occurs in the dissolution of pyrite:

$$4FeS_2(s) + 15O_2(g) + (8 + 2x)H_2O \rightarrow$$
$$2Fe_2O_3 \cdot xH_2O + 8SO_4^{2-}(aq) + 16H^+(aq)$$

or in the following example in which dissolution of an iron(II) mineral is followed by oxidation of iron(II) to iron(III):

$$Fe_2SiO_4(s) + 4CO_2(aq) + 4H_2O \rightarrow 2Fe^{2+} + 4HCO_3^- + H_4SiO_4$$

$$4Fe^{2+} + 8HCO_3^- + O_2(s) \rightarrow 2Fe_2O_3(s) + 8CO_2 + 4H_2O$$

The second of these two reactions may occur at a site some distance from the first, resulting in a net transport of iron from its original location. Iron, manganese, and sulfur are the major elements that undergo oxidation as part of the weathering process.

- **Dissolution with hydrolysis** as occurs with the hydrolysis of carbonate ion when mineral carbonates dissolve:

$$CaCO_3(s) + H_2O \rightarrow Ca^{2+}(aq) + HCO_3^-(aq) + OH^-(aq)$$

Hydrolysis is the major means by which silicates undergo weathering as shown by the following reaction of forsterite:

$$Mg_2SiO_4(s) + 4CO_2 + 4H_2O \rightarrow 2Mg^{2+} + 4HCO_3^- + H_4SiO_4$$

The weathering of silicates yields soluble silicon as species such as H_4SiO_4, and residual silicon-containing minerals (clay minerals).

- **Acid hydrolysis**, which accounts for the dissolution of significant amounts of $CaCO_3$ and $CaCO_3 \cdot MgCO_3$ in the presence CO_2-rich water:

$$CaCO_3(s) + H_2O + CO_2(aq) \rightarrow Ca^{2+}(aq) + 2HCO_3^-(aq)$$

- **Complexation**, as exemplified by the reaction of oxalate ion, $C_2O_4^{2-}$ on muscovite, $K_2(Si_6Al_2)Al_4O_{20}(OH)_4$:

$$K_2(Si_6Al_2)Al_4O_{20}(OH)_4 (s) + 6C_2O_4^{2-} (aq) + 2OH^+ \rightarrow$$
$$6AlC_2O_4^+(aq) + 6Si(OH)_4 + 2K^+$$

Reactions such as these largely determine the kinds and concentrations of solutes in surface water and groundwater. Acid hydrolysis, especially, is the predominant process that releases elements such as Na^+, K^+, and Ca^{2+} from silicate minerals.

14.6. GROUNDWATER IN THE GEOSPHERE

Groundwater (Figure 14.2) is a vital resource in its own right that plays a crucial role in geochemical processes, such as the formation of secondary minerals. The nature, quality, and mobility of groundwater are all strongly dependent upon the rock formations in which the water is held. Physically, an important characteristic of such formations is their **porosity**, which determines the percentage of rock volume available to contain water. A second important physical characteristic is **permeability**, which describes the ease of flow of the water through the rock. High permeability is usually associated with high porosity. However, clays tend to have low permeability even when a large percentage of the volume is filled with water.

The nature of water underground is illustrated by Figure 14.2. Most groundwater originates as **meteoritic** water from precipitation in the form of rain or snow. If water from this source is not lost by evaporation, transpiration, or to stream runoff, it may infiltrate into the ground. Initial amounts of water from precipitation onto dry soil are held very tightly as a film on the surfaces and in the micropores of soil particles in a **belt of soil moisture**. At intermediate levels, the soil particles are covered with films of water, but air is still present in larger voids in the soil. The region in which such water is held is called the **unsaturated zone** or **zone of aeration** and the water present in it is **vadose water**. At lower depths in the presence of adequate amounts of water, all voids are filled to produce a **zone of saturation**, the upper level of which is the **water table**. Water present in a zone of saturation is called **groundwater**. Because of its surface

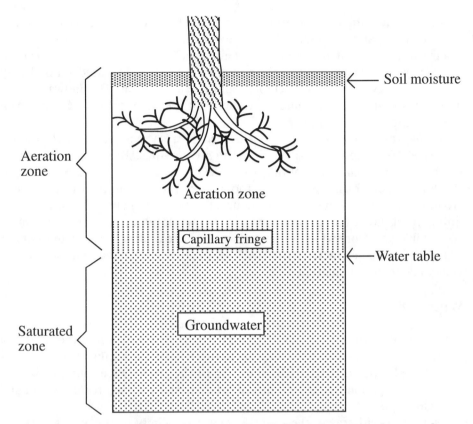

Figure 14.2.   Some major features of the distribution of water underground.

tension, water is drawn somewhat above the water table by capillary-sized passages in soil in a region called the **capillary fringe**.

The water table is crucial in explaining and predicting the flow of wells and springs and the levels of streams and lakes. It is also an important factor in determining the extent to which pollutant and hazardous chemicals underground are likely to be transported by water. The water table can be mapped by observing the equilibrium level of water in wells, which is essentially the same as the top of the saturated zone. The water table is usually not level, but tends to follow the general contours of the surface topography. It also varies with differences in permeability and water infiltration. The water table is at surface level in the vicinity of swamps and frequently above the surface where lakes and streams are encountered. The water level in such bodies may be maintained by the water table. **Influent** streams or reservoirs are located above the water table; they lose water to the underlying aquifer and cause an upward bulge in the water table beneath the surface water.

Groundwater **flow** is an important consideration in determining the accessibility of the water for use and transport of pollutants from underground waste sites. Various parts of a body of groundwater are in hydraulic contact so that a change in pressure at one point will tend to affect the pressure and level at another point. For example, infiltration from a heavy, localized rainfall may affect water level at a point remote from the infiltration. Groundwater flow occurs as the result of the natural tendency of the water table to assume even levels by the action of gravity.

Groundwater flow is strongly influenced by rock permeability. Porous or extensively fractured rock is relatively highly **pervious**. Because water can be extracted from such a formation, it is called an **aquifer**. By contrast, an **aquiclude** is a rock formation that is too impermeable or unfractured to yield groundwater. Impervious rock in the unsaturated zone may retain water infiltrating from the surface to produce a **perched water table** that is above the main water table and from which water may be extracted. However, the amounts of water that can be extracted from such a formation are limited and the water is vulnerable to contamination.

Water Wells

Most groundwater is tapped for use by water wells drilled into the saturated zone. The use and misuse of water from this source has a number of environmental implications. In the U.S. about $2/3$ of the groundwater pumped is consumed for irrigation; lesser amounts of groundwater are used for industrial and municipal applications.

As water is withdrawn, the water table in the vicinity of the well is lowered. This **drawdown** of water creates a **zone of depression**. In extreme cases the groundwater is severely depleted and surface land levels can even subside (which is one reason that Venice, Italy is now very vulnerable to flooding). Heavy drawdown can result in infiltration of pollutants from sources such as septic tanks, municipal refuse sites, and hazardous waste dumps. When soluble iron(II) or manganese(II) are present in groundwater, exposure to air at the well walls can result in the formation of deposits of insoluble iron(III) and manganese(IV) oxides produced by bacterially catalyzed oxidation processes:

$$4Fe^{2+}(aq) + O_2(aq) + 10H_2O \rightarrow 4Fe(OH)_3(s) + 8H^+ \qquad (14.6.1)$$

$$2Mn^{2+}(aq) + O_2(aq) + (2x + 2)H_2O \rightarrow 2MnO_2 \cdot xH_2O(s) + 4H^+ \qquad (14.6.2)$$

Deposits of iron(III) and manganese(IV) that result from the processes outlined above coat the surfaces from which water flows into the well with a coating that is relatively impermeable to water. The deposits fill the spaces that water must traverse to enter the well. As a result, they can seriously impede the flow of water into the well from the water-bearing aquifer. This creates major water source problems for municipalities using groundwater for water supply. As a result of this problem, chemical or mechanical cleaning, drilling of new wells, or even acquisition of new water sources may be required.

CHAPTER SUMMARY

The chapter summary below is presented in a programmed format to review the main points covered in this chapter. It is used most effectively by filling in the blanks, referring back to the chapter as necessary. The correct answers are given at the end of the summary.

Environmental chemistry of the geosphere is most concerned with the (1)_____, which consists of (2)_____.
A (3)_____ is a naturally-occurring inorganic solid with a definite internal structure and chemical composition, and a (4)_____ is a mass of pure mineral or an aggregate of two or more minerals.

Physical properties used to characterize minerals include (5)_____
_____. Most minerals are (6)_____, because the two most abundant elements in the earth's crust are (7)_____, which comprise (8)_____ percent of the earth's crust, respectively. The formula of mineral quartz is (9)_____, that of calcite is (10)_____, and pyrite is (11)_____.
Clays are examples of (12)_____ minerals formed by alteration of (13)_____.

Granite, basalt, and quartz are all examples of (14)_____ rocks. Exposed rocks of this kind are not in chemical equilibrium with their surroundings, so they undergo (15)_____ to produce (16)_____
_____.

Sedimentary materials may be carried by flowing water in streams as (17)__
_____. The reaction $Ca^{2+} + 2HCO_3^- \rightarrow CaCO_3(s) + CO_2(g) + H_2O$ is an example of the formation of sediment from (18)_____ load.

A group of microcrystalline secondary minerals consisting of hydrous aluminum silicates that have sheet-like structures defines (19)_____.
These minerals may attain a net negative charge by (20)_____,

in which Si(IV) and Al(III) ions are replaced by metal ions of similar size but (21)_____. This results in the ability of the mineral to hold (22)_____.

The science that deals with chemical species, reactions, and processes in the lithosphere and their interactions with the atmosphere and hydrosphere is called (23)_____. The branch of this science that explores the complex interactions among the rock/water/air/life systems that determine the chemical characteristics of the surface environment is (24)_____

_____.

Factors determining weathering include (25)_____

_____

_____. As a chemical phenomenon, weathering can be viewed as the result of the tendency of the rock/water/mineral system to attain (26)_____. This occurs through the chemical mechanisms of (27)_____

_____

_____. Probably the most active chemical species in weathering phenomena is (28)_____. Other weathering agents are (29)_____

_____.

Because of its high reactivity toward common minerals, rainwater is said to be (30)_____. The major processes in chemical weathering are (31)_____

_____

_____.

Most groundwater originates as (32)_____ water. Initial amounts of water from precipitation onto dry soil are held very tightly as a film on the surfaces and in the micropores of soil particles in a (33)_____

_____. Water held so that the soil particles are covered with films of water, but air is still present in larger voids in the soil is present as

(34)_____ water in the (35)_____ zone. Water in a region in which all voids are filled with water is held in a zone of (36)_____. A rock formation from which water may be extracted is called (37)_____.

The reactions (38)_____

and (39)_____

result in the formation of harmful deposits of iron(III) and manganese(IV) in water wells.

Answers

1. lithosphere
2. the outer mantle and the crust
3. mineral
4. rock
5. streak, luster, color, crystal form, hardness, cleavage, fracture, specific gravity
6. silicates
7. oxygen and silicon
8. 46.6% and 27.7%
9. SiO_2
10. $CaCO_3$
11. FeS_2
12. secondary minerals
13. parent mineral matter
14. igneous
15. weathering
16. sedimentary rock or soil
17. dissolved load, suspended load, and bed load
18. dissolved
19. clays
20. ion replacement
21. lesser charge
22. exchangeable cations
23. geochemistry
24. environmental geochemistry
25. alternate freezing and thawing, wet periods alternating with severe drying, swelling and shrinking of minerals with hydration and dehydration, growth of roots through cracks in rocks, and temperature
26. equilibrium
27. dissolution/precipitation, acid-base reactions, complexation, hydrolysis, and oxidation-reduction

28. water
29. CO_2, O_2, organic acids, sulfur acids, and nitrogen acids
30. chemically aggressive
31. hydration/dehydration, dissolution, oxidation, dissolution with hydrolysis, acid hydrolysis, and complexation
32. meteoritic
33. belt of soil moisture
34. vadose
35. unsaturated
36. saturation
37. an aquifer
38. $4Fe^{2+}(aq) + O_2(aq) + 10H_2O \rightarrow 4Fe(OH)_3(s) + 8H^+$
39. $2Mn^{2+}(aq) + O_2(aq) + (2x + 2)H_2O \rightarrow 2MnO_2 \cdot xH_2O(s) + 4H^+$

QUESTIONS AND PROBLEMS

1. Of the following, the one that is **not** a manifestation of desertification is (a) declining groundwater tables, (b) salinization of topsoil and water, (c) production of deposits of MnO_2 and $Fe_2O_3 \cdot H_2O$ from anaerobic processes, (d) reduction of surface waters, (e) unnaturally high soil erosion.

2. Give an example of how each of the following chemical or biochemical phenomena in soils operate to reduce the harmful nature of pollutants: (a) oxidation/reduction processes, (b) hydrolysis, (c) acid-base reactions, (d) precipitation, (e) sorption, (f) biochemical degradation.

3. Why do silicates and oxides predominate among Earth's minerals?

4. Match the following:

1. Metamorphic rock	(a) Produced by the precipitation or coagulation of dissolved or colloidal weathering products
2. Chemical sedimentary rocks	
3. Sedimentary rock	
4. Organic sedimentary rocks	(b) Contain residues of plant and animal remains
	(c) Formed from action of heat and pressure on sedimentary rock
	(d) Formed from solid particles eroded from igneous rocks as a consequence of weathering

5. Give the common characteristic of the minerals with the following formulas: $NaCl$, Na_2SO_4, $CaSO_4 \cdot 2H_2O$, $MgCl_2 \cdot 6H_2O$, $MgSO_4 \cdot 7H_2O$, $KMgClSO_4 \cdot 11/4H_2O$, $K_2MgCa_2(SO_4)_4 \cdot 2H_2O$.

6. Explain how the following are related: weathering, igneous rock, sedimentary rock, soil.

7. Where does most flowing water that contains dissolved load originate? Why does it tend to come from this source?

8. What role might be played by water pollutants in the production of dissolved load and in the precipitation of secondary minerals from it?

9. Which three elements are most likely to undergo oxidation as part of the chemical weathering process? Give example reactions of each.

10. Match the following:

1. Groundwater	(a) Water from precipitation in the form of rain or snow
2. Vadose water	
3. Meteoritic water	(b) Water present in a zone of saturation
4. Water in capillary fringe	(c) Water held in the unsaturated zone or zone of aeration
	(d) Water drawn somewhat above the water table by surface tension

11. Describe one important way in which the geosphere interacts with the atmosphere and climate. Describe another important way in which the geosphere affects the hydrosphere and another in which it is affected by the hydrosphere.

12. Why are silicates so abundant among minerals in Earth's crust?

13. In what sense is weathering a physical process? In what ways is weathering a chemical process? How are physical and chemical weathering related?

14. Explain with a biochemical reaction how a common sediment is formed by photosynthesis in hard water.

15. What kind of process is illustrated by the reaction

$$2KAlSi_3O_8(s) + 2H^+ + 9H_2O \rightarrow Al_2Si_2O_5(OH)_4(s) + 2K^+(aq) + 4H_4SiO_4(aq)$$

What kind of mineral does it form?

16. Match the chemical weathering process on the left, below, with the reactions that illustrate it from the right.
   1. Oxidation   (a) $CaCO_3(s) + H_2O + CO_2(aq) \rightarrow$
   2. Hydration/dehydration $Ca^{2+}(aq) + 2HCO_3^-(aq)$

3. Acid hydrolysis
4. Dissolution

(b) $CaSO_4 \cdot 2H_2O(s)$ (water) $\rightarrow Ca^{2+}(aq)$
$+ SO_4^{2-}(aq) + 2H_2O$

(c) $4Fe^{2+} + 8HCO_3^- + O_2(s) \rightarrow 2Fe_2O_3(s)$
$+ 8CO_2 + 4H_2O$

(d) $CaSO_4(s) + 2H_2O \rightarrow CaSO_4 \cdot 2H_2O(s)$
$2Fe(OH)_3 \cdot xH_2O(s) \rightarrow Fe_2O_3(s) +$
$(3 + 2x)H_2O$

17. Explain how the following reactions may cause problems with water wells:

$$4Fe^{2+}(aq) + O_2(aq) + 10H_2O \rightarrow 4Fe(OH)_3(s) + 8H^+$$

$$2Mn^{2+}(aq) + O_2(aq) + (2x + 2)H_2O \rightarrow 2MnO_2 \cdot xH_2O(s) + 4H^+$$

SUPPLEMENTARY REFERENCES

Barney, G. O., *The Global 2000 Report to the President*, 2 vols., U. S. Government Printing Office, Washington, DC, 1980.

Plant, J. A., and R. Raiswell, "Principles of Environmental Geochemistry," Chapter 1 in *Applied Environmental Geochemistry*, I. Thornton, Ed. (New York, Academic Press, 1983), pp. 1–40.

Montgomery, C. W., *Environmental Geology,* 3rd ed. (Dubuque, IA: Wm. C. Brown, Publishers, 1992).

Tarbuck, E. J., and F. K. Lutgens, *The Earth, an Introduction to Physical Geology* (Columbus, OH: Charles E. Merrill Publishing Co., 1984).

Sposito, G., *The Chemistry of Soils* (New York: Oxford University Press, 1989).

Suffet, I. H., and P. MacCarthy, Eds., *Aquatic Humic Substances: Influence on Fate and Treatment of Pollutants*, Advances in Chemistry Series 219. American Chemical Society, Washington, DC, 1989.

Wolf, K., W. J. Van Den Brink, and F. J. Colon, Contaminated Soil '88, Vols. 1 and 2 (Norwell, MA: Kluwer Academic Publishers, 1988).

Brownlow, A. H., *Geochemistry* (Englewood Cliffs, NJ: Prentice-Hall, Inc., 1978).

Craig, P. J., *The Natural Environment and The Biogeochemical Cycles* (New York: Springer-Verlag, Inc., 1980).

Fortescue, A. C., *Environmental Geochemistry* (New York: Springer-Verlag, Inc., 1980).

Krenvolden, K. A., Ed., *Geochemistry of Organic Molecules* (New York: Academic Press, Inc., 1980).

Lerman, A., *Geochemical Processes*, (New York: Wiley-Interscience, 1979).

Lindsay, W. L., *Chemical Equilibria in Soils* (New York: Wiley-Interscience, 1979).

Morrill, L. G., B. C. Mahilum, and S. H. Mohiuddin, *Organic Compounds in Soils: Sorption, Degradation, and Persistence*, (Ann Arbor, MI: Ann Arbor Science Publishers, Inc., 1982).

Nancollas, G. H., Ed., *Biological Mineralization and Demineralization,* (New York: Springer-Verlag, Inc., 1982).

Olson, G. W., *Soils and the Environment*, (New York: Methuen, Inc., 1981).

Paton, T. R., *The Formation of Soil Material* (Winchester, MA: Allen and Unwin, Inc., 1979).

Speidel, G. H., and A. F. Agnew, *The Natural Geochemistry of Our Environment*, (Boulder, CO: Westview Press, 1982).

Stevenson, F. J., *Humus Chemistry: Genesis, Composition, Reactions,* (New York: Wiley-Interscience, 1982).

15 SOIL AND AGRICULTURAL CHEMISTRY

15.1. THE NATURE AND IMPORTANCE OF SOIL

To humans and most terrestrial organisms, soil is the most important part of the geosphere. Though only a tissue-thin layer compared to the Earth's total diameter, soil is the medium that produces most of the food required by living things. In addition to being the site of most food production, soil is the receptor of large quantities of pollutants, such as particulate matter from power plant smokestacks. Fertilizers and some other materials applied to soil often contribute to water and air pollution. Therefore, soil is a key component of environmental chemical cycles.

Soil is formed by the weathering of parent rocks as the result of interactive geological, hydrological, and biological processes and is a variable mixture of minerals, organic matter, and water, capable of supporting plant life on the Earth's surface. Soils are porous and are vertically stratified into horizons as the result of downward-percolating water and biological processes, including the production and decay of biomass. The organic portion of soil consists of plant biomass in various stages of decay. High populations of bacteria, fungi, and animals such as earthworms may be found in soil. Soil contains air spaces and generally has a loose texture (Figure 15.1). Soils are open systems that undergo continual exchange of matter and energy with the atmosphere, hydrosphere, and biosphere.

The solid fraction of typical productive soil is approximately 5% organic matter and 95% inorganic matter. Some soils, such as peat soils, may contain as much as 95% organic material. Other soils contain as little as 1% organic matter.

Typical soils exhibit distinctive layers called **horizons** (Figure 15.2). Horizons form as the result of complex interactions among processes that occur during weathering. Rainwater percolating through soil carries dissolved and colloidal solids to lower horizons where they are deposited. Biological processes, such as bacterial decay of residual plant biomass, produces slightly acidic CO_2, organic acids, and complexing compounds that are carried by rainwater to lower hori-

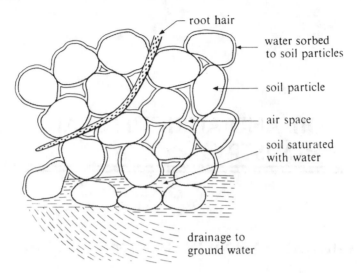

Figure 15.1.   Fine structure of soil, showing solid, water, and air phases.

zons where they interact with clays and other minerals, altering the properties of the minerals. The top layer of soil, typically several inches in thickness, is known as the A horizon, or **topsoil**. This is the layer of maximum biological activity in the soil and contains most of the soil organic matter. Metal ions and clay particles in the A horizon are subject to considerable leaching. The next layer is the B horizon, or **subsoil**. It receives material such as organic matter, salts, and clay

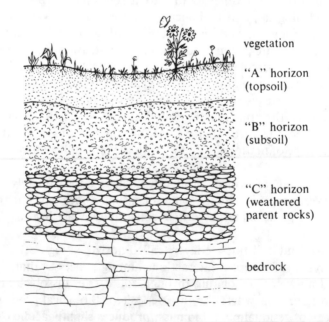

Figure 15.2.   Soil profile showing soil horizons.

particles leached from the topsoil. The C horizon is composed of weathered parent rocks from which the soil originated.

Water, Air, and Particles in Soil

Soil particle size largely determines the physical character of soil. Classified according to size, clays consist of particles less than 0.002 mm, silt particles are 0.002–0.05 mm, sand is 0.05–1 mm, and gravel exceeds 1 mm. **Soil texture** is classified according to relative amounts of clay, silt, and sand. One of the more productive kinds of soils is **loam**, which consists of approximately 40 percent silt, 40 percent sand, and 20 percent clay. Even better for agricultural purposes is **sandy loam**, which is typically 70% sand, 10% clay, and 20% silt.

Large quantities of water are required for the production of most plant materials. For example, several hundred kg of water are required to produce one kg of dry hay. Water is part of the three-phase, solid-liquid-gas system making up soil. It is the basic transport medium for carrying essential plant nutrients from solid soil particles into plant roots and to the farthest reaches of the plant's leaf structure (Figure 15.3). The water enters the atmosphere from the plant's leaves, a process called **transpiration**.

Normally, because of the small size of soil particles and the presence of small capillaries and pores in the soil, the water phase is not totally independent of soil

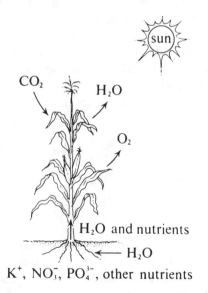

K^+, NO_3^-, PO_4^{3-}, other nutrients

Figure 15.3. Plants transport water from the soil to the atmosphere by transpiration. Nutrients are also carried from the soil to the plant extremities by this process. Plants remove CO_2 from the atmosphere and add O_2 by photosynthesis. The reverse occurs during plant respiration.

solid matter. Access of soil water to plants is governed by gradients arising from capillary and gravitational forces. The availability of nutrient solutes in water depends upon concentration gradients and electrical potential gradients. Water present in larger spaces in soil is relatively more available to plants and readily drains away. Water present in smaller pores, or between the unit layers of clay particles is held much more strongly. Soils high in organic matter may hold appreciably more water than other soils, but it is relatively less available to plants because of physical and chemical sorption of the water by the organic matter.

There is a very strong interaction between clays and water in soil. Water is absorbed on the surfaces of clay particles. Because of the high surface/volume ratio of colloidal clay particles, a great deal of water may be bound in this manner. Water is also held between the unit layers of the expanding clays, such as the montmorillonite clays.

As soil becomes waterlogged (water-saturated) it undergoes drastic changes in physical, chemical, and biological properties. Oxygen in such soil is rapidly used up by the respiration of microorganisms that degrade soil organic matter. In such soils, the bonds holding soil colloidal particles together are broken, which causes disruption of soil structure. Thus, the excess water in such soils is detrimental to plant growth, and the soil does not contain the air required by most plant roots. Most useful crops, with the notable exception of rice, cannot grow on waterlogged soils.

One of the most marked chemical effects of waterlogging is a reduction of pE by the action of organic reducing agents acting through bacterial catalysts. Thus, the redox condition of the soil becomes much more reducing, and the soil pE (see Section 11.8) may drop from that of water in equilibrium with air ($+13.6$ at pH 7) to 1 or less. One of the more significant results of this change is the mobilization of iron and manganese as soluble iron(II) and manganese(II) through reduction of their insoluble higher oxides:

$$MnO_2 + 4H^+ + 2e^- \rightarrow Mn^{2+} + 2H_2O \qquad (15.1.1)$$

$$Fe_2O_3 + 6H^+ + 2e^- \rightarrow 2Fe^{2+} + 3H_2O \qquad (15.1.2)$$

Although soluble manganese generally is found in soil as Mn^{2+} ion, soluble iron(II) frequently occurs as negatively charged iron-organic chelates. Strong chelation of iron(II) by soil fulvic acids (Section 11.9) apparently facilitates reduction of iron(III).

Some soluble metal ions such as Fe^{2+} and Mn^{2+} are toxic to plants at high levels. Their oxidation to insoluble oxides may cause formation of deposits of Fe_2O_3 and MnO_2, which clog tile drains in fields.

The Soil Solution

The **soil solution** is the aqueous portion of soil that contains dissolved matter from soil chemical and biochemical processes and from exchange with the hydrosphere and biosphere. This medium transports chemical species to and from soil particles and provides intimate contact between the solutes and the soil particles. In addition to providing water for plant growth, it is an essential pathway for the exchange of plant nutrients between roots and solid soil.

Air in Soil

Roughly 35% of the volume of typical soil is composed of air-filled pores. Whereas the normal dry atmosphere at sea level contains 21% O_2 and 0.03% CO_2 by volume, these percentages may be quite different in soil air because of the decay of organic matter:

$$\{CH_2O\} + O_2 \rightarrow CO_2 + H_2O \qquad (15.1.3)$$

This process consumes oxygen and produces CO_2. As a result, the oxygen content of air in soil may be as low as 15%, and the carbon dioxide content may be several percent. Thus, the decay of organic matter in soil increases the equilibrium level of dissolved CO_2 in groundwater. This lowers the pH and contributes to weathering of carbonate minerals, particularly calcium carbonate (see Reaction 11.7.6). The presence of CO_2 also shifts the equilibrium of the process by which roots absorb metal ions from soil (see Reaction 15.2.1).

The Inorganic Component of Soil

The weathering of parent rocks and minerals to form the inorganic soil components results ultimately in the formation of inorganic colloids. These colloids are repositories of water and plant nutrients, which may be made available to plants as needed. Inorganic soil colloids often absorb toxic substances in soil, thus playing a role in detoxification of substances that otherwise would harm plants. The abundance and nature of inorganic colloidal material in soil are obviously important factors in determining soil productivity.

The uptake of plant nutrients by roots often involves complex interactions with the water and inorganic phases. For example, a nutrient held by inorganic colloidal material has to traverse the mineral/water interface and then the water/root interface. This process is often strongly influenced by the ionic structure of soil inorganic matter.

As noted in Section 14.2, the most common elements in the Earth's crust are oxygen, silicon, aluminum, iron, calcium, sodium, potassium, and magnesium. Therefore, minerals composed of these elements—particularly silicon and oxygen—constitute most of the mineral fraction of the soil. Common soil mineral constituents are finely divided quartz (SiO_2), orthoclase ($KAlSi_3O_8$), albite ($NaAlSi_3O_8$), epidote ($4CaO \cdot 3(AlFe)_2O_3 \cdot 6SiO_2 \cdot H_2O$), geothite ($FeO(OH)$), magnetite ($Fe_3O_4$), calcium and magnesium carbonates ($CaCO_3$, $CaCO_3 \cdot MgCO_3$), and oxides of manganese and titanium.

Organic Matter in Soil

Though typically comprising less than 5% of a productive soil, organic matter largely determines soil productivity. It serves as a source of food for microorganisms, undergoes chemical reactions such as ion exchange, and influences the physical properties of soil. Some organic compounds even contribute to the weathering of mineral matter, the process by which soil is formed. For example, $C_2O_4^{2-}$, oxalate ion, produced as a soil fungi metabolite, occurs in soil as the calcium salts whewellite and weddelite. Oxalate in soil water dissolves minerals, thus speeding the weathering process and increasing the availability of nutrient ion species. This weathering process involves oxalate complexation of iron or aluminum in minerals, represented by the reaction

$$3H^+ + M(OH)_3(s) + 2CaC_2O_4(s) \rightarrow M(C_2O_4)_2^-(aq) + 2Ca^{2+}(aq) + 3H_2O \qquad (15.1.4)$$

in which M is Al or Fe. Some soil fungi produce citric acid and other chelating organic acids, which react with silicate minerals and release potassium and other nutrient metal ions held by these minerals.

The strong chelating agent 2-ketogluconic acid is produced by some soil bacteria. By solubilizing metal ions, it may contribute to the weathering of minerals. It may also be involved in the release of phosphate from insoluble phosphate compounds.

Biologically active components of the organic soil fraction include polysaccharides, amino sugars, nucleotides, and organic sulfur and phosphorus compounds. Humus, a water-insoluble material that biodegrades very slowly, makes up the bulk of soil organic matter.

The accumulation of organic matter in soil is strongly influenced by temperature and by the availability of oxygen. Since the rate of biodegradation decreases with decreasing temperature, organic matter does not degrade rapidly in colder climates and tends to build up in soil. In water and in waterlogged soils, decaying vegetation does not have easy access to oxygen, and organic matter accumulates. The organic content may reach 90% in areas where plants grow and decay in soil saturated with water.

Of the organic components listed as soil, **soil humus** is by far the most significant. Humus, produced from the partial decay of plant matter and composed of humic and fulvic acids (described in Section 11.9) and an insoluble fraction called humin, is the residue left when bacteria and fungi biodegrade plant material. The bulk of plant biomass consists of relatively degradable cellulose and degradation-resistant lignin, which is a polymeric substance with a higher carbon content than cellulose. Among lignin's prominent chemical components are aromatic rings connected by alkyl chains, methoxyl groups, and hydroxyl groups. Lignin is the precursor of most soil humus.

Humic materials in soil strongly sorb many solutes in soil water and have a particular affinity for heavy polyvalent cations. Soil humic substances may contain levels of uranium more than 10^4 times that of the water with which they are in equilibrium. Thus, water becomes depleted of its cations (or purified) in passing through humic-rich soils. Humic substances in soils also have a strong affinity for organic compounds with low water-solubility such as DDT or Atrazine, a herbicide widely used to kill weeds in corn fields.

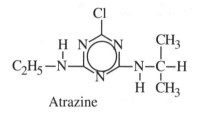

Atrazine

In some cases, there is a strong interaction between the organic and inorganic portions of soil. This is especially true of the strong complexes formed between clays and humic (fulvic) acid compounds. In many soils, 50–100% of soil carbon is complexed with clay. These complexes play a role in determining the physical properties of soil, soil fertility, and stabilization of soil organic matter. The synthesis, chemical reactions, and biodegradation of humic materials are affected by interaction with clays. The lower-molecular-weight fulvic acids may be bound in the spaces in layers in clay particles.

15.2. ACID-BASE AND ION EXCHANGE REACTIONS IN SOILS

One of the more important chemical functions of soils is the exchange of cations. The ability of a sediment or soil to exchange cations is expressed as the cation-exchange capacity (CEC), the quantity of monovalent cations that can be exchanged per 100 g of soil (on a dry-weight basis). The CEC varies with soil conditions such as pE and pH. Both the mineral and organic portions of soils exchange cations. Clay minerals exchange cations because of the presence of negatively charged sites on the mineral, resulting from the substitution of an atom of lower oxidation number for one of higher number; for example, magnesium for aluminum. Organic materials exchange cations because of the presence of the carboxylate group and other basic functional groups. Humus typically has a very high cation-exchange capacity.

Cation exchange in soil is the mechanism by which potassium, calcium, magnesium, and essential trace-level metals are made available to plants. When nutrient metal ions are taken up by plant roots, hydrogen ion is exchanged for the metal ions. This process, plus the leaching of calcium, magnesium, and other metal ions from the soil by water containing carbonic acid, tends to make the soil acidic:

$$\text{Soil}\}Ca^{2+} + 2CO_2 + 2H_2O \rightarrow \text{Soil}\}(H^+)_2 + Ca^{2+}(\text{root}) + 2HCO_3^- \quad (15.2.1)$$

Soil acts as a buffer and resists changes in pH. The buffering capacity depends upon the type of soil.

Production of Mineral Acid in Soil

The oxidation of pyrite in soil causes formation of acid-sulfate soils sometimes called "cat clays":

$$FeS_2 + 7/2O_2 + H_2O \rightarrow Fe^{2+} + 2H^+ + 2SO_4^{2-} \quad (15.2.2)$$

Cat clay soils may have pH values as low as 3.0. These soils, which are commonly found in Delaware, Florida, New Jersey, and North Carolina, are formed when neutral or basic marine sediments containing FeS_2 become acidic upon oxidation of pyrite when exposed to air. For example, soil reclaimed from marshlands and used for citrus groves has developed high acidity detrimental to plant growth. In addition, H_2S released by increased acidity is very toxic to citrus roots.

Pyrite-containing mine spoils (residue left over from mining) also form soils similar to acid-sulfate soils of marine origin. In addition to high acidity and toxic H_2S, a major chemical species limiting plant growth on such soils is Al(III). Aluminum ion liberated in acidic soils is very toxic to plants.

Adjustment of Soil Acidity

Most common plants grow best in soil with a pH near neutrality. If the soil becomes too acidic for optimum plant growth, it may be restored to productivity by liming, ordinarily through the addition of calcium carbonate:

$$\text{Soil}\}(H^+)_2 + CaCO_3 \rightarrow \text{Soil}\}Ca^{2+} + CO_2 + H_2O \quad (15.2.3)$$

In areas of low rainfall, soils may become too basic (alkaline) due to the presence of basic salts such as Na_2CO_3. Alkaline soils may be treated with aluminum or iron sulfate, which release acid on hydrolysis:

$$2Fe^{3+} + 3SO_4^{2-} + 6H_2O \rightarrow 2Fe(OH)_3(s) + 6H^+ + 3SO_4^{2-} \quad (15.2.4)$$

Sulfur added to soils is oxidized by bacterially mediated reactions to sulfuric acid:

$$S + \tfrac{3}{2}O_2 + H_2O \rightarrow 2H^+ + SO_4^{2-} \quad\quad\quad (15.2.5)$$

and sulfur is used, therefore, to acidify alkaline soils. The huge quantities of sulfur now being removed from fossil fuels to prevent air pollution by sulfur dioxide may make the treatment of alkaline soils by sulfur much more attractive economically.

15.3. MACRONUTRIENTS IN SOIL

One of the most important functions of soil in supporting plant growth is to provide essential plant nutrients—macronutrients and micronutrients. **Macronutrients** are those elements that occur in substantial levels in plant materials or in fluids in the plant. Micronutrients (Section 15.5) are elements that are essential only at very low levels and generally are required for the functioning of essential enzymes.

The elements generally recognized as essential macronutrients for plants are carbon, hydrogen, oxygen, nitrogen, phosphorus, potassium calcium, magnesium, and sulfur. Carbon, hydrogen, and oxygen are obtained from the atmosphere. The other essential macronutrients must be obtained from soil. Of these, nitrogen, phosphorus, and potassium are the most likely to be lacking and are commonly added to soil as fertilizers. Because of their importance, these elements are discussed separately in Section 15.4.

Calcium-deficient soils are relatively uncommon. Liming, a process used to treat acid soils (see Section 15.2), provides a more than adequate calcium supply for plants. However, calcium uptake by plants and leaching by carbonic acid (Reaction 15.2.1) may produce a calcium deficiency in soil. Acid soils may still contain an appreciable level of calcium which, because of competition by hydrogen ion, is not available to plants. Treatment of acid soil to restore the pH to near-neutrality generally remedies the calcium deficiency. In alkaline soils, the presence of high levels of sodium, magnesium, and potassium sometimes produces calcium deficiency because these ions compete with calcium for availability to plants.

Although magnesium makes up 2.1% of the Earth's crust, most of it is rather strongly bound in minerals. Generally, exchangeable magnesium is considered available to plants and is held by ion-exchanging organic matter or clays. The availability of magnesium to plants depends upon the calcium/magnesium ratio. If this ratio is too high, magnesium may not be available to plants and magnesium deficiency results. Similarly, excessive levels of potassium or sodium may cause magnesium deficiency.

Sulfur is assimilated by plants as the sulfate ion, SO_4^{2-}. In addition, in areas where the atmosphere is contaminated with SO_2, sulfur may be absorbed as sulfur dioxide by plant leaves. Atmospheric sulfur dioxide levels have been high enough to kill vegetation in some areas (see Section 17.7). However, some experiments designed to show SO_2 toxicity to plants have resulted in increased plant growth where there was an unexpected sulfur deficiency in the soil used for the experiment.

Soils deficient in sulfur do not support plant growth well, largely because sulfur is a component of some essential amino acids and of thiamin and biotin. Sulfate ion is generally present in the soil as immobilized insoluble sulfate minerals or as soluble salts, which are readily leached from the soil and lost as soil water runoff. Unlike the case of nutrient cations such as K^+, little sulfate is adsorbed to the soil (that is, bound by ion exchange binding) so that it would be resistant to leaching while still available for assimilation by plant roots.

Soil sulfur deficiencies have been found in a number of regions of the world. Whereas most fertilizers used to contain sulfur, its use in commercial fertilizers has declined, raising the possibility of its being a limiting nutrient in some cases.

The reaction of FeS_2 with acid in acid-sulfate soils may release H_2S, which is very toxic to plants and which also kills many beneficial microorganisms. Toxic hydrogen sulfide can also be produced by reduction of sulfate ion through microorganism-mediated reactions with organic matter (see Section 11.13). Production of hydrogen sulfide in flooded soils may be inhibited by treatment with oxidizing compounds, one of the most effective of which is KNO_3.

15.4. NITROGEN, PHOSPHORUS, AND POTASSIUM IN SOIL

Nitrogen, phosphorus, and potassium are plant nutrients that are obtained from soil. They are so important for crop productivity that they are commonly added to soil as fertilizers. The environmental chemistry of these elements is discussed here and their production as fertilizers in Section 15.6.

Nitrogen

Figure 15.4 summarizes the primary sinks and pathways of nitrogen in soil. In most soils, over 90% of the nitrogen content is organic. This organic nitrogen is primarily the product of the biodegradation of dead plants and animals. It is eventually hydrolyzed to NH_4^+, which can be oxidized to NO_3^- by the action of bacteria in the soil.

Nitrogen bound to soil humus is especially important in maintaining soil fertility. Unlike potassium or phosphate, nitrogen is not a significant product of

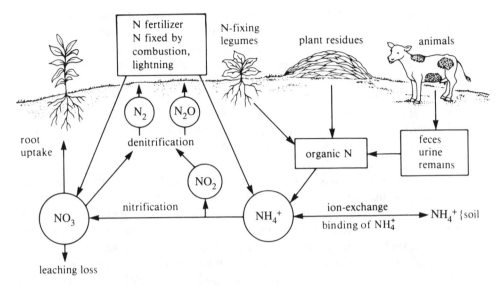

Figure 15.4. Nitrogen sinks and pathways in soil.

mineral weathering. Nitrogen-fixing organisms ordinarily cannot supply sufficient nitrogen to meet peak demand. Inorganic nitrogen from fertilizers and rainwater is often largely lost by leaching. Soil humus, however, serves as a reservoir of nitrogen required by plants. It has the additional advantage that its rate of decay, hence its rate of nitrogen release to plants, roughly parallels plant growth—rapid during the warm growing season, slow during the winter months.

Nitrogen is an essential component of proteins and other constituents of living matter. Plants and cereals grown on nitrogen-rich soils not only provide higher yields, but are often substantially richer in protein and therefore, more nutritious. Nitrogen is most generally available to plants as nitrate ion, NO_3^-. Some plants such as rice may utilize ammonium nitrogen; however, other plants are poisoned by this form of nitrogen. When nitrogen is applied to soils in the ammonium form, nitrifying bacteria perform an essential function in converting it to available nitrate ion.

Nitrogen fixation is the process by which atmospheric N_2 is converted to nitrogen compounds available to plants. Human activities are resulting in the fixation of a great deal more nitrogen than would otherwise be the case. Artificial sources now account for 30–40% of all nitrogen fixed. These include chemical fertilizer manufacture; nitrogen fixed during fuel combustion; combustion of nitrogen-containing fuels; and the increased cultivation of nitrogen-fixing legumes (see the following paragraph). A concern with this increased fixation of nitrogen is the possible effect upon the atmospheric ozone layer by N_2O released during denitrification of fixed nitrogen.

Prior to the widespread introduction of nitrogen fertilizers, soil nitrogen was provided primarily by legumes. These are plants such as soybeans, alfalfa, and clover, which contain on their root structures bacteria capable of fixing atmospheric nitrogen. Leguminous plants have a symbiotic (mutually advantageous) relationship with the bacteria that provide their nitrogen. Legumes may add significant quantities of nitrogen to soil, up to 10 pounds per acre per year, which is comparable to amounts commonly added as synthetic fertilizers. Soil fertility with respect to nitrogen may be maintained by rotating plantings of nitrogen-consuming plants with plantings of legumes, a fact recognized by agriculturists as far back as the Roman era.

The nitrogen-fixing bacteria in legumes exist in special structures on the roots called root nodules (see Figure 15.5). The rod-shaped bacteria that fix nitrogen are members of a special genus called Rhizobium. These bacteria may exist independently, but cannot fix nitrogen except in symbiotic combination with plants. Although all species of Rhizobium appear to be very similar, they exhibit a great deal of specificity in their choice of host plants. Curiously, legume root nodules also contain a form of hemoglobin, which apparently is involved in the nitrogen-fixation process.

Nitrate pollution of some surface waters and groundwater from fertilizers and from the degradation of nitrogenous feedlot wastes has become a major problem in some agricultural areas. Nitrate in farm wells is a common and especially

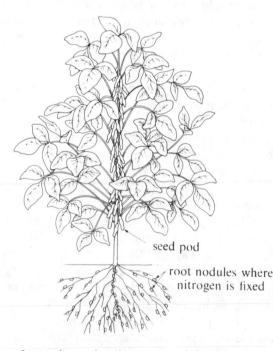

seed pod

root nodules where nitrogen is fixed

Figure 15.5.   A soybean plant, showing root nodules where nitrogen is fixed.

damaging manifestation of nitrogen pollution from feedlots because of the susceptibility of ruminant animals to nitrate poisoning. The stomach contents of ruminant animals such as cattle and sheep constitute a reducing medium (low pE) and contain bacteria capable of reducing nitrate ion to toxic nitrite ion:

$$NO_3^- + 2H^+ + 2e^- \rightarrow NO_2^- + H_2O \qquad (15.4.1)$$

The origin of most nitrate produced from feedlot wastes is amino nitrogen present in nitrogen-containing waste products. Approximately one-half of the nitrogen excreted by cattle is contained in the urine. Part of this nitrogen is proteinaceous and the other part is in the form of urea, NH_2CONH_2. As a first step in the degradation process, the amino nitrogen is probably hydrolyzed to ammonia, or ammonium ion:

$$RNH_2 + H_2O \rightarrow R\text{-}OH + NH_3 (NH_4^+) \qquad (15.4.2)$$

This product is then oxidized through microorganism-catalyzed reactions to nitrate ion:

$$NH_3 + 2O_2 \rightarrow H^+ + NO_3^- + H_2O \qquad (15.4.3)$$

Under some conditions, an appreciable amount of the nitrogen originating from the degradation of feedlot wastes is present as ammonium ion. Ammonium ion is rather strongly bound to soil (recall that soil is a generally good cation exchanger), and a small fraction is fixed as nonexchangeable ammonium ion in the crystal lattice of clay minerals. Because nitrate ion is not strongly bound to soil, it is readily carried through soil formations by water. Many factors, including soil type, moisture, and level of organic matter, affect the production of ammonia and nitrate ion originating from feedlot wastes, and a marked variation is found in the levels and distributions of these materials in feedlot areas.

Phosphorus

Although the percentage of phosphorus in plant material is relatively low, it is an essential component of plants. Phosphorus, like nitrogen, must be present in a simple inorganic form before it can be taken up by plants. In the case of phosphorus, the utilizable species is some form of orthophosphate ion. In the pH range that is present in most soils, $H_2PO_4^-$ and HPO_4^{2-} are the predominant orthophosphate species.

Orthophosphate is most available to plants at pH values near neutrality. It is believed that in relatively acidic soils, orthophosphate ions are precipitated or sorbed by species of Al(III) and Fe(III). In alkaline soils, orthophosphate may react with calcium carbonate to form relatively insoluble hydroxyapatite:

$$3HPO_4^{2-} + 5CaCO_3(s) + 2H_2O \rightarrow Ca_5(PO_4)_3(OH)(s)$$
$$+ 5HCO_3^- + OH^- \qquad (15.4.4)$$

In general, because of sorption and precipitation, little phosphorus applied as fertilizer leaches from the soil. This is important from the standpoint of both water pollution and utilization of phosphate fertilizers.

Potassium

Relatively high levels of potassium are utilized by growing plants. Potassium activates some enzymes and plays a role in the water balance in plants. It is also essential for some carbohydrate transformations. Crop yields are generally greatly reduced in potassium-deficient soils. The higher the productivity of the crop, the more potassium is removed from soil. When nitrogen fertilizers are added to soils to increase productivity, removal of potassium is enhanced. Therefore, potassium may become a limiting nutrient in soils heavily fertilized with other nutrients.

Although potassium is one of the most abundant elements in the Earth's crust, of which it makes up 2.6%, much of this potassium is not easily available to plants. For example, some silicate minerals such as leucite, $K_2O \cdot Al_2O_3 \cdot 4SiO_2$, contain strongly bound potassium. Exchangeable potassium held by clay minerals is relatively more available to plants.

15.5.   MICRONUTRIENTS IN SOIL

Boron, chlorine, copper, iron, manganese, molybdenum (for N-fixation), sodium, vanadium, and zinc are considered essential plant micronutrients. These elements are needed by plants only at very low levels and frequently are toxic at higher levels. There is some chance that other elements will be added to this list as techniques for growing plants in environments free of specific elements improve. Most of these elements function as components of essential enzymes. Manganese, iron, chlorine, zinc, and vanadium may be involved in photosynthesis.

Iron and manganese occur in a number of soil minerals. Sodium and chlorine (as chloride) occur naturally in soil and are transported as atmospheric particulate matter from marine sprays. Some of the other micronutrients and trace elements are found in primary (unweathered) minerals that occur in soil. Boron is substituted for a small percentage of the Si in some micas and is present in tourmaline, a mineral with the formula $NaMg_3Al_6B_3Si_6O_{27}(OH,F)_4$. Low levels of copper are present and substituted for other elements in feldspars, amphiboles, olivines, pyroxenes, and micas; it also occurs as trace levels of copper sulfides in silicate minerals. Molybdenum occurs as molybdenite (MoS_2). Vanadium is substituted for some of the Fe or Al in oxides, pyroxenes, amphiboles, and micas. Zinc is present as the result of substitution for Mg, Fe, and Mn in oxides, amphiboles, olivines, and pyroxenes and as traces of zinc sulfide in silicates. Other trace level elements that occur as specific minerals, sulfide inclusions, or by substitution for other elements in minerals are chromium, cobalt, arsenic, selenium, nickel, lead, and cadmium.

The trace elements listed above may be coprecipitated with secondary minerals (see Section 14.2) that are involved in soil formation. Such secondary minerals include oxides of aluminium, iron, and manganese (precipitation of hydrated oxides of iron and manganese very efficiently removes many trace metal ions from solution); calcium and magnesium carbonates; smectites; vermiculites; and illites.

Some plants accumulate extremely high levels of specific trace metals. Those accumulating more than 1.00 mg/g of dry weight are called **hyperaccumulators**. Hyperaccumulation of nickel and copper has been described; for example, *Aeolanthus biformifolius DeWild* growing in copper-rich regions of Shaba Province, Zaire, contains up to 1.3% copper (dry weight) and is known as a "copper flower".

15.6. FERTILIZERS

Crop fertilizers contain nitrogen, phosphorus, and potassium as major components. Magnesium, sulfate, and micronutrients may also be added. Fertilizers are designated by numbers, such as 6–12–8, showing the respective percentages of nitrogen expressed as N (in this case 6%), phosphorus as P_2O_5 (12%), and potassium as K_2O (8%). Farm manure corresponds to an approximately 0.5–0.24–0.5 fertilizer. The organic fertilizers such as manure must undergo biodegradation to release the simple inorganic species (NO_3^-, $H_xPO_4^{x-3}$, K^+) assimilable by plants.

Most modern nitrogen fertilizers are made by the Haber process, in which N_2 and H_2 are combined over a catalyst at temperatures of approximately 500°C and pressures up to 1000 atm:

$$N_2 + 3H_2 \rightarrow 2NH_3 \qquad (15.6.1)$$

The anhydrous ammonia product has a very high nitrogen content of 82%. It may be added directly to the soil, for which it has a strong affinity because of its water solubility and formation of ammonium ion:

$$NH_3(g) \ (water) \rightarrow NH_3(aq) \qquad (15.6.2)$$

$$NH_3(aq) + H_2O \rightarrow NH_4^+ + OH^- \qquad (15.6.3)$$

Special equipment is required, however, because of the toxicity of ammonia gas. Aqua ammonia, a 30% solution of NH_3 in water, may be used with much greater safety. It is sometimes added directly to irrigation water. It should be pointed out that ammonia vapor is toxic and NH_3 is reactive with some substances. Improperly discarded or stored ammonia can be a hazardous waste.

Ammonium nitrate, NH_4NO_3, is a common solid nitrogen fertilizer. It is made by oxidizing ammonia over a platinum catalyst, converting the nitric oxide product to nitric acid, and reacting the nitric acid with ammonia. The molten ammonium nitrate product is forced through nozzles at the top of a *prilling tower* and solidifies to form small pellets while falling through the tower. The particles are coated with a water repellent. Ammonium nitrate contains 33.5% nitrogen. Although convenient to apply to soil, it requires considerable care during manufacture and storage because it is explosive. Ammonium nitrate also poses some hazards. It is mixed with fuel oil to form an explosive that serves as a substitute for dynamite in quarry blasting and construction.

Urea,

$$H_2N-\overset{\overset{\displaystyle O}{\|}}{C}-NH_2$$

is easier to manufacture and handle than ammonium nitrate. It is now the favored solid nitrogen-containing fertilizer. The overall reaction for urea synthesis is

$$CO_2 + 2NH_3 \rightarrow CO(NH_2)_2 + H_2O \qquad (15.6.4)$$

involving a rather complicated process in which ammonium carbamate, $NH_2CO_2NH_4$, is an intermediate.

Other compounds used as nitrogen fertilizers include sodium nitrate (obtained largely from Chilean deposits, see Section 14.2), calcium nitrate, potassium nitrate, and ammonium phosphates. Ammonium sulfate, a by-product of coke ovens, used to be widely applied as fertilizer. The alkali metal nitrate tends to make soil alkaline, whereas ammonium sulfate leaves an acidic residue.

Phosphate minerals are found in several states, including Idaho, Montana, Utah, Wyoming, North Carolina, South Carolina, Tennessee, and Florida. The principal mineral is fluorapatite, $Ca_5(PO_4)_3F$. The phosphate from fluorapatite is relatively unavailable to plants and is frequently treated with phosphoric or sulfuric acids to produce superphosphates:

$$2Ca_5(PO_4)_3F(s) + 14H_3PO_4 + 10H_2O \rightarrow 2HF(g)$$
$$+ 10CaH_4(PO_4)_2 \cdot H_2O \qquad (15.6.5)$$

$$2Ca_5(PO_4)_3F(s) + 7H_2SO_4 + 3H_2O \rightarrow 2HF(g)$$
$$+ 3CaH_4(PO_4)_2 \cdot H_2O + 7CaSO_4 \qquad (15.6.6)$$

The superphosphate products are much more soluble than the parent phosphate minerals. The HF produced as a by-product of superphosphate production used to cause air pollution problems, but these have been largely controlled.

Phosphate minerals are rich in trace elements required for plant growth, such as boron, copper, manganese, molybdenum, and zinc. Ironically, these elements are lost in processing phosphate for fertilizers and are sometimes added to fertilizers as micronutrients later.

Ammonium phosphates are excellent, highly soluble phosphate fertilizers. Liquid ammonium polyphosphate fertilizers consisting of ammonium salts of pyrophosphate, triphosphate, and small quantities of higher polymeric phosphate anions in aqueous solution can be used as phosphate fertilizers. The polyphosphates are believed to have the additional advantage of chelating iron and other micronutrient metal ions, thus making the metals more available to plants.

Potassium fertilizer components consist of potassium salts, generally KCl. Such salts are found as deposits in the ground or may be obtained from some brines. Very large deposits are found in Saskatchewan, Canada. These salts are all quite soluble in water. One problem encountered with potassium fertilizers is the luxury uptake of potassium by some crops, which absorb more potassium than is really needed for their maximum growth. In a crop where only the grain is harvested, leaving the rest of the plant in the field, luxury uptake does not create much of a problem because most of the potassium is returned to the soil with the dead plant. However, when hay or forage is harvested, potassium contained in the plant as a consequence of luxury uptake is lost from the soil.

15.7. WASTES AND POLLUTANTS IN SOIL

Soil receives large quantities of waste products. Much of the sulfur dioxide emitted in the burning of sulfur-containing fuels ends up on soil as sulfates. Atmospheric nitrogen oxides are converted to nitrates in the atmosphere, and the nitrates eventually are deposited on soil. Soil sorbs NO and NO_2 readily, and these gases are oxidized to nitrate in the soil. Carbon monoxide is converted to CO_2 and possibly to biomass by soil bacteria and fungi (see Section 17.6). Particulate lead from automobile exhausts is found at elevated levels in soil along heavily traveled highways. Elevated levels of lead from lead mines and smelters are found on soil near such facilities.

Soil also receives enormous quantities of pesticides as an inevitable result of their application to crops. The degradation and eventual fate of these pesticides on soil largely determines the ultimate environmental effects of the pesticides, and detailed knowledge of these effects is now required for licensing of a new pesticide (in the U.S. under the Federal Insecticide, Fungicide, and Rodenticide, Act, FIFRA). Among the factors to be considered are the sorption of the pesticide by soil; leaching of the pesticide into water, as related to its potential for water pollution; effects of the pesticide on microorganisms and animal life in the soil; and possible production of relatively more toxic degradation products.

Adsorption by soil is a key step in the degradation of a pesticide. The degree of adsorption and the speed and extent of ultimate degradation are influenced by a number of factors. Some of these, including solubility, volatility, charge, polarity, and molecular structure and size, are properties of the medium. Adsorption of a pesticide by soil components may have several effects. Under some circumstances, it retards degradation by separating the pesticide from the microbial enzymes that degrade it, whereas under other circumstances the reverse is true. Purely chemical degradation reactions may be catalyzed by adsorption. Loss of the pesticide by volatilization or leaching is diminished. The toxicity of a herbicide to plants may be strongly affected by soil sorption.

The forces holding a pesticide to soil particles may be of several types. Physical adsorption involves van der Waals forces arising from dipole-dipole interactions between the pesticide molecule and charged soil particles. Ion exchange is especially effective in holding cationic organic compounds, such as the herbicide paraquat,

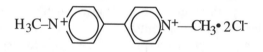

to anionic soil particles. Some neutral pesticides become cationic by binding with H^+ and are bound as the positive form. Hydrogen bonding (Section 7.3) is another mechanism by which some pesticides are held to soil. In some cases, a pesticide may act as a ligand coordinating to metals in soil mineral matter.

The three primary ways in which pesticides are degraded in or on soil are *biodegradation, chemical degradation*, and *photochemical reactions*. Various combinations of these processes may operate in the degradation of a pesticide. Although insects, earthworms, and plants may be involved in the biodegradation of pesticides, microorganisms have the most important role. Despite the importance of biodegradation, chemical degradation of pesticides has been observed experimentally in soils and clays sterilized to remove all microbial activity. Chemical degradation reactions are mostly hydrolysis reactions in which a molecule of pollutant is split with addition of a molecule of H_2O. A number of pesticides have been shown to undergo photochemical reactions; that is, chemical reactions brought about by the absorption of light. Frequently, isomers of the pesticides are produced as products. Many of the studies reported apply to pesticides in water or on thin films, and the photochemical reactions of pesticides on soil and plant surfaces remain largely a matter of speculation.

Soil is the receptor of many hazardous wastes from landfill leachate, lagoons, and other sources. In some cases, land disposal of degradable hazardous organic wastes is practiced as a means of disposal and degradation. The degradable material is worked into the soil, and soil microbial processes bring about its degradation. Sewage and fertilizer-rich sewage sludge may be applied to soil.

15.8.   SOIL EROSION

Soil erosion can occur by the action of both water and wind, although water is the primary source of erosion. Vast quantities of topsoil are swept from the mouth of the Mississippi River. About one-third of U.S. topsoil has been lost since cultivation began on the continent. At the present time approximately one-third of U.S. cultivated land is eroding at a rate sufficient to reduce soil productivity. It is estimated that 48 million acres of land, somewhat more than 10 percent of that under cultivation, is eroding at unacceptable levels, taken to mean a loss of more than 14 tons of topsoil per acre each year. Specific areas in which the greatest erosion is occurring include northern Missouri, southern Iowa, west Texas, western Tennessee, and the Mississippi Basin. Figure 15.6 shows the pattern of soil erosion in the continental U.S.

Intensive cultivation of high-income crops, particularly corn and soybeans, can contribute strongly to soil erosion. With conventional tillage methods, these crops grow in rows with bare soil in between, which tends to wash away with

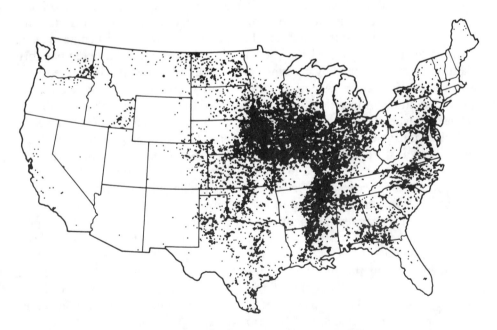

Figure 15.6.   Pattern of soil erosion in the continental U.S. as of 1977. The dark areas indicate locations where the greatest erosion is occurring.

each rainfall. The problem can be aggravated by planting corn and soybeans year after year, without intervening plantings of soil-restoring clover or grass. The problem of decreased productivity due to soil erosion has been masked somewhat by increased use of chemical fertilizers.

Wind erosion, such as occurs on the generally dry, high plains soils of eastern Colorado, poses another threat. After the Dust Bowl days of the 1930s, much of this land was allowed to revert to grassland, and the topsoil was held in place by the strong root systems of the grass cover. However, in an effort to grow more wheat and improve the sale value of the land, much of it has been cultivated in recent years. For example, from 1979 through 1982, more than 450,000 acres of Colorado grasslands were plowed. Much of this was done by speculators who purchased grassland at a low price of $100-$200 per acre, broke it up and sold it as cultivated land at more than double the original purchase price. Although freshly cultivated grassland may yield well for one or two years, the nutrients and soil moisture are rapidly exhausted, and the land becomes very susceptible to wind erosion.

There are a number of solutions to the soil erosion problem. Some are old, well-known agricultural practices, such as terracing, contour plowing, and periodically planting fields with cover crops, such as clover. For some crops **no-till agriculture** greatly reduces erosion. This practice consists of planting a crop among the residue of the previous year's crop, without plowing. Weeds are killed in the newly planted crop row by application of a herbicide prior to planting. The surface residue of plant material left on top of the soil prevents erosion.

Another, more experimental, solution to the soil erosion problem is the cultivation of perennial plants, which develop a large root system and come up each spring after being harvested the previous fall. For example, a perennial corn plant has been developed by crossing corn with a distant, wild relative, teosinte, which grows in Central America. Unfortunately, the resulting plant does not give outstanding grain yields. It should be noted that an annual plant's ability to propagate depends upon producing large quantities of seeds, whereas a perennial plant must develop a strong root system with bulbous growths called rhizomes, which store food for the coming year. However, it is possible that the application of genetic engineering (see Section 15.9) may result in the development of perennial crops with good seed yields. The cultivation of such a crop would cut down on a great deal of soil erosion.

The best known perennial plant is the tree, which is very effective in stopping soil erosion. Wood from trees can be used as biomass fuel, as a source of raw materials, and as food (see below). There is a tremendous unrealized potential for an increase in the production of biomass from trees. For example, the production of biomass from natural forests of loblolly pine trees in South Carolina has been about 3 dry tons per hectare per year. This has now been increased to 11 tons through selection of superior trees, and 30 tons may eventually be possible. In Brazil, experiments have been conducted with a species of Eucalyptus, which has a 7-year growth cycle. With improved selection of trees, the annual yields for three successive cycles of these trees in dry tons per hectare per year has been 23, 33, and 40.

The most important use for wood is, of course, as lumber for construction. This use will remain important as higher energy costs increase the costs of other construction materials, such as steel, aluminum, and cement. Wood is about 50 percent cellulose, which can be hydrolyzed by enzyme processes to yield glucose sugar. The glucose can be used directly as food, fermented to ethyl alcohol for fuel (gasohol), or employed as a carbon and energy source for protein-producing yeasts. Given these and other potential uses, the future of trees as an environmentally desirable and profitable crop is very bright.

15.9. GENETIC ENGINEERING AND AGRICULTURE

The nuclei of living cells contain the genetic instruction for cell reproduction. These instructions are in the form of a special material called deoxyribonucleic acid, DNA (see Section 10.7). In combination with proteins, DNA makes up the cell chromosomes. During the 1970s the ability to manipulate DNA through genetic engineering became a reality and in the 1990s has become the basis of a major industry. Such manipulation falls into the category of recombinant DNA technology. Recombinant DNA gets its name from the fact that it contains DNA from two different organisms, recombined together. This technology promises some exciting developments in agriculture.

The "green revolution" of the mid-1960s used conventional plant-breeding techniques of selective breeding, hybridization, cross-pollination, and backcrossing to develop new strains of rice, wheat, and corn, which, when combined with chemical fertilizers, yielded spectacularly increased crop yields. For example, India's output of grain increased 50 percent. By working at the cellular level, however, it is now possible to greatly accelerate the process of plant breeding. Thus, plants may be developed that resist particular diseases, grow in seawater, or have much higher productivity. The possibility exists for developing entirely new kinds of plants.

One exciting possibility with genetic engineering is the development of plants other than legumes which fix their own nitrogen. For example, if nitrogen-fixing corn could be developed, the savings in fertilizer would be enormous. Furthermore, since the nitrogen is fixed in an organic form in plant root structures, there would be no pollutant runoff of chemical fertilizers.

Another promising possibility with genetic engineering is increased efficiency of photosynthesis. Plants utilize only about 1 percent of the sunlight striking their leaves, so there is appreciable room for improvement in that area.

Cell culture techniques can be applied in which billions of cells are allowed to grow in a medium and develop mutants which, for example, might be resistant to particular viruses or herbicides or have other desirable qualities. Regeneration of these cells into whole plants can enable development of new varieties that might have taken decades using conventional plant-breeding techniques.

Despite the enormous potential of the "green revolution," genetic engineering, and more intensive cultivation of land to produce food and fiber, these technologies cannot be relied upon to support an uncontrolled increase in world population and may even simply postpone an inevitable day of reckoning with the consequences of population growth, resulting in even greater catastrophes than would have otherwise occurred. Changes in climate resulting from global warming, ozone depletion, or natural disasters, such as massive volcanic eruptions or collisions with large meteorites can, and almost certainly will, result in worldwide famine conditions in the future that no agricultural technology will be able to alleviate.

15.10. AGRICULTURE AND HEALTH

Some authorities hold that soil has an appreciable effect upon health. An obvious way in which such an effect might be manifested is the incorporation into food of micronutrient elements essential for human health. One such nutrient (which is toxic at overdose levels) is selenium. It is definitely known that the health of animals is adversely affected in selenium-deficient areas, as it is in areas of selenium excess. Human health might be similarly affected.

There are some striking geographic correlations with the occurrence of cancer. Some of these correlations may be due to soil type. A high incidence of stomach cancer has been shown to occur in areas with certain types of soil in the Netherlands, the United States, France, Wales, and Scandinavia. These soils are high in organic matter content, are acidic, and frequently are waterlogged.

One possible reason for the existence of "stomach cancer-producing soils" is the production of cancer-causing secondary metabolites by plants and microorganisms. Secondary metabolites are biochemical compounds that are of no apparent use to the organism producing them. It is believed that they are formed from the precursors of primary metabolites when the primary metabolites accumulate to excessive levels.

The role of soil in environmental health is not well known, nor has it been extensively studied. The amount of research on the influence of soil in producing foods that are more nutritious and lower in content of naturally occurring toxic substances is quite small compared to research on higher soil productivity. It is to be hoped that the environmental health and nutrition aspects of soil and its products will receive much greater emphasis in the future.

Chemical Contamination

Sometimes human activities contaminate food grown on soil. Most often this occurs through contamination by pesticides. An interesting example of such contamination occurred in Hawaii in early 1982. It was found that milk from several sources on Oahu contained very high levels of the organohalide insecticide heptachlor. This pesticide causes cancer and liver disorders in mice; therefore, it is a suspected human carcinogen. Remarkably, in this case it was not until 57 days after the initial discovery that the public was informed of the contamination by the Department of Health. The source of heptachlor was traced to contaminated "green chop," chopped-up pineapple leaves fed to cattle.

In the late 1980s Alar residues on food caused considerable controversy in the marketplace. **Alar,** daminozide, is a growth regulator that was widely used on apples to bring about uniform ripening of the fruit and to improve firmness and color of the apples. It was discontinued for this purpose after 1988 because of concerns that it might cause cancer, particularly in those children who consume relatively large amounts of apples, apple juice, and other apple products. Dire predictions of apple crop losses and financial devastation caused by discontinuing Alar have not been realized.

CHAPTER SUMMARY

The chapter summary below is presented in a programmed format to review the main points covered in this chapter. It is used most effectively by filling in the blanks, referring back to the chapter as necessary. The correct answers are given at the end of the summary.

Soil is the final product of the (1)_____ action of physical, chemical, and biological processes on rocks. Typical soils exhibit distinctive layers with increasing depth called (2)_____. The top layer of soil is called (3)_____, and the next layer is called (4)_____.

Water is transferred from soil to the atmosphere by plant leaves through a process called (5)_____. Water present in smaller pores, or between the unit layers of clay particles is held (6)_____ than is water in the large pores of soil. One of the most marked chemical effects of waterlogging is a reduction of (7)_____ by organic reducing agents acting through bacterial catalysts. The reaction $Fe_2O_3 + 6H^+ + 2e^- \rightarrow 2Fe^{2+} + 3H_2O$ followed by exposure to oxidizing conditions can result in formation of (8)_____. The (9)_____ is the aqueous portion of soil that contains dissolved matter from soil chemical and biochemical processes and from exchange with the hydrosphere and biosphere.

Soil air contains much higher levels of CO_2 than does atmospheric air because (10)_____.

The inorganic (mineral) content of soil reflects the high abundance of the two elements (11)_____. The weathering of parent rocks and minerals to form the inorganic soil components results ultimately in the formation of inorganic (12)_____.

Some strong effects of organic matter in soil include (13)_____

_____

_____

_____.

Of the organic components of soil, (14)_____, composed of (15)_____ is by far the most significant.

The ability of a sediment or soil to exchange cations is expressed as its (16)_____. When nutrient metal ions are taken up by plant roots, (17)_____ is exchanged for the metal ions. Soil acts as a (18)_____ and resists changes in pH.

Production of acid in soil may result from oxidation of (19)_____ to produce (20)_____. Most common plants grow best in soil with a pH near (21)_____. If the soil becomes too acidic for optimum plant growth, it may be restored to productivity by addition of (22)_____.

Carbon, hydrogen, oxygen, nitrogen, phosphorus, potassium, calcium, magnesium, and sulfur are generally recognized as essential (23)_____ for plants. Of these, nitrogen, phosphorus, and potassium are so important that they are commonly added to soil as (24)_____. The form of nitrogen utilized by plants is (25)_____, phosphorus is utilized as (26)_____

_____, and potassium as (27)_____. Boron, chlorine, copper, iron, manganese, molybdenum, sodium, vanadium, and zinc are considered essential plant (28)_____. Crop fertilizers contain (29)_____

_____ as major components.

One of the most common waste products received by soil consists of (30)___ _____ applied to crops. The three primary ways in which pesticides are degraded in or on soil are (31)_____

_____.

(32)_____ is the primary source of erosion. An effective way of combatting soil erosion is (33)_____, a practice that consists of planting a crop among the residue of the previous year's crop, without plowing.

One exciting possibility with genetic engineering for the improvement of agri-culture is the development of plants other than legumes which (34)_____ _____. Another promising possibility is increased efficiency of photosynthesis because plants utilize only about (35)_____ percent of the sunlight striking their leaves.

Answers

1. weathering
2. horizons
3. topsoil
4. subsoil
5. transpiration
6. much more strongly
7. pE
8. $Fe(OH)_3$
9. soil solution
10. of the biochemical reaction $\{CH_2O\} + O_2 \rightarrow CO_2 + H_2O$
11. oxygen and silicon
12. colloids
13. determination of soil productivity, source of food for microorganisms, chemical reactions such as ion exchange, and influence on the physical properties of soil
14. soil humus
15. humic acids, fulvic acids, and humin
16. cation-exchange capacity
17. hydrogen ion
18. buffer
19. pyrite, FeS_2
20. sulfuric acid

21. neutrality
22. lime
23. macronutrients
24. fertilizers
25. nitrate ion, NO_3^-
26. $H_2PO_4^-$ or HPO_4^{2-}
27. K^+
28. micronutrients
29. nitrogen, phosphorus, and potassium
30. pesticides
31. biodegradation, chemical degradation, and photochemical reactions
32. Water
33. no-till agriculture
34. fix their own nitrogen
35. 1

QUESTIONS AND PROBLEMS

1. Give two examples of reactions involving manganese and iron compounds that may occur in waterlogged soil.

2. What temperature and moisture conditions favor the buildup of organic matter in soil?

3. "Cat clays" are soils containing a high level of iron pyrite, FeS_2. Hydrogen peroxide, H_2O_2, is added to such a soil, producing sulfate, as a test for cat clays. Suggest the chemical reaction involved in this test.

4. What effect upon soil acidity would result from heavy fertilization with ammonium nitrate accompanied by exposure of the soil to air and the action of aerobic bacteria?

5. Prolonged waterlogging of soil does **not** (a) increase NO_3^- production, (b) increase Mn^{2+} concentration, (c) increase Fe^{2+} concentration, (d) have harmful effects upon most plants, (e) increase production of NH_4^+ from NO_3^-.

6. Of the following phenomena, the one that eventually makes soil more basic is (a) removal of metal cations by roots, (b) leaching of soil with CO_2-saturated water, (c) oxidation of soil pyrite, (d) fertilization with $(NH_4)_2SO_4$, (e) fertilization with KNO_3.

7. Suggest how chelating agents that are produced from soil microorganisms could be involved in soil formation? In what sense might such chelating agents promote weathering?

8. What specific compound is both a particular animal waste product and a major fertilizer?

9. Suggest why plants grown on either excessively acidic or excessively basic soils may suffer from calcium deficiency.

10. What are the three major ways in which pesticides are degraded in or on soil?

11. Lime from lead mine tailings containing 0.5% lead was applied at a rate of 10 metric tons per acre of soil and worked in to a depth of 20 cm. The soil density was 2.0 g/cm. To what extent did this add to the burden of lead in the soil?

12. Match the soil or soil-solution constituent in the left column with the soil condition described on the right, below:

    (1) High Mn^{2+} content in   (a) "Cat clays" containing initially high lev-
        soil solution   els of pyrite, FeS_2.
    (2) Excess H^+ (b) Soil in which biodegradation has not
    (3) High H^+ and SO_4^{2-} occurred to a great extent
        content   (c) Waterlogged soil
    (4) High organic content   (d) Soil whose fertility can be improved by
                                       adding limestone

13. What are the processes occurring in soil that operate to reduce the harmful effects of pollutants?

14. Under what conditions do the reactions,

$$MnO_2 + 4H^+ + 2e^- \rightarrow Mn^{2+} + 2H_2O$$

and

$$Fe_2O_3 + 6H^+ + 2e^- \rightarrow 2Fe^{2+} + 3H_2O$$

occur in soil? Name two detrimental effects that can result from these reactions.

15. What are four important effects of organic matter in soil?

16. How might irrigation water treated with fertilizer potassium and ammonia become depleted of these nutrients in passing through humus-rich soil?

17. What biological processes and, therefore, resulting chemical effects occur when soil becomes waterlogged?

18. In what respects is the soil solution particularly important in soil biological and chemical processes?

19. Explain, using a hypothetical chemical reaction, if possible, how chelating agents produced by microorganisms contribute to weathering and soil formation.

20. Using a chemical reaction, show how cation exchange is involved in the uptake of potassium, calcium, or magnesium from soil. Using the same reaction illustrate why the soil from which these metals are taken becomes acidic.

21. Distinguish among macronutrients, micronutrients, and fertilizers in soil. Give examples of each.

22. To what degree is calcium in lime used to treat soil an important nutrient? Is addition of calcium the main objective of lime treatment?

23. Considering the leachability of nitrate, why might it be advantageous to have nitrogen bound to soil humus and other soil organic matter as organic nitrogen as a source of nitrogen in soil?

24. Using chemical reactions as appropriate, illustrate how organically-bound nitrogen originally present as feedlot wastes can be converted to a toxic nitrogen species in the stomachs of ruminant animals.

25. What is the major biochemical function of plant micronutrients, such as copper, manganese, or molybdenum?

26. Estimate the number of kg of farm manure that would be equivalent to 100 kg of a 10–48–10 commercial fertilizer.

27. Using a series of chemical reactions, show how nitrogen from the atmosphere can be converted to a form that can be used by plants by (a) biochemical fixation by legumes and (b) chemical fixation and application as anhydrous liquid ammonia fertilizer.

SUPPLEMENTARY REFERENCES

Conway, G. R., and E. B. Barbier, *After the Green Revolution. Sustainable Agriculture for Development*, (London: Earthscan Publications, 1990).

Bollag, J.-M., and G. Stotzky, Eds., *Soil Biochemistry*, Vol. 6, (New York: Marcel Dekker, 1990).

Heling E., *Sediments and Environmental Geochemistry. Selected Aspects and Case Histories*, (New York: Springer-Verlag, 1990).

Lal, R., and B. A. Stewart, Eds., *Soil Degradation*, (New York: Springer-Verlag, 1990).

Roberts, W. L., T. J. Campbell, and G. R. Rapp, Jr., *Encyclopedia of Minerals*, 2nd ed., (New York: Van Nostrand Reinhold, 1989).

Simkiss, K., and K. M. Wilbur, *Biomineralization. Cell Biology and Mineral Deposition*, (San Diego, CA: Academic Press, 1989).

Sparks, D. L., *Kinetics of Soil Chemical Processes*, (San Diego, CA: Academic Press, 1989).

Paul, E. A., and F. E. Clark, *Soil Microbiology and Biochemistry*, (San Diego, CA: Academic Press, 1988).

Rump, H. H., and H. Krist, *Laboratory Manual for the Examination of Water, Waste Water, and Soil*, (New York: VCH, 1989).

16 THE ATMOSPHERE AND ATMOSPHERIC CHEMISTRY

16.1. IMPORTANCE OF THE ATMOSPHERE

The atmosphere is a protective blanket which nurtures life on the Earth and protects it from the hostile environment of outer space. The atmosphere is the source of carbon dioxide for plant photosynthesis and of oxygen for respiration. It provides the nitrogen that nitrogen-fixing bacteria and ammonia-manufacturing plants use to produce chemically-bound nitrogen, an essential component of life molecules. As a basic part of the hydrologic cycle (Figure 11.1) the atmosphere transports water from the oceans to land, thus acting as the condenser in a vast solar-powered still. Unfortunately, the atmosphere also has been used as a dumping ground for many pollutant materials — ranging from sulfur dioxide to refrigerant Freon — a practice which causes damage to vegetation and materials, shortens human life, and alters the characteristics of the atmosphere itself.

In its vital protective role the atmosphere absorbs most of the cosmic rays from outer space and protects organisms from their effects. It also absorbs most of the electromagnetic radiation from the sun, allowing transmission of significant amounts of radiation only in the regions of 300–2500 namometers (near-ultraviolet, visible, and near-infrared radiation) and 0.01–40 meters (radio waves). By absorbing electromagnetic radiation below 300 nm the atmosphere filters out damaging ultraviolet radiation that would otherwise be very harmful to living organisms. Furthermore, because it reabsorbs much of the infrared radiation by which absorbed solar energy is re-emitted to space, the atmosphere stabilizes the Earth's temperature, preventing the tremendous temperature extremes that occur on planets and moons lacking substantial atmospheres.

16.2. PHYSICAL CHARACTERISTICS OF THE ATMOSPHERE

Atmospheric science deals with the movement of air masses in the atmosphere, atmospheric heat balance, and atmospheric chemical composition and reactions. In order to understand atmospheric chemistry and air pollution, it is important to have an overall appreciation of the atmosphere, its composition, and physical characteristics as discussed in the first parts of this chapter.

Atmospheric Composition

Dry air within several kilometers of ground level consists of two **major components,**

- Nitrogen, 78.08% (by volume)
- Oxygen, 20.95 %

minor components,

- Argon, 0.934%
- Carbon dioxide, 0.035 %

noble gases,

- Neon, 1.818×10^{-3}%
- Krypton, 1.14×10^{-4}%
- Helium, 5.24×10^{-4}%
- Xenon, 8.7×10^{-6}%

and **trace gases** as given in Table 16.1. Atmospheric air may contain 0.1% to 5% water by volume, with a normal range of 1 to 3%.

Variation of Pressure and Density with Altitude

As anyone who has exercised at high altitudes well knows, the density of the atmosphere decreases sharply with increasing altitude as a consequence of the gas laws and gravity. More than 99% of the total mass of the atmosphere is found within approximately 30 km (about 20 miles) of the Earth's surface. Such an altitude is miniscule compared to the Earth's diameter, so it is not an exaggeration to characterize the atmosphere as a "tissue-thin" protective layer. Although the total mass of the global atmosphere is a huge figure, approximately 5.14×10^{15} metric tons, it is still only approximately one millionth of the Earth's total mass.

The fact that atmospheric pressure decreases as an approximately exponential function of altitude largely determines the characteristics of the atmosphere. Plots of pressure in atmospheres at a specified altitude, P_h, and temperature versus altitude are shown in Figure 16.1. The plot of P_h is nonlinear because of variations arising from temperature differences and the mixing of air masses. The plot reflects nonlinear variations in temperature with altitude that are discussed later in this section.

The characteristics of the atmosphere vary widely with altitude, time (season), location (latitude), and even solar activity. Extremes of pressure and tempera-

Table 16.1. Atmospheric Trace Gases in Dry Air Near Ground Level

Gas or Species	Volume Percent[a]	Major Sources	Process for Removal from the Atmosphere
CH_4	1.6×10^{-4}	Biogenic[b]	Photochemical[c]
CO	$\sim 1.2 \times 10^{-5}$	Photochemical, anthropogenic[d]	Photochemical
N_2O	3×10^{-5}	Biogenic	Photochemical
NO_x[e]	10^{-10}–10^{-6}	Photochemical, lightning, anthropogenic	Photochemical
HNO_3	10^{-9}–10^{-7}	Photochemical	Washed out by precipitation
NH_3	10^{-8}–10^{-7}	Biogenic	Photochemical, washed out by precipitation
H_2	5×10^{-5}	Biogenic, photochemical	Photochemical
H_2O_2	10^{-8}–10^{-6}	Photochemical	Washed out by precipitation
$HO\cdot$[g]	10^{-13}–10^{-10}	Photochemical	Photochemical
$HO_2\cdot$[g]	10^{-11}–10^{-9}	Photochemical	Photochemical
H_2CO	10^{-8}–10^{-7}	Photochemical	Photochemical
CS_2	10^{-9}–10^{-8}	Anthropogenic, biogenic	Photochemical
OCS	10^{-8}	Anthropogenic, biogenic, photochemical	Photochemical
SO_2	$\sim 2 \times 10^{-8}$	Anthropogenic, photochemical, volcanic	Photochemical
I_2	0-trace	—	—
CCl_2F_2[g]	2.8×10^{-5}	Anthropogenic	Photochemical
H_3CCCl_3[h]	$\sim 1 \times 10^{-8}$	Anthropogenic	Photochemical

[a] Levels in the absence of gross pollution.
[b] From biological sources.
[c] Reactions induced by the absorption of light energy as described later in this chapter.
[d] Sources arising from human activities.
[e] Sum of NO and NO_2.
[f] Reactive free radical species with one unpaired electron; described later in the chapter; these are transient species whose concentrations become much lower at night.
[g] A chlorofluorocarbon, Freon F-12.
[h] Methyl chloroform.

ture are illustrated in Figure 16.1. At very high altitudes normally reactive species, such as atomic oxygen, O, persist for long periods of time. That occurs because the pressure is very low at these altitudes so that the distance traveled by a reactive species before it collides with a potential reactant—its **mean free path**—is quite high. A particle with a mean free path of 1×10^{-6} cm at sea level has a mean free path greater than 1×10^6 cm at an altitude of 500 km, where the pressure is lower by many orders of magnitude.

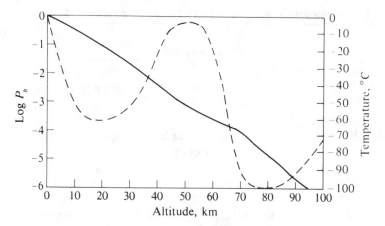

Figure 16.1.   Variation of pressure in atmospheres (solid line) and temperature (dashed line) with altitude.

Stratification of the Atmosphere

As shown in Figure 16.2, the atmosphere is stratified on the basis of temperature/density relationships resulting from interrelationships between physical and photochemical (light-induced chemical phenomena) processes in air.

The lowest layer of the atmosphere extending from sea level to an altitude of 10–16 km is the **troposphere**, characterized by a generally homogeneous composition of major gases other than water and decreasing temperature with increasing altitude from the heat-radiating surface of the earth. The upper limit of the troposphere, which has a temperature minimum of about –56°C, varies in altitude by a kilometer or more with atmospheric temperature, underlying terrestrial surface, and time. The homogeneous composition of the troposphere results from constant mixing by circulating air masses. However, the water vapor content of the troposphere is extremely variable because of cloud formation, precipitation, and evaporation of water from terrestrial water bodies.

The very cold temperature of the **tropopause** layer at the top of the troposphere serves as a barrier that causes water vapor to condense to ice so that it cannot reach altitudes at which it would photodissociate through the action of intense high-energy ultraviolet radiation. If this happened, the hydrogen produced would escape the Earth's atmosphere and be lost. (Much of the hydrogen and helium originally present in the Earth's atmosphere was lost by this process.)

The atmospheric layer directly above the troposphere is the **stratosphere**, in which the temperature rises to a maximum of about –2°C with increasing altitude. This temperature maximum is due to the presence of ozone, O_3, which may reach a level of around 10 ppm by volume in the mid-range of the stratosphere.

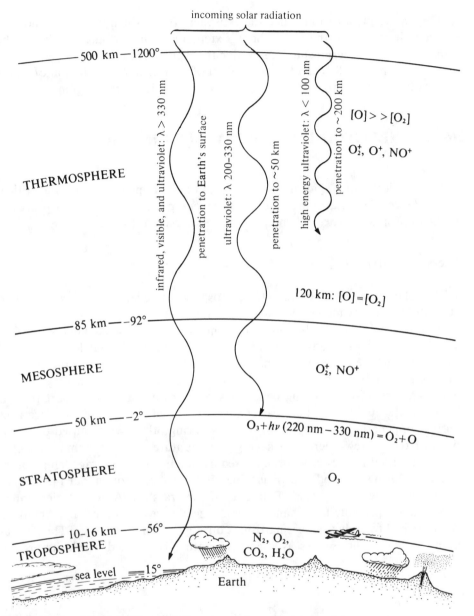

Figure 16.2.   Major regions of the atmosphere (not to scale).

The heating effect is caused by the absorption of ultraviolet radiation energy by ozone, a phenomenon discussed later in this chapter.

The absence of high levels of radiation-absorbing species in the mesosphere immediately above the stratosphere results in a further temperature decrease to about –92°C at an altitude around 85 km. The upper regions of the mesosphere and higher define a region called the exosphere from which molecules and ions can completely escape the atmosphere. Extending to the far outer reaches of the atmosphere is the **thermosphere,** in which the highly rarified gas reaches temperatures as high as 1200°C by the absorption of very energetic radiation of wavelengths less than approximately 200 nm by gas species in this region.

16.3.   ENERGY AND MASS TRANSFER IN THE ATMOSPHERE

The physical and chemical characteristics of the atmosphere and the critical heat balance of the Earth are determined by energy and mass transfer processes in the atmosphere. These phenomena are addressed in this section.

Energy Transfer

The solar energy flux reaching the atmosphere is huge, amounting to 1.34×10^3 watts per square meter (19.2 kcal per minute per square meter) perpendicular to the line of solar flux at the top of the atmosphere, as illustrated in Figure 16.3. This value is the **solar constant.** If all this energy reached the Earth's surface and were retained, the planet would have vaporized long ago. As it is, the complex factors involved in maintaining the Earth's heat balance within very narrow limits are crucial to retaining conditions of climate that will support present levels of life on Earth. The great changes of climate that resulted in ice ages during some periods or tropical conditions during others were caused by variations of only a few degrees in average temperature. Marked climate changes within recorded history have been caused by much smaller average temperature changes. During a period known as "The Little Ice Age" from about 1450 A.D. to 1850, winters were very cold. Paintings of the era show Alpine glaciers much larger than at present. The Thames River froze over several times, and American colonists had to endure severe hardships during the winter months. The mechanisms by which the Earth's average temperature is retained within its present

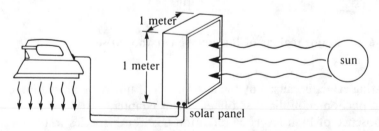

Figure 16.3.   The solar flux at the distance of the Earth from the sun is 1.34×10^3 watts/m².

narrow range are complex and not completely understood, but the main features are explained here.

About half of the solar radiation entering the atmosphere reaches the Earth's surface either directly or after scattering by clouds, atmospheric gases, or particles. The remaining half of the radiation is either reflected directly back or absorbed in the atmosphere and its energy radiated back into space at a later time as infrared radiation. Most of the solar energy reaching the surface is absorbed and it must be returned to space in order to maintain heat balance. In addition, a very small amount of energy (less than 1% of that received from the sun) reaches the Earth's surface by convection and conduction processes from the Earth's hot mantle, and this, too, must be lost.

Energy transport, which is crucial to eventual reradiation of energy from the Earth is accomplished by three major mechanisms — conduction, convection, and radiation. **Conduction** of energy occurs through the interaction of adjacent atoms or molecules without the bulk movement of matter. **Convection** involves the movement of whole masses of air, which may be either relatively warm or cold. It is the mechanism by which abrupt temperature variations occur when large masses of air move across an area. As well as carrying **sensible heat** due to the kinetic energy of molecules, convection carries **latent heat** in the form of water vapor which releases heat as it condenses. An appreciable fraction of the Earth's surface heat is transported to clouds in the atmosphere by conduction and convection before being lost ultimately by radiation.

Radiation of energy occurs through electromagnetic radiation in the infrared region of the spectrum. As the only way in which energy is transmitted through a vacuum, radiation is the means by which all energy lost from the planet to maintain its heat balance is ultimately returned to space. The electromagnetic radiation that carries energy away from the Earth is of a much longer wavelength than the sunlight that brings energy to the Earth. This is a crucial factor in maintaining the Earth's heat balance and one susceptible to upset by human activities. The maximum intensity of incoming radiation occurs at 0.5 micrometers (500 nanometers) in the visible region, with essentially none outside the range of 0.2 μm to 3 μm. This range encompasses the whole visible region and small parts of the ultraviolet and infrared adjacent to it. Outgoing radiation is in the infrared region, with maximum intensity at about 10 μm, primarily between 2 μm and 40 μm. Thus, the Earth loses energy by electromagnetic radiation of a much lower wavelength (lower energy per photon) than the radiation by which it receives energy.

Earth's Radiation Budget

The Earth's radiation budget is illustrated in Figure 16.4. The average surface temperature is maintained at a relatively comfortable 15°C because of an atmospheric "greenhouse effect" in which water vapor and, to a lesser extent carbon

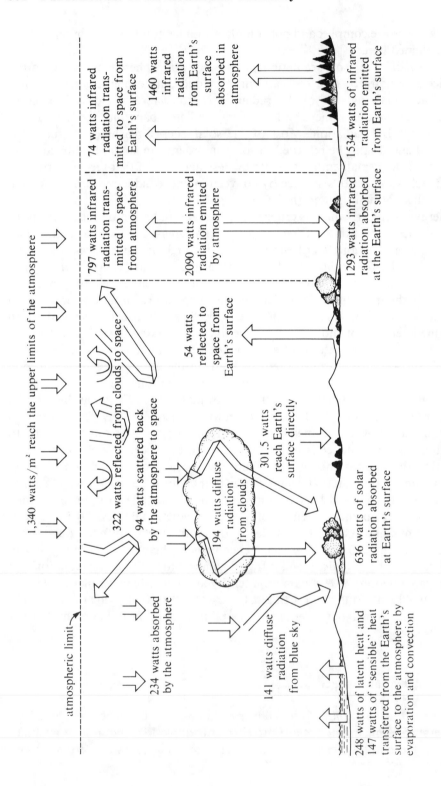

Figure 16.4.   Earth's radiation budget expressed on the basis of portions of the 1,340 watts/m² composing the solar flux.

dioxide, reabsorb much of the outgoing radiation and reradiate about half of it back to the surface. Were this not the case, the surface temperature would average around −18°C. Most of the absorption of infrared radiation is done by water molecules in the atmosphere. Absorption is weak in the regions 7–8.5 μm and 11–14 μm and nonexistent between 8.5 μm and 11 μm, leaving a "hole" in the infrared absorption spectrum through which radiation may escape. Carbon dioxide, though present at a much lower concentration than water vapor, absorbs strongly between 12 μm and 16.3 μm, and plays a key role in maintaining the heat balance. There is concern that an increase in the carbon dioxide level in the atmosphere could prevent sufficient energy loss to cause a perceptible and damaging increase in the Earth's temperature. This phenomenon, discussed in more detail in Sections 16.7 and 17.6, is popularly known as the **greenhouse effect** and may occur from elevated CO_2 levels caused by increased use of fossil fuels and the destruction of massive quantities of forests.

Mass Transfer and Meteorology

Meteorology is the science of atmospheric phenomena, encompassing the study of the movement of air masses as well as physical forces in the atmosphere such as heat, wind, and transitions of water, primarily liquid to vapor, or vice versa. Meteorological phenomena affect, and in turn are affected by, the chemical properties of the atmosphere. For example, meteorological phenomena determine whether or not power plant stack gas heavily laced with sulfur dioxide is dispersed high in the atmosphere, with little direct effect upon human health, or settles as a choking chemical blanket in the vicinity of the power plant. Los Angeles largely owes its susceptibility to smog to the meteorology of the Los Angeles basin, which holds hydrocarbons and nitrogen oxides long enough to cook up an unpleasant brew of damaging chemicals under the intense rays of the sun (see Sections 18.9 to 18.12).

Short-term variations in the state of the atmosphere are described as **weather**. The weather is defined in terms of seven major factors: temperature, clouds, winds, humidity, horizontal visibility (as affected by fog, etc.), type and quantity of precipitation, and atmospheric pressure. All of these factors are closely interrelated.

Horizontally moving air is called wind, whereas vertically moving air is referred to as an air current. Wind and air currents are strongly involved with air pollution phenomena. Wind carries and disperses air pollutants. Prevailing wind direction is an important factor in determining the areas most affected by an air pollution source.

Condensation of water vapor, which forms clouds, must occur prior to the formation of precipitation in the form of rain or snow. For this condensation to happen, air must be cooled below the dew point, and nuclei of condensation must be present. These nuclei are hygroscopic substances such as salts, sulfuric acid droplets, and some organic materials, including bacterial cells. Air pollution in some forms is now an important source of condensation nuclei.

The complicated movement of air across the Earth's surface is a crucial factor in the creation and dispersal of air pollution phenomena. When air movement ceases, air stagnation can occur with a resultant build-up of air pollutants in localized regions. Although the temperature of air relatively near the Earth's surface normally decreases with increasing altitude, certain atmospheric conditions can result in the opposite condition—increasing temperature with increasing altitude. Such conditions are characterized by high atmospheric stability and are known as **temperature inversions**. Typically, an inversion occurs by the collision of a warm air mass (warm front) with a cold air mass (cold front). The warm air mass overrides the cold air mass in the frontal area, producing the inversion. Because they limit the vertical circulation of air, temperature inversions result in air stagnation and the trapping of air pollutants in localized areas.

Human activities have succeeded in changing the meteorology of whole cities, a localized so-called mesometeorological effect. By paving over large areas of a city with nonreflecting asphalt, and by activities that generate heat, humans have created conditions under which the center of a city may be as much as 5°C warmer than the surrounding area. In such a case, the warmer air rises, bringing in a breeze from the surrounding area. Large cities have been described as "heat islands." Pollutants and carbon dioxide given off from cities may absorb emitted infrared radiation, causing a local greenhouse effect that probably is largely counterbalanced by reflection of incoming solar energy by particulate matter above cities.

16.4.   CHEMICAL AND PHOTOCHEMICAL REACTIONS IN THE ATMOSPHERE

The study of atmospheric chemical reactions is difficult. One of the primary obstacles encountered in studying atmospheric chemistry is that the chemist generally must deal with incredibly low concentrations, so that the detection and analysis of reaction products is quite difficult. Simulating high-altitude conditions in the laboratory can be extremely hard because of interferences, such as those from species given off from container walls under conditions of very low pressure. Many chemical reactions that require a third body to absorb excess energy occur very slowly in the upper atmosphere, where there is a sparse concentration of third bodies, but occur readily in a container whose walls effectively absorb energy. Container walls may serve as catalysts for some important reactions, or they may absorb important species and react chemically with the more reactive ones.

Photochemical Processes

The absorption of light by chemical species can bring about reactions, called **photochemical reactions**, which do not otherwise occur under the conditions (particularly the temperature) of the medium in the absence of light. Thus, photochemical reactions, even in the absence of a chemical catalyst, occur at temperatures much lower than those which otherwise would be required. Photochemical reactions, which are induced by intense solar radiation, play a very important role in determining the nature and ultimate fate of a chemical species in the atmosphere.

Nitrogen dioxide, NO_2, is one of the most photochemically active species found in a polluted atmosphere and is an essential participant in the smog-formation process. A species such as NO_2 may absorb light of energy $h\nu$, producing an **electronically excited molecule**,

$$NO_2 + h\nu \rightarrow NO_2^* \qquad (16.4.1)$$

designated in the reaction above by an asterisk, *. The photochemistry of nitrogen dioxide is discussed in greater detail in Section 17.8 and in Chapter 18.

Electronically excited molecules are one of the three relatively reactive and unstable species that are encountered in the atmosphere and are strongly involved with atmospheric chemical processes. The other two species are atoms or molecular fragments with unshared electrons, called **free radicals**, and **ions** consisting of charged atoms or molecular fragments.

Electronically excited molecules are produced when stable molecules absorb energetic electromagnetic radiation in the ultraviolet or visible regions of the spectrum. A molecule may possess several possible excited states, but generally ultraviolet or visible radiation is energetic enough to excite molecules only to several of the lowest energy levels. The nature of the excited state may be understood by considering the disposition of electrons in a molecule. Most molecules have an even number of electrons. The electrons occupy orbitals, with a maximum of two electrons with opposite spin occupying the same orbital. The absorption of light may promote one of these electrons to a vacant orbital of higher energy. In some cases the electron thus promoted retains a spin opposite to that of its former partner, giving rise to an **excited singlet state**. In other cases the spin of the promoted electron is reversed, so that it has the same spin as its former partner; this gives rise to an **excited triplet state**.

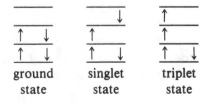

ground   singlet   triplet
state   state   state

These excited states are relatively energized compared to the ground state; therefore, excited chemical species are reactive. Their participation in atmospheric chemical reactions, such as those involved in smog formation are discussed later in this chapter and in Chapters 17 and 18.

In order for a photochemical reaction to occur, electromagnetic radiation, usually from the ultraviolet region of the spectrum, must be absorbed by the reacting species. If visible light is absorbed, the absorbing species is colored. Colored NO_2 is a common example of such a species in the atmosphere. Normally, the first step in a photochemical process is the activation of the molecule by the absorption of a single unit of photochemical energy characteristic of the frequency of the light called a **quantum** of light. The energy of one quantum is equal to the product $h\nu$, where h is Planck's constant, 6.62×10^{-27} erg sec, and ν is the frequency of the absorbed light in sec^{-1} (inversely proportional to its wavelength, λ).

The reactions that occur following absorption of a photon of light to produce an electronically excited species are largely determined by the way in which the excited species loses its excess energy. This may occur by one of several processes, including transfer of energy to other molecules, transfer of energy within the absorbing molecule, or luminescence processes in which energy is lost by emission of photons. From the standpoint of photochemistry, the most important ways in which excited molecules lose energy are the following:

- **Dissociation** of the excited molecule (the process responsible for the predominance of atomic oxygen in the upper atmosphere)

$$O_2^* \rightarrow O + O \tag{16.4.2}$$

- **Direct reaction** with another species

$$O_2^* + O_3 \rightarrow 2O_2 + O \tag{16.4.3}$$

- **Photoionization** through loss of an electron

$$N_2^* \rightarrow N_2^+ + e^- \tag{16.4.4}$$

Electromagnetic radiation absorbed in the infrared region is not sufficiently energetic to break chemical bonds, but does cause the receptor molecules to gain vibrational and rotational energy. The energy absorbed as infrared radiation ultimately is dissipated as heat and raises the temperature of the whole atmosphere. As noted in Section 16.7, the absorption of infrared radiation is very important in the Earth's acquiring heat from the sun and in the retention of energy radiated from the Earth's surface.

Ions and Radicals in the Atmosphere

One of the characteristics of the upper atmosphere which is difficult to duplicate under laboratory conditions is the presence of significant levels of electrons and positive ions. Because of the rarefied conditions, these ions may exist in the upper atmosphere for long periods before recombining to form neutral species.

At altitudes of approximately 50 km and up, ions are so prevalent that the region is called the **ionosphere**. The presence of the ionosphere has been known since about 1901, when it was discovered that radio waves could be transmitted over long distances, where the curvature of the Earth makes line-of-sight transmission impossible. These radio waves bounce off the ionosphere.

Ultraviolet light is the primary producer of ions in the ionosphere. In darkness, the positive ions slowly recombine with free electrons. The process is especially rapid in the lower regions of the ionosphere, where the concentration of species is relatively high. Thus, the lower limit of the ionosphere lifts at night and makes possible the transmission of radio waves over much greater distances.

Free Radicals

In addition to forming ions by photoionization, energetic electromagnetic radiation in the atmosphere may produce atoms or groups of atoms with unpaired electrons called **free radicals**:

$$\overset{\displaystyle O}{\underset{\displaystyle H_3C-C-H}{\|}} + h\nu \rightarrow H_3C\cdot + H\dot{C}O \qquad (16.4.5)$$

Free radicals are involved with most significant atmospheric chemical phenomena and are of the utmost importance in the atmosphere. Because of their unpaired electrons and the strong pairing tendencies of electrons under most circumstances, free radicals are highly reactive. The upper atmosphere is so rarefied, however, that at very high altitudes radicals may have half-lives of several minutes, or even longer. Radicals can take part in chain reactions in which one of the products of each reaction is a radical. Eventually, through processes such as reaction with another radical, one of the radicals in a chain is destroyed and the chain ends:

$$H_3C\cdot + H_3C\cdot \rightarrow C_2H_6 \qquad (16.4.6)$$

This process is a **chain-terminating reaction**. Reactions involving free radicals are responsible for smog formation, discussed in Chapter 18.

Free radicals are quite reactive; therefore, they generally have short lifetimes.

It is important to distinguish between high reactivity and instability. A totally isolated free radical or atom would be quite stable. Therefore, free radicals and single atoms from diatomic gases tend to persist under the rarefied conditions of very high altitudes because they can travel long distances before colliding with another reactive species. However, electronically excited species have a finite, generally very short, lifetime because they can lose energy through radiation without having to react with another species.

Hydroxyl and Hydroperoxyl Radicals in the Atmosphere

The hydroxyl radical, $HO\cdot$, is the single most important reactive intermediate species in atmospheric chemical processes. It is formed by several mechanisms. At higher altitudes it is produced by photolysis of water:

$$H_2O + h\nu \rightarrow HO\cdot + H \tag{16.4.7}$$

In the presence of organic matter, hydroxyl radical is produced in abundant quantities as an intermediate in the formation of photochemical smog (see Section 18.10). To a certain extent in the atmosphere, and for laboratory experimentation, $HO\cdot$ is made by the photolysis of nitrous acid vapor:

$$HONO + h\nu \rightarrow HO\cdot + NO \tag{16.4.8}$$

In the relatively unpolluted troposphere, hydroxyl radical is produced as the result of the photolysis of ozone,

$$O_3 + h\nu(\lambda < 315 \text{ nm}) \rightarrow O^* + O_2 \tag{16.4.9}$$

followed by the reaction of a fraction of the excited oxygen atoms with water molecules:

$$O^* + H_2O \rightarrow 2HO\cdot \tag{16.4.10}$$

Among the important atmospheric trace species that react with hydroxyl radical are carbon monoxide, sulfur dioxide, hydrogen sulfide, methane, and nitric oxide.

Hydroxyl radical is most frequently removed from the troposphere by reaction with carbon monoxide,

$$CO + HO \cdot \rightarrow CO_2 + H \qquad (16.4.11)$$

or with methane:

$$CH_4 + HO \cdot \rightarrow H_3C \cdot + H_2O \qquad (16.4.12)$$

The highly reactive methyl radical, $H_3C \cdot$, reacts with O_2,

$$H_3C \cdot + O_2 \rightarrow H_3COO \cdot \qquad (16.4.13)$$

to form **methylperoxyl radical**, $H_3COO \cdot$. (Further reactions of this species are discussed in Chapter 18.) Hydrogen atoms produced in Reactions 16.4.7 and 16.4.11 react with O_2 to produce **hydroperoxyl radical**, an intermediate in some important chemical reactions:

$$H + O_2 \rightarrow HOO \cdot \qquad (16.4.14)$$

The hydroperoxyl radical can undergo chain termination reactions, such as

$$HOO \cdot + HO \cdot \rightarrow H_2O + O_2 \qquad (16.4.15)$$

$$HOO \cdot + HOO \cdot \rightarrow H_2O_2 + O_2 \qquad (16.4.16)$$

or reactions that regenerate hydroxyl radical:

$$HOO \cdot + NO \rightarrow NO_2 + HO \cdot \qquad (16.4.17)$$

$$HOO \cdot + O_3 \rightarrow 2O_2 + HO \cdot \qquad (16.4.18)$$

The global concentration of hydroxyl radical, averaged diurnally and seasonally, is estimated to range from 2×10^5 to 1×10^6 radicals per cm^3 in the troposphere. Because of the higher humidity and higher incident sunlight, which result in elevated O^* levels, the concentration of $HO \cdot$ is higher in tropical regions. The southern hemisphere probably has about a 20% higher level of $HO \cdot$ than does the northern hemisphere because of greater production of anthropogenic $HO \cdot$-consuming CO in the northern hemisphere.

Chemical and Biochemical Processes in Evolution of the Atmosphere

It is now widely believed that the Earth's atmosphere originally was very different from its present state and that the changes were brought about by biological activity and accompanying chemical changes. Approximately 3.5 billion years ago, when the first primitive life molecules were formed, the atmosphere was chemically reducing, consisting primarily of methane, ammonia, water vapor, and hydrogen. The atmosphere was bombarded by intense, bond-breaking ultraviolet light, which, along with lightning and radiation from radio-nuclides, provided the energy to bring about chemical reactions that resulted in the production of relatively complicated molecules, including even amino acids and sugars. From the rich chemical mixture in the sea, life molecules evolved. Initially, these very primitive life forms derived their energy from fermentation of organic matter formed by chemical and photochemical processes, but eventually they gained the capability to produce organic matter, "$\{CH_2O\}$," by photosynthesis,

$$CO_2 + H_2O + h\nu \rightarrow \{CH_2O\} + O_2(g) \qquad (16.4.18)$$

and the stage was set for the massive biochemical transformation that resulted in the production of almost all the atmosphere's oxygen.

The oxygen initially produced by photosynthesis was probably quite toxic to primitive life forms. However, much of this oxygen was converted to iron oxides by reaction with soluble iron(II):

$$4Fe^{2+} + O_2 + 4H_2O \rightarrow 2Fe_2O_3 + 8H^+ \qquad (16.4.19)$$

This resulted in the formation of enormous deposits of iron oxides, the existence of which provides major evidence for the liberation of free oxygen in the primitive atmosphere.

Eventually, enzyme systems developed that enabled organisms to mediate the reaction of waste-product oxygen with oxidizable organic matter in the sea. Later, this mode of waste-product disposal was utilized by organisms to produce energy by respiration, which is now the mechanism by which nonphotosynthetic organisms obtain energy.

In time, O_2 accumulated in the atmosphere, providing an abundant source of oxygen for respiration. It had an additional benefit in that it enabled the formation of an ozone shield (see Section 16.5). The ozone shield absorbs bond-rupturing ultraviolet light. With the ozone shield protecting tissue from destruction by high-energy ultraviolet radiation, the Earth became a much more hospitable environment for life, and life forms were enabled to move from the sea to land.

16.5. REACTIONS OF ATMOSPHERIC OXYGEN

Some of the primary features of the exchange of oxygen among the atmosphere, lithosphere, hydrosphere, and biosphere are summarized in Figure 16.5. The oxygen cycle is critically important in atmospheric chemistry, geochemical transformations, and life processes.

Oxygen in the troposphere plays a strong role in processes that occur on the

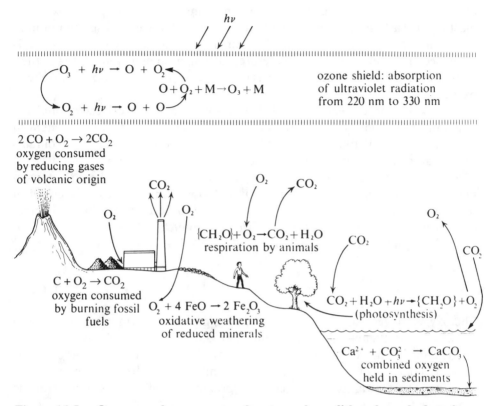

Figure 16.5.   Oxygen exchange among the atmosphere, lithosphere, hydrosphere, and biosphere.

Earth's surface. Atmospheric oxygen takes part in energy-producing reactions, such as the burning of fossil fuels:

$$CH_4 \text{(in natural gas)} + 2O_2 \rightarrow CO_2 + 2H_2O \qquad (16.5.1)$$

Atmospheric oxygen is utilized by aerobic organisms in the degradation of organic material. Some oxidative weathering processes consume oxygen, such as

$$4FeO + O_2 \rightarrow 2Fe_2O_3 \qquad (16.5.2)$$

Oxygen is returned to the atmosphere through plant photosynthesis:

$$CO_2 + H_2O + h\nu \rightarrow \{CH_2O\} + O_2 \qquad (16.5.3)$$

All molecular oxygen now in the atmosphere is thought to have originated through the action of photosynthetic organisms, which shows the importance of photosynthesis in the oxygen balance of the atmosphere. It can be shown that most of the carbon fixed by these photosynthetic processes is dispersed in mineral formations as humic material (Section 11.9); only a very small fraction is deposited in fossil fuel beds. Therefore, although combustion of fossil fuels consumes large amounts of O_2, there is no danger of running out of atmospheric oxygen.

Because of the extremely rarefied atmosphere and the effects of ionizing radiation, elemental oxygen in the upper atmosphere exists to a large extent in forms other than diatomic O_2. In addition to O_2, the upper atmosphere contains oxygen atoms, O; excited oxygen molecules, O_2^*, and ozone, O_3.

Atomic oxygen, O, is stable primarily in the thermosphere, where the atmosphere is so rarefied that the three-body collisions necessary for the chemical reaction of atomic oxygen seldom occur (the third body in this kind of three-body reaction absorbs energy to stabilize the products). Atomic oxygen is produced by a photochemical reaction:

$$O_2 + h\nu \rightarrow O + O \qquad (16.5.4)$$

The oxygen-oxygen bond is strong (120 kcal/mole) and ultraviolet radiation in the wavelength regions 135–176 nm and 240–260 nm is most effective in causing dissociation of molecular oxygen. Because of photochemical dissociation, O_2 is virtually nonexistent at very high altitudes and less than 10% of the oxygen in the atmosphere at altitudes exceeding approximately 400 km is present in the molecular form.

Oxygen atoms in the atmosphere can exist in the ground state (O) and in excited states (O*). Excited oxygen atoms are produced by the photolysis of ozone, which has a relatively weak bond energy of 26 kcal/mole, at wavelengths below 308 nm,

$$O_3 + h\nu(\lambda < 308 \text{ nm}) \rightarrow O^* + O_2 \qquad (16.5.5)$$

or by highly energetic chemical reactions such as

$$O + O + O \rightarrow O_2 + O^* \qquad (16.5.6)$$

Excited atomic oxygen emits visible light at wavelengths of 636 nm, 630 nm, and 558 nm. This emitted light is partially responsible for **airglow**, a very faint electromagnetic radiation continuously emitted by the Earth's atmosphere. Although its visible component is extremely weak, airglow is quite intense in the infrared region of the spectrum.

Oxygen ion, O^+, which may be produced by ultraviolet radiation acting upon oxygen atoms,

$$O + h\nu \rightarrow O^+ + e^- \qquad (16.5.7)$$

is the predominant positive ion in some regions of the ionosphere.

Ozone, O_3, has an essential protective function because it absorbs harmful ultraviolet radiation in the stratosphere and serves as a radiation shield, protecting living beings on the Earth from the effects of excessive amounts of such radiation. It is produced by a photochemical reaction,

$$O_2 + h\nu \rightarrow O + O \qquad (16.5.8)$$

(where the wavelength of the exciting radiation must be less than 242.4 nm), followed by a three-body reaction,

$$O + O_2 + M \rightarrow O_3 + M(\text{increased energy}) \qquad (16.5.9)$$

in which M is another species, such as a molecule of N_2 or O_2, which absorbs the excess energy given off by the reaction and enables the ozone molecule to stay together. The region of maximum ozone concentration is found within the range of 25–30 km high in the stratosphere where it may reach 10 ppm.

Ozone absorbs ultraviolet light very strongly in the region 220–330 nm. If this light were not absorbed by ozone, severe damage would result to exposed forms of life on the Earth. Absorption of electromagnetic radiation by ozone converts the radiation's energy to heat and is responsible for the temperature maximum encountered at the boundary between the stratosphere and the mesosphere at an altitude of approximately 50 km. The reason that the temperature maximum occurs at a higher altitude than that of the maximum ozone concentration arises from the fact that ozone is such an effective absorber of ultraviolet radiation. Therefore, most of this radiation is absorbed in the upper stratosphere, where it generates heat, and only a small fraction reaches the lower altitudes, which remain relatively cool.

Thermodynamically, the overall reaction,

$$2O_3 \rightarrow 3O_2 \qquad (16.5.10)$$

is favored so that ozone is inherently unstable. Its decomposition in the stratosphere is catalyzed by a number of natural and pollutant trace constituents, including NO, NO_2, H, HO·, HOO·, ClO, Cl, Br, and BrO. Ozone decomposition also occurs on solid surfaces, such as metal oxides and salts produced by rocket exhausts.

Ozone is an undesirable pollutant in the troposphere. It is toxic, and a mild overdose causes labored breathing, a feeling of chest pressure, cough, and irritated eyes. In addition to its toxicological effects, which are discussed in Section 21.7, ozone damages materials, such as rubber.

16.6.   REACTIONS OF ATMOSPHERIC NITROGEN

The 78% by volume of nitrogen contained in the atmosphere constitutes an inexhaustible reservoir of that essential element. The nitrogen cycle and nitrogen fixation by microorganisms were discussed in Section 11.13. A small amount of nitrogen is thought to be fixed in the atmosphere by lightning, and some is also fixed by combustion processes, as in the internal combustion engine.

Before the use of synthetic fertilizers reached its current high levels, chemists were concerned that denitrification processes (see Figure 11.13) in the soil would lead to nitrogen depletion on the Earth. Now, with millions of tons of synthetically fixed nitrogen being added to the soil each year, major concern has shifted to possible excess accumulation of nitrogen in soil, fresh water, and the oceans.

Unlike oxygen, which is almost completely dissociated to the monatomic form in higher regions of the thermosphere, molecular nitrogen is not readily dissociated by ultraviolet radiation. However, at altitudes exceeding approximately 100 km, atomic nitrogen is produced by photochemical reactions:

$$N_2 + h\nu \rightarrow N + N \qquad (16.6.1)$$

Most stratospheric ozone is probably removed by the action of nitric oxide, which reacts as follows:

$$O_3 + NO \rightarrow NO_2 + O_2 \qquad (16.6.2)$$

$$NO_2 + O \rightarrow NO + O_2 \text{ (regeneration of NO from NO}_2) \qquad (16.6.3)$$

Pollutant oxides of nitrogen, particularly NO_2, are key species involved in air pollution and the formation of photochemical smog. The most important photochemical process involving NO_2 is its facile photochemical dissociation to NO and reactive atomic oxygen:

$$NO_2 + h\nu \rightarrow NO + O \qquad (16.6.4)$$

This reaction is the most important primary photochemical process that initiates smog formation. The roles played by nitrogen oxides in smog formation and other forms of air pollution are discussed in Chapter 18.

16.7.   ATMOSPHERIC CARBON DIOXIDE

Although only about 0.035% (350 ppm) of air consists of carbon dioxide, it is the atmospheric "nonpollutant" species of most concern. As mentioned in Section 16.3, carbon dioxide, along with water vapor, is primarily responsible for the absorption of infrared energy re-emitted by the Earth so that some of this energy is reradiated back to the Earth's surface. Current evidence suggests that changes in the atmospheric carbon dioxide level will substantially alter the Earth's climate through the greenhouse effect.

Valid measurements of overall atmospheric CO_2 can only be taken in areas remote from industrial activity. Such areas include Antarctica and the top of Mauna Loa Mountain in Hawaii. Measurements of carbon dioxide levels in these locations over the last 40 years suggest an annual increase in CO_2 of about 1 ppm per year.

The most obvious factor contributing to increased atmospheric carbon dioxide is consumption of carbon-containing fossil fuels. In addition, release of CO_2 from the biodegradation of biomass and uptake by photosynthesis are important factors determining overall CO_2 levels in the atmosphere. The role of photosynthesis is illustrated by the seasonal cycle in carbon dioxide levels in the northern hemisphere. Maximum values occur in April and minimum values in late September or October. These oscillations are due to the "photosynthetic pulse," influenced most strongly by forests in middle latitudes. Forests have a much greater influence than other vegetation because in general forest trees carry out more photosynthesis than other kinds of plants, such as prairie grasses. Furthermore, forests store enough fixed, but readily oxidizable carbon in the form of wood and humus to have a marked influence on atmospheric CO_2 content. Thus, during the summer months, forests carry out enough photosynthesis to reduce the atmospheric carbon dioxide content markedly. During the winter, metabolism of biota, such as bacterial decay of humus, releases a significant amount of CO_2. Therefore, the current worldwide trend toward destruction of forests and conversion of forest lands to agricultural uses will contribute substantially to a greater overall increase in atmospheric CO_2 levels.

With current trends, it is likely that global CO_2 levels will double by the middle of the next century, which may well raise the Earth's mean surface temperature by 1.5°C to 4.5°C. Such a change might have more potential to cause massive irreversible environmental changes than any other disaster short of global nuclear war.

Chemically and photochemically, carbon dioxide is a comparatively insignificant species because of its relatively low concentrations and low photochemical reactivity. However, calculations based on known photochemical reactions, carbon dioxide levels, and ultraviolet radiation intensity indicate that photodissociation of CO_2 by solar ultraviolet radiation should occur in the upper atmosphere:

$$CO_2 + h\nu \rightarrow CO + O \qquad (16.7.1)$$

This reaction could be a major source of CO at higher altitudes. The infrared radiation absorbed by carbon dioxide is not energetic enough to cause photochemical reactions to occur.

16.8. ATMOSPHERIC WATER

The water vapor content of the troposphere is normally within a range of 1-3% by volume with a global average of about 1%. However, air can contain as little as 0.1% or as much as 5% water. The percentage of water in the atmosphere decreases rapidly with increasing altitude. Water circulates through the atmosphere in the hydrologic cycle, as shown in Figure 11.1.

Water vapor absorbs infrared radiation even more strongly than does carbon dioxide, thus greatly influencing the Earth's heat balance. Clouds formed from water vapor reflect light from the sun and have a temperature-lowering effect. On the other hand, water vapor in the atmosphere acts as a kind of "blanket" at night, retaining heat from the Earth's surface by absorption of infrared radiation.

Gaseous water in the upper atmosphere is involved in the formation of hydroxyl and hydroperoxyl radicals as discussed in Section 16.4. Condensed water vapor in the form of very small droplets is of considerable concern in atmospheric chemistry. The harmful effects of some air pollutants—for instance, the corrosion of metals by acid-forming gases—requires the presence of water which may come from the atmosphere. Atmospheric water vapor has an important influence upon pollution-induced fog formation under some circumstances. Water vapor interacting with pollutant particulate matter in the atmosphere may reduce visibility to undesirable levels through the formation of aerosol particles (see Section 16.9).

When ice particles in the atmosphere change to liquid droplets or when these droplets evaporate, heat is absorbed from the surrounding air. Reversal of these processes results in heat release to the air (as latent heat). This may occur many miles from the place where heat was absorbed and is a major mode of energy transport in the atmosphere. It is the predominant type of energy transition involved in thunderstorms, hurricanes, and tornadoes.

On a global basis, rivers drain only about one-third of the precipitation that falls on the Earth's continents. This means that two-thirds of the precipitation is lost as combined evaporation and transpiration (**evapotranspiration**, Figure 11.1). During the summer, evapotranspiration may exceed precipitation because of the large quantities of water stored in the root zone of the soil. In some cases, evapotranspiration furnishes atmospheric water vapor necessary for cloud formation and precipitation. It is probable, therefore, that large-scale deforestation, soil damage (such as by plowing up grasslands in semi-arid areas), and irrigation could have an effect on regional climate and rainfall. As noted in Section 16.2, the cold tropopause serves as a barrier to the movement of water into the stratosphere. The main source of water in the stratosphere is the photochemical oxidation of methane:

$$CH_4 + 2O_2 + h\nu \xrightarrow[\text{(several steps)}]{} CO_2 + 2H_2O \qquad (16.8.1)$$

The water thus produced serves as a source of stratospheric hydroxyl radical as shown by the following reaction:

$$H_2O + h\nu \rightarrow HO\cdot + H \qquad (16.8.2)$$

16.9.   ATMOSPHERIC PARTICLES

Particles are common significant components of the atmosphere, particularly the troposphere. Colloidal-sized particles in the atmosphere are called **aerosols**. Most aerosols from natural sources have a diameter of less than 0.1 μm. These particles originate in nature from sea sprays, smokes, dusts, and the evaporation of organic materials from vegetation. Other typical particles of natural origin in the atmosphere are bacteria, fog, pollen grains, and volcanic ash.

As shown in Figure 16.6, atmospheric particles undergo a number of processes in the atmosphere. Small colloidal particles are subject to *diffusion processes*. Smaller particles *coagulate* together to form larger particles. *Sedimentation* and *scavenging* by raindrops and other forms of precipitation are the major mechanisms for particle removal from the atmosphere. Particles also react with atmospheric gases.

Many important atmospheric phenomena involve aerosol particles, including electrification phenomena, cloud formation, and fog formation. Particles help determine the heat balance of the Earth's atmosphere by reflecting light. Probably the most important function of particles in the atmosphere is their action as nuclei for the formation of ice crystals and water droplets. Current efforts at rain-making are centered around the addition of condensing particles to atmospheres supersaturated with water vapor. Dry ice was used in early attempts; now silver iodide, which forms huge numbers of very small particles, is used.

Particles are involved in many chemical reactions in the atmosphere. Neutralization reactions, which occur most readily in solution, may take place in water

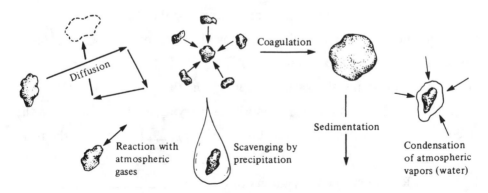

Figure 16.6.   Processes that particles undergo in the atmosphere.

droplets suspended in the atmosphere. Small particles of metal oxides and carbon have a catalytic effect on oxidation reactions. Particles may also participate in oxidation reactions induced by light.

CHAPTER SUMMARY

The chapter summary below is presented in a programmed format to review the main points covered in this chapter. It is used most effectively by filling in the blanks, referring back to the chapter as necessary. The correct answers are given at the end of the summary.

Dry air consists of (1)_____ percent nitrogen, (2)_____ percent oxygen, (3)_____ percent argon, and (4)_____ percent carbon dioxide. The three major sources of trace gases in the atmosphere are (5)_____ _____.

More than (6)_____ percent of the total mass of the atmosphere is found within approximately 30 km of the Earth's surface. This mass is approximately (7)_____ metric tons. With higher altitude the mean free path of a molecule or atom in the atmosphere becomes (8)_____.

The atmosphere is stratified on the basis of (9)_____. The lowest layer of the atmosphere is the (10)_____, characterized by (11)_____

_____, and above it is the (12)_____ containing the essential ultraviolet-absorbing (13)_____ layer.

The value of the solar constant is (14)_____. Energy transport in the atmosphere occurs by the three processes of (15)_____ _____. As well as carrying sensible heat due to the kinetic energy of molecules, convection carries (16)_____ heat in the form of (17)_____

_____. Radiation of energy occurs through (18)_____, which can travel through a vacuum in space.

The average surface temperature of Earth is maintained at a relatively comfortable 15°C because of (19)_____,
in which water vapor and, to a lesser extent (20)_____,
reabsorb much of the outgoing radiation.

The science of atmospheric phenomena, encompassing the study of the movement of air masses as well as physical forces in the atmosphere such as heat, wind, and transitions of water, is called (21)_____. The two things required for condensation of water vapor from air are (22)_____

_____. Stagnant air masses leading to air pollution incidents, such as formation of photochemical smog, are produced by (23)_____.

In order for photochemical reactions to occur, light must be (24)_____
_____ which results in production of (25)_____
_____. In addition to this highly reactive species, two other kinds of active species formed when light is absorbed are (26)_____
_____. When an electron excited to a higher orbital by absorption of light retains a spin opposite to that of its former partner, it is said to be in (27)_____, whereas if it has the same spin as its former partner, it is in (28)_____. Some of the ways by which an excited species loses excess energy are (29)_____

_____

_____.

A layer of ions in the atmosphere produced primarily by the action of (30)____
_____ is known as the (31)_____.
Free radicals are highly reactive species that contain (32)_____.
Radicals can take part in (33)_____ in which one of the products of each reaction is a radical.

The single most important reactive intermediate species in atmospheric chemi-

cal processes is (34)_____. The species $H_3COO\cdot$ is the
(35)_____ and $HOO\cdot$ is the (36)_____.

Atmospheric molecular oxygen, O_2, is consumed by (37)_____
and is produced by (38)_____.
A crucially important species formed by oxygen in the stratosphere is
(39)_____.

Molecular nitrogen is (40)_____ dissociated by ultraviolet
radiation. As air pollutants, nitrogen oxides play an important role in the formation of (41)_____.

Atmospheric carbon dioxide levels are increasing at a rate of about
(42)_____ per year, of which the most obvious
contributor is (43)_____. In
addition, carbon dioxide is released biologically by (44)_____
and taken up from the atmosphere by (45)_____.

The water vapor content of the troposphere is normally within a range of
(46)_____ percent by volume with a global average of about
(47)_____ percent. Water vapor (48)_____ infrared radiation, but clouds
formed from it (49)_____ light from the sun. Energy taken up
and released when water evaporates and condenses in the atmosphere is particu-
lary important in (50)_____

_____.

Colloidal-sized particles in the atmosphere are called (51)_____.
Atmospheric particles undergo a number of processes including (52)_____

_____

_____. Among the important
atmospheric phenomena involving atmospheric aerosol particles are (53)_____
_____. Among the
chemical reactions involving atmospheric particles are (54)_____

_____

_____.

Answers

1. 78.08
2. 20.95
3. 0.934
4. 0.035
5. biogenic, photochemical, and anthropogenic
6. 99
7. 5.14×10^{15}
8. longer
9. temperature/density relationships
10. troposphere
11. a generally homogeneous composition of major gases other than water and decreasing temperature with increasing altitude
12. stratosphere
13. ozone
14. 1.34×10^3 watts per square meter
15. conduction, convection, and radiation
16. latent
17. water vapor which releases heat as it condenses
18. electromagnetic radiation
19. an atmospheric "greenhouse effect"
20. carbon dioxide
21. meteorology
22. air must be cooled below the dew point, and nuclei of condensation must be present
23. temperature inversions
24. absorbed
25. an electronically excited molecule
26. free radicals and ions
27. an excited singlet state
28. an exicted triplet state
29. transfer of energy to other molecules, transfer of energy within the absorbing molecule, luminescence processes, dissociation, direct reaction, and photoionization
30. ultraviolet radiation
31. ionosphere
32. unpaired electrons
33. chain reactions
34. the hydroxyl radical, HO·
35. methylperoxyl radical
36. hydroperoxyl radical
37. burning of fossil fuels and respiration of organisms
38. photosynthesis

39. ozone, O_3
40. not readily
41. photochemical smog
42. 1 part per million
43. combustion of carbon-containing fossil fuels
44. biodegradation processes
45. photosynthesis
46. 1–3
47. 1
48. absorbs
49. reflect
50. weather phenomena involving energy
51. aerosols
52. diffusion, coagulation, sedimentation, scavenging, and reaction with atmospheric gases
53. electrification phenomena, cloud formation, and fog formation
54. neutralization reactions in water droplets and catalysis of oxidation reactions by particles of metal oxides and carbon

QUESTIONS AND PROBLEMS

1. What phenomenon is responsible for the temperature maximum at the boundary of the stratosphere and the mesosphere?

2. What function does a third body serve in an atmospheric chemical reaction?

3. Why does the lower boundary of the ionosphere lift at night?

4. Why might it be expected that the reaction of a free radical with NO_2 is a chain-terminating reaction (consider the total number of electrons in NO_2).

5. Suppose that 22.4 liters of air at STP is used to burn 1.50 g of carbon to form CO_2, and that the gaseous product is adjusted to STP. What is the volume and the average molecular mass of the resulting mixture?

6. Measured in μm, what are the lower wavelength limits of solar radiation reaching the Earth; the wavelength at which maximum solar radiation reaches the earth; and the wavelength at which maximum energy is radiated back into space?

7. Of the species O, $HO\cdot$, NO_2^*, $H_3C\cdot$, and N^+, which could most readily revert to a nonreactive, "normal" species in total isolation?

8. Of the gases neon, sulfur dioxide, helium, oxygen, and nitrogen, which shows the most variation in its atmospheric concentration?

9. A 12.0-liter sample of air at 25°C and 1.00 atm pressure was collected and dried. After drying, the volume of the sample was exactly 11.50 L. What was the percentage *by mass* of water in the original air sample?

10. The sunlight incident upon a 1 square meter area perpendicular to the line of transmission of the solar flux just above the Earth's atmosphere provides energy at a rate most closely equivalent to: (a) that required to power a pocket calculator, (b) that required to provide a moderate level of lighting for a 40-person capacity classroom illuminated with fluorescent lights, (c) that required to propel a 2500 pound automobile at 55 mph, (d) that required to power a 100-watt incandescent light bulb, (e) that required to heat a 40-person classroom to 70°F when the outside temperature is –10°F.

11. At an altitude of 50 km, the average atmospheric temperature is essentially 0°C. Given the information in Figure 16.1, estimate the average number of air molecules per cubic centimeter of air at this altitude.

12. What two types of condensation nuclei originate with bursting sea-foam bubbles?

13. State two factors that make the stratosphere particularly important in terms of acting as a region where atmospheric trace contaminants are converted to other, chemically less reactive, forms.

14. What two chemical species are most generally responsible for the removal of hydroxyl radical from the unpolluted troposphere?

15. What is the distinction between the symbols * and · in discussing chemically active species in the atmosphere?

16. Given the total mass of Earth's atmosphere and the percentage by volume that is carbon dioxide, attempt to calculate the mass of CO_2 in the atmosphere. This exercise will require a review of the gas laws from Chapter 2. It may also be useful to calculate the average molecular mass of air assuming that it is composed only of oxygen and nitrogen.

17. After completing the calculation above, calculate the mass of pure carbon that would have to be burned to double the carbon dioxide content of the atmosphere.

18. From Figure 16.1 estimate the change in altitude required for atmospheric pressure to be reduced in half. What increase in altitude is required for a 10-fold reduction in pressure?

19. Discuss in which sense the troposphere is homogenous. In regard to which important species is it **not** homogeneous? Why are regions of the atmosphere above the troposphere relatively less homogeneous?

20. What vital protective function is served by the stratosphere? Illustrate the answer with chemical and photochemical reactions.

21. What vital protective function is served by the tropopause?

22. What is the area in square kilometers that receives solar energy undiminished by atmospheric absorption equivalent to the output of a 1,000 megawatt power plant?

23. Distinguish among the atmospheric energy transport processes of conduction, convection, and radiation.

24. Distinguish between sensible and latent heat. Why is the latter particularly important in atmospheric energy exchange processes?

25. What areas are covered by meteorology? How are meteorologic phenomena involved in air pollution phenomena? In this respect, why are temperature inversions particularly important?

26. Explain why the study of atmospheric chemical reactions is particularly difficult.

27. Explain what causes photochemical reactions and how they can lead to chain reactions.

28. Distinguish between the two hypothetical chemical species, X^* and $X\cdot$.

29. What does $h\nu$ represent in photochemical terms? How may it lead to the occurrence of chemical reactions?

30. Distinguish between excited singlet states and excited triplet states. In what sense are they both "excited"?

31. Define the photochemical phenomena represented by each of the following:

    (a) $O_2^* \rightarrow O + O$

    (b) $O_2^* + O_3 \rightarrow 2O_2 + O$

    (c) $N_2^* \rightarrow N_2^+ + e^-$

32. Why are free radicals so highly reactive? Despite their high reactivity, why do free radicals tend to persist for significant lengths of time at high altitudes?

33. What is the kind of reaction below called? Explain.

$$H_3C\cdot \ + \ H_3C\cdot \ \rightarrow \ C_2H_6$$

34. What is the "single most important reactive intermediate species in atmospheric chemical processes"? Explain.

35. Show the reactions by which stratospheric ozone is formed. Why is it not so formed at lower altitudes?

36. Justify or refute the statement that, "Although elemental nitrogen is the most abundant species in the atmosphere, it is one of the least significant atmospheric constituents in atmospheric chemical phenomena."

37. Why is it true that chemically and photochemically, carbon dioxide is a comparatively insignificant species? If this is true, however, why is an appropriate level of atmospheric carbon dioxide important for life on Earth?

38. Give possible examples of neutralization reactions and oxidation reactions catalyzed by particles in the atmosphere. What kinds of particles would be required for each of these types of reactions?

SUPPLEMENTARY REFERENCES

Seinfeld, J. H., *Atmospheric Chemistry and Physics of Air Pollution*, (New York: John Wiley and Sons, Inc., 1986).

Warneck, P. *Chemistry of the Natural Atmosphere*, (San Diego, CA: Academic Press, 1988).

Regens, J. L., and R. W. Rycroft, *The Acid Rain Controversy*, (Pittsburgh, PA: University of Pittsburgh Press, 1988).

Ferraudi, G. J., *Elements of Inorganic Photochemistry*, (New York, NY: John Wiley and Sons, 1988).

Lodge, J. P. Jr., Ed., *Methods of Air Sampling and Analysis*, 3rd ed., (Chelsea, MI: Lewis Publishers, Inc., 1989).

Brown, L. R., *State of the World 1990*, (Washington, DC: Worldwatch Institute, 1989).

Botkin, D. B., Ed., *Changing the Global Environment: Perspectives on Human Involvement*, (San Diego, CA: Academic Press, 1989).

Jorgensen, S. E., and I. Johnsen, *Principles of Environmental Science and Technology*, 2nd ed., (Amsterdam, The Netherlands: Elsevier Science Publishers, 1989).

Miller, E. W., and R. M. Miller, *Environmental Hazards: Air Pollution,* (Santa Barbara, CA: ABC-CLIO, 1989).

Knap, A. H., and M.-S. Kaiser, Eds., *The Long-Range Atmospheric Transport of Natural and Contaminant Substances*, (Norwell, MA: Kluwer Academic Publishers, 1989).

Bryce-Smith, D., and A. Gilbert, *Photochemistry*, Vol. 20, (Letchworth, England: Royal Society of Chemistry, 1989).

17   INORGANIC AIR POLLUTANTS

17.1.   INTRODUCTION

This chapter addresses inorganic air pollutants of various kinds. Organic air pollutants and the photochemical smog formed from them are discussed in Chapter 18.

Inorganic air pollutants consist of many kinds of substances. The first to be addressed here are particulate pollutants. Many solid and liquid substances may become particulate air contaminants. Another important class of inorganic air pollutants consists of oxides of carbon, sulfur, and nitrogen. Carbon monoxide is a directly toxic material that is fatal at relatively small doses. Carbon dioxide is a natural and essential constituent of the atmosphere, and it is required for plants to use during photosynthesis. However, CO_2 may turn out to be the most deadly air pollutant of all because of its potential as a greenhouse gas that might cause devastating global warming. Oxides of sulfur and nitrogen are acid-forming gases that can cause acid precipitation. Several other inorganic air pollutants, such as ammonia, hydrogen chloride, and hydrogen sulfide are also discussed in this chapter.

A number of gaseous inorganic pollutants enter the atmosphere as the result of human activities. Those added in the greatest quantities are CO, SO_2, NO, and NO_2. (These quantities are relatively small compared to the amount of CO_2 in the atmosphere. The possible environmental effects of increased atmospheric CO_2 levels are discussed later in this chapter.) Other inorganic pollutant gases include NH_3, N_2O, N_2O_5, H_2S, Cl_2, HCl, and HF. Substantial quantities of some of these gases are added to the atmosphere each year by human activities. Globally, atmospheric emissions of carbon monoxide, sulfur oxides, and nitrogen are of the order of one to several hundred million tons per year.

As with most aspects of chemistry in the real world, it is somewhat artificial and arbitrary to divide air pollutants between the inorganic and organic realms. For example, inorganic NO_2 undergoes photodissociation to start the processes that convert organic vapors to aldehydes, oxidants including inorganic O_3, and other substances characteristic of photochemical smog. Oxidants generated in such smog convert inorganic SO_2 to much more acidic sulfuric acid, the major contributor to acid precipitation. Numerous other examples could be cited to illustrate the interrelationships among air pollutants of various kinds.

17.2.   PARTICLES IN THE ATMOSPHERE

Particles in the atmosphere, which range in size from about one-half millimeter (the size of sand or drizzle) down to molecular dimensions, are made up of an amazing variety of materials and discrete objects that may consist of either solids or liquid droplets. **Particulates** is a term that has come to stand for particles in the atmosphere, although *particulate matter* or simply *particles*, is preferred usage. Particulate matter makes up the most visible and obvious form of air pollution. Atmospheric **aerosols** are solid or liquid particles smaller than 100 μm in diameter. Pollutant particles in the 0.001 to 10 μm range are commonly suspended in the air near sources of pollution, such as the urban atmosphere, industrial plants, highways, and power plants. Some of the terms commonly used to describe atmospheric particles are summarized in Table 17.1.

Very small, solid particles include carbon black, silver iodide, combustion nuclei, and sea-salt nuclei formed by the loss of water from droplets of seawater. Larger particles include cement dust, wind-blown soil dust, foundry dust, and pulverized coal. Liquid particulate matter, **mist**, includes raindrops, fog, and sulfuric acid mist. Some particles are of biological origin, such as viruses, bacteria, bacterial spores, fungal spores, and pollen. Particulate matter may be organic or inorganic; both types are very important atmospheric contaminants.

Chemical Processes for Inorganic Particle Formation

Metal oxides constitute a major class of inorganic particles in the atmosphere. These are formed whenever fuels containing metals are burned. For example, particulate iron oxide is formed during the combustion of pyrite-containing coal:

Table 17.1.   Important Terms Describing Atmospheric Particles

Term	Meaning
Aerosol	Colloidal-sized atmospheric particle
Condensation aerosol	Formed by condensation of vapors or reactions of gases
Dispersion aerosol	Formed by grinding of solids, atomization of liquids, or dispersion of dusts
Fog	Term denoting high level of water droplets
Haze	Denotes decreased visibility due to the presence of particles
Mists	Liquid particles
Smoke	Particles formed by incomplete combustion of fuel

$$3FeS_2 + 8O_2 \rightarrow Fe_3O_4 + 6SO_2 \tag{17.2.1}$$

Organic vanadium in residual fuel oil is converted to particulate vanadium oxide. Part of the calcium carbonate in the ash fraction of coal is converted to calcium oxide and is emitted to the atmosphere through the stack:

$$CaCO_3 + heat \rightarrow CaO + CO_2 \tag{17.2.2}$$

A common process for the formation of aerosol mists involves the oxidation of atmospheric sulfur dioxide to sulfuric acid, a hygroscopic substance that accumulates atmospheric water to form small liquid droplets:

$$2SO_2 + O_2 + 2H_2O \rightarrow 2H_2SO_4 \tag{17.2.3}$$

In the presence of basic air pollutants, such as ammonia or calcium oxide, the sulfuric acid reacts to form salts:

$$H_2SO_4(droplet) + 2NH_3(g) \rightarrow (NH_4)_2SO_4 \text{ (droplet)} \tag{17.2.4}$$

$$H_2SO_4(droplet) + CaO(s) \rightarrow CaSO_4(droplet) + H_2O \tag{17.2.5}$$

Under low-humidity conditions water is lost from these droplets and a solid aerosol is formed.

The preceding examples show several ways in which solid or liquid inorganic aerosols are formed by chemical reactions. Such reactions constitute an important general process for the formation of aerosols, particularly the smaller particles.

17.3.   THE COMPOSITION OF INORGANIC PARTICLES

Figure 17.1 illustrates the basic factors responsible for the composition of inorganic particulate matter. In general, the proportions of elements in atmospheric particulate matter reflect relative abundances of elements in the parent material. The source of particulate matter is reflected in its elemental composition, taking into consideration chemical reactions that may change the composition. For example, particulate matter largely from ocean spray origin in a coastal area receiving sulfur dioxide pollution may show anomalously high sulfate and corresponding low chloride content. The sulfate comes from atmospheric oxidation of sulfur dioxide to form nonvolatile ionic sulfate, whereas some chloride originally from the NaCl in the seawater may be lost from the solid aerosol as volatile HCl:

$$2SO_2 + O_2 + 2H_2O \rightarrow 2H_2SO_4 \tag{17.3.1}$$

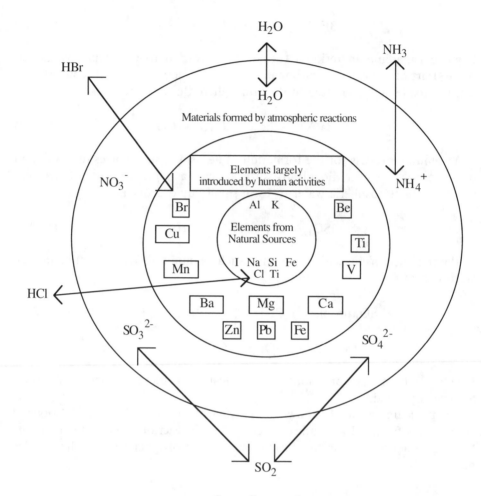

Figure 17.1. Some of the components of inorganic particulate matter and their origins.

$$H_2SO_4 + 2NaCl(particulate) \rightarrow Na_2SO_4(particulate) + 2HCl \quad (17.3.2)$$

The chemical composition of atmospheric particulate matter is quite diverse. Among the constituents of inorganic particulate matter found in polluted atmospheres are salts, oxides, nitrogen compounds, sulfur compounds, various metals, and radionuclides. In coastal areas, sodium and chlorine get into atmospheric particles as sodium chloride from sea spray. The major trace elements that typically occur at levels above 1 $\mu g/m^3$ in particulate matter are aluminum, calcium, carbon, iron, potassium, sodium, and silicon; note that most of these tend to originate from terrestrial sources. Lesser quantities of copper, lead, titanium, and zinc and even lower levels of antimony, beryllium, bismuth, cadmium, cobalt, chromium, cesium, lithium, manganese, nickel, rubidium, selenium, strontium, and vanadium are commonly observed. The likely sources of some of these elements are given below:

- **Al, Fe, Ca, Si**: Soil erosion, rock dust, coal combustion
- **C**: Incomplete combustion of carbonaceous fuels
- **Na, Cl**: Marine aerosols, chloride from incineration of organohalide polymer wastes
- **Sb, Se**: Very volatile elements, possibly from the combustion of oil, coal, or refuse
- **V**: Combustion of residual petroleum (present at very high levels in residues from Venezuelan crude oil)
- **Zn**: Tends to occur in small particles, probably from combustion
- **Pb**: Combustion of leaded fuels and wastes containing lead

Particulate carbon as soot, carbon black, coke, and graphite originates from auto and truck exhausts, heating furnaces, incinerators, power plants, and steel and foundry operations and composes one of the more visible and troublesome particulate air pollutants. Because of its good adsorbent properties, carbon can be a carrier of gaseous and other particulate pollutants. Particulate carbon surfaces may catalyze some heterogeneous atmospheric reactions, including the important conversion of SO_2 to sulfate.

Fly Ash

Much of the mineral particulate matter in a polluted atmosphere is in the form of oxides and other compounds produced during the combustion of high-ash fossil fuel. Much of the mineral matter in fossil fuels such as coal or lignite is converted during combustion to a fused, glassy bottom ash which presents no air pollution problems. Smaller particles of **fly ash** enter furnace flues and are efficiently collected in a properly equipped stack system. However, some fly ash escapes through the stack and enters the atmosphere. Unfortunately, the fly ash thus released tends to consist of smaller particles that do the most damage to human health, plants, and visibility.

The composition of fly ash varies widely, depending upon the source of fuel. The predominant constituents are oxides of aluminum, calcium, iron, and silicon. Other elements that occur in fly ash are magnesium, sulfur, titanium, phosphorus, potassium, and sodium. Elemental carbon (soot, carbon black) is a significant fly ash constituent.

Asbestos

Asbestos is the name given to a group of fibrous silicate minerals, typically those of the serpentine group, for which the approximate formula is $Mg_3P(Si_2O_5)(OH)_4$. The tensile strength, flexibility, and nonflammability of asbestos have led to many uses in the past including structural materials, brake linings, insulation, and pipe manufacture.

Asbestos is of concern as an air pollutant because when inhaled it may cause asbestosis (a pneumonia condition), mesothelioma (tumor of the mesothelial tissue lining the chest cavity adjacent to the lungs), and bronchogenic carcinoma (cancer originating with the air passages in the lungs). Therefore, uses of asbestos have been severely curtailed and widespread programs have been undertaken to remove the material from buildings.

Toxic Metals

Some of the metals found predominantly as particulate matter in polluted atmospheres are known to be hazardous to human health. All of these except beryllium are so-called "heavy metals." Lead is the toxic metal of greatest concern in the urban atmosphere because it comes closest to being present at a toxic level; mercury ranks second. Others include beryllium, cadmium, chromium, vanadium, nickel, and arsenic (a metalloid).

Atmospheric mercury is of concern because of its toxicity, volatility, and mobility. Some atmospheric mercury is associated with particulate matter. Much of the mercury entering the atmosphere does so as volatile elemental mercury from coal combustion and volcanoes. Volatile organomercury compounds such as dimethylmercury, $(CH_3)_2Hg$, and monomethylmercury salts, such as $(CH_3)HgBr$, are also encountered in the atmosphere.

With the reduction of leaded fuels, atmospheric lead is of less concern than it used to be. However, during the decades that leaded gasoline containing tetraethyllead was the predominant automotive fuel, particulate lead halides were emitted in large quantities. Lead halides are emitted from engines burning leaded gasoline through the action of dichloroethane and dibromoethane added as halogenated scavengers to form volatile lead chloride, lead bromide and lead chlorobromide, thereby preventing the accumulation of lead oxides inside engines.

Beryllium is used for the formulation of specialty alloys employed in electrical equipment, electronic instrumentation, space gear, and nuclear reactor components, so that distribution of beryllium is by no means comparable to that of other toxic metals such as lead or mercury. However, because of its "high tech" applications, consumption of beryllium may increase in the future. Because of its high toxicity beryllium has the lowest allowable limit in the atmosphere of all the elements. One of the main results of the recognition of beryllium toxicity hazards was the elimination of this element from phosphors (coatings which produce visible light from ultraviolet light) in fluorescent lamps.

Radioactive Particles

A significant natural source of radionuclides in the atmosphere is **radon**, a noble gas product of radium decay. Radon may enter the atmosphere as either of two isotopes, ^{222}Rn (half-life 3.8 days) and ^{220}Rn (half-life 54.5 seconds). (As discussed in Section 3.4, the superscript numbers on the preceding symbols denote mass numbers, the sum of protons and neutrons in the isotope nucleus.) Both ^{220}Rn and ^{222}Rn are alpha emitters in decay chains that terminate with stable isotopes of lead. The initial decay products, ^{218}Po and ^{216}Po are nongaseous and adhere readily to atmospheric particulate matter. Therefore, some of the radio-activity detected in these particles is of natural origin. Furthermore, cosmic rays act on nuclei in the atmosphere to produce other radionuclides, including ^{7}Be, ^{10}Be, ^{14}C, ^{39}Cl, ^{3}H, ^{22}Na, ^{32}P, and ^{33}P.

The combustion of fossil fuels introduces radioactivity into the atmosphere in the form of radionuclides contained in fly ash. Large coal-fired power plants lacking ash-control equipment may introduce up to several hundred millicuries of radionuclides into the atmosphere each year, far more than either an equivalent nuclear or oil-fired power plant.

The aboveground detonation of nuclear weapons can add large amounts of radioactive particulate matter to the atmosphere. Among the radioisotopes that can be detected in rainfall falling after atmospheric nuclear weapon detonation are ^{91}Y, ^{141}Ce, ^{144}Ce, ^{147}Nd, ^{147}Pm, ^{149}Pm, ^{151}Sm, ^{153}Sm, ^{155}Eu, ^{156}Eu, ^{89}Sr, ^{90}Sr, ^{115m}Cd, ^{129m}Te, ^{131}I, ^{132}Te, and ^{140}Ba. (Note that "^{m}" denotes a metastable state that decays by gamma-ray emission to an isotope of the same element.)

17.4. EFFECTS OF PARTICLES

Atmospheric particles have numerous effects. The most obvious of these is reduction and distortion of visibility. They provide active surfaces upon which heterogeneous atmospheric chemical reactions can occur and nucleation bodies for the condensation of atmospheric water vapor, thereby exerting a significant influence upon weather and air pollution phenomena.

The most visible influence that aerosol particles have upon air quality results from their optical effects. Particles smaller than about 0.1 μm in diameter scatter light much like molecules; that is, Rayleigh scattering. Generally, such particles have an insignificant effect upon visibility in the atmosphere. The light-scattering and intercepting properties of particles larger than 1 μm are approximately proportional to the particle's cross-sectional area. Particles of 0.1 μm–1 μm cause interference phenomena because they are about the same dimensions as the wavelengths of visible light, so their light-scattering properties are especially significant.

Atmospheric particles inhaled through the respiratory tract may damage health. Relatively large particles are likely to be retained in the nasal cavity and in the pharynx, whereas very small particles are likely to reach the lungs and be retained by them. The respiratory system possesses mechanisms for the expulsion of inhaled particles. In the ciliated region of the respiratory system, particles are carried as far as the entrance to the gastrointestinal tract by a flow of mucus. Macrophages in the nonciliated pulmonary regions carry particles to the ciliated region.

The respiratory system may be damaged directly by particulate matter that enters the blood system or lymph system through the lungs. In addition, the soluble components of the particulate material may be transported to organs some distance from the lungs and have a detrimental effect on these organs. Particles cleared from the respiratory tract are to a large extent swallowed into the gastrointestinal tract.

A strong correlation has been found between increases in the daily mortality rate and acute episodes of air pollution. In such cases, high levels of particulate matter are accompanied by elevated concentrations of SO_2 and other pollutants, so that any conclusions must be drawn with caution.

17.5.   CONTROL OF PARTICULATE EMISSIONS

The removal of particulate matter from gas streams is the most widely practiced means of air pollution control. A number of devices have been developed for this purpose, which differ widely in effectiveness, complexity, and cost. The selection of a particle removal system for a gaseous waste stream depends upon the particle loading, nature of particles (size distribution), and type of gas scrubbing system used.

Particle Removal by Sedimentation and Inertia

The simplest means of particulate matter removal is **sedimentation**, a phenomenon that occurs continuously in nature. Gravitational settling chambers may be employed for the removal of particles from gas streams by simply settling under the influence of gravity. These chambers take up large amounts of space and have low collection efficiencies, particularly for small particles.

Gravitational settling of particles is enhanced by increased particle size which occurs spontaneously by coagulation. Thus, over time, the size of particles increases and the number of particles decreases in a mass of air that contains particles. Brownian motion of particles less than about 0.1 μm in size is primarily responsible for their contact, enabling coagulation to occur. Particles greater than about 0.3 μm in radius do not diffuse appreciably and serve primarily as receptors of smaller particles.

Inertial mechanisms are effective for particle removal. These depend upon the fact that the radius of the path of a particle in a rapidly moving, curving air stream is larger than the path of the stream as a whole. Therefore, when a gas stream is spun by vanes, a fan, or a tangential gas inlet, the particulate matter may be collected on a separator wall because the particles are forced outward by centrifugal force. Devices utilizing this mode of operation are called **dry centrifugal collectors**.

Particle filtration

Fabric filters, as their name implies, consist of fabrics that allow the passage of gas but retain particulate matter. These are used to collect dust in bags contained in structures called *baghouses*. Periodically, the fabric composing the filter is shaken to remove the particles and to reduce back-pressure to acceptable levels. Typically, the bag is in a tubular configuration as shown in Figure 17.2. Numerous other configurations are possible. Collected particulate matter is removed from bags by mechanical agitation, blowing air on the fabric, or rapid expansion and contraction of the bags.

Although simple, baghouses are generally effective in removing particles from exhaust gas. Particles as small as 0.01 μm in diameter are removed, and removal efficiency is relatively high for particles down to 0.5 μm in diameter.

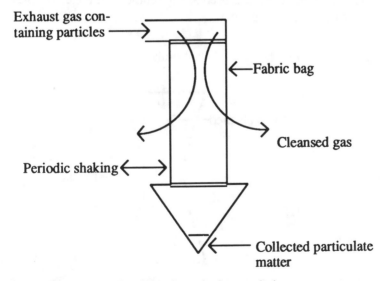

Figure 17.2.   Baghouse collection of particulate emissions.

Scrubbers

A venturi scrubber passes gas through a converging section, throat, and diverging section, as shown in Figure 17.3. Injection of the scrubbing liquid at right angles to incoming gas breaks the liquid into very small droplets, which are ideal for scavenging particles from the gas stream. In the reduced-pressure (expanding) region of the venturi, some condensation can occur, adding to the scrubbing efficiency. In addition to removing particles, venturis may serve as quenchers to cool exhaust gas and as scrubbers for pollutant gases.

Electrostatic Removal

Aerosol particles may acquire electrical charges. In an electric field, such particles are subjected to a force, F (dynes) given by

$$F = Eq \qquad (17.5.1)$$

where E is the voltage gradient (statvolt/cm) and q is the electrostatic charge charge on the particle (in esu). This phenomenon has been widely used in highly efficient **electrostatic precipitators**, as shown in Figure 17.4. The particles acquire a charge when the gas stream is passed through a high-voltage, direct-current corona. Because of the charge, the particles are attracted to a grounded surface from which they may be later removed. Ozone may be produced by the corona discharge. Similar devices used as household dust collectors may produce toxic ozone if not operated properly.

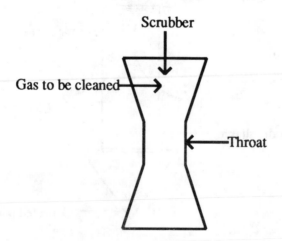

Figure 17.3.   Venturi scrubber.

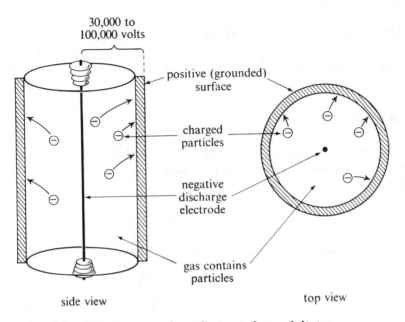

30,000 to
100,000 volts

positive (grounded)
surface

charged
particles

negative
discharge
electrode

gas contains
particles

side view

top view

Figure 17.4. Schematic diagram of an electrostatic precipitator

17.6. CARBON OXIDES

Carbon Monoxide

Carbon monoxide, CO, causes problems in cases of locally high concentrations. The toxicity of carbon monoxide is discussed in Section 21.8. The overall atmospheric concentration of carbon monoxide is about 0.1 ppm. Much of this CO is present as an intermediate in the oxidation of methane by hydroxyl radical.

Because of carbon monoxide emissions from internal combustion engines, highest levels of this toxic gas tend to occur in congested urban areas at times when the maximum number of people are exposed, such as during rush hours. At such times, carbon monoxide levels in the atmosphere may become as high as 50–100 ppm.

Control of Carbon Monoxide Emissions

Since the internal combustion engine is the primary source of localized pollutant carbon monoxide emissions, control measures have been concentrated on the automobile. Carbon monoxide emissions may be lowered by employing a leaner air-fuel mixture, that is, one in which the weight ratio of air to fuel is relatively high. At air-fuel (weight:weight) ratios exceeding approximately 16:1, an internal combustion engine emits virtually no carbon monoxide.

Modern automobiles use catalytic exhaust reactors to cut down on carbon monoxide emissions. Excess air is pumped into the exhaust gas, and the mixture is passed through a catalytic converter in the exhaust system, resulting in oxidation of CO to CO_2.

Fate of Atmospheric CO

The residence time of carbon monoxide in the atmosphere is of the order of 4 months. It is generally agreed that carbon monoxide is removed from the atmosphere by reaction with hydroxyl radical, $HO\cdot$:

$$CO + HO\cdot \rightarrow CO_2 + H \qquad (17.6.1)$$

The reaction produces hydroperoxyl radical as a product:

$$(17.6.2)$$
$$O_2 + H + M \rightarrow HOO\cdot + M \text{ (M is an energy-absorbing third body, usually a molecule of } O_2 \text{ or } N_2)$$

$HO\cdot$ is regenerated from $HOO\cdot$ by the following reactions:

$$HOO\cdot + NO \rightarrow HO\cdot + NO_2 \qquad (17.6.3)$$

$$HOO\cdot + HOO\cdot \rightarrow H_2O_2 + O_2 \qquad (17.6.4)$$

The latter reaction is followed by photochemical dissociation of H_2O_2 to regenerate $HO\cdot$:

$$H_2O_2 + h\nu \rightarrow 2HO\cdot \qquad (17.6.5)$$

Methane is also involved through the atmospheric CO-$HO\cdot$-CH_4 cycle.

Soil microorganisms act to remove CO from the atmosphere. Therefore, soil is a sink for carbon monoxide.

Carbon Dioxide and Global Warming

Carbon dioxide and other infrared-absorbing trace gases in the atmosphere contribute to global warming—the "greenhouse effect"—by allowing incoming solar radiant energy to penetrate to the Earth's surface while reabsorbing infrared radiation emanating from it. Levels of these "greenhouse gases" have increased at a rapid rate during recent decades and are continuing to do so. Concern over this phenomenon has intensified since about 1980. This is because ever since accurate temperature records have been kept, the 1980s have been the warmest 10-year period recorded. On an annual basis, 1988 was the warmest year ever recorded, 1987 was second and 1981 third. To a degree, perhaps masked in part by the cooling effect of at least one major volcanic eruption, the global trend seemed to be continuing in the 1990s.

There are many uncertainties surrounding the issue of greenhouse warming. However, several things about the phenomenon are certain. It is known that CO_2 and other greenhouse gases, such as CH_4, absorb infrared radiation by which Earth loses heat. The levels of these gases have increased markedly since about 1850 as nations have become industrialized and as forest lands and grasslands have been converted to agriculture. Chlorofluorocarbons, which also are greenhouse gases, were not even introduced into the atmosphere until the 1930s. Although trends in levels of these gases are well known, their effects on global temperature and climate are much less certain. The phenomenon has been the subject of much computer modeling. Most models predict global warming of 1.5 to 5°C, about as much again as has occurred since the last ice age. Such warming would have profound effects on rainfall, plant growth, and sea levels, which might rise as much as 0.5 to 1.5 meters.

Carbon dioxide is the gas most commonly thought of as a greenhouse gas; it is responsible for about half of the atmospheric heat retained by trace gases. It is produced primarily by burning of fossil fuels and deforestation accompanied by burning and biodegradation of biomass. On a molecule-for-molecule basis, methane, CH_4, is 20–30 times more effective in trapping heat than is CO_2. Other trace gases that contribute are chlorofluorocarbons and N_2O.

Analyses of gases trapped in polar ice samples indicate that pre-industrial levels of CO_2 and CH_4 in the atmosphere were approximately 260 parts per million and 0.70 ppm, respectively. Over the last 300 years these levels have increased to current values of around 350 ppm, and 1.7 ppm, respectively; most of the increase by far has taken place at an accelerating pace over the last 100 years. (A note of interest is the observation based upon analyses of gases trapped in ice cores that the atmospheric level of CO_2 at the peak of the last ice age about 18,000 years past was 25 percent *below* preindustrial levels.) About half of the increase in carbon dioxide in the last 300 years can be attributed to deforestation, which still accounts for approximately 20 percent of the annual increase in this gas. Carbon dioxide is increasing by about 1 ppm per year. Methane is going up at a rate of almost 0.02 ppm/year. The comparatively very rapid increase in methane levels is attributed to a number of factors resulting from human activities. Among these are direct leakage of natural gas, by-product emissions from coal mining and petroleum recovery, and release from the burning of savannas and tropical forests. Biogenic sources resulting from human activities produce large amounts of atmospheric methane. These include methane from bacteria degrading organic matter, such as municipal refuse in landfills; methane evolved from anaerobic biodegradation of organic matter in rice paddies; and methane emitted as the result of bacterial action in the digestive tracts of ruminant animals.

Both positive and negative feedback mechanisms may be involved in determining the rates at which carbon dioxide and methane build up in the atmosphere. Laboratory studies indicate that increased CO_2 levels in the atmosphere cause accelerated uptake of this gas by plants undergoing photosynthesis, which tends to slow buildup of atmospheric CO_2. Given adequate rainfall, plants living in a warmer climate that would result from the greenhouse effect would grow faster and take up more CO_2. This could be an especially significant effect of forests, which have a high CO_2-fixing ability. However, the projected rate of increase in carbon dioxide levels is so rapid that forests would lag behind in their ability to fix additional CO_2. Similarly, higher atmospheric CO_2 concentrations will result in accelerated sorption of the gas by oceans. The amount of dissolved CO_2 in the oceans is about 60 times the amount of CO_2 gas in the atmosphere. However, the times for transfer of carbon dioxide from the atmosphere to the ocean are of the order of years. Because of low mixing rates, the times for transfer of oxygen from the upper approximately 100-meter layer of the oceans to ocean depths is much longer, of the order of decades. Therefore, like the uptake of CO_2 by forests, increased absorption by oceans will lag behind the emissions of CO_2. Severe drought conditions resulting from climatic warming could cut down substantially on CO_2 uptake by plants. Warmer conditions would accelerate release of both CO_2 and CH_4 by microbial degradation of organic matter. (It is important to realize that about twice as much carbon is held in soil in dead organic matter — necrocarbon — potentially degradable to CO_2 and CH_4 as is present in the atmosphere.) Global warming might speed up the rates at which biodegradation adds these gases to the atmosphere.

It is certain that atmospheric CO_2 levels will continue to increase significantly. The degree to which this occurs depends upon future levels of CO_2 production and the fraction of that production that remains in the atmosphere. Given plausible projections of CO_2 production and a reasonable estimate that half of that amount will remain in the atmosphere, projections can be made that indicate that sometime during the middle part of the next century the concentration of this gas will reach 600 ppm in the atmosphere. This is well over twice the levels estimated for pre-industrial times. Much less certain are the effects that this change will have on climate. It is virtually impossible for the elaborate computer models used to estimate these effects to accurately take account of all variables, such as the degree and nature of cloud cover. Clouds both reflect incoming light radiation and absorb outgoing infrared radiation, with the former effect tending to predominate. The magnitude of these effects depends upon the degree of cloud cover, brightness, altitude, and thickness. In the case of clouds, too, feedback phenomena occur; for example, warming induces formation of more clouds, which reflect more incoming energy. Most computer models predict global warming of at least 3.0°C and as much as 5.5°C occurring over a period of just a few decades. These estimates are sobering because they correspond to the approximate temperature increase since the last ice age 18,000 years past, which took place at a much slower pace of only about 1 or 2°C per 1,000 years.

Drought is one of the most serious problems that could arise from major climatic change resulting from greenhouse warming. Typically, a 3–degree warming would be accompanied by a 10 percent decrease in precipitation. Water shortages would be aggravated, not just from decreased rainfall, but from increased evaporation, as well. Increased evaporation results in decreased run-off, thereby reducing water available for agricultural, municipal, and industrial use. Water shortages, in turn, lead to increased demand for irrigation and to the production of lower quality, higher salinity runoff water and wastewater. In the U. S., such a problem would be especially intense in the Colorado River basin, which supplies much of the water used in the rapidly growing U. S. Southwest.

A variety of other problems, some of them unforeseen as of now, could result from global warming. An example is the effect of warming on plant and animal pests—insects, weeds, diseases, and rodents. Many of these would certainly thrive much better under warmer conditions.

Interestingly, another air pollutant, acid-rain-forming sulfur dioxide (see Section 17.7), may have a counteracting effect on greenhouse gases. This is because sulfur dioxide is oxidized in the atmosphere to sulfuric acid, forming a light-reflecting haze. Furthermore, the sulfuric acid and resulting sulfates act as condensation nuclei (Section 16.9) that increase the extent, density, and brightness of light-reflecting cloud cover.

17.7. SULFUR DIOXIDE SOURCES AND THE SULFUR CYCLE

Figure 17.5 shows the main aspects of the global sulfur cycle. This cycle involves primarily H_2S, SO_2, SO_3, and sulfates. There are many uncertainties regarding the sources, reactions, and fates of these atmospheric sulfur species. On a global basis, sulfur compounds enter the atmosphere to a very large extent through human activities. Approximately 100 million metric tons of sulfur per year enter the global atmosphere through anthropogenic activities, primarily as SO_2 from the combustion of coal and residual fuel oil. The greatest uncertainties in the cycle have to do with nonanthropogenic sulfur, which enters the atmosphere largely as H_2S from volcanoes and from the biological decay of organic matter and reduction of sulfate. The quantity added from biological processes may be as low as 1 million metric tons per year. Any H_2S that does get into the atmosphere is converted rapidly to SO_2 by processes that involve several intermediate steps, including reactions with hydroxyl radical.

Sulfur Dioxide Reactions in the Atmosphere

Many factors, including temperature, humidity, light intensity, atmospheric transport, and surface characteristics of particulate matter, may influence the atmospheric chemical reactions of sulfur dioxide. Like many other gaseous pollutants, sulfur dioxide undergoes chemical reactions resulting in the formation of particulate matter. Whatever the processes involved, much of the sulfur

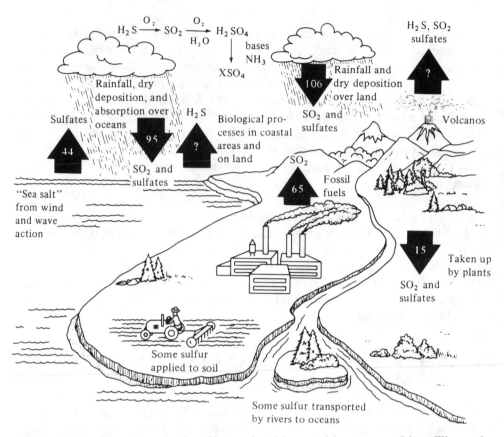

Figure 17.5. **The atmospheric sulfur cycle with quantities expressed in millions of metric tons per year.**

dioxide in the atmosphere ultimately is oxidized to sulfuric acid and sulfate salts, particularly ammonium sulfate and ammonium hydrogen sulfate.

Effects of Atmospheric Sulfur Dioxide

Though not terribly toxic to most people, low levels of sulfur dioxide in air do have some health effects. Sulfur dioxide's primary effect is upon the respiratory tract, producing irritation and increasing airway resistance, especially to people with respiratory weaknesses and sensitized asthmatics. Therefore, exposure to the gas may increase the effort required to breathe. Mucus secretion is also stimulated by exposure to air contaminated by sulfur dioxide.

Atmospheric sulfur dioxide is harmful to plants. Acute exposure to high levels of the gas kills leaf tissue (leaf necrosis). The edges of the leaves and the areas between the leaf veins are particularly damaged. Chronic exposure of plants to sulfur dioxide causes chlorosis, a bleaching or yellowing of the normally green portions of the leaf. Sulfur dioxide in the atmosphere is converted to sulfuric acid, so that in areas with high levels of sulfur dioxide pollution, plants may be damaged by sulfuric acid aerosols. Such damage appears as small spots where sulfuric acid droplets have impinged in leaves.

Sulfur Dioxide Removal

A number of processes are being used to remove sulfur and sulfur oxides from fuel before combustion and from stack gas after combustion. Most of these efforts concentrate on coal, since it is the major source of sulfur oxides pollution. Physical separation techniques may be used to remove discrete particles of pyritic sulfur from coal. Chemical methods may also be employed for removal of sulfur from coal. Fluidized bed combustion of coal promises to eliminate SO_2 emissions at the point of combustion. The process consists of burning granular coal in a bed of finely divided limestone or dolomite maintained in a fluid-like condition by air injection. Heat calcines the limestone,

$$CaCO_3 \rightarrow CaO + CO_2 \qquad (17.7.1)$$

and the lime produced absorbs SO_2:

$$CaO + SO_2 + \frac{1}{2}O_2 \rightarrow CaSO_4 \qquad (17.7.2)$$

Many processes have been proposed or studied for the removal of sulfur dioxide from stack gas. Table 17.2 summarizes major stack gas scrubbing systems. These include throwaway and recovery systems as well as wet and dry systems.

Current practice with lime and limestone scrubber systems often uses injection of the slurry into the scrubber loop beyond the boilers. Experience to date has shown that these scrubbers remove well over 90% of both SO_2 and fly ash when operating properly. In addition to corrosion and scaling problems, disposal of lime sludge poses formidable obstacles. The quantity of this sludge may be appreciated by considering that approximately 1 ton of limestone is required for each 5 tons of coal.

Recovery systems in which sulfur dioxide or elemental sulfur are removed from the spent sorbing material, which is recycled, are much more desirable from an environmental viewpoint than are throwaway systems. Many kinds of recovery processes have been investigated, including those that involve scrubbing with magnesium oxide slurry, sodium sulfite solution, ammonia solution, or sodium citrate solution.

Sulfur dioxide trapped in a stack-gas-scrubbing process can be converted to hydrogen sulfide by reaction with synthesis gas (H_2, CO, CH_4),

Table 17.2. Major Stack Gas Scrubbing Systems[a]

Process	Chemical Reactions	Major Advantages or Disadvantages
Lime slurry scrubbing[b]	$Ca(OH)_2 + SO_2 \rightarrow CaSO_3 + H_2O$	Up to 200 kg of lime are needed per metric ton of coal, producing huge quantities of waste product.
Limestone slurry scrubbing[b]	$CaCO_3 + SO_2 \rightarrow CaSO_3 + CO_2(g)$	Lower pH than lime slurry, and not so efficient.
Magnesium oxide scrubbing	$Mg(OH)_2(slurry) + SO_2 \rightarrow MgSO_3 + H_2O$	The sorbent can be regenerated, and this need not be done on site.
Sodium-base scrubbing	$Na_2SO_3 + H_2O + SO_2 \rightarrow 2\,NaHSO_3$ $2\,NaHSO_3 + heat \rightarrow Na_2SO_3 + H_2O + SO_2$ (regeneration)	There are no major technological limitations. Annual costs are relatively high.
Double alkali[b]	$2\,NaOH + SO_2 \rightarrow Na_2SO_3 + H_2O$ $Ca(OH)_2 + Na_2SO_3 \rightarrow CaSO_3(s) + 2\,NaOH$ (regeneration of NaOH)	Allows for regeneration of expensive sodium alkali solution with inexpensive lime.

[a]For details regarding these and more advanced processes, see the following two books. Satriana, M. 1981. *New developments in flue gas desulfurization technology.* Park Ridge, N.J.: Noyes Data Corp. Hudson, J.D., and Rochelle G.T., eds. 1982. *Flue gas desulfurization.* Washington, D.C.: American Chemical Society.
[b]These processes have also been adapted to produce a gypsum product by oxidation of $CaSO_3$ in the spent scrubber medium:

$$CaSO_3 + \tfrac{1}{2}O_2 + 2\,H_2O \rightarrow CaSO_4 \cdot 2\,H_2(s)$$

Gypsum has some commercial value, such as in the manufacture of plasterboard, and makes a relatively settleable waste product.

$$SO_2 + (H_2, CO, CH_4) \rightleftharpoons H_2S + CO_2 \qquad (17.7.3)$$

The Claus reaction is then employed to produce elemental sulfur:

$$2H_2S + SO_2 \rightleftharpoons 2H_2O + 3S \qquad (17.7.4)$$

17.8. NITROGEN OXIDES IN THE ATMOSPHERE

The three oxides of nitrogen normally encountered in the atmosphere are nitrous oxide (N_2O), nitric oxide (NO), and nitrogen dioxide (NO_2). Microbially generated nitrous oxide is relatively unreactive and probably does not significantly influence important chemical reactions in the lower atmosphere. Its concentration decreases rapidly with altitude in the stratosphere due to the photochemical reaction

$$N_2O + h\nu \rightarrow N_2 + O \qquad (17.8.1)$$

and some reaction with singlet atomic oxygen:

$$N_2O + O \rightarrow N_2 + O_2 \qquad (17.8.2)$$

$$N_2O + O \rightarrow NO + NO \qquad (17.8.3)$$

These reactions are significant in terms of depletion of the ozone layer. Increased global fixation of nitrogen, accompanied by increased microbial production of N_2O, could contribute to ozone layer depletion.

Colorless, odorless nitric oxide (NO) and pungent red-brown nitrogen dioxide (NO_2) are very important in polluted air. Collectively designated NO_x, these gases enter the atmosphere from natural sources, such as lightning and biological processes, and from pollutant sources. The latter are much more significant because of regionally high NO_2 concentrations, which can cause severe air quality deterioration. Practically all anthropogenic NO_2 enters the atmosphere as a result of the combustion of fossil fuels in both stationary and mobile sources. The contribution of automobiles to nitric oxide production in the U.S. has become somewhat lower in the last decade as newer automobiles with nitrogen oxide pollution controls have become more common.

Most NO_2 entering the atmosphere from pollution sources does so as NO generated from internal combustion engines. At the very high temperatures in an automobile combustion chamber, the following reaction occurs:

$$N_2 + O_2 \rightarrow 2NO \qquad (17.8.4)$$

High temperatures favor both a high equilibrium concentration and a rapid rate of formation of NO. Rapid cooling of the exhaust gas from combustion "freezes" NO at a relatively high concentration because equilibrium is not maintained. Thus, by its very nature, the combustion process both in the internal combustion engine and in furnaces produces high levels of NO in the combustion products.

Atmospheric Reactions of NO_x

Atmospheric chemical reactions convert NO_x to nitric acid, inorganic nitrate salts, organic nitrates, and peroxyacetyl nitrate (see Chapter 18). The principal reactive nitrogen oxide species in the troposphere are NO, NO_2, and HNO_3. These species cycle among each other, as shown in Figure 17.6. Although NO is the primary form in which NO_x is released to the atmosphere, the conversion of NO to NO_2 is relatively rapid in the troposphere.

Nitrogen dioxide is a very reactive and significant species in the atmosphere. It absorbs light throughout the ultraviolet and visible spectrum penetrating the troposphere. At wavelengths below 398 nm, photodissociation to oxygen atoms occurs,

$$NO_2 + h\nu \rightarrow NO + O \qquad (17.8.5)$$

giving rise to several significant inorganic reactions, in addition to a host of atmospheric reactions involving organic species. The reactivity of NO_2 to photodissociation is shown clearly by the fact that in direct sunlight the half-life of NO_2 is much shorter than that of any other atmospheric component (only 1 or 2 minutes). The significance of NO_2 in initiating photochemical smog formation is discussed in Chapter 18.

Harmful Effects of Nitrogen Oxides

Nitric oxide, NO, is biochemically less active and less toxic than NO_2. Acute exposure to NO_2 can be quite harmful to human health. For exposures ranging from several minutes to one hour, a level of 50–100 ppm of NO_2 causes inflammation of lung tissue for a period of 6–8 weeks, after which time the subject normally recovers. Exposure of the subject to 150–200 ppm of NO_2 causes

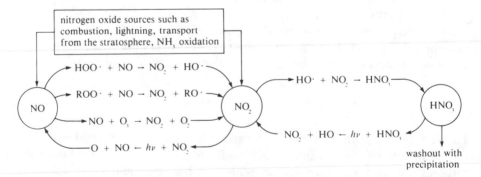

Figure 17.6.   Principal reactions among NO, NO_2, and HNO_3 in the atmosphere. $ROO\cdot$ represents an organic peroxyl radical, such as the methylperoxyl radical, $CH_3OO\cdot$.

bronchiolitis fibrosa obliterans, a condition fatal within 3–5 weeks after exposure. Death generally results within 2–10 days after exposure to 500 ppm or more of NO_2. "Silo-filler's disease," caused by NO_2 generated by the fermentation of ensilage (fermented corn plants used for cattle feed) containing nitrate, is a particularly striking example of nitrogen dioxide poisoning. Deaths have resulted from the inhalation of NO_2-containing gases from burning celluloid and nitrocellulose film and from leakage of NO_2 oxidant from missile rocket motors.

Although extensive damage to plants is observed in areas receiving heavy exposure to NO_2, most of this damage probably comes from secondary products of nitrogen oxides, such as PAN formed in smog (see Chapter 18). Exposure of plants to several parts per million of NO_2 in the laboratory causes leaf spotting and breakdown of plant tissue. Exposure to 10 ppm of NO causes a reversible decrease in the rate of photosynthesis.

Nitrogen oxides are known to cause fading of dyes and inks used in some textiles. This has been observed in gas clothes dryers and is due to NO_x formed in the dryer flame. Much of the damage to materials caused by NO_x, such as stress-corrosion cracking of electrical apparatus, comes from secondary nitrates and nitric acid.

Control of Nitrogen Oxides

The level of NO_x emitted from stationary sources such as power plant furnaces generally falls within the range of 50–1000 ppm. NO production is favored both kinetically and thermodynamically by high temperatures and by high excess oxygen concentrations. These factors must be considered in reducing NO emissions from stationary sources. Reduction of flame temperature to prevent NO formation is accomplished by adding recirculated exhaust gas, cool air, or inert gases.

Low-excess-air firing is effective in reducing NO_x emissions during the combustion of fossil fuels. As the term implies, low-excess-air firing uses the minimum amount of excess air required for oxidation of the fuel, so that less oxygen is available for the reaction

$$N_2 + O_2 \rightarrow 2NO \qquad (17.8.6)$$

in the high temperature region of the flame. Incomplete fuel burnout, with the emission of hydrocarbons, soot, and CO, is an obvious problem with low-excess-air firing. This may be overcome by a two-stage combustion process. In the first stage fuel is fired at a relatively high temperature with a substoichiometric amount of air, and NO formation is limited by the absence of excess oxygen. In the second stage fuel burnout is completed at a relatively low temperature in excess air; the low temperature prevents formation of NO.

Removal of NO_x from stack gas presents some formidable problems. Possible approaches to NO_x removal are catalytic decomposition of nitrogen oxides, catalytic reduction of nitrogen oxides, and sorption of NO_x by liquids or solids.

Ammonia in the Atmosphere

Ammonia is present even in unpolluted air as a result of natural biochemical and chemical processes. Among the various sources of atmospheric ammonia are microorganisms, decay of animal wastes, sewage treatment, coke manufacture, ammonia manufacture, and leakage from ammonia-based refrigeration systems. High concentrations of ammonia gas in the atmosphere are generally indicative of accidental release of the gas.

Ammonia is removed from the atmosphere by its affinity for water and by its action as a base. It is a key species in the formation and neutralization of nitrate and sulfate aerosols in polluted atmospheres. Ammonia reacts with these acidic aerosols to form ammonium salts:

$$NH_3 + HNO_3 \rightarrow NH_4NO_3 \tag{17.8.7}$$

$$NH_3 + H_2SO_4 \rightarrow NH_4HSO_4 \tag{17.8.8}$$

Ammonium salts are among the more corrosive salts in atmospheric aerosols.

17.9. ACID RAIN

As discussed in this chapter, much of the sulfur and nitrogen oxides entering the atmosphere are converted to sulfuric and nitric acids, respectively. When combined with hydrochloric acid arising from hydrogen chloride emissions, these acids cause acidic precipitation that is now a major pollution problem in some areas. Precipitation made acidic by the presence of acids stronger than $CO_2(aq)$ is commonly called **acid rain**; the term applies to all kinds of acidic aqueous precipitation, including fog, dew, snow, and sleet. In a more general sense, **acid deposition** refers to the deposition on the Earth's surface of aqueous acids, acid gases (such as SO_2), and acidic salts (such as NH_4HSO_4). According to this definition, deposition in solution form is *acid precipitation*, and deposition of dry gases and compounds is *dry deposition*. Sulfur dioxide, SO_2, contributes more to the acidity of precipitation than does CO_2 present at higher levels in the atmosphere because SO_2 is a stronger acid and it is more water soluble

Although acid rain can originate from the direct emission of strong acids, such as HCl gas or sulfuric acid mist, most of it is a secondary air pollutant produced by the atmospheric oxidation of acid-forming gases such as the following:

$$SO_2 + \frac{1}{2}O_2 + H_2O \xrightarrow[\text{consisting of several steps}]{\text{Overall reaction}} \{2H^+ + SO_4^{2-}\}(aq) \quad (17.9.1)$$

$$2NO_2 + \frac{1}{2}O_2 + H_2O \xrightarrow[\text{consisting of several steps}]{\text{Overall reaction}} 2\{H^+ + NO_3^-\}(aq) \quad (17.9.2)$$

Chemical reactions such as these play a dominant role in determining the nature, transport, and fate of acid precipitation. As the result of such reactions the chemical properties (acidity, ability to react with other substances) and physical properties (volatility, solubility) of acidic atmospheric pollutants are altered drastically. For example, even the small fraction of NO that does dissolve in water does not react significantly. However, its ultimate oxidation product, HNO_3, though volatile, is highly water-soluble, strongly acidic, and very reactive with other materials. Therefore, it tends to be removed readily from the atmosphere and to do a great deal of harm to plants, corrodable materials, and other things that it contacts.

Although emissions from industrial operations and fossil fuel combustion are the major sources of acid-forming gases, acid rain has also been encountered in areas far from such sources. This is due in part to the fact that acid-forming gases are oxidized to acidic constituents and deposited over several days, during which time the air mass containing the gas may have moved as much as several thousand km. It is likely that the burning of biomass, such as is employed in "slash-and-burn" agriculture, evolves the gases that lead to acid formation in more remote areas. In arid regions, dry acid gases or acids sorbed to particles may be deposited, with effects similar to those of acid rain deposition.

Acid rain spreads out over areas of several hundred to several thousand kilometers, so it is classified as a *regional* air pollution problem compared to a *local* air pollution problem for smog and a *global* one for ozone-destroying chlorofluorocarbons and greenhouse gases. Analyses of the movements of air masses have shown a correlation between acid precipitation and prior movement of an air mass over major sources of anthropogenic sulfur and nitrogen oxides emissions. This is particularly obvious in southern Scandinavia, which receives a heavy burden of air pollution from densely populated, heavily industrialized areas in Europe. The strong geographic dependence of acid precipitation is illustrated in Figure 17.7, representing the pH of precipitation in the continental U.S. The preponderance of acidic rainfall in the northeastern U.S. is obvious.

Acid rain is not a new phenomenon; it has been observed for well over a century, with many of the older observations from Great Britain. The first manifestations of this phenomenon were elevated levels of SO_4^{2-} in precipitation collected in industrialized areas. More modern evidence was obtained from analyses of precipitation in Sweden in the 1950s and of U.S. precipitation a decade or so later. A vast research effort on acid rain was conducted in North

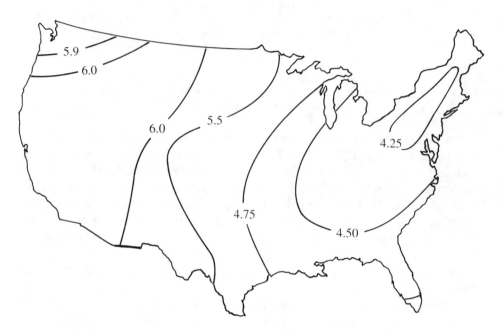

Figure 17.7.   Hypothetical precipitation-pH pattern in the continental United States. Actual values found may vary with the time of year and climatic conditions.

America by the National Acid Precipitation Assessment Program, which resulted from the U.S. Acid Precipitation Act of 1980.

Ample evidence exists of the damaging effects of acid rain. The major such effects are the following:

- Direct phytotoxicity to plants from excessive acid concentrations. (Evidence of direct or indirect phytoxicity of acid rain is provided by the declining health of Eastern U.S. and Scandinavian forests and especially by damage to Germany's Black Forest.)
- Phytotoxicity from acid-forming gases, particularly SO_2 and NO_2, that accompany acid rain
- Indirect phytotoxicity, such as from Al^{3+} liberated from acidified soil
- Destruction of sensitive forests
- Respiratory effects on humans and other animals
- Acidification of lake water with toxic effects to lake flora and fauna, especially fish fingerlings
- Corrosion to exposed structures, electrical relays, equipment, and ornamental materials. Because of the effect of hydrogen ion,

$$2H^+ + CaCO_3(s) \rightarrow Ca^{2+} + CO_2(g) + H_2O$$

limestone, $CaCO_3$, is especially susceptible to damage from acid rain
- Associated effects, such as reduction of visibility by sulfate aerosols and the influence of sulfate aerosols on physical and optical properties of clouds.

17.10. FLUORINE, CHLORINE, AND THEIR GASEOUS INORGANIC COMPOUNDS

Fluorine, hydrogen fluoride, and other volatile fluorides are produced in the manufacture of aluminum, and hydrogen fluoride is a by-product in the conversion of fluorapatite (rock phosphate) to phosphoric acid, superphosphate fertilizers, and other phosphorus products. Hydrogen fluoride gas is a dangerous substance that is so corrosive that it even reacts with glass. It is irritating to body tissues, and the respiratory tract is very sensitive to it. Brief exposure to HF vapors at the part-per-thousand level may be fatal. The acute toxicity of F_2 is even higher than that of HF. Chronic exposure to high levels of fluorides causes fluorosis, the symptoms of which include mottled teeth and pathological bone conditions.

Plants are particularly susceptible to the effects of gaseous fluorides. Fluorides from the atmosphere appear to enter the leaf tissue through the stomata. Fluoride is a cumulative poison in plants, and exposure of sensitive plants to even very low levels of fluorides for prolonged periods results in damage. Characteristic symptoms of fluoride poisoning are chlorosis (fading of green color due to conditions other than the absence of light), edge burn, and tip burn. Conifers (such as pine trees) afflicted with fluoride poisoning may have reddish-brown necrotic needle tips as far as 10 miles from the plant.

Silicon tetrafluoride gas, SiF_4, is a gaseous fluoride pollutant produced during some steel and metal smelting operations that employ CaF_2, fluorspar. Fluorspar reacts with silicon dioxide (sand), releasing SiF_4 gas:

$$2CaF_2 + 3SiO_2 \rightarrow 2CaSiO_3 + SiF_4 \qquad (17.10.1)$$

Another gaseous fluorine compound, sulfur hexafluoride, SiF_6, occurs in the atmosphere at levels of about 0.3 parts per trillion. It is extremely unreactive and is used as an atmospheric tracer. It does not absorb ultraviolet light in either the troposphere or stratosphere and is probably destroyed above 50 km by reactions beginning with its capture of free electrons.

Chlorine and Hydrogen Chloride

Chlorine gas, Cl_2, does not occur as an air pollutant on a large scale but can be quite damaging on a local scale. Chlorine was the first poisonous gas deployed in World War I. It is widely used as a manufacturing chemical in the plastics industry, for example, as well as for water treatment and as a bleach. Therefore, possibilities for its release exist in a number of locations. Chlorine is quite toxic and is a mucous-membrane irritant. It is very reactive and a powerful oxidizing agent. Chlorine dissolves in atmospheric water droplets, yielding hydrochloric acid and hypochlorous acid, an oxidizing agent:

$$H_2O + Cl_2 \rightarrow H^+ + Cl^- + HOCl \qquad (17.10.2)$$

Spills of chlorine gas have caused fatalities among exposed persons.

Hydrogen chloride, HCl, is emitted from a number of sources. Incineration of chlorinated plastics, such as polyvinylchloride, releases HCl as a combustion product.

```
     Cl  H   H   H   Cl  H   H   H   Cl  H
      |  |   |   |   |   |   |   |   |   |
·····C – C – C – C – C – C – C – C – C – C·····Polyvinylchloride
      |  |   |   |   |   |   |   |   |   |
     H   H   Cl  H   H   H   Cl  H   H   H
```

Some compounds released to the atmosphere as air pollutants hydrolyze to form HCl. One such incident occurred on April 26, 1974, when a storage tank containing 750,000 gallons of liquid silicon tetrachloride, $SiCl_4$, began to leak in South Chicago, Illinois. This compound reacted with water in the atmosphere to form a choking fog of hydrochloric acid droplets:

$$SiCl_4 + 2H_2O \rightarrow SiO_2 + 4HCl \qquad (17.10.3)$$

Many people became ill from inhaling the vapor.

In February, 1981, in Stroudsburg, Pennsylvania, a wrecked truck dumped 12 tons of powdered aluminum chloride during a rainstorm. This compound produces HCl gas when wet,

$$AlCl_3 + 3H_2O \rightarrow Al(OH)_3 + 3HCl \qquad (17.10.4)$$

and more than 1,200 residents had to be evacuated from their homes because of the fumes generated.

17.11.   HYDROGEN SULFIDE, CARBONYL SULFIDE, AND CARBON DISULFIDE

Hydrogen sulfide is produced by microbial decay of sulfur compounds and microbial reduction of sulfate (see Section 11.13), from geothermal steam, from wood pulping, and from a number of miscellaneous natural and anthropogenic sources. Most atmospheric hydrogen sulfide is rapidly converted to SO_2 and to sulfates by oxidizing processes in the atmosphere. The organic homologs of hydrogen sulfide, the mercaptans or thiols, enter the atmosphere from decaying organic matter and have particularly objectionable odors.

Hydrogen sulfide pollution from artificial sources is not as much of an overall air pollution problem as sulfur dioxide pollution. However, there have been several acute incidents of hydrogen sulfide emissions resulting in damage to human health and even fatalities. The most notorious such incident occurred in Poza Rica, Mexico, in 1950. Accidental release of hydrogen sulfide from a plant used for the removal of sulfur from natural gas caused the deaths of 22 people and the hospitalization of over 300.

Hydrogen sulfide at levels well above ambient concentrations destroys immature plant tissue. This type of plant injury is readily distinguished from that due to other phytotoxins. More sensitive species are killed by continuous exposure to around 3000 ppb H_2S, whereas other species exhibit reduced growth, leaf lesions, and defoliation.

Damage to certain kinds of materials is a very expensive effect of hydrogen sulfide pollution. Paints containing lead pigments, $2PbCO_3 \cdot Pb(OH)_2$ (no longer used), were particularly susceptible to darkening by H_2S. A black layer of copper sulfide forms on copper metal exposed to H_2S. Eventually, this layer is replaced by a green coating of basic copper sulfate such as $CuSO_4 \cdot 3Cu(OH)_2$. The green "patina," as it is called, is very resistant to further corrosion. Such layers of corrosion can seriously impair the function of copper contacts on electrical equipment. Hydrogen sulfide also forms a black sulfide coating on silver.

Carbonyl sulfide, COS, is now recognized as a component of the atmosphere at a tropospheric concentration of approximately 500 parts per trillion by volume, corresponding to a global burden of about 2.4 teragrams. It is, therefore, a significant sulfur species in the atmosphere.

Both COS and CS_2 are oxidized in the atmosphere by reactions initiated by the hydroxyl radical. The initial reactions are

$$HO\cdot + COS \rightarrow CO_2 + HS\cdot \qquad (17.11.1)$$

$$HO\cdot + CS_2 \rightarrow COS + HS\cdot \qquad (17.11.2)$$

The sulfur-containing products undergo further reactions to sulfur dioxide and, eventually, to sulfate species.

CHAPTER SUMMARY

The chapter summary below is presented in a programmed format to review the main points covered in this chapter. It is used most effectively by filling in the blanks, referring back to the chapter as necessary. The correct answers are given at the end of the summary.

Atmospheric aerosols are (1)_____

_____. Condensation aerosols are formed by

(2)_____

_____ and dispersion aerosols are formed by (3)_____

_____.

As applied to the formation of atmospheric aerosols, the reaction $2SO_2 + O_2 +$

$2H_2O \rightarrow 2H_2SO_4$ shows (4)_____

_____.

In general, the proportions of elements in atmospheric particulate matter reflect

(5)_____. The

two reactions $2SO_2 + O_2 + 2H_2O \rightarrow 2H_2SO_4$ and $H_2SO_4 + 2NaCl$(particulate)

Na_2SO_4(particulate) $+ 2HCl$ show why particulate matter largely from ocean

spray origin in a coastal area receiving sulfur dioxide pollution may show

(6)_____. Insofar as the origin

of elements in particulate matter is concerned, soil erosion, rock dust, and coal

combustion produce (7)_____; incomplete combustion of

carbonaceous fuels produces (8)_____; marine aerosols and

incineration of organohalide polymer wastes produce (9)_____; and

combustion of residual petroleum produces (10)_____. Mineral

particulate matter in the form of oxides and other compounds produced during

the combustion of high-ash fossil fuel that can be collected in furnace flues is

called (11)_____. Three adverse respiratory health effects

caused by inhalation of asbestos are (12)_____

_____. The toxic metal of greatest concern in the urban atmosphere is

(13)_____ because it (14)_____.

A significant natural source of radionuclides in the atmosphere is (15)_____

for which the initial decay products are (16)_____. Two

significant anthropogenic sources of radioactivity in the atmosphere are

(17)_____

_____.

Three effects of atmospheric particles are (18)_____

_____

_____.

The most common health effects of atmospheric particles are on the (19)_____.

Some of the most common ways to control particulate emissions are (20)__

_____

_____

Carbon monoxide causes pollution problems in cases of (21)_____

_____. It is of most concern because of its (22)_____. Automobiles are equipped with (23)_____

_____ to cut down carbon monoxide emissions. It is generally agreed that carbon monoxide is removed from the atmosphere by (24)_____. The greatest pollution concern with carbon dioxide is its role as (25)_____

_____. The two major sources of atmospheric carbon dioxide are (26)_____.

Most sulfates enter the atmosphere from (27)_____ and most sulfur dioxide from (28)_____. The fate of any H_2S that does get into the atmosphere is that it is (29)_____

_____. Sulfur dioxide's primary health effect is upon the (30)_____. Chronic exposure of plants to sulfur dioxide causes (31)_____. Sulfur dioxide can be removed from stack gas by (32)_____ processes, of which the major types are (33)_____ as well as (34)_____

_____.

The three oxides of nitrogen normally encountered in the atmosphere are (35)_____. Nitric oxide (NO) and nitrogen dioxide (NO_2) are collectively designated (36)_____, most of which enters the atmosphere as (37)_____. The most important photochemical reaction of NO_2, and one that can lead to photochemical smog formation is (38)_____. Pulmonary exposure to 150–200 ppm of NO_2 causes (39)_____, a condition fatal within 3–5 weeks after exposure. During the combustion of fossil fuels, (40)_____ is effective in reducing NO_x emissions. Sources of atmospheric ammonia are (41)_____

_____

_____.

   Much of the sulfur and nitrogen oxides entering the atmosphere are converted to (42)_____, which when combined with hydrochloric acid arising from hydrogen chloride emissions cause (43)_____

_____. The major damaging effects of acid precipitation are (44)_____

_____

_____

_____

_____

_____.

   Two specific gaseous fluorine-containing air pollutants are (45)_____

_____. Though not directly toxic, the fluorine-containing air pollutants with the greatest potential for damage to the atmosphere are the (46)_____. Halons, which are related to chlorofluorocarbons, are compounds that contain (47)_____ and are used in (48)_____. The concern with chlorofluorocarbon air pollutants is (49)_____.

The first reaction in this harmful process is (50)_____,

the Cl atoms from which undergo the reaction (51)_____

with ozone. The most prominent instance of ozone layer destruction that has been documented in recent years is (52)_____. The highly damaging consequence of stratospheric ozone destruction is that it would allow

(53)_____, which would result in (54)_____

_____.

Atmospheric chlorine dissolves in atmospheric water droplets, yielding

(55)_____.

Natural sources of hydrogen sulfide are (56)_____

_____.

Answers

1. solid or liquid particles smaller than 100 μm in diameter
2. condensation of vapors or reactions of gases
3. grinding of solids, atomization of liquids, aerosol or dispersion of dusts
4. oxidation of atmospheric sulfur dioxide to sulfuric acid, a hygroscopic substance that accumulates atmospheric water to form small liquid droplets
5. relative abundances of elements in the parent material
6. anomalously high sulfate and corresponding low chloride content
7. Al, Fe, Ca, Si
8. C
9. Na, Cl
10. V
11. fly ash
12. asbestosis, mesothelioma, and bronchogenic carcinoma
13. lead
14. comes closest to being present at a toxic level
15. radon
16. ^{218}Po and ^{216}Po
17. the combustion of fossil fuels and the aboveground detonation of nuclear weapons
18. reduction and distortion of visibility, provision of active surfaces upon which heterogeneous atmospheric chemical reactions can occur, and nucleation bodies for the condensation of atmospheric water vapor
19. respiratory tract
20. sedimentation, inertial mechanisms, fabric filters, venturi scrubbers, and electrostatic precipitators
21. locally high concentrations
22. toxicity
23. catalytic exhaust reactors
24. reaction with hydroxyl radical
25. a greenhouse gas

26. fossil fuel combustion and deforestation
27. evaporated sea salt
28. combustion of fossil fuels
29. converted rapidly to SO_2
30. respiratory tract
31. chlorosis
32. scrubbing
33. throwaway and recovery systems
34. wet and dry systems
35. nitrous oxide (N_2O), nitric oxide (NO), and nitrogen dioxide (NO_2)
36. NO_x
37. NO
38. $NO_2 + h\nu \rightarrow NO + O$
39. *bronchiolitis fibrosa obliterans*
40. low-excess-air firing
41. microorganisms, decay of animal wastes, sewage treatment, coke manufacture, ammonia manufacture, and leakage from ammonia-based refrigeration systems
42. sulfuric and nitric acids
43. acidic precipitation
44. direct phytotoxicity to plants from excessive acid concentrations; phytotoxicity from acid-forming gases; indirect phytotoxicity, such as from Al^{3+} liberated from acidified soil; destruction of sensitive forests, respiratory effects on humans and other animals; acidification of lake water with toxic effects to lake flora and fauna; corrosion to exposed structures, electrical relays, equipment, and ornamental materials; and associated effects, such as reduction of visibility by sulfate aerosols and the influence of sulfate aerosols on physical and optical properties of clouds.
45. fluorine gas and hydrogen fluoride
46. chlorofluorocarbons (CFC)
47. bromine
48. fire extinguisher systems
49. their potential to destroy stratospheric ozone
50. $Cl_2CF_2 + h\nu \rightarrow Cl\cdot + ClCF_2\cdot$
51. $Cl + O_3 \rightarrow ClO + O_2$
52. the Antarctic ozone hole
53. penetration of high-energy ultraviolet radiation
54. adverse biological effects, such as increased skin cancer
55. hydrochloric acid and hypochlorous acid
56. microbial decay of sulfur compounds and microbial reduction of sulfate

QUESTIONS AND PROBLEMS

1. A freight train that included a tank car containing anhydrous NH_3 and one containing concentrated HCl was wrecked, causing both of the tank cars to leak. In the region between the cars a white aerosol formed. What was it, and how was it produced?

2. What two vapor forms of mercury might be found in the atmosphere?

3. Analysis of particulate matter collected in the atmosphere near a seashore shows considerably more Na than Cl on a molar basis. What does this indicate?

4. What type of process results in the formation of very small aerosol particles?

5. Which size range encompasses most of the particulate matter mass in the atmosphere?

6. Why are aerosols in the 0.1–1 μm size range especially effective in scattering light?

7. Per unit mass, why are smaller particles more effective catalysts for atmospheric chemical reactions than larger ones?

8. What is the rationale for classifying most acid rain as a secondary pollutant?

9. Why is it that "highest levels of carbon monoxide tend to occur in congested urban areas at times when the maximum number of people are exposed?"

10. Which unstable, reactive species is responsible for the removal of CO from the atmosphere?

11. Which of the following fluxes in the atmospheric sulfur cycle is smallest: (a) sulfur species washed out in rainfall over land, (b) sulfates entering the atmosphere as "sea salt," (c) sulfur species entering the atmosphere from volcanoes, (d) sulfur species entering the atmosphere from fossil fuels, (e) hydrogen sulfide entering the atmosphere from biological processes in coastal areas and on land.

12. Of the following agents, the one that would not favor conversion of sulfur dioxide to sulfate species in the atmosphere is: (a) ammonia, (b) water, (c) contaminant-reducing agents, (d) ions of transition metals such as manganese, (e) sunlight.

13. The air inside a garage was found to contain 10 ppm CO by volume at standard temperature and pressure (STP). What is the concentration of CO in mg/L and in ppm by mass?

14. Assume that an incorrectly adjusted lawn mower is operated in a garage so that the combustion reaction in the engine is

$$C_8H_{18} + \ ^{17}/_2O_2 \rightarrow 8CO + 9H_2O$$

If the dimensions of the garage are $5 \times 3 \times 3$ meters, how many grams of gasoline must be burned to raise the level of CO in the air to 1000 ppm by volume at STP?

15. A 12.0-L sample of waste air from a smelter process was collected at 25°C and 1.00 atm pressure, and the sulfur dioxide was removed. After SO_2 removal, the volume of the air sample was 11.50 L. What was the percentage by weight of SO_2 in the original sample?

16. What is the oxidant in the Claus reaction?

17. How many metric tons of 5%–S coal would be needed to yield the H_2SO_4 required to produce a 3.00–cm rainfall of pH 2.00 over a 100 km² area?

18. In what major respect is NO_2 a more significant species than SO_2 in terms of participation in atmospheric chemical reactions?

19. Assume that the wet limestone process requires 1 metric ton of $CaCO_3$ to remove 90% of the sulfur from 4 metric tons of coal containing 2% S. Assume that the sulfur product is $CaSO_4$. Calculate the percentage of the limestone converted to calcium sulfate.

SUPPLEMENTARY REFERENCES

Seinfeld, J. H., "Urban Air Pollution: State of the Science," *Science*, 243, February 10, 1989, pp. 745–752.

Schwartz, S. E., "Acid Deposition: Unraveling a Regional Phenomenon," *Science*, 243, February 10, 1989, pp. 753–763.

McElroy, M. B., and R. J. Salawitch, "Changing Composition of the Global Stratosphere," *Science*, 243, February 10, 1989, pp. 763–770.

Schneider, S. H., "The Changing Climate," in *Managing Planet Earth*, special issue of *Scientific American*, September, 1989, pp. 70–79.

Sagan, C., and R. Turco, *A Path Where No Man Thought: Nuclear Winter and the End of the Arms Race*, (New York: Random House, 1990).

Fisher, G. L., and M. A. Gallo, Eds., *Asbestos Toxicity*, (New York: Marcel Dekker, 1988).

Nazaroff, W. W., and A. V. Nero, Jr., *Radon and its Decay Products in Indoor Air*, (New York: John Wiley and Sons, 1988).

Finlayson-Pitts, B. J., and J. N. Pitts, Jr., *Atmospheric Chemistry: Fundamentals and Experimental Techniques*, (New York: John Wiley and Sons, 1986).

Seinfeld, J. H., *Atmospheric Chemistry and Physics of Air Pollution*, (Somerset, N. J.: John Wiley and Sons, Inc., 1986).

Botkin, D. B., Ed., *Changing the Global Environment: Perspectives on Human Involvement*, (San Diego, CA: Academic Press, 1989).

Regens, J. L., and R. W. Rycroft, *The Acid Rain Controversy*, (Pittsburgh, PA: University of Pittsburgh Press, 1988).

White, J. C., Ed., *Acid Rain: The Relationship Between Sources and Receptors*, (New York, NY: Elsevier, 1988).

Singer, F. S., Ed., *Global Climate Change: Human and Natural Influences*, (New York, NY: Paragon House, 1989).

Moomaw, W. R., and I. M. Mintzer, *Strategies for Limiting Global Climate Change*, (Washington, DC: World Resources Institute Publications, 1989).

Schneider, S. H., *Global Warming: Are We Entering The Greenhouse Century?* (San Francisco, CA: Sierra Club Books, 1989).

Oppenheimer, M., and R. H. Boyle, *Dead Heat: The Race Against the Greenhouse Effect*, (New York, NY: Basic Books, 1990).

18   ORGANIC AIR POLLUTANTS

18.1. ORGANIC COMPOUNDS IN THE ATMOSPHERE

Organic pollutants are common atmospheric contaminants that may have a strong effect upon atmospheric quality. Such pollutants may come from both natural and artificial sources. In some cases, contaminants from both kinds of sources interact to produce a pollution effect. This occurs, for example, when terpene hydrocarbons evolved from citrus and conifer trees interact with nitrogen oxides from automobiles to produce photochemical smog.

The effects of organic pollutants in the atmosphere may be divided into two major categories. The first consists of **direct effects**, such as cancer caused by exposure to vinyl chloride. The second is the formation of **secondary pollutants**, especially photochemical smog (discussed in Sections 18.9–18.12). In the case of pollutant hydrocarbons in the atmosphere, the latter is the more important effect. In some localized situations, particularly the workplace, direct effects of organic air pollutants may be equally important.

This chapter discusses the nature and distribution of organic compounds in the atmosphere; it deals with photochemical smog and addresses the mechanisms by which organic compounds undergo photochemical reactions in the atmosphere. Although **smog** is the term used in this book to denote a photochemically oxidizing atmosphere, the word originally was used to describe the unpleasant combination of smoke and fog laced with sulfur dioxide, which was formerly prevalent in London when high-sulfur coal was the primary fuel used in that city. This mixture is characterized by the presence of sulfur dioxide, a reducing compound; therefore, it is a **reducing smog** or **sulfurous smog**. In fact, readily oxidized sulfur dioxide has a short lifetime in an atmosphere where oxidizing photochemical smog is present.

Smog has a long history. Exploring what is now southern California in 1542, Juan Rodriguez Cabrillo named San Pedro Bay "The Bay of Smokes" because of the heavy haze that covered the area. Complaints of eye irritation from anthropogenically polluted air in Los Angeles were recorded as far back as 1868. Characterized by reduced visibility, eye irritation, cracking of rubber, and deterioration of materials, smog became a serious nuisance in the Los Angeles area during the 1940s. It is now recognized as a major air pollution problem in many areas of the world.

A smoggy day may be regarded as one having moderate to severe eye irritation or visibility below 3 miles when the relative humidity is below 60%. The formation of oxidants in the air, particularly ozone, is indicative of smog formation. The three ingredients required to generate photochemical smog are ultraviolet light, hydrocarbons, and nitrogen oxides.

18.2.   ORGANIC COMPOUNDS FROM NATURAL SOURCES

Most organics in the atmosphere come from natural sources, with only about $1/7$ of the total atmospheric hydrocarbons originating from human activities. This ratio is primarily the result of the huge quantities of methane produced by anaerobic bacteria in the decomposition of organic matter in water, sediments, and soil:

$$2\{CH_2O\} \text{ (bacterial action)} \rightarrow CO_2(g) + CH_4(g) \qquad (18.2.1)$$

Flatulent emissions from domesticated animals, arising from bacterial decomposition of food in their digestive tracts, add about 85 million metric tons of methane to the atmosphere each year. Methane is a natural constituent of the atmosphere and is present at a level of about 1.4 parts per million (ppm) in the troposphere.

Methane in the troposphere contributes to the photochemical production of carbon monoxide and ozone. The photochemical oxidation of methane is a major source of water vapor in the stratosphere.

Vegetation is the most important natural source of atmospheric hydrocarbons. Ethylene, C_2H_4, is released to the atmosphere by a variety of plants. Most of the hydrocarbons emitted by plants (predominantly trees) are **terpenes**. Others are α-pinene, a principal component of turpentine; limonene, which evolves from citrus fruit and pine needles and is encountered in the atmosphere around these sources; β-pinene; myrcene; ocimene; α-terpinene; and isoprene (2-methyl-1,3–butadiene, a hemiterpene).

As exemplified by the structures of α-pinene, isoprene, and limonene shown on p. 615, terpenes contain alkenyl (olefinic) bonds, usually two or more per molecule. Therefore, terpenes are among the most reactive compounds in the atmosphere. Terpenes react very rapidly with hydroxyl radical, HO·, and with other oxidizing agents in the atmosphere, particularly ozone, O_3. Such reactions form aerosols, which cause much of the blue haze in the atmosphere above some heavy growths of vegetation.

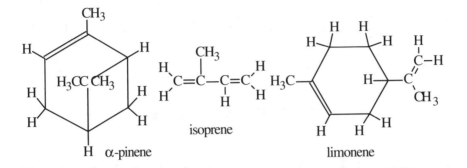

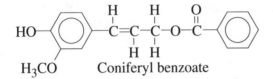
Coniferyl benzoate

Perhaps the greatest variety of compounds emitted by plants consist of **esters**, such as coniferyl benzoate below:

However, they are released in such small quantities that they have little influence upon atmospheric chemistry. Esters are primarily responsible for the fragrances associated with much vegetation.

18.3. POLLUTANT HYDROCARBONS

Ethylene and terpenes, which were discussed in the preceding section, are **hydrocarbons**, organic compounds containing only hydrogen and carbon. As discussed in Section 9.2, hydrocarbons may be **alkanes, alkenes,** such as ethylene; **alkynes** (compounds with triple bonds), and **aromatic compounds,** such as naphthalene, which have characteristic benzene rings.

Because of their widespread use in fuels, hydrocarbons predominate among organic atmospheric pollutants. Petroleum products, primarily gasoline, are the source of most of the anthropogenic pollutant hydrocarbons found in the atmosphere. Hydrocarbons from fuel may enter the atmosphere either directly or as by-products of the partial combustion of other hydrocarbons. The latter are particularly important because they tend to be unsaturated and relatively reactive.

Atmosphere alkenes come from a variety of processes, including emissions from internal combustion engines and turbines, foundry operations, and petroleum refining. Several alkenes, including the ones shown below, are among the top 50 chemicals produced each year, with worldwide production of several billion kg/year:

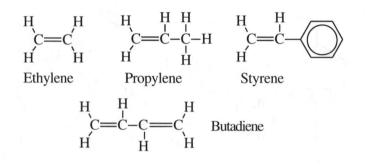

Ethylene   Propylene   Styrene

Butadiene

These compounds are used primarily as monomers, which are polymerized to create polymers for plastics, synthetic rubber, latex paints, and other applications. All of these compounds, as well as others manufactured in lesser quantities, are released to the atmosphere. In addition to the direct release of alkenes, these hydrocarbons are commonly produced by the partial combustion and "cracking" at high temperatures of alkanes, particularly in the internal combustion engine.

Aromatic Hydrocarbons

Aromatic hydrocarbons may be divided into the two major classes of those that have only one benzene ring and those with multiple rings. As discussed in Section 18.8, the latter are **polycyclic aromatic hydrocarbons, PAH.** The aromatic hydrocarbons shown below are among the top 50 chemicals manufactured. Single-ring aromatic compounds are important constituents of lead-free gasoline, which has largely replaced leaded gasoline. Aromatic solvents are widely used in industry. Aromatic hydrocarbons are raw materials for the manufacture of monomers and plasticizers in polymers. Styrene is a monomer in plastics and synthetic rubber. Cumene is oxidized to produce phenol and acetone as a by-product. With all of these applications, plus production of aromatic combustion by-products, it is no surprise that aromatic compounds are common atmospheric pollutants.

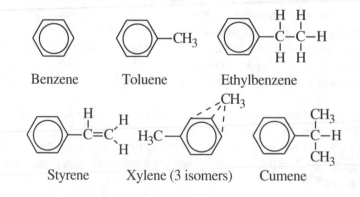

Benzene   Toluene   Ethylbenzene

Styrene   Xylene (3 isomers)   Cumene

18.4. OXYGEN-CONTAINING ORGANIC COMPOUNDS

Aldehydes and Ketones

Carbonyl compounds, consisting of aldehydes and ketones, are often the first species formed, other than unstable reaction intermediates, in the photochemical oxidation of atmospheric hydrocarbons. The simplest and most widely produced of the carbonyl compounds is **formaldehyde**:

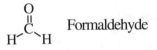

The structures of some other important aldehydes and ketones are shown below:

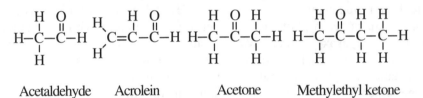

Acetaldehyde   Acrolein   Acetone   Methylethyl ketone

Acetaldehyde is a widely produced organic chemical used in the manufacture of acetic acid, plastics, and raw materials. On the order of a billion kg of acetone are produced each year as a solvent and for applications in the rubber, leather, and plastics industries. Methylethyl ketone is employed as a low-boiling solvent for coatings and adhesives and for the synthesis of other chemicals.

In addition to their production from hydrocarbons by photochemical oxidation, carbonyl compounds enter the atmosphere from a large number of sources and processes. These include direct emissions from internal combustion engine exhausts, incinerator emissions, spray painting, polymer manufacture, printing, petrochemicals manufacture, and lacquer manufacture. Formaldehyde and acetaldehyde are produced by microorganisms, and acetaldehyde is emitted by some kinds of vegetation.

Aldehydes are second only to NO_2 as atmospheric sources of free radicals produced by the absorption of light. This is because the carbonyl group is a **chromophore**, a molecular group that readily absorbs light. It absorbs well in the near-ultraviolet region of the spectrum, and the excited species resulting therefrom may initiate chain reactions involved in photochemical smog formation.

Because of the presence of both double bonds and carbonyl groups, alkenyl aldehydes are especially reactive in the atmosphere. The most common of these found in the atmosphere is acrolein,

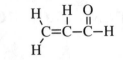

a powerful lachrymator (tear producer) which is used as an industrial chemical and produced as a combustion by-product.

Alcohols

Of the alcohols, methanol, ethanol, isopropanol, and ethylene glycol rank among the top 50 chemicals with annual worldwide production of the order of a billion kg or more. A number of aliphatic alcohols have been reported in the atmosphere. Because of their volatility, the lower alcohols, especially methanol and ethanol, predominate as atmospheric pollutants. Among the other alcohols released to the atmosphere are 1-propanol, 2-propanol, propylene glycol, 1-butanol, and even octadecanol, $CH_3(CH_2)_{16}CH_2OH$, which is evolved by plants. Alcohols can undergo photochemical reactions, beginning with abstraction of hydrogen by hydroxyl radical. Mechanisms for scavenging alcohols from the atmosphere are relatively efficient because the lower alcohols are quite water soluble and the higher ones have low vapor pressures.

Some alkenyl alcohols have been found in the atmosphere, largely as by-products of combustion. Typical of these is 2-buten-1-ol, below, which has been detected in automobile exhausts:

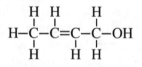

Phenols

Phenols, which are aryl alcohols, are more noted as water pollutants than as air pollutants. Some typical phenols that have been reported as atmospheric contaminants are the following:

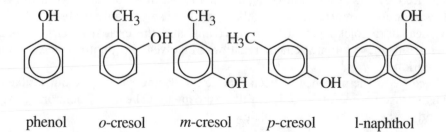

phenol   o-cresol   m-cresol   p-cresol   1-naphthol

Phenol is among the top 50 chemicals produced. It is most commonly used in the manufacture of resins and polymers, such as Bakelite, a phenol-formaldehyde copolymer. Phenols are produced by the pyrolysis of coal and are major by-products of coking. Thus, in local situations involving coal coking and similar operations, phenols can be troublesome air pollutants.

Ethers

Ethers are relatively uncommon atmospheric pollutants; however, the flammability hazard of diethyl ether vapor in an enclosed work space is well known. In addition to aliphatic ethers, such as dimethyl ether and diethyl ether, several alkenyl ethers, including vinylethyl ether are produced by internal combustion engines. A cyclic ether and important industrial solvent, tetrahydrofuran, occurs as an air contaminant. As the most widely used gasoline octane booster, methyl-tertiarybutyl ether, MTBE, has the potential to be an air pollutant, although its hazard is limited by its low vapor pressure. The structural formulas of the ethers mentioned above are given below:

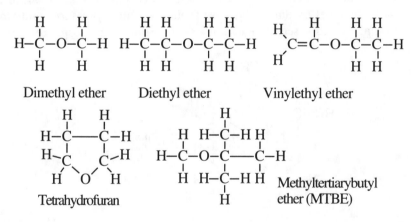

Dimethyl ether   Diethyl ether   Vinylethyl ether

Tetrahydrofuran   Methyltertiarybutyl ether (MTBE)

Oxides

Ethylene oxide and propylene oxide,

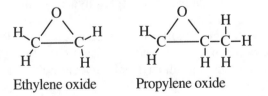

Ethylene oxide   Propylene oxide

rank among the 50 most widely produced industrial chemicals and have a limited potential to enter the atmosphere as pollutants. Ethylene oxide is a moderately to highly toxic sweet-smelling, colorless, flammable, explosive gas used as a chemical intermediate, sterilant, and fumigant. It is a mutagen and a carcinogen to experimental animals. It is classified as hazardous for both its toxicity and ignitability.

Carboxylic Acids

Carboxylic acids have at least one of the functional groups,

attached to an alkane, alkene, or aromatic hydrocarbon moiety. A carboxylic acid, pinonic acid (structure below), is produced by the photochemical oxidation of naturally produced α-pinene. Most of the many carboxylic acids found in the atmosphere are probably the result of the photochemical oxidation of other organic compounds through gas-phase reactions or by reactions of other organic compounds dissolved in aqueous aerosols. These acids are often the end products of photochemical oxidation because their low vapor pressures and water solubilities make them susceptible to scavenging from the atmosphere.

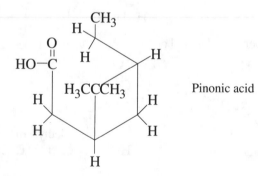

Pinonic acid

18.5. ORGANOHALIDE COMPOUNDS

Organohalides consist of halogen-substituted hydrocarbon molecules, each of which contains at least one atom of F, Cl, Br, or I. They may be saturated (**alkyl halides**), unsaturated (**alkenyl halides**), or aromatic (**aryl halides**). Organohalides exhibit a wide range of physical and chemical properties. Structural formulas of several alkyl halides and alkenyl halides commonly encountered in the atmosphere are given in Figure 18.1.

Volatile **chloromethane** (methyl chloride) is consumed in the manufacture of silicones. **Dichloromethane** is a volatile liquid with excellent solvent properties for nonpolar organic solutes. It has been used as a solvent for the decaffeination of coffee, in paint strippers, as a blowing agent in urethane polymer manufac-

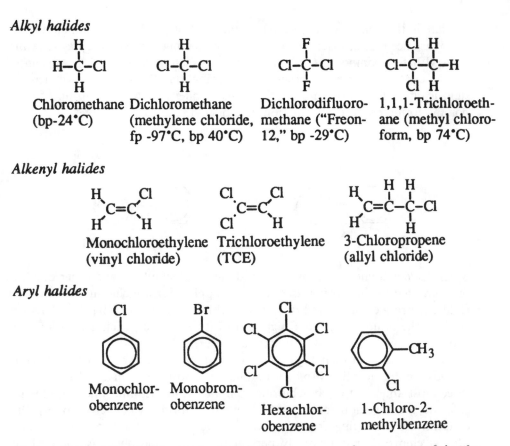

Figure 18.1. Alkyl halides and alkenyl halides commonly encountered in the atmosphere.

ture, and to depress vapor pressure in aerosol formulations. **Dichlorodifluoromethane** is one of the chlorofluorocarbon compounds used as a refrigerant and involved in stratospheric ozone depletion. One of the more common industrial chlorinated solvents is **1,1,1-trichloroethane**.

Viewed as halogen-substituted derivatives of alkenes, the **alkenyl** or **olefinic organohalides** contain at least one halogen atom and at least one carbon-carbon double bond. **Vinyl chloride** is consumed in large quantities as a raw material to manufacture pipe, hose, wrapping, and other products fabricated from polyvinylchloride plastic. This highly flammable, volatile, sweet-smelling gas is known to cause angiosarcoma, a rare form of liver cancer. **Trichloroethylene** is a clear, colorless, nonflammable, volatile liquid. It is an excellent degreasing and dry cleaning solvent and has been used as a household solvent and for food extraction (for example, in decaffeination of coffee). **Allyl chloride** is an intermediate in the manufacture of allyl alcohol and other allyl compounds, including pharmaceuticals, insecticides, and thermosetting varnish and plastic resins.

Aryl halide compounds have many uses, which have resulted in substantial human exposure and environmental contamination. The most environmentally significant aryl halides are polychlorinated biphenyls, PCBs, a group of compounds formed by the chlorination of biphenyl,

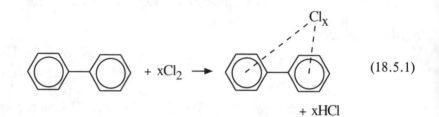

$$+ \text{xHCl} \qquad (18.5.1)$$

which have extremely high physical and chemical stabilities and other qualities that have led to their being used in many applications, including heat transfer fluids, hydraulic fluids, and dielectrics. Although not very volatile, PCBs can get into the atmosphere from high temperature sources, such as incinerators, and be transported with atmospheric particles.

As expected from their high vapor pressures, the lighter organohalide compounds are the most likely to be found in the atmosphere. On a global basis, the three most abundant organochlorine compounds in the atmosphere are methyl chloride, methyl chloroform, and carbon tetrachloride, which have tropospheric concentrations ranging from a tenth to several tenths of a part per billion. Methyl chloroform is relatively persistent in the atmosphere, with residence times of several years. Therefore, it may pose a threat to the stratospheric ozone layer in the same way as chlorofluorocarbons. Also found are methylene chloride; methyl bromide, CH_3Br; bromform, $CHBr_3$; assorted chlorofluorocarbons; and halogen-substituted ethylene compounds, such as trichloroethylene, vinyl chloride, perchloroethylene, $(CCl_2 = CCl_2)$ and solvent ethylene dibromide $(CHBr = CHBR)$.

Chlorofluorocarbons and Halons

The fluorine-containing air pollutants with the greatest potential for damage to the atmosphere are the **chlorofluorocarbons (CFC)**, commonly called Freons, which are used as fluids in refrigeration mechanisms, as blowing agents in the fabrication of flexible and rigid foams, and, until several years ago, as propellants in spray cans containing deodorants, hair spray, and many other products. Chlorofluorocarbons are volatile 1- and 2-carbon compounds that contain Cl and F bonded to carbon. These compounds are notably stable and nontoxic. The most widely manufactured of these compounds have been CCl_3F (CFC-11, bp

24°C), CCl_2F_2 (CFC-12, bp 28°C), $C_2Cl_3F_3$ (CFC-113), $C_2Cl_2F_4$ (CFC-114), and C_2ClF_5 (CFC-115).

Halons are related compounds that contain bromine and are used in fire extinguisher systems. The major commercial halons are $CBrClF_2$ (Halon-1211), $CBrF_3$ (Halon-1301), and $C_2Br_2F_4$ (Halon-2402), where the sequence of numbers denotes the number of carbon, fluorine, chlorine, and bromine atoms, respectively, per molecule. Halons are particularly effective fire extinguishing agents.

Stratospheric Ozone Depletion

All of the chlorofluorocarbons and halons discussed above have been implicated in the halogen-atom-catalyzed destruction of atmospheric ozone, which filters out cancer-causing ultraviolet radiation from the sun. Although quite inert in the lower atmosphere, CFCs undergo photodecomposition by the action of high-energy ultraviolet radiation in the stratosphere through reactions such as

$$CCl_2F_2 + h\nu \rightarrow Cl\cdot + \cdot CClF_2 \qquad (18.5.2)$$

which releases Cl· atoms, denoted below simply as Cl. These atoms react with ozone, destroying it and producing ClO:

$$Cl + O_3 \rightarrow ClO + O_2 \qquad (18.5.3)$$

In this region of the atmosphere, there is an appreciable concentration of atomic oxygen, by virtue of the reaction

$$O_3 + h\nu \rightarrow O_2 + O \qquad (18.5.4)$$

Nitric oxide, NO, is also present. The ClO species may react with either O or NO, regenerating Cl atoms and resulting in chain reactions that cause the net destruction of ozone:

$$ClO + O \rightarrow Cl + O_2 \qquad (18.5.5)$$

$$\underline{Cl + O_3 \rightarrow ClO + O_2} \qquad (18.5.3)$$

$$O + O_3 \rightarrow 2O_2 \qquad (18.5.6)$$

$$ClO + NO \rightarrow Cl + NO_2 \qquad (18.5.7)$$

$$\underline{Cl + O_3 \rightarrow ClO + O_2} \qquad (18.5.3)$$

$$O_3 + NO \rightarrow NO_2 + O_2 \qquad (18.5.8)$$

The most prominent instance of ozone layer destruction is the so-called "Antarctic ozone hole" that has been documented in recent years. This phenomenon is manifested by the appearance during the Antarctic's late winter and early spring of severely depleted stratospheric ozone (up to 50%) over the polar region. The reasons why this occurs are related to the formation of chlorine nitrate, $ClONO_2$, and species held by freezing in polar stratospheric clouds at very low temperatures.

Consequences of Ozone Layer Destruction

The concentration of ozone in the stratosphere is a steady-state concentration resulting from the balance of ozone production and destruction. The quantities of ozone involved are interesting. A total of about 350,000 metric tons of ozone are formed and destroyed daily. Ozone never makes up more than a small fraction of the gases in the ozone layer. In fact, if all the atmosphere's ozone were in a layer at 273 K and 1 atm, it would be only 3 mm thick!

Despite its miniscule atmospheric concentration, ozone has a vital protective function. Ozone serves its protective function because it absorbs ultraviolet radiation very strongly in the region 220–330 nm. Therefore, it is effective in filtering out dangerous UV-B radiation, 290 nm $< \lambda <$ 320 nm. (UV-A radiation, 320 nm-400 nm, is relatively less harmful and UV-C radiation, $<$ 290 nm does not penetrate to the troposphere.)

Increased intensities of ground-level ultraviolet radiation caused by stratospheric ozone destruction would have some significant adverse consequences. One effect would be on plants, including crops used for food. The destruction of microscopic phytoplankton plants that are the basis of the ocean's food chain could severely reduce the productivity of the world's seas. Human exposure would result in more cataracts. The effect of most concern to humans is the elevated occurrence of skin cancer in individuals exposed to ultraviolet radiation. This is because UV-B radiation is absorbed by cellular DNA (see Chapter 21) resulting in photochemical reactions that alter the function of DNA so that the genetic code is improperly translated during cell division. This can result in uncontrolled cell division leading to skin cancer. People with light complexions lack protective melanin, which absorbs UV-B radiation, and are especially susceptible to its effects.

The effects of CFCs on the ozone layer may be the single greatest threat to the global atmosphere. U. S. Environmental Protection Agency regulations, imposed in accordance with the 1986 Montreal Protocol on Substances that Deplete the Ozone Layer, curtailed production of CFCs and halocarbons in the U. S. starting in 1989. The most likely substitutes for these halocarbons are hydrogen-containing chlorofluorocarbons (HCFCs) and hydrogen-containing fluorocarbons (HFCs). The substitute compounds include CH_2FCF_3 (HFC-134a, a substitute for CFC-12 in automobile air conditioners and refrigeration equipment), $CHCl_2CF_3$ (HCFC-123, substitute for CFC-11 in plastic foam-blowing), CH_3CCl_2F (HCFC-141b, substitute for CFC-11 in plastic foam-blowing), CHClF₂ (HCFC-22, to be used in air conditioners and manufacture of plastic foam food containers). Because of the more readily broken H-C bonds that they contain, these compounds are more easily destroyed by atmospheric chemical reactions (particularly with hydroxyl radical, HO·) before they reach the stratosphere. The Du Pont Company, which introduced chlorofluorocarbons in the 1930s and is the largest manufacturer of them, has announced that it intends to cease production shortly after the year 2000. In 1991, Du Pont began marketing Freon substitutes for refrigeration applications under the trade name Suva.

18.6. ORGANOSULFUR COMPOUNDS

Substitution of alkyl or aryl hydrocarbon groups such as phenyl and methyl for H on hydrogen sulfide, H₂S, leads to a number of different organosulfur compounds. Substitution for one H yields thiols, or mercaptans, R–SH; substitution for two H's yields sulfides, also called thioethers, R–S–R. Structural formulas of examples of these compounds are shown below:

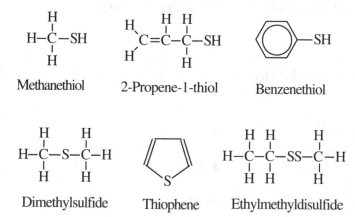

Methanethiol   2-Propene-1-thiol   Benzenethiol

Dimethylsulfide   Thiophene   Ethylmethyldisulfide

Methanethiol and other lighter alkyl thiols are fairly common air pollutants that make their presence known by their very strong odors. These can be described as "ultragarlic" odors, and both 1- and 2-butanethiol are associated with skunk odor. Gaseous methanethiol and volatile liquid ethanethiol are used as odorant leak-detecting additives for natural gas, propane, and butane and are also employed as intermediates in pesticide synthesis. Allyl mercaptan (2-propene-1-thiol) is a toxic, irritating volatile liquid with a strong garlic odor. Benzenethiol (phenyl mercaptan), is the simplest of the aryl thiols. It is a toxic liquid with a severely "repulsive" odor.

Alkyl sulfides or thioethers contain the C-S-C functional group. The lightest of these compounds is dimethyl sulfide, a volatile liquid (bp 38°C) that is moderately toxic by ingestion. Cyclic sulfides contain the C-S-C group in a ring structure. The most common of these compounds is thiophene, a heat-stable liquid (bp 84°C) with a solvent action much like that of benzene, that is used in the manufacture of pharmaceuticals, dyes, and resins.

Although not highly significant as atmospheric contaminants on a large scale – organosulfur compounds effects on atmospheric chemistry is minimal in areas such as aerosol formation or production of acid precipitation components – these compounds are the worst of all in producing odor. Therefore, organosulfur compounds can cause local air pollution problems because of their bad odors. Major sources of organosulfur compounds in the atmosphere include microbial degradation, wood pulping, volatile matter evolved from plants, animal wastes, packing house and rendering plant wastes, starch manufacture, sewage treatment, and petroleum refining.

As with all H-containing organic species in the atmosphere, reaction of organosulfur compounds with hydroxyl radical is a first step in their atmospheric photochemical reactions. The sulfur from both mercaptans and sulfides ends up as SO_2.

18.7. ORGANONITROGEN COMPOUNDS

Organic nitrogen compounds that may be found as atmospheric contaminants may be classified as **amines, amides, nitriles, nitro compounds**, or **heterocyclic nitrogen compounds**. Structures of common examples of each of these five classes of compounds reported as atmospheric contaminants are:

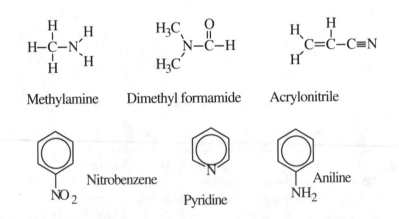

Lower-molecular-mass amines are volatile. These are prominent among the compounds giving rotten fish their characteristic odor – an obvious reason why air contamination by amines is undesirable. The simplest and most important aromatic amine is aniline, used in the manufacture of dyes, amides, photographic chemicals, and drugs. A number of amines are widely used industrial chemicals and solvents, so that industrial sources have the potential to contaminate the atmosphere with these chemicals. Decaying organic matter, especially protein wastes, produces amines, so that rendering plants, packing houses, and sewage treatment plants are important sources of these substances.

Aromatic amines are of particular concern as atmospheric pollutants, particularly in the workplace, because some are known to cause urethral tract cancer (particularly of the bladder) in exposed individuals. Aromatic amines are widely used as chemical intermediates, antioxidants, and curing agents in the manufacture of polymers (rubber and plastics), drugs, pesticides, dyes, pigments, and inks. In addition to aniline, some aromatic amines of potential concern are the following:

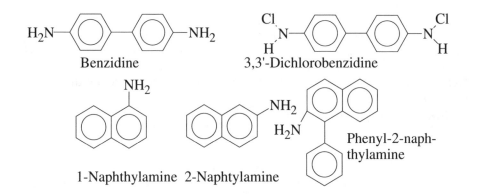

In the atmosphere, amines can be attacked by hydroxyl radical and undergo further reactions. Amines are bases (electron-pair donors). Therefore, their acid-base chemistry in the atmosphere may be important, particularly in the presence of acids in acidic precipitation.

The amide most likely to be encountered as an atmospheric pollutant is dimethylformamide. It is widely used commercially as a solvent for the synthetic polymer, polyacrylonitrile (Orlon, Dacron). Most amides have relatively low vapor pressures, which limits their entry into the atmosphere.

Nitriles, which are characterized by the $-C\equiv N$ group, have been reported as air contaminants, particularly from industrial sources. Both acrylonitrile and acetonitrile, CH_3CN, have been reported in the atmosphere as a result of synthetic rubber manufacture. As expected from their volatilities and levels of industrial production, most of the nitriles reported as atmospheric contaminants are low-molecular-mass aliphatic or alkenyl nitriles, or aromatic nitriles with only one benzene ring. Acrylonitrile, used to make polyacrylonitrile polymer, is the only nitrogen-containing organic chemical among the top 50 chemicals, with annual worldwide production exceeding 1 billion kg.

Among the nitro compounds (RNO_2) reported as air contaminants are nitromethane, nitroethane, and nitrobenzene, all produced from industrial sources. Highly oxygenated compounds containing the NO_2 group, particularly peroxyacetyl nitrate (PAN, discussed in Section 18.10), are end products of the photo-chemical oxidation of hydrocarbons in urban atmospheres.

A large number of **heterocyclic nitrogen compounds** have been reported in tobacco smoke, and it is inferred that many of these compounds can enter the atmosphere from burning vegetation. Coke ovens are another major source of these compounds. In addition to the derivatives of pyridine, some of the hetero-cyclic nitrogen compounds are derivatives of pyrrole:

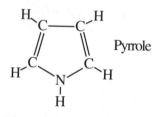

Heterocyclic nitrogen compounds occur almost entirely in association with aero-sols in the atmosphere.

Nitrosamines, general formula,

deserve special mention as atmospheric contaminants because some are known carcinogens. Both N-nitrosodimethylamine and N-nitrosodiethylamine have been detected in the atmosphere.

18.8.   ORGANIC PARTICLES IN THE ATMOSPHERE

A significant portion of organic particulate matter is produced by internal combustion engines in complicated processes that involve pyrosynthesis and nitrogenous compounds. These products may include nitrogen-containing com-pounds and oxidized hydrocarbon polymers. Lubricating oil and its additives may also contribute to organic particulate matter.

PAH Synthesis

The most significant atmospheric organic particles are polycyclic aromatic hydrocarbons (PAH), which consist of condensed ring aromatic molecules. As discussed in Section 21.9, some PAHs can be metabolized in the body to species that bond to DNA and cause cancer. The most often cited example of a PAH compound is benzo(a)pyrene, a compound that the body can metabolize to a carcinogenic form:

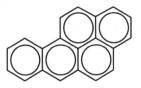   Benzo(a)pyrene

PAHs may be synthesized from saturated hydrocarbons under oxygen-deficient conditions. Hydrocarbons with very low molecular masses, including even methane, may act as precursors for the polycyclic aromatic compounds. Low-molecular-mass hydrocarbons form PAHs by **pyrosynthesis**. This happens at temperatures exceeding approximately 500°C at which carbon-hydrogen and carbon-carbon bonds are broken to form free radicals. These radicals undergo dehydrogenation and combine chemically to form aromatic ring structures, which are resistant to thermal degradation. Polycyclic aromatic compounds may also be formed from higher alkanes present in fuels and plant materials by the process of **pyrolysis,** the "cracking" of organic compounds to form smaller and less stable molecules and radicals.

Elevated levels of PAH compounds of up to about 20 $\mu g/m^3$ are found in the atmosphere. Elevated levels of PAHs are most likely to be encountered in polluted urban atmospheres and in the vicinity of natural fires, such as forest and prairie fires. Coal furnace stack gas may contain over 1000 $\mu g/m^3$ of PAH compounds and cigarette smoke almost 100 $\mu g/m^3$.

Atmospheric polycyclic aromatic hydrocarbons are found almost exclusively in the solid phase, largely sorbed to soot particles. Soot itself is a highly condensed product of PAHs. Soot contains 1–3% hydrogen and 5–10% oxygen, the latter due to partial surface oxidation. Benzo(a)pyrene adsorbed on soot disappears very rapidly in the presence of light-yielding oxygenated products; the large surface area of the particle contributes to the high rate of reaction.

18.9.   OVERVIEW OF SMOG FORMATION

Mentioned in the introduction to this chapter, photochemical smog is a secondary air pollutant formed when pollutant hydrocarbons, nitrogen oxides, and atmospheric oxygen interact under the influence of ultraviolet radiation from the sun under stagnant air conditions. This section addresses the conditions that are characteristic of a smoggy atmosphere and the overall processes involved in smog formation. In atmospheres that receive hydrocarbon and NO pollution accompanied by intense sunlight and stagnant air masses, oxidants tend to form. In air-pollution parlance, **gross photochemical oxidant** is a substance in the atmosphere capable of oxidizing iodide ion to elemental iodine. The primary oxidant in the atmosphere is ozone. Other atmospheric oxidants include H_2O_2, organic peroxides (ROOR′), organic hydroperoxides (ROOH), and peroxyacyl nitrates, such as PAN (see pages 630 and 633).

The formation of peroxyacetyl nitrate, PAN, is shown in Equation 18.10.8. PAN and its homologs, such as peroxybenzoyl nitrate (PBN),

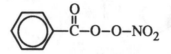

a powerful eye irritant and lachrymator, are produced photochemically in atmospheres containing hydrocarbons and NO_x. PAN, especially, is a notorious organic oxidant. In addition to PAN and PBN, some other specific organic oxidants that may be important in polluted atmospheres are peroxypropionyl nitrate (PPN); peracetic acid, $CH_3(CO)OOH$; acetylperoxide, $CH_3(CO)OO(CO)CH_3$; m-butyl hydroperoxide, CH_3CH_2OOH; and tert-butylhydroperoxide, $(CH_3)_3COOH$.

As shown in Figure 18.2, smoggy atmospheres show characteristic variations with time of day in levels of NO, NO_2, hydrocarbons, aldehydes, and oxidants. Examination of the figure shows that, shortly after sunrise, the level of NO in the atmosphere decreases markedly, a decrease that is accompanied by a peak in the concentration of NO_2. During midday (significantly, after the concentration of NO has fallen to a very low level), the levels of aldehydes and oxidants become relatively high. The concentration of total hydrocarbons in the atmosphere peaks sharply in the morning, then decreases during the remaining daylight hours.

An overview of the processes responsible for the behavior just discussed is

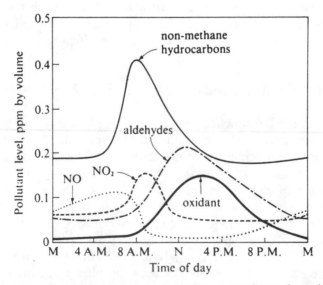

Figure 18.2.   Generalized plot of atmospheric concentrations of species involved in smog formation as a function of time of day.

summarized in Figure 18.3. The chemical basis for the processes illustrated in this figure is explained in Section 18.10.

Smog-Forming Emissions

The production of nitrogen oxides was discussed in Section 17.8. At the high temperature and pressure conditions in an internal combustion engine, products of incompletely burned gasoline undergo chemical reactions which produce several hundred different hydrocarbons. Many of these are highly reactive in forming photochemical smog. In addition to hydrocarbons from automobile exhausts, automobile fuel systems and emissions from the processing and transfer of gasoline produce pollutant hydrocarbons in the atmosphere. Therefore, much of the effort to control photochemical smog has concentrated on reducing and controlling hydrocarbon emissions from automobiles. Catalytic converters are now used to destroy pollutants in exhaust gases. A reduction catalyst is employed to reduce NO in the exhaust gas and, after injection of air, an oxida-

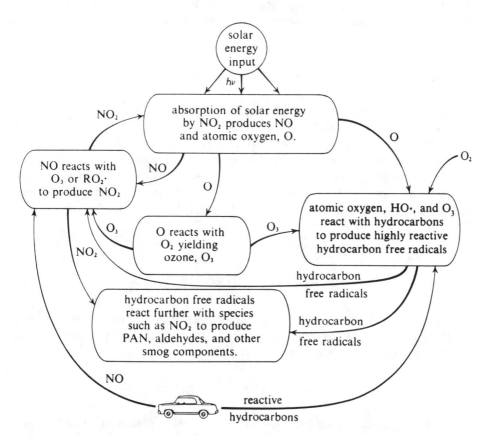

Figure 18.3.   Generalized scheme for the formation of photochemical smog.

tion catalyst oxidizes hydrocarbons and CO. Exhaust catalysts and other control measures, particularly computerized control of engine operating parameters, have reduced automotive emissions remarkably during the last two decades, while fuel efficiency has been improved.

18.10.   SMOG-FORMING REACTIONS OF ORGANIC COMPOUNDS IN THE ATMOSPHERE

Hydrocarbons are eliminated from the atmosphere by a number of chemical and photochemical reactions. These reactions are responsible for the formation of many noxious secondary pollutant products and intermediates from relatively innocuous hydrocarbon precursors. These pollutant products and intermediates make up photochemical smog.

Hydrocarbons and most other organic compounds in the atmosphere are thermodynamically unstable toward oxidation and tend to be oxidized through a series of steps. The oxidation process terminates with formation of CO_2, solid organic particulate matter which settles from the atmosphere, or water-soluble products (for example, acids, aldehydes), which are removed by rain. Inorganic species such as ozone or nitric acid are by-products of these reactions.

Hydrocarbon Reactions Leading to Smog Formation

Some of the major reactions involved in the oxidation of atmospheric hydrocarbons may be understood by considering the oxidation of methane, the most common and widely dispersed atmospheric hydrocarbon. Although methane is the least reactive of the simple hydrocarbons involved in smog formation, it provides a simple molecule to show example reactions that likewise occur with other hydrocarbons. Like other hydrocarbons, methane reacts with oxygen atoms generally produced by the photochemical dissociation of NO_2:

$$NO_2 + h\nu \rightarrow NO + O \qquad (18.10.1)$$

$$CH_4 + O \rightarrow H_3C\cdot + HO\cdot \qquad (18.10.2)$$

The latter reaction generates the all-important hydroxyl radical and an alkyl (methyl) radical. The methyl radical produced reacts rapidly with molecular oxygen to form very reactive peroxyl radicals,

$$H_3C\cdot + O_2 + M \rightarrow H_3COO\cdot + M \qquad (18.10.3)$$

which with methane as the hydrocarbon is the methoxyl radical, $H_3COO\cdot$. Such radicals participate in a variety of subsequent chain reactions, including those leading to smog formation. The hydroxyl radical from Reaction 18.10.2 reacts rapidly with hydrocarbons to form reactive hydrocarbon (alkyl) radicals,

$$CH_4 + HO\cdot \rightarrow H_3C\cdot + H_2O \qquad (18.10.4)$$

in this case, the methyl radical, $H_3C\cdot$. The following are additional reactions that are involved in the overall oxidation of methane:

$$H_3COO\cdot + NO \rightarrow H_3CO\cdot + NO_2 \qquad (18.10.5)$$

(This is a very important kind of reaction in smog formation because the oxidation of NO by peroxyl radicals is the predominant means of regenerating NO_2 in the atmosphere, after it has been photochemically dissociated to NO.)

$$H_3CO\cdot + O_3 \rightarrow \text{various products} \qquad (18.10.6)$$

$$H_3CO\cdot + O_2 \rightarrow CH_2O + HOO\cdot \qquad (18.10.7)$$

$$H_3COO\cdot + NO_2 + M \rightarrow CH_3OONO_2 + M \qquad (18.10.8)$$

(The species CH_3OONO_2 is peroxyacetyl nitrate, PAN, a very strong oxidant.)

$$CH_2O + h\nu \rightarrow \text{photodissociation products} \qquad (18.10.9)$$

Hydroxyl radical, $HO\cdot$, and hydroperoxyl radical, $HOO\cdot$, are ubiquitous intermediates in photochemical chain-reaction processes. These two species are known collectively as odd hydrogen radicals.

Reactions such as (18.10.2) and (18.10.4) are **abstraction reactions** involving the removal of an atom, usually hydrogen, by reaction with an active species. **Addition reactions** of unsaturated organic compounds are also common. Typically, hydroxyl radical reacts with an alkene such as propene to form another reactive free radical:

$$(18.10.10)$$

Ozone adds to unsaturated compounds to form reactive ozonides:

$$(18.10.11)$$

Organic compounds (in the troposphere, almost exclusively carbonyls) can undergo primary photochemical reactions resulting in the direct formation of free radicals. By far the most important of these is the photochemical dissociation of aldehydes:

$$H_3C-\overset{\displaystyle O}{\overset{\|}{C}}-H + h\nu \longrightarrow H_3C\bullet + H\overset{\bullet}{C}O \qquad (18.10.12)$$

Organic free radicals undergo a number of chemical reactions. Hydroxyl radicals may be generated from organic peroxyl reactions such as,

$$H_3C-\underset{\displaystyle H}{\overset{\displaystyle \overset{\bullet}{O}}{\overset{\displaystyle |}{\underset{\displaystyle |}{C}}}}-CH_3 \longrightarrow H_3C-\overset{\displaystyle O}{\overset{\|}{C}}-CH_3 + HO\bullet \qquad (18.10.13)$$

producing aldehydes (among the most noxious smog products) or ketones. The hydroxyl radical may react with other organic compounds, maintaining the chain reaction. Gas-phase reaction chains commonly have many steps. Furthermore, chain-branching reactions take place in which a free radical reacts with an excited molecule, causing it to produce two new radicals. Chain termination may occur in several ways, including reaction of two free radicals,

$$2HO\bullet \rightarrow H_2 + O_2 \qquad (18.10.14)$$

adduct formation with nitric oxide or nitrogen dioxide (which, because of their odd numbers of electrons, are themselves stable free radicals),

$$HO\bullet + NO_2 + M \rightarrow HNO_3 + M \qquad (18.10.15)$$

or reaction of the radical with a solid particle surface.

Hydrocarbons may undergo heterogeneous reactions on particles in the atmosphere. Dusts composed of metal oxides or charcoal have a catalytic effect upon the oxidation of organic compounds. Metal oxides may enter into photochemical reactions. For example, zinc oxide photosensitized by exposure to light promotes oxidation of organic compounds.

As noted earlier, ozone is a characteristic species in smog. It can be formed by the following reaction:

$$O_2 + O + M \rightarrow O_3 + M \qquad (18.10.16)$$

Alkyl nitrates and alkyl nitrites may be formed by the reaction of alkoxyl radicals (such as methoxyl radical, $H_3CO\bullet$) with nitrogen dioxide and nitric oxide, respectively:

$$H_3CO\cdot + NO_2 \rightarrow H_3CONO_2 \text{ (methyl nitrate)} \qquad (18.10.17)$$

$$H_3CO\cdot + NO \rightarrow H_3CONO \text{ (methyl nitrite)} \qquad (18.10.18)$$

Addition reactions with NO_2 such as these are important in terminating the reaction chains involved in smog formation.

18.11. INORGANIC PRODUCTS FROM SMOG

Two major classes of inorganic products from smog are sulfates and nitrates. Inorganic sulfates and nitrates, along with sulfur and nitrogen oxides can contribute to acidic precipitation, corrosion, reduced visibility, and adverse health effects.

Oxidation of SO_2 and H_2S

Although the oxidation of SO_2 to sulfate species is relatively slow in a clean atmosphere, it is much faster under smoggy conditons. During severe photochemical smog conditions, oxidation rates of 5–10% per hour may occur, as compared to only a fraction of a percent per hour under normal atmospheric conditions. Thus, sulfur dioxide exposed to smog can produce very high local concentrations of sulfate, which can aggravate already bad atmospheric conditions.

Several oxidant species in smog can oxidize SO_2. Among the oxidants are molecular species, including O_3, NO_3, and N_2O_5, as well as reactive radical species, particularly $HO\cdot$, $HOO\cdot$, O, $RO\cdot$, and $ROO\cdot$. The two major primary reactions are oxygen transfer,

$$SO_2 + O \text{ (from } O, RO\cdot, RO_2\cdot,) \rightarrow SO_3 \rightarrow H_2O_4, \text{ sulfates} \quad (18.11.1)$$

or addition. As an example of the latter, $HO\cdot$ adds to SO_2 to form a reactive species which can further react with oxygen, nitrogen oxides, or other species to yield sulfates, other sulfur compounds, or compounds of nitrogen:

$$HO\cdot + SO_2 \rightarrow HOSOO\cdot \qquad (18.11.2)$$

The presence of $HO\cdot$ (typically at a level of 3×10^6 radicals/cm^3, but appreciably higher in smoggy atmosphere), makes this a likely route. Addition of SO_2 to $RO\cdot$ or $ROO\cdot$ can yield organic sulfur compounds.

It should be noted that the reaction of H_2S with $HO\cdot$ is quite rapid. As a result, the normal atmospheric half-life of H_2S of about one-half day becomes much shorter in the presence of photochemical smog.

Formation of Nitrates and Nitric Acid

Inorganic nitrates or nitric acid are formed by several reactions in smog. Among the important reactions forming nitric acid are the reaction of N_2O_5 with water and the addition of hydroxyl radical to NO_2. The oxidation of NO or NO_2 to nitrate species may occur after absorption of gas by an aerosol droplet. Nitric acid formed by these reactions reacts with ammonia in the atmosphere to form ammonium nitrate:

$$NH_3 + HNO_3 \rightarrow NH_4NO_3 \qquad (18.11.3)$$

Other nitrate salts may also be formed.

Nitric acid and nitrates are among the more damaging end-products of smog. In addition to possible adverse effects on plants and animals, they cause severe corrosion problems. Electrical relay contacts and small springs associated with electrical switches are especially susceptible to damage from nitrate-induced corrosion.

18.12.   EFFECTS OF SMOG

The harmful effects of smog occur mainly in the areas of (1) human health and comfort, (2) damage to materials, (3) effects on the atmosphere, and (4) toxicity to plants. The exact degree to which exposure to smog affects human health is not known, although substantial adverse effects are suspected. Pungent-smelling, smog-produced ozone is known to be toxic. Ozone at 0.15 ppm causes coughing, wheezing, bronchial constriction, and irritation to the respiratory mucous system in healthy, exercising individuals. Peroxyacyl nitrates and aldehydes found in smog are eye irritants. Materials are adversely affected by some smog components. Rubber has a high affinity for ozone and is cracked and aged by it. Indeed, the cracking of rubber used to be employed as a test for the presence of ozone.

Ozone attacks natural rubber and similar materials by oxidizing and breaking double bonds in the polymer according to the following:

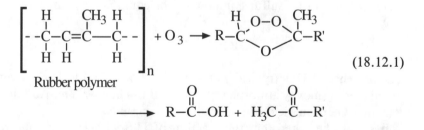

$$(18.12.1)$$

This oxidative scission type of reaction causes bonds in the polymer structure to break and results in deterioration of the polymer.

Aerosol particles that reduce visibility are formed by the polymerization of the smaller molecules produced in smog-forming reactions. Since these reactions largely involve the oxidation of hydrocarbons, it is not suprising that oxygen-containing organics make up the bulk of the particulate matter produced from smog. Ether-soluble aerosols collected from the Los Angeles atmosphere have shown an empirical formula of approximately CH_2O. Among the specific kinds of compounds identified in organic smog aerosols are oxygen-containing alcohols, aldehydes, ketones, organic acids, esters, and organic nitrates.

In view of worldwide shortages of food, the known harmful effects of smog on plants is of particular concern. These effects are largely due to oxidants in the smoggy atmosphere. The three major oxidants involved are ozone, PAN, and nitrogen oxides. Of these, PAN has the highest toxicity to plants, attacking younger leaves and causing "bronzing" and "glazing" of their surfaces. Exposure for several hours to an atmosphere containing PAN at a level of only 0.02–0.05 ppm will damage vegetation. The sulfhydryl group of proteins in organisms is susceptible to damage by PAN. Fortunately, PAN is usually present at only low levels. Nitrogen oxides occur at relatively high concentrations during smoggy conditions, but their toxicity to plants is relatively low. The low toxicity of nitrogen oxides and the usually low levels of PAN leave ozone as the greatest smog-produced threat to plant life.

Typical of the phytotoxicity of O_3, ozone damage to a lemon leaf is often manifested by chlorotic stippling (characteristic yellow spots on a green leaf), as represented in Figure 18.4. Reduction in plant growth may occur without visible lesions on the plant.

Brief exposure to approximately 0.06 ppm of ozone may temporarily cut photosynthesis rates in some plants in half. Crop damage from ozone and other photochemical air pollutants in California alone is estimated to cost millions of dollars each year. The geographic distribution of damage to plants in California is illustrated in Figure 18.5.

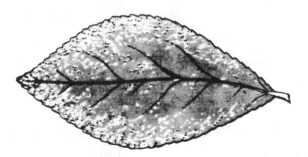

Figure 18.4.   Representation of ozone damage to a lemon leaf. In color, the spots appear as yellow chlorotic stippling on the green upper surface caused by ozone exposure.

Figure 18.5.   Geographic distribution of plant damage from smog in California.

CHAPTER SUMMARY

The chapter summary below is presented in a programmed format to review the main points covered in this chapter. It is used most effectively by filling in the blanks, referring back to the chapter as necessary. The correct answers are given at the end of the summary.

The effects of organic pollutants in the atmosphere may be divided into the two major categories of (1)_____ and (2)_____ _____. The kind of smog discussed in this chapter is (3)_____ _____ smog, the other kind of smog is (4)_____ _____. A smoggy day may be regarded as one having (5)_____

_____

_____

and is characterized by the formation of (6)_____ in the air, particularly ozone.

Most organics in the atmosphere come from (7)_____, largely because of the (8)_____ of (9)_____. Most of the hydrocarbons emitted by plants (predominantly trees) are (10)_____, which are characterized by usually two or more (11)_____ in their molecules. The greatest variety (but not quantity) of organic compounds emitted by plants probably consists of (12)_____.

The source of most of the anthropogenic pollutant hydrocarbons found in the atmosphere consists of (13)_____. The most reactive hydrocarbons introduced into the atmosphere by human activities are those that are (14)_____. Their high reactivity is due to the fact that they (15)_____

_____. Low-molecular mass alkenes have significant potential to be air pollutants because of their widespread use to make (16)_____. Alkenes are also produced from alkanes by (17)_____. Aryl (aromatic) compounds with multiple rings are called (18)_____,

and occur in the atmosphere in the form of (19)_____.

Of the oxygen-containing organic compounds found in the atmosphere, (20)_____ are often the first species formed, other than unstable reaction intermediates, in the photochemical oxidation of atmospheric hydrocarbons. Aldehydes and ketones are known collectively as (21)_____

_____ of which the simplest and most widely produced is (22)_____. Aldehydes are second only to NO_2 as atmospheric sources of (23)_____

_____. Alkenyl aldehydes are especially reactive in the atmosphere because they have (24)_____

_____.

The two alcohols most likely to be found as atmospheric pollutants are (25)_____, because of their (26)_____. Mechanisms for scavenging alcohols from the atmosphere are relatively efficient because (27)_____

_____.

Phenols are (28)_____ alcohols, which are more noted as (29)_____ _____ than as air pollutants. Although ethers are relatively uncommon atmospheric pollutants, diethyl ether is of concern in enclosed atmospheres because of its (30)_____. An ether with a significant potential to be an air pollutant despite its low vapor pressure is (31)_____, widely used as (32)_____. A low-molecular mass oxygen-containing compound that is a mutagen and a carcinogen to experimental animals and is classified as hazardous for both its toxicity and ignitability that has some limited potential to be an air pollutant is (33)_____. Compounds with the functional group,

are called (34)_____, and are produced from atmospheric hydrocarbons as the end products of (35)_____. Their (36)_____ make them susceptible to scavenging from the atmosphere.

The three major classes of halogen-substituted hydrocarbon molecules that may be found in the atmosphere are (37)_____. Alkenyl organohalides contain (38)_____

_____. Of the alkenyl organohalides, one that is highly volatile and a known human carcinogen is (39)_____. Although not very volatile, organohalide (40)_____ can get into the atmosphere from high temperature sources, such as incinerators, and be transported with atmospheric particles. On a global basis, the three most abundant organochlorine compounds in the atmosphere are (41)_____

_____. Of these, the compound that is persistent enough in the atmosphere to constitute a threat to atmospheric ozone is (42)_____

_____.

The two major classes of organohalide compounds that are thought to pose threats to stratospheric ozone are (43)_____. Of these compounds, those that contain bromine are used in (44)_____

_____. In the destruction of stratospheric ozone, these compounds produce (45)_____ by photochemical processes, and this product reacts with O_3 to produce (46)_____. The most prominent instance of ozone layer destruction documented in recent years is the (47)_____. Stratospheric ozone serves its protective function because it (48)_____. Some effects of stratospheric ozone destruction are (49)_____ _____. Chlorofluorocarbon substitutes contain (50)_____, which because of (51)_____ _____ are more easily destroyed by atmospheric chemical reactions.

The most significant air pollution problem with organosulfur compounds is their (52)_____. Major sources of organosulfur compounds in the atmosphere include (53)_____ _____ _____.

Organic nitrogen compounds that may be found as atmospheric contaminants may be classified as (54)_____ _____.

Aromatic amines are of high concern as atmospheric pollutants, particularly in the workplace, because (55)_____ _____.

The amide most likely to be encountered as an atmospheric pollutant is (56)__ _____. Air pollutant nitriles are characterized by the presence of the (57)_____ group. Organonitrogen compounds with the general formula,

$$O{=}N{-}N\underset{\textstyle R'}{\overset{\textstyle R}{\diagup}}$$

are of particular concern as atmospheric contaminants because (58)_____ _____.

The organic particles of greatest concern are (59)_____ _____, which consist of (60)_____ _____. The most often cited example of a PAH compound is (61)_____, a compound that the body can metabolize to (62)_____. Low-molecular-mass hydrocarbons form PAHs by (63)_____ under (64)_____ _____ conditions, and higher alkanes form PAHs by (65)_____ processes.

Photochemical smog is a (66)_____ air pollutant formed when (67)_____ interact under the influence of ultraviolet radiation from the sun under stagnant air conditions. The general type of species characteristic of photochemical smog consists of (68)_____, the most common inorganic example of which is (69)_____. The daily cycle of chemical species in a smoggy atmosphere shows that, shortly after sunrise, the level of (70)_____ in the atmosphere decreases markedly, a decrease that is accompanied by a peak in the concentration of (71)_____. During midday, the levels of (72)_____ become relatively high. Automobile exhausts produce both (73)_____ _____, which are involved in forming photochemical smog.

The first (photochemical) reaction that leads to smog formation is (74)_____ _____. The (75)_____ product of this initial reaction reacts with a hydrocarbon to produce (76)_____ _____. These processes set off a series of (77)_____, including those leading to smog formation. The hydroxyl radical can undergo (78)_____ reactions with alkanes and (79)_____ reactions with alkenes. The reaction $H_3COO\cdot + NO \rightarrow H_3CO\cdot + NO_2$ is especially important in smog formation because it (80)_____.

Two major classes of inorganic products from smog are (81)_____,
which, along with sulfur and nitrogen oxides can contribute to (82)_____
_____. Although the oxidation of SO_2 to
sulfate species is relatively (83)_____ in a clean atmosphere, it is much
(84)_____ in a smoggy atmosphere. This is due to the presence of
(85)_____ under smoggy conditions. Nitric acid formed under conditions
of photochemical smog reacts with (86)_____ in the atmosphere
to form (87)_____.

The harmful effects of smog occur mainly in the areas of (88)_____
_____
_____.

(89)_____ has a high affinity for ozone and is cracked and aged by it.
The three major oxidants involved in damaging plants are (90)_____
_____, of
which (91)_____ has the highest toxicity to plants.

Answers

1. direct effects
2. effects from formation of secondary pollutants
3. photochemical or oxidizing
4. reducing or sulfurous smog
5. moderate to severe eye irritation or visibility below 3 miles when the relative
   humidity is below 60%
6. oxidants
7. natural sources
8. microbial production
9. methane
10. terpenes
11. alkenyl bonds
12. esters
13. petroleum products, primarily gasoline
14. by-products of the partial combustion of other hydrocarbons
15. tend to be unsaturated and relatively reactive
16. polymers
17. high temperature cracking processes in engines
18. polycyclic aromatic hydrocarbons, PAH

19. particulate matter
20. aldehydes and ketones
21. carbonyl compounds
22. formaldehyde
23. free radicals produced by the absorption of light
24. both double bonds and carbonyl groups
25. methanol and ethanol
26. high volatility
27. the lower alcohols are quite water-soluble and the higher ones have low vapor pressures
28. aryl
29. water pollutants
30. flammability hazard
31. methyltertiarybutyl ether, MTBE
32. a gasoline octane booster
33. ethylene oxide
34. carboxylic acids
35. photochemical oxidation
36. low vapor pressures and water solubilities
37. alkyl halides, alkenyl halides, and aryl halides
38. at least one halogen atom and at least one carbon-carbon double bond
39. vinyl chloride
40. PCBs
41. methyl chloride, methyl chloroform, and carbon tetrachloride
42. methyl chloroform
43. chlorofluorocarbons and halons
44. fire extinguishing systems
45. Cl atoms
46. ClO radical
47. Antarctic ozone hole
48. absorbs ultraviolet radiation very strongly in the region 220–330 nm
49. adverse effects on plants, destruction of phytoplankton, cataracts, skin cancer
50. H atoms
51. their more readily broken H-C bonds
52. foul odors
53. microbial degradation, wood pulping, volatile matter evolved from plants, animal wastes, packing house and rendering plant wastes, starch manufacture, sewage treatment, and petroleum refining.
54. amines, amides, nitriles, nitro compounds, or heterocyclic nitogen compounds
55. some are known to cause urethral tract cancer (particularly of the bladder) in exposed individuals
56. dimethylformamide

57. $-C\equiv N$
58. some are known carcinogens
59. polycyclic aromatic hydrocarbons (PAHs)
60. condensed ring aromatic molecules
61. benzo(a)pyrene
62. a carcinogenic form
63. pyrosynthesis
64. oxygen-deficient
65. pyrolysis
66. secondary
67. pollutant hydrocarbons, nitrogen oxides, and atmospheric oxygen
68. oxidants
69. ozone
70. NO
71. NO_2
72. aldehydes and oxidants
73. hydrocarbons and nitrogen oxides
74. $NO_2 + h\nu \rightarrow NO + O$
75. O atom
76. an alkyl radical and a hydroxyl radical
77. chain reactions
78. abstraction
79. addition
80. regenerates NO_2 from NO
81. sulfates and nitrates
82. acidic precipitation, corrosion, reduced visibility, and adverse health effects
83. slow
84. faster
85. oxidants
86. ammonia
87. ammonium nitrate
88. (1) human health and comfort, (2) damage to materials, (3) effects on the atmosphere, and (4) toxicity to plants
89. Rubber
90. ozone, PAN, and nitrogen oxides
91. PAN

QUESTIONS AND PROBLEMS

1. Of the following species, the one most likely to be found in reducing smogs is: ozone, relatively high levels of atomic oxygen, SO_2, PAN, PBN.

2. Why are automotive exhaust pollutant hydrocarbons even more damaging to the environment than their quantities would indicate?

3. At what point in the smog-producing chain reaction is PAN formed?

4. Which of the following species reaches its peak value last on a smog-forming day: NO, oxidants, hydrocarbons, NO_2?

5. What is the main species responsible for the oxidation of NO to NO_2 in a smoggy atmosphere?

6. Some atmospheric chemical reactions are abstraction reactions and others are addition reactions. Which of these applies to the reaction of hydroxyl radical with propane? With propene (propylene)?

7. How might oxidants be detected in the atmosphere?

8. Why is ozone especially damaging to rubber?

9. Show how hydroxyl radical, $HO\cdot$, might react differently with ethylene, $H_2C=CH_2$, and methane, CH_4.

10. Why are hydrocarbon emissions from uncontrolled automobile exhaust particularly reactive?

11. What important photochemical property do carbonyl compounds share with NO_2?

SUPPLEMENTARY REFERENCES

Finlayson-Pitts, B. J., and J. N. Pitts, *Atmospheric Chemistry*, (New York: John Wiley and Sons, Inc., 1986), p. 478.

Seinfeld, J. H., *Atmospheric Chemistry and Physics of Air Pollution*, (New York: John Wiley and Sons, Inc., 1986).

Fawell, J. K., and S. Hunt, *Environmental Toxicology: Organic Pollutants*, (New York, NY: John Wiley and Sons, 1988).

Watson, A. Y., R. R. Bates, D. Kennedy, Eds., *Air Pollution, the Automobile, and Public Health*, (Washington, DC: National Academy Press, 1988).

Warneck, P., *Chemistry of the Natural Atmosphere*, (San Diego, CA: Academic Press, 1988).

Schneider, T., Ed., *Atmospheric Ozone Research and its Policy Implications,* (New York, NY: Elsevier Science Publishing Co., 1989).

IARC, *Diesel and Gasoline Engine Exhausts and Some Nitroarenes*, World Health Organization. (Albany, NY: WHO Publications Center USA, 1989).

Howard, P. H., *Handbook of Environmental Fate and Exposure Data for Organic Chemicals*, Vol. 1: *Large Production and Priority Pollutants*, (Chelsea, MI: Lewis Publishers, 1989.)

Bryce-Smith, D., and A. Gilbert, *Photochemistry*, Vol. 20, (Letchworth, England: Royal Society of Chemistry, 1989).

Chameides, W. L., and D. D. Davis, "Chemistry in the Troposphere," *Chemical and Engineering News*, October 4, 1982, pp. 38–52.

Warneck, P., *Chemistry of the Natural Atmosphere*, (San Diego, CA: Academic Press, 1988).

Seinfeld, J. H., "Urban Air Pollution: State of the Science," *Science*, 243, February 10, 1989, pp. 745–752.

Khalil, M. A. K., and R. A. Rasmussen, "Atmospheric Methane: Recent Global Trends," *Environmental Science and Technology*, 24, 549–553 (1991).

19 THE NATURE, SOURCES, AND ENVIRONMENTAL CHEMISTRY OF HAZARDOUS SUBSTANCES

19.1. INTRODUCTION

A hazardous substance is a material that may pose a danger to living organisms, materials, structures, or the environment by explosion or fire hazards, corrosion, toxicity to organisms, or other detrimental effects. What then is a hazardous waste? Although it has been stated that,[1] "The discussion on this question is as long as it is fruitless," a simple definition of a **hazardous waste** is that it is a hazardous substance that has been discarded, abandoned, neglected, released or designated as a waste material. In a simple sense a hazardous waste is a material that has been left somewhere such that it may cause harm if encountered.

History of Hazardous Substances

Humans have always been exposed to hazardous substances going back to prehistoric times when they inhaled noxious volcanic gases or succumbed to carbon monoxide from inadequately vented fires in cave dwellings sealed too well against Ice-Age cold. Slaves in Ancient Greece developed lung disease from weaving mineral asbestos fibers into cloth to make it more degradation-resistant. Some archaeological and historical studies have concluded that lead wine containers were a leading cause of lead poisoning in the more affluent ruling class of the Roman Empire leading to erratic behavior such as fixation on spectacular sporting events, chronic unmanageable budget deficits, poorly regulated financial institutions, and ill-conceived, overly ambitious military ventures in foreign lands. Alchemists who worked during the Middle Ages often suffered debilitating injuries and illnesses resulting from the hazards of their explosive and toxic chemicals. During the 1700s, runoff from mine spoils or tailing piles began to create serious contamination problems in Europe. As the production of dyes and other organic chemicals developed from the coal tar industry in Germany during the 1800s, pollution and poisoning from coal tar by-products was observed. By around 1900 the quantity and variety of chemical wastes produced each year was increasing sharply with the addition of wastes such as spent steel and iron pickling liquor, lead battery wastes, chromic wastes, petroleum refinery wastes, radium wastes, and fluoride wastes from aluminum ore refining. As the century progressed into the World War II era, the wastes and hazardous by-products of manufacturing increased markedly from sources such as chlorinated solvents manufacture, pesticides synthesis, polymers manufacture, plastics, paints, and wood preservatives.

The Love Canal affair of the 1970s and 1980s brought hazardous wastes to the public attention as a major political issue in the U.S. Starting around 1940, this site in Niagara Falls, New York, had received about 20,000 metric tons of chemical wastes containing at least 80 different chemicals. By 1992, state and federal governments had spent well over $100 million to clean up the site and relocate residents.

Other areas containing hazardous wastes that received attention included an industrial site in Woburn, Massachusetts, that had been contaminated by wastes from tanneries, glue-making factories, and chemical companies dating back to about 1850; the Stringfellow Acid Pits near Riverside, California; the Valley of the Drums in Kentucky; and Times Beach, Missouri, an entire town that was abandoned because of contamination by TCDD (dioxin).

Legislation

Governments in a number of nations have passed legislation to deal with hazardous substances and wastes. In the U.S. such legislation has included the following:

- Toxic Substances Control Act of 1976
- Resource, Conservation and Recovery Act (RCRA) of 1976 [amended and strengthened by the Hazardous and Solid Wastes Amendments (HSWA) of 1984]
- Comprehensive Environmental Response, Compensation, and Liability Act (CERCLA) of 1980.

RCRA legislation charged the U.S. Environmental Protection Agency (EPA) with protecting human health and the environment from improper management and disposal of hazardous wastes by issuing and enforcing regulations pertaining to such wastes. RCRA requires that hazardous wastes and their characteristics be identified and managed from the time of their generation until their proper disposal or destruction.

CERCLA (Superfund) legislation deals with actual or potential releases of hazardous materials that have the potential to endanger people or the surrounding environment at uncontrolled or abandoned hazardous waste sites in the U.S.

The act requires responsible parties or the U.S. government to clean up waste sites.

Hazardous Wastes and Air and Water Pollution Control

Somewhat paradoxically, measures taken to reduce air and water pollution (Figure 19.1) have had a tendency to increase production of hazardous wastes. Most water treatment processes yield sludges or concentrated liquors that require stabilization and disposal. Hazardous substances, such as heavy metals, removed by the pollution control equipment into the sludges or concentrated liquors may render them hazardous. Air scrubbing processes likewise produce sludges that may be hazardous. Additives required for efficient scrubbing may

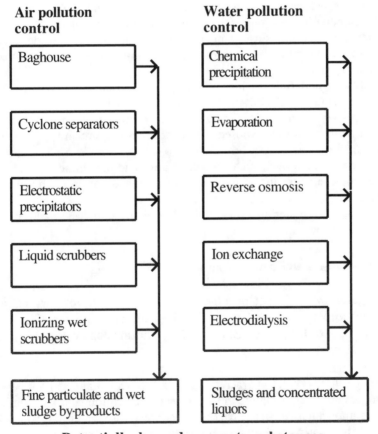

Figure 19.1.   Potential contributions of air and water pollution control measures to hazardous wastes production.

increase the quantities of sludges to volumes considerably larger than those of the original hazardous substances. Baghouses and precipitators used to control air pollution all yield significant quantities of solids, some of which are hazardous.

19.2   CLASSIFICATION OF HAZARDOUS SUBSTANCES AND WASTES

Many specific chemicals in widespread use are hazardous because of their chemical reactivities, fire hazards, toxicities, and other properties. There are numerous kinds of hazardous substances, usually consisting of mixtures of specific chemicals. These include the following:

- **Explosives,** such as dynamite, or ammunition
- **Compressed gases,** such as hydrogen and sulfur dioxide
- **Flammable liquids,** such as gasoline and aluminum alkyls
- **Flammable solids,** such as magnesium metal, sodium hydride, and calcium carbide that burn readily, are water-reactive, or spontaneously combustible
- **Oxidizing materials,** such as lithium peroxide, that supply oxygen for the combustion of normally nonflammable materials
- **Corrosive materials,** including oleum, sulfuric acid and caustic soda, which may injure exposed flesh or cause disintegration of metal containers
- **Poisonous materials,** such as hydrocyanic acid or aniline
- **Etiologic agents,** including causative agents of anthrax, botulism, or tetanus
- **Radioactive materials,** including plutonium, cobalt-60, and uranium hexafluoride.

Hazardous Wastes Defined by Characteristics

For regulatory and legal purposes in the U.S. hazardous wastes are listed specifically and are defined according to general characteristics. Under the authority of RCRA, the EPA defines hazardous wastes in terms of **characteristics:**

- **Ignitability,** (D001) characteristic of substances that are liquids whose vapors are likely to ignite in the presence of ignition sources, nonliquids that may catch fire from friction or contact with water and which burn vigorously or persistently, ignitable compressed gases, and oxidizers.
- **Corrosivity,** (D002) characteristic of substances that exhibit extremes of acidity or basicity or a tendency to corrode steel.
- **Reactivity,** (D003) characteristic of substances that have a tendency to

undergo violent chemical change (an explosive substance is an obvious example).
- **Toxicity**, (D004–D043) defined in terms of a standard extraction procedure followed by chemical analysis for specific substances.

Listed Hazardous Wastes

In addition to classification by characteristics, EPA designates more than 450 **listed hazardous wastes** which are specific substances or classes of substances determined to be hazardous. Each such substance is assigned an EPA **hazardous waste number** in the format of a letter followed by 3 numerals, where a different letter is assigned to substances from each of the four following lists:

- **F list hazardous wastes from nonspecific sources**: For example, quenching waste water treatment sludges from metal heat treating operations where cyanides are used in the process (F012).
- **K list hazardous wastes from specific sources**: For example, heavy ends from the distillation of ethylene dichloride in ethylene dichloride production (K019).
- **P list acute hazardous wastes**: These are mostly specific chemical species such as fluorine (P056) or 3–chloropropane nitrile (P027).
- **U list general hazardous wastes**: These are predominantly specific compounds such as calcium chromate (U032) or phthalic anhydride (U190).

19.3. FLAMMABLE AND COMBUSTIBLE SUBSTANCES

In a broad sense a **flammable substance** is something that will burn readily, whereas a **combustible substance** requires relatively more persuasion to burn. Before trying to sort out these definitions it is necessary to define several other terms. Most chemicals that are likely to catch fire accidentally and burn are liquids. Liquids form **vapors** which are usually more dense than air, and thus tend to settle. The tendency of a liquid to ignite is measured by a test in which the liquid is heated and periodically exposed to a flame until the mixture of vapor and air ignites at the liquid's surface. The temperature at which this occurs is called the **flash point**.

With these definitions in mind it is possible to divide ignitable materials into four major classes. A **flammable solid** is one that can ignite from friction or from heat remaining from its manufacture, or which may cause a serious hazard if ignited. Explosive materials are not included in this classification. A **flammable liquid** is one having a flash point below 37.8°C (100°F). A **combustible liquid** has a flash point in excess of 37.8°C, but below 93.3°C. Gases are sub-

Table 19.1. Flammabilities of Some Common Organic Liquids

Liquid	Flash Point (°C)[a]	Volume Percent in Air	
		LFL[b]	UFL[b]
Diethyl ether	−43	1.9	36
Pentane	−40	1.5	7.8
Acetone	−20	2.6	13
Toluene	4	1.27	7.1
Methanol	12	6.0	37
Gasoline (2,2,4-tri-methylpentane)	−	1.4	7.6
Naphthalene	157	0.9	5.9

[a]Closed-cup flash point test.
[b]LFL, lower flammability limit; UFL, upper flammability limit at 25°C.

stances that exist entirely in the gaseous phase at 0°C and 1 atm pressure. A **flammable compressed gas** meets specified criteria for lower flammability limit, flammability range (see below), and flame projection.

In considering the ignition of vapors, two important concepts are those of flammability limit and flammability range. Values of the vapor/air ratio below which ignition cannot occur because of insufficient fuel define the lower **flammability limit**. Similarly, values of the vapor/air ratio above which ignition cannot occur because of insufficient air define the upper flammability limit. The difference between upper and lower flammability limits at a specified temperature is the **flammability range**. Table 19.1 gives some examples of these values for common liquid chemicals.

Combustion of Finely Divided Particles

Finely divided particles of combustible materials are somewhat analogous to vapors in respect to flammability. One such example is a spray or mist of hydrocarbon liquid in which oxygen has the opportunity for intimate contact with the liquid particles; the liquid may ignite at a temperature below its flash point.

Dust explosions can occur with a large variety of solids that have been ground to a finely divided state. Many metal dusts, particularly those of magnesium and its alloys, zirconium, titanium, and aluminum can burn explosively in air. In the case of aluminum, for example, the highly exothermic (heat-releasing) reaction is the following:

$$4Al(powder) + 3O_2(from\ air) \rightarrow 2Al_2O_3 \qquad (19.3.1)$$

Coal dust and grain dusts have caused many fatal fires and explosions in coal mines and grain elevators, respectively. Dusts of polymers such as cellulose acetate, polyethylene, and polystyrene can also be explosive.

Oxidizers

Combustible substances are reducing agents that react with **oxidizers** (oxidizing agents or oxidants) to produce heat. Diatomic oxygen, O_2, from air is the most common oxidizer. Many oxidizers are chemical compounds that contain oxygen in their formulas. The halogens and many of their compounds are oxidizers. Some examples of oxidizers are given in Table 19.2.

An example of a reaction of an oxidizer is that of concentrated HNO_3 with copper metal, which gives toxic NO_2 gas as a product:

$$4HNO_3 + Cu \rightarrow Cu(NO_3)_2 + 2H_2O + 2NO_2 \qquad (19.3.2)$$

The toxic effects of some oxidizers are due to their ability to oxidize biomolecules in living systems.

Whether or not a substance acts as an oxidizer depends upon the reducing strength of the material that it contacts. For example, carbon dioxide is a common fire extinguishing material that can be sprayed onto a burning substance to keep air away. However, aluminum is such a strong reducing agent that carbon dioxide in contact with hot, burning aluminum reacts as an oxidizing agent to give off toxic combustible carbon monoxide gas:

Table 19.2.   Examples of Some Oxidizers

Name	Formula	State of Matter
Ammonium nitrate	NH_4NO_3	Solid
Ammonium perchlorate	NH_4ClO_4	Solid
Bromine	Br_2	Liquid
Chlorine	Cl_2	Gas (stored as liquid)
Fluorine	F_2	Gas
Hydrogen peroxide	H_2O_2	Solution in water
Nitric acid	HNO_3	Concentrated solution
Nitrous oxide	N_2O	Gas (stored as liquid)
Ozone	O_3	Gas
Perchloric acid	$HClO_4$	Concentrated solution
Potassium permanganate	$KMnO_4$	Solid
Sodium dichromate	$Na_2Cr_2O_7$	Solid

$$2Al + 3CO_2 \rightarrow Al_2O_3 + 3CO \qquad (19.3.3)$$

Oxidizers can contribute strongly to fire hazards because fuels may burn explosively in contact with an oxidizer.

Spontaneous Ignition

Substances that catch fire spontaneously in air without an ignition source are called **pyrophoric**. These include several elements—white phosphorus, the alkali metals (group 1A), and powdered forms of magnesium, calcium, cobalt, manganese, iron, zirconium and aluminum. Also included are some organometallic compounds, such as lithium ethyl (LiC_2H_4) and lithium phenyl (LiC_6H_5), and some metal carbonyl compounds, such as iron pentacarbonyl, $Fe(CO)_5$. Another major class of pyrophoric compounds consists of metal and metalloid hydrides, including lithium hydride, LiH; pentaborane, B_5H_9; and arsine, AsH_3. Moisture in air is often a factor in spontaneous ignition. For example, lithium hydride undergoes the following reaction with water from moist air:

$$LiH + H_2O \rightarrow LiOH + H_2 + heat \qquad (19.3.4)$$

The heat generated from this reaction can be sufficient to ignite the hydride so that it burns in air:

$$2LiH + O_2 \rightarrow Li_2O + H_2O \qquad (19.3.5)$$

Many mixtures of oxidizers and oxidizable chemicals catch fire spontaneously and are called **hypergolic mixtures**. Nitric acid and phenol form such a mixture as does potassium permanganate mixed with glycerin.

Toxic Products of Combustion

Some of the greater dangers of fires are from toxic products and by-products of combustion. The most obvious of these is carbon monoxide, CO, which can cause serious illness or death because it forms carboxyhemoglobin with hemoglobin in the blood so that the blood no longer carries oxygen to body tissues. Toxic SO_2, P_4O_{10}, and HCl are formed by the combustion of sulfur, phosphorus, and organochloride compounds, respectively. A large number of noxious organic compounds such as aldehydes are generated as by-products of combustion. In addition to forming carbon monoxide, combustion under oxygen-deficient conditions produces polycyclic aromatic hydrocarbons consisting of fused ring structures. Some of these compounds, such as benzo(a)pyrene, below, are precarcinogens that are acted upon by enzymes in the body to yield cancer-producing metabolites.

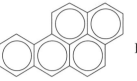

Benzo(a)pyrene

19.4.   REACTIVE SUBSTANCES

Reactive substances are those that tend to undergo rapid or violent reactions under certain conditions. Such substances include those that react violently or form potentially explosive mixtures with water. An example is sodium metal which reacts strongly with water as follows:

$$2Na + 2H_2O \rightarrow 2NaOH + H_2 + heat \qquad (19.4.1)$$

This reaction usually generates enough heat to ignite the sodium. Explosives constitute another class of reactive substances. For regulatory purposes hazardous wastes are also classified as reactive if they react with water, acid, or base to produce toxic fumes, particularly those of hydrogen sulfide or hydrogen cyanide.

Heat and temperature are usually very important factors in reactivity. Many reactions require energy of activation to get them started. The rates of most reactions tend to increase sharply with increasing temperature, and most chemical reactions give off heat. Therefore, once a reaction is started in a reactive mixture lacking an effective means of heat dissipation, the rate may increase exponentially with time, leading to an uncontrollable event. Other factors that may affect reaction rate include physical form of reactants (for example, a finely divided metal powder that reacts explosively with oxygen, whereas a single mass of metal barely reacts), rate and degree of mixing of reactants, degree of dilution with nonreactive media (solvent), presence of a catalyst, and pressure.

Chemical Structure and Reactivity

As shown in Table 19.3, some chemical structures are associated with high reactivity. High reactivity in some organic compounds results from unsaturated bonds in the carbon skeleton, particularly where multiple bonds are adjacent (allenes, $C=C=C$) or separated by only one carbon-carbon single bond (dienes, $C=C-C=C$). Some organic structures involving oxygen are very reactive. Examples are oxiranes, such as ethylene oxide,

Table 19.3.   Examples of Reactive Compounds and Structures

Name	Structure or Formula
Organic	
Allenes	$C=C=C$
Dienes	$C=C-C=C$
Azo compounds	$C=N-N=C$
Triazenes	$C-N=N-N$
Hydroperoxides	$R-OOH$
Peroxides	$R-OO-R'$
Alkyl nitrates	$R-O-NO_2$
Nitro compounds	$R-NO_2$
Inorganic	
Nitrous oxide	N_2O
Nitrogen halides	NCl_3, NI_3
Interhalogen compounds	$BrCl$
Halogen oxides	ClO_2
Halogen azides	ClN_3
Hypohalites	$NaClO$

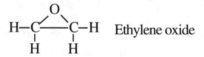

hydroperoxides (ROOH), and peroxides (ROOR′), where R and R′ stand for hydrocarbon moieties such as the methyl group, –CH₃. Many organic compounds containing nitrogen, along with carbon and hydrogen are very reactive. Included are triazenes (R–N=N–N), some azo compounds (R–N=N–R′), and some nitriles:

$$R-C\equiv N \quad \text{Nitrile}$$

Functional groups containing both oxygen and nitrogen tend to impart reactivity to an organic compound. Examples are alkyl nitrates (R–O–NO₂), alkyl nitrites (R–O–N=O), nitroso compounds (R–N=O), and nitro compounds (R–NO₂).

   Many different classes of inorganic compounds are reactive. These include some of the halogen compounds of nitrogen (shock-sensitive nitrogen triiodide, NI₃, is an outstanding example), compounds with metal-nitrogen bonds, halogen oxides (ClO₂), and compounds with oxyanions of the halogens. An example of the last group of compounds is ammonium perchlorate, NH₄ClO₄.

   Explosives such as nitroglycerin or TNT that are single compounds containing both oxidizing and reducing functions in the same molecule are called **redox compounds**. Some redox compounds have more oxygen than is needed for a complete reaction and are said to have a positive balance of oxygen, some have exactly the stoichiometric quantity of oxygen required (zero balance, maximum energy release), and others have a negative balance and require oxygen from outside sources to completely oxidize all components.

19.5. CORROSIVE SUBSTANCES

Conventionally, **corrosive substances** are regarded as those that dissolve metals or cause oxidized material to form on the surface of metals—rusted iron is a prime example. In a broader sense corrosives cause deterioration of materials, including living tissue, that they contact. Most corrosives belong to at least one of the four following chemical classes: (1) strong acids, (2) strong bases, (3) oxidants, (4) dehydrating agents. Table 19.4 lists some of the major corrosive substances and their effects.

Table 19.4.   Examples of Some Corrosive Substances

Name and formula	Properties and Effects
Sulfuric acid, H_2SO_4	Strong acid, dehydrating agent, and oxidant, tremendous affinity for water, including water in exposed flesh
Nitric acid, HNO_3	Strong acid and strong oxidizer, corrodes metal, reacts with protein in tissue to form yellow xanthoproteic acid, lesions are slow to heal
Hydrochloric acid, HCl vapor	Strong acid, corrodes metals, gives off HCl gas vapor, which can damage respiratory tract tissue
Hydrofluoric acid, HF	Corrodes metals, dissolves glass, causes particularly bad burns to flesh
Alkali metal hydroxides, KOH	Strong bases, corrodes zinc, lead, and NaOH and aluminum, caustic substances that dissolve tissue and cause severe burns
Hydrogen peroxide, H_2O_2	Oxidizer, all but very dilute solutions cause severe burns
Interhalogen compounds such as ClF, BrF_3	Powerful corrosive irritants that acidify, oxidize, and dehydrate tissue
Halogen oxides such as OF_2, Cl_2O, Cl_2O_7	Powerful corrosive irritants that acidify, oxidize, and dehydrate tissue
Elemental fluorine, chlorine, bromine $(F_2, Cl_2, Br_2,)$	Very corrosive to mucous membranes and moist tissue, strong irritants

19.6. TOXIC SUBSTANCES

Toxicity is of the utmost concern in dealing with hazardous substances. This includes both long-term chronic effects from continual or periodic exposures to low levels of toxicants and acute effects from a single large exposure. Toxic substances are covered in greater detail in Chapter 21.

Toxicity Characteristic Leaching Procedure

For regulatory and remediation purposes a standard test is needed to measure the likelihood of toxic substances getting into the environment and causing harm to organisms. The test required by the U.S. EPA is the **Toxicity Characteristic Leaching Procedure** (TCLP) which is based on the mobility of both organic and inorganic contaminants present in liquid, solid, and multiphasic hazardous wastes. For analysis of toxic species a test solution is leached or filtered from a test sample of the waste and is designated as the TCLP extract. After the TCLP extract is separated from the solids, it is analyzed for a number of specified volatile organic compounds, semivolatile organic compounds, and elements to determine if their concentrations exceed specified levels for these contaminants.

19.7. PHYSICAL FORMS AND SEGREGATION OF WASTES

Three major categories of wastes based upon their physical forms are **organic materials, aqueous materials,** and **sludges**. These forms largely determine the course of action taken in treating and disposing of the wastes. The **level of segregation**, a concept illustrated in Figure 19.2, is very important in treating, storing, and disposing of different kinds of wastes. It is relatively easy to deal with wastes that are not mixed with other kinds of wastes; that is, those that are highly segregated. For example, spent hydrocarbon solvents can be used as fuel in boilers. However, if these solvents are mixed with spent organochloride solvents, the production of contaminant hydrogen chloride during combustion may prevent fuel use and require disposal in special hazardous waste incinerators. Further mixing with inorganic sludges adds mineral matter and water. These impurities complicate the treatment processes required by producing mineral ash in incineration or lowering the heating value of the material incinerated because of the presence of water. Among the most difficult types of wastes to handle and treat are those with the least segregation, of which a "worst case scenario" would be "dilute sludge consisting of mixed organic and inorganic wastes," as shown in Figure 19.2.

Concentration of wastes is an important factor in their management. A waste that has been concentrated or preferably never diluted is generally much easier and more economical to handle than one that is dispersed in a large quantity of water or soil. Dealing with hazardous wastes is greatly facilitated when the

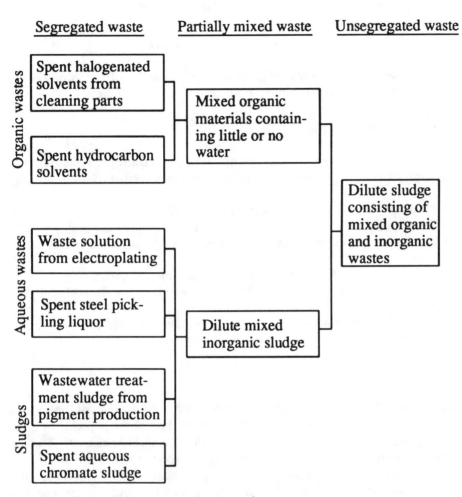

Figure 19.2.   Illustration of waste segregation.

original quantities of wastes are minimized and the wastes remain separated and concentrated insofar as possible.

19.8.   ENVIRONMENTAL CHEMISTRY OF HAZARDOUS WASTES

Having outlined the nature and sources of hazardous substances and hazardous wastes earlier in this chapter, it is now possible to discuss their environmental chemistry. The environmental chemistry of hazardous wastes in the environment may be considered on the basis of the definition of environmental chemistry (Section 1.1) according to the following factors:

- Origin   • Transport   • Reactions
    • Effects   • Ultimate fate

In addition, consideration must be given to the distribution of hazardous wastes among the geosphere, hydrosphere, atmosphere, and biosphere.

Origin of Hazardous Wastes

For purposes of discussion in this chapter, *origin* of hazardous wastes refers to their points of entry into the environment. These may consist of the following:

- Deliberate addition to soil, water, or air by humans
- Evaporation or wind erosion from waste dumps into the atmosphere
- Leaching from waste dumps into groundwater, streams, and bodies of water
- Leakage, such as from underground storage tanks or pipelines
- Evolution and subsequent deposition by accidents, such as fire or explosion
- Release from improperly operated waste treatment, storage, and disposal facilities

Transport of Hazardous Wastes

The transport (sometimes called chemodynamic transport) of hazardous wastes in the environment is largely a function of their physical properties, the physical properties of the surrounding matrix, the physical conditions to which they are subjected, and chemical factors. Highly volatile hazardous wastes are obviously more likely to be transported through the atmosphere and more soluble ones to be carried by water. Hazardous wastes will move farther, faster in porous sandy formations than in denser soils. Volatile hazardous wastes are more mobile under hot, windy conditions, and soluble ones during periods of heavy rainfall. Hazardous wastes that are more chemically and biochemically reactive will not move as far as less reactive hazardous wastes before breaking down.

Physical Factors

The major physical properties of hazardous wastes that determine their amenability to transport are volatility, solubility, and the degree to which they are sorbed to solids, including soil and sediments.

The distribution of hazardous waste constituents between the atmosphere and the geosphere or hydrosphere is largely a function of compound volatility. Compound volatilities are usually measured by vapor pressures, which vary over a wide range.

Usually, in the hydrosphere, and often in soil, hazardous waste constituents are dissolved in water; therefore, the tendency of water to hold the constituent is a factor in its mobility. For example, although ethyl alcohol has a higher evaporation rate and lower boiling temperature than toluene, vapor of the latter compound is more readily evolved from soil because of its limited solubility in water compared to ethanol, which is miscible with water.

Chemical Factors

As an illustration of chemical factors involved in transport of wastes, consider largely cationic inorganic species. Inorganic species can be divided into three groups based upon their retention by clay minerals. Elements that tend to be highly retained by clay include cadmium, mercury, lead, and zinc. Potassium, magnesium, iron, silicon, and NH_4^+ are moderately retained by clay, whereas sodium, chloride, calcium, manganese, and boron are poorly retained. The retention of the last three elements is probably biased in that they are leached from clay, so that negative retention (elution) is often observed. It should be noted, however, that the retention of iron and manganese is a strong function of oxidation state in that the reduced forms of Mn and Fe are relatively poorly retained, whereas the oxidized forms of $Fe_2O_3 \cdot xH_4O$ and MnO_2 are very insoluble and stay on soil as solids.

Reactions of Hazardous Wastes

A myriad of environmental chemical and environmental biochemical processes operate on hazardous wastes in water, air, and soil. One of the important results of this is that, in addition to primary pollutants added directly to the environment, there are secondary pollutants as well, which may be even more harmful than their precursor species. Reactions that result in the formation of secondary pollutants have been discussed extensively in Chapters 11–18. An example is the oxidation of primary pollutant sulfur dioxide,

$$2SO_2 + O_2 + 2H_2O \rightarrow 2H_2SO_4 \qquad (19.8.1)$$
(several steps and intermediates)

to yield secondary pollutant sulfuric acid, the prime ingredient of acid rain.

Effects of Hazardous Wastes

The effects of hazardous wastes in the environment may be divided among effects on organisms, effects on materials, and effects on the environment. These are addressed briefly here and in greater detail in later sections.

The ultimate concern with hazardous wastes has to do with their toxic effects to animals, plants, and microbes. Virtually all hazardous wastes are poisonous to a degree, some extremely so. The toxicity of a waste is a function of many factors, including its chemical nature, the matrix in which it is contained, circumstances of exposure, the species exposed, manner of exposure, degree of exposure, and time of exposure. The toxicities of hazardous wastes are discussed in more detail in Chapter 21, "Toxicological Chemistry."

Many hazardous wastes are *corrosive* (Section 19.5) to materials, usually because of extremes of pH or because of dissolved salt content. Oxidant wastes can cause ignitable substances to burn uncontrollably. Highly reactive hazardous wastes can explode, causing damage to materials and structures. Contamination by hazardous substances, such as by toxic pesticides in grain, can result in products becoming unfit for use.

In addition to their toxic effects in the biosphere, hazardous wastes can damage air, water, and soil. Hazardous wastes that get into air can cause deterioration of air quality, either directly, or by the formation of secondary pollutants. Hazardous waste constituents dissolved in, suspended in, or as surface films on, the surface of water can render it unfit for use and for sustenance of aquatic organisms.

Soil exposed to hazardous wastes can be severely damaged by alteration of its physical and chemical properties and ability to support plants. For example, soil exposed to concentrated brines from petroleum production may become unable to support plant growth so that the soil becomes extremely susceptible to erosion.

Fates of Hazardous Wastes

The fates of hazardous waste substances are addressed in more detail in subsequent sections. As with all environmental pollutants, they eventually reach a state of physical and chemical stability, although that may take many centuries to occur. In some cases, the fate of a hazardous waste is a simple function of its physical properties and surroundings.

The fate of a hazardous waste in water is a function of its solubility, density, biodegradability, and chemical reactivity. Dense, water-immiscible liquids may simply sink to the bottoms of bodies of water or aquifers and accumulate there as "blobs" of liquid. This has happened, for example, with hundreds of tons of PCB wastes that have accumulated in sediments in the Hudson River in New York State. Biodegradable compounds are broken down by bacteria, a process for which the availability of oxygen is an important variable. Substances that readily undergo bioaccumulation are taken up by organisms, exchangeable cationic materials become bound to sediments, and organophilic constituents may be sorbed by organic matter in sediments.

The fates of hazardous waste substances in the atmosphere are often determined by photochemical reactions. Ultimately, such substances may be converted to nonvolatile, insoluble matter and precipitate from the atmosphere onto soil or plants.

19.9.   HAZARDOUS WASTES IN THE GEOSPHERE

The sources, transport, interactions, and fates of the constituents in the geosphere involve a complex scheme, some aspects of which are illustrated in Figure 19.3. The primary environmental concern regarding hazardous wastes in the geosphere is the possible contamination of groundwater aquifers by waste leachates and leakage from wastes. As the figure shows, there are a number of possible contamination sources. The most obvious one is leachate from landfills containing hazardous wastes. In some cases, liquid hazardous materials are placed in lagoons, which can leak into aquifers. Leaking sewers can also result in contamination, as can the discharge from septic tanks. Hazardous wastes spread on land can result in aquifer contamination by leachate. Hazardous wastes are sometimes deliberately disposed underground in waste disposal wells. This means of disposal can result in interchange of contaminated water between surface water and groundwater at discharge and recharge points.

The transport of contaminants in the geosphere depends largely upon the hydrologic factors governing the movement of water underground and the interactions of hazardous waste constituents with geological strata, particularly unconsolidated earth materials. As shown in Figure 19.4, groundwater contaminated with hazardous wastes tends to flow as a relatively undiluted plug trailing an increasingly diluted plume along with the groundwater in an aquifer. The

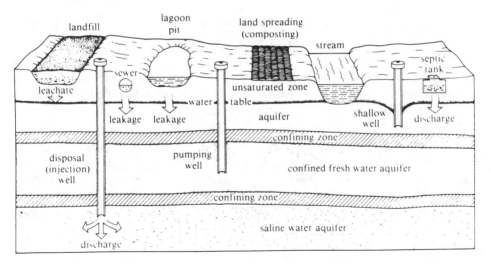

Figure 19.3.   Sources, disposal, and movement of hazardous wastes in the geosphere.

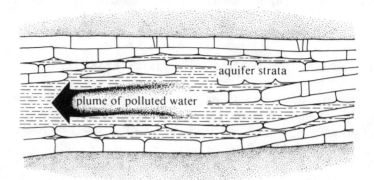

Figure 19.4.   Plug-flow of hazardous wastes in groundwater.

groundwater flow rate depends upon the water gradient and aquifer characteristics, such as permeability and cross-section area. The rate of flow is generally relatively slow; 1 meter per day would be considered fast. Contaminated groundwater can result in contamination of a surface water source. This can occur at a discharge area where the groundwater flows into a lake or stream.

As discussed in the preceding section, hazardous waste dissolved in groundwater can be attenuated by soil or rock by means of various sorption mechanisms. Mathematically, the distribution of a solute between groundwater or leachate water and soil is expressed by a **distribution coefficient, K_d,**

$$K_d = \frac{C_S}{C_W} \qquad (19.9.1)$$

where C_S is the concentration of the species in the solid phase and C_W is its concentration in water. This equation assumes that the relative degree of sorption is independent of C_W.

The degree of attenuation depends upon the surface properties of the solid, particularly its surface area. The chemical nature of the attenuating solid is also important because attenuation is a function of the organic matter (humus) content, presence of hydrous metal oxides, and the content and types of clays present. The chemical characteristics of the leachate also affect attenuation greatly. For example, attenuation of metals is very poor in acidic leachate because precipitation reactions such as,

$$M^{2+} + 2OH^- \rightarrow M(OH)_2(s) \qquad (19.9.2)$$

are reversed in acid:

$$M(OH)_2(s) + 2H^+ \rightarrow M^{2+} + 2H_2O \qquad (19.9.3)$$

Organic solvents in leachates tend to prevent attenuation of organic hazardous waste constituents.

The degree of attenuation of a pollutant by soil depends upon the water content of the soil. As shown in Figure 14.2, above the water table there is an unsaturated zone of soil in which attenuation is more highly favored. Normally, soil has a greater surface area at liquid-solid interfaces in this zone so that absorption and ion-exchange processes are favored. Aerobic degradation (see Chapter 11) is possible in the unsaturated zone, enabling more rapid and complete degradation of biodegradable hazardous wastes.

Codisposal of chelating agents with heavy metals can have a strong effect upon the mobility of metal ions in soil. This effect was observed resulting from codisposal of intermediate-level nuclear wastes with chelating agents during the period 1951–1965 at Oak Ridge National Laboratory. The presence of chelating agents resulted from the use of salts of chelating ethylenediaminetetraacetic acid (EDTA, see Section 11.9) in decontaminating facilities exposed to nuclear wastes. Whereas metal cations are readily held by ion exchange processes and precipitation on soil,

$$2Soil\}^-H^+ + Co^{2+} \rightarrow (Soil\}^-)_2Co^{2+} + 2H^+ \qquad (19.9.4)$$

$$Co^{2+} + 2OH^- \rightarrow Co(OH)_2(s) \qquad (19.9.4)$$

chelated anionic species, such as CoY^{2-} (where Y^{4-} is the chelating EDTA anion), are not strongly retained by the negatively charged functional groups in soil.

Radionuclides have been buried in shallow trenches on the grounds of Oak Ridge National Laboratory since 1944, so ample time has elapsed to observe the effects of this means of radioactive waste disposal. It has been found that chelating agents used for decontamination, as well as naturally occurring humic substance chelators, are responsible for migration in excess of that expected. Most notably, ^{60}Co has been found outside the disposal trenches.

19.10. HAZARDOUS WASTES IN THE HYDROSPHERE

Figure 19.5 illustrates a typical pathway for the entry of hazardous waste constituents into the hydrosphere. Other sources consist of precipitation from the atmosphere with rainfall, deliberate release to streams and bodies of water, runoff from soil, and mobilization from sediments. Once in an aquatic system,

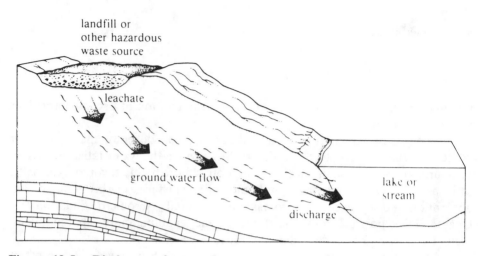

Figure 19.5. Discharge of groundwater contaminated from hazardous waste landfill into a body of water.

hazardous waste species are subject to a number of chemical and biochemical processes, including acid-base, oxidation-reduction, precipitation-dissolution, and hydrolysis reactions, as well as biodegradation.

The presence of organic matter in water has a tendency to increase the solubility of hazardous organic substances. Typically, the solubility of hexachlorobenzene is 1.8 μg/L in pure water at 25°C whereas it is 2.3 μg/L in creek water containing organic solutes and 4–4.5 μg/L in landfill leachate.

In considering the processes that hazardous wastes undergo in water, it is important to recall the nature of aquatic systems and the unique properties of water discussed in detail in Chapter 11. Water in the environment is far from pure. Just as the atmosphere is a constantly changing mass of bodies of moving air with different temperatures, pressures, and humidities, bodies of water are highly dynamic systems. Rivers, impoundments, and groundwater aquifers are subject to the input and loss of a variety of materials from both natural and anthropogenic sources. These materials may be gases, liquids, or solids. They interact chemically with each other and with living organisms—particularly bacteria—in the water. They are subject to dispersion and transport by stream flow, convection currents, and other physical phenomena. Hazardous substances or their by-products in water may undergo bioaccumulation through food chains involving aquatic organisms.

Several physical, chemical, and biochemical processes are particularly important in determining the transformations and ultimate fates of hazardous chemical species in the hydrosphere. These include **hydrolysis reactions**, through which a molecule is cleaved with the addition of H_2O; **precipitation reactions**, generally accompanied by **aggregation** of colloidal particles suspended in water; **oxidation-reduction reactions**, generally mediated by microorganisms; **sorption** of hazardous solutes by sediments and by suspended mineral and organic matter; **biochemical processes**, often involving hydrolysis and oxidation-reduction reactions; **photolysis reactions**; and miscellaneous chemical phenomena.

The hydrolysis of hazardous waste acetic anhydride is illustrated by the following reaction:

$$\underset{\overset{|}{H}}{\overset{\overset{H}{|}}{H-C}}-\overset{\overset{O}{\|}}{C}-O-\overset{\overset{O}{\|}}{C}-\underset{\overset{|}{H}}{\overset{\overset{H}{|}}{C}}-H \ +HOH \ \longrightarrow \ 2\,H-\underset{\overset{|}{H}}{\overset{\overset{H}{|}}{C}}-\overset{\overset{O}{\|}}{C}-OH \qquad (19.10.1)$$

The rates at which compounds hydrolyze in water vary widely. Acetic anhydride hydrolyzes very rapidly. In fact, the great affinity of this compound for water (including water in skin) is one of the reasons that it is hazardous. Once in the aquatic environment, though, acetic anhydride is converted very rapidly to essentially harmless acetic acid. Many ethers, esters, and other compounds formed originally by the joining together of two or more molecules with the loss of water hydrolyze very slowly, although the rate may be greatly increased by the action of enzymes in micoorganisms (biochemical processes).

As discussed in Section 20.5, the formation of precipitates in the form of sludges is one of the most common means of isolating hazardous components from an unsegregated waste. Although solid inorganic ionic compounds are often discussed in terms of very simple formulas, such as $PbCO_3$ for lead carbonate, much more complicated species [for example, $2PbCO_3 \cdot Pb(OH)_2$] generally result when precipitates are formed in the aquatic environment. For example, a hazardous heavy metal ion in the hydrosphere may be precipitated as a relatively complicated compound, coprecipitated as a minor constituent of some other compound, or be sorbed by the surface of another solid.

The major anions present in natural waters and wastewaters are OH^-, CO_3^{2-}, and SO_4^{2-} Since these anions are all capable of forming precipitates with cationic impurities, such pollutants tend to precipitate as hydroxides, carbonates, and sulfates. Sometimes a distinction can be made between hydroxides and hydrated oxides with similar, or identical, empirical formulas. For example, iron(III) hydroxide, $Fe(OH)_3$, is a relatively uncommon species; iron(III) usually is precipitated from water as hydrated iron(III) oxides, such as β ferric oxide monohydrate, $Fe_2O_3 \cdot H_2O$. Basic salts containing OH^- ion along with some other anions are very common in solids formed by precipitation from water. A typical example is azurite, $2CuCO_3 \cdot Cu(OH)_2$. Two or more metal ions may be present in a compound, as is the case with chalcopyrite, $CuFeS_2$.

Sorption processes are particularly common methods for the removal of low level hazardous materials from water. Many heavy metals are sorbed by or coprecipitated with hydrated iron(III) oxide ($Fe_2O_3 \cdot xH_2O$) or manganese(IV) oxide ($MnO_2 \cdot xH_2O$). Oxidation-reduction reactions are very important means of transformation of hazardous wastes in water.

Under many circumstances, biochemical processes largely determine the fates of hazardous chemical species in the hydrosphere. The most important such processes are those mediated by microorganisms, as discussed in Chapter 11. In particular, the oxidation of biodegradable hazardous organic wastes in water generally occurs by means of microorganism-mediated biochemical reactions. Bacteria produce organic acids and chelating agents, such as citrate, which have the effect of solubilizing hazardous heavy metal ions. Some mobile methylated forms, such as compounds of methylated arsenic and mercury, are produced by bacterial action.

As discussed in Chapter 16, photolysis reactions are those initiated by the absorption of light. The effect of photolytic processes on the destruction of hazardous wastes in the hydrosphere is minimal, although some photochemical reactions of hazardous waste compounds can occur when the compounds are present as surface films on water exposed to sunlight.

Groundwater is the part of the hydrosphere most vulnerable to damage from hazardous wastes. Although surface water supplies are subject to contamination, groundwater can become almost irreversibly contaminated by the improper land disposal of hazardous wastes.

19.11.   HAZARDOUS WASTES IN THE ATMOSPHERE

Some chemicals found in hazardous waste sites may enter the atmosphere by evaporation or even as windblown particles. Three major areas of interest in respect to hazardous waste compounds in the atmosphere are their **pollution potential, atmospheric fate**, and **residence time**. These strongly interrelated factors are discussed in this section.

Air Pollution Potential of Hazardous Waste Compounds

The pollution potential of hazardous wastes in the atmosphere depends upon whether they are *primary pollutants* that have a direct effect or *secondary pollutants* that are converted to harmful substances by atmospheric chemical processes. Hazardous waste sites do not usually evolve sufficient quantities of pollutants to give significant amounts of secondary pollutants, so primary air pollutants are the greater concern. Examples of primary air pollutants include toxic organic vapors (vinyl chloride), corrosive acid gases (HCl), and toxic inorganic gases, such as H_2S released by the accidental mixing of waste acid (HCl from waste steel pickling liquor) and waste metal sulfides:

$$2HCl + FeS \rightarrow FeCl_2 + H_2S(g) \qquad (19.11.1)$$

Primary air pollutants are most dangerous in the immediate vicinity of a site, usually posing hazards to workers involved in disposal or cleanup or people living adjacent to the site. Quantities are rarely sufficient to cause any kind of regional air pollution problems.

The two major kinds of secondary air pollutants from hazardous wastes are those that are oxidized in the atmosphere to corrosive substances and organic substances that undergo photochemical oxidation. Plausible examples of the former are sulfur dioxide released from the action of waste strong acids on sulfites and subsequently oxidized in the atmosphere to corrosive sulfuric acid,

$$SO_2 + \frac{1}{2}O_2 + H_2O \rightarrow H_2SO_4(aerosol) \qquad (19.11.2)$$

and nitrogen dioxide (itself a toxic primary air pollutant) produced by the reaction of waste nitric acid with reducing agents such as metals and oxidized to corrosive nitric acid or converted to corrosive nitrate salts:

$$4HNO_3 + Cu \rightarrow Cu(NO_3)_2 + 2NO_2(g) + 2H_2O \qquad (19.11.3)$$

$$2NO_2(g) + \frac{1}{2}O_2 + H_2O \rightarrow 2HNO_3(aerosol) \qquad (19.11.4)$$

$$HNO_3(aerosol) + NH_3(g) \rightarrow NH_4NO_3(aerosol) \qquad (19.11.5)$$

Organic species that produce secondary air pollutants are those that form photochemical smog (see Chapter 18). The more reactive of these are unsaturated compounds that react with atomic oxygen or hydroxyl radical in air,

$$R\text{-}CH = CH_2 + HO\cdot \rightarrow RCH_2CH_2O\cdot \qquad (19.11.6)$$

to yield reactive radicals that participate in chain reactions to eventually yield ozone, organic oxidants, noxious aldehydes, and other products characteristic of photochemical smog.

Fate and Residence Times of Hazardous Waste Compounds in the Atmosphere

An obvious means by which hazardous waste constituents may be removed from the atmosphere is by **dissolution** in water in the form of cloud or rain droplets. Inorganic acid, base, and salt compounds, such as H_2SO_4, HNO_3, and NH_4NO_3 mentioned above, are readily removed from the atmosphere by dissolution. For vapors of compounds that are not highly soluble in water, solubility information combined with information about rainfall amounts and mixing in the atmosphere can be used to estimate the atmospheric half-life, $\tau_{1/2}$, of the species. Solubility rates may be used to estimate half-lives for substances that are more miscible in water. For poorly water-soluble compounds, such calculations tend to drastically underestimate lifetimes, which indicates that other removal mechanisms must predominate.

The lifetimes of vaporized hazardous waste species removed from the atmosphere through **adsorption by aerosol particles** is limited to that of the sorbing aerosol particles (typically about 7 days) plus the time spent in the vapor phase before adsorption. This mechanism appears to be viable only for highly nonvolatile constituents such as benzo(a)pyrene.

Sorptive removal by soil, water, or plants on the Earth's surface, called **dry deposition**, is another means for physical removal of hazardous substances from the atmosphere. Predictions of dry deposition rate vary greatly with type of compound, type of surface, and weather conditions. For highly volatile organic compounds, such as low molecular mass organohalide compounds, predicted rates of dry deposition give atmospheric lifetimes many-fold higher than those actually observed so, for such compounds, dry deposition is probably not a common removal mechanism.

Predicted rates of physical removal of a number of volatile organic compounds that are not very soluble in water are far too slow to account for the loss of such compounds from the atmosphere, so chemical processes must predominate. As discussed in Chapter 18, the most important of these processes is reaction with hydroxyl radical, HO·, in the troposphere. Ozone can react with compounds having a double bond. Other oxidant species that might react with hazardous waste compounds in the troposphere and stratosphere are atomic oxygen (O), peroxyl radicals (HOO·), alkylperoxyl radicals (ROO·), and NO_3.

Despite the fact that its concentration in the troposphere is relatively low, HO· is so reactive that it tends to initiate most of the reactions leading to the chemical removal of most refractory organic compounds from the atmosphere. As noted in Section 18.10, hydroxyl radical undergoes *abstraction reactions* to remove H atoms from organic compounds containing R-H,

$$R\text{-}H + HO\cdot \rightarrow R\cdot + H_2O \tag{19.11.7}$$

and may react with those containing unsaturated bonds by addition as illustrated in Reaction 19.11.6. In both cases, reactive free radicals are formed that undergo further reactions, leading to nonvolatile and/or water-soluble species, which are scavenged from the atmosphere by physical means. These scavengeable species tend to be aldehydes, ketones, or acids. Halogenated organic compounds may lose halogen atoms in the form of halo-oxy radicals and undergo further reactions to form scavengeable species.

In general, reactions with species other than HO· or O_3 are not considered significant in the removal of hazardous organic waste compounds from the troposphere. Perhaps, in some cases, such reactions do contribute to a very slow removal of such contaminants compounds.

Photolytic transformations involve direct cleavage (photodissociation) of compounds by reactions with light and ultraviolet radiation:

$$R\text{-}X + h\nu \rightarrow R\cdot + X\cdot \qquad (19.11.8)$$

The extent of these reactions varies greatly with light intensity, quantum yields (chemical reactions per quantum absorbed) and other factors. In order for photolysis to be an important process for its removal from the atmosphere, a molecule must have a **chromophore** (light-absorbing group) that absorbs light in a wavelength region of significant intensity in the impinging light spectrum. This requirement limits the importance of photolysis as a removal mechanism to only a few classes of compounds, including conjugated alkenes, carbonyl compounds, some halides, and some nitrogen compounds, particularly nitro compounds. However, these do include a number of the more important hazardous waste compounds.

19.12. HAZARDOUS WASTES IN THE BIOSPHERE

One of the most crucial aspects of the fates and toxic effects of hazardous waste constituents is their accumulation by organisms from their surroundings, including bioaccumulation and biomagnification phenomena. **Biodegradation** of wastes is their conversion by biological processes to simple inorganic molecules and, to a certain extent, to biological materials. The complete bioconversion of an organic compound to inorganic species such as CO_2, NH_3, and phosphate is called **mineralization**. **Detoxification** refers to the biological conversion of a toxic compound to a less toxic species, which may still be relatively complex, or biological conversion to an even more complex material. An example of detoxification is illustrated below for the enzymatic conversion of paraoxon (a highly toxic organophosphate insecticide) to *p*-nitrophenol, which has only about 1/200 the toxicity of the parent compound:

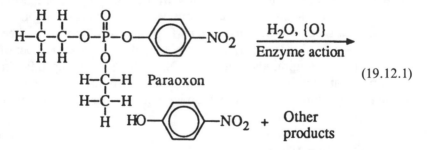

$$(19.12.1)$$

Usually the products of biodegradation are molecular forms that tend to occur in nature. Because the organisms that carry out biodegradation do so as a means of extracting free energy for their metabolic and growth needs (see Chapter 6) they form products that are in greater thermodynamic equilibrium with their surroundings. The definition of biodegradation is illustrated by an example in Figure 19.6. Biodegradation is usually carried out by the action of microorganisms, particularly bacteria and fungi.

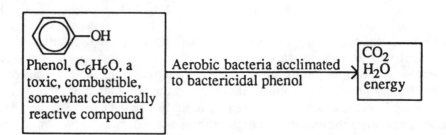

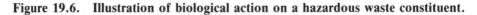

Figure 19.6.   Illustration of biological action on a hazardous waste constituent.

Biodegradation Processes

The biotransformations of environmental hazardous waste constituents, including pesticides and industrial chemicals, in vertebrates (birds, mammals, fish, reptiles) can be of the utmost importance in determining their fates and effects. **Biotransformation** is what happens to any substance that is metabolized and thereby altered by biochemical processes in an organism. **Metabolism** is divided into the two general categories of **catabolism**, which is the breaking down of more complex molecules, and **anabolism**, which is the building up of life molecules from simpler materials. The substances subjected to biotransformation may be naturally occurring or *anthropogenic* (made by human activities). They may consist of *xenobiotic* molecules that are foreign to living systems.

An important biochemical process that occurs in the biodegradation of many hazardous waste compounds is **cometabolism**. Cometabolism does not serve a useful purpose to an organism in terms of providing energy or raw material to build biomass, but occurs concurrently with normal metabolic processes. An example of cometabolism of hazardous wastes is provided by the white rot fungus, *Phanerochaete chrysosporium*. This organism, which has been investigated for its hazardous waste treatment potential, degrades a number of kinds of organochlorine compounds—including DDT, PCBs, and chlorodioxins—under the appropriate conditions. The enzyme system responsible for this degradation is one that the fungus uses to break down lignin in plant material under normal conditions.

Enzymes in Waste Degradation

Enzyme systems hold the key to biodegradation of hazardous wastes. For most biological treatment processes currently in use, enzymes are present in living organisms in contact with the wastes. However, in some cases it is possible to use cell-free extracts of enzymes removed from bacterial or fungal cells to treat hazardous wastes. For this application the enzymes may be present in solution or, more commonly, immobilized in biochemical reactors.

Biodegradation of municipal wastewater and solid wastes in landfills occurs by design. Biodegradation of any kind of waste that can be metabolized takes place whenever the wastes are subjected to conditions conducive to biological processes. The most common type of biodegradation is that of organic compounds in the presence of air; that is, **aerobic processes**. However, in the absence of air, **anaerobic biodegradation** may also take place. Furthermore, inorganic species are subject to both aerobic and anaerobic biological processes.

Although biological treatment of wastes is normally regarded as degradation to simple inorganic species such as carbon dioxide, water, sulfates, and phosphates, the possibility must always be considered of forming more complex or more hazardous chemical species. An example of the latter is the production of volatile, soluble, toxic methylated forms of arsenic and mercury from inorganic species of these elements by bacteria under anaerobic conditions.

For the most part, anthropogenic compounds resist biodegradation much more strongly than do naturally occurring compounds. This is generally due to the absence of enzymes that can bring about an initial attack on the compound (see Phase I reactions, Section 21.3). A number of physical and chemical characteristics of a compound are involved in its amenability to biodegradation. Such characteristics include hydrophobicity, solubility, volatility, and affinity for lipids. Some organic structural groups impart particular resistance to biodegradation. These include branched carbon chains, ether linkages, meta-substituted benzene rings, chlorine, amines, methoxy groups, sulfonates, and nitro groups.

Several groups of microorganisms are capable of partial or complete degradation of hazardous organic compounds. Among the aerobic bacteria, those of the *Pseudomonas* family are the most widespread and most adaptable to the degradation of synthetic compounds. These bacteria degrade biphenyl, naphthalene, DDT, and many other compounds. Anaerobic bacteria are very fastidious and difficult to study in the laboratory because they require oxygen-free (anoxic) conditions and pE values of less than –3.4 in order to survive. These bacteria catabolize biomass through hydrolytic processes, breaking down proteins, lipids, and saccharides. They are also known to reduce nitro compounds to amines, degrade nitrosamines, promote reductive dechlorination, reduce epoxide groups to alkenes, and break down aromatic structures. **Actinomycetes** are microorganisms that are morphologically similar to both bacteria and fungi. They are involved in the degradation of a variety of organic compounds, including degradation-resistant alkanes, and lignocellulose. Other compounds attacked include pyridines, phenols, nonchlorinated aromatics, and chlorinated aromatics. Fungi are particularly noted for their ability to attack long-chain and complex hydrocarbons and are more successful than bacteria in the initial attack on PCB compounds. Phototrophic microorganisms, which include algae, photosynthetic bacteria, and cyanobacteria (blue-green algae) tend to concentrate organophilic compounds in their lipid stores and induce photochemical degradation of the stored compounds. For example, *Oscillatoria* can initiate the biodegradation of naphthalene by the attachment of –OH groups.

Practically all classes of synthetic organic compounds can be at least partially degraded by various microorganisms. These classes include nonhalogenated alkanes, halogenated alkanes (trichloroethane, dichloromethane) nonhalogenated aromatic compounds (benzene, naphthalene, benzo(a)pyrene), halogenated aromatic compounds (hexachlorobenzene, pentachlorophenol), phenols (phenol, cresols), polychlorinated biphenyls, phthalate esters, and pesticides (chlordane, parathion).

CHAPTER SUMMARY

The chapter summary below is presented in a programmed format to review the main points covered in this chapter. It is used most effectively by filling in the blanks, referring back to the chapter as necessary. The correct answers are given at the end of the summary.

Three legislative acts passed in the U.S. to deal with hazardous substances and wastes are (1)_____

_____

_____

_____

_____.

The four characteristics by which hazardous substances are defined are (2)___

_____.

The four kinds of listed wastes are (3)_____

_____

_____

_____.

Measurement of the temperature at which a hot liquid ignites is called its (4)_____. Values of the vapor/air ratio below which ignition cannot occur because of insufficient fuel define the (5)_____

_____. The difference between upper and lower flammability limits at a specified temperature is the (6)_____.

Ammonium nitrate, ammonium perchlorate, hydrogen peroxide, and sodium dichromate are all examples of (7)_____. Such substances can contribute strongly to fire hazards because (8)_____

_____. White phosphorus; the alkali metals; powdered forms of magnesium, calcium, cobalt, manganese, iron, zirconium and aluminum; some organometallic compounds, such as lithium ethyl (LiC_2H_4) and lithium phenyl (LiC_6H_5); and some metal carbonyl compounds, such as iron pentacarbonyl, $Fe(CO)_5$ are all examples of (9)_____, which (10)_____

_____.

The most obvious toxic product of combustion is (11)_____

_____. Other products include (12)_____

_____.

Reactive substances are those that (13)_____

_____.

One such substance is sodium metal which undergoes the following reaction with water: (14)_____.

High reactivity in some organic compounds results from (15)_____

_____, and some organic structures involving oxygen are very reactive. Functional groups containing both oxygen and nitrogen tend to impart (16)_____ to an organic compound. Explosives such as nitroglycerin or TNT are single compounds containing (17)_____ in the same molecule. Corrosive substances as the definition is applied to interaction with metals are regarded as (18)_____

_____. The major chemical classes of corrosive substances are (19)_____

_____. A measurement of toxicity applied to hazardous wastes for regulatory purposes is (20)_____.

Three major categories of wastes based upon their physical forms are

(21)_____. In consideration of mixtures and impurities, wastes are best treated that have a high degree of (22)_____ and (23)_____.

Major points of origin of hazardous wastes into the environment are (24)_____

_____

_____

_____

_____

_____.

The major physical properties of wastes that determine their amenability to transport are (25)_____

_____.

Two major subdivisions of hazardous waste pollutants based largely upon their environmental chemical behavior are (26)_____ pollutants and (27)_____

_____ pollutants. The effects of hazardous wastes in the environment may be divided among effects on (28)_____

_____. The fates of hazardous waste constituents are largely determined by the tendencies of the compounds to attain (29)_____. The fate of a hazardous waste substance in water is a function of the substance's (30)_____.

The primary environmental concern regarding hazardous wastes in the geo-sphere is (31)_____.

The transport of contaminants in the geosphere depends largely upon (32)_____

_____ that govern the movement of water underground. Mathe-matically, the distribution of a solute between groundwater or leachate water and soil is expressed by a (33)_____. Codisposal of (34)_____

_____ with heavy metals, including radioactive metals, can have a strong effect upon the mobility of metal ions in soil.

Among the chemical and biochemical processes to which a hazardous waste species is subjected in an aquatic system are (35)_____

_____

_____. The presence of organic matter in water has a tendency to increase the (36)_____ of hazardous organic substances. Among the physical, chemical, and biochemical processes that are particularly important in determining the transformations and ultimate fates of hazardous chemical species in the hydrosphere are (37)_____

_____

_____.

The most important biochemical processes that determine fates of hazardous waste constituents in water are those that are (38)_____.

Three major areas of interest in respect to hazardous waste compounds in the atmosphere are their (39)_____

_____. The reaction,

$$SO_2 + 1/2 O_2 + H_2O \rightarrow H_2SO_4 (aerosol) \qquad (19.12.2)$$

is an example of conversion of a (40)_____ hazardous waste pollutant to a (41)_____ hazardous waste pollutant. Organic species that produce secondary air pollutants are those that form (42)_____

_____ (see Chapter 18).

The conversion by biological processes of waste constituents to simple inorganic molecules and, to a certain extent, to biological materials is known as (43)_____, whereas the complete bioconversion of an organic compound to inorganic species such as CO_2, NH_3, and phosphate is called (44)_____, and (45)_____ refers to the biological conversion of a toxic substance to a less toxic species. An important biochemical process that occurs in the biodegradation of many synthetic and hazardous waste materials that does not serve a useful purpose to an organism in terms of providing energy or raw material to build biomass, but occurs concurrently with normal metabolic processes is called (46)_____.

Microorganisms that are morphologically similar to both bacteria and fungi and that are involved in the degradation of a variety of organic compounds, including degradation-resistant alkanes, and lignocellulose are (47)_____.

Answers

1. Toxic Substances Control Act of 1976, Resource, Conservation and Recovery Act (RCRA) of 1976, and Comprehensive Environmental Response, Compensation, and Liability Act (CERCLA)
2. ignitability, corrosivity, reactivity, toxicity
3. F-type wastes from nonspecific sources, K-type wastes from specific sources, P-type acute hazardous wastes, U-Type generally hazardous wastes
4. flash point
5. lower flammability limit
6. flammability range
7. oxidizers
8. fuels may burn explosively in contact with an oxidizer
9. pyrophoric compounds
10. catch fire spontaneously in air without an ignition source
11. carbon monoxide
12. SO_2, P_4O_{10}, HCl, polycyclic aromatic hydrocarbons
13. tend to undergo rapid or violent reactions under certain conditions
14. $2Na + 2H_2O \rightarrow 2NaOH + H_2 + heat$
15. unsaturated bonds in the carbon skeleton
16. reactivity
17. both oxidizing and reducing functions
18. those that dissolve metals or cause oxidized material to form on the surface of metals
19. (1) strong acids, (2) strong bases, (3) oxidants, (4) dehydrating agents
20. Toxicity Characteristic Leaching Procedure (TCLP)
21. organic materials, aqueous wastes, and sludges
22. segregation
23. concentration
24. deliberate addition to soil, water, or air by humans; evaporation or wind erosion from waste dumps into the atmosphere; leaching from waste dumps into groundwater, streams, and bodies of water; leakage, such as from underground storage tanks or pipelines; evolution and subsequent deposi-

tion by accidents, such as fire or explosion; release from improperly operated waste treatment or storage facilities

25. volatility, solubility, and the degree to which they are sorbed to solids, including soil and sediments
26. primary
27. secondary
28. organisms, materials, and the environment
29. a state of physical and chemical stability
30. solubility, density, biodegradability, and chemical reactivity
31. the possible contamination of groundwater aquifers by waste leachates and leakage from wastes
32. hydrologic factors
33. distribution coefficient
34. chelating agents
35. acid-base, oxidation-reduction, precipitation-dissolution, and hydrolysis reactions, as well as biodegradation
36. solubility
37. hydrolysis reactions, precipitation reactions, oxidation-reduction reactions, sorption, biochemical processes, and photolysis reactions
38. mediated by microorganisms
39. pollution potential, atmospheric fate, and residence time
40. primary
41. secondary
42. photochemical smog
43. biodegradation
44. mineralization
45. detoxification
46. cometabolism
47. actinomycetes

QUESTIONS AND PROBLEMS

1. Match the following kinds of hazardous substances on the left with a specific example of each from the right, below:
   1. Explosives   (a) Oleum, sulfuric acid, caustic soda
   2. Compressed gases   (b) Magnesium metal, sodium hydride
   3. Radioactive materials (c) Lithium peroxide
   4. Flammable solids   (d) Hydrogen, sulfur dioxide
   5. Oxidizing materials   (e) Dynamite, ammunition
   6. Corrosive materials   (f) Plutonium, cobalt-60

2. Of the following, the property that is **not** a member of the same group as the other properties listed is (a) substances that are liquids whose vapors are

likely to ignite in the presence of ignition sources, (b) nonliquids that may catch fire from friction or contact with water and which burn vigorously or persistently, (c) ignitable compressed gases, (d) oxidizers, (e) substances that exhibit extremes of acidity or basicity.

3. In what respects may it be said that measures taken to alleviate air and water pollution tend to aggravate hazardous waste problems?

4. Discuss the significance of LFL, UFL, and flammability range in determining the flammability hazards of organic liquids.

5. Concentrated HNO_3 and its reaction products pose several kinds of hazards. What are these?

6. What are substances called that catch fire spontaneously in air without an ignition source?

7. Name four or five hazardous products of combustion and specify the hazards posed by these materials.

8. What kind of property tends to be imparted to a functional group of an organic compound containing both oxygen and nitrogen?

9. Match the corrosive substance from the column on the left, below, with one of its major properties from the right column:

   1. Alkali metal hydroxides   (a) Reacts with protein in tissue to form yel-
   2. Hydrogen peroxide   low xanthoproteic acid
   3. Hydrofluoric acid, HF   (b) Dissolves glass
   4. Nitric acid, HNO_3 (c) Strong bases
                                  (d) Oxidizer

10. Rank the following wastes in increasing order of segregation (a) mixed halogenated and hydrocarbon solvents containing little water, (b) spent steel pickling liquor (c) dilute sludge consisting of mixed organic and inorganic wastes, (d) spent hydrocarbon solvents free of halogenated materials, (e) dilute mixed inorganic sludge.

11. What is the role of a POTW in the treatment of hazardous wastes?

12. What are three major properties of hazardous wastes that determine their amenability to transport in the environment?

13. What is the influence of organic solvents in leachates upon attenuation of organic hazardous waste constituents?

14. Match the following physical, chemical, and biochemical processes dealing with the transformations and ultimate fates of hazardous chemical species in the hydrosphere on the left with the description of the process on the right, below:

    1. Precipitation reactions
    2. Biochemical processes
    3. Hydrolysis reactions
    4. Sorption
    5. Oxidation-reduction

    (a) Molecule is cleaved with the addition of H_2O
    (b) Generally accompanied by aggregation of colloidal particles suspended in water
    (c) Generally mediated by microorganisms
    (d) By sediments and by suspended matter
    (e) Often involve hydrolysis and oxidation-reduction

15. Describe the particular danger posed by codisposal of strong chelating agents with radioactive wastes. What may be said about the chemical nature of the latter in regard to this danger?

16. Describe a beneficial effect that might result from the precipitation of either $Fe_2O_3 \cdot xH_2O$ or $MnO_2 \cdot xH_2O$ from hazardous wastes in water.

17. Why are secondary air pollutants from hazardous waste sites usually of only limited concern as compared to primary air pollutants? What is the distinction between the two?

18. What are the major means by which hazardous waste species may be removed from the atmosphere? What is meant by $\tau_{1/2}$?

19. What may be said about the relative rates of reaction of hazardous waste compounds in the atmosphere with $HO\cdot$ and O_3?

20. As applied to hazardous wastes in the biosphere distinguish among biodegradation, biotransformation, detoxification, and mineralization.

21. What is the potential role of *Phanerochaete chrysosporium* in treatment of hazardous waste compounds? For which kinds of compounds might it be most useful?

22. What is a specific example of the formation of relatively more hazardous materials by the action of biological processes on hazardous wastes?

23. Several physical and chemical characteristics are involved in determining the amenability of a hazardous waste compound to biodegradation. These include hydrophobicity, solubility, volatility, and affinity for lipids. Suggest and discuss ways in which each one of these factors might affect biodegradability.

24. List and discuss some of the important processes determining the transformations and ultimate fates of hazardous waste constituents in the hydrosphere.

25. Which part of the hydrosphere is most subject to long-term, largely irreversible contamination from the improper disposal of hazardous wastes in the environment?

26. What features or characteristics should a compound possess in order for direct photolysis to be a significant factor in its removal from the atmosphere?

27. List and discuss the significance of major sources for the origin of hazardous wastes; that is, their main modes of entry into the environment. What are the relative dangers posed by each of these? Which part of the environment would each be most likely to contaminate?

28. In what form would a large quantity of hazardous waste PCB be likely to be found in the hydrosphere?

29. Why is attenuation of metals likely to be very poor in acidic leachate?

LITERATURE CITED

1. Wolbeck, B., "Political Dimensions and Implications of Hazardous Waste Disposal," in *Hazardous Waste Disposal*, Lehman, J. P., Ed., (New York: Plenum Press, 1982), pp. 7–18.

SUPPLEMENTARY REFERENCES

Manahan, S. E., *Hazardous Waste Chemistry, Toxicology and Treatment*, (Chelsea, MI: Lewis Publishers, 1990).
Nott, S., C. Arnstein, S. Ramsey, and M. Crough, *Superfund Handbook*, 2nd ed., (Chicago, IL: Sidley and Austin, 1987).
Weiss, G., Ed., *Hazardous Chemicals Data Book*, 2nd ed., (Park Ridge, NJ: Noyes Data Corporation, 1986).
U. S. Environmental Protection Agency, *Extremely Hazardous Substances*, (Park Ridge, NJ: Noyes Data Corporation, 1988).
Standard Handbook of Hazardous Waste Treatment and Disposal, Harry M. Freeman, Ed., (New York: McGraw-Hill, 1989).
Manahan, S. E., *Toxicological Chemistry*, 2nd ed., (Chelsea, MI: Lewis Publishers, 1992).
Extremely Hazardous Substances: Superfund Chemical Profiles, Vols. 1 & 2,

U.S. Environmental Protection Agency, (Park Ridge, NJ: Noyes Publications, 1989).

Wolf, K., W. J. Van Den Brink, and F. J. Colon, *Contaminated Soil '88*. Vols. 1 & 2, (Norwell, MA: Kluwer Academic Publishers, 1988).

Goldman, B. A., J. A. Hulme, and C. Johnson, *Hazardous Waste Management: Reducing the Risk*, Council on Economic Priorities, New York, 1986.

20   REDUCTION, TREATMENT, AND DISPOSAL OF HAZARDOUS WASTE

20.1.   INTRODUCTION

This chapter discusses how environmental chemistry can be applied to hazard-ous waste management to develop measures by which chemical wastes can be minimized, recycled, treated, and disposed. In descending order of desirability, hazardous waste management attempts to accomplish the following:

- Do not produce it
- If making it cannot be avoided, produce only minimum quantities
- Recycle it
- If it is produced and cannot be recycled, treat it, preferably in a way that makes it nonhazardous
- If it cannot be rendered nonhazardous, dispose of it in a safe manner
- Once it is disposed, monitor it for leaching and other adverse effects

20.2.   WASTE REDUCTION AND MINIMIZATION

Many hazardous waste problems can be avoided at early stages by **waste reduction** (cutting down quantities of wastes from their sources) and **waste minimization** (utilization of treatment processes which reduce the quantities of wastes requiring ultimate disposal). This section outlines basic approaches to waste minimization and reduction.

There are several ways in which quantities of wastes can be reduced, including source reduction, waste separation and concentration, resource recovery, and waste recycling. The most effective approaches to minimizing wastes center around careful control of manufacturing processes, taking into consideration discharges and the potential for waste minimization at every step of manufactur-ing. Viewing the process as a whole (as outlined for a generalized chemical manufacturing process in Figure 20.1) often enables crucial identification of the source of a waste, such as a raw material impurity, catalyst, or process solvent.

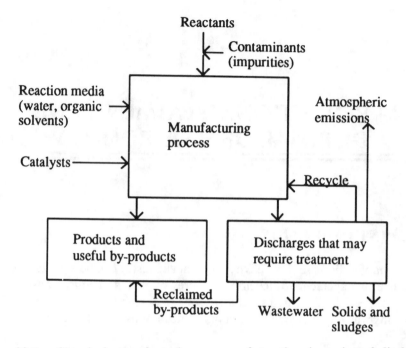

Figure 20.1.   Chemical manufacturing process frôm the viewpoint of discharges and waste minimization.

Once a source is identified, it is much easier to take measures to eliminate or reduce the waste.

Modifications of the manufacturing process can yield substantial waste reduction. Some such modifications are of a chemical nature. Changes in chemical reaction conditions can minimize production of by-product hazardous substances. In some cases potentially hazardous catalysts, such as those formulated from toxic substances, can be replaced by catalysts that are nonhazardous or that can be recycled rather than discarded. Wastes can be minimized by volume reduction; for example, through dewatering and drying sludge.

20.3.   RECYCLING

Wherever possible, recycling and reuse should be accomplished onsite because it avoids having to move wastes and because a process that produces recyclable materials is often the most likely to have use for them. The four broad areas in which something of value may be obtained from wastes are the following:

- Direct recycle as raw material to the generator, as with the return to feedstock of raw materials not completely consumed in a synthesis process

- Transfer as a raw material to another process; a substance that is a waste product from one process may serve as a raw material for another, sometimes in an entirely different industry
- Utilization for pollution control or waste treatment, such as use of waste alkali to neutralize waste acid
- Recovery of energy; for example, from the incineration of combustible hazardous wastes

Examples of Recycling

Recycling of scrap industrial impurities and products occurs on a large scale with a number of different materials. Most of these materials are not hazardous, but, as with most large-scale industrial operations, their recycle may involve the use or production of hazardous substances. Some of the more important examples are the following:

- **Ferrous metals** composed primarily of iron and used largely as feedstock for electric-arc furnaces
- **Nonferrous metals**, including aluminum (which ranks next to iron in terms of quantities recycled), copper and copper alloys, zinc, lead, cadmium, tin, silver, and mercury
- **Metal compounds**, such as metal salts
- **Inorganic substances**, including alkaline compounds (such as sodium hydroxide used to remove sulfur compounds from petroleum products), acids (steel pickling liquor where impurities permit reuse), and salts (for example, ammonium sulfate from coal coking used as fertilizer)
- **Glass**, which makes up about 10 percent of municipal refuse
- **Paper**, commonly recycled from municipal refuse
- **Plastic**, consisting of a variety of moldable polymeric materials and composing a major constituent of municipal wastes
- **Rubber**
- **Organic substances**, especially solvents and oils, such as hydraulic and lubricating oils
- **Catalysts** from chemical synthesis or petroleum processing
- Materials with **agricultural uses**, such as waste lime or phosphate-containing sludges used to treat and fertilize acidic soils.

Waste Oil Utilization and Recovery

Waste oil generated from lubricants and hydraulic fluids is one of the more commonly recycled materials. Annual production of waste oil in the U.S. is of the order of 4 billion liters per year. Around half of this amount is burned as fuel and lesser quantities are recycled or disposed of as waste. The collection, recycling, treatment, and disposal of waste oil are all complicated by the fact that it comes from diverse, widely dispersed sources and contains several classes of potentially hazardous contaminants. These are divided between organic constituents (polycyclic aromatic hydrocarbons, chlorinated hydrocarbons) and inorganic constituents (aluminum, chromium, and iron from wear of metal parts; barium and zinc from oil additives; lead from leaded gasoline).

Recycling Waste Oil

The processes used to convert waste oil to a feedstock hydrocarbon liquid for lubricant formulation are illustrated in Figure 20.2. The first of these uses distillation to remove water and light ends that have come from condensation and contaminant fuel. The second, or processing, step may be a vacuum distillation in which the three products are oil for further processing, a fuel oil cut, and a heavy residue. The processing step may also employ treatment with a mixture of solvents including isopropyl and butyl alcohols and methylethyl ketone to dissolve the oil and leave contaminants as a sludge; or contact with sulfuric acid to remove inorganic contaminants followed by treatment with clay to take out acid and contaminants that cause odor and color. The third step shown in Figure 20.2 employs vacuum distillation to separate lubricating oil stocks from a fuel fraction and heavy residue. This phase of treatment may also involve hydrofinishing, treatment with clay, and filtration.

Waste Oil Fuel

For economic reasons, waste oil that is to be used for fuel is given minimal treatment of a physical nature, including settling, removal of water, and filtration. Metals in waste fuel oil become highly concentrated in its fly ash, which may be hazardous.

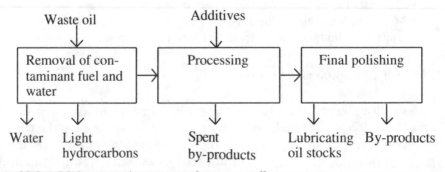

Figure 20.2.   Major steps in reprocessing waste oil.

Waste Solvent Recovery and Recycle

The recovery and recycling of waste solvents is similar in many respects to the recycling of waste oil. Among the many solvents listed as hazardous wastes and recoverable from wastes are dichloromethane, tetrachloroethylene, trichloro-ethylene, 1,1,1-trichloroethane, benzene, liquid alkanes, 2-nitropropane, methylisobutyl ketone, and cyclohexanone. For reasons of both economics and pollution control, many industrial processes that use solvents are equipped for solvent recycle. The basic scheme for solvent reclamation and reuse is shown in Figure 20.3.

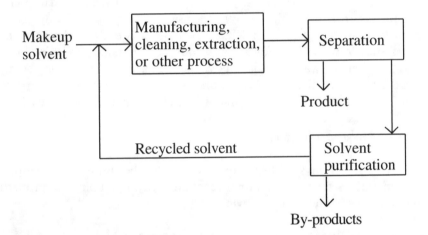

Figure 20.3. Overall process for recycling solvents.

A number of operations are used in solvent purification. Entrained solids are removed by settling, filtration, or centrifugation. Drying agents may be used to remove water from solvents and various adsorption techniques and chemical treatment may be required to free the solvent from specific impurities. Fractional distillation, often requiring several distillation steps, is the most important operation in solvent purification and recycle. It is used to separate solvents from impurities, water, and other solvents.

20.4.   PHYSICAL METHODS OF WASTE TREATMENT

This section addresses predominantly physical methods for waste treatment and the following section addresses methods that utilize chemical processes. It should be kept in mind that most waste treatment measures have both physical and chemical aspects. The appropriate treatment technology for hazardous wastes obviously depends upon the nature of the wastes. These may consist of volatile wastes (gases, volatile solutes in water, gases or volatile liquids held by solids), liquid wastes (wastewater, organic solvents), dissolved or soluble wastes (water-soluble inorganic species, water-soluble organic species, compounds soluble in organic solvents) semisolids (sludges, greases), and solids (dry solids, including granular solids with a significant water content, such as dewatered sludges, as well as solids suspended in liquids). The type of physical treatment to be applied to wastes depends strongly upon the physical properties of the material treated, including state of matter, solubility in water and organic solvents, density, volatility, boiling point, and melting point.

As shown in Figure 20.4, waste treatment may occur at three major levels — **primary, secondary,** and **polishing** — somewhat analogous to the treatment of wastewater (see Chapter 13). Primary treatment is generally regarded as preparation for further treatment, although it can result in the removal of by-products and reduction of the quantity and hazard of the waste. Secondary treatment detoxifies, destroys, and removes hazardous constituents. Polishing usually refers to treatment of water that is removed from wastes so that it may be safely discharged. However, the term can be broadened to apply to the treatment of other products as well so that they may be safely discharged or recycled.

Methods of Physical Treatment

Knowledge of the physical behavior of wastes has been used to develop various unit operations for waste treatment that are based upon physical properties. These operations include the following:

- Phase separation
    Filtration
- Phase transition
    Distillation
    Evaporation
    Physical precipitation

- Phase transfer
    Extraction
    Sorption
- Membrane separations
    Reverse osmosis
    Hyper- and ultrafiltration

Phase Separations

The most straightforward means of physical treatment involves separation of components of a mixture that are already in two different phases. **Sedimentation** and **decanting** are easily accomplished with simple equipment. In many cases the separation must be aided by mechanical means, particularly **filtration** or **centrifugation. Flotation** is used to bring suspended organic matter or finely divided particles to the surface of a suspension. In the process of **dissolved air flotation** (DAF), air is dissolved in the suspending medium under pressure and comes out of solution when the pressure is released as minute air bubbles attached to suspended particles, which causes the particles to float to the surface.

An important and often difficult waste treatment step is **emulsion breaking** in

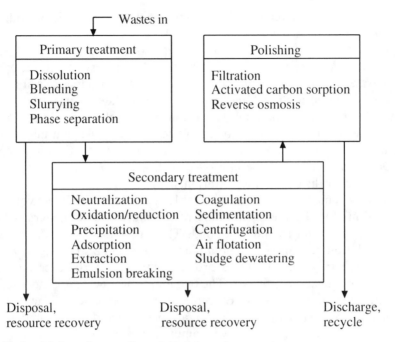

Figure 20.4. Major phases of waste treatment.

which colloidal-sized **emulsions** are caused to aggregate and settle from suspension. Agitation, heat, acid, and the addition of **coagulants** consisting of organic polyelectrolytes or inorganic substances, such as an aluminum salt, may be used for this purpose. The chemical additive acts as a flocculating agent to cause the particles to stick together and settle out.

Phase Transition

A second major class of physical separation is that of **phase transition** in which a material changes from one physical phase to another. It is best exemplified by **distillation**, which is used in treating and recycling solvents, waste oil, aqueous phenolic wastes, xylene contaminated with paraffin from histological laboratories, and mixtures of ethylbenzene and styrene. Distillation produces **distillation bottoms** (still bottoms), which are often hazardous and polluting. These consist of unevaporated solids, semisolid tars, and sludges from distillation. Specific examples are distillation bottoms from the production of acetaldehyde from ethylene, and still bottoms from toluene reclamation distillation in the production of disulfoton. The landfill disposal of these and other hazardous distillation bottoms used to be widely practiced but is now severely limited.

Evaporation is usually employed to remove water from an aqueous waste to concentrate it. A special case of this technique is **thin-film evaporation** in which volatile constituents are removed by heating a thin layer of liquid or sludge waste spread on a heated surface.

Drying — removal of solvent or water from a solid or semisolid (sludge) or the removal of solvent from a liquid or suspension — is a very important operation because water is often the major constituent of waste products, such as sludges obtained from emulsion breaking. In **freeze drying**, the solvent, usually water, is sublimed from a frozen material. Hazardous waste solids and sludges are dried to reduce the quantity of waste, to remove solvent or water that might interfere with subsequent treatment processes, and to remove hazardous volatile constituents. Dewatering can often be improved with addition of a filter aid, such as diatomaceous earth, during the filtration step.

Stripping is a means of separating volatile components from less volatile ones in a liquid mixture by the partitioning of the more volatile materials to a gas phase of air or steam (steam stripping). The gas phase is introduced into the aqueous solution or suspension containing the waste in a stripping tower that is equipped with trays or packed to provide maximum turbulence and contact between the liquid and gas phases. The two major products are condensed vapor and a stripped bottoms residue. Examples of two volatile components that can be removed from water by air stripping are benzene and dichloromethane. Air stripping can also be used to remove ammonia from water that has been treated with a base to convert ammonium ion to volatile ammonia.

Physical precipitation is used here as a term to describe processes in which a solid forms from a solute in solution as a result of a physical change in the solution, as compared to chemical precipitation (see Section 20.5) in which a chemical reaction in solution produces an insoluble material. The major changes that can cause physical precipitation are cooling the solution, evaporation of solvent, or alteration of solvent composition. The most common type of physical precipitation by alteration of solvent composition occurs when a water-miscible organic solvent is added to an aqueous solution, so that the solubility of a salt is lowered below its concentration in the solution.

Phase Transfer

Phase transfer consists of the transfer of a solute in a mixture from one phase to another. An important type of phase transfer process is **solvent extraction**, a process in which a substance is transferred from solution in one solvent (usually water) to another (usually an organic solvent) without any chemical change taking place. When solvents are used to leach substances from solids or sludges, the process is called **leaching**. Solvent extraction and the major terms applicable to it are summarized in Figure 20.5. The same terms and general principles apply to leaching. The major application of solvent extraction to waste treatment has been in the removal of phenol from by-product water produced in coal coking, petroleum refining, and chemical syntheses that involve phenol.

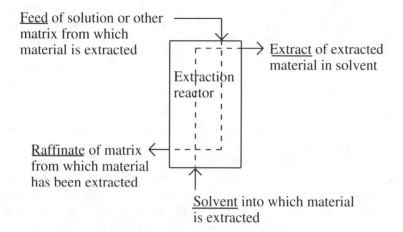

Figure 20.5.   Outline of solvent extraction/leaching process with important terms underlined.

One of the more promising approaches to solvent extraction and leaching of hazardous wastes is the use of **supercritical fluids**, most commonly CO_2, as extraction solvents. A supercritical fluid is one that has characteristics of both liquid and gas and consists of a substance above its supercritical temperature and pressure (31.1°C and 73.8 atm, respectively, for CO_2). After a substance has been extracted from a waste into a supercritical fluid at high pressure, the pressure can be released, resulting in separation of the substance extracted. The fluid can then be compressed again and recirculated through the extraction system. Some possibilities for treatment of hazardous wastes by extraction with supercritical CO_2 include removal of organic contaminants from wastewater, extraction of organohalide pesticides from soil, extraction of oil from emulsions used in aluminum and steel processing, and regeneration of spent activated carbon. Waste oils contaminated with PCBs, metals, and water can be purified using supercritical ethane.

Transfer of a substance from a solution to a solid phase is called **sorption**. The most important sorbent is **activated carbon** used for several purposes in waste treatment; in some cases it is adequate for complete treatment. It can also be applied to pretreatment of waste streams going into processes such as reverse osmosis to improve treatment efficiency and reduce fouling. Effluents from other treatment processes, such as biological treatment of degradable organic solutes in water can be polished with activated carbon. Activated carbon sorption is most effective for removing from water those hazardous waste materials that are poorly water-soluble and that have high molecular masses, such as xylene, naphthalene, cyclohexane, chlorinated hydrocarbons, phenol, aniline, dyes, and surfactants. Activated carbon does not work well for organic compounds that are highly water-soluble or polar.

Solids other than activated carbon can be used for sorption of contaminants from liquid wastes. These include synthetic resins composed of organic polymers and mineral substances. Of the latter, clay is employed to remove impurities from waste lubricating oils in some oil recycling processes.

Molecular Separation

A third major class of physical separation is **molecular separation**, often based upon **membrane processes** in which dissolved contaminants or solvent pass through a size-selective membrane under pressure. The products are a relatively pure solvent phase (usually water) and a concentrate enriched in the solute impurities. **Hyperfiltration** allows passage of species with molecular masses of about 100 to 500, whereas **ultrafiltration**, is used for the separation of organic solutes with molecular masses of 500 to 1,000,000. With both of these techniques, water and lower molecular mass solutes under pressure pass through the membrane as a stream of purified **permeate**, leaving behind a stream of **concentrate** containing impurities in solution or suspension. Ultrafiltration and hyperfiltration are especially useful for concentrating suspended oil, grease, and fine solids in water. They also serve to concentrate solutions of large organic molecules and heavy metal ion complexes.

Reverse osmosis is the most widely used of the membrane techniques. Although superficially similar to ultrafiltration and hyperfiltration, it operates on a different principle in that the membrane is selectively permeable to water and excludes ionic solutes. Reverse osmosis uses high pressures to force permeate through the membrane, producing a concentrate containing high levels of dissolved salts.

Electrodialysis, sometimes used to concentrate plating wastes, employs membranes alternately permeable to cations and to anions. The driving force for the separation is provided by electrolysis with a direct current between two electrodes. Alternate layers between the membranes contain concentrate (brine) and purified water.

20.5.   CHEMICAL TREATMENT: AN OVERVIEW

The applicability of chemical treatment to wastes depends upon the chemical properties of the waste constituents, particularly acid-base, oxidation-reduction, precipitation, and complexation behavior; reactivity; flammability/combustibility; corrosivity; and compatibility with other wastes. The chemical behavior of wastes translates to various unit operations for waste treatment that are based upon chemical properties and reactions. These include the following:

- Acid/base neutralization
- Chemical extraction and leaching
- Hydrolysis
- Chemical precipitation
- Oxidation/reduction
- Ion exchange

Some of the more sophisticated means available for treatment of wastes have been developed for pesticide disposal.

Acid/Base Neutralization

Waste acids and bases are treated by **neutralization**:

$$H^+ + OH^- \rightarrow H_2O \qquad (20.5.1)$$

Although simple in principle, neutralization can present some problems in practice. These include evolution of volatile contaminants, mobilization of soluble substances, excessive heat generated by the neutralization reaction, and corrosion to apparatus. By adding too much or too little of the neutralizing agent, it is possible to get a product that is too acidic or basic.

Lime, $Ca(OH)_2$, is the least expensive source of base and is widely used for treating acidic wastes. Because of lime's limited solubility, solutions of excess lime do not reach extremely high pH values. Sulfuric acid, H_2SO_4, is a relatively inexpensive acid for treating alkaline wastes. However, addition of too much sulfuric acid can produce highly acidic products; for some applications, acetic acid, CH_3COOH, is preferable. As noted above, acetic acid is a weak acid and an excess of it does little harm. It is also a nontoxic natural product and biodegradable.

Neutralization, or pH adjustment, is often required prior to the application of other waste treatment processes. Processes that may require neutralization include oxidation/reduction, activated carbon sorption, wet air oxidation, stripping, and ion exchange. Microorganisms usually require a pH in the range of 6–9, so neutralization may be required prior to biochemical treatment.

Chemical Precipitation

Chemical precipitation is used in hazardous waste treatment primarily for the removal of heavy metal ions from water as shown below for the chemical precipitation of cadmium:

$$Cd^{2+}(aq) + HS^-(aq) \rightarrow CdS(s) + H^+(aq) \qquad (20.5.2)$$

Precipitation of Metals

The most widely used means of precipitating metal ions is by the formation of hydroxides such as chromium(III) hydroxide:

$$Cr^{3+} + 3OH^- \rightarrow Cr(OH)_3 \qquad (20.5.3)$$

The source of hydroxide ion is a base (alkali), such as lime ($Ca(OH)_2$), sodium hydroxide ($NaOH$), or sodium carbonate (Na_2CO_3). Most metal ions tend to produce basic salt precipitates, such as basic copper(II) sulfate, $CuSO_4 \cdot 3Cu(OH)_2$, formed as a solid when hydroxide is added to a solution containing Cu^{2+} and SO_4^{2-} ions. The solubilities of many heavy metal hydroxides reach a minimum value, often at a pH in the range of 9–11, then increase with increasing pH values due to the formation of soluble hydroxo complexes, as illustrated by the following reaction:

$$Zn(OH)_2(s) \ + \ OH^-(aq) \ \rightarrow \ Zn(OH)_3^-(aq) \qquad (20.5.4)$$

The chemical precipitation method that is used most is precipitation of metals as hydroxides and basic salts with lime. Sodium carbonate can be used to precipitate hydroxides ($Fe(OH)_3 \cdot xH_2O$) carbonates ($CdCO_3$), or basic carbonate salts ($2PbCO_3 \cdot Pb(OH)_2$). The carbonate anion produces hydroxide by virtue of its hydrolysis reaction with water:

$$CO_3^{2-} \ + \ H_2O \ \rightarrow \ HCO_3^- \ + \ OH^- \qquad (20.5.5)$$

Carbonate, alone, does not give as high a pH as do alkali metal hydroxides, which may have to be used to precipitate metals that form hydroxides only at relatively high pH values.

The solubilities of some heavy metal sulfides are extremely low, so precipitation by H_2S or other sulfides (see Reaction 20.5.2) can be a very effective means of treatment. Hydrogen sulfide is a toxic gas that is itself considered to be a hazardous waste. Iron(II) sulfide (ferrous sulfide), one of the less insoluble metal sulfides, can be used as a safe source of sulfide ion to produce sulfide precipitates with other metals that are less soluble than FeS. However, toxic H_2S can be produced when metal sulfide wastes contact acid:

$$MS \ + \ 2H^+ \ \rightarrow \ M^{2+} \ + \ H_2S \qquad (20.5.6)$$

Some metals can be precipitated from solution in the elemental metal form by the action of a reducing agent, such as sodium borohydride,

$$4Cu^{2+} \ + \ NaBH_4 \ + \ 2H_2O \ \rightarrow \ 4Cu \ + \ NaBO_2 \ + \ 8H^+ \qquad (20.5.7)$$

or with more active metals in a process called **cementation**:

$$Cd^{2+} \ + \ Zn \ \rightarrow \ Cd \ + \ Zn^{2+} \qquad (20.5.8)$$

Oxidation/Reduction

As shown by the reactions in Table 20.1, **oxidation** and **reduction** can be used for the treatment and removal of a variety of inorganic and organic wastes. Some waste oxidants are employed to treat oxidizable wastes in water and cyanides.

Ozone, O_3, is a strong oxidant that can be generated onsite by an electrical discharge through dry air or oxygen. Ozone employed as an oxidant gas at levels of 1–2 wt % in air and 2–5 wt % in oxygen has been used to treat a large variety of oxidizable contaminants, effluents, and wastes including wastewater and sludges containing oxidizable constituents.

Electrolysis

As shown in Figure 20.6, **electrolysis** is a process in which one species in solution (usually a metal ion) is reduced by electrons at the **cathode** and another gives up electrons to the **anode** and is oxidized there. In hazardous waste applications electrolysis is most widely used in the recovery of cadmium, copper, gold, lead, silver, and zinc. Metal recovery by electrolysis is made more difficult by the presence of cyanide ion, which stabilizes metals in solution as the cyanide complexes, such as $Ni(CN)_4^{2-}$.

Table 20.1.   Oxidation/Reduction Reactions Used to Treat Wastes

Waste Substance	Reaction with Oxidant or Reductant
Oxidation of Organics	
Organic matter, $\{CH_2O\}$	$\{CH_2O\} + 2\{O\}^a \rightarrow CO_2 + H_2O$
Aldehyde	$CH_3CHO + \{O\} \rightarrow CH_3COOH(acid)$
Oxidation of Inorganics	
Cyanide	$2CN^- + 5OCl^- + H_2O \rightarrow N_2 + 2HCO_3^- + 5\ Cl^-$
Iron(II)	$4Fe^{2+} + O_2 + 10H_2O \rightarrow 4Fe(OH)_3 + 8H^+$
Sulfur dioxide	$2SO_2 + O_2 + 2H_2O \rightarrow 2H_2SO_4$
Reduction of Inorganics	
Chromate	$2CrO_4^{2-} + 3SO_2 + 4H^+ \rightarrow Cr_2(SO_4)_3 + 2H_2O$
Permanganate	$MnO_4^- + 3Fe^{2+} + 7H_2O \rightarrow$ $MnO_2(s) + 3Fe(OH)_3(s) + 5H^+$

[a]The symbol $\{O\}$ is used to denote a source of oxygen in an oxidizing species.

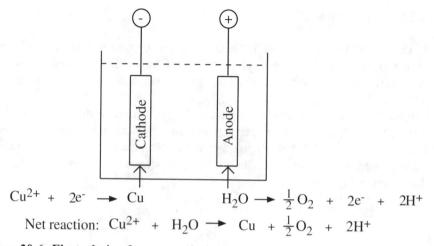

$$Cu^{2+} + 2e^- \rightarrow Cu \qquad H_2O \rightarrow \tfrac{1}{2}O_2 + 2e^- + 2H^+$$

$$\text{Net reaction: } Cu^{2+} + H_2O \rightarrow Cu + \tfrac{1}{2}O_2 + 2H^+$$

Figure 20.6. Electrolysis of copper solution.

Hydrolysis

One of the ways to dispose of chemicals that are reactive with water is to allow them to react with water under controlled conditions, a process called hydrolysis. Inorganic chemicals that can be treated by hydrolysis include metals that react with water; metal carbides, hydrides, amides, alkoxides, and halides; and nonmetal oxyhalides and sulfides. Examples of the treatment of these classes of inorganic species are given in Table 20.2.

Organic chemicals may also be treated by hydrolysis. For example, toxic acetic anhydride is hydrolyzed to relatively safe acetic acid:

Table 20.2.   Inorganic Chemicals That May Be Treated by Hydrolysis

Class of Chemical	Reaction with Water
Active metals (calcium)	$Ca + 2H_2O \rightarrow H_2 + Ca(OH)_2$
Hydrides (sodium aluminum hydride)	$NaAlH_4 + 4H_2O \rightarrow 4H_2 + NaOH + Al(OH)_3$
Carbides (calcium carbide)	$CaC_2 + 2H_2O \rightarrow Ca(OH)_2 + C_2H_2$
Amides (sodium amide)	$NaNH_2 + H_2O \rightarrow NaOH + NH_3$
Halides (silicon tetrachloride)	$SiCl_4 + 2H_2O \rightarrow SiO_2 + 4HCl$
Alkoxides (sodium ethoxide)	$NaOC_2H_5 + H_2O \rightarrow NaOH + C_2H_5OH$

$$H-\underset{\underset{H}{|}}{\overset{\overset{H}{|}}{C}}-\underset{}{\overset{\overset{O}{||}}{C}}-O-\underset{}{\overset{\overset{O}{||}}{C}}-\underset{\underset{H}{|}}{\overset{\overset{H}{|}}{C}}-H + H_2O \longrightarrow 2H-\underset{\underset{H}{|}}{\overset{\overset{H}{|}}{C}}-\overset{\overset{O}{||}}{C}-OH$$

(20.5.9)

Acetic anhydride (an
acid anhydride)

Chemical Extraction and Leaching

Chemical extraction or **leaching** in hazardous waste treatment is the removal of a hazardous constituent by chemical reaction with an extractant in solution. Poorly soluble heavy metal salts can be extracted by reaction of the salt anions with H^+ as illustrated by the following:

$$PbCO_3 + H^+ \rightarrow Pb^{2+} + HCO_3^-$$ (20.5.10)

Acids also dissolve basic organic compounds such as amines and aniline. Extraction with acids should be avoided if cyanides or sulfides are present to prevent formation of toxic hydrogen cyanide or hydrogen sulfide. Nontoxic weak acids are usually the safest to use. These include acetic acid, CH_3COOH, and the acid salt, NaH_2PO_4.

Chelating agents, such as dissolved ethylenedinitrilotetraacetate (EDTA, HY^{3-}), dissolve insoluble metal salts by forming soluble species with metal ions:

$$FeS + HY^{3-} \rightarrow FeY^{2-} + HS^-$$ (20.5.11)

Heavy metal ions in soil contaminated by hazardous wastes may be present in a coprecipitated form with insoluble iron(III) and manganese(IV) oxides, Fe_2O_3 and MnO_2, respectively. These oxides can be dissolved by reducing agents, such as solutions of sodium dithionate/citrate or hydroxylamine. This results in the production of soluble Fe^{2+} and Mn^{2+} and the release of heavy metal ions, such as Cd^{2+} or Ni^{2+}, which are removed with the water.

Ion Exchange

Ion exchange is a means of removing cations or anions from solution onto a solid resin, which can be regenerated by treatment with acids, bases or salts. The greatest use of ion exchange in hazardous waste treatment is for the removal of low levels of heavy metal ions from wastewater:

$$2H^+\{^-CatExchr\} + Cd^{2+} \rightarrow Cd^{2+}\{^-CatExchr\}_2 + 2H^+$$ (20.5.12)

Ion exchange is employed in the metal plating industry to purify rinsewater and spent plating bath solutions. Cation exchangers are used to remove cationic metal species, such as Cu^{2+}, from such solutions. Anion exchangers remove anionic cyanide metal complexes (for example, $Ni(CN)_4^{2-}$) and chromium(VI) species, such as CrO_4^{2-}. Radionuclides may be removed from radioactive wastes and mixed waste by ion exchange resins.

20.6. PHOTOLYTIC REACTIONS

Photolytic reactions were discussed in Chapter 16. *Photolysis* can be used to destroy a number of kinds of hazardous wastes. In such applications it is most useful in breaking chemical bonds in refractory organic compounds. TCDD (see Section 12.9), one of the most troublesome and refractory of wastes, can be treated by ultraviolet light in the presence of hydrogen atom donors {H} resulting in reactions such as the following:

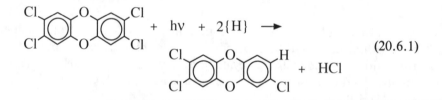

$$+ \text{ h}\nu \text{ } + \text{ 2\{H\}} \longrightarrow \text{ + HCl} \tag{20.6.1}$$

As photolysis proceeds, more H-C bonds are broken, the C-O bonds are broken, and the final product is a harmless organic polymer.

An initial photolysis reaction can result in the generation of reactive intermediates that participate in **chain reactions** that lead to the destruction of a compound. One of the most important reactive intermediates is free radical HO·. In some cases **sensitizers** are added to the reaction mixture to absorb radiation and generate reactive species that destroy wastes.

Hazardous waste substances other than TCDD that have been destroyed by photolysis are herbicides (atrazine), 2,4,6-trinitrotoluene (TNT), and polychlorinated biphenyls (PCBs). The addition of a chemical oxidant, such as potassium peroxidisulfate, $K_2S_2O_8$, enhances destruction by oxidizing active photolytic products.

20.7. THERMAL TREATMENT METHODS

Thermal treatment of hazardous wastes can be used to accomplish most of the common objectives of waste treatment — volume reduction; removal of volatile, combustible, mobile organic matter; and destruction of toxic and pathogenic materials. The most widely applied means of thermal treatment of hazardous wastes is **incineration**. Incineration utilizes high temperatures, an oxidizing atmosphere, and often turbulent combustion conditions to destroy wastes. Methods other than incineration that make use of high temperatures to destroy or neutralize hazardous wastes include pyrolysis, wet oxidation, and gasification.

Incineration

Hazardous waste incineration will be defined here as a process that involves exposure of the waste materials to oxidizing conditions at a high temperature, usually in excess of 900°C. Normally, the heat required for incineration comes from the oxidation of organically bound carbon and hydrogen contained in the waste material or in supplemental fuel:

$$C(organic) + O_2 \rightarrow CO_2 + heat \qquad (20.7.1)$$

$$4H(organic) + O_2 \rightarrow 2H_2O + heat \qquad (20.7.2)$$

These reactions destroy organic matter and generate heat required for endothermic reactions, such as the breaking of C–Cl bonds in organochlorine compounds.

Incinerable Wastes

Ideally, incinerable wastes are predominantly organic materials that will burn with a heating value of at least 5,000 Btu/lb and preferably over 8,000 Btu/lb. Such heating values are readily attained with wastes having high contents of the most commonly incinerated waste organic substances, including methanol, acetonitrile, toluene, ethanol, amyl acetate, acetone, xylene, methylethyl ketone, adipic acid, and ethyl acetate. In some cases, however, it is desirable to incinerate wastes that will not burn alone and which require **supplemental fuel**, such as methane and petroleum liquids. Examples of such wastes are nonflammable organochloride wastes, some aqueous wastes, or soil in which the elimination of a particularly troublesome contaminant is worth the expense and trouble of incinerating it. Inorganic matter, water, and organic hetero element contents of liquid wastes are important in determining their incinerability.

Hazardous Waste Fuel

Many industrial wastes, including hazardous wastes, are burned as **hazardous waste fuel** for energy recovery in industrial furnaces and boilers and in incinerators for nonhazardous wastes, such as sewage sludge incinerators. This process is called **coincineration,** and more combustible wastes are utilized by it than are burned solely for the purpose of waste destruction. In addition to heat recovery from combustible wastes, it is a major advantage to use an existing onsite facility for waste disposal rather than a separate hazardous waste incinerator.

Incineration Systems

The four major components of hazardous waste incineration systems are shown in Figure 20.7.

Waste preparation for liquid wastes may require filtration, settling to remove solid material and water, blending to obtain the optimum incinerable mixture, or heating to decrease viscosity. Solids may require shredding and screening. Atomization is commonly used to feed liquid wastes. Several mechanical devices, such as rams and augers, are used to introduce solids into the incinerator.

The most common kinds of **combustion chambers** are liquid injection, fixed hearth, rotary kiln, and fluidized bed. These types are discussed in more detail later in this section.

Often the most complex part of a hazardous waste incineration system is the **air pollution control system**, which involves several operations. The most common operations in air pollution control from hazardous waste incinerators are combustion gas cooling, heat recovery, quenching, particulate matter removal, acid gas removal, and treatment and handling of by-product solids, sludges, and liquids.

Hot ash is often quenched in water. Prior to disposal it may require dewatering and chemical stabilization. A major consideration with hazardous waste

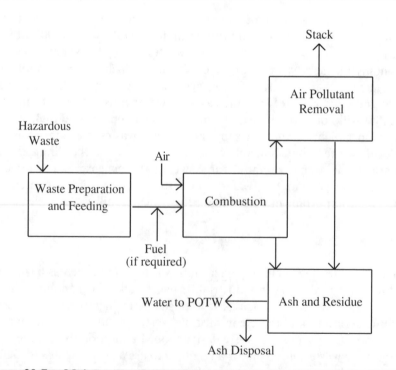

Figure 20.7.   Major components of a hazardous waste incinerator system.

incinerators and the types of wastes that are incinerated is the disposal problem posed by the ash, especially in respect to potential leaching of heavy metals.

Types of Incinerators

Hazardous waste incinerators may be divided among the following, based upon type of combustion chamber:

- **Rotary kiln** (about 40% of U.S. hazardous waste incinerator capacity) in which the primary combustion chamber is a rotating cylinder lined with refractory materials and an afterburner downstream from the kiln to complete destruction of the wastes
- **Liquid injection incinerators** (also about 40% of U.S. hazardous waste incinerator capacity) that burn pumpable liquid wastes dispersed as small droplets
- **Fixed-hearth incinerators** with single or multiple hearths upon which combustion of liquid or solid wastes occurs
- **Fluidized-bed incinerators** that have a bed of granular solid (such as sand) maintained in a suspended state by injection of air to remove pollutant acid gas and ash products
- **Advanced design incinerators** including **plasma incinerators** that make use of an extremely hot plasma of ionized air injected through an electrical arc; **electric reactors** that use resistance-heated incinerator walls at around 2,200°C to heat and pyrolyze wastes by radiative heat transfer; **infrared systems**, which generate intense infrared radiation by passing electricity through resistance-heating elements composed of silicon carbide; **molten salt combustion** that uses a bed of molten sodium carbonate at about 900°C to destroy the wastes and retain gaseous pollutants; and **molten glass processes** that use a pool of molten glass to transfer heat to the waste and to retain products in a poorly leachable glass.

Combustion Conditions

The key to effective incineration of hazardous wastes lies in the combustion conditions. These require (1) sufficient free oxygen in the combustion zone; (2) turbulence for thorough mixing of waste, oxidant, and (where used) supplemental fuel; (3) high combustion temperatures above about 900°C to ensure that thermally resistant compounds do react; and (4) sufficient residence time (at least 2 seconds) to allow reactions to occur.

Effectiveness of Incineration

EPA standards for hazardous waste incineration are based upon the effectiveness of destruction of the **principal organic hazardous constituents** (POHC).

Measurement of these compounds before and after incineration gives the **destruction removal efficiency** (DRE) according to the formula,

$$\text{DRE} = \frac{W_{in} - W_{out}}{W_{in}} \times 100 \qquad (20.7.3)$$

where W_{in} and W_{out} are the mass flow rates of the principal organic hazardous constituent (POHC) input and output (at the stack downstream from emission controls), respectively.

Wet Air Oxidation

Organic compounds and oxidizable inorganic species can be oxidized by oxygen in aqueous solution. The source of oxygen usually is air. Rather extreme conditions of temperature and pressure are required, with a temperature range of 175–327°C and a pressure range of 300–3,000 psig (2070–20,700 kPa). The high pressures allow a high concentration of oxygen to be dissolved in the water and the high temperatures enable the reaction to occur.

Wet air oxidation has been applied to the destruction of cyanides in electroplating wastewaters. The oxidation reaction for sodium cyanide is the following:

$$2Na^+ + 2CN^- + O_2 + 4H_2O \rightarrow 2Na^+ + 2HCO_3^- + 2NH_3 \qquad (20.7.4)$$

UV-Enhanced Wet Oxidation

Hydrogen peroxide (H_2O_2) can be used as an oxidant in solution assisted by ultraviolet radiation ($h\nu$). For the oxidation of organic species represented in general as $\{CH_2O\}$, the overall reaction is

$$2H_2O_2 + \{CH_2O\} + h\nu \rightarrow CO_2 + 3H_2O \qquad (20.7.5)$$

The ultraviolet radiation breaks chemical bonds and serves to form reactive oxidant species, such as HO·.

20.8. BIODEGRADATION OF WASTES

Biodegradation of wastes is their conversion by biological processes to simple inorganic molecules (mineralization) and, to a certain extent, to biological materials. Usually the products of biodegradation are molecular forms that tend to occur in nature and that are in greater thermodynamic equilibrium with their surroundings than are the starting materials. **Detoxification** refers to the biological conversion of a toxic substance to a less toxic species. Microbial bacteria and fungi possessing enzyme systems required for biodegradation of wastes are usually best obtained from populations of indigenous microorganisms at a hazardous waste site where they have developed the ability to degrade particular kinds of molecules. Although it has some shortcomings in the degradation of complex chemical mixtures, biological treatment offers a number of significant advantages and has considerable potential for the degradation of hazardous wastes, even in situ.

Biodegradability

The **biodegradability** of a compound is influenced by its physical characteristics, such as solubility in water and vapor pressure, and by its chemical properties, including molecular mass, molecular structure, and presence of various kinds of functional groups, some of which provide a "biochemical handle" for the initiation of biodegradation. With the appropriate organisms and under the right conditions, even substances such as phenol that are considered to be biocidal to most microorganisms can undergo biodegradation.

Recalcitrant or **biorefractory** substances are those that resist biodegradation and tend to persist and accumulate in the environment. Such materials are not necessarily toxic to organisms, but simply resist their metabolic attack. However, even some compounds regarded as biorefractory may be degraded by microorganisms adapted to their biodegradation; for example DDT is degraded by properly acclimated *Pseudomonas*. As mentioned in Section 12.9, relatively highly chlorinated PCBs are degraded under anaerobic conditions by anaerobic bacteria to less highly chlorinated congeners. These products can be further degraded by aerobic bacteria upon subsequent exposure to air. Chemical pretreatment, especially by partial oxidation, can make some kinds of recalcitrant wastes much more biodegradable.

Properties of hazardous wastes and their media can be changed to increase biodegradability. This can be accomplished by adjustment of conditions to optimum temperature, pH (usually in the range of 6–9), stirring, oxygen level, and material load. Biodegradation can be aided by removal of toxic organic and inorganic substances, such as heavy metal ions.

Aerobic Treatment

Aerobic waste treatment processes utilize aerobic bacteria and fungi that require molecular oxygen, O_2. These processes are often favored by microorganisms, in part because of the high energy yield obtained when molecular oxygen reacts with organic matter. Aerobic waste treatment is well adapted to the use of an activated sludge process. It can be applied to hazardous wastes such as chemical process wastes and landfill leachates. Some systems use powdered activated carbon as an additive to absorb nonbiodegradable organic wastes.

Contaminated soils can be mixed with water and treated in a bioreactor to eliminate biodegradable contaminants in the soil. It is possible in principle to treat contaminated soils biologically in place by pumping oxygenated, nutrient-enriched water through the soil in a recirculating system.

Anaerobic Treatment

Anaerobic waste treatment in which microorganisms degrade wastes in the absence of oxygen can be practiced on a variety of organic hazardous wastes. Compared to the aerated activated sludge process, anaerobic digestion requires less energy; yields less sludge by-product; generates sulfide (H_2S), which precipitates toxic heavy metal ions; and produces methane gas, CH_4, which can be used as an energy source.

The overall process for anaerobic digestion is a fermentation process in which organic matter is both oxidized and reduced. The simplified reaction for the anaerobic fermentation of a hypothetical organic substance, "$\{CH_2O\}$", is the following:

$$2\{CH_2O\} \rightarrow CO_2 + CH_4 \qquad (20.8.1)$$

In practice, the microbial processes involved are quite complex. Most of the wastes for which anaerobic digestion is suitable consist of oxygenated compounds, such as acetaldehyde or methylethyl ketone.

20.9.   LAND TREATMENT AND COMPOSTING

Land Treatment

Soil may be viewed as a natural filter for wastes. Soil has physical, chemical, and biological characteristics that can enable waste detoxification, biodegradation, chemical decomposition, and physical and chemical fixation. Therefore, **land treatment** of wastes may be accompished by mixing the wastes with soil under appropriate conditions.

Soil is a natural medium for a number of living organisms that may have an effect upon biodegradation of hazardous wastes. Of these, the most important are bacteria, including those from the genera *Agrobacterium, Arthrobacteri, Bacillus, Flavobacterium,* and *Pseudomonas. Actinomycetes* and fungi are important organisms in decay of vegetable matter and may be involved in biodegradation of wastes.

Wastes that are amenable to land treatment are biodegradable organic substances. However, in soil contaminated with hazardous wastes, bacterial cultures may develop that are effective in degrading normally recalcitrant compounds through acclimation over a long period of time. Land treatment is most used for petroleum refining wastes and is applicable to the treatment of fuels and wastes from leaking underground storage tanks. It can also be applied to biodegradable organic chemical wastes, including some organohalide compounds. Land treatment is not suitable for the treatment of wastes containing acids, bases, toxic inorganic compounds, salts, heavy metals, and organic compounds that are excessively soluble, volatile, or flammable.

Composting

Composting of hazardous wastes is the biodegradation of solid or solidified materials in a medium other than soil. Bulking material, such as plant residue, paper, municipal refuse, or sawdust may be added to retain water and enable air to penetrate to the waste material. Successful composting of hazardous waste depends upon a number of factors, including those discussed above under land treatment. The first of these is the selection of the appropriate microorganism or **inoculum**. Once a successful composting operation is underway, a good inoculum is maintained by recirculating spent compost to each new batch. Other parameters that must be controlled include oxygen supply, moisture content (which should be maintained at a minimum of about 40%), pH (usually around neutral), and temperature. The composting process generates heat so, if the mass of the compost pile is sufficiently high, it can be self-heating under most conditions. Some wastes are deficient in nutrients, such as nitrogen, which must be supplied from commercial sources or from other wastes.

20.10. PREPARATION OF WASTES FOR DISPOSAL

Immobilization, stabilization, fixation, and solidification are terms that describe techniques whereby hazardous wastes are placed in a form suitable for long-term disposal. These aspects of hazardous waste management are addressed in detail in reference works on the subject.

Immobilization

Immobilization includes physical and chemical processes that reduce surface areas of wastes to minimize leaching. It isolates the wastes from their environment, especially groundwater, so that they have the least possible tendency to migrate. This is accomplished by physically isolating the waste, reducing its solubility, and decreasing its surface area. Immobilization usually improves the handling and physical characteristics of wastes.

Stabilization

Stabilization means the conversion of a waste from its original form to a physically and chemically more stable material. Stabilization may include chemical reactions that produce products that are less volatile, soluble, and reactive. Solidification, which is discussed below, is one of the most common means of stabilization. Stabilization is required for land disposal of wastes. **Fixation** is a process that binds a hazardous waste in a less mobile and less toxic form; it means much the same thing as stabilization.

Solidification

Solidification may involve chemical reaction of the waste with the solidification agent, mechanical isolation in a protective binding matrix, or a combination of chemical and physical processes. It can be accomplished by evaporation of water from aqueous wastes or sludges, sorption onto solid material, reaction with cement, reaction with silicates, encapsulation, or imbedding in polymers or thermoplastic materials.

In many solidification processes, such as reaction with portland cement, water is an important ingredient of the hydrated solid matrix. Therefore, the solid should not be heated excessively or exposed to extremely dry conditions, which could result in diminished structural integrity from loss of water. In some cases, however, heating a solidified waste is an essential part of the overall solidification procedure. For example, an iron hydroxide matrix can be converted to highly insoluble, refractory iron oxide by heating, and organics in solidified wastes may be converted to inert carbon by pyrolytic heating. Heating is an integral part of the process of vitrification (see below).

Sorption to a Solid Matrix Material

Hazardous waste liquids, emulsions, sludges, and free liquids in contact with sludges may be solidified and stabilized by fixing onto solid **sorbents,** including activated carbon (for organics), fly ash, kiln dust, clays, vermiculite, and various proprietary materials. Sorption may be done to convert liquids and semisolids to dry solids, improve waste handling, and reduce solubility of waste constituents. Sorption can also be used to improve waste compatibility with substances such as portland cement used for solidification and setting. Specific sorbents may also be used to stabilize pH and pE (a measure of the tendency of a medium to be oxidizing or reducing, see Section 11.8).

The action of sorbents can include simple mechanical retention of wastes, physical sorption, and chemical reactions. It is important to match the sorbent to the waste. A substance with a strong affinity for water should be employed for wastes containing excess water and one with a strong affinity for organic materials should be used for wastes with excess organic solvents.

Thermoplastics and Organic Polymers

Thermoplastics are solids or semisolids that become liquified at elevated temperatures. Hazardous waste materials may be mixed with hot thermoplastic liquids and solidified in the cooled thermoplastic matrix, which is rigid but deformable. The thermoplastic material most used for this purpose is asphalt bitumen. Other thermoplastics, such as paraffin and polyethylene, have also been used to immobilize hazardous wastes.

Among the wastes that can be immobilized with thermoplastics are those containing heavy metals, such as electroplating wastes. Organic thermoplastics repel water and reduce the tendency toward leaching in contact with groundwater. Compared to cement, thermoplastics add relatively less material to the waste.

A technique similar to that described above uses **organic polymers** produced in contact with solid wastes to imbed the wastes in a polymer matrix. Three kinds of polymers that have been used for this purpose include polybutadiene, urea-formaldehyde, and vinyl ester-styrene polymers. This procedure is more complicated than is the use of thermoplastics but, in favorable cases, yields a product in which the waste is held more strongly.

Vitrification

Vitrification or **glassification** consists of imbedding wastes in a glass material. In this application, glass may be regarded as a high-melting-temperature inorganic thermoplastic. Molten glass can be used, or glass can be synthesized in contact with the waste by mixing and heating with glass constituents — silicon dioxide (SiO_2), sodium carbonate (Na_2CO_3), and calcium oxide (CaO). Other constituents may include boric oxide, B_2O_3, which yields a borosilicate glass that is especially resistant to changes in temperature and chemical attack. In some cases, glass is used in conjunction with thermal waste destruction processes, serving to immobilize hazardous waste ash constituents. Some wastes are detrimental to the quality of the glass. Aluminum oxide, for example, may prevent glass from fusing.

Vitrification is relatively complicated and expensive, the latter because of the energy consumed in fusing glass. Despite these disadvantages, it is the best immobilization technique for some special wastes and has been promoted for solidification of radionuclear wastes because glass is chemically inert and resistant to leaching. However, high levels of radioactivity can cause deterioration of glass and lower its resistance to leaching.

Solidification with Cement

Portland cement is widely used for solidification of hazardous wastes. In this application, portland cement provides a solid matrix for isolation of the wastes, chemically binds water from sludge wastes, and may react chemically with wastes (for example, the calcium and base in portland cement react chemically with inorganic arsenic sulfide wastes to reduce their solubilities). However, most wastes are held physically in the rigid portland cement matrix and are subject to leaching.

As a solidification matrix, portland cement is most applicable to inorganic sludges containing heavy metal ions that form insoluble hydroxides and carbonates in the basic carbonate medium provided by the cement. The success of solidification with portland cement strongly depends upon whether or not the waste adversely affects the strength and stability of the concrete product. A number of substances—organic matter such as petroleum or coal; some silts and clays; sodium salts of arsenate, borate, phosphate, iodate, and sulfide; and salts of copper, lead, magnesium, tin, and zinc—are incompatible with portland cement because they interfere with its set and cure and cause deterioration of the cement matrix with time. However, a reasonably good disposal form can be obtained by absorbing organic wastes with a solid material, which in turn is set in portland cement. This approach has been used with hydrocarbon wastes sorbed by an activated coal char matrix.

Solidification with Silicate Materials

Water-insoluble **silicates**, (pozzolanic substances) containing oxyanionic silicon such as SiO_3^{2-} are used for waste solidification. These substances include fly ash, flue dust, clay, calcium silicates, and ground-up slag from blast furnaces. Soluble silicates, such as sodium silicate, may also be used. Silicate solidification usually requires a setting agent, which may be portland cement (see above), gypsum (hydrated $CaSO_4$), lime, or compounds of aluminum, magnesium, or iron. The product may vary from a granular material to a concrete-like solid. In some cases, the product is improved by additives, such as emulsifiers, surfactants, activators, calcium chloride, clays, carbon, zeolites, and various proprietary materials.

Success has been reported for the solidification of both inorganic wastes and organic wastes (including oily sludges) with silicates. The advantages and disadvantages of silicate solidification are similar to those of portland cement discussed above. One consideration that is especially applicable to fly ash is the presence in some silicate materials of leachable hazardous substances, which may include arsenic and selenium.

Encapsulation

As the name implies, **encapsulation** is used to coat wastes with an impervious material so that they do not contact their surroundings. For example, a water-soluble waste salt encapsulated in asphalt would not dissolve, so long as the asphalt layer remains intact. A common means of encapsulation uses heated, molten thermoplastics, asphalt, and waxes that solidify when cooled. A more sophisticated approach to encapsulation is to form polymeric resins from monomeric substances in the presence of the waste.

Chemical Fixation

Chemical fixation is a process that binds a hazardous waste substance in a less mobile, less toxic form by a chemical reaction that alters the waste chemically. Physical and chemical fixation often occur together. Polymeric inorganic silicates containing some calcium and often some aluminum are the inorganic materials most widely used as a fixation matrix. Many kinds of heavy metals are chemically bound in such a matrix, as well as being held physically by it. Similarly, some organic wastes are bound by reactions with matrix constituents.

20.11.   ULTIMATE DISPOSAL OF WASTES

Regardless of the destruction, treatment, and immobilization techniques used, there will always remain from hazardous wastes some material that has to be put somewhere. This section briefly addresses the ultimate disposal of ash, salts, liquids, solidified liquids, and other residues that must be placed where their potential to do harm is minimized.

Disposal Aboveground

In some important respects disposal aboveground, essentially in a pile designed to prevent erosion and water infiltration, is the best way to store solid wastes. Perhaps its most important advantage is that it avoids infiltration by groundwater that can result in leaching and groundwater contamination common to storage in pits and landfills. In a properly designed aboveground disposal facility, any leachate that is produced drains quickly by gravity to the leachate collection system, where it can be detected and treated.

Aboveground disposal can be accomplished with a storage mound deposited on a layer of compacted clay covered with impermeable membrane liners laid somewhat above the original soil surface and shaped to allow leachate flow and collection. The slopes around the edges of the storage mound should be sufficiently great to allow good drainage of precipitation, but gentle enough to deter erosion.

Landfill

Landfill historically has been the most common way of disposing of solid hazardous wastes and some liquids, although it is being severely limited in many nations by new regulations and high land costs. Landfill involves disposal that is at least partially underground in excavated cells, quarries, or natural depressions. Usually, fill is continued aboveground to most efficiently utilize space and provide a grade for drainage of precipitation.

The greatest environmental concern with landfill of hazardous wastes is the generation of leachate from infiltrating surface water and groundwater with resultant contamination of groundwater supplies. Modern hazardous waste landfills provide elaborate systems to contain, collect, and control such leachate.

There are several components to a modern landfill. A landfill should be placed on a compacted low-permeability medium, preferably clay, which is covered by a flexible-membrane liner consisting of water-tight impermeable material. This liner is covered with granular material in which is installed a secondary drainage system. Next is another flexible-membrane liner above which is installed a primary drainage system for the removal of leachate. This drainage system is covered with a layer of granular filter medium, upon which the wastes are placed. In the landfill, wastes of different kinds are separated by berms consisting of clay or soil covered with liner material. When the fill is complete, the waste is capped to prevent surface water infiltration and covered with compacted soil. In addition to leachate collection, provision may be made for a system to treat evolved gases, particularly when methane-generating biodegradable materials are disposed in the landfill.

The flexible-membrane liner made of rubber (including chlorosulfonated polyethylene) or plastic (including chlorinated polyethylene, high-density polyethylene, and polyvinylchloride), is a key component of state-of-the-art landfills. It controls seepage out of, and infiltration into the landfill. Liners have to meet stringent standards to serve their intended purpose. In addition to being impermeable, the liner material must be strongly resistant to biodegradation, chemical attack, and tearing.

Capping is done to cover the wastes, prevent infiltration of excessive amounts of surface water, and prevent release of wastes to overlying soil and the atmosphere. Caps come in a variety of forms and are often multilayered. Some of the problems that may occur with caps are settling, erosion, ponding, damage by rodents and penetration by plant roots.

Surface Impoundment of Liquids

Many liquid hazardous wastes, slurries, and sludges are placed in **surface impoundments**, which usually serve for treatment and often are designed to be filled in eventually as a landfill disposal site. Most liquid hazardous wastes and a significant fraction of solids are placed in surface impoundments in some stage of treatment, storage, or disposal.

A surface impoundment may consist of an excavated "pit", a structure formed with dikes, or a combination thereof. The construction is similar to that discussed above for landfills in that the bottom and walls should be impermeable to liquids and provision must be made for leachate collection. The chemical and mechanical challenges to liner materials in surface impoundments are severe so that proper geological siting and construction with floors and walls composed of low-permeability soil and clay are important in preventing pollution from these installations.

Deep-well Disposal of Liquids

Deep-well disposal of liquids consists of their injection under pressure to underground strata isolated by impermeable rock strata from aquifers. Early experience with this method was gained in the petroleum industry where disposal is required of large quantities of saline wastewater coproduced with crude oil The method was later extended to the chemical industry for the disposal c brines, acids, heavy metal solutions, organic liquids, and other liquids.

A number of factors must be considered in deep-well disposal. Wastes ar injected into a region of elevated temperature and pressure, which may caus chemical reactions to occur involving the waste constituents and the minera strata. Oils, solids, and gases in the liquid wastes can cause problems such a: clogging. Corrosion may be severe. Microorganisms may have some effects. Most problems from these causes can be mitigated by proper waste pretreatment.

The most serious consideration involving deep-well disposal is the potential contamination of groundwater. Although injection is made into permeable saltwater aquifers presumably isolated from aquifers that contain potable water, contamination may occur. Major routes of contamination include fractures, faults, and other wells. The disposal well itself can act as a route for contamination if it is not properly constructed and cased or if it is damaged.

20.12.   LEACHATE AND GAS EMISSIONS

Leachate

The production of contaminated leachate is a possibility with most disposal sites. Therefore, new hazardous waste landfills require leachate collection/ treatment systems and many older sites are required to have such systems retrofitted to them. Modern hazardous waste landfills typically have dual leachate collection systems, one located between the two impermeable liners required for the bottom and sides of the landfill and another just above the top liner of the double-liner system. The upper leachate collection system is called the primary leachate collection system, and the bottom is called the secondary leachate collection system. Leachate is collected in perforated pipes that are imbedded in granular drain material.

Chemical and biochemical processes have the potential to cause some problems for leachate collection systems. One such problem is clogging by insoluble manganese(IV) and iron(III) hydrated oxides upon exposure to air as described for water wells in Section 14.6.

Leachate consists of water that has become contaminated by wastes as it passes through a waste disposal site. It contains waste constituents that are soluble, not retained by soil, and not degraded chemically or biochemically. Some potentially harmful leachate constituents are products of chemical or biochemical transformations of wastes.

The best approach to leachate management is to prevent its production by limiting infiltration of water into the site. Rates of leachate production may be very low when sites are selected, designed, and constructed with minimal production of leachate as a major objective. A well-maintained, low-permeability cap over the landfill is very important for leachate minimization.

Hazardous Waste Leachate Treatment

The first step in treating leachate is to characterize it fully, particularly with a thorough chemical analysis of possible waste constituents and their chemical and metabolic products. The biodegradability of leachate constituents should also be determined.

The options available for the treatment of hazardous waste leachate are generally those that can be used for industrial wastewaters. These include biological treatment by an activated sludge, or related process, and sorption by activated carbon usually in columns of granular activated carbon. Hazardous waste leachate can be treated by a variety of chemical processes, including acid/base neutralization, precipitation, and oxidation/reduction. In some cases, these treatment steps must precede biological treatment; for example, leachate exhibiting extremes of pH must be neutralized in order for microorganisms to thrive in it. Cyanide in the leachate may be oxidized with chlorine and organics with ozone, hydrogen peroxide promoted with ultraviolet radiation, or dissolved oxygen at high temperatures and pressures. Heavy metals may be precipitated with base, carbonate, or sulfide.

Leachate can be treated by a variety of physical processes. In some cases, simple density separation and sedimentation can be used to remove water-immiscible liquids and solids. Filtration is frequently required and flotation may be useful. Leachate solutes can be concentrated by evaporation, distillation, and membrane processes, including reverse osmosis, hyperfiltration, and ultrafiltration. Organic constituents can be removed by solvent extraction, air stripping, or steam stripping.

Gas Emissions

In the presence of biodegradable wastes, methane and carbon dioxide gases are produced in landfills by anaerobic degradation (see Reaction 20.8.1). Gases may also be produced by chemical processes with improperly pretreated wastes, as would occur in the hydrolysis of calcium carbide to produce acetylene:

$$CaC_2 + 2H_2O \rightarrow C_2H_2 + Ca(OH)_2 \qquad (20.12.1)$$

Odorous and toxic hydrogen sulfide, H_2S, may be generated by the chemical reaction of sulfides with acids or by the biochemical reduction of sulfate by anaerobic bacteria (*Desulfovibrio*) in the presence of biodegradable organic matter:

$$SO_4^{2-} + 2\{CH_2O\} + 2H^+ \xrightarrow[\text{bacteria}]{\text{Anaerobic}} H_2S + 2CO_2 + 2H_2O \quad (20.12.2)$$

Gases such as these may be toxic, they may burn, or they may explode. Furthermore, gases permeating through landfilled hazardous waste may carry along waste vapors, such as those of volatile aryl compounds and low-molecular-mass chlorinated hydrocarbons. Of these, the ones of most concern are benzene, carbon tetrachloride, chloroform, 1,2-dibromoethane, 1,2-dichloroethane, dichloromethane, tetrachloroethane, 1,1,1-trichloroethane, tri-chloroethylene, and vinyl chloride. Because of the hazards from these and other volatile species, it is important to minimize production of gases and, if significant amounts of gases are produced, they should be vented or treated by activated carbon sorption or flaring.

20.13.   IN-SITU TREATMENT

In-situ treatment refers to waste treatment processes that can be applied to wastes in a disposal site by direct application of treatment processes and reagents to the wastes. Where possible, in-situ treatment is highly desirable as a waste site remediation option.

In-Situ Immobilization

In-situ immobilization is used to convert wastes to insoluble forms that will not leach from the disposal site. Heavy metal contaminants, including lead, cadmium, zinc, and mercury, can be immobilized by chemical precipitation as the sulfides by treatment with gaseous H_2S or alkaline Na_2S solution. Disadvantages include the high toxicity of H_2S and the contamination potential of soluble sulfide. Although precipitated metal sulfides should remain as solids in the anaerobic conditions of a landfill, unintentional exposure to air can result in oxidation of the sulfide and remobilization of the metals as soluble sulfate salts.

Oxidation and reduction reactions can be used to immobilize heavy metals in-situ. Oxidation of soluble Fe^{2+} and Mn^{2+} to their insoluble hydrous oxides, $Fe_2O_3 \cdot xH_2O$ and $MnO_2 \cdot xH_2O$, respectively, can precipitate these metal ions and coprecipitate other heavy metal ions. However, subsurface reducing conditions could later result in reformation of soluble reduced species. Reduction can be used in-situ to convert soluble, toxic chromate to insoluble chromium(III).

Chelation may convert metal ions to less mobile forms, although with most agents, chelation has the opposite effect. A chelating agent called Tetran is supposed to form metal chelates that are strongly bound to clay minerals. The humin fraction of soil humic substances likewise immobilizes metal ions.

Solidification In-Situ

In-situ solidification can be used as a remedial measure at hazardous waste sites. One approach is to inject soluble silicates followed by reagents that cause them to solidify. For example, injection of soluble sodium silicate followed by calcium chloride or lime forms solid calcium silicate.

Detoxification In-Situ

When only one, or a limited number of harmful constituents is present in a waste disposal site, it may be practical to consider detoxification in-situ. This approach is most practical for organic contaminants including pesticides (organophosphate esters and carbamates), amides, and esters. Among the chemical and biochemical processes that can detoxify such materials are chemical and enzymatic oxidation, reduction, and hydrolysis. Chemical oxidants that have been proposed for this purpose include hydrogen peroxide, ozone, and hypochlorite.

Enzyme extracts collected from microbial cultures and purified have been considered for in-situ detoxification. One cell-free enzyme that has been used for detoxification of organophosphate insecticides is parathion hydrolase. The hostile environment of a chemical waste landfill, including the presence of enzyme-inhibiting heavy metal ions, is detrimental to many biochemical approaches to in-situ treatment. Furthermore, most sites contain a mixture of hazardous constituents, which might require several different enzymes for their detoxification.

Permeable Bed Treatment

Some groundwater plumes contaminated by dissolved wastes can be treated by a permeable bed of material placed in a trench through which the groundwater must flow. For example, limestone contained in a permeable bed neutralizes acid and precipitates some kinds of heavy metals as hydroxides or carbonates. Synthetic ion exchange resins can be used in a permeable bed to retain heavy metals and even some anionic species, although competition with ionic species present naturally in the groundwater can cause some problems with the use of ion exchangers. Activated carbon in a permeable bed will remove some organics, especially less soluble, higher molecular mass organic compounds.

Permeable bed treatment requires relatively large quantities of reagent, which argues against the use of activated carbon and ion exchange resins. In such an application, it is unlikely that either of these materials could be reclaimed and regenerated as is done when they are used in columns to treat wastewater. Furthermore, ions taken up by ion exchangers and organic species retained by activated carbon may be released at a later time, causing subsequent problems. Finally, a permeable bed that has been truly effective in collecting waste materials may, itself, be considered a hazardous waste requiring special treatment and disposal.

In-Situ Thermal Processes

Heating of wastes in-situ can be used to remove or destroy some kinds of hazardous substances. Both steam injection and radio frequency heating have been proposed for this purpose. Volatile wastes brought to the surface by heating can be collected and held as condensed liquids or by activated carbon.

One approach to immobilizing wastes in-situ is high temperature vitrification using electrical heating. This process involves placing conducting graphite between two electrodes poured on the surface and passing an electrical current between the electrodes. In principle, the graphite becomes very hot and "melts" into the soil leaving a glassy slag in its path. Volatile species evolved are collected and, if the operation is successful, a nonleachable slag is left in place. It is easy to imagine problems that might occur with this "molten glob" approach to treating wastes in situ, including difficulties in getting a uniform melt, problems from groundwater infiltration, and astoundingly high consumption of electricity per unit mass of waste treated.

Soil Washing and Flushing

Extraction with water containing various additives can be used to cleanse soil contaminated with hazardous wastes. When the soil is left in place and the water pumped into and out of it, the process is called flushing; when soil is removed and contacted with liquid the process is referred to as washing. Here, washing is used as a term applied to both processes.

The composition of the fluid used for soil washing depends upon the contaminants to be removed. The washing medium may consist of pure water or it may contain acids (to leach out metals or neutralize alkaline soil contaminants), bases (to neutralize contaminant acids), chelating agents (to solubilize heavy metals), surfactants (to enhance the removal of organic contaminants from soil and improve the ability of the water to emulsify insoluble organic species), or reducing agents (to reduce oxidized species). Soil contaminants may dissolve, form emulsions, or react chemically. Inorganic species commonly removed from soil by washing include heavy metals salts; lighter aromatic hydrocarbons, such as toluene and xylenes; lighter organohalides, such as trichloro- or tetrachloroethylene; and light-to-medium molecular mass aldehydes and ketones.

CHAPTER SUMMARY

The chapter summary below is presented in a programmed format to review the main points covered in this chapter. It is used most effectively by filling in the blanks, referring back to the chapter as necessary. The correct answers are given at the end of the summary.

In descending order of desirability, hazardous waste management attempts to accomplish the following goals (1)_____

_____

_____

_____

_____.

Cutting down quantities of wastes from their sources is called (2)_____

_____ and utilization of treatment processes which reduce the quantities of wastes requiring ultimate disposal is called (3)_____

_____. The four broad areas in which something of value may be obtained from recycling wastes are (4)_____

_____

_____.

The two main classes of potentially hazardous contaminants in motor oil are (5)_____. The three major steps in reprocessing waste lubricating oil are (6)_____

_____. Ash from waste fuel oil may be hazardous because of (7)_____.

Among the operations used in solvent purification for recycling are
(8)_____

_____

_____.

Primary treatment of wastes is generally regarded as (9)_____

_____, secondary treatment (10)_____

_____, and polishing (11)_____

_____.

The main methods of physical treatment can be divided into the categories of
(12)_____

_____.

A waste treatment process by which colloidal-sized particles are caused to aggregate and settle from suspension is called (13)_____ and it is aided by
(14)_____. A phase
transistion process which is used in treating and recycling solvents, waste oil,
aqueous phenolic wastes, xylene contaminated with paraffin from histological
laboratories, and mixtures of ethylbenzene and styrene is (15)_____.
This process produces by-product (16)_____, which are often
hazardous and polluting. A means of separating volatile components from less
volatile ones in a liquid mixture by the partitioning of the more volatile materials
to a gas phase of air or steam is called (17)_____.
The major changes in a solution that can cause physical precipitation are
(18)_____.

Solvent extraction and leaching are examples of (19)_____
processes. A solvent extraction process that makes use of CO_2 at very high
pressures is (20)_____ extraction. Transfer of a substance from a
solution to a solid phase is called (21)_____, for which the most
important solid for organics removal is (22)_____.

In membrane processes, such as hyperfiltration and ultrafiltration, water and lower molecular mass solutes under pressure pass through the membrane as a stream of purified (23)_____, leaving behind a stream of (24)_____ containing (25)_____. Reverse osmosis uses a membrane that is (26)_____.

The major kinds of chemical phenomena used in chemical purification of wastes are (27)_____

_____.

The most widely used means of precipitating metal ions is by (28)_____

_____. The reaction $Cd^{2+} + Zn \rightarrow Cd + Zn^{2+}$ illustrates a process called (29)_____. A process that can be used to treat metal carbides, hydrides, amides, alkoxides, and halides; nonmetal oxyhalides and sulfides; and some organic species, such as acetic anhydride is (30)_____.

The most widely applied means of thermal treatment of hazardous wastes is (31)_____, which utilizes (32)_____

_____ to destroy wastes. The four major components of hazardous waste incineration systems are (33)_____

_____.

The two most common types of hazardous waste incinerators in the U. S. are (34)_____

_____. The type of thermal treatment devices that make use of an extremely hot plasma of ionized air injected through an electrical arc are (35)_____.

Biodegradation of wastes is (36)_____

_____.

Detoxification refers to (37)_____

_____. Recalcitrant or biorefractory substances are (38)_____

_____. Some of

the ways in which anaerobic digestion differs from aerobic digestion in the treatment of wastes is (39)_____ _____. Immobilization of wastes includes physical and chemical processes that (40)_____ _____ whereas stabilization means (41)_____ _____. Some of the sorbents used to solidify and stabilize hazardous wastes are (42)_____ _____. Vitrification consists of (43)_____. Pozzolanic substances used to solidify wastes include (44)_____ _____.

Dissolved iron and manganese in hazardous waste leachate can cause problems because of (45)_____ _____.

The best approach to leachate management is (46)_____ _____. Two gases, one explosive and the other highly toxic, produced by anaerobic degradation in a landfill are (47)_____.

Waste treatment processes that can be applied to wastes in a disposal site by direct application of treatment processes and reagents to the wastes are called (48)_____ processes. Where possible, in-situ treatment is highly desirable as a waste site remediation option.

When contaminated soil is left in place and purified by water pumped into and out of it, the process is called (49)_____ and when soil is removed and contacted with liquid the process is referred to as (50)_____.

Answers

1. Do not produce waste, produce only minimum quantities, recycle, treat in a way that makes the waste nonhazardous, dispose of the waste in a safe manner, monitor disposed wastes for leaching and other adverse effects

2. waste reduction
3. waste minimization
4. direct recycle as raw material to the generator, transfer as a raw material to another process, utilization for pollution control or waste treatment, recovery of energy
5. organic constituents and inorganic constituents
6. removal of contaminant fuel and water, processing, and final polishing
7. its heavy metal content
8. settling, filtration, centrifugation, application of drying agents, adsorption, chemical treatment, and fractional distillation
9. preparation for further treatment
10. detoxifies, destroys, and removes hazardous constituents
11. usually refers to treatment of water that is removed from wastes
12. phase separation, phase transfer, phase transition, and membrane separations
13. coagulation
14. agitation, heat, acid, and addition of coagulants
15. distillation
16. distillation bottoms
17. stripping
18. cooling the solution, evaporation of solvent, or alteration of solvent composition
19. phase transfer
20. supercritical fluid
21. sorption
22. activated carbon
23. permeate
24. concentrate
25. impurities
26. selectively permeable to water
27. acid/base neutralization, chemical precipitation, chemical extraction and leaching, oxidation/reduction, hydrolysis, and ion exchange
28. the formation of hydroxides
29. cementation
30. hydrolysis
31. incineration
32. high temperatures, an oxidizing atmosphere, and often turbulent combustion conditions
33. waste preparation and feeding, combustion, air pollutant removal, and ash and residue handling
34. rotary kiln and liquid injection incinerators
35. plasma incinerators
36. their conversion by biological processes to simple inorganic molecules and to biological materials

37. the biological conversion of a toxic substance to a less toxic species
38. those that resist biodegradation and tend to persist and accumulate in the environment
39. it requires less energy, yields less sludge by-product, generates sulfide (H_2S) and produces methane gas
40. that reduce surface areas of wastes to minimize leaching
41. the conversion of a waste from its original form to a physically and chemically more stable material
42. activated carbon, fly ash, kiln dust, clays, vermiculite
43. imbedding wastes in a glass material
44. fly ash, flue dust, clay, calcium silicates, and ground-up slag from blast furnaces
45. clogging by insoluble manganese(IV) and iron(III) hydrated oxides upon exposure to air
46. to prevent its production by limiting infiltration of water into the site
47. methane and carbon dioxide
48. in-situ
49. flushing
50. washing

QUESTIONS AND PROBLEMS

1. Place the following hazardous waste management options in order of increasing desirability: (a) treat the waste to make it nonhazardous, (b) do not produce waste, (c) minimize quantities of waste produced, (d) dispose of the waste in a safe manner, (e) recycle the waste.

2. Match the waste recycling process or industry from the column on the left with the kind of material that can be recycled from the list on the right, below:

   1. Recycle as raw material to the generator
   2. Utilization for pollution control or waste treatment
   3. Materials with agricultural uses
   4. Organic substances
   5. Energy production

   (a) Waste alkali
   (b) Hydraulic and lubricating oils
   (c) Incinerable materials
   (d) Incompletely consumed feedstock material
   (e) Waste lime or phosphate-containing sludges

3. What material is recycled using hydrofinishing, treatment with clay, and filtration?

4. What is the "most important operation in solvent purification and recycle" that is used to separate solvents from impurities, water, and other solvents?

5. Dissolved air flotation (DAF) is used in the secondary treatment of wastes. What is the principle of this technique? For what kinds of hazardous waste substances is it most applicable?

6. Match the process or industry from the column on the left with its "phase of waste treatment" from the list on the right, below:

   1. Activated carbon sorption
   2. Precipitation
   3. Reverse osmosis
   4. Emulsion breaking
   5. Slurrying

   (a) Primary treatment
   (b) Secondary treatment
   (c) Polishing

7. Distillation is used in treating and recycling a variety of wastes, including solvents, waste oil, aqueous phenolic wastes, and mixtures of ethylbenzene and styrene. What is the major hazardous waste problem that arises from the use of distillation for waste treatment?

8. Supercritical fluid technology has a great deal of potential for the treatment of hazardous wastes. What are the principles involved with the use of supercritical fluids for waste treatment? Why is this technique especially advantageous? Which substance is most likely to be used as a supercritical fluid in this application? For which kinds of wastes are supercritical fluids most useful?

9. What are some advantages of using acetic acid, compared, for example, to sulfuric acid, as a neutralizing agent for treating waste alkaline materials?

10. Which of the following would be **least likely** to be produced by, or used as a reagent for the removal of heavy metals by their precipitation from solution? (a) Na_2CO_3, (b) CdS, (c) $Cr(OH)_3$, (d) KNO_3, (e) $Ca(OH)_2$

11. Both $NaBH_4$ and Zn are used to remove metals from solution. How do these substances remove metals? What are the forms of the metal products?

12. Of the following, thermal treatment of wastes is **not** useful for (a) volume reduction, (b) destruction of heavy metals, (c) removal of volatile, combustible, mobile organic matter, (d) destruction of pathogenic materials, (e) destruction of toxic substances.

13. From the following, choose the waste liquid that is least amenable to incineration and explain why it is not readily incinerated: (a) methanol, (b) tetrachloroethylene, (c) acetonitrile, (d) toluene, (e) ethanol, (f) acetone.

14. Name and give the advantages of the process that is used to destroy more hazardous wastes by thermal means than are burned solely for the purpose of waste destruction.

15. What is the major advantage of fluidized-bed incinerators from the standpoint of controlling pollutant by-products?

16. What is the best way to obtain microorganisms to be used in the treatment of hazardous wastes by biodegradation?

17. What are the principles of composting? How is it used to treat hazardous wastes?

18. How is portland cement used in the treatment of hazardous wastes for disposal? What might be some disadvantages of such a use?

19. What are the advantages of aboveground disposal of hazardous wastes as opposed to burying wastes in landfills?

20. Describe and explain the best approach to managing leachate from hazardous waste disposal sites.

SUPPLEMENTARY REFERENCES

Manahan, S. E., *Hazardous Waste Chemistry, Toxicology and Treatment,* (Chelsea, MI: Lewis Publishers, 1990).

Freeman, H. M., Ed., *Hazardous Waste Minimization*, (New York: McGraw-Hill Publishing Co., 1990).

Mueller Associates, Inc., *Waste Oil Reclaiming Technology, Utilization and Disposal*, (Park Ridge, NJ: Noyes Data Corporation, 1989).

U.S. Environmental Protection Agency, ICF Consulting Associates, Inc., *Solvent Waste Reduction,* (Park Ridge, NJ: Noyes Data Corporation, 1990).

Donahue, B. A., *Reclamation and Reprocessing of Spent Solvents*, (Park Ridge, NJ: Noyes Data Corporation, 1989).

Oppelt, E. T., "Hazardous Waste Destruction," *Environmental Science and Technology*, 22, 403–404 (1988).

Brunner, C. R., "Incineration: Today's Hot Option for Waste Disposal," *Chemical Engineering*, October 12, 1987, pp. 96–106.

Torpy, M. F., Ed., *Anaerobic Treatment of Industrial Wastewaters*, (Park Ridge, NJ: Noyes Data Corporation, 1988).

Côté, P., and M. Gilliam, Eds., *Environmental Aspect of Stabilization and Solidification of Hazardous and Radioactive Wastes*, (Philadelphia, PA: American Society for Testing and Materials, 1989).

Conner, J. R., *Chemical Fixation and Solidification of Hazardous Wastes*, (New York: Van Nostrand Reinhold, 1990).

Raghavan, R., E. Coles, and D. Dietz, "Cleaning Excavated Soil Using Extraction Agents: A State-of-the-Art Review," EPA/600/S2–89/034, United States Environmental Protection Agency, Risk Reduction Engineering Laboratory, Cincinnati, OH, 1990.

Lankford, P. W., and W. W. Eckenfelder, Jr., Eds., *Toxicity Reduction in Industrial Effluents*, (New York: Van Nostrand Reinhold, 1990).

Allegri, T. C., *Handling and Management of Hazardous Materials and Wastes,* (New York, NY: Chapman and Hall, 1986).

Krauskopf, K. B., *Radioactive Waste Disposal and Geology*, (New York, NY: Chapman and Hall, 1988).

21 TOXICOLOGICAL CHEMISTRY

21.1. BIOCHEMISTRY AND THE CELL

Ultimately, most pollutants and hazardous substances are of concern because of their toxic effects. **Biochemistry**, discussed in some detail in Chapter 10, is the science that deals with chemical processes and materials in living systems.[1] It is summarized very briefly here preparatory to discussing the toxic effects of environmental pollutants.

The Cell

Figure 21.1 shows the major features of the eukaryotic cell, which is the basic structure in which biochemical processes occur in animals. These features are the following:

- **Cytoplasm**, which fills the cell and in which several important kinds of cell structures are contained
- **Mitochondria**, which mediate energy conversion and utilization in the cell
- **Ribosomes**, which participate in protein synthesis

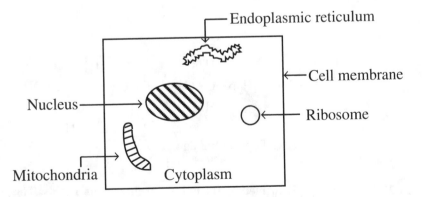

Figure 21.1. Some major features of the eukaryotic cell.

- **Endoplasmic reticulum**, which is involved in the metabolism of some toxicants by enzymatic processes
- **Cell membrane,** which encloses the cell and regulates the passage of ions, nutrients, metabolic products, toxicants, and toxicant metabolites into and out of the cell interior; when its membrane is damaged by toxic sustances, a cell may not function properly and the organism may be harmed
- **Cell nucleus,** the "control center" of the cell which contains **deoxyribo-nucleic acid** through which the nucleus regulates cell division

21.2.   PROTEINS, CARBOHYDRATES, AND LIPIDS

Proteins are the basic building blocks of biological material that constitute enzymes and most of the cytoplasm inside the cell. Proteins are composed primarily of biopolymers of **amino acids**, which have the general structure shown in Figure 21.2, where R represents a group ranging from H atom (in the amino acid glycine) to moderately complex structures. The amino acids in proteins are joined at **peptide linkages** outlined by dashed lines in Figure 21.2.

The **structures** of protein molecules determine the behavior of proteins in crucial areas such as the processes by which the body's immune system recognizes substances that are foreign to the body. Proteinaceous enzymes depend upon their structures for the very specific functions of the enzymes. The order of amino acids in the protein molecule determines its primary structure. Secondary and tertiary protein structures depend upon the ways in which the polypeptide molecules are bent and folded and by hydrogen bonding (see Section 7.3). The loss of a protein's secondary and tertiary structure is called denaturation and may be caused by heat or the action of foreign chemicals. Some corrosive poisons act by denaturing proteins.

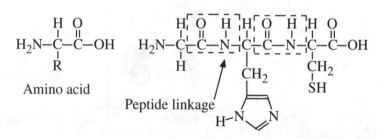

Figure 21.2.   A tripeptide formed by the linking of three amino acids. The peptide linkages with which the amino acids are joined are outlined by dashed lines.

Carbohydrates have the approximate simple formula CH_2O and include a diverse range of substances composed of simple sugars such as glucose:

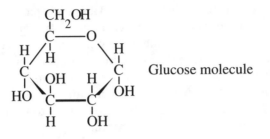

Glucose molecule

High-molecular-mass **polysaccharides,** such as starch and glycogen ("animal starch"), are biopolymers of simple sugars. The major functions of carbohydrates are to store and transfer energy, processes with which toxic substances may interfere.

Lipids are substances that can be extracted from plant or animal matter by organic solvents, such as chloroform, diethyl ether, or toluene. The most common lipids are fats and oils composed of **triglycerides** formed from the alcohol glycerol, $CH_2(OH)CH(OH)CH_2(OH)$ and a long-chain fatty acid such as stearic acid, $CH_3(CH_2)_{16}C(O)OH$ (Figure 21.3). Numerous other biological materials, including waxes, cholesterol, and some vitamins and hormones, are classified as lipids.

Lipids are toxicologically important for several reasons. Some toxic substances interfere with lipid metabolism, leading to detrimental accumulation of lipids. Many toxic organic compounds are poorly soluble in water, but are lipid-soluble, so that bodies of lipids in organisms serve to dissolve and store toxicants.

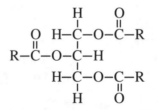

Figure 21.3.   General formula of triglycerides, which make up fats and oils. The R group is from a fatty acid and is a hydrocarbon chain, such as $-(CH_2)_{16}CH_3$**.**

21.3.   ENZYMES

Enzymes are proteinaceous substances with highly specific structures that interact with particular substances or classes of substances called **substrates.** Enzymes act as catalysts to enable biochemical reactions to occur, after which they are regenerated intact to take part in additional reactions:

$$\text{Enzyme} + \text{Substrate} \rightarrow \text{Enzyme-substrate complex} \rightarrow$$
$$\text{Products} + \text{Regenerated enzyme} \qquad (21.3.1)$$

Enzymes are named for what they do. As examples, **lipase** enzymes cause lipid triglycerides to dissociate and form glycerol and fatty acids.

Some substances, such as cyanide, heavy metals, or insecticidal parathion, alter or destroy enzymes so that they function improperly or not at all. Toxicants can affect enzymes in several ways. Parathion, for example, bonds covalently to the nerve enzyme acetylcholinesterase, which can then no longer serve to stop nerve impulses. Heavy metals tend to bind to sulfur atoms in enzymes, thereby altering the shape and function of the enzyme.

Enzyme-Catalyzed Reactions of Xenobiotic Substances

The processes by which organisms metabolize xenobiotic species (synthetic substances that are foreign to living systems) are enzyme-catalyzed phase I and phase II reactions,[2] which are described briefly here.

Phase I Reactions

Lipophilic xenobiotic species in the body tend to undergo **phase I reactions** that make them more water-soluble and reactive by the attachment of polar functional groups, such as $-OH$ (Figure 21.4). Most phase I processes are "microsomal mixed-function oxidase" reactions catalyzed by the cytochrome P-450 enzyme system associated with the **endoplasmic reticulum** of the cell and occurring most abundantly in the livers of vertebrates.

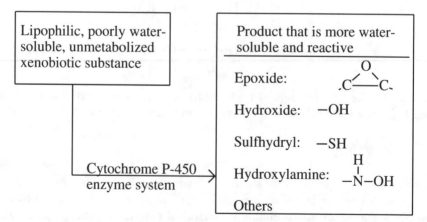

Figure 21.4.   Illustration of phase I reactions.

Phase II Reactions

The polar functional groups attached to a xenobiotic compound in a phase I reaction provide reaction sites for **phase II** reactions. These are **conjugation reactions** in which enzymes attach **conjugating agents** to xenobiotics, their phase I reaction products, and nonxenobiotic compounds (Figure 21.5). The **conjugation product** of such a reaction is usually less toxic than the original xenobiotic compound, less lipid-soluble, more water-soluble, and more readily eliminated from the body. The major conjugating agents and the enzymes that catalyze their phase II reactions are glucuronide (UDP glucuronyltransferase enzyme), glutathione (glutathionetransferase enzyme), sulfate (sulfotransferase enzyme), and acetyl (acetylation by acetyltransferase enzymes). The most abundant conjugation products are glucuronides. A glucuronide conjugate is illustrated in Figure 21.6, where -X-R represents a xenobiotic species conjugated to glucuronide and R is an organic moiety. For example, if the xenobiotic compound

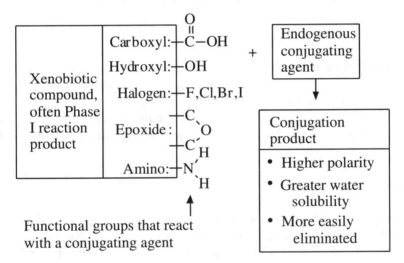

Figure 21.5.   Illustration of phase II reactions.

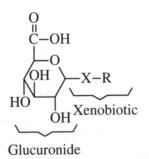

Glucuronide

Figure 21.6.   Glucuronide conjugate formed from a xenobiotic, HX-R.

conjugated is phenol, HXR is HOC_6H_5, X is the O atom, and R represents the phenyl group, C_6H_5.

21.4. TOXICOLOGY AND TOXICOLOGICAL CHEMISTRY

Toxicology is the science that deals with the effects of poisons upon living organisms. A **poison** or **toxicant** is a substance that, above a certain level of exposure or dose, has detrimental effects on tissues, organs, or biological processes. Many toxicants are xenobiotic materials, which were defined and discussed in the preceding section. Among the important aspects of toxicology are the relationship between the demonstrated presence of a chemical or its metabolites in the body and observed symptoms of poisoning, mechanisms by which toxicants are transformed to other species by biochemical processes, the processes by which toxicants and their metabolites are eliminated from an organism, and treatment of poisoning with antidotes.

Toxicological Chemistry

Toxicological chemistry is the science that deals with the chemical nature and reactions of toxic substances, including their origins, uses, and chemical aspects of exposure, fates, and disposal.[3] Toxicological chemistry addresses the relationships between the chemical properties and molecular structures of molecules and their toxicological effects. Figure 21.7 outlines the terms discussed above and the relationships among them.

Toxicities

The major variables in the ways in which toxicants affect organisms are (1) toxicity of the same substance to different organisms, (2) toxicity of different substances to the same organism, (3) minimum levels for observable toxic effects, (4) sensitivity to small increments of toxicant, (5) levels at which most organisms experience the ultimate effect, particularly death, (6) reversibility of toxic effect, and (7) acute or chronic effect.

When a toxicant leaves no permanent effect, either through the action of the organism's natural defense mechanisms or the administration of substances to

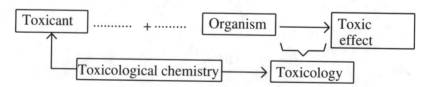

Figure 21.7.   Toxicology is the science of poisons. Toxicological chemistry relates toxicology to the chemical nature of toxicants.

counteract the toxicant's action (antidotes), it is said to act in a **reversible** manner. Effects that last after the toxicant is eliminated, such as the scar from a sulfuric acid burn on the skin, are termed **irreversible**.

Acute toxicity refers to responses that are observed soon after exposure to a toxic substance. **Chronic toxicity** deals with effects that take a long time to be manifested. Chronic responses to toxicants may have latency periods as long as several decades in humans. Acute effects normally result from brief exposures to relatively high levels of toxicants and are comparatively easy to observe and relate to exposure to a poison. Chronic effects are often obscured by normal background maladies and tend to result from low exposures to a toxicant over relatively long periods of time. Chronic effects are much more difficult to study, but are of greater importance in dealing with hazardous wastes and pollutants.

Dose-Response Relationship

Dose is defined as the degree of exposure of an organism to a toxicant, commonly in units of mass of toxicant per unit of body mass of the organism. The observed effect of the toxicant is the **response**. A plot of the percentage of organisms that exhibit a particular response as a function of dose is a **dose-response** curve.[4] The statistical estimate of the dose that would cause death in 50 percent of the subjects is the inflection point of the S-shaped dose-response curve, shown in Figure 21.8, and is designated as LD_{50}. The figure shows that LD_{50} values and the slopes of the dose-response curves may differ substantially.

A response to a very low level of toxicant is known as **hypersensitivity**, an allergic reaction, which is an exaggerated response of the body's immune system, whereas response to only extremely high levels is called **hyposensitivity**. In some

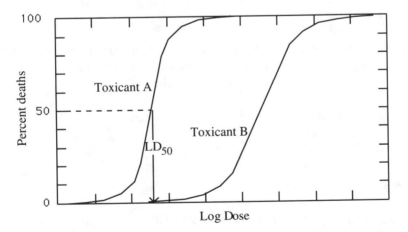

Figure 21.8.   Illustration of a dose-response curve in which the response is the death of the organism. The cumulative percentage of deaths of organisms is plotted on the Y axis.

cases hypersensitivity develops as an extreme reaction to a toxicant after one or more doses of it.

Toxicity Ratings

Substances may be assigned toxicity ratings ranging from 1 to 6 as follows: (1) practically nontoxic, >15 g/kg mass of toxicant per unit body mass; (2) slightly toxic, 5–15 g/kg; (3) moderately toxic, 0.5–5 g/kg; (4) very toxic, 50–500 mg/kg; (5) extremely toxic, 5–50 mg/kg; (6) supertoxic, <5 mg/kg.

Toxicants in the Body

The major routes and sites of absorption, metabolism, binding, and excretion of toxic substances in the body are illustrated in Figure 21.9. Toxicants in the body are metabolized, transported, and excreted; they have adverse biochemical effects; and they cause manifestations of poisoning. It is convenient to divide these processes into two major phases, a kinetic phase and a dynamic phase.

Kinetic Phase

In the **kinetic phase** a toxicant or the metabolic precursor of a toxic substance (**protoxicant**) may undergo absorption, metabolism, temporary storage, distribution, and excretion, as illustrated in Figure 21.10. A toxicant that is absorbed may be passed through the kinetic phase unchanged as an **active parent compound**, metabolized to a **detoxified metabolite** that is excreted, or converted to a toxic **active metabolite**.

Dynamic Phase

In the **dynamic phase** (Figure 21.11) a toxicant or toxic metabolite interacts with cells, tissues, or organs in the body to cause some toxic response. The three major subdivisions of the dynamic phase are the following:

- **Primary reaction** with a receptor or target organ
- A **biochemical response**
- Observable effects

Biochemical Effects in the Dynamic Phase

A toxicant or an active metabolite reacts with a receptor to cause a toxic response. Such a reaction occurs, for example, when benzene epoxide (see Section 21.12), forms an adduct with a nucleic acid unit in DNA resulting in alteration of the DNA. This is an example of an **irreversible** reaction between a

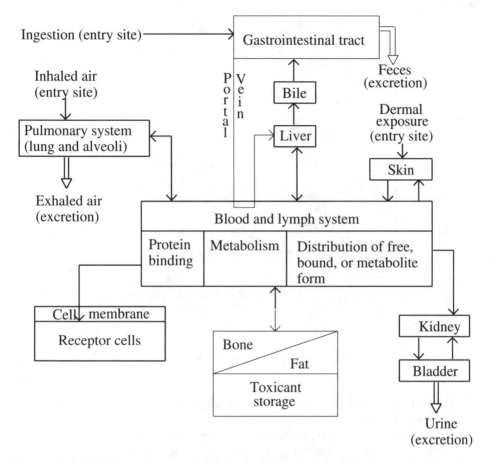

Figure 21.9.   Major routes and sites of absorption, metabolism, binding, and excretion of toxic substances in the body.

toxicant and a receptor. A **reversible** reaction that can result in a toxic response is illustrated by the binding between carbon monoxide and oxygen-transporting hemoglobin in blood:

$$O_2Hb + CO \rightleftharpoons COHb + O_2 \qquad (21.4.1)$$

The binding of a toxicant to a receptor may result in some kind of biochemical effect. The major ones are the following:

- Impairment of enzyme function by binding to the enzyme, coenzymes, metal activators of enzymes, or enzyme substrates
- Alteration of cell membrane or carriers in cell membranes
- Interference with carbohydrate metabolism

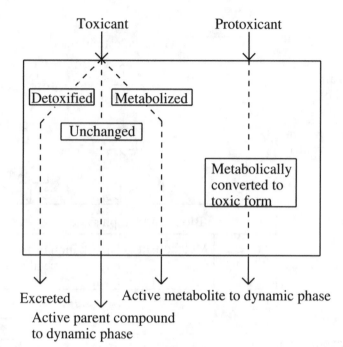

Figure 21.10.   Processes involving toxicants or protoxicants in the kinetic phase.

- Interference with lipid metabolism resulting in excess lipid accumulation ("fatty liver")
- Interference with respiration, the overall process by which electrons are transferred to molecular oxygen in the biological oxidation of energy-yielding substrates
- Stopping or interfering with protein biosynthesis by the action of toxicants on DNA
- Interference with regulatory processes mediated by hormones or enzymes.

Responses to Toxicants

Prominent among the more chronic responses to toxicant exposure are mutations, cancer, and birth defects and effects on the immune system. Other observable effects, some of which may occur soon after exposure, include gastrointestinal illness, cardiovascular disease, hepatic (liver) disease, renal (kidney) malfunction, neurologic symptoms (central and peripheral nervous systems), skin abnormalities (rash, dermatitis).

Among the more immediate and readily observed manifestations of poisoning are alterations in the **vital signs** of **temperature, pulse rate, respiratory rate,** and **blood pressure.** Poisoning by some substances may cause an abnormal skin

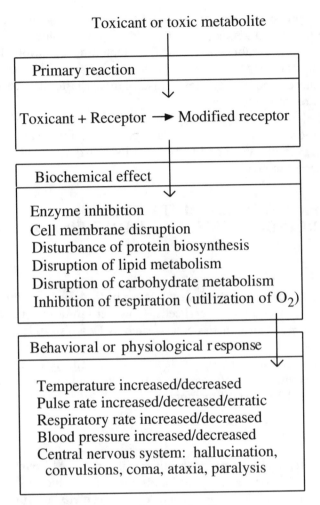

Figure 21.11.   **The dynamic phase of toxicant action.**

color (jaundiced yellow skin from CCl_4 poisoning) or excessively moist or dry skin. Toxic levels of some materials or their metabolites cause the body to have unnatural **odors,** such as the bitter almond odor of HCN in tissues of victims of cyanide poisoning. Symptoms of poisoning manifested in the eye include **miosis** (excessive or prolonged contraction of the eye pupil), **mydriasis** (excessive pupil dilation), **conjunctivitis** (inflammation of the mucus membrane that covers the front part of the eyeball and the inner lining of the eyelids) and **nystagmus** (involuntary movement of the eyeballs). Some poisons cause a moist condition of the mouth, whereas others cause a dry mouth. Gastrointestinal tract effects including pain, vomiting, or paralytic ileus (stoppage of the normal peristalsis movement of the intestines) occur as a result of poisoning by a number of toxic substances.

Central nervous system poisoning may be manifested by **convulsions, paralysis, hallucinations,** and **ataxia** (lack of coordination of voluntary movements of the body), as well as abnormal behavior, including agitation, hyperactivity, disorientation, and delirium. Severe poisoning by some substances, including organophosphates and carbamates, causes **coma,** the term used to describe a lowered level of consciousness.

Often the effects of toxicant exposure are subclinical in nature. The most common of these are some kinds of damage to immune system, chromosomal abnormalities, modification of functions of liver enzymes, and slowing of conduction of nerve impulses.

21.5. TERATOGENESIS, MUTAGENESIS, CARCINOGENESIS, AND IMMUNE SYSTEM EFFECTS

Teratogenesis

Teratogens are chemical species that cause birth defects. These usually arise from damage to embryonic or fetal cells. However, mutations in germ cells (egg or sperm cells) may cause birth defects, such as Down's syndrome.

The biochemical mechanisms of teratogenesis are varied. These include enzyme inhibition by xenobiotics; deprivation of the fetus of essential substrates, such as vitamins; interference with energy supply; or alteration of the permeability of the placental membrane.

Mutagenesis

Mutagens alter DNA to produce inheritable traits. Although mutation is a natural process that occurs even in the absence of xenobiotic substances, most mutations are harmful. The mechanisms of mutagenicity are similar to those of carcinogenicity, and mutagens often cause birth defects as well. Therefore, mutagenic hazardous substances are of major toxicological concern.

Carcinogenesis

Chemical carcinogenesis is the term that applies to the role of substances foreign to the body in causing the uncontrolled cell replication commonly known as cancer. In the public eye chemical carcinogenesis is the aspect of toxicology most commonly associated with hazardous substances.

The two major steps in the overall processes by which xenobiotic chemicals cause cancer are an **initiation stage** followed by a **promotional stage.**[5] Chemical carcinogens usually have the ability to form covalent bonds with macromolecular life molecules, especially DNA.[6] This can alter the DNA in a manner such that the cells replicate uncontrollably and form cancerous tissue. Many chemical carcinogens are **alkylating agents**, which act to attach alkyl groups—such as methyl (CH_3) or ethyl (C_2H_5)—or **arylating agents**, which act to attach aryl moieties, such as the phenyl group, C_6H_5, to DNA. Attachment is made through the N and O atoms in the nitrogenous bases that are contained in DNA.

Primary Carcinogens and Procarcinogens

Chemical substances that cause cancer directly are called **primary** or **direct-acting carcinogens**. Most xenobiotics involved in causing cancer are **precarcinogens** or **procarcinogens**. These species require metabolic activation by phase I or phase II reactions to produce **ultimate carcinogens**, which are actually responsible for carcinogenesis.[7]

Only a few chemicals have definitely been established as human carcinogens. A well-documented example is vinyl chloride, $CH_2 = CHCl$, which is known to have caused a rare form of liver cancer (angiosarcoma) in individuals who cleaned autoclaves in the polyvinylchloride fabrication industry. Animal tests are used to infer chemical carcinogenicity in humans.

Bruce Ames Test

Mutagenicity used to infer carcinogenicity is the basis of the **Bruce Ames** test, in which observations are made of the reversion of mutant histidine-requiring *Salmonella* bacteria back to a form that can synthesize its own histidine.[8] The test makes use of enzymes in homogenized liver tissue to convert potential procarcinogens to ultimate carcinogens. Histidine-requiring *Salmonella* bacteria are inoculated onto a medium that does not contain histidine, and those that mutate back to a form that can synthesize histidine establish visible colonies that are assayed to indicate mutagenicity.

According to Bruce Ames, the developer of the test by the same name, animal tests for carcinogens that make use of massive doses of chemicals may give results that cannot be accurately extrapolated to assess cancer risks from smaller doses of chemicals.[9] This is because of the huge doses of chemicals used kill large numbers of cells, which the organism's body attempts to replace with new cells. Rapidly dividing cells greatly increase the likelihood of mutations that result in cancer simply as the result of rapid cell proliferation, not genotoxicity.

Immune System Response

The **immune system** acts as the body's natural defense system to protect it from xenobiotic chemicals; infectious agents, such as viruses or bacteria; and neoplastic cells, which give rise to cancerous tissue. Adverse effects on the body's immune system are being increasingly recognized as important consequences of exposure to hazardous substances. Toxicants can cause **immunosup-**

pression, which is the impairment of the body's natural defense mechanisms. Xenobiotics can also cause the immune system to lose its ability to control cell proliferation, resulting in leukemia or lymphoma.

Another major toxic response of the immune system is **allergy** or **hypersensitivity**. This kind of condition results when the immune system overreacts to the presence of a foreign agent or its metabolites in a self-destructive manner. Among the xenobiotic materials that can cause such reactions are beryllium, chromium, nickel, formaldehyde, pesticides, resins, and plasticizers.

21.6.   HEALTH HAZARDS

In recent years, attention in toxicology has shifted from readily recognized, usually severe, acute maladies that developed on a short time scale as a result of brief, intense exposure to toxicants to delayed, chronic, often less severe illnesses caused by long-term exposure to low levels of toxicants. Although the total impact of the latter kinds of health effects may be substantial, their assessment is very difficult because of factors such as uncertainties in exposure, low occurrence above background levels of disease, and long latency periods.

Assessment of Potential Exposure

A critical step in assessing exposure to toxic substances, such as those from environmental pollutants or hazardous waste sites, is evaluation of potentially exposed populations. The most direct approach to this is to determine chemicals or their metabolic products in organisms. For inorganic species, this is most readily done for heavy metals, radionuclides, and some minerals, such as asbestos. Symptoms associated with exposure to particular chemicals may also be evaluated. Examples of such effects include skin rashes or subclinical effects, such as chromosomal damage.

Epidemiologic Evidence

Epidemiologic studies applied to toxic environmental pollutants attempt to correlate observations of particular illnesses with probable exposure to such substances. There are two major approaches to such studies. One approach is to look for diseases known to be caused by particular agents in areas where exposure is likely from such agents. A second approach is to look for **clusters** consisting of an abnormally large number of cases of a particular disease in a limited geographic area, then attempt to locate sources of exposure to hazardous wastes that may be responsible. The most common types of maladies observed in clusters are spontaneous abortions, birth defects, and particular types of cancer.

Epidemiologic studies are complicated by long latency periods from exposure to onset of disease, lack of specificity in the correlation between exposure to a particular waste and the occurrence of a disease, and background levels of a disease in the absence of exposure to a hazardous waste capable of causing the disease.

Estimation of Health Effects Risks

An important part of estimating the risks of adverse health effects from exposure to toxicants involves extrapolation from experimentally observable data. Usually, the end result needed is an estimate of a low occurrence of a disease in humans after a long latency period resulting from low-level exposure to a toxicant for a long period of time. The data available are almost always taken from animals exposed at high levels of the substance for a relatively short period of time. Extrapolation is then made using linear or curvilinear projections to estimate the risk to human populations. There are, of course, very substantial uncertainties in this kind of approach.

Risk Assessment

Toxicological considerations are very important in estimating potential dangers of pollutants and hazardous waste chemicals. One of the major ways in which toxicology interfaces with the area of hazardous wastes is in **health risk assessment**, providing guidance for risk management, cleanup, or regulation needed at a hazardous waste site based upon knowledge about the site and the chemical and toxicological properties of wastes in it. Risk assessment includes the factors of site characteristics; substances present, including indicator species; potential receptors; potential exposure pathways; and uncertainty analysis. It may be divided into the following major components:[10]

- Identification of hazard
- Dose-response assessment
- Exposure assessment
- Risk characterization

21.7.   TOXIC ELEMENTS AND ELEMENTAL FORMS

This section discusses toxicological aspects of elements (particularly heavy metals) whose presence in a compound frequently means that the compound is toxic, as well as the toxicities of some commonly used elemental forms, such as the chemically uncombined elemental halogens.

Ozone

Ozone (O_3, see Chapters 16–18) has several toxic effects. Air containing 1 ppm by volume ozone has a distinct odor. Inhalation of ozone at this level causes severe irritation and headache. Ozone irritates the eyes, upper respiratory system, and lungs. Inhalation of ozone can cause sometimes fatal pulmonary edema. Chromosomal damage has been observed in subjects exposed to ozone.

Ozone generates free radicals in tissue. These reactive species can cause lipid peroxidation, oxidation of sulfhydryl ($-SH$) groups, and other destructive oxidation processes. Compounds that protect organisms from the effects of ozone include radical scavengers, antioxidants, and compounds containing sulfhydryl groups.

White Phosphorus

Elemental white phosphorus can enter the body by inhalation, by skin contact, or orally. It is a systemic poison; that is, one that is transported through the body to sites remote from its entry site. White phosphorus causes anemia, gastrointestinal system dysfunction, bone brittleness, and eye damage. Exposure also causes **phossy jaw**, a condition in which the jawbone deteriorates and becomes fractured.

Elemental Halogens

Elemental **fluorine** (F_2) is a pale yellow, highly reactive gas that is a strong oxidant. It is a toxic irritant and attacks skin and the mucous membranes of the nose and eyes.

Chlorine (Cl_2) gas reacts in water to produce a strongly oxidizing solution. This reaction is responsible for some of the damage caused to the moist tissue lining the respiratory tract when the tissue is exposed to chlorine. The respiratory tract is rapidly irritated by exposure to 10–20 ppm of chlorine gas in air, causing acute discomfort that warns of the presence of the toxicant. Even brief exposure to 1,000 ppm of Cl_2 can be fatal.

Bromine (Br_2) is a volatile dark red liquid that is toxic when inhaled or ingested. Like chlorine and fluorine, it is strongly irritating to the mucous tissue of the respiratory tract and eyes and may cause pulmonary edema. The toxicological hazard of bromine is limited somewhat because its irritating odor elicits a withdrawal response.

Elemental **iodine** (I_2), a solid, is irritating to the lungs much like bromine or chlorine. However, the relatively low vapor pressure of iodine limits exposure to I_2 vapor.

Heavy Metals

Heavy metals, discussed as water pollutants in Section 12.2, are particularly toxic in their chemically combined forms and some, notably mercury, are toxic in the elemental form. The toxic properties of some of the most hazardous heavy metals and metalloids are discussed here.

Although not truly a *heavy* metal, **beryllium** (atomic mass 9.01) is one of the more hazardous toxic elements. Its most serious toxic effect is **berylliosis**, a condition manifested by lung fibrosis and pneumonitis, which may develop after a latency period of 5–20 years. Beryllium exposure also causes skin granulomas and ulcerated skin and is a hypersensitizing agent.

Cadmium adversely affects several important enzymes; it can also cause painful osteomalacia (bone disease) and kidney damage. Inhalation of cadmium oxide dusts and fumes results in cadmium pneumonitis characterized by edema and pulmonary epithelium necrosis.

Lead, widely distributed as metallic lead, inorganic compounds, and organometallic compounds, has a number of toxic effects, including inhibition of the synthesis of hemoglobin. It also adversely affects the central and peripheral nervous systems and the kidneys.

Arsenic is a metalloid which forms a number of toxic compounds. The toxic $+3$ oxide, As_2O_3, is absorbed through the lungs and intestines. Biochemically, arsenic acts to coagulate proteins, forms complexes with coenzymes, and inhibits the production of adenosine triphosphate (ATP) in essential metabolic processes.

Elemental **mercury** vapor can enter the body through inhalation and be carried by the bloodstream to the brain where it penetrates the blood-brain barrier. It disrupts metabolic processes in the brain causing tremor and psychopathological symptoms such as shyness, insomnia, depression, and irritability. Divalent ionic mercury, Hg^{2+}, damages the kidney. Organometallic mercury compounds such as dimethylmercury, $Hg(CH_3)_2$, are also very toxic.

21.8. TOXIC INORGANIC COMPOUNDS

Cyanide

Both **hydrogen cyanide** (HCN) and **cyanide salts** (which contain CN^- ion) are rapidly acting poisons; a dose of only 60–90 mg is sufficient to kill a human. Uses of hydrogen cyanide as a pesticidal fumigant and cyanide salt solutions in chemical synthesis and metal processing pose risks of human exposure.

Metabolically, cyanide bonds to iron(III) in iron-containing ferricytochrome oxidase enzyme (see enzymes, Section 21.3), preventing its reduction to iron(II) in the oxidative phosphorylation process by which the body utilizes O_2. The crucial enzyme is inhibited because ferrous cytochrome oxidase, which is required to react with O_2, is not formed and utilization of oxygen in cells is prevented so that metabolic processes cease.

Carbon Monoxide

Carbon monoxide, CO, is a common cause of accidental poisonings. At CO levels in air of 10 parts per million (ppm) impairment of judgment and visual perception occur; exposure to 100 ppm causes dizziness, headache, and weariness; loss of consciousness occurs at 250 ppm; and inhalation of 1,000 ppm results in rapid death. Chronic long-term exposures to low-levels of carbon monoxide are suspected of causing disorders of the respiratory system and the heart.

After entering the blood stream through the lungs, carbon monoxide reacts with hemoglobin (Hb) to convert oxyhemoglobin (O_2Hb) to carboxyhemoglobin (COHb):

$$O_2Hb + CO \rightarrow COHb + O_2 \qquad (21.8.1)$$

Carboxyhemoglobin is much more stable than oxyhemoglobin so that its formation prevents hemoglobin from carrying oxygen to body tissues.

Nitrogen Oxides

The two most common toxic oxides of nitrogen are NO and NO_2, of which the latter is regarded as the more toxic. Nitrogen dioxide causes severe irritation of the innermost parts of the lungs resulting in pulmonary edema. In cases of severe exposures, fatal bronchiolitis fibrosa obliterans may develop approximately three weeks after exposure to NO_2. Fatalities may result from even brief periods of inhalation of air containing 200–700 ppm of NO_2. Biochemically, NO_2 disrupts lactic dehydrogenase and some other enzyme systems, possibly acting much like ozone, a stronger oxidant discussed in Section 21.7. Free radicals, particularly HO, are likely formed in the body by the action of nitrogen dioxide, and the compound probably causes lipid peroxidation in which the $C = C$ double bonds in unsaturated body lipids are attacked by free radicals and undergo chain reactions in the presence of O_2, resulting in their oxidative destruction.

Nitrous oxide, N_2O is used as an oxidant gas and in dental surgery as a general anesthetic. This gas was once known as "laughing gas," and was used in the late 1800s as a "recreational gas" at parties held by some of our not-so-staid Victorian ancestors. Nitrous oxide is a central nervous system depressant and can act as an asphyxiant.

Hydrogen Halides

Hydrogen halides (general formula HX, where X is F, Cl, Br, or I) are relatively toxic gases. The most widely used of these gases are HF and HCl; their toxicities are discussed here.

Hydrogen Fluoride

Hydrogen fluoride, (HF, mp –83.1°C, bp 19.5C) is used as a clear, colorless liquid or gas or as a 30–60% aqueous solution of **hydrofluoric acid**, both referred to here as HF. Both are extreme irritants to any part of the body that they contact, causing ulcers in affected areas of the upper respiratory tract. Lesions caused by contact with HF heal poorly, and tend to develop gangrene.

Fluoride ion, F, is toxic in soluble fluoride salts, such as NaF, causing **fluorosis**, a condition characterized by bone abnormalities and mottled, soft teeth. Livestock is especially susceptible to poisoning from fluoride fallout on grazing land; severely afflicted animals become lame and even die. Industrial pollution has been a common source of toxic levels of fluoride. However, about 1 ppm of fluoride used in some drinking water supplies prevents tooth decay.

Hydrogen Chloride

Gaseous **hydrogen chloride** and its aqueous solution, called **hydrochloric acid**, both denoted as HCl, are much less toxic than HF. Hydrochloric acid is a natural physiological fluid present as a dilute solution in the stomachs of humans and other animals. However, inhalation of HCl vapor can cause spasms of the larynx as well as pulmonary edema and even death at high levels. The high water affinity of hydrogen chloride vapor tends to dehydrate eye and respiratory tract tissue.

Interhalogen Compounds and Halogen Oxides

Interhalogen compounds, including ClF, BrCl, and BrF_3, are extremely reactive and are potent oxidants. They react with water to produce hydrohalic acid solutions (HF, HCl) and nascent oxygen {O}. Too reactive to enter biological systems in their original chemical state, interhalogen compounds tend to be powerful corrosive irritants that acidify, oxidize, and dehydrate tissue, much like those of the elemental forms of the elements from which they are composed. Because of these effects skin, eyes, and mucous membranes of the mouth, throat, and pulmonary systems are especially susceptible to attack.

Major halogen oxides, including fluorine monoxide (OF_2), chlorine monoxide (Cl_2O), chlorine dioxide (ClO_2), chlorine heptoxide (Cl_2O_7), and bromine monoxide (Br_2O), tend to be unstable, highly reactive, and toxic compounds that pose hazards similar to those of the interhalogen compounds discussed previously in this section. Chlorine dioxide, the most commonly used halogen oxide is employed for odor control and bleaching wood pulp. As a substitute for chlorine in water disinfection, it produces fewer undesirable chemical by-products, particularly trihalomethanes.

The most important of the oxyacids and their salts formed by halogens are hypochlorous acid, HOCl, and hypochlorites, such as NaOCl, used for bleaching and disinfection. The hypochlorites irritate eye, skin, and mucous membrane tissue because they react to produce active (nascent) oxygen ($\{O\}$) and acid as shown by the reaction below:

$$HClO \rightarrow H^+ + Cl^- + \{O\} \qquad (21.8.2)$$

Inorganic Compounds of Silicon

Silica (SiO_2, quartz) occurs in a variety of minerals such as sand, sandstone, and diatomaceous earth. **Silicosis** resulting from human exposure to silica dust from construction materials, sand blasting, and other sources has been a common occupational disease. A type of pulmonary fibrosis that causes lung nodules and makes victims more susceptible to pneumonia and other lung diseases, silicosis is one of the most common disabling conditions resulting from industrial exposure to hazardous substances. It can cause death from insufficient oxygen or from heart failure in severe cases.

Silane, SiH_4, and disilane, H_3SiSiH_3, are examples of inorganic **silanes**, which have H-Si bonds. Numerous organic ("organometallic") silanes exist in which alkyl moieties are substituted for H. Little information is available regarding the toxicities of silanes.

Silicon tetrachloride, $SiCl_4$, is the only industrially significant of the **silicon tetrahalides**, a group of compounds with the general formula SiX_4, where X is a halogen. The two commercially produced **silicon halohydrides**, general formula $H_{4-x}SiX_x$, are dichlorosilane (SiH_2Cl_2) and trichlorosilane, ($SiHCl_3$). These compounds are used as intermediates in the synthesis of organosilicon compounds and in the production of high-purity silicon for semiconductors. Silicon tetrachloride and trichlorosilane, fuming liquids which react with water to give off HCl vapor, have suffocating odors and are irritants to eye, nasal, and lung tissue.

Asbestos

Asbestos is the name given to a group of fibrous silicate minerals, typically those of the serpentine group, for which the approximate formula is $Mg_3P(Si_2O_5)(OH)_4$. Asbestos has been widely used in structural materials, brake linings, insulation, and pipe manufacture.[11] Inhalation of asbestos may cause asbestosis (a pneumonia condition), mesothelioma (tumor of the mesothelial tissue lining the chest cavity adjacent to the lungs), and broncho-genic carcinoma (cancer originating with the air passages in the lungs)[12] so that uses of asbestos have been severely curtailed and widespread programs have been undertaken to remove the material from buildings.

Inorganic Phosphorus Compounds

Phosphine (PH_3), a colorless gas that undergoes autoignition at 100°C, is used for the synthesis of organophosphorus compounds and is sometimes inadvertently produced in chemical syntheses involving other phosphorus compounds. It is a potential hazard in industrial processes and in the laboratory. Symptoms of poisoning from potentially fatal phosphine gas include pulmonary tract irritation, central nervous system depression, fatigue, vomiting, and difficult, painful breathing.

Phosphorus pentoxide, P_4O_{10}, is produced as a fluffy white powder from the combustion of elemental phosphorus and reacts with water from air to form syrupy orthophosphoric acid, Because of the formation of acid by this reaction and its dehydrating action, P_4O_{10} is a corrosive irritant to skin, eyes and mucous membranes.

The most important of the **phosphorus halides**, general formulas PX_3 and PX_5, is phosphorus pentachloride used as a catalyst in organic synthesis, as a chlorinating agent and as a raw material to make phosphorus oxychloride ($POCl_3$). Because they react violently with water to produce the corresponding hydrogen halides and oxo phosphorus acids,

$$PCl_5 + 4H_2O \rightarrow H_3PO_4 + 5HCl \qquad (21.8.3)$$

the phosphorus halides are strong irritants to eyes, skin, and mucous membranes.

The major **phosphorus oxyhalide** in commercial use is phosphorus oxychloride ($POCl_3$) a faintly yellow fuming liquid. Reacting with water to form toxic vapors of hydrochloric acid and phosphonic acid (H_3PO_3), phosphorus oxyhalide is a strong irritant to the eyes, skin, and mucous membranes.

Inorganic Compounds of Sulfur

A colorless gas with a foul rotten-egg odor, **hydrogen sulfide** is very toxic. In some cases, inhalation of H_2S kills faster than even hydrogen cyanide; rapid death ensues from exposure to air containing more than about 1000 ppm H_2S due to asphyxiation from respiratory system paralysis. Lower doses cause symptoms that include headache, dizziness, and excitement because of damage to the central nervous system. General debility is one of the numerous effects of chronic H_2S poisoning.

Sulfur dioxide, SO_2, dissolves in water, to produce sulfurous acid, H_2SO_3; hydrogen sulfite ion, HSO_3^-; and sulfite ion, SO_3^{2-}. Because of its water solubility, sulfur dioxide is largely removed in the upper respiratory tract. It is an irritant to the eyes, skin, mucous membranes and respiratory tract. Some individuals are hypersensitive to sodium sulfite (Na_2SO_3), which has been used as a chemical food preservative. These uses were further severely restricted in the U.S. in early 1990.

Number one in synthetic chemical production, sulfuric acid (H_2SO_4) is a severely corrosive poison and dehydrating agent in the concentrated liquid form; it readily penetrates skin to reach subcutaneous tissue causing tissue necrosis with effects resembling those of severe thermal burns. Sulfuric acid fumes and mists irritate eye and respiratory tract tissue and industrial exposure has caused tooth erosion in workers.

The more important halides, oxides and oxyhalides of sulfur are listed in Table 21.1. The major toxic effects of these compounds are given in the table.

Organometallic Compounds

The toxicological properties of some organometallic compounds — pharmaceutical organoarsenicals, organomercury fungicides, and tetraethyllead antiknock gasoline additives — that have been used for many years are well known. However, toxicological experience is lacking for many relatively new organometallic compounds that are now being used in semiconductors, as catalysis, and for chemical synthesis, so they should be treated with great caution until proved safe.

Organometallic compounds often behave in the body in ways totally unlike the inorganic forms of the metals that they contain. This is due in large part to the fact that, compared to inorganic forms, organometallic compounds have an organic nature and higher lipid solubility.

Organolead Compounds

Perhaps the most notable toxic organometallic compound is tetraethyllead, $Pb(C_2H_5)_4$, a colorless, oily liquid that was widely used as an octane-boosting gasoline additive. Tetraethyllead has a strong affinity for lipids and can enter the body by inhalation, ingestion, and absorption through the skin. Acting differently from inorganic compounds in the body, it affects the central nervous system with symptoms such as fatigue, weakness, restlessness, ataxia, psychosis, and convulsions. Recovery from severe lead poisoning tends to be slow. In cases of fatal tetraethyllead poisoning, death has occurred as soon as one or two days after exposure.

Table 21.1. Inorganic Sulfur Compounds

Compound Name	Formula	Properties
Sulfur		
Monofluoride	S_2F_2	Colorless gas, mp –104°C, bp 99°C, toxicity similar to HF
Tetrafluoride	SF_4	Gas, bp –40°C, mp –124°C, powerful irritant
Hexafluoride	SF_6	Colorless gas, mp –51°C, surprisingly nontoxic when pure, but often contaminated with toxic lower fluorides
Monochloride	S_2Cl_2	Oily, fuming orange liquid, mp –80°C, bp 138°C, strong irritant to eyes, skin, and lungs
Tetrachloride	SCl_4	Brownish/yellow liquid/gas, mp –30°C, Decom. below 0°C, irritant
Trioxide	SO_3	Solid anhydride of sulfuric acid reacts with moisture or steam to produce sulfuric acid
Sulfuryl chloride	SO_2Cl_2	Colorless liquid, mp –54°C, bp 69°C, used for organic synthesis, corrosive toxic irritant
Thionyl chloride	$SOCl_2$	Colorless-to-orange fuming liquid, mp –105°C, bp 79°C, toxic corrosive irritant
Carbon oxysulfide	COS	Volatile liquid by-product of natural gas or petroleum refining, toxic narcotic
Carbon disulfide	CS_2	Colorless liquid, industrial chemical, narcotic and central nervous system anesthetic

Organotin Compounds

The greatest number of organometallic compounds in commercial use are those of tin—tributyltin chloride and related tributyltin (TBT) compounds. These compounds have bactericidal, fungicidal, and insecticidal properties and have particular environmental significance because of their increasing applications as industrial biocides.[13,14] Organotin compounds are readily absorbed through the skin, sometimes causing a skin rash. They probably bind with sulfur groups on proteins and appear to interfere with mitochondrial function.

Carbonyls

Metal carbonyls regarded as extremely hazardous because of their toxicities include nickel carbonyl ($Ni(CO)_4$), cobalt carbonyl, and iron pentacarbonyl. Some of the hazardous carbonyls are volatile and readily taken into the body through the respiratory tract or through the skin. The carbonyls affect tissue directly and they break down to toxic carbon monoxide and products of the metal, which have additional toxic effects.

Reaction Products of Organometallic Compounds

An example of the production of a toxic substance from the burning of an organometallic compound is provided by the oxidation of diethylzinc:

$$Zn(C_2H_5)_2 + 7O_2 \rightarrow ZnO(s) + 5H_2O(g) + 4CO_2(g) \qquad (21.8.4)$$

Zinc oxide is used as a healing agent and food additive. However, inhalation of zinc oxide fume particles produced by the combustion of zinc organometallic compounds causes zinc **metal fume fever**. This is an uncomfortable condition characterized by elevated temperature and "chills."

21.9. TOXICOLOGY OF ORGANIC COMPOUNDS

Alkane Hydrocarbons

Methane, ethane, n-butane, and isobutane (both C_4H_{10}) are regarded as **simple asphyxiants** which deprive air of sufficient oxygen to support respiration. The most common toxicological occupational problem associated with the use of hydrocarbon liquids in the workplace is dermatitis caused by dissolution of the fat portions of the skin and characterized by inflamed, dry, scaly skin. Inhalation of volatile liquid hydrocarbon n-alkanes and branched-chain alkanes may cause central nervous system depression manifested by dizziness and loss of coordination. Exposure to n-hexane and cyclohexane results in loss of myelin (a fatty substance constituting a sheath around certain nerve fibers) and degeneration of axons (part of a nerve cell through which nerve impulses are transferred out of the cell). This has resulted in multiple disorders of the nervous system (**polyneuropathy**), including muscle weakness and impaired sensory function of the hands and feet.

Alkene and Alkyne Hydrocarbons

Ethylene, a widely used colorless gas with a somewhat sweet odor, acts as a simple asphyxiant and anesthetic to animals and is phytotoxic (toxic to plants). The toxicological properties of propylene (C_3H_6) are very similar to those of ethylene. Colorless, odorless gaseous 1,3–butadiene is an irritant to eyes and respiratory system mucous membranes; at higher levels, it can cause unconsciousness and even death. Acetylene, H–C≡C–H, is a colorless gas with an odor resembling garlic. It acts as an asphyxiant and narcotic, causing headache, dizziness, and gastric disturbances. Some of these effects may be due to the presence of impurities in the commercial product.

Benzene and Aromatic Hydrocarbons

Inhaled benzene is readily absorbed by blood, from which it is strongly taken up by fatty tissues. For the nonmetabolized compound, the process is reversible and benzene is excreted through the lungs. As shown in Figure 21.12, benzene is converted to phenol by a phase I oxidation reaction (see Section 21.3) in the liver. The benzene epoxide intermediate in this reaction is probably responsible for the unique toxicity of benzene, which involves damage to bone marrow.

Benzene is a skin irritant, and progressively higher local exposures can cause skin redness (erythema), burning sensations, fluid accumulation (edema) and blistering. Inhalation of air containing about 7 g/m³ of benzene causes acute poisoning within an hour, because of a narcotic effect upon the central nervous system manifested progressively by excitation, depression, respiratory system failure, and death. Inhalation of air containing more than about 60 g/m³ of benzene can be fatal within a few minutes.

Long-term exposures to lower levels of benzene cause nonspecific symptoms, including fatigue, headache, and appetite loss. Chronic benzene poisoning causes blood abnormalities, including a lowered white cell count, an abnormal increase in blood lymphocytes (colorless corpuscles introduced to the blood from the lymph glands), anemia, a decrease in the number of blood platelets required for clotting (thrombocytopenia), and damage to bone marrow. It is thought that preleukemia, leukemia, or cancer may result.

Toluene

Toluene, a colorless liquid boiling at 101.4°C, is classified as moderately toxic through inhalation or ingestion; it has a low toxicity by dermal exposure. Toluene can be tolerated without noticeable ill effects in ambient air up to 200 ppm. Exposure to 500 ppm may cause headache, nausea, lassitude, and impaired coordination without detectable physiological effects. Massive expo-

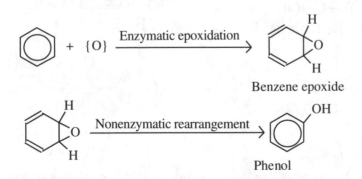

Figure 21.12.   Conversion of benzene to phenol in the body.

sure to toluene has a narcotic effect, which can lead to coma. Because it possesses an aliphatic side chain that can be oxidized enzymatically leading to products that are readily excreted from the body (see the metabolic reaction scheme in Figure 21.13), toluene is much less toxic than benzene.

Naphthalene

As is the case with benzene, **naphthalene** undergoes a phase I oxidation reaction that places an epoxide group on the aromatic ring. This process is followed by phase II conjugation reactions to yield products that can be eliminated from the body.

Exposure to naphthalene can cause anemia and marked reductions in red cell count, hemoglobin, and hematocrit in genetically susceptible individuals. Naphthalene causes skin irritation or severe dermatitis in sensitized individuals. Headaches, confusion, and vomiting may result from inhalation or ingestion of naphthalene. Death from kidney failure occurs in severe instances of poisoning.

Polycyclic Aromatic Hydrocarbons

Benzo(a)pyrene (see Section 18.8) is the most studied of the polycyclic aromatic hydrocarbons (PAHs). Some metabolites of PAH compounds, particularly the 7,8-diol-9,10 epoxide of benzo(a)pyrene shown in Figure 21.14 are

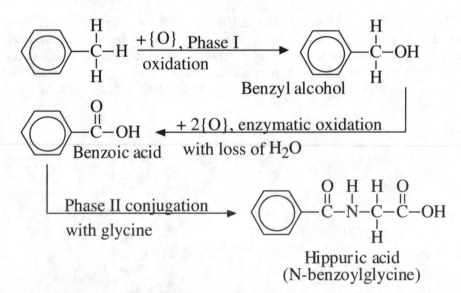

Figure 21.13.   Metabolic oxidation of toluene with conjugation to hippuric acid, which is excreted with urine.

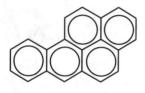 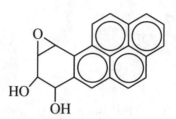

Benzo(a)pyrene   7,8-Diol-9,10-epoxide
of benzo(a)pyrene

Figure 21.14.   Benzo(a)pyrene and its carcinogenic metabolic product.

known to cause cancer. There are two stereoisomers of this metabolite, both of
which are known to be potent mutagens and presumably can cause cancer.

Oxygen-Containing Organic Compounds

Oxides

Hydrocarbon **oxides** such as ethylene oxide and propylene oxide,

$$\underset{\underset{H}{|}}{H-C}\overset{\overset{O}{\diagdown}}{-}\underset{\underset{H}{|}}{C-H} \quad \text{Ethylene oxide} \quad \underset{\underset{H}{|}}{H-C}\overset{\overset{O}{\diagdown}}{-}\underset{\underset{H}{|}}{C}\overset{\overset{H}{|}}{\underset{\underset{H}{|}}{-C}}-H \quad \text{Propylene oxide}$$

which are characterized by an **epoxide** functional group bridging oxygen
between two adjacent C atoms, are significant for both their uses and their toxic
effects. Ethylene oxide, a gaseous colorless, sweet-smelling, flammable, explosive gas used as a chemical intermediate, sterilant, and fumigant, has a moderate
to high toxicity, is a mutagen, and is carcinogenic to experimental animals.
Inhalation of relatively low levels of this gas results in respiratory tract irritation,
headache, drowsiness, and dyspnea, whereas exposure to higher levels causes
cyanosis, pulmonary edema, kidney damage, peripheral nerve damage, and even
death. Propylene oxide is a colorless, reactive, volatile liquid (bp 34°C) with uses
similar to those of ethylene oxide and similar, though less severe, toxic effects.
The toxicity of 1,2,3,4-butadiene epoxide, the oxidation product of 1,3–butadiene, is notable in that it is a direct acting (primary) carcinogen.

Alcohols

Human exposure to the three light alcohols shown in Figure 21.15 is common
because they are widely used industrially and in consumer products.

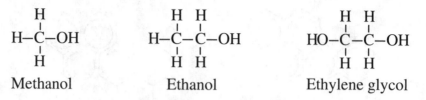

Figure 21.15.   Alcohols such as these three compounds are oxygenated compounds in which the hydroxyl functional group is attached to an aliphatic or olefinic hydrocarbon skeleton.

Methanol, which has caused many fatalities when ingested accidentally or consumed as a substitute for beverage ethanol, is metabolically oxidized to formaldehyde and formic acid. In addition to causing acidosis, these products affect the central nervous system and the optic nerve. Acute exposure to lethal doses causes an initially mild inebriation, followed in about 10–20 hours by unconsciousness, cardiac depression, and death. Sublethal exposures can cause blindness from deterioration of the optic nerve and retinal ganglion cells. Inhalation of methanol fumes may result in chronic, low level exposure.

Ethanol is usually ingested through the gastrointestinal tract, but can be absorbed as vapor by the alveoli of the lungs. Ethanol is oxidized metabolically more rapidly than methanol, first to acetaldehyde (discussed later in this section), then to CO_2. Ethanol has numerous acute effects resulting from central nervous system depression. These range from decreased inhibitions and slowed reaction times at 0.05% blood ethanol, through intoxication, stupor and — at more than 0.5% blood ethanol — death.

Despite its widespread use in automobile cooling systems, exposure to ethylene glycol is limited by its low vapor pressure. However, inhalation of droplets of ethylene glycol can be very dangerous. In the body, ethylene glycol initially stimulates the central nervous system, then depresses it. Glycolic acid, $HOCH_2CO_2H$, formed as an intermediate metabolite in the metabolism of ethylene glycol, may cause acedemia and oxalic acid produced by further oxidation may precipitate as kidney-damaging calcium oxalate, CaC_2O_4.

Of the higher alcohols, 1-butanol is an irritant, but its toxicity is limited by its low vapor pressure. Unsaturated (alkenyl) allyl alcohol, $CH_2 = CHCH_2OH$, has a pungent odor and is strongly irritating to eyes, mouth, and lungs.

Phenols

Figure 21.16 shows some of the more important phenolic compounds, aryl analogs of alcohols which have properties much different from those of the aliphatic and olefinic alcohols. Nitro groups (NO_2) and halogen atoms (particularly Cl) bonded to the aromatic rings strongly affect the chemical and toxicological behavior of phenolic compounds.

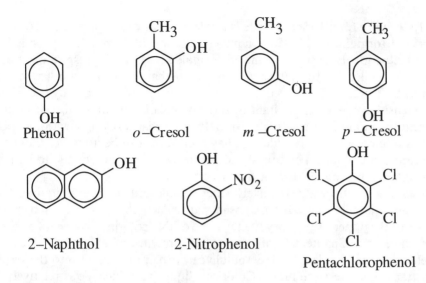

Phenol   o–Cresol   m–Cresol   p–Cresol

2–Naphthol   2-Nitrophenol   Pentachlorophenol

Figure 21.16. Some phenols and phenolic compounds.

Although the first antiseptic used on wounds and in surgery, phenol is a protoplasmic poison that damages all kinds of cells and is alleged to have caused "an astonishing number of poisonings" since it came into general use.[15] The acute toxicological effects of phenol are predominantly upon the central nervous system and death can occur as soon as one-half hour after exposure. Acute poisoning by phenol can cause severe gastrointestinal disturbances, kidney malfunction, circulatory system failure, lung edema, and convulsions. Fatal doses of phenol may be absorbed through the skin. Key organs damaged by chronic exposure to phenol include the spleen, pancreas, and kidneys. The toxic effects of other phenols resemble those of phenol.

Aldehydes and Ketones

Aldehydes and ketones are compounds that contain the carbonyl (C=O) group, as shown by the examples in Figure 21.17.

Formaldehyde is uniquely important because of its widespread use and toxic-

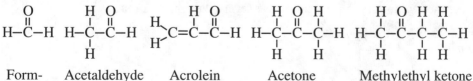

Form-aldehyde   Acetaldehyde   Acrolein   Acetone   Methylethyl ketone

Figure 21.17. Commercially and toxicologically significant aldehydes and ketones.

ity. In the pure form formaldehyde is a colorless gas with a pungent, suffocating odor and **formalin** is a 37–50% aqueous solution of formaldehyde containing some methanol. Exposure to inhaled formaldehyde via the respiratory tract is usually to molecular formaldehyde vapor, whereas exposure by other routes is usually to formalin. Prolonged, continuous exposure to formaldehyde can cause hypersensitivity. A severe irritant to the mucous membrane linings of both the respiratory and alimentary tracts, formaldehyde reacts strongly with functional groups in molecules. Formaldehyde has been shown to be a lung carcinogen in experimental animals. The toxicity of formaldehyde is largely due to its metabolic oxidation product, formic acid (see below).

The lower aldehydes are relatively water-soluble and intensely irritating. These compounds attack exposed moist tissue, particularly the eyes and mucous membranes of the upper respiratory tract. (Some of the irritating properties of photochemical smog, Chapter 18, are due to the presence of aldehydes.) However, aldehydes that are relatively less soluble can penetrate further into the respiratory tract and affect the lungs. Colorless, liquid acetaldehyde is relatively less toxic than acrolein and acts as an irritant, and systemically, as a narcotic to the central nervous system. Extremely irritating, lachrimating acrolein vapor has a choking odor, and inhalation of it can cause severe damage to respiratory tract membranes. Tissue exposed to acrolein may undergo severe necrosis, and direct contact with the eye can be especially hazardous.

The ketones shown in Figure 21.17 are relatively less toxic than the aldehydes. Pleasant-smelling acetone can act as a narcotic and causes dermatitis by dissolving fats from skin. Not many toxic effects have been attributed to methylethyl ketone exposure. It is suspected of having caused neuropathic disorders in shoe factory workers.

Carboxylic Acids

Formic acid, HCO_2H, is a relatively strong acid that is corrosive to tissue. In Europe, decalcifier formulations for removing mineral scale that contain about 75% formic acid are sold and children ingesting these solutions have suffered corrosive lesions to mouth and esophageal tissue. Although acetic acid as a 4–6% solution in vinegar is an ingredient of many foods, pure acetic acid (glacial acetic acid) is extremely corrosive to tissue that it contacts. Ingestion of, or skin contact with, acrylic acid can cause severe damage to tissues.

Ethers

The common ethers have relatively low toxicities because of the low reactivity of the C–O–C functional group, which has very strong carbon-oxygen bonds. Exposure to volatile diethyl ether is usually by inhalation, and about 80% of this compound that gets into the body is eliminated unmetabolized as the vapor through the lungs. Diethyl ether depresses the central nervous system and is a depressant widely used as an anesthetic for surgery. Low doses of diethyl ether cause drowsiness, intoxication, and stupor, whereas higher exposures cause unconsciousness and even death.

Acid Anhydrides

Strong smelling, intensely lachrimating **acetic anhydride**,

$$\underset{\underset{\displaystyle H}{|}}{\overset{\overset{\displaystyle H}{|}}{H-C}}-\overset{\overset{\displaystyle O}{\|}}{C}-O-\overset{\overset{\displaystyle O}{\|}}{C}-\underset{\underset{\displaystyle H}{|}}{\overset{\overset{\displaystyle H}{|}}{C}}-H \quad \text{Acetic anhydride}$$

is a systemic poison. It is especially corrosive to the skin, eyes, and upper respiratory tract, causing blisters and burns that heal only slowly. Levels in the air should not exceed 0.04 mg/m³, and adverse effects to the eyes have been observed at about 0.4 mg/m³.

Esters

Many esters (Figure 21.18) have relatively high volatilities, so that the pulmonary system is a major route of exposure. Because of their generally good solvent properties, esters penetrate tissues and tend to dissolve body lipids. For example, vinyl acetate acts as a skin defatting agent. Because they hydrolyze in water, ester toxicities tend to be the same as the toxicities of the acids and alcohols from which they were formed. Many volatile esters exhibit asphyxiant and narcotic action. Whereas many of the naturally occurring esters have insignificant toxicities at low doses, allyl acetate and some of the other synthetic esters are relatively toxic.

Organonitrogen Compounds

Organonitrogen compounds constitute a large group of compounds with diverse toxicities. Examples of several of the kinds of organonitrogen compounds discussed here are given in Figure 21.19.

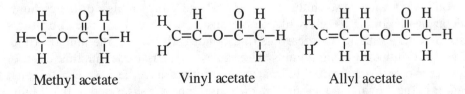

| Methyl acetate | Vinyl acetate | Allyl acetate |

Figure 21.18. Examples of esters.

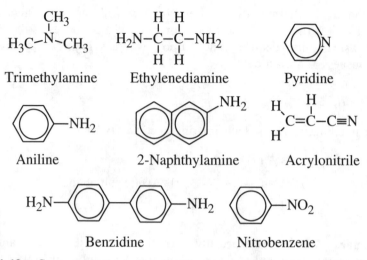

Figure 21.19.   **Some toxicologically significant organonitrogen compounds.**

Aliphatic Amines

The lower amines, such as the methylamines, are rapidly and easily taken into the body by all common exposure routes. They are basic and react with water in tissue,

$$R_3N + H_2O \rightarrow R_3NH^+ + OH^- \tag{21.9.1}$$

raising the pH of the tissue to harmful levels, acting as corrosive poisons (especially to sensitive eye tissue), and causing tissue necrosis at the point of contact. Among the systemic effects of amines are necrosis of the liver and kidneys, lung hemorrhage and edema, and sensitization of the immune system. The lower amines are among the more toxic substances in routine, large-scale use.

Ethylenediamine is the most common of the alkyl polyamines, compounds in which two or more amino groups are bonded to alkane moieties. Its toxicity rating is only 3, but it is a strong skin sensitizer and can damage eye tissue.

Carbocyclic Aromatic Amines

Aniline is a widely used industrial chemical and is the simplest of the **carbocyclic aromatic amines**, a class of compounds in which at least one substituent group is an aromatic hydrocarbon ring bonded directly to the amino group. There are numerous compounds with many industrial uses in this class of amines. Some of the carbocyclic aromatic amines have been shown to cause cancer in the human bladder, ureter, and pelvis, and are suspected of being lung, liver, and prostate carcinogens. A very toxic colorless liquid with an oily consistency and distinct odor, aniline readily enters the body by inhalation, ingestion, and through the skin. Metabolically, aniline converts iron(II) in hemoglobin to iron(III). This causes a condition called **methemoglobinemia**, characterized by cyanosis and a brown-black color of the blood, in which the hemoglobin can no longer transport oxygen in the body. This condition is not reversed by oxygen therapy.

Both **1-naphthylamine** (alpha-naphthylamine) and **2-naphthylamine** (beta-naphthylamine) are proven human bladder carcinogens. In addition to being a proven human carcinogen, **benzidine**, *p*-aminodiphenyl, is highly toxic and has systemic effects that include blood hemolysis, bone marrow depression, and kidney and liver damage. It can be taken into the body orally, by inhalation, and by skin sorption.

Pyridine

Pyridine, a colorless liquid with a sharp, penetrating, "terrible" odor, is an aromatic amine in which an N atom is part of a 6-membered ring. This widely used industrial chemical is only moderately toxic with a toxicity rating of 3. Symptoms of pyridine poisoning include anorexia, nausea, fatigue, and, in cases of chronic poisoning, mental depression. In a few rare cases, pyridine poisoning has been fatal.

Nitriles

Nitriles contain the $-C\equiv N$ functional group. Colorless, liquid **acetonitrile**, CH_3CN, is widely used in the chemical industry. With a toxicity rating of 3-4, acetonitrile is considered relatively safe, although it has caused human deaths, perhaps by metabolic release of cyanide. **Acrylonitrile**, a colorless liquid with a peach-seed (cyanide) odor, is highly reactive because it contains both nitrile and $C = C$ groups. Ingested, absorbed through the skin, or inhaled as vapor, acrylonitrile metabolizes to release deadly HCN, which it resembles toxicologically.

Nitro Compounds

The simplest of the **nitro compounds, nitromethane** H_3CNO_2, is an oily liquid that causes anorexia, diarrhea, nausea, and vomiting and damages the kidneys and liver. **Nitrobenzene**, a pale yellow oily liquid with an odor of bitter almonds or shoe polish, can enter the body by all routes. It has a toxic action much like that of aniline, converting hemoglobin to methemoglobin, which cannot carry oxygen to body tissue. Nitrobenzene poisoning is manifested by cyanosis.

Nitrosamines

N-nitroso compounds (**nitrosamines**), which are characterized by the $>N-N=O$ functional group, have been found in a variety of materials to which humans may be exposed, including beer, whiskey, and cutting oils used in machining. Cancer may result from exposure to a single large dose or from chronic exposure to relatively small doses of some nitrosamines. Once widely used as an industrial solvent and known to cause liver damage and jaundice in exposed workers, dimethylnitrosamine was shown to be carcinogenic from studies starting in the 1950s.

$$\begin{array}{l} CH_3 \\ | \\ N-N=O \\ | \\ CH_3 \end{array}$$ Dimethylnitrosamine
(N-nitrosodimethylamine)

Isocyanates and Methyl Isocyanate

Compounds with the general formula $R-N=C=O$, **isocyanates** are widely used industrial chemicals noted for the high chemical and metabolic reactivity of their characteristic functional group. **Methyl isocyanate**, $H_3C-N=C=O$, was the toxic agent involved in the catastrophic industrial poisoning in Bhopal, India on December 2, 1984, the worst industrial accident in history. In this incident, several tons of methyl isocyanate were released, killing 2,000 people and affecting about 100,000. The lungs of victims were attacked; survivors suffered long-term shortness of breath and weakness from lung damage as well as numerous other toxic effects including nausea and bodily pain.[16]

Organonitrogen Pesticides

Pesticidal *carbamates* are characterized by the structural skeleton of carbamic acid outlined by the dashed box in the structural formula of carbaryl in Figure 21.20. Widely used on lawns and gardens, insecticidal **carbaryl** has a low toxicity to mammals. Highly water-soluble **carbofuran** is taken up by the roots and leaves of plants and poisons insects that feed on the leaves. The toxic effects to animals of carbamates are due to the fact that they inhibit acetylcholinesterase directly without the need to first undergo biotransformation. This effect is relatively reversible because of metabolic hydrolysis of the carbamate ester.

Reputed to have "been responsible for hundreds of human deaths,"[17] herbicidal **paraquat** has a toxicity rating of 5. Dangerous or even fatal acute exposures can occur by all pathways, including inhalation of spray, skin contact, and ingestion. Paraquat is a systemic poison that affects enzyme activity and is

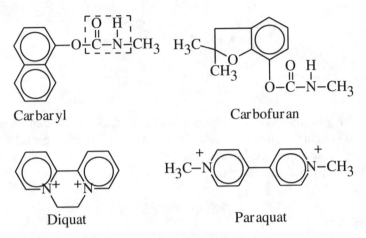

Figure 21.20.   Examples of organonitrogen pesticides.

devastating to a number of organs. Pulmonary fibrosis results in animals that have inhaled paraquat aerosols, and the lungs are also adversely affected by nonpulmonary exposure. Acute exposure may cause variations in the levels of catecholamine, glucose, and insulin. The most prominent initial symptom of poisoning is vomiting, followed within a few days by dyspnea, cyanosis, and evidence of impairment of the kidneys, liver, and heart. Pulmonary fibrosis, often accompanied by pulmonary edema and hemorrhaging, is observed in fatal cases.

Organohalide Compounds

Alkyl Halides

The toxicities of alkyl halides, such as carbon tetrachloride, CCl_4, vary a great deal with the compound. Most of these compounds cause depression of the central nervous system, and individual compounds exhibit specific toxic effects.

During its many years of use as a consumer product, carbon tetrachloride compiled a grim record of toxic effects which led the U. S. Food and Drug Administration (FDA) to prohibit its household use in 1970. It is a systemic poison that affects the nervous system when inhaled, and the gastrointestinal tract, liver, and kidneys when ingested. The biochemical mechanism of carbon tetrachloride toxicity involves reactive radical species including,

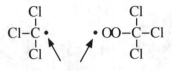

Unpaired
electrons

that react with biomolecules, such as proteins and DNA. The most damaging such reaction occurs in the liver as **lipid peroxidation**, consisting of the attack of free radicals on unsaturated lipid molecules, followed by oxidation of the lipids through a free radical mechanism.

Alkenyl Halides

The most significant **alkenyl** or **olefinic organohalides** are the lighter chlorinated compounds, such as vinyl chloride and tetrachloroethylene:

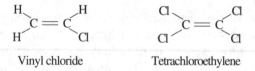

Vinyl chloride   Tetrachloroethylene

Because of their widespread use and disposal in the environment, the numerous acute and chronic toxic effects of the alkenyl halides are of considerable concern.

The central nervous system, respiratory system, liver, and blood and lymph systems are all affected by vinyl chloride exposure, which has been widespread because of this compound's use in polyvinylchloride manufacture. Most notably, vinyl chloride is carcinogenic, causing a rare angiosarcoma of the liver. This deadly form of cancer has been observed in workers chronically exposed to vinyl chloride while cleaning autoclaves in the polyvinylchloride fabrication industry. The alkenyl organohalide, 1,1-dichloroethylene, is a suspect human carcinogen based upon animal studies and its structural similarity to vinyl chloride The toxicities of both 1,2-dichloroethylene isomers are relatively low. These compounds act in different ways in that the *cis* isomer is an irritant and narcotic, whereas the *trans* isomer affects both the central nervous system and the gastrointestinal tract, causing weakness, tremors, cramps, and nausea. A suspect human carcinogen, trichloroethylene has caused liver carcinoma in experimental animals and is known to affect numerous body organs. Like other organohalide solvents, trichloroethylene causes skin dermatitis from dissolution of skin lipids and it can affect the central nervous and respiratory systems, liver, kidneys, and heart. Symptoms of exposure include disturbed vision, headaches, nausea, cardiac arrhythmias, and burning/tingling sensations in the nerves (paresthesia).

Tetrachloroethylene damages the liver, kidneys, and central nervous system. It is a suspect human carcinogen.

Aryl Halides

Individuals exposed to irritant monochlorobenzene by inhalation or skin contact suffer symptoms to the respiratory system, liver, skin, and eyes. Ingestion of this compound causes effects similar to those of toxic aniline, including incoordination, pallor, cyanosis, and eventual collapse.

The dichlorobenzenes are irritants that affect the same organs as monochlorobenzene. The 1,2- isomer is more toxic than the 1,4-isomer. In addition to being a local irritant, it is toxic to the liver and kidney and depresses the central nervous system. *Para*-dichlorobenzene (1,4-dichlorobenzene), a chemical used in air fresheners and mothballs, has become the center of a controversy regarding the evaluation of carcinogenicity. Based upon animal tests that involved subjecting rats to large amounts of the chemical, then extrapolating to humans with adjustments for differences in body size and many orders of magnitude in dose, the U. S. Department of Health and Human Service's National Toxicology Program has classified 1,2-dichlorobenzene as a potential cancer-causing substance. Some industry groups, including the Synthetic Organic Chemicals Association, have filed suit contending that the procedures used are inaccurate and out of date. The suit is viewed as a general challenge to the current use of animal studies to establish possible carcinogenicity of chemicals.[18] Specifically, the suit asks that the report acknowledge that the protein through which the chemical acts in rats does not exist in humans. Many authorities contend that more sophisticated alternatives to animal studies should be used to predict human carcinogenicity of chemicals, especially biochemical investigations of the interactions of suspect chemicals and their metabolites in the body, pathways through the body, and chemical (structural and functional) similarities to known carcinogens.

Because of their once widespread use in electrical equipment, as hydraulic fluids, and in many other applications, polychlorinated biphenyls (PCBs, see Section 12.9) became widespread, extremely persistent environmental pollutants.[19] PCBs have a strong tendency to undergo bioaccumulation in lipid tissue. Polybrominated biphenyl analogs (PBBs) were much less widely used and distributed. However, PBBs were involved in one major incident that resulted in catastrophic agricultural losses when livestock feed contaminated with PBB flame retardant caused massive livestock poisoning in Michigan in 1973.

Organohalide Pesticides

Exhibiting a wide range of kind and degree of toxic effects, many organohalide insecticides affect the central nervous system, causing symptoms such as tremor, irregular jerking of the eyes, changes in personality, and loss of memory. Such symptoms are characteristic of acute DDT poisoning. However, the acute toxicity of DDT to humans is very low and it was used for the control of typhus and malaria in World War II by large-scale direct application to people. The chlorinated cyclodiene insecticides — aldrin, dieldrin, endrin, chlordane, hepta-chlor, endosulfan, and isodrin — act on the brain, releasing betaine esters and causing headaches, dizziness, nausea, vomiting, jerking muscles, and convulsions. Dieldrin, chlordane, and heptachlor have caused liver cancer in test animals and some chlorinated cyclodiene insecticides are teratogenic or fetotoxic. Because of these effects, aldrin, dieldrin, heptachlor, and — more recently — chlordane have been prohibited from use in the U. S.

The major **chlorophenoxy** herbicides are 2,4-dichlorophenoxyacetic acid (2,4-D), 2,4,5-trichlorophenoxyacetic acid (2,4,5-T or Agent Orange), and Silvex. Large doses of 2,4-dichlorophenoxyacetic acid have been shown to cause nerve damage, (peripheral neuropathy), convulsions, and brain damage. According to a National Cancer Institute study,[20] Kansas farmers who had handled 2,4-D extensively have suffered 6 to 8 times the incidence of non-Hodgkins lymphoma as comparable unexposed populations. With a toxicity somewhat less than that of 2,4-D, Silvex is largely excreted unchanged in the urine. The toxic effects of 2,4,5-T (used as a herbicidal warfare chemical called "Agent Orange") have resulted from the presence of 2,3,7,8-tetrachloro-p-dioxin (TCDD, commonly known as "dioxin," discussed below), a manufacturing by-product. Autopsied carcasses of sheep poisoned by this herbicide have exhibited nephritis, hepatitis, and enteritis.

TCDD

Polychlorinated dibenzodioxins, which have the same basic structure as that of TCDD (2,3,7,8-tetrachlorodibenzo-*p*-dioxin),

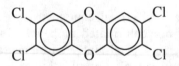

TCDD (2,3,7,8-tetrachloro-dibenzo-p-dioxin)

but different numbers and arrangements of chlorine atoms on the ring structure. Extremely toxic to some animals, the toxicity of TCDD to humans is rather uncertain; it is known to cause a skin condition called chloracne. TCDD has been a manufacturing by-product of some commercial products (see the discussion of 2,4,5-T, above), contaminant identified in some municipal incineration emissions, and widespread environmental pollutant from improper waste disposal. This compound has been released in a number of industrial accidents, the most massive of which exposed several tens of thousands of people to a cloud of chemical emissions spread over an approximately 3–square-mile area at the

Givaudan-La Roche Icmesa manufacturing plant near Seveso, Italy, in 1976. On an encouraging note from a toxicological perspective, no abnormal occurrences of major malformations were found in a study of 15,291 children born in the area within 6 years after the release.[21]

Chlorinated Phenols

The chlorinated phenols used in largest quantities have been **penta-chlorophenol** and the trichlorophenol isomers used as wood preservatives. Although exposure to these compounds has been correlated with liver malfunction and dermatitis, contaminant polychlorinated dibenzodioxins may have caused some of the observed effects.

Organosulfur Compounds

Despite the high toxicity of H_2S, not all organosulfur compounds are particularly toxic. Their hazards are often reduced by their strong, offensive odors that warn of their presence.

Inhalation of even very low concentrations of the alkyl **thiols**, such as methanethiol, H_3CSH, can cause nausea and headaches; higher levels can cause increased pulse rate, cold hands and feet, and cyanosis. In extreme cases, unconsciousness, coma, and death occur. Like H_2S, the alkyl thiols are precursors to cytochrome oxidase poisons.

An oily water-soluble liquid, **methylsulfuric acid** is a strong irritant to skin, eyes, and mucous tissue. Colorless, odorless **dimethylsulfate** is highly toxic and

$$H_3C-O-\overset{\overset{O}{\|}}{\underset{\underset{O}{\|}}{S}}-OH \quad \text{Methylsulfuric acid} \qquad H_3C-O-\overset{\overset{O}{\|}}{\underset{\underset{O}{\|}}{S}}-O-CH_3 \quad \text{Dimethyl-sulfate}$$

is a primary carcinogen. Skin or mucous membranes exposed to dimethylsulfate develop conjunctivitis and inflammation of nasal tissue and respiratory tract mucous membranes following an initial latent period during which few symptoms are observed. Damage to the liver and kidney, pulmonary edema, cloudiness of the cornea, and death within 3–4 days can result from heavier exposures.

Sulfur Mustards

A typical example of deadly **sulfur mustards**, compounds used as military poisons, or "poison gases," is mustard oil (bis(2-chloroethyl)sulfide):[22]

$$\underset{\substack{| \ \ | \quad\quad | \ \ | \\ \text{H H} \quad\quad \text{H H}}}{\overset{\substack{\text{H H} \quad\quad \text{H H} \\ | \ \ | \quad\quad | \ \ |}}{\text{Cl--C--C--S--C--C--Cl}}} \quad \text{Mustard oil}$$

An experimental mutagen and primary carcinogen, mustard oil produces vapors that penetrate deep within tissue, resulting in destruction and damage at some depth from the point of contact; penetration is very rapid, so that efforts to remove the toxic agent from the exposed area are ineffective after 30 minutes. This military "blistering gas" poison, causes tissue to become severely inflamed with lesions that often become infected. These lesions in the lung can cause death.

Organophosphorous Compounds

Organophosphorus compounds have varying degrees of toxicity. Some of these compounds, such as the "nerve gases" produced as industrial poisons, are deadly in minute quantities. The toxicities of major classes of organophosphate compounds are discussed in this section.

Organophosphate Esters

Some organophosphate esters are shown in Figure 21.21. **Trimethylphosphate** is probably moderately toxic when ingested or absorbed through the skin, whereas moderately toxic **triethylphosphate**, $(C_2H_5O)_3PO$, damages nerves and inhibits acetylcholinesterase. Notoriously toxic **tri-o-cresylphosphate, TOCP,** apparently is metabolized to products that inhibit acetylcholinesterase. Exposure to TOCP causes degeneration of the neurons in the body's central and peripheral nervous systems with early symptoms of nausea, vomiting, and diarrhea accompanied by severe abdominal pain. About 1–3 weeks after these symptoms have subsided, peripheral paralysis develops manifested by "wrist drop" and "foot drop," followed by slow recovery, which may be complete or leave a permanent partial paralysis.

Briefly used in Germany as a substitute for insecticidal nicotine, **tetraethylpyrophosphate, TEPP**, is a very potent acetylcholinesterase inhibitor. With a toxicity rating of 6 (supertoxic), TEPP is deadly to humans and other mammals.

Phosphorothionate and Phosphorodithioate Ester Insecticides

Because esters containing the $P=S$ (thiono) group are resistant to nonenzymatic hydrolysis and are not as effective as $P=O$ compounds in inhibiting acetylcholinesterase, they exhibit higher insect:mammal toxicity ratios than their nonsulfur analogs. Therefore, **phosphorothionate** and **phosphorodithioate** esters (Figure 21.22) are widely used as insecticides. The insecticidal activity of

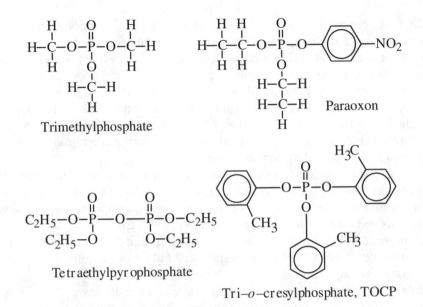

Figure 21.21. Some organophosphate esters.

these compounds requires metabolic conversion of P=S to P=O (oxidative desulfuration). Environmentally, organophosphate insecticides are superior to many of the organochlorine insecticides because the organophosphates readily undergo biodegradation and do not bioaccumulate.

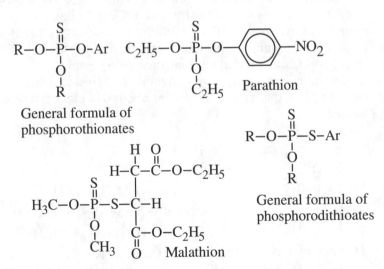

Figure 21.22. Phosphorothionate and phosphorodithioate ester insecticides. Malathion contains hydrolyzable carboxyester linkages.

The first commercially successful phosphorothionate/phosphorodithioate ester insecticide was **parathion**, *O,O*-diethyl-*O*-*p*-nitrophenylphosphorothionate, first licensed for use in 1944. This insecticide has a toxicity rating of 6 (supertoxic). Since its use began, several hundred people have been killed by parathion, including 17 of 79 people exposed to contaminated flour in Jamaica in 1976. As little as 120 mg of parathion has been known to kill an adult human and a dose of 2 mg has been fatal to a child. Most accidental poisonings have occurred by absorption through the skin. Methylparathion (a closely related compound with methyl groups instead of ethyl groups) is regarded as extremely toxic.

In order for parathion to have a toxic effect, it must be converted metabolically to paraoxon (Figure 21.21), which is a potent inhibitor of acetylcholinesterase. Because of the time required for this conversion, symptoms develop several hours after exposure, whereas the toxic effects of TEPP or paraoxon develop much more rapidly. Humans poisoned by parathion exhibit skin twitching and respiratory distress. In fatal cases, respiratory failure occurs due to central nervous system paralysis.

Malathion is the best known of the phosphorodithioate insecticides. It has a relatively high insect:mammal toxicity ratio because of its two carboxyester linkages which are hydrolyzable by carboxylase enzymes (possessed by mammals, but not insects) to relatively nontoxic products. For example, although malathion is a very effective insecticide, its LD_{50} for adult male rats is about 100 times that of parathion.

Organophosphorus Military Poisons

Powerful inhibitors of acetylcholinesterase enzyme, organophosphorus "nerve gas" military poisons such as **Sarin** and **VX** are among the most toxic synthetic compounds ever made. (The possibility that military poisons such as these might be used in war was a major concern during the 1991 mid-East conflict, which, fortunately, ended without their being employed.) A systemic poison to the central nervous system that is readily absorbed as a liquid through the skin, Sarin may be lethal at doses as low as about 0.01 mg/kg; a single drop can kill a human.

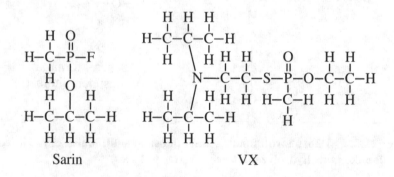

Sarin   VX

CHAPTER SUMMARY

The chapter summary below is presented in a programmed format to review the main points covered in this chapter. It is used most effectively by filling in the blanks, referring back to the chapter as necessary. The correct answers are given at the end of the summary.

The kind of cell found in humans is the (1)_____, of which the main features are (2)_____

_____.

Proteins are composed of (3)_____, which are joined by (4)_____. The loss of a protein's secondary and tertiary structure, which may be caused by exposure to foreign chemicals is called (5)_____. Carbohydrates, of which a simple example is (6)_____, have the approximate simple formula (7)_____. Lipids are substances that can be (8)_____. The most common lipids are fats and oils composed of (9)_____ formed from (10)_____. Lipids are toxicologically important because (11)_____

_____

_____.

Enzymes are (12)_____

_____ that act as (13)_____

_____, after which (14)_____

_____.

As an example of the ways in which toxic substances affect enzymes, parathion is toxic because it (15)_____.

The processes by which organisms metabolize xenobiotic species can be divided between the two major categories of (16)_____

_____. Lipophilic xenobiotic species in the body tend to undergo phase I reactions that (17)_____

_____

_____.

Phase II reactions are (18)_____ reactions in which enzymes attach (19)_____ to xenobiotics, their phase I reaction products, and nonxenobiotic compounds. The product of such a reaction is usually less (20)_____ and (21)_____ water-soluble than is the original xenobiotic compound. The most abundant conjugation products are (22)_____.

A (23)_____ is a substance that, above a certain level of exposure or dose, has detrimental effects on tissues, organs, or biological processes. Toxicological chemistry is defined as (24)_____

_____

_____.

(25)_____ refers to responses that are observed soon after exposure to a toxic substance. (26)_____ deals with effects that take a long time to be manifested.

Dose is defined as (27)_____

_____, whereas the observed effect of the toxicant is the (28)_____. A plot of the percentage of organisms that exhibit a particular response as a function of dose is a (29)____

_____. When death is the response measured in such a curve, the inflection point is designated as (30)_____. A response of a subject to an abnormally low level of toxicant is known as (31)_____.

The two major phases into which the processes that toxicants undergo in the body may be divided into (32)_____. In the first of these, a toxicant or a protoxicant may undergo (33)_____

_____. In the dynamic phase, a toxicant or toxic metabolite interacts with (34)_____

_____ to cause (35)_____. The three major subdivisions of the dynamic phase are (36)_____

_____.

The major kinds of biochemical effects caused by binding of a toxicant to a receptor are (37)_____

_____

_____

_____

_____.

Vital signs affected by toxicants are (38)_____

_____.

Teratogens are chemical species that cause (39)_____.
Mutagens alter (40)_____ to produce (41)_____. The term that applies to the role of substances foreign to the body in causing the uncontrolled cell replication commonly known as cancer is (42)_____.
The two major steps in the overall processes by which xenobiotic chemicals cause cancer are (43)_____. Many chemical carcinogens are (44)_____, which act to attach alkyl groups. Chemical substances that cause cancer directly are called (45)_____ _____. Most xenobiotics involved in causing cancer are (46)_____ _____. Mutagenicity used to infer carcinogenicity is the basis of the (47)_____ test, which makes use of the organism (48)_____.

Toxicants can cause (49)_____, which is the impairment of the body's natural defense mechanisms. Two other adverse responses of the immune system that can be caused by exposure to xenobiotic compounds are (50)_____ and (51)_____ _____.

Epidemiologic studies applied to toxic environmental pollutants attempt to correlate observations of particular illnesses with probable exposure to such wastes, or to look for (52)_____

_____.

In tissue, ozone generates (53)_____, which cause (54)_____

_____. Exposure to white phosphorus causes a devastating bone condition known as (55)_____. Exposure to chlorine gas causes irritation of (56)_____.

Important toxic heavy metals are (57)_____

_____.

Elemental mercury vapor can enter the bloodstream through (58)_____ where it is especially dangerous because of its ability to penetrate (59)_____

_____.

A highly toxic inorganic species that inhibits the utilization of oxygen by the body is (60)_____. A toxic inorganic species that prevents transport of oxygen by hemoglobin is (61)_____. Bronchiolitis fibrosa obliterans is a fatal condition of the (62)_____ caused by exposure to (63)_____. Hydrogen fluoride is an extreme (64)_____ to any part of the body that it contacts. Fluorosis is caused by exposure to (65)_____ and is characterized by (66)_____.

Silicosis results from human exposure to (67)_____. A fibrous silicate mineral known to cause certain kinds of cancer associated with the lungs is (68)_____. The phosphorus(IV) oxide, P_4O_{10}, is toxic because of its (69)_____ action and because it forms (70)_____. Of two simple inorganic sulfur-containing gases, (71)_____ is very toxic and (72)_____ is an irritant to the eyes, skin, mucous membranes and respiratory tract.

The most notable toxic organometallic compound is (73)_____

_____. The most widely used organometallic compounds are (74)_____

_____, which have (75)_____

_____ properties.

Toxicologically, methane, ethane, n-butane, and isobutane are regarded as (76)_____ which act by (77)_____

_____.

Exposure to n-hexane and cyclohexane can cause disorders of the nervous system called (78)_____. The greatest toxicological concern with chronic benzene exposure is (79)_____. Some metabolites of PAH compounds, particularly the 7,8-diol-9,10 epoxide of benzo(a)pyrene, are known to cause (80)_____.

Methanol is metabolically oxidized to (81)_____. Sublethal exposures to methanol can be very serious because they can cause (82)_____. Ethanol is oxidized metabolically to (83)_____ then to (84)_____.

Phenol is a (85)_____ poison. Of the carbonyl compounds, formaldehyde is uniquely important toxicologically because of (86)_____ _____. The lower aldehydes are intensely irritating and attack (87)_____. Acetone causes dermatitis by (88)_____. The common ethers have relatively low toxicities because of (89)_____ _____.

The lower amines act as (90)_____ and cause tissue (91)_____. The simplest member of a class of compounds in which at least one substituent group is an aromatic hydrocarbon ring bonded directly to an amino group is (92)_____. Toxicologically, both 1-naphthylamine and 2-naphthylamine (beta-naphthylamine) are proven (93)_____ _____. N-nitroso compounds are commonly known as (94)_____, and some are known (95)_____.

An organohalide solvent that was widely used as a consumer product until banned because of its toxicity in 1970 is (96)_____. In the body, this compound forms reactive (97)_____. Vinyl chloride is notable toxicologically because it is known to be (98)_____. A class of aryl halides containing two benzene rings per molecule and noted as environmental pollutants consists of (99)_____. Pesticidal 2,4-dichlorophenoxyacetic acid (2,4-D) and 2,4,5-trichlorophenoxy-acetic acid (2,4,5-T or Agent Orange) are examples of a class of pesticides known as (100)_____. The compound 2,3,7,8-tetrachloro-p-dioxin is commonly known as (101)_____.

A chlorinated phenolic compound once widely used as a wood preservative is (102)_____.

Replacement of an H atom on H_2S with an alkyl group, such as CH_3 results in the formation of a (103)_____ compound, most of which are noted for their (104)_____. Toxicologically, dimethylsulfate is noted for being a (105)_____. Sulfur mustards have been used as (106)_____.

Trimethylphosphate, tri-o-cresylphosphate, and tetraethylpyrophosphate are all examples of (107)_____. Esters containing the $P=S$ (thiono) group are favored as insecticides because they exhibit (108)____ _____ than their nonsulfur analogs. The first commercially successful phosphorothionate/phosphorodithioate ester insecticide, now banned for most uses, was (109)_____.

Answers

1. eukaryotic cell
2. cytoplasm, mitochondria, ribosomes, endoplasmic reticulum, cell membrane, cell nucleus
3. amino acids
4. peptide linkages
5. denaturation
6. glucose
7. CH_2O
8. extracted from plant or animal matter by organic solvents
9. triglycerides
10. glycerol and a long-chain fatty acid
11. some toxic substances interfere with lipid metabolism and many toxic organic compounds are lipid-soluble, so that bodies of lipids in organisms serve to dissolve and store toxicants
12. proteinaceous substances with highly specific structures
13. catalysts to enable biochemical reactions to occur
14. they are regenerated intact to take part in additional reactions
15. bonds covalently to the nerve enzyme acetylcholinesterase
16. enzyme-catalyzed phase I and phase II reactions
17. make them more water-soluble and reactive by the attachment of polar functional groups, such as –OH
18. conjugation

19. conjugating agents
20. polar
21. more
22. glucuronides
23. poison or toxicant
24. the science that deals with the chemical nature and reactions of toxic substances, including their origins, uses, and chemical aspects of exposure, fates, and disposal
25. Acute toxicity
26. Chronic toxicity
27. the degree of exposure of an organism to a toxicant
28. response
29. dose-response curve
30. LD_{50}
31. hypersensitivity
32. the kinetic phase and the dynamic phase
33. absorption, metabolism, temporary storage, distribution, and excretion
34. cells, tissues, or organs in the body
35. some toxic response
36. primary reaction, biochemical response, observable effects
37. impairment of enzyme function, alteration of cell membrane, interference with carbohydrate metabolism, interference with lipid metabolism, interference with respiration, stopping or interfering with protein biosynthesis, interference with regulatory processes mediated by hormones or enzymes
38. temperature, pulse rate, respiratory rate, and blood pressure
39. birth defects
40. DNA
41. inheritable traits
42. chemical carcinogenesis
43. an initiation stage followed by a promotional stage
44. alkylating agents
45. primary or direct-acting carcinogens
46. precarcinogens or procarcinogens
47. Bruce Ames
48. *Salmonella* bacteria
49. immunosuppression
50. loss of ability to control cell proliferation
51. hypersensitivity
52. clusters consisting of an abnormally large number of cases of a particular disease in a limited geographic area
53. free radicals
54. destructive oxidation processes
55. phossy jaw
56. the respiratory tract

57. lead, cadmium, mercury, arsenic (a metalloid), and beryllium (not truly a "heavy" metal)
58. inhalation
59. the blood-brain barrier
60. cyanide
61. CO
62. lungs
63. NO_2
64. irritant
65. F^-
66. bone abnormalities and mottled, soft teeth
67. silica dust
68. asbestos
69. dehydrating
70. acid
71. H_2S
72. SO_2
73. tetraethyllead, $Pb(C_2H_5)$
74. tributyltin chloride and related tributyltin (TBT) compounds
75. bactericidal, fungicidal, and insecticidal
76. simple asphyxiants
77. depriving air of sufficient oxygen to support respiration
78. polyneuropathy
79. blood abnormalities
80. cancer
81. formaldehyde and formic acid
82. blindness
83. aetaldehyde
84. CO_2
85. protoplasmic
86. its widespread use and toxocity
87. exposed moist tissue
88. dissolving fats from skin.
89. the low reactivity of the C–O–C functional group
90. corrosive poisons
91. necrosis
92. aniline
93. human bladder carcinogens
94. nitrosamines
95. carcinogens
96. carbon tetrachloride
97. radical species
98. a human carcinogen
99. polychlorinated biphenyls, PCBs

100. chlorophenoxy herbicides
101. TCDD or "dioxin"
102. pentachlorophenol
103. thiol
104. bad odors
105. primary carcinogen
106. military poisons
107. organophosphate esters
108. higher insect:mammal toxicity ratios
109. parathion

QUESTIONS AND PROBLEMS

1. How are conjugating agents and phase II reactions involved with some toxicants?

2. What is the toxicological importance of proteins, particularly as related to protein structure?

3. What is the toxicological importance of lipids? How do lipids relate to hydrophobic ("water-disliking") pollutants and toxicants?

4. What is the function of a hydrolase enzyme?

5. What are phase I reactions? What enzyme system carries them out? Where is this enzyme system located in the cell?

6. Match the cell structure on the left with its function on the right, below:

   1. Mitochondria
   2. Endoplasmic reticulum
   3. Cell membrane
   4. Cytoplasm
   5. Cell nucleus

   (a) Toxicant metabolism
   (b) Fills the cell
   (c) Deoxyribonucleic acid
   (d) Mediate energy conversion and utilization
   (e) Encloses the cell and regulates the passage of materials into and out of the cell interior

7. Name and describe the science that deals with the chemical nature and reactions of toxic substances, including their origins, uses, and chemical aspects of exposure, fates, and disposal.

8. What is a dose-response curve?

9. What is meant by a toxicity rating of 6?

10. What are the three major subdivisions of the *dynamic phase* of toxicity, and what happens in each?

11. Characterize the toxic effect of carbon monoxide in the body. Is its effect reversible or irreversible? Does it act on an enzyme system?

12. Of the following, choose the one that is **not** a biochemical effect of a toxic substance: (a) impairment of enzyme function by binding to the enzyme, (b) alteration of cell membrane or carriers in cell membranes, (c) change in vital signs, (d) interference with lipid metabolism, (e) interference with respiration.

13. Distinguish among teratogenesis, mutagenesis, carcinogenesis, and immune system effects. Are there ways in which they are related?

14. As far as environmental toxicants are concerned, compare the relative importance of acute and chronic toxic effects and discuss the difficulties and uncertainties involved in studying each.

15. What are some of the factors that complicate epidemiologic studies of toxicants?

16. List and discuss two elements that are invariably toxic in their elemental forms. For another element, list and discuss two elemental forms, one of which is quite toxic and the other of which is essential for the body. In what sense is even the toxic form of this element "essential for life?"

17. What is a toxic substance that bonds to iron(III) in iron-containing ferricytochrome oxidase enzyme, preventing its reduction to iron(II) in the oxidative phosphorylation process by which the body utilizes O_2?

18. What are interhalogen compounds, and which elemental forms do their toxic effects most closely resemble?

19. Name and describe the three health conditions that may be caused by inhalation of asbestos.

20. Why might tetraethyllead be classified as "the most notable toxic organometallic compound"?

21. What is the most common toxic effect commonly attributed to low-molecular-mass alkanes?

22. Although benzene and toluene have a number of chemical similarities, their metabolisms and toxic effects are quite different. Explain.

23. Information about the toxicities of many substances to humans is lacking because of limited data on direct human exposure. (Volunteers to study human health effects of toxicants are in notably short supply.) However, there is a great deal of information available about human exposure to phenol and the adverse effects of such exposure. Explain.

24. What are neuropathic disorders? Why are organic solvents frequently the cause of such disorders?

25. What is a major metabolic effect of aniline? What is this effect called? How is it manifested?

26. What are the organic compounds characterized by the $>N-N=O$ functional group? What is their major health effect?

27. What structural group is characteristic of carbamates? For what purpose are these compounds commonly used? What are their major advantages in such an application?

28. What is lipid peroxidation? Which common toxic substance is known to cause lipid peroxidation?

29. Biochemically, what do organophosphate esters such as parathion do that could classify them as "nerve poisons"?

30. Comment on the toxicity of the compound below:

$$\begin{array}{c} O \\ \parallel \\ H_3C-P-F \\ | \\ O \\ | \\ H_3C-C-CH_3 \\ | \\ H \end{array}$$

LITERATURE CITED

1. Feigl, D. M., J. W. Hill, and E. Boschman, *Foundations of Life: An Introduction to General, Organic, and Biological Chemistry*, 3rd Ed., (New York: Macmillan Publishing Co., 1991).
2. Hodgson, E., and W. C. Dauterman, "Metabolism of Toxicants—Phase I

Reactions," Chapter 4, and Walter C. Dauterman, "Metabolism of Toxicants," Chapter 5, in *Introduction to Biochemical Toxicology*, E. Hodgson and F. E. Guthrie, Eds., (New York: Elsevier, 1980), pp. 67–105.

3. Manahan, S. E., *Toxicological Chemistry*, 2nd ed. (Chelsea, MI: Lewis Publishers, Inc., 1992).

4. James, R. C., "General Principles of Toxicology," Chapter 2 in *Industrial Toxicology*, P. L. Williams and J. L. Burson, Eds., (New York: Van Nostrand Reinhold Company, 1985).

5. Diamond, L., "Tumor Promoters and Cell Transformation," Chapter 3 in *Mechanisms of Cellular Transformation by Carcinogenic Agents*, D. Grunberger and S. P. Goff, Eds., (New York: Pergamon Press, 1987), pp. 73–132.

6. Singer, B., and D. Grunberger *Molecular Biology of Mutagens and Carcinogens*, (New York: Plenum Press, 1983).

7. Levi, P. E., "Toxic Action," Chapter 6 in *Modern Toxicology*, E. Hodgson and P. E. Levi, Eds., (New York, Elsevier, 1987), pp. 133–184.

8. "The Detection of Environmental Mutagens and Potential Carcinogens," Bruce N. Ames, *Cancer*, 53, 1034–1040 (1984).

9. "Tests of Chemicals on Animals are Unreliable as Predictors of Cancer in Humans," *Environmental Science and Technology* 24, 1990 (1990).

10. Paustenbach, D. J., Ed., *The Risk Assessment of Environmental and Human Health Hazards: A Textbook of Case Studies*, (New York: John Wiley and Sons, 1988).

11. Steinway, D. M., "Scope and Numbers of Regulations for Asbestos-Containing Materials, Abatement Continue to Grow," *Hazmat World*, April, 1990, pp. 32–58.

12. Fisher, G. L., and M. A. Gallo, Eds., *Asbestos Toxicity*, (New York: Marcel Dekker, 1988).

13. Seligman, P. F., et al., "Distribution and Fate of Tributyltin in the Marine Environment," *American Chemical Society Division of Environmental Chemistry Preprint Extended Abstracts*, 28, 573–579 (1988).

14. Clark, E. M., R. M. Sterritt, and J. N. Lester, "The Fate of Tributyltin in the Aquatic Environment," *Environmental Science and Technology*, 22, 600–604 (1988).

15. Gosselin, R. E., R. P. Smith, and H. C. Hodge, "Phenol," in *Clinical Toxicology of Commercial Products*, 5th ed., (Baltimore/London: Williams and Wilkins, 1984), pp. III-344 to III-348.

16. Lepowski, W., "Methyl Isocyanate: Studies Point to Systemic Effects," *Chemical and Engineering News*, June 13, 1988, p. 6.

17. Gosselin, R. E., R. P. Smith, and H. C. Hodge, "Paraquat," in *Clinical Toxicology of Commercial Products*, 5th ed., (Baltimore/London: Williams and Wilkins, 1984), pp. III-328 to III-336.

18. Shabecoff, P., "Industry Fights Use of Animal Tests to Assess Cancer Risk," *New York Times*, July 25, 1989, p. 21.

19. Safe, S., Ed., *Polychlorinated Biphenyls (PCBs)*: *Mammalian and Environmental Toxicology*, (New York: Springer-Verlag, 1987).
20. Silberner, J., "Common Herbicide Linked to Cancer," *Science News*, 130(11), 167–174 (1986).
21. "Dioxin is Found Not to Increase Birth Defects," *New York Times*, March 18, 1988, p. 12.
22. "Global Experts Offer Advice on Chemical Weapons Treaty," *Chemical and Engineering News*, July 27, 1987, pp. 16–27.

SUPPLEMENTARY REFERENCES

Hallenbeck, W. H., and K. M. Cunningham-Burns, *Quantitative Risk Assessment for Environmental and Occupational Health*, (Chelsea, MI: Lewis Publishers, 1986).

Fawell, J. K., and S. Hunt, *Environmental Toxicology: Organic Pollutants,* (New York, NY: John Wiley & Sons, 1988).

Kneip, T. J., and J. V. Crable, Eds., *Methods for Biological Monitoring: A Manual for Assessing Human Exposure to Hazardous Substances*, (Washington, DC: American Public Health Association, 1988).

Elements of Toxicology and Chemical Risk Assessment, Revised Ed., Environ Corp., Washington, DC, 1988.

Proctor, N. H., J. P. Hughes, and M. L. Fischman, *Chemical Hazards of the Workplace*, 2nd Ed., (Philadelphia, PA: J. B. Lippincott Co., 1988).

Lioy, P. J., and J. M. Daisey, Eds., *Toxic Air Pollution: A Comprehensive Study of Non-Criteria Air Pollutants*, (Chelsea, MI: Lewis Publishers, 1987).

Ram, N. M., E. J. Calabrese, and R. F. Christman, Eds., *Organic Carcinogens in Drinking Water: Detection, Treatment, and Risk Assessment*, (New York, NY: John Wiley and Sons, 1986).

Tardiff, R. G., and J. V. Rodricks, Eds., *Toxic Substances and Human Risk,* (New York, NY: Plenum Press, 1987).

Worobec, M. D., and G. Ordway, *Toxic Substances Controls Guide*, (Edison, NJ: BNA Books Distribution Center, 1989).

Tardiff, R. G., and J. V. Rodricks, Eds., *Toxic Substances and Human Risk: Principles of Data Interpretation*, (New York, NY: Plenum Publishing, 1988).

Paddock, T., *Dioxins and Furans: Questions and Answers*, (Philadelphia, PA: Academy of Natural Sciences, 1989).

Garner, R. C., and J. Hradec, Eds., *Biochemistry of Chemical Carcinogenesis,* (New York: Plenum, 1990).

Cohen, G., and J. O'Connor, *Fighting Toxics. A Manual for Protecting Your Family, Community, and Workplace*, (Washington, DC: Island Press, 1990).

Poole, A., and G. B. Leslie, *A Practical Approach to Toxicological Investigations*, (New York: Cambridge University Press, 1990).

Cooper, C. S., and P. L. Gover, Eds., *Chemical Carcinogenesis and Mutagensis II*, (New York: Springer-Verlag, 1990).

Nichol, J., *Bites and Stings. The World of Venomous Animals*, (New York: Facts on File, 1989).

Baker, S. R., and C. F. Wilkinson, Eds., *The Effects of Pesticides on Human Health*, (Princeton, NJ: Princeton Scientific, 1990).

Sandhu, S. S., Ed., *In Situ Evaluation of Biological Hazards of Environmental Pollutants*, (New York: Plenum Press, 1990).

Clarke, L., *Acceptable Risk? Making Decisions in a Toxic Environment*, (Berkeley: University of California Press, 1989).

Liberman, D. F., and J. G. Gordon, Eds., *Biohazards Management Handbook,* (New York: Dekker, 1989).

22   RESOURCES AND ENERGY

22.1. THE NATURAL RESOURCES-ENERGY-ENVIRONMENT TRIANGLE

Natural resources, energy, and the environment are intimately related (Fig. 22.1). Perturbations in one usually cause perturbations in the other two. For example, reductions in automotive exhaust pollutant levels with the use of catalytic devices have resulted in increased demand for platinum metal, a scarce natural resource. The availability of many metals depends upon the quantity of energy used and the amount of environmental damage tolerated in the extraction of low-grade ores. Many other such examples could be cited. Because of these intimate interrelationships, resources and energy must be discussed along with environmental chemistry.

Problems of energy, natural resources, and environment have been more

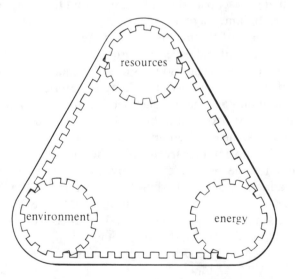

Figure 22.1. Strong connections exist among natural resources, energy, and environment.

apparent since the mid-1970s, particularly following the 1973–74 "energy crisis." This crisis was accompanied by worldwide shortages of some foods and minerals, followed in some cases by surpluses, such as the surplus wheat resulting from increased planting and a copper surplus resulting from the efforts of copper-producing nations to acquire foreign currency by copper export. Even petroleum prices have fallen in inflation-adjusted terms. However, as long as pressures of unrestrained population continue to place demands on limited resources, shortages such as those just described will reoccur and will worsen.

It is beyond the scope of one chapter to discuss natural resources and energy in anything like a comprehensive manner. Instead, this chapter touches upon the major aspects of natural resources and energy resources as they relate to environmental chemistry.

In discussing minerals and fossil fuels, two terms related to available quantities are used which should be defined. The first of these is **resources**, which refers to quantities that are estimated to be *ultimately* available. The second term is **reserves**, which refers to well-identified resources that can be profitably utilized with existing technology.

22.2.   METALS

With an adequate supply of all of the important elements and energy, almost any needed material can be manufactured. Most of the elements, including practically all of those likely to be in short supply, are metals. Some of these are virtually unavailable in the U.S., which imports almost all its aluminum, chromium, cobalt, manganese, palladium, platinum, and titanium, and the majority of its bismuth, cadmium, mercury, nickel, tungsten and zinc. Some countries produce virtually none of the metals they require.

Some metals are considered especially crucial because of their importance to industrialized societies, uncertain sources of supply, and price volatility in world markets. One of these is antimony, used in auto batteries, fire-resistant fabrics, and rubber. Chromium, another crucial metal, is used to manufacture stainless steel (especially for parts exposed to high temperatures and corrosive gases), jet aircraft, automobiles, hospital equipment, and mining equipment. The U.S. imports 91% of its chromium, largely from the Republic of South Africa and Russia, and obtains the remainder by recycling. The platinum-group metals (platinum, palladium, iridium, rhodium) are used as catalysts in the chemical industry, in petroleum refining, and in automobile exhaust antipollution devices. The U.S. imports 87% of these metals and receives the remainder from recycling. The Republic of South Africa and Russia are the major import sources. Substitutes in the electrical and electronic industries include gold, silver, and tungsten. Nickel, vanadium, titanium, and rare earths can substitute in catalytic applications.

Mining and processing of metal ores involve major environmental concerns, including disturbance of land, air pollution from dust and smelter emissions, and water pollution from disrupted aquifers. This problem is aggravated by the fact that the general trend in mining involves utilization of less rich ores. This is illustrated in Figure 22.2, showing the average percentage of copper in copper ore mined since 1900. The average percentage of copper in ore mined in 1900 was about 4%, but by 1982 it was about 0.6% in domestic ores and 1.4% in richer foreign ores. Ores as low as 0.1% copper may eventually be processed. Increased demand for a particular metal, coupled with the necessity to utilize lower grade ores, has a vicious multiplying effect upon the amount of ore that must be mined and processed, and accompanying environmental consequences.

Discussion of all the major environmental manifestations of the mining of each metal would require a chapter much longer than this one. Instead, these are summarized in Table 22.1.

22.3   NONMETAL MINERAL RESOURCES

A number of minerals other than those used to produce metals are important resources. The major ones are mentioned here. As with metals, the environmental aspects of mining many of these minerals are quite important. Typically, even the extraction of ordinary rock and gravel can have important environmental effects.

About 1.2 million metric tons of barite, composed of finely ground $BaSO_4$, are employed annually in the U.S. to make drilling "mud", which is used in rotary drilling rigs to seal the walls, cool the drill bit, and lubricate the drill stem. World resources of barite exceed 300 million metric tons with about 100 million metric tons in the U.S.

Clays are abundant in suspended and sedimentary matter in water and as secondary minerals in the geosphere (Section 14.4). Various clays are also used for clarifying oils, as catalysts in petroleum processing, as fillers and coatings for paper, and in the manufacture of firebrick, pottery, sewer pipe, and floor tile. Major types of clays that have industrial uses are shown in Table 22.2. U.S.

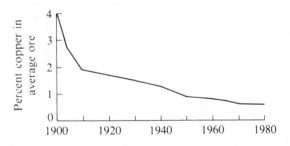

Figure 22.2.   Average percentage of copper in ore that has been mined.

Table 22.1.   Worldwide and Domestic Metal Resources

Metals	Properties[a]	Major Uses	Ores and Aspects of Resources[b]
Aluminum	mp 660°C, bp 2467°C, sg 2.70, malleable, ductile	Metal products, including autos, aircraft, electrical equipment. Conducts electricity better than copper per unit weight and is used in electrical transmission lines.	From bauxite ore containing 35–55% Al_2O_3. About 60 million metric tons of bauxite produced worldwide annually, about 27% used in the U.S., which produces about 1.5 million metric tons of bauxite per year. U.S. resources of bauxite are 40 million metric tons, world resources are 15 billion metric tons.
Chromium	mp 1903°C, bp 2642°C, sg 7.14, hard, silvery color	Metal plating, stainless steel, wear-resistant and cutting-tool alloys, chromium chemicals, including chromate used as an anticorrosive and cooling-water additive.	From chromite having the general formula $[Mg(II), Fe(II)][Cr(III), Al(III), Fe(III)]_2O_4$. Resources of 1 billion metric tons in South Africa and Rhodesia, large deposits in Russia, virtually none in the U.S.
Cobalt	mp 1495°C, bp 2880°C, sg 8.71, bright silvery	Manufacture of hard, heat-resistant alloys such as stellite, permanent magnet alloys, driers, pigments glazes, catalysts, animal-feed additive.	From a variety of minerals, such as linnaeite, Co_3S_4, and as a by-product of other metals. World consumption is 25,000 metric tons Co per year, 30% used in the U.S. U.S. imports 80% of its cobalt. Abundant global and U.S. resources.
Copper	mp 1083°C, bp 2582°C, sg 8.96, dense, ductile, malleable	Electrical conductors, alloys, chemicals. Many uses.	Occurs in low percentages (see Fig. 22.2) as sulfides, oxides, and carbonates in other minerals. U.S. consumption is 1.5 million metric tons per year. World resources of 344 million metric tons, including 78 million in U.S.
Gold	mp 1063°C, bp 2660°C, sg 19.3	Jewelry, basis of currency, electronics, increasing industrial uses.	In various minerals at a very low 10 ppm for ores currently being processed in the U.S.; by-product of copper refining. World resources of 1 billion oz, 80 million oz in U.S.

Table 22.1. Worldwide and Domestic Metal Resources (continued)

Metals	Properties[a]	Major Uses	Ores and Aspects of Resources[b]
Iron	mp 1535°C, bp 2885°C, sg 7.86, silvery metal in (rare) pure form	By far the most widely produced metal, usually as steel, a high-tensile-strength material containing 0.3–1.7% C. Made into many alloys for special purposes.	Occurs as hematite (Fe_2O_3) goethite ($Fe_2O_3 \cdot H_2O$), and magnetite (Fe_3O_4) in abundant supply globally and in the U.S.
Lead	mp 327°C, bp 1750°C, sg 11.35, silvery color	Fifth most widely used metal. Storage batteries, gasoline additives, pigments, ammunition.	Major source is galena, PbS. Worldwide consumption of lead is 3.5 million metric tons, about 1/3 in U.S. (not including recycled scrap constituting 40% of use). Global reserves of 140 million metric tons, 39 million metric tons in U.S.
Manganese	mp 1244°C, bp 2040°C, sg 7.3, hard, brittle, gray-white	Sulfur and oxygen scavenger in steel, manufacture of alloys, dry cells, chemicals.	Found in a variety of minerals, primarily oxides, in manganese nodules on the ocean floor. About 20 million metric tons of manganese ore produced globally each year, 2 million tons consumed in the U.S., no domestic production. World reserves of manganese are 6.5 billion tons.
Mercury	mp –38°C, bp 357°C, sg 13.6 shiny liquid metal	Instruments, electronic apparatus, electrodes, chemical compounds (such as fungicides and slimicides).	From cinnabar, HgS. Annual world production of 11,500 metric tons, 1/3 of which is used in the U.S. World resources of 275,000 metric tons, only 6600 metric tons in the U.S.
Molybdenum	mp 2620°C, bp 4825°C, sg 9.01, ductile, silvery-gray	Alloys, pigments, catalysts, chemicals, and lubricants.	Molybdenite (MoS_2) and wulfenite ($PbMoO_4$) are major ores. About two-thirds of global molybdenum production is in the U.S.; global resources are quite large.
Nickel	mp 1455°C, bp 2835°C, sg 8.90, silvery color	Alloys, coins, storage batteries, catalysts (e.g., for hydrogenation of vegetable oil).	Found in ores associated with iron. U.S. consumption of nickel is 150,000 metric tons per year, of which less than 10% is produced domestically. Large domestic reserves of low-grade ore are available.

Element	Properties	Uses	Occurrence and resources
Silver	mp 961°C, bp 2193°, sg 10.5, shiny metal	Photographic materials, electronics, sterling ware, jewelry, bearings, dentistry.	Found with sulfide minerals, by-product of copper, lead, and zinc smelting. Annual U.S. consumption of 150 million troy ounces will soon exhaust known resources.
Tin	mp 232°C, bp 2687°C, sg 7.31	Coatings, solders, bearing alloys, bronze, chemicals.	Found in many compounds associated with granitic rocks and chrysolites. World consumption of 190,000 metric tons/year, U.S. consumption of 60,000 metric tons/year, world resources of 10 million metric tons.
Titanium	mp 1677°C, bp 3277°C, sg 4.5, silvery color	Strong, corrosion-resistant, used in aircraft and their engines, valves, pumps, paint pigments.	Ranks ninth in elemental abundance, commonly as TiO_2; no shortages of titanium are likely.
Tungsten	mp 3380°C, bp 5530°C, sg 19.3 gray	Very strong, high boiling point, used in alloys, drill bits, turbines, nuclear reactors, tungsten carbide.	Found as tungstates, such as scheelite ($CaWO_4$); U.S. has 7% of world reserves, China 60%. Could be in short supply in U.S. by year 2000.
Vanadium	mp 1917°C, bp 3375°C, sg 5.87, gray	Used to make strong steel alloys.	Occurs in igneous rocks, primarily as V(III), primarily a by-product of other metals, U.S. consumption of 5,000 metric tons/year equals production.
Zinc	mp 420°C, bp 907°C, sg 7.14, bluish white	Widely used in alloys (brass), galvanized, paint pigments, chemicals. Fourth in metal production worldwide.	Found in many ore minerals, including sulfides, oxides, and silicates. World production is 5 million metric tons/year (10% from U.S.) and annual U.S. consumption is 1.5 million metric tons. World resources are 235 million metric tons, 20% in the U.S.

[a] Mp, melting point; bp, boiling point; sg, specific gravity.
[b] All figures are approximate; quantities of minerals considered available depend upon price, technology, recent discoveries, and other factors, so that these quantities are subject to fluctuation.

Table 22.2.   Major Types of Clays and Their Uses in the U.S.

Type of Clay	Percent Use	Composition	Uses
Miscellaneous	72	variable	filler, brick, tile, portland cement, many others
Fireclay	12	variable; can be fired at high temperatures without warping	refractories, pottery, sewer pipe, tile, brick
Kaolin	8	$Al_2(OH)_4Si_2O_5$; is white and can be fired without losing shape or color	paper filler, refractories, pottery, dinnerware, petroleum-cracking catalyst
Bentonite and fuller's earth	7	variable	drilling muds, petroleum catalyst, carriers for pesticides, sealers, clarifying oils
Ball clay	1	variable, very plastic	refractories, tile, whiteware

production of clay is about 60 million metric tons per year, and global and domestic resources are abundant.

Fluorine compounds are widely used in industry, and large quantities of fluorspar, CaF_2, are consumed as a flux in steel manufacture. Synthetic and natural cryolite, Na_3AlF_6, is used as a solvent for aluminum oxide in the electrolytic preparation of aluminum metal. Freon-12, difluorodichloromethane, has been employed extensively as a refrigeration and air-conditioning fluid and is considered to be a major stratospheric pollutant (see Section 18.5). Sodium fluoride is used for water fluoridation. World reserves of high-grade fluorspar are around 190 million metric tons, about 13% of which is in the United States. This is sufficient for several decades at projected rates of use. A great deal of by-product fluorine is recovered from the processing of fluorapatite, $Ca_5(PO_4)_3F$, used as a source of phosphorus.

Phosphorus, along with nitrogen and potassium, is one of the major fertilizer elements (see Section 15.6). Many soils are deficient in phosphate. Its non-fertilizer applications include supplementation of animal feeds, synthesis of detergent builders, and preparation of chemicals such as pesticides and medicines.

The major phosphate minerals are the apatites, having the general formula $Ca_5(PO_4)_3(F, OH, Cl)$. The most common of these are fluorapatite, $Ca_5(PO_4)_3F$

and hydroxyapatite, $Ca_5(PO_4)_3(OH)$. Ions of Na, Sr, Th, and U are found substituted for calcium in apatite minerals. Small amounts of PO_4^{3-} may be replaced by SO_4^{2-}, VO_4^{3-}, or AsO_4^{3-}. Because of the latter ion, traces of arsenic are found in phosphate products. By-product vanadium in phosphate minerals may have economic value.

Approximately 17% of world phosphate production is from igneous minerals, primarily fluorapatites. About three-fourths of world phosphate production is from sedimentary deposits, generally of marine origin. Vast deposits of phosphate, accounting for approximately 5% of world phosphate production, are derived from guano droppings of seabirds and bats.

Current U.S. production of phosphate rock is around 40 million metric tons per year, most of it from Florida. Tennessee and several of the western states are also major producers of phosphate. Reserves of phosphate minerals in the United States amount to 10.5 billion metric tons, containing approximately 1.4 billion metric tons of phosphorus. Identified world reserves of phosphate rock are approximately 6 billion metric tons.

Pigments and fillers of various kinds are used in large quantities. The only naturally occurring pigments still in wide use are those containing iron. These minerals are colored by limonite, an amorphous brown-yellow compound with the formula $2\ Fe_2O_3 \cdot 3H_2O$, and hematite, composed of gray-black Fe_2O_3. Along with varying quantities of clay and manganese oxides, these compounds are found in ocher, sienna, and umber. Manufactured pigments include carbon black, titanium dioxide, and zinc pigments. About 1.5 million metric tons of carbon black, manufactured by the partial combustion of natural gas, are used in the U.S. each year, primarily as a reinforcing agent in tire rubber.

Over 7 million metric tons of minerals are used in the U.S. each year as fillers for paper, rubber, roofing, battery boxes, and many other products. Among the minerals used as fillers are carbon black, diatomite, barite, fuller's earth, kaolin, mica, limestone, pyrophyllite, and wollastonite ($CaSiO_3$).

Although sand and gravel are the cheapest of mineral commodities per ton, the average annual dollar value of these materials is greater than all but a few mineral products because of the huge quantities involved. In tonnage, sand and gravel production is exceeded only by that of fossil fuels.

At present, old river channels and glacial deposits are used as sources of sand and gravel. Many valuable deposits of sand and gravel are covered by construction and lost to development. Transportation and distance from source to use are especially crucial for this resource. Environmental problems involved with defacing land can be severe, although bodies of water used for fishing and other recreational activities frequently are formed by removal of sand and gravel.

The biggest single use for sulfur is in the manufacture of sulfuric acid. However, the element is employed in a wide variety of other industrial and agricultural products. Current consumption of sulfur amounts to approximately 10 million metric tons per year in the United States.

Sulfur can exist in many forms and undergoes a complicated geobiochemical cycle (Figure 22.3). Sulfur may be expelled from volcanoes as sulfur dioxide or as hydrogen sulfide, which is oxidized to sulfur dioxide and sulfates in the atmosphere. Highly insoluble metal sulfides are oxidized upon exposure to atmospheric oxygen to relatively soluble metal sulfates. In the ground and in water, sulfate is converted to organic sulfur by plants and bacteria. Bacteria mediate transitions among sulfate, elemental sulfur, organic sulfur, and hydrogen sulfide. Sulfur(-II) may either escape to the atmosphere as H_2S or volatile organosulfur compounds, or it may precipitate as sulfides of metals, primarily iron.

The four most important sources of sulfur are (in decreasing order): deposits of elemental sulfur; H_2S recovered from sour natural gas; organic sulfur recovered from petroleum; and pyrite (FeS_2). Supply of sulfur is no problem either in the United States or worldwide. The United States has abundant deposits of elemental sulfur, and sulfur recovery from fossil fuels as a pollution control measure adds to the supply.

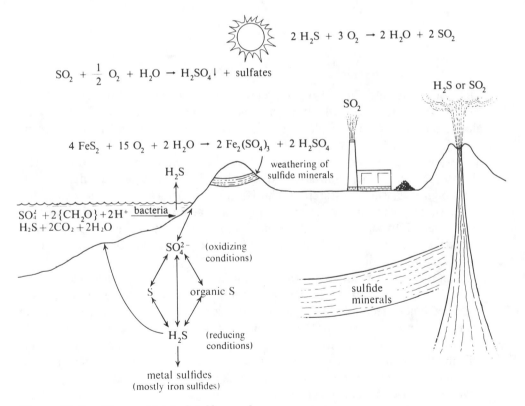

Figure 22.3.   Biogeochemical sulfur cycles.

22.4. WOOD – A MAJOR RENEWABLE RESOURCE

Fortunately, one of the major natural resources in the world, wood, is a renewable resource. Production of wood and wood products is the fifth largest industry in the United States and forests cover one-third of the United States surface area. Wood ranks first worldwide as a raw material for the manufacture of other products, including lumber, plywood, particle board, cellophane, rayon, paper, methanol, plastics, and turpentine.

Chemically, wood is a complicated substance consisting of long cells having thick walls composed of polysaccharides such as cellulose,

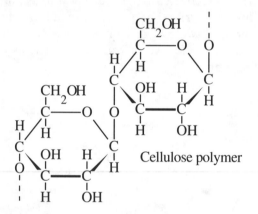

Cellulose polymer

The polysaccharides in cell walls account for approximately three-fourths of *solid wood*, wood from which extractable materials have been removed by an alcohol-benzene mixture. Wood typically contains a few tenths of a percent ash.

A wide variety of organic compounds can be extracted from wood by water, alcohol-benzene, ether, and steam distillation. These compounds include tannins, pigments, sugars, starch, cyclitols, gums, mucilages, pectins, galactans, terpenes, hydrocarbons, acids, esters, fats, fatty acids, aldehydes, resins, sterols, and waxes. Substantial amounts of methanol (sometimes called *wood alcohol*) are extracted from wood. Methanol, once a major source of liquid fuel, is again being considered for that use.

A major use of wood is in paper manufacture, and the widespread use of paper is a mark of an industrialized society. The manufacture of paper is a highly advanced technology. Paper consists essentially of cellulosic fibers tightly pressed together. The lignin fraction must first be removed from the wood, leaving the cellulosic fraction. Both the sulfite and alkaline processes for accomplishing this separation have resulted in severe water and air pollution problems, although substantial progress has been made in alleviating these.

Wood fibers and particles can be used for making fiberboard, paper-base laminates (layers of paper held together by a resin and formed into the desired structures at high temperatures and pressures), particle board (consisting of wood particles bonded together by a phenol-formaldehyde or urea-formaldehyde resin) and nonwoven textile substitutes consisting of wood fibers held together by adhesives. Chemical processing of wood enables the manufacture of many useful products, including methanol (a gasoline substitute) and sugar. Sugar and methanol are potential major products from the 60 million metric tons of wood wastes produced in the U.S. each year.

22.5. THE ENERGY PROBLEM

Since the first "energy crisis" of 1973–74, much has been said and written, many learned predictions have gone awry, and some concrete action has even taken place. Prophecies of catastrophic economic disruption, people "freezing in the dark," and freeways given over to bicycles (though perhaps not a bad idea) have not been fulfilled. However, the several-fold increase in crude oil prices since 1973 has exacted a toll. In the U.S. and other industrialized nations, the economy has been plagued by inflation, recession, unemployment, and obsolescence of industrial equipment. The economies of some petroleum-deficient developing countries have been devastated by energy prices.

The solutions to energy problems are strongly tied to environmental considerations. For example, a massive shift of the energy base to coal in nations that now rely largely on petroleum for energy would involve much more strip mining, potential production of acid mine water, use of scrubbers, and release of greenhouse gases (carbon dioxide from coal combustion and methane from coal mining). Similar examples could be cited for most other energy alternatives.

Clearly, chemists must be involved in developing alternative energy sources. Chemical processes are used in the conversion of coal to gaseous and liquid fuels. New materials developed through the applications of chemistry will be employed to capture solar energy and convert it to electricity. The environmental chemist has a key role to play in making alternative energy sources environmentally acceptable.

22.6. WORLD ENERGY RESOURCES

At present, most of the energy consumed by humans is produced from fossil fuels. Estimates of the amounts of fossil fuels available differ. Because of undiscovered deposits of petroleum and natural gas, there may be considerably more fossil fuel available than is commonly realized (or less than the more optimistic estimates).

Estimates of the quantities of recoverable fossil fuels in the world before 1800 are given in Figure 22.4. By far the greatest recoverable fossil fuel is in the form of coal and lignite. Furthermore, only a small percentage of this energy source has been utilized to date, whereas much of the recoverable petroleum and natu-

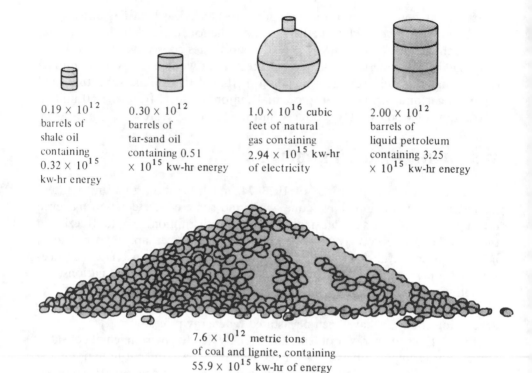

0.19 × 10^{12}
barrels of
shale oil
containing
0.32 × 10^{15}
kw-hr energy

0.30 × 10^{12}
barrels of
tar-sand oil
containing 0.51
× 10^{15} kw-hr energy

1.0 × 10^{16} cubic
feet of natural
gas containing
2.94 × 10^{15} kw-hr
of electricity

2.00 × 10^{12}
barrels of
liquid petroleum
containing 3.25
× 10^{15} kw-hr energy

7.6 × 10^{12} metric tons
of coal and lignite, containing
55.9 × 10^{15} kw-hr of energy

Figure 22.4.   Original amounts of the world's recoverable fossil fuels (quantities in thermal kilowatt hours of energy based upon data taken from M. K. Hubbert, "The Energy Resources of the Earth," in *Energy and Power*, W. H. Freeman and Co., San Francisco, 1971).

ral gas has already been consumed. Projected use of these latter resources indicates rapid depletion.

Although it is not the purpose of this chapter to give a detailed discussion of energy resources and projected rates of use, some of the statistics in these areas are interesting as well as sobering. In the 48 coterminous United States, the peak of petroleum production has been reached. Alaskan oil can help the petroleum supply only temporarily. Worldwide petroleum resources are approximately 10 times those of the United States, and peak world production will be reached within a few years. Despite the vast amounts of oil that remain in tar sand deposits and oil shales, the economic and environmental costs of their utilization may be prohibitive.

Although world coal resources are enormous and potentially can fill energy needs for a century or two, their utilization is limited by environmental disruption from mining and emissions of carbon dioxide and sulfur dioxide. These would become intolerable long before coal resources were exhausted. Assuming only uranium-235 as a fission fuel source, total recoverable reserves of nuclear fuel are roughly about the same as fossil fuel reserves. These are many orders of magnitude higher if the use of breeder reactors is assumed. Extraction of only 2% of the deuterium present in the Earth's oceans would yield about a billion times as much energy by controlled nuclear fusion as was originally present in fossil fuels! This prospect is tempered by the lack of success in developing a controlled nuclear fusion reactor. Hopes for such a process soared with the announcement of a "cold fusion" process in 1989, but the world still awaits an operational cold fusion reactor. Geothermal power, currently utilized in northern California, Italy, and New Zealand, has the potential for providing an appreciable percentage of energy worldwide. The same limited potential is characteristic of several renewable energy resources, including hydroelectric energy, tidal energy, and wind power. All of these will continue to contribute significant, but relatively small, amounts of energy. Renewable, nonpolluting solar energy comes as close to being an ideal energy source as any available. It certainly has a bright future in those areas where the sun shines intensely and consistently.

Globally, therefore, the energy picture is both hopeful and grim. In the U.S., certainly, liquid petroleum and natural gas are on the way out as major producers of energy. The prospects for massive utilization of shale oil deposits appear to be very dim. Nuclear power may provide an increasing share of energy production, though problems of a predominantly political nature will continue to delay its rate of development. Miscellaneous sources such as wind, hydroelectric, tidal, and geothermal power will make useful but limited contributions to total energy production. Were it not for the associated environmental problems, coal, an abundant and versatile fossil fuel, could fill the energy gap until more exotic sources can be developed. Within a few decades, widespread utilization may be made of nuclear fusion or solar power, each of which promises to be an abundant, relatively nonpolluting source of energy.

22.7 ENERGY CONSERVATION

From about 1960 until 1973, energy production increased in the United States slightly over 4% per year, a growth rate with a doubling time of approximately 16 years. Such a growth rate can be used or misused to support many different theories. It can be projected to show, for example, that within a few decades facilities for the production of power must be greatly expanded to provide for energy needs. It can also be used to show that devastating environmental damage will occur as a result of the exploitation of energy resources. It can even be used to calculate the exact date upon which the whole nation will glow in the dark from exponentially increasing waste heat radiated out into space. What is wrong with such projections is that they need not, and indeed cannot, come true.

Any consideration of energy needs and production must take energy conservation into consideration. This does not have to mean cold classrooms with thermostats set at 60°F in mid-winter, nor swelteringly hot homes with no air-conditioning, nor total reliance on the bicycle for transportation, although these, and even more severe, conditions are routine in many countries. The fact remains that the United States has wasted energy at a deplorable rate. For example, U.S. energy consumption is significantly higher per capita than that of some other countries that have equal, or significantly better, living standards. Obviously, a great deal of potential exists for energy conservation that will ease the energy problem.

Transportation is the economic sector with the greatest potential for increased efficiencies. The private auto and airplane are only about one-third as efficient as buses or trains for transportation. Transportation of freight by truck requires about 3800 Btu/ton-mile, compared to only 670 Btu/ton-mile for a train. It is terribly inefficient compared to rail transport (as well as dangerous, labor-intensive, and environmentally disruptive). Major shifts in current modes of transportation in the U.S. will not come without anguish, but energy conservation dictates that they be made.

Household and commercial uses of energy are relatively efficient. Here again, appreciable savings can be made. The all-electric home requires much more energy (considering the percentage wasted in generating electricity) than a home heated with fossil fuels. The sprawling ranch-house style home uses much more energy per person than does an apartment unit or row house. Improved insulation, sealing around the windows, and other measures can conserve a great deal of energy. Electric generating plants centrally located in cities can provide waste heat for commercial and residential heating and cooling and, with proper pollution control equipment, can use refuse for a significant fraction of fuel

As scientists and engineers undertake the crucial task of developing alternative energy sources to replace dwindling petroleum and natural gas supplies, energy conservation must receive proper emphasis. In fact, zero energy-use growth, at least on a per capita basis, is a worthwhile and achievable goal. Such a policy would go a long way toward solving many environmental problems. With ingenuity, planning, and proper management, it could be achieved while increasing the standard of living and quality of life.

22.8.   ENERGY CONVERSION PROCESSES

As shown in Figure 22.5, energy occurs in several forms and must be converted to other forms. The efficiencies of conversion vary over a wide range. Conversion of electrical energy to radiant energy by incandescent light bulbs is very inefficient — less than 5% of the energy is converted to visible light and the remainder is wasted as heat. At the other end of the scale, a large electrical generator is around 80% efficient in converting mechanical energy to electrical energy. The once much-publicized Wankel rotary engine converts chemical to

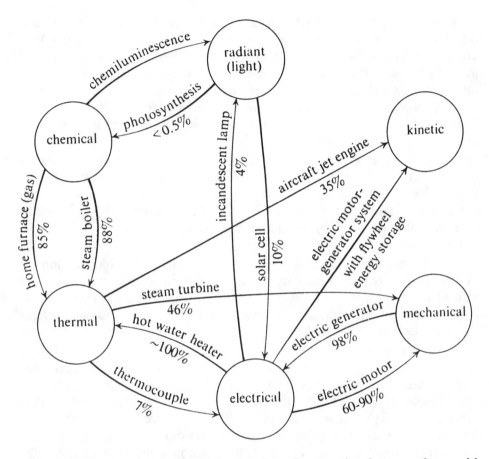

Figure 22.5.   Kinds of energy and examples of conversion between them, with conversion efficiency percentages.

mechanical energy with an efficiency of about 18%, compared to 25% for a gasoline-powered piston engine and about 37% for a diesel engine. A modern coal-fired steam-generating power plant converts chemical energy to electrical energy with an overall efficiency of about 40%.

One of the most significant energy conversion processes is that of thermal energy to mechanical energy in a heat engine such as a steam turbine. The Carnot equation,

$$\text{Percent efficiency} = \frac{T_1 - T_2}{T_1} \times 100 \qquad (22.8.1)$$

states that the percent efficiency is given by a fraction involving the inlet temperature (for example, of steam), T_1, and the outlet temperature, T_2. These temperatures are expressed in Kelvin (°C + 273). Typically, a steam turbine engine operates with approximately 810 K inlet temperature and 330 K outlet temperature. These temperatures substituted into Equation 22.8.1 give a maximum theoretical efficiency of 59%. However, because it is not possible to maintain the incoming steam at the maximum temperature and because mechanical energy losses occur, overall efficiency of conversion of thermal energy to mechanical energy in a modern steam power plant is approximately 47%. Taking into account losses from conversion of chemical to thermal energy in the boiler, the total efficiency is about 40%.

Some of the greatest efficiency advances in the conversion of chemical to mechanical or electrical energy have been made by increasing the peak inlet temperature in heat engines. The use of superheated steam has raised T_1 in a steam power plant from around 550 K in 1900 to about 850 K at present. Improved materials and engineering design, therefore, have resulted in large energy savings.

The efficiency of nuclear power plants is limited by the maximum temperatures attainable. Reactor cores would be damaged by the high temperatures used in fossil-fuel-fired boilers and have a maximum temperature of approximately 620 K. Because of this limitation, the overall efficiency of conversion of nuclear energy to electricity is about 30%.

Most of the 60% of energy from fossil-fuel-fired power plants and 70% of energy from nuclear power plants that is not converted to electricity is dissipated as heat, either to the atmosphere or to bodies of water and streams. The latter is thermal pollution, which may either harm aquatic life or, in some cases, actually increase bioactivity in the water to the benefit of some species. This waste heat is potentially very useful in applications like home heating, water desalination, and aquaculture (growth of plants in water). Increased production of waste heat by human activity could eventually result in marked changes in climate.

Some devices for the conversion of energy are shown in Figure 22.6. Substantial advances have been made in energy conversion technology over many decades and more can be projected for the future. Through the use of higher temperatures and larger generating units, the overall efficiency of fossil-fueled electrical power generation has increased approximately ten-fold since 1900, from less than 4% to a maximum of around 40%. An approximately four-fold increase in the energy-use efficiency of rail transport occurred during the 1940s and 1950s with the replacement of steam locomotives with diesel locomotives. During the coming decades, increased efficiency can be anticipated from the application of sophisticated techniques linking two or more energy conversion processes. For example, combined power cycles in connection with magnetohydrodynamics (Figure 22.8) could provide a very efficient energy source used in combination with conventional steam generation. Entirely new devices such as thermonuclear reactors for the direct conversion of nuclear fusion energy to electricity will very likely be developed.

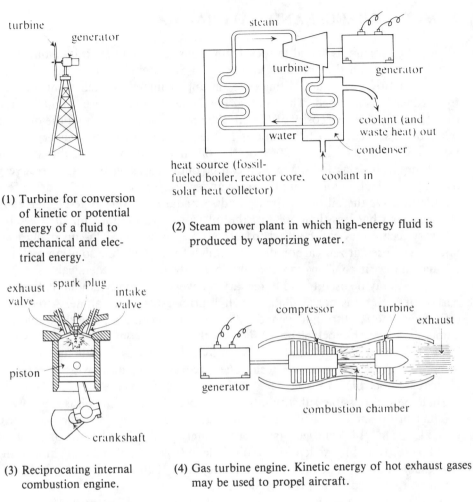

(1) Turbine for conversion of kinetic or potential energy of a fluid to mechanical and electrical energy.

(2) Steam power plant in which high-energy fluid is produced by vaporizing water.

(3) Reciprocating internal combustion engine.

(4) Gas turbine engine. Kinetic energy of hot exhaust gases may be used to propel aircraft.

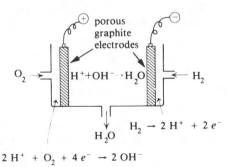

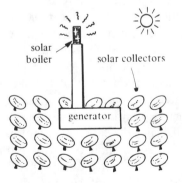

(5) Fuel cell for the direct conversion of chemical energy to electrical energy.

(6) Solar thermal electric conversion.

Figure 22.6.   Some energy conversion devices.

22.9.   PETROLEUM AND NATURAL GAS

Liquid petroleum is found in rock formations ranging in porosity from 10 to 30%. Up to half of the pore space is occupied by water. The oil in these formations must flow over long distances to an approximately 6-inch diameter well from which it is pumped. The rate of flow depends on the permeability of the rock formation, the viscosity of the oil, the driving pressure behind the oil, and other factors. Because of limitations in these factors **primary recovery** of oil yields an average of about 30% of the oil in the formation, although it is sometimes as little as 15%. More oil can be obtained using **secondary recovery** techniques, which involve forcing water under pressure into the oil-bearing formation to drive the oil out. Primary and secondary recovery together typically extract somewhat less than 50% of the oil from a formation. Finally, **tertiary recovery** techniques can be used to extract even more oil. This normally uses injection of pressurized carbon dioxide, which forms a mobile solution with the oil and allows it to flow more easily to the well. Other chemicals, such as detergents, may be used to aid in tertiary recovery. Currently, about 300 billion barrels of U.S. oil is not available through primary recovery alone. A recovery efficiency of 60% through secondary or tertiary techniques could double the amount of available petroleum. Much of this would come from fields which have already been abandoned or essentially exhausted using primary recovery techniques. Advanced recovery techniques are now being more widely practiced as increased petroleum prices have made them more feasible.

Shale oil is a possible substitute for liquid petroleum. Shale oil is a pyrolysis product of oil shale, a rock containing organic carbon in a complex structure called kerogen. It is believed that approximately 1.8. trillion barrels of shale oil could be recovered from deposits of oil shale in Colorado, Wyoming, and Utah. In the Colorado Piceance Creek basin alone, more than 100 billion barrels of oil could be recovered from prime shale deposits.

Shale oil may be recovered from the parent mineral by retorting the mined shale in a surface retort or by burning the shale underground with an in situ process. Both processes present environmental problems. Surface retorting requires the mining of enormous quantities of mineral and disposal of the spent shale, which has a volume greater than the original mineral. In situ retorting limits the control available over infiltration of underground water and resulting water pollution. Water passing through spent shale becomes quite saline, so there is major potential for saltwater pollution.

During the late 1970s and early 1980s, several corporations began building facilities for shale oil extraction in northwestern Colorado. The most ambitious of these was the Colony project, a joint construction project of Exxon and Tosco. This plant, situated about 15 miles north of Parachute, Colorado, was to produce 50,000 barrels of synthetic crude oil per day starting around 1987. The amounts of money involved were staggering. Exxon spent $300 million to purchase a 60% share of the project from Tosco, and the two companies spent an additional $400 million on the project. By 1982 the total cost of the project was estimated to be $6 billion or more. The Synthetic Fuels Corporation, an agency set up by the U.S. government, had lent $80 million on the project. In the face of escalating construction costs and a "soft" crude oil market, Exxon withdrew from the project in May, 1982, stopping construction on the plant, and shale oil extraction is still not practiced commercially.

An even larger synthetic fuels project that failed was the $13.1 billion (Canadian) Alsands oil sands project located in Alberta. This project would have produced 137,000 barrels of oil per day from the mining of tar sand and extraction of heavy oil from it. In April, 1982, both Shell Canada Ltd. and Gulf Canada Ltd. withdrew from the project, the cost of which had escalated from an initial estimate of $5 billion.

Natural gas, consisting almost entirely of methane, has become more attractive as an energy source because of the discovery and development of truly enormous new sources of this premium fuel. Two decades ago, the outlook for natural gas as a fuel was rather poor. In 1968, discoveries of natural gas deposits in the U.S. fell below annual consumption for the first time. This trend continued during the following decade, except for 1970, when gas was discovered in Alaska. During the record-cold winter of 1976–77, severe natural gas shortages occurred in parts of the U.S. Price incentives resulting from the passage of the 1978 Gas Policy Act have tended to reverse the adverse trend of natural gas discoveries versus consumption.

In addition to its use as a fuel, natural gas can be converted to many other hydrocarbon materials. It can be used as a raw material for the Fischer-Tropsch synthesis of gasoline. The discovery and development of truly massive sources of natural gas, such as may exist in geopressurized zones, could provide abundant energy reserves for the U.S., though at substantially increased prices.

22.10. COAL

From Civil War times until World War II, coal was the dominant energy source behind U.S. industrial expansion. However, the greater convenience of lower-cost petroleum resulted in a decrease in the use of coal for U.S. energy requirements after World War II. Annual coal production fell by about one-third, reaching a low of approximately 400 million tons in 1958. Since that time U.S. production has increased substantially to about 1 billion tons per year.

The general term *coal* describes a large range of solid fossil fuels derived from partial degradation of plants. Table 22.3 shows the characteristics of the major classes of coal found in the U.S., differentiated largely by percentage of fixed carbon, percentage of volatile matter, and heating value (*coal rank*). Chemically, coal is a very complex material and is by no means pure carbon. For example, a chemical formula for Illinois No. 6 coal, a type of bituminous coal, would be something like $C_{100}H_{85}S_{2.1}N_{1.5}O_{9.5}$.

Table 22.3.   Major Types of Coal Found in the United States

| Type of Coal | Proximate Analysis, Percent[a] | | | | Range of Heating |
	Fixed Carbon	Volatile Matter	Moisture	Ash	Value (Btu/pound)
Anthracite	82	5	4	9	13,000–16,000
Bituminous					
Low-volatile	66	20	2	12	11,000–15,000
Medium-volatile	64	23	3	10	11,000–15,000
High-volatile	46	44	6	4	11,000–15,000
Subbituminous	40	32	19	9	8,000–12,000
Lignite	30	28	37	5	5,500–8,000

[a]These values may vary considerably with the source of coal.

Anthracite, a hard, clean-burning, low-sulfur coal, is the most desirable of all coals. Approximately half of the anthracite originally present in the United States has been mined. Bituminous coal found in the Appalachian and north central coal fields is widely used. It is an excellent fuel with a high heating value. Unfortunately, most bituminous coals have a high percentage of sulfur (an average of 2–3%), so the use of this fuel presents environmental problems. Large quantities of subbituminous and lignite coals are found in the Rocky Mountain states and in the northern plains. These fuels have the advantage of being low in sulfur content and are finding increasing use in power plants that have to meet SO_2 emission standards. They have a number of disadvantages, often having a low heat content and high moisture contents. Also, they are generally found great distances from areas having the greatest need for fossil fuels. Lignite, in particular, tends to lose moisture and crumble when transported, forming a powdery "bug dust." Despite these disadvantages, the low sulfur content and ease of mining these low-grade fuels is resulting in a rapid increase in their use.

Some geographical areas have very large coal resources. Huge reserves of subbituminous coal are found in the Rocky Mountain states, particularly Montana and Wyoming. An especially striking example is the approximately 350 billion tons of virtually untouched lignite in North Dakota, of which 16–20 billion tons may be mined at present prices. Figure 22.7 shows areas in the U.S. with major coal reserves.

The extent to which coal can be used as a fuel depends upon solutions to several problems, including (1) minimizing the environmental impact of coal mining; (2) removing ash and sulfur from coal prior to combustion; (3) removing ash and sulfur dioxide from stack gas after combustion; (4) conversion of coal to liquid and gaseous fuels free of ash and sulfur (see Section 22.11); and, most important, (5) whether or not the impact of increased carbon dioxide emissions upon global climate can be tolerated. Progress is being made on minimizing the environmental impact of mining. Proper mining practice can

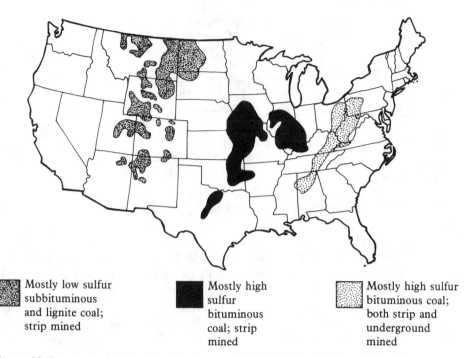

| Mostly low sulfur subbituminous and lignite coal; strip mined | Mostly high sulfur bituminous coal; strip mined | Mostly high sulfur bituminous coal; both strip and underground mined |

Figure 22.7.   Areas with major coal reserves in the coterminous United States.

minimize production of acid mine water. Particularly on flatter lands, strip-mined areas can be reclaimed with relative success. Inevitably, some environmental damage will result from increased coal mining, but the environmental impact can be reduced by various control measures. Washing, flotation, and chemical processes can be used to remove some of the ash and sulfur prior to burning. Approximately half of the sulfur in the average coal occurs as pyrite, FeS_2, and half as organic sulfur. Although little can be done to remove the latter, much of the pyrite can be separated from most coals by physical and chemical processes.

The maintenance of air pollution emission standards requires the removal of sulfur dioxide from stack gas in coal-fired power plants. Stack gas desulfurization presents some economic and technological problems; the major processes available for it are summarized in Section 17.7 and Table 17.1.

Magnetohydrodynamic power combined with conventional steam generating units has the potential for a major breakthrough in the efficiency of coal utilization. A schematic diagram of a magnetohydrodynamic (MHD) generator is shown in Figure 22.8. This device uses a plasma of ionized gas at around 2400°C blasting through a very strong magnetic field of at least 50,000 gauss to generate direct current. The ionization of the gas is accomplished by injecting a "seed" of cesium or potassium salts. In a coal-fired generator, the ultra-high-temperature

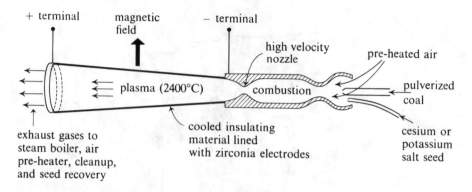

Figure 22.8.   A magnetohydrodynamic power generator.

gas issuing through a supersonic nozzle contains ash, sulfur dioxide, and nitrogen oxides which severely erode and corrode the materials used. This hot gas is used to generate steam for a conventional steam power plant, thus increasing the overall efficiency of the process. The seed salts combine with sulfur dioxide and are recovered along with ash in the exhaust. Pollutant emissions are low. Despite some severe technological difficulties, there is a chance that MHD power could become feasible on a large scale, and an experimental MHD generator was tied to a working power grid in Russia for several years. Most important, the overall efficiency of combined MHD-steam power plants should reach 60%, one and one-half times the maximum of present steam-only plants.

22.11.   COAL CONVERSION

Coal can be converted to gaseous, liquid, or low-sulfur, low-ash fuels. All of these are less polluting than coal, and the gases and liquids can be used with distribution systems and equipment designed for natural gas or petroleum. A major advantage of coal conversion is that it enables use of high-sulfur coal which otherwise could not be burned without intolerable pollution or expensive stack gas cleanup.

Coal conversion is an old idea; a house belonging to William Murdock at Redruth, Cornwall, England, was illuminated with coal gas in 1792. The first municipal coal gas system was employed to light Pall Mall in London in 1807. The coal-gas industry began in the U.S. in 1816. The early coal-gas plants used coal pyrolysis (heating in the absence of air) to produce a hydrocarbon-rich product particularly useful for illumination. Later in the 1800s, the water-gas process was developed, in which steam was added to hot coal to produce a mixture consisting primarily of H_2 and CO. It was necessary to add volatile hydrocarbons to this "carbureted" water-gas to bring its illuminating power up to that of gas prepared by coal pyrolysis. The U.S. had 11,000 coal gasifiers operating in the 1920s. At the peak of its use in 1947, the water-gas method accounted for 57% of U.S.-manufactured gas. The gas was made in low-pressure, low-capacity gasifiers which by today's standards would be inefficient and environmentally unacceptable (several locations of these old plants have been designated as hazardous waste sites because of residues of coal tar and other wastes). During World War II, Germany developed a major synthetic petroleum industry based on coal, which reached a peak capacity of 100,000 barrels per day in 1944. A plant now operating in Sasol, South Africa, converts several tens of thousands tons of coal per day to synthetic petroleum.

The major routes for coal conversion are shown in Figure 22.9. The more promising coal-derived fuels are the following:

1. Solvent-refined coal (SRC), a solid high-Btu product with low sulfur, ash, and water content
2. Low-sulfur boiler fuels that are liquid at elevated temperatures
3. Liquid hydrocarbon fuels, including gasoline, diesel fuel, and naphtha
4. Synthetic natural gas (SNG), essentially pure methane
5. Low-sulfur, low-Btu gas for industrial use

High-heat-content (high-Btu) synthetic natural gas, SNG, can be produced by any of several processes. The steps in a typical process for SNG production are:

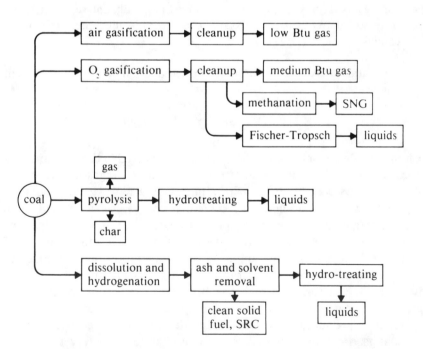

Figure 22.9.   Routes to coal conversion.

1. Steam and oxygen react with coal in a gasifier, producing a low-Btu gas consisting of carbon monoxide, carbon dioxide, hydrogen, and some methane. The major reactions, simplifying coal as C, are:

$$2C + O_2 \rightarrow 2CO \qquad \Delta H = -94.1 \text{ kcal} \qquad (22.11.1)$$
$$\text{(per mole of carbon at } 25°C)$$

$$2CO + O_2 \rightarrow 2CO_2 \qquad\qquad\qquad\qquad (22.11.2)$$

$$C + H_2O \rightarrow CO + H_2 \qquad \Delta H = +31.4 \text{ kcal} \qquad (22.11.3)$$

$$C + 2H_2 \rightarrow CH_4 \qquad \Delta H = -17.9 \text{ kcal} \qquad (22.11.4)$$

2. The gas product is freed from tar and dust by scrubbing with water.

3. The shift reaction,

$$H_2O + CO \rightarrow H_2 + CO_2 \qquad (22.11.5)$$

over a shift catalyst increases the molar ratio of H_2 to CO to an optimum value of approximately 3.5:1.

4. Acid gases (H_2S, COS, CO_2) are removed in an alkaline scrubber. Complete sulfur removal is necessary to avoid poisoning the methanation catalyst.

5. The catalytic hydrogenation of CO (methanation step) produces methane gas:

$$CO + 3H_2 \rightarrow CH_4 + H_2O \qquad (22.11.6)$$

A number of environmental implications are involved in the widespread use of coal conversion. To produce significant quantities of gas from coal would require increased coal mining. Strip mining would have to increase and large quantities of water would be required. Scarcity of water is a particularly severe constraint on gasification processes in the Powder River basin of Wyoming and Montana, which is located in a region where the average rainfall is only 15–30 cm per year. Indigenous water supplies could not support massive coal gasification operations. Another consideration involves the thermal efficiency of the conversion of coal to gas, which has been estimated at approximately 65%. Therefore, large quantities of heat and by-product carbon dioxide "greenhouse gas" would be produced by widespread gasification, causing thermal pollution. Since coal gasification is carried out in closed systems, however, there is little potential for air pollution from particle emissions and acid gases.

Methanol, CH_3OH, is a convenient liquid fuel which can be produced from coal. Methanol is produced on a small scale by the destructive distillation of wood (see Section 22.4). It was widely used for heating and lighting in mid-19th century France and for automobile fuel in gasoline-short countries during World War II. On a large scale, methanol is produced by the reaction of carbon monoxide and hydrogen at a lower H_2/CO ratio than that required for the production of methane:

$$CO + 2H_2 \rightarrow CH_3OH \qquad (22.11.7)$$

This reaction can be carried out using a copper-based catalyst at 50 atm and 250°C. The carbon monoxide and hydrogen used for methanol production are both produced from coal, oxygen, and steam. At levels up to 15%, methanol makes an excellent additive for gasoline. It has a high octane number (106) and improves fuel economy and acceleration time in automobiles. It also reduces the emissions of practically all automotive pollutants. A significant advantage of methanol is that it can be substituted for present liquid fuels with minimum disruption in existing processing, transportation, and end-use facilities.

22.12. NUCLEAR FISSION POWER

The awesome power of the atom revealed at the end of World War II held out enormous promise for the production of abundant, cheap energy. This promise has never really come to fruition, although nuclear energy currently provides a significant percentage of electric energy in many countries.

Nuclear power reactors currently in use depend upon the fission of uranium-235 nuclei by reactions such as

$$^{235}_{92}U + {}^{1}_{0}n \rightarrow {}^{133}_{51}Sb + {}^{99}_{41}Nb + 4\,{}^{1}_{0}n \qquad (22.12.1)$$

to produce two radioactive fission products, an average of 2.5 neutrons, and an average of 200 MeV of energy per fission. The neutrons, initially released as fast-moving, highly energetic particles, are slowed to thermal energies in a moderator medium. For a reactor operating at a steady state, exactly one of the neutron products from each fission is used to induce another fission reaction in a chain reaction:

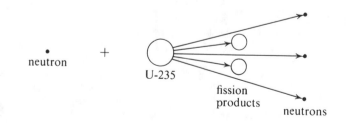

The energy from these nuclear reactions is used to heat water in the reactor core and produce steam to drive a steam turbine, as shown in Figure 22.10.

Because of limitations of structures and materials, nuclear reactors operate at a maximum temperature of around 620 K, compared to approximately 800 K in a fossil-fuel power plant. The thermal efficiency of nuclear power generation is limited by the Carnot relationship (Equation 22.8.1) and is therefore inherently low. The overall efficiency for production of electricity in a nuclear-fission plant does not exceed 30%, leaving 70% of the energy to be disposed of in the environment.

The course of nuclear power development was permanently altered by the now-famous reactor accident at Three Mile Island (TMI) in Pennsylvania. The incident began on March 28, 1979 with a partial loss of coolant water from the Metropolitan Edison Company's nuclear reactor located on Three Mile Island in the Susquehanna River, 28 miles outside of Harrisburg, Pennsylvania. The loss resulted in loss of control, overheating, and partial disintegration of the reactor core. Some radioactive xenon and krypton gas were released and some radioactive water was dumped into the Susquehanna River. A much worse accident occurred at Chernobyl in the Soviet Union in April of 1986 when a reactor blew up, spreading radioactive debris over a wide area and killing a number of people (officially 31, but probably many more). Thousands of people were evacuated and the entire reactor structure had to be entombed in concrete. Food was seriously contaminated as far away as northern Scandinavia.

A limitation of fission reactors is the fact that only 0.71% of natural uranium is fissionable uranium-235. This situation could be improved by the development of **breeder reactors**, which convert uranium-238 (natural abundance 99.28%) to fissionable plutonium-239.

A major consideration in the widespread use of nuclear fission power is the production of large quantities of highly radioactive waste products. These

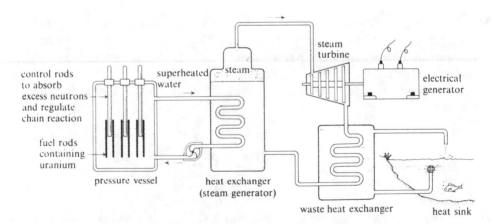

Figure 22.10. A typical nuclear fission power plant.

remain lethal for thousands of years. They must either be stored in a safe place or disposed of permanently in a safe manner. At the present time, spent fuel elements are being stored under water at the reactor sites. Eventually, the wastes from this fuel will have to be buried.

Another problem to be faced with nuclear fission reactors is their eventual decommissioning. There are three possible solutions. One is dismantling soon after shutdown, in which the fuel elements are removed, various components are flushed with cleaning fluids, and the reactor is cut up by remote control and buried. "Safe storage" involves letting the reactor stand 30–100 years to allow for radioactive decay, followed by dismantling. The third alternative is entombment, encasing the reactor in a concrete structure.

22.13. NUCLEAR FUSION POWER

The two main reactions by which energy can be produced from the fusion of two light nuclei into a heavier nucleus are the deuterium-deuterium reaction,

$$\, _1^2\text{H} + \, _1^2\text{H} \rightarrow \, _2^3\text{He} + \, _0^1\text{n} = 1\text{MeV} \qquad (22.13.1)$$

and the deuterium-tritium reaction:

$$\, _1^2\text{H} + \, _1^3\text{H} \rightarrow \, _2^4\text{He} + \, _0^1\text{n} = 17\text{MeV} \qquad (22.13.2)$$

The second reaction is more feasible because less energy is required to fuse the two nuclei than to fuse two deuterium nuclei. The total energy from deuterium-tritium fusion is limited by the availability of tritium, which is made from nuclear reactions of lithium-6 (natural abundance, 7.4%). The supply of deuterium, however, is essentially unlimited; one out of every 6700 atoms of hydrogen is the deuterium isotope. The ^{3}He by-product of Reaction 22.13.1. reacts with neutrons, which are abundant in a nuclear fusion reactor, to produce tritium required for Reaction 22.13.2.

The power of nuclear fusion has not yet been harnessed in a sustained, controlled reaction of appreciable duration. Most approaches have emphasized "squeezing" a plasma (ionized gas) of fusionable nuclei in a strong magnetic field. Another approach to obtaining the high temperatures required for fusion involves the use of a number of high-powered lasers focused on a pellet of fusionable material. A great flurry of excitement over the discovery of a cheap, safe, simple fusion power source was generated by an announcement from the University of Utah in 1989 of the attainment of "cold fusion" in the electrolysis of deuterium oxide (heavy water). Funding was appropriated and laboratories around the world were thrown into frenetic activity in an effort to duplicate the reported results. Some investigators reported evidence, particularly the generation of anomalously large amounts of heat, to support the idea of cold fusion, whereas others scoffed at the idea. To date, the promise of cold fusion has not been realized.

Controlled nuclear fusion processes would produce almost no radioactive waste products. However, tritium is very difficult to contain, and some release of the isotope would occur. The deuterium-deuterium reaction promises an unlimited source of energy. Either of these reactions would be preferable to fission in terms of environmental considerations. Therefore, despite the possibility of insurmountable technical problems involved in harnessing fusion energy, the promise of this abundant, relatively nonpolluting energy source makes its pursuit well worth a continuing persistent effort.

22.14.   GEOTHERMAL ENERGY

Underground heat in the form of steam, hot water, or hot rock used to produce steam is already being used as an energy resource. This energy was first harnessed for the generation of electricity at Larderello, Italy, in 1904, and has since been developed in Japan, Russia, New Zealand, the Philippines, and at the Geysers in northern California.

Underground dry steam is relatively rare but is the most desirable from the standpoint of power generation. More commonly, energy reaches the surface as superheated water and steam. In some cases, the water is so pure that it can be used for irrigation and livestock; in other cases, it is loaded with corrosive, scale-forming salts. Utilization of the heat from contaminated geothermal water generally requires that the water be reinjected into the hot formation after heat removal to prevent contamination of surface water.

The utilization of hot rocks for energy requires fracturing of the hot formation, followed by injection of water and withdrawal of steam. This technology is still in the experimental state, but promises approximately ten times as much energy production as steam and hot water sources.

Land subsidence and seismic effects are environmental factors that may hinder the development of geothermal power. However, this energy source holds considerable promise, and its development continues.

22.15.   THE SUN: AN IDEAL ENERGY SOURCE

The recipe for an ideal energy source calls for one that is unlimited in supply, widely available, and inexpensive; it should not add to the Earth's total heat burden or produce chemical air and water pollutants. Solar energy fulfills all of these criteria. Solar energy does not add excess heat to that which must be radiated from the Earth. On a global basis, utilization of only a small fraction of solar energy reaching the Earth could provide for all energy needs. In the United States, for example, with conversion efficiencies ranging from 10–30%, it would only require collectors ranging in area from one-tenth down to one-thirtieth that of the state of Arizona to satisfy present U.S. energy needs. (This is still an enormous amount of land, and there are economic and environmental problems related to the use of even a fraction of this amount of land for solar energy collection. Certainly, many residents of Arizona would not be pleased at having so much of the state devoted to solar collectors, and some environmental groups would protest the resultant shading of rattlesnake habitat.)

Solar power cells for the direct conversion of sunlight to electricity have been developed and are widely used for energy in space vehicles. With present technology, however, they remain too expensive for large-scale generation of electricity. Therefore, most schemes for the utilization of solar power depend upon the collection of thermal energy, followed by conversion to electrical energy. The simplest such approach involves focusing sunlight on a steam-generating boiler. Parabolic reflectors can be used to focus sunlight on pipes containing heat-transporting fluids. Selective coatings on these pipes can be used so that only a small percentage of the incident energy is reradiated from the pipes.

A major disadvantage of solar energy is its intermittent nature. However, flexibility inherent in an electric power grid would enable it to accept up to 15% of its total power input from solar energy units without special provision for energy storage. Existing hydroelectric facilities may be used for pumped-water energy storage in conjunction with solar electricity generation. Heat or cold can be stored in water, in a latent form in water (ice) or eutectic salts, or in beds of rock. Enormous amounts of heat can be stored in water as a supercritical fluid contained at high temperatures and very high pressures deep underground. Mechanical energy can be stored with compressed air or flywheels.

Hydrogen gas, H_2, is an ideal chemical fuel that may serve as a storage medium for solar energy. Electricity generated by solar means can be used to electrolyze a salt solution containing an anion that is very difficult to oxidize, so that oxygen is released at the anode and hydrogen is produced at the cathode. The net reaction is

$$2H_2O + \text{electrical energy} \rightarrow 2H_2(g) + O_2(g) \qquad (22.15.1)$$

Hydrogen, and even oxygen, can be piped some distance and the hydrogen burned without pollution or used in a fuel cell (Figure 22.6). This may, in fact, make possible a "hydrogen economy." Disadvantages of hydrogen use as a fuel include the fact that it has a heating value per unit volume of about one-third that of natural gas and that it is explosive over a wide range of mixtures with air.

No really insurmountable barriers exist to block the development of solar energy, such as might be the case with fusion power. In fact, the installation of solar space and water heaters became widespread in the late 1970s, and research on solar energy was well supported in the U.S. until after 1980, when it became fashionable to believe that free market forces had solved the "energy crisis." With the installation of more heating devices and the probable development of some cheap, direct solar electrical generating capacity, it is likely that during the coming century solar energy will be providing an appreciable percentage of energy needs in areas receiving abundant sunlight.

22.16.   ENERGY FROM BIOMASS

All fossil fuels originally came from photosynthetic processes. Photosynthesis does hold some promise of producing combustible chemicals to be used for energy production and could certainly produce all needed organic raw materials. It suffers from the disadvantage of being a very inefficient means of solar energy collection (a collection efficiency of only several hundredths of a percent by photosynthesis is typical of most common plants). However, the overall energy conversion efficiency of several plants, such as sugarcane, is around 0.6%. Furthermore, some plants, such as *Euphorbia lathyrus* (gopher plant), a small bush growing wild in California, produce hydrocarbon emulsions directly. The fruit of the Philippine plant, *Pittsosporum reiniferum*, can be burned for illumination due to its high content of hydrocarbon terpenes (see Section 18.2), primarily α-pinene and myrcene. Conversion of agricultural plant residues to energy could be employed to provide some of the energy required for agricultural production. Indeed, until about 70 years ago, virtually all of the energy required in agriculture—hay and oats for horses, home-grown food for laborers, and wood for home heating—originated from plant materials produced on the land. (An interesting exercise is to calculate the number of horses required to provide the energy used for transportation at the present time in the Los Angeles basin. It can be shown that such a large number of horses would fill the entire basin with manure at a rate of several feet per day.)

Annual world production of biomass is estimated at 146 billion metric tons, mostly from uncontrolled plant growth. Many farm crops and trees can produce 10–20 metric tons per acre per year of dry biomass, and some algae and grasses can produce as much as 50 metric tons per acre per year. The heating value of this biomass is 5000–8000 Btu/lb for a fuel having virtually no ash or sulfur (compare heating values of various coals in Table 22.3). Current world demand for oil and gas could be met with about 6% of the global production of biomass. Meeting U.S. demands for oil and gas would require that about 6–8% of the land area of the coterminous 48 states be cultivated intensively for biomass production. A significant advantage of biomass energy is that its use would not add any net carbon dioxide to the atmosphere.

As it has been throughout history, biomass is significant as heating fuel and, in some parts of the world, is the fuel most widely used for cooking. Air pollution from wood burning stoves and furnaces is a growing problem in some areas. Currently, wood provides about 8% of world energy needs. This percentage could increase through the development of "energy plantations" consisting of trees grown solely for their energy content.

Seed oils show promise as fuels, particularly for use in diesel engines. The most common plants producing seed oils are sunflowers and peanuts. More exotic species include the buffalo gourd, cucurbits, and Chinese tallow tree.

Biomass could be used to replace much of the 50–60 million metric tons of petroleum and natural gas currently consumed in the manufacture of primary chemicals in the U.S. each year. Among the sources of biomass that could be used for chemical production are grains and sugar crops (for ethanol manufacture), oilseeds, animal by-products, manure, and sewage (the last two for methane generation). The biggest potential source of chemicals is the lignocellulose making up the bulk of most plant material. For example, both phenol and benzene might be produced directly from lignin. Brazil has maintained a program for the production of chemicals from fermentation-produced ethanol.

22.17. GASOHOL

A major option for converting photosynthetically-produced biochemical energy to a form suitable for internal combustion engines is the production of either methanol or ethanol. Methanol is created by the destructive distillation of wood (Section 22.4) or from synthesis gas manufactured from coal or natural gas (Section 22.11). Ethanol is most commonly manufactured by fermentation of carbohydrates. Either one of these chemicals can be used by itself as fuel in a suitably designed internal combustion engine. More commonly, these alcohols are blended in proportions of up to 20% with gasoline to give **gasohol**, a fuel that can be used in existing internal combustion engines with little or no adjustment.

Gasohol offers a number of advantages. A high octane rating is obtained without adding tetraethyl lead. The reduction in exhaust emissions can be substantial, up to 50% for carbon monoxide and NO_x. Most important, because of its photosynthetic origin, alcohol may be considered a renewable resource rather than a depletable fossil fuel. The manufacture of alcohol can be accomplished by the fermentation of sugar obtained from the hydrolysis of cellulose in wood wastes (see Section 22.4) and crop wastes. Fermentation of these waste products offers an excellent opportunity for recycling.

Brazil has been a leader in the manufacture of ethanol for fuel uses, with 4 billion liters produced in 1982. At one time Brazil had over 450,000 automobiles that could run on pure alcohol, although many of these were converted back to gasoline during the era of relatively low petroleum prices in the 1980s. The extensive involvement of Brazil with ethanol production stems from the fact that it has few fossil resources but does have ideal conditions for the growth of large quantities of biomass. To date, most of Brazil's alcohol has come from the fermentation of sugarcane. However, a potentially much greater source of fermentable biomass is casava, or manioc, a root crop growing abundantly throughout the country. Significant amounts of gasoline in the United States are

Table 22.4.   Possible Future Sources of Energy

Source	Principles
Coal conversion	Manufacture of gas, hydrocarbon liquids, alcohol, or solvent-refined coal from coal.
Oil shale	Retorting petroleum-like fuel from oil shale.
Geothermal	Utilization of underground heat.
Gas-turbine topping cycle	Utilization of hot combustion gases in a turbine, followed by steam generation.
MHD	Electrical generation by passing a hot gas plasma through a magnetic field.
Thermionics	Electricity generated across a thermal gradient.
Fuel cells	Conversion of chemical to electrical energy.
Solar heating and cooling	Direct use of solar energy for heating and cooling through the application of solar collectors.
Solar cells	Use of silicon semiconductor sheets for the direct generation of electricity from sunlight.
Solar thermal electric	Conversion of solar energy to heat followed by conversion to electricity.
Wind	Conversion of wind energy to electricity.
Ocean thermal electric	Use of ocean thermal gradients to convert heat energy to electricity.
Nuclear fission	Conversion of energy released from fission of heavy nuclei to electricity.
Breeder reactors	Nuclear fission combined with conversion of nonfissionable nuclei to fissionable nuclei.
Nuclear fusion	Conversion of energy released by the fusion of light nuclei to electricity.
Bottoming cycles	Utilization of waste heat from power generation for various purposes.
Solid waste	Combustion of trash to produce heat and electricity.
Photosynthesis	Use of plants for the conversion of solar energy to other forms by a biomass intermediate.
Hydrogen	Generation of H_2 by thermochemical means for use as an energy-transporting medium.

supplemented with ethanol, more as an octane-ratings booster than as a fuel supplement.

Methanol, another potential ingredient of gasohol, can also be made from biomass. This can be accomplished by converting biomass, such as wood, to CO and H_2, and synthesizing methanol from these gases.

22.18. FUTURE ENERGY SOURCES

As discussed in this chapter, a number of options are available for the supply of energy in the future. The major possibilities are summarized in Table 22.4.

CHAPTER SUMMARY

The chapter summary below is presented in a programmed format to review the main points covered in this chapter. It is used most effectively by filling in the blanks, referring back to the chapter as necessary. The correct answers are given at the end of the summary.

Mineral and fossil fuel resources refer to (1)_____

_____ whereas reserves refer to

(2)_____

_____.

Some of the major environmental concerns associated with mining and processing of metal ores are (3)_____

_____. The figure below,

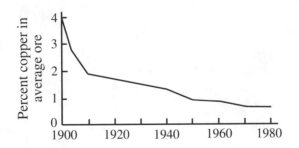

illustrates a problem in mining in that the general trend (4)_____

_____.

Clays are used for (5)_____

_____. Fluorspar, CaF_2, is used as

(6)_____

_____ and cryolite, Na_3AlF_6, is used as (7)_____

_____. The

major phosphate minerals are (8)_____

_____,

widely used in agriculture to manufacture (9)_____.

The biggest single use for sulfur is in the manufacture of (10)_____.

The four most important sources of sulfur are (11)_____

_____.

The most common raw material for the manufacture of other products worldwide is (12)_____. Two products that can be obtained by the chemical
processing of wood are (13)_____.

At present, most of the energy consumed by humans is produced from
(14)_____. Although the most abundant of these is (15)_____,
its utilization is limited by environmental disruption from mining and emissions
of (16)_____.
Reserves of fissionable nuclear fuels are much higher than those of fossil fuels
assuming use of (17)_____.

The Carnot equation,

$$\text{Percent efficiency} = \frac{T_1 - T_2}{T_1} \times 100 \qquad (22.8.1)$$

expresses a limitation to (18)_____

_____. About

(19)_____ percent of energy from fossil-fuel-fired power plants and (20)_____
of energy from nuclear power plants is not converted to electricity and is
(21)_____.

Up to half of the pore space in petroleum-bearing formations is occupied by (22)_____. The three levels of petroleum recovery are (23)_____ _____, the last of which uses (24)_____ _____. Other than crude oil pumped directly from underground, two other potential sources of liquid petroleum are (25)_____.

The general term coal describes (26)_____ _____. Major classes of coal are differentiated largely by the three factors (27)_____ _____

which are used to differentiate coals according to (28)_____. Coal conversion refers to production of (29)_____. Five different fuels formed from coal conversion are (30)_____ _____ _____. In coal gasification, the reaction between hot carbon from coal and water (steam) is (31)_____ and the shift reaction is (32)_____. As related to energy, the reaction,

$$^{235}_{92}U + ^{1}_{0}n \rightarrow ^{133}_{51}Sb + ^{99}_{41}Nb + 4\,^{1}_{0}n \qquad (22.12.8)$$

illustrates (33)_____ _____. The thermal efficiency of nuclear reactors is inherently low because of (34)_____ _____ _____.

The fissionable isotope of uranium is (35)_____, which is (36)_____ percent of natural uranium.

The reaction,

$$^{2}_{1}H + ^{3}_{1}H \rightarrow ^{4}_{2}He + ^{1}_{0}n = 17MeV$$

is an example of (37)_____.

The most desirable form of geothermal energy is (38)_____

_____. An energy source that is "unlimited in

supply, widely available, and inexpensive" is (39)_____. The

efficiency of collection of solar energy by photosynthesis is of the order of

(40)_____

_____.

Answers

1. quantities that are estimated to be ultimately available
2. well-identified resources that can be profitably utilized with existing technology
3. disturbance of land, air pollution from dust and smelter emissions, and water pollution from disrupted aquifers
4. involves utilization of less rich ores
5. clarifying oils, catalysts in petroleum processing, fillers and coatings for paper, and in the manufacture of firebrick, pottery, sewer pipe, and floor tile
6. a flux in steel manufacture
7. a solvent for aluminum oxide in the electrolytic preparation of aluminum metal
8. the apatites, having the general formula $Ca_5(PO_4,CO_3)_3(F,OH,Cl)$
9. phosphate fertilizers
10. sulfuric acid
11. deposits of elemental sulfur; H_2S recovered from sour natural gas; organic sulfur recovered from petroleum; and pyrite (FeS_2)
12. wood
13. methanol and sugar
14. fossil fuels
15. coal
16. carbon dioxide and sulfur dioxide
17. breeder reactors
18. the percentage of energy from a source that can actually be utilized
19. 60
20. 70
21. dissipated as heat
22. water
23. primary, secondary, and tertiary
24. injection of pressurized carbon dioxide

25. oil shale and tar sands
26. a large range of solid fossil fuels derived from partial degradation of plants
27. percentage of fixed carbon, percentage of volatile matter, and heating value
28. coal rank
29. gaseous, liquid, or low-sulfur, low-ash fuels
30. Solvent-refined coal, low-sulfur boiler fuels, liquid hydrocarbon fuels, synthetic natural gas, low-sulfur, low-Btu gas for industrial use
31. $C + H_2O \rightarrow CO + H_2$
32. $H_2O + CO \rightarrow H_2 + CO_2$
33. Production of energy by nuclear fission
34. low operating temperatures that must be maintained because of limitations of structures and materials so that conversion of heat to useful energy as expressed by the Carnot relationship is low
35. uranium-235
36. 0.71
37. nuclear fusion
38. underground dry steam
39. solar energy
40. several hundreths of a percent

QUESTIONS AND PROBLEMS

1. What pollution control measures may produce a shortage of platinum metals?

2. List and discuss some of the major environmental concerns related to the mining and utilization of metal ores.

3. What are the major phosphate minerals?

4. Arrange the following energy conversion processes in order from the least to the most efficient: (a) electric hot water heater, (b) photosynthesis, (c) solar cell, (d) electric generator, (e) aircraft jet engine.

5. Considering the Carnot equation and common means for energy conversion, what might be the role of improved materials (metal alloys, ceramics) in increasing energy conversion efficiency?

6. Why is shale oil, a possible substitute for petroleum in some parts of the world, considered to be a pyrolysis product?

7. List some coal ranks, and describe what is meant by coal rank.

8. Why was it necessary to add hydrocarbons to gas produced by reacting steam with hot carbon from coal in order to make a useful gas product?

9. Consider the production of hydrocarbon liquids from coal, Figure 22.9. The approximate empirical formula of coal is $CH_{0.8}$ and of fuel hydrocarbon liquids is CH_2. Using chemical reactions to illustrate, discuss what is meant by hydrotreating shown in Figure 22.9.

10. As it is now used, what is the principle or basis for the production of energy from uranium by nuclear fission? Is this process actually used for energy production? What are some of its environmental disadvantages? What is one major advantage?

11. What would be at least two highly desirable features of nuclear fusion power if it could ever be achieved in a controllable fashion on a large scale?

12. Justify describing the sun as "an ideal energy source." What are two big disadvantages of solar energy?

13. What are some of the greater implications of the use of biomass for energy? How might such widespread use affect greenhouse warming? How might it affect agricultural production of food?

14. Describe how gasohol is related to energy from biomass.

15. Using some specific examples, describe what is meant by the "resources-energy-environment triangle."

16. How does the trend toward utilization of less rich ores affect the environment? What does it have to do with energy utilization?

17. Of the resources listed in this chapter, list and discuss those that are largely from by-product sources.

18. Why is the total dollar value of "cheap" sand and gravel so high? What does this fact imply for environmental protection?

LITERATURE CITED

1. Lilienthal, D. E., *Atomic Energy: A New Start*, (New York: Harper and Row Publishers, Inc., 1980).
2. "Trash-to-Electricity Disposal Runs into Problems in Maine," *New York Times*, January 22, 1991, p. A13.
3. "Coal Output Near a Record," *New York Times*, December 17, 1990, p. C-2.

SUPPLEMENTARY REFERENCES

Medvedev, Z. A., *The Legacy of Chernobyl,* (New York: Norton, 1990).

Ogden, J. M., and R. H. Williams, *Solar Hydrogen. Moving Beyond Fossil Fuels*, (Washington, DC: World Resources Institute, 1989).

World Resources, 1990–91, (New York: Oxford University Press, 1990).

Cohen, B. L., *The Nuclear Energy Option. An Alternative for the 90s,* (New York: Plenum Press, 1990).

Cassedy, E. S., and P. Z. Grossman, *Introduction to Energy. Resources, Technology, and Society*, (New York: Cambridge University Press, 1990).

Van der Voort, E., and G. Grassi, Eds., *Wind Energy–1*, (New York: Harwood, 1988).

Van der Voort, E., and G. Grassi, Eds., *Wind Energy–2*, (New York: Harwood, 1988).

World Resources Institute and International Institute for Environment and Development, *World Resources, 1988–89*, (New York: Basic Books, 1988).

Berlin, R. E., and C. C. Stanton, *Radioactive Waste Management*, (New York: Wiley-Interscience, 1989).

Rush to Burn. Solving America's Garbage Crisis?, (Washington, DC: Island Press, 1989).

Oppenheimer, E.J., *Natural Gas, the Best Energy Choice*, (New York, NY: Pen & Podium, Inc., 1989).

Sperling, D., *New Transportation Fuels: A Strategic Approach to Technological Change*, (Berkeley, CA: University of California Press, 1989).

Carter, L. J., *Nuclear Imperatives and Public Trust: Dealing with Radioactive Waste*, (Baltimore, MD: Resources for the Future In., 1987).

Klass, D. L., *Energy From Biomass and Wastes IX*, (Chicago, IL: Institute of Gas Technology, 1986).

Krauskopf, K. B., *Radioactive Waste Disposal and Geology*, (New York: Chapman and Hall, 1988).

Burns, M. E., Ed., *Low-Level Radioactive Waste Regulation: Science, Politics, and Fear*, (Chelsea, MI: Lewis Publishers, 1988).

Bibby, D. M., *Methane Conversion*, (New York, NY: Elsevier Science Publishers, 1988).

Kittel, H., Ed., *Radioactive Waste Management Handbook, Vol. I: Near-Surface Land Disposal*, (New York, NY: J. Harwood Academic Publishers, 1989).

Park, C. C., *Chernobyl: The Long Shadow*, (London and New York: Routledge, 1989).

Speight, J. G., Ed., *Fuel Science and Technology Handbook*, (New York, NY: Marcel Dekker, 1990).

INDEX